Mechatronics and Control of Electromechanical Systems

Mechatronics and Control of Electromechanical Systems

Sergey Edward Lyshevski

CRC Press
Taylor & Francis Group
Boca Raton London New York

CRC Press is an imprint of the
Taylor & Francis Group, an **informa** business

CRC Press
Taylor & Francis Group
6000 Broken Sound Parkway NW, Suite 300
Boca Raton, FL 33487-2742

Printed on acid-free paper

International Standard Book Number-13: 978-1-4987-8239-5 (Hardback)

Library of Congress Cataloging-in-Publication Data

Names: Lyshevski, Sergey Edward, author.
Title: Mechatronics and Control of Electromechanical Systems/Sergey Lyshevski.
Description: Boca Raton : CRC Press, 2017. | Includes bibliographical references and index.
Identifiers: LCCN 2016024616 | ISBN 9781498782395 (hardback : acid-free paper)
Subjects: LCSH: Microelectromechanical systems. | Mechatronics. | Power electronics. | Electromechanical devices.
Classification: LCC TK7875 .L959 2017 | DDC 621--dc23
LC record available at https://lccn.loc.gov/2016024616

Visit the Taylor & Francis Web site at
http://www.taylorandfrancis.com

and the CRC Press Web site at
http://www.crcpress.com

Printed and bound in the United States of America by
Edwards Brothers Malloy on sustainably sourced paper

Perseverance and Indebtedness:

With a sincere gratitude and deepest thanks to all who have empowered us with scientific knowledge, scholarly learning, engineering developments, and transformative technologies. With a deep appreciation of the contributions and discoveries by Michael Faraday and Nikola Tesla who enlighten us with knowledge, solutions, and technologies.

Sincerely, Sergey Lyshevski

Quotes:

"A pessimist sees the difficulty in every opportunity; an optimist sees the opportunity in every difficulty.

However beautiful the strategy, you should occasionally look at the results."

Sir Winston S. Churchill

"Nothing is too wonderful to be true, if it be consistent with the laws of nature; and in such things as these, experiment is the best test of such consistency."

Michael Faraday

"Today's scientists have substituted mathematics for experiments, and they wander off through equation after equation, and eventually build a structure which has no relation to reality."

Nicola Tesla

Contents

Preface

ROADMAPPING OF MECHATRONICS AND ELECTROMECHANICAL SYSTEMS

With an emergence of new technologies, a mechatronic paradigm has become increasingly important in the design of electromechanical and mechanical systems [1–9]. These systems are used in cars, consumer electronic devices, energy and power systems, flight vehicles, robots, medical devices, etc. Electromechanical systems comprise electromechanical motion devices, microelectromechanical systems (MEMS), sensors, transducers, microelectronic and electronic components, etc. Devices, components, modules, and subsystems are designed, fabricated, and integrated using different technologies. Enabling mechatronic-centric technologies and advanced electromechanical systems are critical in various applications.

The high-performance electromechanical systems are designed and fabricated using new paradigms, such as information technology, MEMS, nanoscale microelectronics, and nanotechnology. A spectrum of challenges and open problems in system and technology integrations may be solved using mechatronics. These grand challenges, as well as a drastic increase in user-configured electromechanical systems, result in the need for basic, applied, and experimental findings. Hence, basic engineering physics, underlying engineering design, application-specific findings, new technologies, and physical implementation are covered in this textbook. Applied and experimental results, control premises, and enabling hardware solutions are reported.

EDUCATIONAL AND TECHNOLOGICAL OBJECTIVES

The major objective of this book is to empower engineering design and enable a deep understanding of engineering underpinnings and integrated technologies. The modern description of electromechanics, electromechanical motion devices, electronics, MEMS, and control are provided. This book covers the frontiers of electromechanical engineering and science by applying basic theory, emerging technologies, advanced software, and enabling hardware. We demonstrate the application of the underlying fundamentals in designing systems. This book is aimed to: (1) Consistently cover engineering science and engineering design; (2) Educate and help one to develop strong problem-solving skills and design proficiency; (3) Ensure an in-depth presentation and consistent coverage; (4) Empower the end user with the adequate knowledge in concurrent engineering. Recent innovations and discoveries are reported. The emerging technologies and enabling hardware further enable mechatronics, and the recent developments are empowered by mechatronics.

This textbook fosters adequate knowledge and expertise generation, retention, and use. A wide range of worked-out problems, examples, and solutions are treated thoroughly. This bridges the gap between the theory, practical problems, and hardware–software codesign. Step by step, one is guided from theoretical foundation to advanced application and implementation. To enable analysis, MATLAB® and Simulink® with various application-specific environments and toolboxes are used. The book demonstrates the MATLAB capabilities, helps one to master this environment, studies examples, and helps increase designer productivity by showing how to apply the advanced software. MATLAB offers a set of capabilities to effectively solve a variety of problems. One can modify the studied problems and apply the reported results to application-specific practical problems. Our results provide the solutions for various modeling, simulation, control, optimization, and other problems.

MECHATRONICS AND ENGINEERING CURRICULUM

This book can be used in core and elective courses in electrical, mechanical, industrial, chemical, systems, control, and biomedical engineering programs. By using this textbook, the following undergraduate and graduate courses can be taught:

1. Mechatronics;
2. Electromechanical motion devices and electric machines;
3. Electromechanical systems;
4. Power and energy systems;
5. Energy conversion;
6. Clean and renewable energy.

ROLE OF MECHATRONICS IN ENGINEERING EDUCATION

Our nation's success in developing practical engineering solutions, discovering technology-centric platforms, and deploying new technologies requires consistent strategies in how academia educates and trains students. Recent economic, geopolitical, ecological, and societal developments have culminated to awareness on needs of sustainable scientific, engineering, and technological developments in electromechanical systems, energy sources, power systems, energy management, etc. Discoveries and innovations can be accomplished by highly educated and trained researchers, engineers, practitioners, and students. Enabling discoveries, comparable to those made by Michael Faraday, James Watt, Nicola Tesla, Nikolaus Otto, or Rudolf Diesel, may emerge. Many engineering disciplines are tangentially related to the solution of extremely important and challenging problems of national importance, such as the following:

1. Design of high-performance electromechanical systems with enabling functionality;
2. Design of safe, affordable, and sustainable highly efficient *clean* energy and power systems;
3. Discovery and use of novel and alternative energy sources to enable sustainable, *clean*, pollution-free, and ecology-friendly energy.

Mechatronics directly contributes to these problems. New technologies, energy independence, affordability, sustainability, manufacturing advancements, technological superiority, innovations, security, and safety are enabled by mechatronics. The integration challenges can be solved. Core information technologies, electromechanics, electronics, control, and energy areas are at the forefront, not at the periphery. The *contemporary modernism* may be inadequate to cope with economic, manufacturing, security, technological, and other challenges. Therefore, the need for the technology-centric engineering science and engineering design is significantly strengthened. The author hopes that readers will enjoy this book and provide valuable suggestions.

MATLAB® is a registered trademark of The MathWorks, Inc. For product information, please contact:

The MathWorks, Inc.
3 Apple Hill Drive
Natick, MA 01760-2098 USA
Tel: 508-647-7000
Fax: 508-647-7001
E-mail: info@mathworks.com
Web: www.mathworks.com

Acknowledgments

The author expresses his sincere acknowledgments and gratitude to all his colleagues, peers, and students who provided valuable suggestions. He sincerely acknowledges the advanced hardware provided by the Analog Devices, Inc. (http://www.analog.com), Maxon Motor (http://www.maxonmotorusa.com), STMicroelectronics (http://www.st.com), and Texas Instruments (http://www.ti.com). The author also gratefully acknowledges the assistance from MathWorks, Inc. (http://www.mathworks.com) for supplying the MATLAB® environment. It is with a great pleasure that he acknowledges the assistance and help from the outstanding team at CRC Press/Taylor & Francis Group, especially Nora Konopka (acquisitions editor, electrical engineering), Jessica Vakili (project coordinator), and Todd Perry (project editor). Many thanks to all of you.

Sergey Edward Lyshevski, PhD
Professor of Electrical Engineering
Department of Electrical and Microelectronic Engineering
Rochester Institute of Technology
Rochester, New York
E-mail: Sergey.Lyshevski@mail.rit.edu
URLs: http://people.rit.edu/seleee/, www.rit.edu/kgcoe/staff/sergey-lyshevski

Author

Dr. Sergey Edward Lyshevski received his MS and PhD in electrical engineering from Kiev Polytechnic Institute, Ukraine, in 1980 and 1987, respectively. From 1980 to 1993, he held research and faculty positions at the Department of Electrical Engineering, Kiev Polytechnic Institute, and the National Academy of Sciences of Ukraine, Kiev, Ukraine. From 1989 to 1992, he was the Microelectronic and Electromechanical Systems Division Head at the National Academy of Sciences of Ukraine. From 1993 to 2002, Dr. Lyshevski was an Associate Professor of Electrical and Computer Engineering at the Purdue University. In 2002, he joined Rochester Institute of Technology as a Professor of Electrical Engineering. He served as a Professor of Electrical and Computer Engineering in the U.S. Department of State Fulbright Program. Dr. Lyshevski is a full professor faculty fellow at the Air Force Research Laboratory, U.S. Naval Surface Warfare Center and U.S. Naval Undersea Warfare Center.

Dr. Lyshevski has authored and coauthored 16 books, 14 handbook chapters, 80 journal articles, and more than 300 refereed conference papers. He has edited encyclopedias and handbooks. Dr. Lyshevski conducted more than 75 invited tutorials, workshops, and keynote talks. As a principal investigator (project director), he performed contracts and grants for high-technology industry (Allison Transmission, Cummins, Delco, Delphi, Harris, Lockheed Martin, Raytheon, General Dynamics, General Motors, etc.), the U.S. Department of Defense (Air Force Research Laboratory, Air Force Office of Scientific Research, Defense Advanced Research Projects Agency, Office of Naval Research, and Air Force), and government agencies (Department of Energy, Department of Transportation, and National Science Foundation). He conducts research and technology developments in cyber-physical systems, microsystems, microelectromechanical systems (MEMS), mechatronics, control, and electromechanical systems. Dr. Lyshevski made a significant contribution in the design, deployment, and commercialization of advanced aerospace, automotive, and naval systems.

1 Mechatronic and Electromechanical Systems

1.1 INTRODUCTION AND EXAMPLES

The mechatronic paradigm addresses and empowers a synergy of electronics, electromagnetics, electromechanics, microelectronics, MEMS, and information technologies in the design, fabrication, integration, and implementation of electromechanical systems. The knowledge-base multidisciplinary designs of electromechanical systems have been demonstrated for various consumer electronic devices, robots, energy systems, etc. The mechatronic paradigm is aimed to guarantee design consistency, empower new technologies and discoveries, ensure multidisciplinary descriptive advancements, enable compliant hardware and software solutions, as well as deploying and commercializing products. Multidisciplinary engineering underpinnings and hardware–software codesign are examined. New science frontiers and technologies contribute to engineering advancements. Optoelectronics, photonics, microelectromechanical systems (MEMS), and other advanced hardware are widely used. There are numerous electromechanical systems in passenger cars, consumer electronics, flight vehicles, industrial robots, power generation systems, etc. Microelectronics, advanced actuators, and sensors enable synergy, foster new solutions, advance products, and empower commercialization.

Mechatronics and electromechanical systems: With an emphasis on integrated systems design, the ultimate objectives are to achieve safety, affordability, technology compliance, enhanced functionality, superior capabilities, and optimal performance. In electromechanical systems, one must ensure a consistent integration of advanced kinematics, actuators, sensors, power electronics, ICs, microprocessors, digital signal processors (DSPs), and MEMS. A concurrent design at the device, module, and system levels should provide consistency and complementarity. These features are achieved by integration of various electronic, electromagnetic, mechanical, and MEMS subsystems, modules, components, and devices.

A phenomenal growth of electromechanical systems has been accomplished due to:

1. Raising industrial and societal needs with strong, growing market and demands;
2. Affordability, safety, practicality, and overall superiority of electromechanical systems;
3. Advances in high-performance actuators, sensors, power electronics, ICs, and MEMS;
4. Cost-effective, high-yield mass production;
5. Sustainable environment-friendly fabrication, manufacturing, and assembly.

The electromechanical systems performance is measured by safety, functionality, efficiency, stability, robustness, sensitivity, accuracy, etc. In *scalable* and *modular* designs, one uses proven advanced technology devices, components, and modules. Solutions are application dependent and applications specific. The existing actuation, electronics, and sensing technologies guarantee safety, durability, and affordability. To meet the rising demands on the system performance, the front-end solutions, technologies, and devices are used and examined. Figure 1.1a illustrates the functional diagram of the electromechanical system, which comprises different components and modules. Various specifications and requirements are imposed. System functionality is examined by studying the physical quantities and variables. The input-output behavior is studied considering references $r(t)$, outputs $y(t)$, and states $x(t)$. The efficiency, accuracy, bandwidth, and other quantitative estimates, metrics, and measures may be represented using the system performance tuple (R, X, Y). The structural and systems complexity of electromechanical systems increase due to hardware advancements and stringent requirements. The *optimum-performance* design implies the use of multidisciplinary science, engineering, and technologies, as reported in Figure 1.1b.

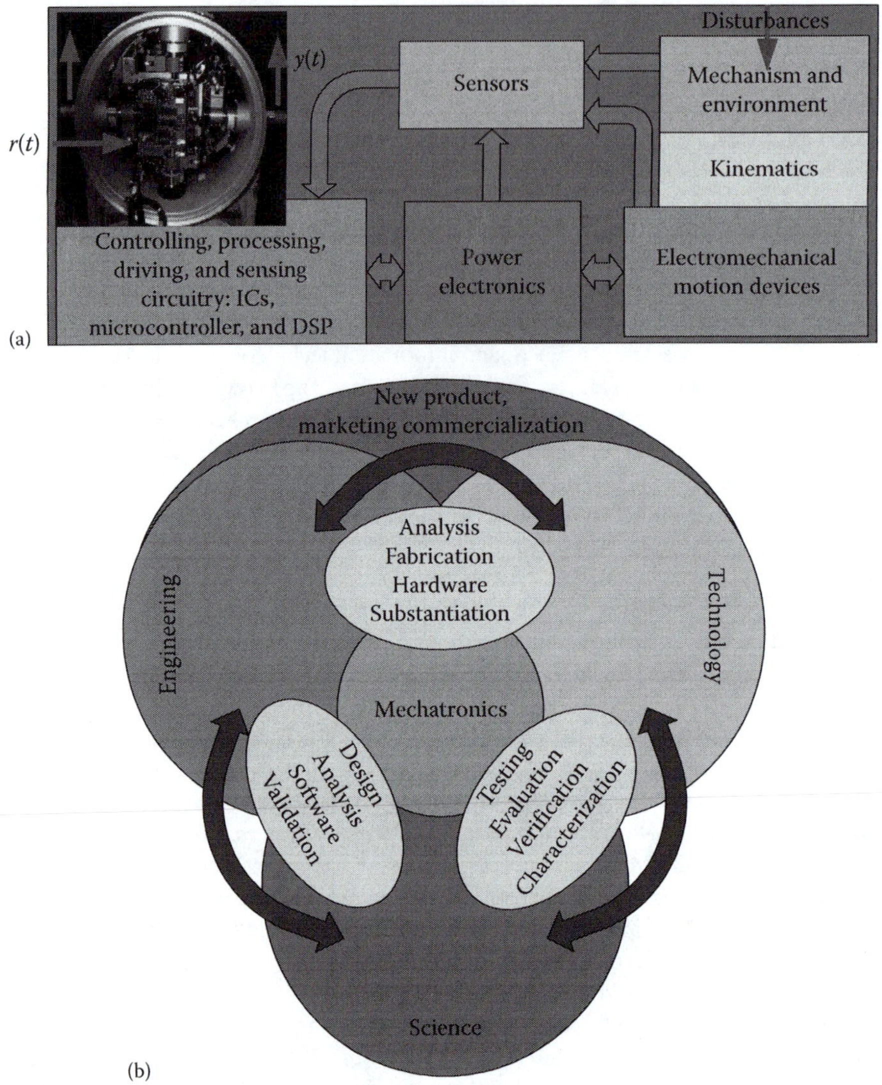

FIGURE 1.1 (a) High-level functional diagram of an electromechanical system; (b) Multidisciplinary engineering, science, and technology within a mechatronics paradigm.

Electromechanical systems comprise electromechanical motion devices, sensors, MEMS, power electronics, controlling/processing and driving/sensing circuitry (microcontrollers, DSPs, DC–DC regulators, converters, etc.), energy sources, and mechanics (kinematics). The examples and images of electromechanical systems are documented in Figures 1.2.

Applications: The mechatronic paradigm is used in automotive, aerospace, energy, naval, and other systems. Various closed-loop actuators with sensors (accelerometers, gyroscopes, current, magnetic field, noise, pressure, temperature, etc.) are used in primary and secondary energy sources, engines, propulsion systems, etc. Figure 1.3a depicts the application of mechatronics in servos and underwater vehicles. The hull with the fins actuators and power electronics is reported. The underwater vehicle path and velocity are controlled by displacing the control surfaces and by changing thrust by varying the angular velocity of the motor propulsor or propeller. Electromechanical systems are core components of propulsion, manufacturing, power generation (wind, hydropower, etc.), and other systems, as reported in Figures 1.3.

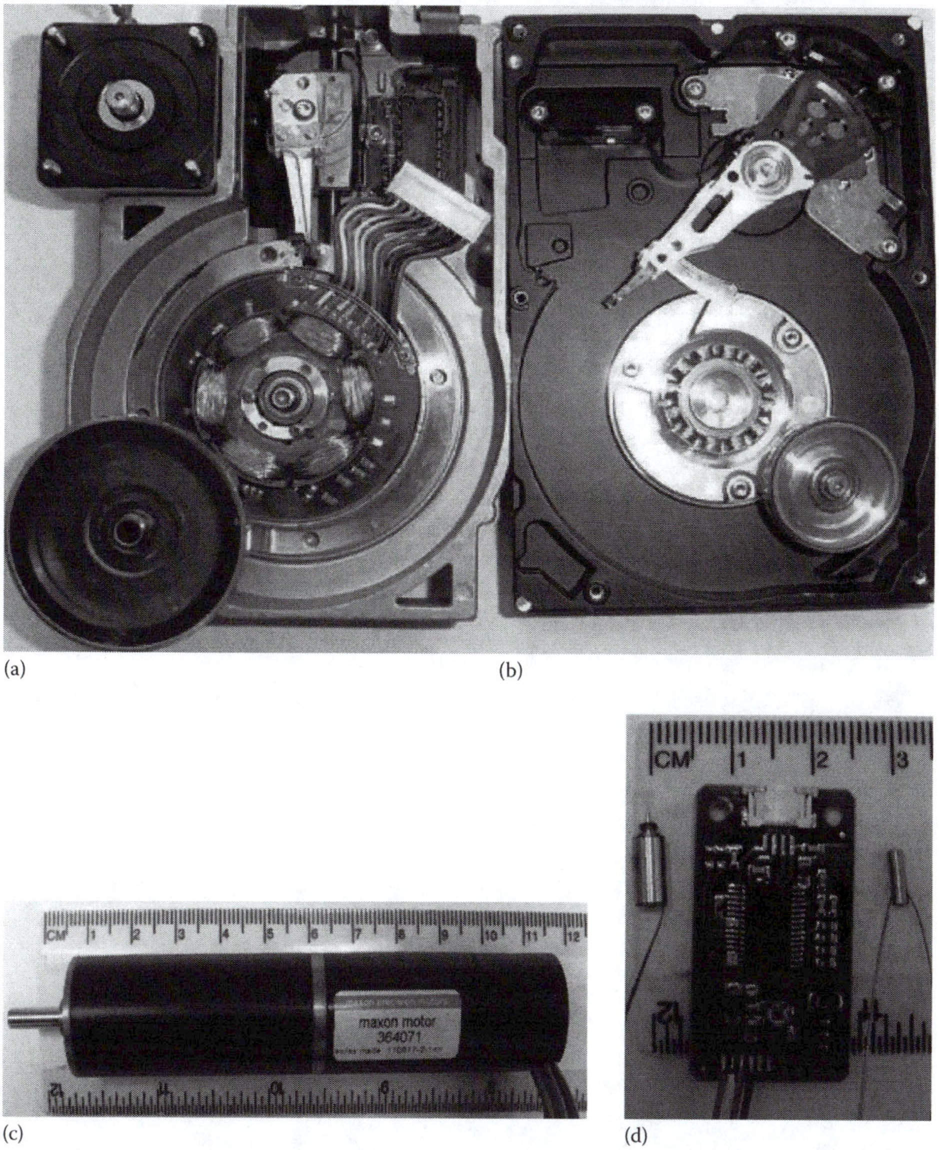

FIGURE 1.2 Electromechanical systems: (a) The 1990s closed-loop hard disk drive: Medium torque density permanent-magnet synchronous motor (at the center) with ferrite magnets rotating the disk. A stepper motor (at the top left) repositions a pointer; (b) The 2010s closed-loop hard disk drive: high torque density permanent-magnet synchronous motor with SmCo magnets rotating the disk. A synchronous motor (at the center of hard drive) is controlled by low-power electronics. The pointer is repositioned by the direct-drive axial-topology actuator (planar winding is on the rotor, and the segmented nickel-plated $Nd_2Fe_{14}B$ magnets are on the stationary member); (c) Permanent-magnet DC Maxon motor (22 mm diameter, graphite brushes, ball bearings, 8 W, 24 V, 720 rad/sec, 11.8 mN-m) with a spur five-stage planetarey ceramic gearhead (1014:1, 2 N-m rated). PWM controller-driver with position sensor is built in to control a closed-loop system; (d) Slotless permanent-magnet synchronous motors (*brushless DC motors*) with high-frequency PWM driver and sensorless controller.

FIGURE 1.3 (a) High-performance servos: Maxon motors with sensors, electronics, and controllers. Underwater vehicle hull with actuators to displace fins. The fins' actuators are controlled by PWM amplifiers. The fins displacement is measured by sensors; (b) Torpedo with control surfaces (fins) and propeller—Gerald R. Ford class nuclear-powered aircraft carrier (337 m length, 76 m height, 78 m flight deck beam, 41 m waterline beam, and 100,000 tons displacement) powered by two A4B nuclear reactors. In naval propulsion systems, ~20 to 36.5 MW electric motors can rotate ~30 ton propellers. Northrop Grumman designed and tested the 36.5 MW, 120 rpm high-temperature superconductor electric motor for the *series* ship propulsion system; (c) Robotic and manufacturing systems use a large number of electromechanical systems; (d) Wind turbines rotate permanent-magnet synchronous generators (1 kW to ~1 MW), which generate three-phase AC voltage. The blade pitch control is ensured by servo-systems.

Mechatronics and energy systems: The mechatronic paradigm has been applied in power generation systems. Automation, diagnostics, control, monitoring, and other tasks must be ensured for a broad class of electric machines (motors, generators, actuators, servos, etc.) and electromechanical systems. In power generation systems, the light-, medium-, and heavy-duty electric machines (generators from 1 kW to 1000 MW) ensure controlled energy conversion by converting the mechanical energy into electric energy. The wind power generation system is shown in Figure 1.3d.

In addition to primary generators, there are auxiliary and secondary electromechanical systems, such as actuators, motors, and transducers with power electronics, microelectronics, and sensors. Electric generators and auxiliary electromechanical systems are used in fossil fuel, natural gas, nuclear, hydro, and other power stations. Per the U.S. Energy Information Administration (www.eia.gov), the energy sources and contributing percentage in total U.S. electricity generation in 2015 are as follows: 39% coal, 27% natural gas, 19% nuclear, 6% hydropower, and 7% renewables (1.7% biomass, 0.4% geothermal, 0.4% solar, and 4.4% wind). Turbines, electric machines, and other major components are significantly different by functionality, design, etc. For example, the Bruce Nuclear Generating Station has two 772 MW, two 730 MW, and four 817 MW units with the total capacity 6.23 GW, ensuring 45 TWh electricity generation annually. Synchronous generators are used to convert energy and produce power. In 2015, hydropower generated ~3500 TWh with ~1000 GW world capacity (single unit capacity is up to 1 GW). The total wind power systems capacity is ~350 GW, while a single unit capacity is up to 10 MW. For 2015, the world final energy consumption estimate was ~125,000 TWh out of which ~20% are electricity generated. The *clean* and renewable, sustainable energy is of great importance. Annual investments in renewable and *clean* technologies, including energy harvesting, electric machines, and electronics, amount to more than $200 billion worldwide (www.gwec.net). The major spending and investments are in nuclear-, wind-, and hydropower technologies with more than $140 billion in new developments.

Clean and renewable sustainable energy production is of a great importance for US because this will decrease negative effect of unclean energy sources, as well as:

1. Ensures national energy security, economic security, and energy independence;
2. Reduces import of energy sources and eases energy dependence;
3. Ensures high-technology developments and U.S. technological superiority;
4. Significantly reduces pollution and contamination.

Air, soil, and water pollution, as well as to some extent environment variations, are caused by greenhouse gases (carbon monoxide (CO) and dioxide (CO_2), methane (CH_4), nitrous oxide (N_2O), hydrofluorocarbons, perfluorocarbons, sulfur hexafluoride, nitrogen trifluoride, etc.), nitrogen monoxide and dioxide, sulfur oxides, radicals, and toxic metals (such as lead and mercury, and their compounds). The greenhouse gases released to the atmosphere affect the irradiation exchange, result in dimming, cause the ozone depletion, etc. Pollution is very harmful to health, environment, agriculture, etc.

Note: Per the U.S. Energy Information Administration (www.eia.gov), in 2014, the CO_2 emissions from U.S. electricity generation by coal, natural gas, and petroleum were 1562, 444, and 23 million metric tons. The amount of carbon dioxide released by wildfire is greater than from all vehicles. Volcanoes release ~150 million tons of CO_2 into the atmosphere every year. The explosive eruption of the Mount Pinatubo volcano on June 15, 1991, injected ~80 and 20 million metric tons of CO_2 and SO_2 into the stratosphere resulting in significant cooling of the Earth's surface.

1.2 ROLE OF MECHATRONICS: FROM DESIGN TO COMMERCIALIZATION AND DEPLOYMENT

A mechatronic paradigm focuses on consistent analysis, synergetic design, and integration of motion devices, sensors, power electronics, controllers (analog, microcontrollers, DSPs, etc.), MEMS, and other components and modules. The advanced hardware, compliant software, and concurrent design tools have been developed using multidisciplinary engineering, science, and technology. Mechatronics, electromechanics, and emerging technologies signify the need for basic engineering science and engineering design. Reflecting a broad spectrum of recent discoveries and technologies, one emphasizes:

1. Advanced-technology devices, components, and modules;
2. Device-, component-, module-, and system-level integrity, functionality, and hierarchy;
3. System-to-device and device-to-system matching and compliance;
4. Application of advanced microelectronics, microactuation, MEMS, and microsensing solutions;
5. Microfabrication technologies such as microelectronics, photonics, *bulk* and *surface* micromachining, etc.

The *modular* functional and organizational designs of high-performance electromechanical systems frequently imply the devices, components, and modules selection and integration. Electromechanical motion devices (actuators, generators, and sensors) are one of the major components. The following tasks and problems are usually emphasized:

- Selection, design, and optimization of devices according to their applications within the system kinematics, specifications, and requirements;
- Integration of high-performance electromechanical motion devices with sensors, power electronics, and microelectronics. One emphasizes integrity, regularity, modularity, compliance, matching, and completeness;
- Control of electromechanical motion devices, testing, data acquisition, and characterization.

One strives to achieve a synergistic combination of advanced kinematical–electromechanical–electronic hardware and software solutions. This ensures adequate designs. Actuators and sensors must be integrated with power electronics. Matching and kinematics–actuators–sensors–ICs–power electronics compliance must be ensured. Analog controllers, microprocessors, and DSPs are used. In aerospace and naval vehicles, control surfaces must be displaced, see Figures 1.3a and b. The functional diagram of the closed-loop system for an aircraft is shown in Figure 1.4.

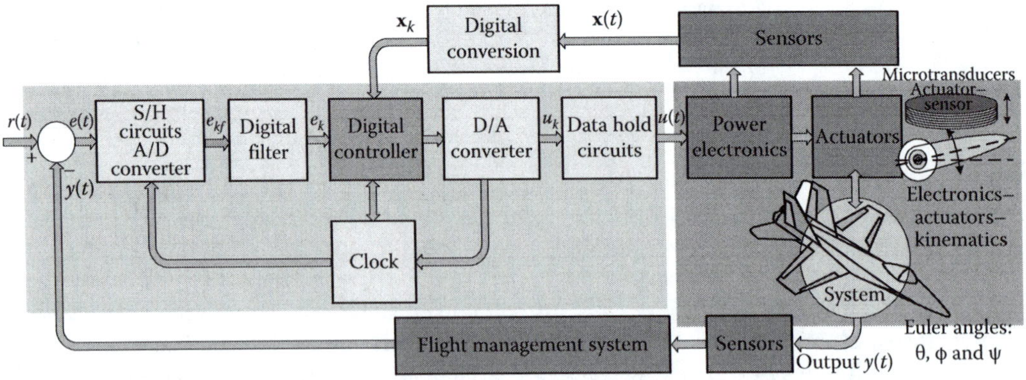

FIGURE 1.4 Functional diagram: Flight actuators displace control surfaces. The supervisory control is accomplished by a flight management system, while control of actuators is performed by microcontrollers.

As reported in Figure 1.4, aircraft and actuators are regulated by using the state variables **x**(*t*), as well as the difference between the desired reference input *r*(*t*) and the output *y*(*t*). The tracking error *e*(*t*) = *r*(*t*) − *y*(*t*) is used. For robots, aircraft, submarines, and other systems, the Euler angles (θ, ϕ, and ψ) and other variables (angle of attack, sideslip angle, acceleration, velocity, altitude, etc.) may be considered as the outputs. Let the aircraft's output vector be $\mathbf{y} = [\theta \quad \phi \quad \psi]^T$. The reference inputs are the desired Euler angles r_θ, r_ϕ, and r_ψ, and $\mathbf{r} = [r_\theta \quad r_\phi \quad r_\psi]^T$. To control the aircraft, control surfaces are deflected using the control surface servos. By applying the voltage to the actuator, one changes the angular or linear displacement. For each rotational or translational actuator, the desired (reference) deflection *r*(*t*) is compared with the actual displacement *y*(*t*). One finds the error *e*(*t*). Advanced aircraft is controlled by varying the thrust as well as displacing hundreds of rotational and translational actuators with the corresponding outputs y_i.

Microcontrollers and DSPs are widely used to control electromechanical systems. Digital systems: (1) Implement control algorithms; (2) Perform data acquisition; (3) Implement filters; (4) Perform decision-making and managements; (5) Generate control signals; etc. For an analog error signal *e*(*t*) = *r*(*t*) − *y*(*t*), the *e*(*t*) is converted to digital e_k to perform digital filtering and control. As illustrated in Figure 1.4, the sample-and-hold circuit (S/H circuit) receives the analog signal and holds this signal at the constant value for the specified period of time depending on the sampling period. An analog-to-digital (A/D) converter converts this piecewise or continuous-time signal to digital. The conversion of continuous-time signals to discrete-time is called sampling or discretization. The input signal to the filter is the sampled version of analog *e*(*t*). The input of a digital controller is the filter output. At each sampling, the discretized value of the error e_k in binary form is used by a digital controller to generate signals to control a power converter. The digital-to-analog (D/A) conversion (decoding) is performed by the D/A converter and the data-hold circuit. Coding and decoding are synchronized by clock. There are various signal conversions, such as multiplexing, demultiplexing, sample and hold, A/D (quantizing and encoding), D/A (decoding), etc.

An electromechanical system is shown in Figure 1.5a. A pointing system kinematics, geared motor (coupled within the kinematics using the *torque limiter*), PWM amplifier, ICs, and DSP are shown. Using the reference *r* and the measured angular displacement θ, the DSP develops the PWM signals to drive high-frequency MOSFETs. The number of the PWM outputs depends on the converter topology. A three-phase permanent-magnet synchronous motor is used. Six PWM outputs drive six transistors to vary the phase voltages u_{as}, u_{bs}, and u_{cs}. The magnitude of the output voltage of the PWM amplifier is controlled by changing the transistors duty cycle. The Hall-effect sensors measure the rotor angular displacement θ_r to generate the balanced phase voltages u_{as}, u_{bs}, and u_{cs}. A fully integrated electromechanical system hardware includes actuator–mechanism coupling with a *torque limiter*, sensors, microelectronics, and power electronics. The operating principles and basic foundations of conventional, mini and microscale electromechanical motion devices are based on classical electromagnetics and mechanics, as depicted in Figure 1.5b.

The two most challenging problems in the electromechanical systems design are as follows: (1) The development and integration of advanced hardware components, ensuring modular synergy of actuators, sensors, power electronics, ICs, microcontrollers, DSPs, and MEMS; (2) Device-, module-, and system-level analyses, design, and optimization. The software developments focus on environments, tools, and algorithms to perform design, control, data acquisition, simulation, visualization, prototyping, evaluation, and other tasks. Designing of high-performance electromechanical systems can be performed using the best practices, existing technologies, and proven designs at the device and system levels. The electromechanical systems must be tested, evaluated, and characterized to validate and substantiate the results. The proven designs with a sufficient technology-readiness level ensure a path toward marketing and commercialization of electromechanical systems.

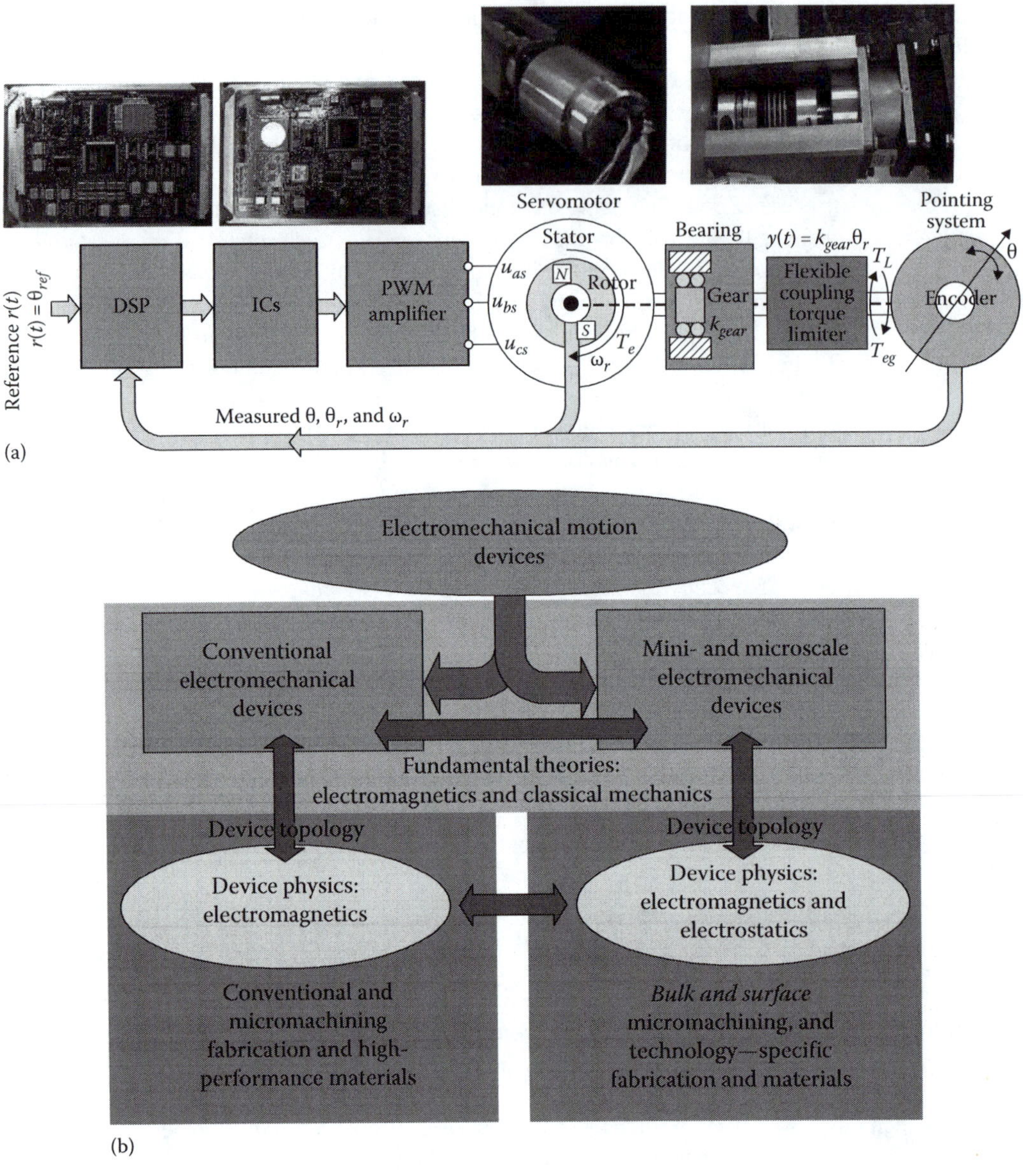

FIGURE 1.5 (a) Functional diagram and images of an electromechanical system; (b) Electromechanical and MEMS motion devices: There are technological and fabrication differences. Conventional electromechanical systems use high-power, high-torque-density electromagnetic transducers. Microsystems may use electrostatic transducers and actuators. MEMS are used as sensors and transducers.

1.3 ELECTROMECHANICAL SYSTEMS SYNTHESIS

The design process starts with a given set of requirements. Using specifications, high-level functional design is performed by applying available technologies, using high-performance devices, examining consistent solutions, and estimating the *achievable* performance. The designer examines alternative engineering solutions, evaluates and advances (if applicable) technologies, and assesses system capabilities. At each level of the design flow and hierarchy, there are specified supporting activities. The bidirectional design flow taxonomy, which ends with the evaluation and commercialization, is illustrated in Figure 1.6a. The performance requirements cannot exceed the *achievable* capabilities defined by the physical, cost, and technological limits. Performance estimates, indexes, metrics, and measures are used. One examines safety, affordability, efficiency, robustness, redundancy, power density, accuracy, and other quantitative characteristics assessing the experimentally substantiated and tested critical components and modules. From the physics-consistent scientific findings, applied research, engineering solutions, and technology developments, one progresses to proven prototype, implementation, commercialization, and deployment.

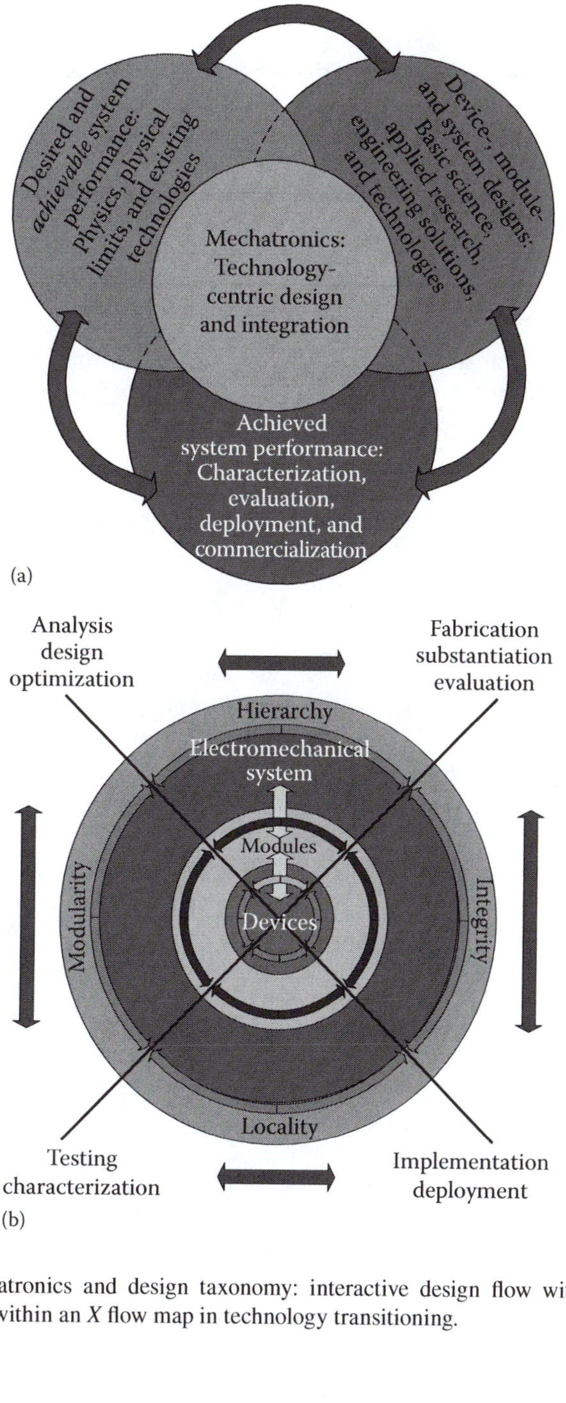

FIGURE 1.6 (a) Mechatronics and design taxonomy: interactive design flow with taxonomy domains; (b) Interactive synthesis within an *X* flow map in technology transitioning.

Mechatronics and systems design: The mechatronic paradigm, which was articulated in Sections 1.1 and 1.2, is fundamentally distinct compared with the *systems design*. Inadequateness and failure of abstract *systems design* fostered alternative inroads. The application of highly-abstract hypothetically-centric methodologies under projected assumptions, countless hypotheses, and assumed conjectures may be impractical and inadequate. One should have a great deal of expertise on high-performance hardware, design concepts, and engineering solutions to carry out design. The application of conjectural hypotheses, without the use of the *first principles* and technologies, commonly results in catastrophic failures. There is only a hope that quantitative synthesis and technology-centric designs can be accomplished by means of abstract descriptions, *intelligent searches*, *knowledge-base* libraries, *virtual designs*, *evolutionary learning*, and other *intelligent* tools. Usually, qualitative conjectural tools may not provide descriptive features and practical prospects. In contrast, experienced practitioners design near-optimal systems.

The design of electromechanical systems starts from the specifications. It progressively moves to a functional design, which is refined through a sequence of steps. The device-to-system and system-to-device hardware and hardware–software integrations must be guaranteed. One ensures a consistency between various tasks. The system performance and capabilities are predefined by engineering solutions and hardware. Analysis, design, and optimization are complementary activities. Analysis starts by deriving and examining system organization, studying modular compliance, and analyzing system functionality. A flow map of the different tasks is depicted in Figure 1.6b. The reported taxonomy is built on a multiple hierarchy, modularity, locality, and integrity. The application of mechatronics reflects:

- Progress in fundamental, applied, and experimental research in response to long-standing challenges and needs in order to advance electromechanical systems;
- Engineering and technological enterprises and entreaties of steady evolutionary demands;
- Evolving technologies and enabling solutions.

Consider the following examples, which portray the need for consistent engineering solutions.

Example 1.1: Applications of Electromechanics and Mechatronics in Drivetrains

Electric drives are widely used in drivetrains to accomplish propulsion of ships, submarines, trains, heavy trucks, etc. There are *parallel*, *power-split*, *parallel-hybrid*, and *series* drivetrain configurations. The *series* drivetrains are used in ships, heavy trucks, etc. Their organization is as follows: engine (combustion, turbine, or nuclear)—synchronous generator—traction motors. Using a synchronous generator (from ~50 to 200 kW for a full-size sedan or heavy truck) and traction motors, one eliminates complex kinematics. There is no mechanical transmission. High-number-of-poles, low-speed traction motors direct-drive rotate the wheels. A simple and robust direct-drive motor-wheel kinematics reduces cost, guarantees robustness, and increases efficiency. Other advantages of the *series* drivetrains are: (1) Fuel efficiency, because the engine is *decoupled* from the load and operates at near-steady-state torque-speed regimes in the minimum fuel consumption envelope; (2) Emission and pollution reduction; (3) Optimal engine operating envelope; (4) Mass reduction; (5) Affordability and cost-effectiveness; (6) Safety; etc. Starters/alternators (~1 kW) are used in all cars. The reported solution may imply the use of an electric machine to be used as a starter/alternator and generator. The *series* configuration has a number of advantages and benefits when compared with the *parallel* drivetrains. ■

Example 1.2: Efficiency Estimates and Feasibility Assessment of Electric Drivetrains and Energy Sources

Approximately 50 to 200 miles range may be ensured for a sedan-class passenger car if expensive, high-energy-density and high specific power batteries are used. The specific energy density of the most advanced rechargeable batteries reaches ~600 kJ/kg, while for conventional lead-acid batteries it reaches ~100 kJ/kg. Current battery technologies impose significant limits. Worldwide, power plants, many of which use the ecology-adverse fossil fuels, produce the energy which is delivered by power lines. The discharge and charge efficiencies of batteries are ~80% or less. The high-current and high-voltage battery charger efficiency is from 40% up to 75% for high-end expensive chargers. The overall efficiency of the charger-battery system is ~50%. The maximum number of charge/discharge battery cycles is ~1000. Then, new batteries must be installed, while highly toxic used batteries must be recycled. Safe, practical, affordable, and cost-effective solutions must be developed.

A fuel cell car uses a fuel cell to power traction electric motors. The oxygen and highly flammable hydrogen, compressed to ~700 bar, are used. A fuel cell emits water and heat. However, in "zero-emission vehicles," the flammable compressed gaseous-state hydrogen is used. The volumetric and mass energy densities of hydrogen is 10.1 MJ/L (at 20 K) and 142 MJ/kg. For a regular 87 gasoline, one has 34.8 MJ/L and 44.4 MJ/kg. The power density for a technology-proven commercialized fuel cell is unknown, but it is less than that of gasoline engines. Considering the alternative fuels, the energy densities are: (1) 21.2 MJ/L and 26.8 MJ/kg for ethyl alcohol (ethanol fuel E100); (2) 34.8 MJ/L and 44.4 MJ/kg for regular 87 gasoline; (3) 38.6 MJ/L and 45.4 MJ/kg for diesel fuel. Recall that 1 MJ is 0.278 kWh. The degradable ethanol-blended gasoline results in metal corrosion, deterioration of fuel system components, clogging, and other adverse effects. Correspondingly, many auto, marine, motorcycle, and other internal combustion engine manufacturers have issued warnings and precautions on the use of the ethanol-blended gasoline of any type. The benefits and effectiveness of ethanol as fuel have been debated. There are arguments that ethanol and ethanol-blended fuels may be economically, environmentally, and energy adverse. Similar to gasoline and diesel, ethanol-based fuels are a source of toxic contamination and pollution. ■

This book covers and delivers the *first principles*, basic fundamentals, and practical solutions. No matter how well the individual components of electromechanical systems perform, the overall performance can be degraded if a consistency is not achieved. While the component-centric *divide-and-solve* approach is applicable in a preliminary design phase, system integration must be accomplished in the context of consistent physics, general and specific objectives, specifications, requirements, hardware limits, existing solutions, etc. These tasks are not within the scope of the *divide-and-solve* concept.

Example 1.3: Energy Harvesting and Storage

Various energy harvesting, energy conversion, transmission, energy storage, and other system modules exist. Medium and large power systems were discussed. The functional diagram of light-duty *modular* power system is depicted in Figure 1.7. The studied systems may include: (1) Electromagnetic radiation (photovoltaic panel or solar cell) or electromechanical energy generation (DC or synchronous generator) module; (2) Power electronic module with controlled PWM regulators, output power stage filters, PWM charger, switches, and other electronic components; (3) Energy storage elements, such as rechargeable batteries or electric double-layer capacitors; (4) Sensors and transducers; (5) Energy management and data acquisition systems. For example, the solar energy can be harvested by solar cells. Figure 1.7 illustrates the flexible amorphous silicon and cadmium telluride solar cells as well as crystalline Si solar cells [10]. The lightweight flexible and thin-film solar cells commonly used are: (1) Crystalline Si solar cells and photovoltaic modules (brittle, 100 μm thickness wafer-based cells with the efficiency η up to ~15% at the standard irradiation and temperature); (2) Amorphous silicon (highly durable, stable, with η varying from ~4% to 8% under standard irradiation); (3) Cadmium telluride with ~15% efficiency (in this case

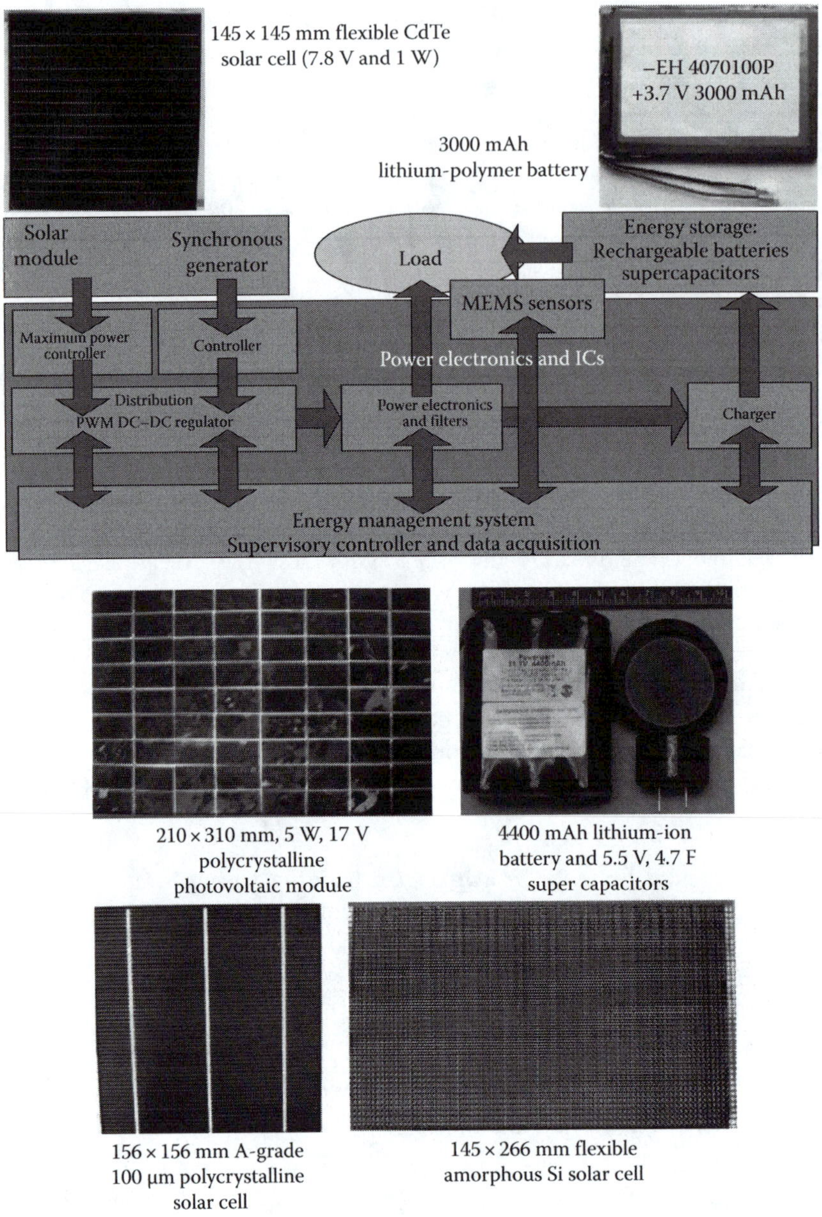

FIGURE 1.7 Modular organization of an energy system with the energy harvesting and energy storage components.

packaging is required because cadmium is highly toxic); (4) Copper–indium–gallium–selenide (η is ~15%, humidity-sensitive, and must be hermetically sealed); (5) Gallium arsenide with ~30% efficiency; (6) III–V multijunction and multilayer solar cells; (7) Polymer solar cells with ~5% efficiency. High-energy-density rechargeable batteries or other energy storage solutions are required. The energy can be stored in: (1) Lithium-ion and lithium-ion-polymer rechargeable batteries with the specific energy of ~0.3 to 1 MJ/kg and energy density from ~200 to 500 Wh/kg, 1000 cycles capability; (2) Lithium-sulfur rechargeable battery (2 V, ~1 MJ/kg, 500 Wh/kg, specific capacity ~700 mAh/g, 1000 cycles); (3) Electric double-layer supercapacitors (the specific energy and power densities of existing commercial supercapacitors and hybrid capacitors are ~5 Wh/kg and ~5000 W/kg, with an efficiency of ~95%). The lifetime for lithium-ion batteries and supercapacitors, shown in Figure 1.7, is ~5 years. High-performance, low-power electronics, microelectronics, sensors, battery chargers, and energy management systems are commercialized. Enabling technologies are emerging, and active research and development has been conducted to find and commercialize new solutions. ■

Example 1.4: Systems Design, System Performance, and Capabilities

One quantitatively evaluates a system design using various metrics, estimates, and measures. This problem may be formulated by considering a tuple $(d, p, c) \equiv (D, P, C)$, where $d \in D$, $p \in P$, and $c \in C$ are the development (technology, engineering solutions, market, financial, cost, effectiveness, quality, yield, sustainability, and other factors), performance (efficiency, robustness, stability, disturbance rejection, etc.), and capabilities (safety, security, reliability, quality, etc.) metrics. One may examine $D \times P \times C$ attempting to solve a mini-max nonlinear optimization problem by minimizing $J = \min_{d\in D, p\in P, c\in C} \max W(d,p,c)$ subject to constraints $V_{\min} \le V(d,p,c) \le V_{\max}$, $J: D \times P \times C \to \mathbb{R}$, with the objective function $W(d, p, c)$. The specified development, performance, and capability levels γ_d, γ_p, and γ_c can be imposed. In practice, this problem is mathematically unsolved. However, by means of technologies developed and engineering solutions, experienced practitioners solve this problem. The mechatronics paradigm contributes to a synergetic quantitative analysis, technology-centric designs, and product developments. ■

MATLAB® has become a standard software tool. It allows the end-user to accelerate design, apply consistent tools, gain productivity, facilitate creative solutions, accelerate prototyping, generate real-time C code, visualize results, perform data acquisition, and ensure data-intensive analysis. The following toolboxes can be applied: Real-Time Workshop, Control System, Nonlinear Control Design, Optimization, Robust Control, Symbolic Math, System Identification, Partial Differential Equations, etc. We will demonstrate MATLAB and Simulink® by solving practical examples, thereby empowering the user's competence and expertise. MATLAB offers a set of capabilities to effectively solve complex problems. These examples will ensure practice, enable educational experience, and train one within the highest degree of comprehensiveness and coverage.

The IEEE code of ethics (www.ieee.org) and ASME principles and canons (www.asme.org), emphasize the importance of advanced scientific, engineering, and technological developments that enhance the quality of life and welfare. Engineers, researchers, scientists, and instructors are accepting various responsibilities and commitments to: (1) Make decisions; (2) Contribute to technological improvements; (3) Serve in the areas of competence. Engineers and professionals who perform research and development in the fields of electromechanics and mechatronics can meet these standards and expectations through adequate education and training. This book is written to enable learning and, thereby, educate and train a new generation of knowledgeable, creative, and exceptionally skillful professionals.

HOMEWORK PROBLEMS

1.1 Provide examples of electromechanical systems and electromechanical motion devices. You may consider clean and renewable energy sources and power generation systems, robotics, drivetrains, etc.

1.2 What is the difference between electromechanical systems and electromechanical motion devices?

1.3 What is the difference between mechatronics and electromechanical system?

1.4 Choose a specific electromechanical system or device (propulsion system, traction electric drive, control surface actuator, loudspeaker, microphone, etc.) and explicitly define problems needed to be addressed, studied, and solved. Formulate and report the specifications and requirements imposed on the system of your interest. Develop the high-level functional diagram with major components.

1.5 Explain why systems and devices must be examined using a consistent taxonomy (*systems engineering*).

REFERENCES

1. D. M. Auslander and C. J. Kempf, *Mechatronics: Mechanical System Interfacing*, Prentice Hall, Upper Saddle River, NJ, 1996.
2. W. Bolton, *Mechatronics: Electronic Control Systems in Mechanical Engineering*, Addison Wesley Logman Publishing, New York, 2013.
3. C. Fraser and J. Milne, *Electro-Mechanical Engineering*, IEEE Press, New York, 1994.
4. M. B. Histand and D. G. Alciatore, *Introduction to Mechatronics and Measurement Systems*, McGraw-Hill, New York, 2011.
5. J. L. Kamm, *Understanding Electro-Mechanical Engineering: An Introduction to Mechatronics*, IEEE Press, New York, 1996.
6. S. E. Lyshevski, *Electromechanical Systems, Electric Machines, and Applied Mechatronics*, CRC Press, Boca Raton, FL, 1999.
7. D. Shetty and R. A. Kolk, *Mechatronics System Design*, CL Engineering, New York, 2010.
8. C. W. de Silva, *Mechatronics: A Foundation Course*, CRC Press, Boca Raton, FL, 2010.
9. C. W. de Silva, *Mechatronics: An Integrated Approach*, CRC Press, Boca Raton, FL, 2004.
10. R. Messenger and J. Ventre, *Photovoltaic Systems Engineering*, CRC Press, Boca Raton, FL, 2010.

2 Mechanics and Electromagnetics

Analysis, Modeling, and Simulation

2.1 INTRODUCTION AND BASELINE PRINCIPLES

Electromechanical systems are designed, analyzed, and optimized using distinct compliant technologies and paradigms. Market needs, practicality, affordability, safety, and other requirements are emphasized. The ultimate objective is to guarantee functionality, best performance, and *achievable* capabilities under technological and cost constraints. The development of electromechanical systems starts from analysis and progresses to optimization, control, testing, and characterization. The design and analysis may imply different meanings. The engineering design implies development of electromechanical devices and systems, hardware, and software. The evolving design taxonomy focuses on:

1. Devising and assessing system organization to ensure overall system functionality to meet the specified requirements;
2. Developing devices and systems focusing on the *first principles* and the laws of physics;
3. Device, components, and modules matching, compliance, and completeness;
4. Device- and system-level data-intensive electromagnetic and mechanical analyses;
5. Development of advanced software and hardware to attain the highest degree of synergy, integration, efficiency, and performance;
6. Coherent assessment and experimental evaluation with redesign;
7. Hardware and software testing, characterization, and evaluation;
8. Marketing, commercialization, and deployment.

To reach the sufficient technology-readiness levels, the emphases are placed on:

- Designing high-performance systems by applying proven technologies and consistent engineering solutions at the device, component, and system levels;
- Analysis and integration of electromechanical motion devices (actuators, motors, sensors, transducers, etc.), high-performance power electronics, low-power microelectronics (signal conditioning, processing, controlling, and other ICs), MEMS, and other components;
- Synthesis and implementation of optimal control and management strategies;
- Testing, evaluation, redesign, substantiation, technology transfer, and implementation.

One studies different components, such as ICs, power electronics, actuators, motion devices, kinematics, etc. The complexity of electromechanical systems has increased drastically due to hardware and software advancements. With stringent performance requirements, the analysis of such systems requires multidisciplinary studies of electromagnetics, mechanics, electronics, microelectronics, MEMS, software, etc. All components are very important. For example, one cannot achieve the specified angular acceleration rate if the actuator does ensure sufficient torque or power electronics cannot ensure the required current. The required accuracy cannot be met if the sensor resolution (the number of pulses per revolution in resolver or optical encoder) is not adequate. The overall system performance and capabilities are largely defined by hardware. The actuation, sensing, and electronic solutions predefine the system performance.

Analysis implies explicit system organization assessment, device physics evaluation, modeling, performance evaluation, capabilities assessment, experimental substantiation, characterization, etc. Complex electromagnetic, mechanical, thermodynamic, vibroacoustic, and other phenomena are examined. Deterministic, stochastic, probabilistic, continuous- and discrete-time, *hybrid*, and other mathematical models and quantitative descriptions are developed by applying physical laws pertaining to the devices used. Mathematical modeling implies deviation of governing and descriptive equations that describe the electromagnetic, mechanical, thermal, and other phenomena and transitions. High-fidelity modeling, such as three-dimensional Maxwell's equations and tensor calculus, may be applied to ensure data-intensive analysis. The complexity may be relaxed and adequacy may be ensured by applying consistent Newtonian dynamics, Kirchhoff and Lagrange equations, Faraday's law and other key concepts without loss of generality. Any descriptive physics-consistent models are the idealization of physical systems, phenomena, effects, and processes. Mathematical models are never absolutely accurate. However, adequate models can be found to solve engineering problems.

For moderate-complexity systems, the experienced designers may accomplish a near-optimal design without overcomplicated analyses by using the experience, expertise, and knowledge gained. Some performance features, such as efficiency and power density, can be estimated avoiding high-fidelity modeling. Over-simplification and over-complexity may result in an overall failure. The overall system performance and capabilities (safety, affordability, efficiency, stability, robustness, error, accuracy, etc.) are predetermined by: (1) hardware and software used; (2) technologies applied; (3) physical and technological limits and constraints. Fundamental and analytic results cannot replace testing, characterization, and experimental validation. The *generic* approaches, *model-free* concepts, *linguistic* models, *descriptive* techniques, and other *abstract* conjectural approaches may be impractical. Usually, these are not based on the underlying device physics, but focus on the narrative/descriptive features with limited applicability and hypothetic practicality. We apply the laws of physics to examine systems and devices, use technology-centric proven solutions, derive governing equations, analyze efficiency, define practical control strategies, assess performance, evaluate capabilities, etc.

2.2 ENERGY CONVERSION AND FORCE PRODUCTION IN ELECTROMECHANICAL DEVICES

This book focuses on high-performance electromagnetic electromechanical motion devices. The electrostatic devices are also covered. Depending on applications, the designer may use electromagnetic actuators that are robust and affordable and guarantee highest power and force/torque densities and high efficiency. Alternatively, electrostatic, piezoelectric, thermal, and hydraulic transducers can be used. We examine motion devices applying classical electromagnetic and mechanics. Energy conversion takes place in electromechanical motion devices that convert electrical energy to mechanical energy and vice versa [1–8]. The device physics define the phenomena exhibited and utilized. The key principle of energy conversion is as follows: *For a lossless electromechanical motion device (in the conservative system, no energy is lost through friction, heat, or other irreversible energy conversion), the sum of the instantaneous kinetic and potential energies of the system remains constant.* The energy conversion is represented in Figure 2.1.

The energy conversion is represented as

$$\underbrace{\mathbf{E}_E}_{\text{Electrical energy input}} - \underbrace{\mathbf{L}_E}_{\text{Ohmic losses}} - \underbrace{\mathbf{L}_M}_{\text{Magnetic losses}} = \underbrace{\mathbf{E}_M}_{\text{Mechanical energy}} + \underbrace{\mathbf{L}_M}_{\text{Friction losses}} + \underbrace{\mathbf{E}_S}_{\text{Stored energy}}.$$

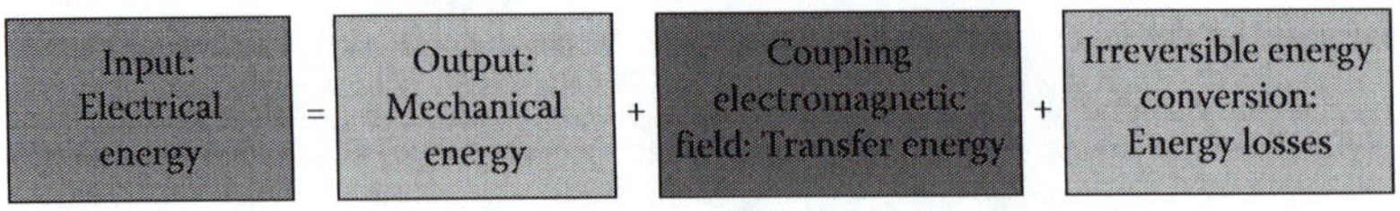

FIGURE 2.1 Energy conversion and transfer in an electromechanical device.

For conservative (lossless) energy conversion

$$\underbrace{\Delta \mathbf{W}_E}_{\text{Change in electrical energy input}} = \underbrace{\Delta \mathbf{W}_M}_{\text{Change in mechanical energy}} + \underbrace{\Delta \mathbf{W}_m}_{\text{Change in electromagnetic energy}}.$$

The electrical energy, mechanical energy, and energy losses are examined. The electromagnetic motion devices ensure superior performance. High power and force densities are achieved by using permanent magnets and electromagnets, which establish a stationary magnetic field. The total energy stored in the magnetic field is expressed using the magnetic field density B and intensity H, $W_m = \frac{1}{2}\int_V \vec{B} \cdot \vec{H}\, dV$. Using the dimensionless magnetic susceptibility χ_m and relative permeability μ_r, one has

$$B = \mu H = \mu_0 \mu_r H = \mu_0(1 + \chi_m)H, \quad \mu_0 = 4\pi \times 10^{-7}\ \text{H/m}.$$

The magnetization is

$$M = \chi_m H, \quad \mu_r = (1 + \chi_m).$$

The materials are classified as follows:

- Nonmagnetic if $\chi_m = 0$, which implies $\mu_r = 1$;
- Diamagnetic (copper, gold, silver, etc.) if χ_m is -1×10^{-5};
- Paramagnetic (aluminum, palladium, etc.) if χ_m is $\sim 1 \times 10^{-4}$;
- Ferrimagnetic (yttrium iron garnet, ferrites composed of iron oxides, etc.) if $|\chi_m| > 1$;
- Ferromagnetic if $|\chi_m| \gg 1$. Ferromagnetic materials classified as *hard* (rare-earth elements, copper-nickel, and other alloys) and *soft* materials.

The relative permeability μ_r of some bulk materials is shown in Table 2.1.

TABLE 2.1
Relative Permeability of Diamagnetic, Paramagnetic, Ferrimagnetic, and Ferromagnetic Materials

Media	Materials	Relative Permeability, μ_r
Diamagnetic	Silver	0.9999736
	Copper	0.9999905
Paramagnetic	Aluminum	1.000021
	Tungsten	1.00008
	Platinum	1.0003
	Manganese	1.001
Ferrimagnetic	Nickel–zinc ferrite	600–1,000
	Manganese–zinc ferrite	700–1,500
Ferromagnetic	Purified iron ($Fe_{99.96\%}$)	$\mu_{r\,\max}$ 280,000
	Electric steel ($Fe_{99.6\%}$)	$\mu_{r\,\max}$ 5,000
	Permalloy ($Ni_{78.5\%}Fe_{21.5\%}$)	$\mu_{r\,\max}$ 70,000
	Superpermalloy ($Ni_{79\%}Fe_{15\%}Mo_{5\%}Mn_{0.5\%}$)	$\mu_{r\,\max}$ 1,000,000

Note and Declaimer: The permeability, *B–H* characteristics, and other descriptive material quantities significantly vary and depend on electromagnetic loading and are affected by composition, fabrication, geometry, dimensionality, temperature, defects, degradation, aging, demagnetization, etc. For example, μ_r of superpermalloy may be much lower than 1,000,000, and the energy product $(BH)_{\max}$ of the $Nd_2Fe_{14}B$ may be much less than 400 kJ/m^3. The data reported in various reputable sources are quite different.

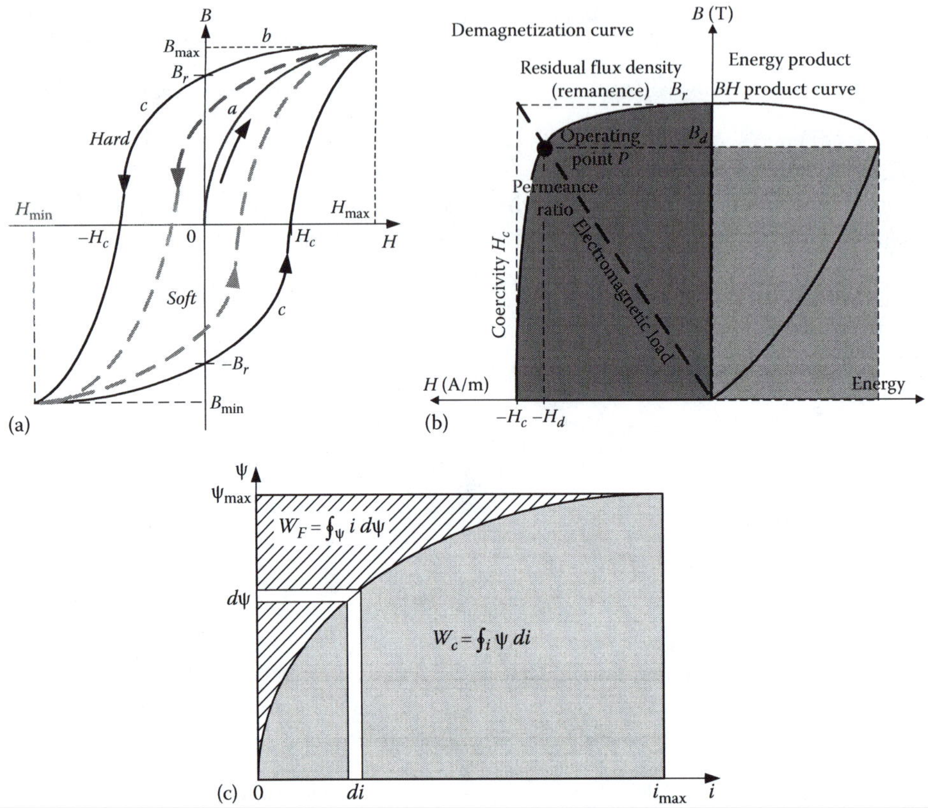

FIGURE 2.2 (a) *B–H* curves for *hard* and *soft* ferromagnetic materials; (b) *B–H* demagnetization and energy product curves; (c) Magnetization curve and energies.

The magnetization behavior of the ferromagnetic materials is described by the magnetization curve, where H is the externally applied magnetic field and B is the total magnetic flux density. Typical $B - H$ curves for *hard* and *soft* ferromagnetic materials are shown in Figure 2.2a.

Assume that initially $B_0 = 0$ and $H_0 = 0$. If H increases from $H_0 = 0$ to H_{max}, B increases from $B_0 = 0$ until the maximum value B_{max} is reached. Then, if H decreases to H_{min}, the flux density B decreases to B_{min} through the remanent value B_r (residual magnetic flux density) along the different curve, as illustrated in Figure 2.2a. For variations of H, $H \in \left[H_{min} \quad H_{max} \right]$, B changes within the *hysteresis loop*, and $B \in \left[B_{min} \quad B_{max} \right]$. Figure 2.2a reports typical curves representing the dependence of magnetic induction B on magnetic field H for ferromagnetic materials. When H is first applied, B follows curve a as the favorably oriented magnetic domains grow. This curve reaches the saturation. When H is then reduced, B follows curve b, but retains a finite value (the remanence B_r) at $H = 0$. In order to demagnetize the material, a negative field $-H_c$ must be applied. This H_c is called the coercive field or coercivity. As H is further decreased and then increased to complete the cycle (curve c), a hysteresis loop is formed.

The area within this loop is a measure of the energy loss per cycle for a unit volume of the material. The *B–H* curve yields the energy analysis. In the per-unit volume, the applied field energy is $W_F = \oint_B HdB$, while the stored energy is $W_c = \oint_H BdH$. The equations for field and stored energies represent the areas enclosed by the corresponding curve. For volume V, the expressions for the field and stored energies are

$$W_F = V\oint_B H dB \quad \text{and} \quad W_c = V\oint_H B dH.$$

In ferromagnetic materials, time-varying magnetic flux produces core losses, which consist of: (1) Hysteresis losses due to the hysteresis loop of the B–H curve; (2) Eddy current losses, which are proportional to the current frequency, lamination thickness, etc. The area of the hysteresis loop is related to the hysteresis losses. *Soft* ferromagnetic materials have narrow hysteresis loop and they are easily magnetized and demagnetized. This results in lower hysteresis losses and *hard* ferromagnetic materials.

The *soft* and *hard* ferromagnetic materials are characterized by the B–H curves, maximum energy product $(BH)_{max}$ [kJ/m^3], Curi temperature, and mechanical properties. The term *soft* implies high saturation magnetization, low coercivity (narrow B–H curve), and low magnetostriction. The *soft* materials are used in magnetic recording heads. The *hard* magnets have wide B–H curves (high coercivity) ensuring high-energy storage capacity. These magnets (ceramic ferrite, neodymium iron boron NdFeB, samarium cobalt SmCo, Alnico, and others) are widely used in electric machines and electromagnetic actuators to attain high force, torque, and power densities. The energy density is given as the area enclosed by the B–H curve. The magnetic volume energy density is $w_m = \frac{1}{2}B \cdot H$ [J/m^3]. Most *hard* magnets are fabricated using the metallurgical processes, e.g., sintering (creating a solid but porous material from a powder), pressure bonding, injection molding, casting, and extruding.

Permanent magnets store energy, contriving to energy exchange, energy conversion, *emf* and force generation, etc. The stationary magnetic field is produced without external energy sources. Magnets operate on the demagnetization curve of the hysteresis loop, e.g., on the second quadrant of the B–H curve, as shown in Figure 2.2b, which illustrates the energy product curve and the demagnetization curve. Using the flux linkages ψ, one may analyze the ψ–i curve. The energy stored in the magnetic field is $W_F = \oint_\psi i d\psi$, and the coenergy is $W_c = \oint_i \psi\, di$ as shown in Figure 2.2c. In actuators and electric machines, almost all the energy is stored in the air gap. The air is a conservative medium, implying that the coupling field is lossless. Figure 2.2c illustrates the nonlinear magnetizing characteristic (normal magnetization curve). The total energy is

$$W_F + W_c = \oint_\psi i\, d\psi + \oint_i \psi\, di = \psi i.$$

The flux linkage ψ is a function of the current i and position x for translational motion, or angular displacement θ for rotational motion. That is, $\psi = f(i, x)$ or $\psi = f(i, \theta)$. Assuming that the coupling field is lossless, the differential change in the mechanical energy is found using the differential displacement $d\vec{l}$,

$$dW_{mec} = \vec{F}_m \cdot d\vec{l},$$

where W_{mec} is related to the differential change of the coenergy. For the displacement dx at constant current, $dW_{mec} = dW_c$. Hence, the electromagnetic force is $F_e(i, \mathbf{r}) = \nabla W_c(i, \mathbf{r})$.

For a one-dimensional case, the electromagnetic force and torque, which are the vectors, are

$$F_e(i,x) = \frac{\partial W_c(i,x)}{\partial x}, \quad T_e(i,\theta) = \frac{\partial W_c(i,\theta)}{\partial \theta}.$$

Using advanced electric machine topologies, enabling magnets, and optimal design, one can:

1. Guarantee safety, affordability, effectiveness, and robustness;
2. Ensure hazard-free manufacturability, automated assembly, and sustainability;
3. Assure optimal energy production and conversion with maximum power, force, and torque densities;
4. Maximize efficiency and minimize losses;
5. Enable scalability, integrity, and modularity.

Figure 2.3 documents the Alnico rotor assembly and permanent-magnet synchronous machine (motor and generator) with the SmCo magnets on the rotor. This book centers on electromagnetic actuators and high-performance, permanent-magnet electric machines, which are widely commercialized in automotive, aerospace, naval, and other applications. For example, autonomous aerial vehicles, auxiliary power units, milliwatt to 10 MW power generation systems, propulsion systems, and robots.

With a focus on advanced technologies and high-performance electromechanical motion devices, we examine advanced magnets. The second quadrant demagnetization B–H curves and images of Alnico, rare-earth magnets (samarium cobalt SmCo and neodymium iron boron NdFeB), and ceramic ferrite magnets are shown in Figures 2.4.

Alnico magnets: Affordable Alnico magnets (iron alloy with 8%–12% of Al, 15%–26% of Ni, 5%–24% of Co, up to 6% of Cu, up to 1% of Ti, and Fe) guarantee high coercivity, high energy product $(BH)_{max}$, robustness, corrosion resistance, superior temperature stability up to ~525°C, and high Curie temperature, ~850°C. Their cobalt content varies from zero in isotropic Alnico 3 to 24% and 35% in Alnico 5 and 8. In commonly used anisotropic Alnico 5, 5/7, 6, 8, and 9, the orientation is achieved by applying a directed magnetic field to ensure the preferred direction of magnetization, and heat treatment by cooling from ~1090°C at a controlled rate. Alnico 5 and 5/7 are widely used due to their high energy density, high coercive force, excellent durability, low reversible temperature coefficient, and high operating temperature, as shown in Table 2.2.

Cast and sintered Alnico magnets ensure robust positioning, mounting, bonding, and assembly with adequate protection and robustness. Alnico magnets can be premagnetized, or magnetized after assembling. The magnetic saturation requires the application of a magnetizing force about five times greater than the coercive force. For Alnico 5, an impulse magnetizing force (capacitor discharge or direct current magnetizer) should be higher than 250 kA/m.

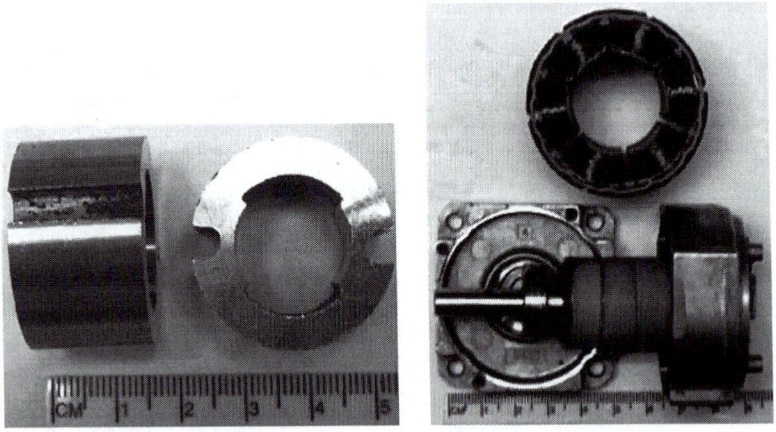

FIGURE 2.3 Images of Alnico rotor and synchronous generator with SmCo magnets on the rotor.

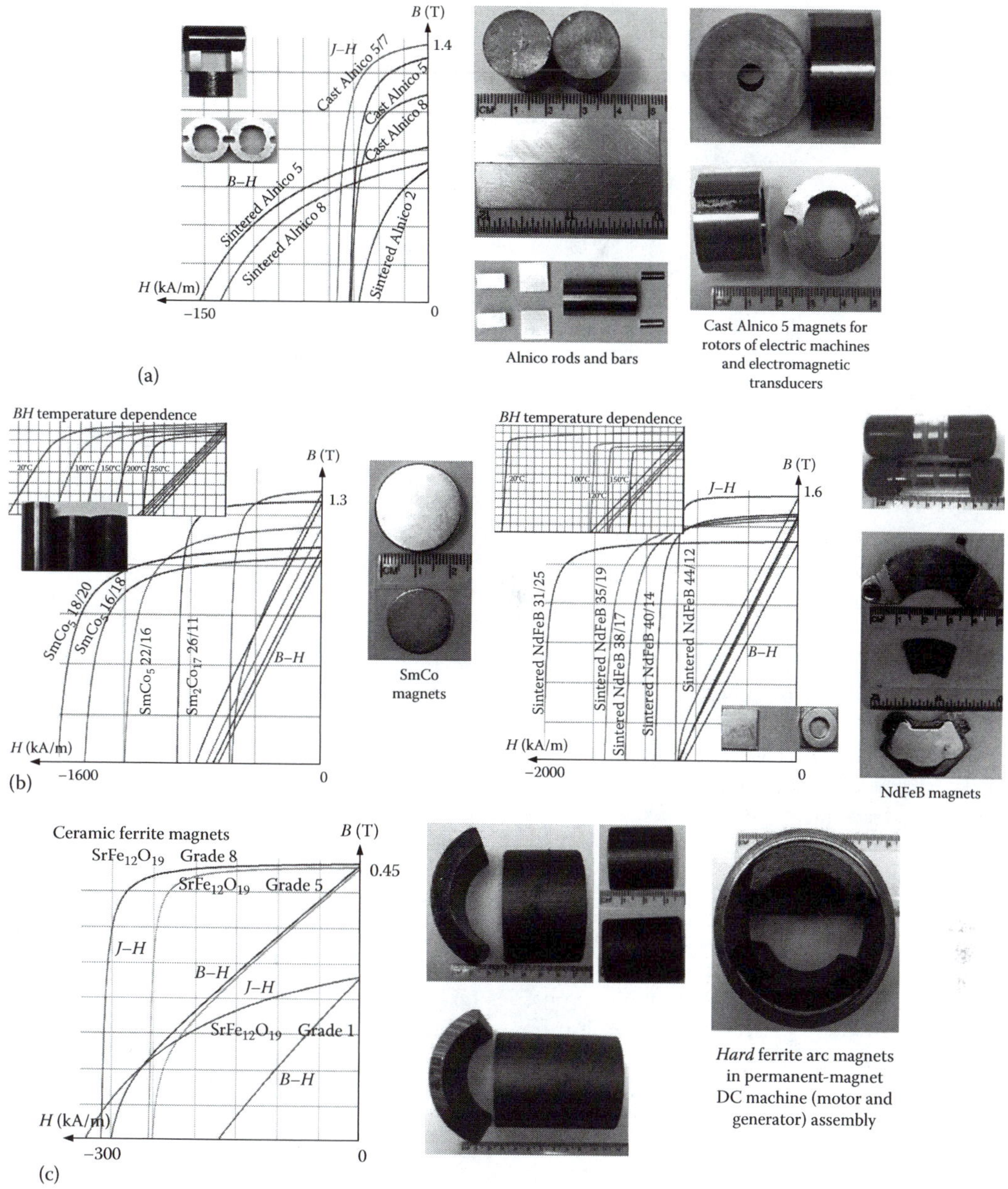

FIGURE 2.4 The intrinsic *J–H* demagnetization curves, extrinsic *B–H* demagnetization curves and images of: (a) Cast and sintered Alnico magnets; (b) SmCo and NdFeB magnets at room temperature (300 K). The demagnetization curves of SmCo and NdFeB magnets vary if temperature changes; (c) Sintered ceramic hard $SrFe_{12}O_{19}$ ferrite magnets at room temperature (300 K).

Note**:** The National Electrical Manufacturers Association (NEMA) specifies the insulation classes and insulation materials. The A (enamel wire polyamide resins), B (inorganic materials hardened with high-temperature binders and adhesives), F, and H (inorganic material with silicone resin or adhesives) insulation classes withstand a temperature up to 105°, 130°, 155°, and 180°C, respectively.

TABLE 2.2
Properties of Magnetic Materials at Room Temperature (300 K)

Permanent Magnets	J_s (T)	H_A (kA/m)	H_{Ic} (kA/m)	$(BH)_{max}$ (kJ/m^3)	T_C (K)
Alnico 5: 8Al, 14Ni, 24Co, 3Cu	1.3	400	130	45	870
$SmCo_5$	1.1	23,000	1600	280	1020
Sm_2Co_{17}	1.3	5,200	960	360	1195
$Sm_2Fe_{17}N_3$	1.5	11,200	2240	380	740
$Nd_2Fe_{14}B$	1.6	5,680	1200	420	580
$Nd_{12.6}(Fe,Co,Zr)_{81.4}$-$B_6$/α-Fe	1.63	5,100	505	180	668
$SrFe_{12}O_{19}$ Grade 8	0.45	1,500	280	35	720

Note: Here, J_s is the saturation magnetic polarization; H_A is the anisotropy field; H_{Ic} is the intrinsic coercivity.

Rare-earth magnets: The rare-earth magnets are used in high-energy-density electromagnetic transducers (synchronous machines, hard disk drives, loudspeakers, MEMS, etc.). Samarium cobalt (SmCo) and neodymium iron boron (NdFeB) magnets are alloys of the Lanthanide group of elements. Rare-earth magnets are available in a number of different grades with a wide range of properties and application requirements. The anisotropic SmCo and NdFeB magnets guarantee the highest energy density and maximum energy product $(BH)_{max}$. However, the rare-earth magnets may not ensure expected thermal stability, must be coated to prevent corrosion, are brittle, are sensitive to mechanical impacts, and are prone to chipping and cracking.

Ceramic magnets: The ceramic *hard* ferrite magnets are used in AC and DC electric machines, loudspeakers, etc. The *hard* ferrite magnets have a high coercivity and remanence, sufficient power density, and adequate energy product. The common affordable and corrosion-resistant ceramic ferrites are the sintered strontium ($SrFe_{12}O_{19}$) and barium ($BaFe_{12}O_{19}$) *hard* ferrites.

The remanence, coercive field, energy density, and other descriptive quantities depend on the electromagnetic load and temperature. The demagnetizing field H_d is induced. The operating point $P(H_d, B_d)$ is located on the demagnetization curve at the intersection of the demagnetization curve and the electromagnetic load line, see Figure 2.2b. The slope of the electromagnetic load line depends on the magnets' magnetic properties and geometry. The relation between the field in the gap H_g is $H_g^2V_g = B_mH_mV_m$, where V is the volume and subscripts g and m correspond to the air gap and magnet. In an air gap with V_g, the produced magnetic field intensity H_g depends on the product B_mH_m. The maximum energy product $(BH)_{max}$ may be ensured. For a given air gap, for the magnetic flux, according to Ampere's law $\oint_l H \cdot dl = 0$. Hence, $H_ml_m = H_gl_g$, where l_m is the length of the magnet; l_g is the length of the air gap parallel to the flux lines. Using the cross-sectional areas of the air gap A_g and magnet A_m, for uniform stationary field, $\Phi = B_gA_g = B_mA_m$. Hence, the magnet cross-sectional area is $A_m = B_gA_g/B_d$, where B_g is the flux density in the air gap. In a lossless system, an informative energy equation for the air gap is

$$V_gB_gH_g = A_gl_g\frac{A_ml_mB_dH_d}{A_gl_g} = A_ml_mB_dH_d = \psi i.$$

The flux density at position **r** can be derived. In one-dimensional case for cylindrical magnets (length l_c and radius r_m) which have *near-linear* demagnetization curves, the flux density at a distance x is

$$B = \frac{B_r}{2}\left(\frac{l_c + x}{\sqrt{r_m^2 + (l_c + x)^2}} - \frac{x}{\sqrt{r_m^2 + x^2}}\right).$$

From the Biot–Savart law, the magnetic flux density on the axis for a uniformly magnetized circular magnet with constant magnetization M is

$$\vec{B} = \frac{1}{2}\mu_0 M\left(\frac{z}{\sqrt{z^2 + r_m^2}} - \frac{z - l_c}{\sqrt{(z - l_c)^2 + r_m^2}}\right)\vec{a}_z.$$

Table 2.3 reports the initial permeability μ_i, maximum relative permeability $\mu_{r\,\max}$, coercivity H_c, saturation polarization J_s, hysteresis loss per cycle W_h, and Curie temperature T_C for high-permeability bulk metals and alloys. Table 2.4 reports the remanence B_r, flux coercivity H_{Fc}, intrinsic coercivity H_{Ic}, maximum energy product $(BH)_{\max}$, Curie temperature, and the maximum operating temperature $T_{\max}$.

The provided data and constants are strongly affected by the dimensions, temperature, and fabrication. These parameters significantly vary for specific applications and devices. Therefore, there are discrepancies in the magnetic properties reported in different sources.

TABLE 2.3
Magnetic Properties of High-Permeability Soft Metals and Alloys at Room Temperature (300 K)

Material	Composition (%)	μ_i	$\mu_{r\,\max}$	H_c (A/m)	J_s (T)	W_h (J/m³)	T_C (K)
Iron	$Fe_{99\%}$	200	6,000	70	2.16	500	1043
Iron	$Fe_{99.9\%}$	25,000	350,000	0.8	2.16	60	1043
Silicon–iron	$Fe_{96\%}Si_{4\%}$	500	7,000	40	1.95	50–150	1008
Silicon–iron (110) [001]	$Fe_{97\%}Si_{3\%}$	9,000	40,000	12	2.01	35–140	1015
Silicon–iron {100} <100>	$Fe_{97\%}Si_{3\%}$		100,000	6	2.01		1015
Steel	$Fe_{99.4\%}C_{0.1\%}Si_{0.1\%}Mn_{0.4\%}$	800	1,100	200			
Hypernik	$Fe_{50\%}Ni_{50\%}$	4,000	7,0000	4	1.60	22	753
Deltamax {100} <100>	$Fe_{50\%}Ni_{50\%}$	500	20,0000	16	1.55		773
Isoperm {100} <100>	$Fe_{50\%}Ni_{50\%}$	90	100	480	1.60		
78 Permalloy	$Ni_{78\%}Fe_{22\%}$	4,000	100,000	4	1.05	50	651
Supermalloy	$Ni_{79\%}Fe_{16\%}Mo_{5\%}$	100,000	1,000,000	0.15	0.79	2	673
Mumetal	$Ni_{77\%}Fe_{16\%}Cu_{5\%}Cr_{2\%}$	20,000	100,000	4	0.75	20	673
Hyperco	$Fe_{64\%}Co_{35\%}Cr_{0.5\%}$	650	10,000	80	2.42	300	1243
Permendur	$Fe_{50\%}Co_{50\%}$	500	6,000	160	2.46	1200	1253
2V Permendur	$Fe_{49\%}Co_{49\%}V_{2\%}$	800	4,000	160	2.45	600	1253
Supermendur	$Fe_{49\%}Co_{49\%}V_{2\%}$		60,000	16	2.40	1150	1253
25 Perminvar	$Ni_{45\%}Fe_{30\%}Co_{25\%}$	400	2,000	100	1.55		
7 Perminvar	$Ni_{70\%}Fe_{23\%}Co_{7\%}$	850	4,000	50	1.25		
Perminvar (magnetically annealed)	$Ni_{43\%}Fe_{34\%}Co_{23\%}$		400,000	2.4	1.50		
Alfenol (Alperm)	$Fe_{84\%}Al_{16\%}$	3,000	55,000	3.2	0.8		723
Alfer	$Fe_{87\%}Al_{13\%}$	700	3,700	53	1.20		673
Aluminum–Iron	$Fe_{96.5\%}Al_{3.5\%}$	500	19,000	24	1.90		
Sendust	$Fe_{85\%}Si_{10\%}Al_{5\%}$	36,000	120,000	1.6	0.89		753

Sources: Data from Lide, D.R., *Handbook of Chemistry and Physics*, 83rd edn., CRC Press, Boca Raton, FL, 2002; Dorf, R.C., *Handbook of Engineering Tables*, CRC Press, Boca Raton, FL, 2003; Lyshevski, S.E., *Nano- and Micro-Electromechanical Systems: Fundamentals of Nano- and Microengineering*, CRC Press, Boca Raton, FL, 2004.

TABLE 2.4
Magnetic Properties of High-Permeability Hard Metals and Alloys at Room Temperature (300 K)

Magnet Composite and Composition	B_r (T)	H_{Fc} (A/m)	H_{Ic} (A/m)	$(BH)_{max}$ (kJ/m³)	T_C (°C)	T_{max} (°C)
Alnico 1: 12Al,20Ni,5Co	0.71		38	11	760	
Alnico 2: 10Al,17Ni,12.5Co,3–6Cu	0.75		45	13.5	800	
Alnico 3: 12–14Al,24–30Ni, 0–3Cu	0.6		45	10.8	750	
Alnico 4: 11–13Al,21–28Ni,3–5Co,2–4Cu	0.55		50	11.5	760	
Alnico 5: 8Al,14Ni,24Co,3Cu	1.25	53	54	45	870	520
Alnico 5/7: 8Al,14Ni,24Co,3Cu	1.28		59	57	860	
Alnico 6: 8Al,16Ni,24Co,3Cu,2Ti	1.05		70	41	860	
Alnico 8: 7Al,15Ni,35Co,4Cu,5Ti	0.83		140	43	860	
Alnico 9: 7Al,5Ni,35Co,4Cu,5Ti	1.10		120	73	850	520
Alnico 12: 8Al,3.5Ni,24.5Co,2Nb	1.20		64	76.8	850	
Ferroxdur: $BaFe_{12}O_{19}$	0.4	1.6	192	29	450	400
$SrFe_{12}O_{19}$	0.4	2.95	3.3	30	450	400
$LaCo_5$	0.91			164	567	
$CeCo_5$	0.77			117	380	
$PrCo_5$	1.20			286	620	
$NdCo_5$	1.22			295	637	
$SmCo_5$	1.00	7.9	696	196	700	250
$Sm(Co_{0.76}Fe_{0.10}Cu_{0.14})_{6.8}$	1.04	4.8	5	212	800	300
$Sm(Co_{0.65}Fe_{0.28}Cu_{0.05}Zr_{0.02})_{7.7}$	1.2	10	16	264	800	300
$Nd_2Fe_{14}B$ (sintered)	1.22	8.4	1120	280	300	100
Vicalloy II: Fe,52Co,14V	1.0	42		28	700	500
Fe,24Cr,15Co,3Mo (anisotropic)	1.54	67		76	630	500
Chromindur II: Fe,28Cr,10.5Co	0.98	32		16	630	500
Fe,23Cr,15Co,3V,2Ti	1.35	4		44	630	500
Fe,36Co	1.04		18	8		
Co (rare-earth)	0.87		638	144		
Cunife: Cu,20Ni,20Fe	0.55	4		12	410	350
Cunico: Cu,21Ni,29Fe	0.34	0.5		8		
Pt,23Co	0.64	4		76	480	350
Mn,29.5Al,0.5C (anisotropic)	0.61	2.16	2.4	56	300	120

Sources: Data from Lide, D.R., *Handbook of Chemistry and Physics*, 83rd edn., CRC Press, Boca Raton, FL, 2002; Dorf, R.C., *Handbook of Engineering Tables*, CRC Press, Boca Raton, FL, 2003; Lyshevski, S.E., *Nano- and Micro-Electromechanical Systems: Fundamentals of Nano- and Microengineering*, CRC Press, Boca Raton, FL, 2004.

2.3 FUNDAMENTALS OF ELECTROMAGNETICS

The fundamental laws of electromagnetics are applied to study various field quantities. The Ohm law for circuits is $V = RI$. For a media, the Ohm law relates the current density $\vec{J}$ and electric field intensity $\vec{E}$. Using the rank-2 tensors (3 × 3 matrices) of conductivity and resistivity σ and ρ,

$$\vec{J} = \sigma\vec{E}, \quad \text{and} \quad \vec{E} = \rho\vec{J}.$$

The resistance r of the conductor is related to the resistivity ρ and conductivity σ,

$$r = \frac{\rho l}{A} = \frac{l}{\sigma A},$$

where l and A are the length and cross-sectional area.

TABLE 2.5
Equations of Electrostatic and Magnetostatic Fields

	Electrostatic Equations	Magnetostatic Equations
Governing equations	$\nabla\times\vec{E}(x,y,z)=0$	$\nabla\times\vec{H}(x,y,z)=\vec{J}(x,y,z)$
	$\nabla\cdot\vec{D}(x,y,z)=\rho_v(x,y,z)$	$\nabla\cdot\vec{B}(x,y,z)=0$
Constitutive equations	$\vec{D}=\varepsilon\vec{E}$	$\vec{B}=\mu\vec{H}$

For copper and aluminum, $\sigma = 5.96\times10^7$ A/V-m and $\sigma = 3.5\times10^7$ A/V-m. The resistivity depends on temperature T, and

$$\rho(T)=\rho_0\left[1+\alpha_{\rho 1}(T-T_0)+\alpha_{\rho 2}(T-T_0)^2+\cdots\right],$$

where $\alpha_{\rho i}$ are the coefficients.

For copper, for temperatures up to 160°C, $\rho(T)=1.7\times10^{-8}\left[1+0.0039\left(T_0-20\right)\right]$, T_0 = 20°C.

Electromagnetic theory and classical mechanics form the basis to examine the device physics, study the inherent phenomena exhibited and utilized, and derive the governing equations of motion. The electrostatic and magnetostatic equations in linear isotropic media are found using the vectors of the electric field intensity $\vec{E}$, electric flux density $\vec{D}$, magnetic field intensity $\vec{H}$, and magnetic flux density $\vec{B}$ The governing and constitutive equations in the Cartesian coordinate system are reported in Table 2.5.

In the steady-state (time-invariant) fields, electric and magnetic field vectors form separate and independent pairs. That is, $\vec{E}$ and $\vec{D}$ are not related to $\vec{H}$ and $\vec{B}$, and vice versa. In electromechanical motion devices, the electric and magnetic fields are time varying. The changes of magnetic field affect the electric field, and vice versa. The four Maxwell's equations in the differential form for time-varying fields are:

1. Faraday's law $\nabla\times\vec{E}(x,y,z,t)=-\dfrac{\partial\vec{B}(x,y,z,t)}{\partial t}$;
2. Ampere's law $\nabla\times\vec{H}(x,y,z,t)=\vec{J}(x,y,z,t)+\dfrac{\partial\vec{D}(x,y,z,t)}{\partial t}, \vec{J}=\sigma\vec{E}$;
3. Gauss's law for electric field $\nabla\cdot\vec{D}(x,y,z,t)=\rho_v(x,y,z,t)$;
4. Gauss's law for magnetic field $\nabla\cdot\vec{B}(x,y,z,t)=0$.

Here, ρ_v is the volume charge density, and the total electric flux through a closed surface is $\Phi=\oint_s\vec{D}\cdot d\vec{s}=\oint_v\rho_v dv=Q$ (Gauss's law), while the magnetic flux through a closed surface is $\Phi=\oint_s\vec{B}\cdot d\vec{s}$; $d\vec{s}$ is the vector surface area, $d\vec{s}=ds\vec{a}_n$; $\vec{a}_n$ is the unit vector which is normal to the surface; Q_s is the total charge enclosed by the surface.

The constitutive equations are given using the permittivity ε, permeability μ, and conductivity σ tensors. One has $\vec{D}=\varepsilon\vec{E}$ or $\vec{D}=\varepsilon\vec{E}+\vec{P}$, $\vec{B}=\mu\vec{H}$ or $\vec{B}=\mu(\vec{H}+\vec{M})$, $\vec{J}=\sigma\vec{E}$ or $\vec{J}=\rho_v\vec{v}$.

Examining actuators and electromechanical motion devices, one derives the expressions for force F, torque T, *electromotive* and *magnetomotive* forces, etc. The Lorenz force, which relates the electromagnetic and mechanical variables, is

$$\vec{F}=\rho_v(\vec{E}+\vec{v}\times\vec{B})=\rho_v\vec{E}+\vec{J}\times\vec{B}.$$

The total potential energy stored in the electrostatic field is found using the potential difference V, $W_e = \frac{1}{2}\int_v \rho_v V dv$. Using the volume charge density $\rho_v = \vec{\nabla}\cdot\vec{D}$ and $\vec{E} = -\vec{\nabla}V$, one obtains the energy stored in the electrostatic field $W_e = \frac{1}{2}\int_v \vec{D}\cdot\vec{E} dv$. The electrostatic volume energy density is $\frac{1}{2}\vec{D}\cdot\vec{E}$. For a linear isotropic medium

$$W_e = \frac{1}{2}\int_v \varepsilon\left|\vec{E}\right|^2 dv = \frac{1}{2}\int_v \varepsilon^{-1}\left|\vec{D}\right|^2 dv.$$

From

$$W_e = \frac{1}{2}\int_v \rho_v V dv,$$

the potential energy that is stored in the electric field between two surfaces (for example, in capacitor) is $W_e = \frac{1}{2}QV = \frac{1}{2}CV^2$.

Using the concept of virtual work, for the lossless conservative system, the differential change of the electrostatic energy dW_e is equal to the differential change of mechanical energy dW_{mec}. That is, $dW_e = dW_{mec}$. For translational motion, $dW_{mec} = \vec{F}_e\cdot d\vec{l}$, where $d\vec{l}$ is the differential displacement.

From $dW_e = \vec{\nabla}W_e\cdot d\vec{l}$, one concludes that the force is the gradient of the stored electrostatic energy,

$$\vec{F}_e = \vec{\nabla}W_e \quad \text{or} \quad \vec{F}_e = -\vec{\nabla}W_e.$$

In the Cartesian coordinates,

$$F_{ex} = \frac{\partial W_e}{\partial x}, \quad F_{ey} = \frac{\partial W_e}{\partial y} \quad \text{and} \quad F_{ez} = \frac{\partial W_e}{\partial z}.$$

The stored energy in the magnetostatic field is

$$W_m = \frac{1}{2}\int_v \vec{B}\cdot\vec{H}\, dv$$

or

$$W_m = \frac{1}{2}\int_v \mu\left|\vec{H}\right|^2 dv = \frac{1}{2}\int_v \mu^{-1}\left|\vec{B}\right|^2 dv.$$

The energy of a magnetic moment $\vec{m}$ in an externally produced $\vec{B}$ is characterized by the potential energy $\Pi = -\vec{m}\cdot\vec{B}$. Using the magnetization, $\Pi = -\frac{1}{2}\int \vec{M}\cdot\vec{B}\, dv$. The work done by a conservative force is $W = -\Delta\Pi$, where $\Delta\Pi$ is the change in the potential energy associated with the force. The negative sign indicates that work performed against a force field increases potential energy, while work done by the force field decreases potential energy.

The energy stored in the magnetic field can be expressed using the inductance as follows: $W_m = \frac{1}{2} i_i L_{ij} i_j$.

Using the current vector $\mathbf{i} = [i_1, \ldots, i_n]^T$ and the inductance mapping $\mathbf{L}(\mathbf{r}) \in \mathbb{R}^{n \times n}$, we have $W_m = \frac{1}{2}\mathbf{i}^T\mathbf{L}(\mathbf{r})\mathbf{i}$, where T denotes the transpose symbol. The magnetic energy stored in the inductor with a single winding is $W_m = \frac{1}{2}Li^2$. The force is the gradient of the stored magnetic energy, $\vec{F}_m = \vec{\nabla} W_m$. In the XYZ coordinate system for the translational motion,

$$F_{mx} = \frac{\partial W_m}{\partial x}, \quad F_{my} = \frac{\partial W_m}{\partial y}, \quad \text{and} \quad F_{mz} = \frac{\partial W_m}{\partial z}.$$

For the rotational motion, using the differential change in the mechanical energy as a function of the angular displacement θ, one finds the torque T as

$$\vec{T}_e = \nabla W_m \quad \text{or} \quad \vec{T}_e = -\nabla W_m.$$

For the rigid body rotor that is constrained to rotate around the z-axis, $dW_{mec} = T_e\, d\theta$, where T_e is the z-component of the electromagnetic torque. For a lossless system, the electromagnetic torque is $T_e = \partial W_m / \partial \theta$.

Applying the Maxwell–Faraday equation $\nabla \times \vec{E} = -\frac{\partial \vec{B}}{\partial t}$, the *electromotive* and *magnetomotive* forces (*emf* and *mmf*) are

$$emf = \oint_l \vec{E} \cdot d\vec{l} = -\int_s \frac{\partial \vec{B}}{\partial t} \cdot d\vec{s} \quad \text{or} \quad emf = \oint_l (\vec{v} \times \vec{B}) \cdot d\vec{l} \quad \text{(Faraday s law of induction)},$$

$$mmf = \oint_l \vec{H} \cdot d\vec{l} = \int_s \vec{J} \cdot d\vec{s} + \int_s \frac{\partial \vec{D}}{\partial t} \cdot d\vec{s}.$$

The motional *emf* $\mathscr{E}$ is a function of the velocity and the magnetic flux density. The *emf* induced in a stationary closed circuit is equal to the negative rate of increase of the magnetic flux. The induced *mmf* is the sum of the induced current and the rate of change of the flux penetrating the surface bounded by the contour. Various transducers (sensors and actuators) are designed applying the reported physical principles.

Variations in induction are used to sense and measure the displacement, frequency, and magnitude of oscillations, acceleration, and motion. A magnetic pickup consists of a permanent magnet (Alnico of ferrite), while a humbucker uses a coil wound around ferromagnetic core. The inductance varies as a function of frequency and amplitude of motion. The image of the Texas Instrument LDC1614 Evaluation Module is documented in Figure 2.5a. The inductive sensing technology is used to measure the displacement and motion of a ferromagnetic material. As illustrated, there are planar spiral coils on the printed circuit board. Processing and data acquisition are accomplished using microcontroller and interfacing capabilities.

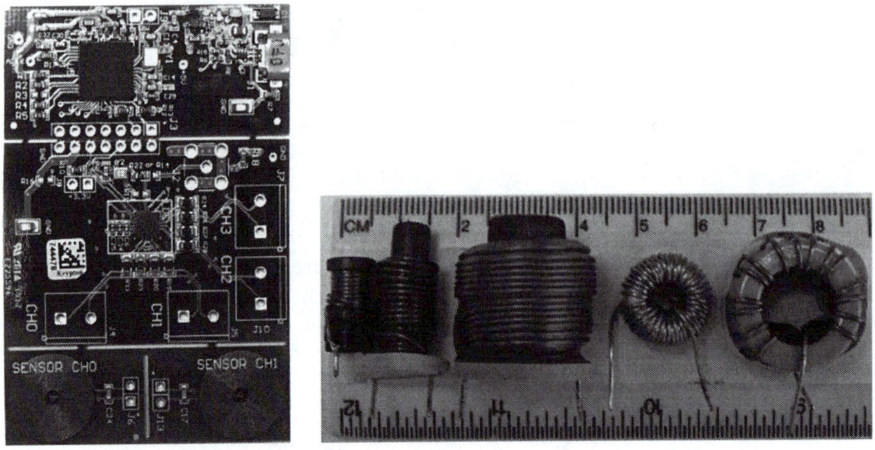

(a) (b)

FIGURE 2.5 (a) Inductive-based Texas Instrument LDC1614 Evaluation Module with spiral planar coils; (b) Ferrite rod and toroidal core inductors.

Example 2.1: Ferrite Core Inductors

Consider the ferrite rod and toroidal inductors illustrated in Figure 2.5b. We find the inductance of a ferrite rod core inductor with N turns, length l, and radius R, which is documented in Figure 2.5b. Ignoring eddy currents and applying the Ampere law $\oint_l \vec{H}\cdot d\vec{l} = Ni$, the magnetic field inside a solenoid is $\vec{B} = \dfrac{\mu Ni}{l}\vec{k}$. The magnetic flux through each turn is $\Phi = BA = \dfrac{\mu Ni}{l}\pi R^2$. Hence, the inductance is a function of the geometry (l and R), and,

$$L = \frac{\psi}{i} = \frac{N\Phi}{i} = \mu\frac{N^2}{l}\pi R^2. \qquad \blacksquare$$

Example 2.2: Ferrite Toroidal Core Inductors

From $\oint_l \vec{H}\cdot d\vec{l} = Ni$, one obtains the relationship between H and current. The unit for H is A/m or A-turn/m. Consider a ferrite-core toroidal inductors, shown in Figure 2.5b. Using the *effective* length l_e of the magnetic path, we have $H = Ni/l_e$. With the inner and outer radii r_{in} and r_{out}, one obtains

$$l_e = 2\pi\frac{r_{out}r_{in}}{r_{out}-r_{in}}\ln\frac{r_{out}}{r_{in}}.$$

The *effective* length is smaller than the mean length $l_{mean} = \pi(r_{in} + r_{out})$.

For a toroid $N = 50$, let $r_{in} = 1$ cm, $r_{out} = 1.5$ cm, and the thickness $h = 1$ cm. For $i = 1$ A, one finds that $H = 574$ A-turn/m. The MATLAB® statement to perform calculations is

```
>> N=50; rout=0.02; rin=0.01; i=1; le=2*pi*(rout*rin/(rout-rin))*log(rout/rin); H=N*i/le; le,H
le = 0.0871
H = 574.0301
```

We find the inductance of an N-turn rectangular cross section toroidal inductor with the inner and outer radii r_{in} and r_{out}. From the Ampere law in integral form $\oint_s B ds = B\oint_s ds = B2\pi r = \mu Ni$, the magnetic field inside toroid is $B = \frac{\mu Ni}{2\pi r}$. The magnetic flux through each turn is

$$\Phi = \iint_s B ds = \int_{r_{in}}^{r_{out}} Bh dr = \mu \frac{Nih}{2\pi} \int_{r_{in}}^{r_{out}} \frac{1}{r} dr = \mu \frac{Nih}{2\pi} \ln \frac{r_{out}}{r_{in}},$$

where the differential area element is the rectangular cross section $ds = hdr$. The inductance is

$$L = \frac{\psi}{i} = \frac{N\Phi}{i} = \mu \frac{N^2 h}{2\pi} \ln \frac{r_{out}}{r_{in}}.$$

From $\oint_l \vec{H} \cdot d\vec{l} = Ni$, one finds the relationship between H and current. The magnetic field intensity is $H_\phi = \frac{Ni}{l_e(r)}$, $r_{in} \le r \le r_{out}$. Assuming the field uniformity,

$$H_\phi = \frac{Ni}{l_e}, \quad H_\phi \approx \frac{Ni}{\pi(r_{in} + r_{out})}.$$

From $\mu = \mu_0 \mu_r$, one has

$$\mu_r\left(H, \frac{dH}{dt}\right) = \frac{1}{\mu_0} \frac{dB\left(H, \frac{dH}{dt}\right)}{dH}.$$

Hence,

$$L = \mu_0 \mu_r \frac{N^2 h}{2\pi} \ln \frac{r_{out}}{r_{in}}.$$

For the ferrite cores,

$$B\left(H, \frac{dH}{dt}\right) = B_{\max} \tanh\left(aH - bH_c \operatorname{sgn}\left(\frac{dH}{dt}\right)\right).$$

From $\mu_r = \frac{1}{\mu_0} \frac{dB(H)}{dH}$, one obtains $\mu_r\left(H, \frac{dH}{dt}\right) = \frac{1}{\mu_0} B_{\max} a \operatorname{sech}^2\left(aH - bH_c \operatorname{sgn}\frac{dH}{dt}\right)$.

The magnetic field intensity H and field density B depend on current i. Using the nonlinear BH curve of the ferromagnetic material, one may apply the following approximation

$$B\left(H, \frac{dH}{dt}\right) = B_{\max} \tanh\left(aH - bH_c \operatorname{sgn}\left(\frac{dH}{dt}\right)\right), \quad H = \frac{N}{l_e} i = ki, \quad l_e = 2\pi \frac{r_{out} r_{in}}{r_{out} - r_{in}} \ln \frac{r_{out}}{r_{in}},$$

where $B_{\max}$, a and b are the known constants.

One obtains

$$\mu_r\left(i,\frac{di}{dt}\right)=\frac{1}{\mu_0}B_{\max}c\left[1-\tanh^2\left(ci-d\,\mathrm{sgn}\left(\frac{di}{dt}\right)\right)\right],\quad c=ak,\quad d=bH_ck,\quad \text{or}$$

$$\mu_r\left(i,\frac{di}{dt}\right)=\begin{cases}\frac{1}{\mu_0}B_{\max}c\left[1-\tanh^2(ci-d)\right] & \text{if } \frac{di}{dt}>0\\ \frac{1}{\mu_0}B_{\max}c\left[1-\tanh^2(ci+d)\right] & \text{if } \frac{di}{dt}<0\end{cases}.$$

Hence,

$$L\left(i,\frac{di}{dt}\right)=\mu_0\mu_r\left(i,\frac{di}{dt}\right)\frac{N^2h}{2\pi}\ln\frac{r_{out}}{r_{in}}=B_{\max}c\left[1-\tanh^2\left(ci-d\,\mathrm{sgn}\left(\frac{di}{dt}\right)\right)\right]\frac{N^2h}{2\pi}\ln\frac{r_{out}}{r_{in}}.\quad \blacksquare$$

Example 2.3

Consider the toroidal inductor with N = 50, r_{in} = 1 cm, r_{out} = 2 cm, h = 1 cm, and the B–H curve $B\left(H,\frac{dH}{dt}\right)=B_{\max}\tanh\left(aH-bH_c\,\mathrm{sgn}\left(\frac{dH}{dt}\right)\right)$, $B_{\max}$ = 10 T, a = 0.0001, b = 0.00002, and H_c = 5000 A/m. Example 2.2 reports the resulting equation for $L\left(i,\frac{di}{dt}\right)$. The numeric solution can be found. The MATLAB script and Simulink® model. The resulting plots are documented in Figures 2.6 within the current envelope $i\in\left[-50\quad 50\right]$ A.

```
N=50; rin=0.01; rout=0.02; h=0.01; le=2*pi*(rout*rin/(rout-rin))*log(rout/rin);
Bmax=50; a=0.0001; b=0.00002; Hc=5000; mu_0=4*pi*1e-7; i=-50:0.1:50; H=N*i./le;
BD=Bmax*tanh(a*H+b*Hc); BI=Bmax*tanh(a*H-b*Hc);
plot(H,BD,H,BI,'LineWidth',2.5);
xlabel('Magnetic Field Intensity, {\itH} [A/m]','FontSize',14);
ylabel('Magnetic Field Density, {\itB} [T]','FontSize',14);
title('{\itBH} Magnetization Curve','FontSize',14); pause;
mu_rD=diff(BD)./diff(H)/mu_0; mu_rI=diff(BI)./diff(H)/mu_0;
LD=mu_0*mu_rD.*(N^2*h/(2*pi))*log(rout/rin);
LI=mu_0*mu_rI.*(N^2*h/(2*pi))*log(rout/rin);
k=length(i)-1; plot(i(1:k(1)),LI,i(1:k(1)),LD,'LineWidth',2.5);
xlabel('Current, {\iti} [A]','FontSize',14); ylabel('Inductance, {\itL} [H]','FontSize',14);
title('Inductance, {\itL}({\iti},{\itdi/dt}) [H]','FontSize',14);
```
■

2.4 CLASSICAL MECHANICS WITH APPLICATIONS

2.4.1 Newtonian Mechanics

2.4.1.1 Translational Motion

Newtonian mechanics, the Lagrange equations, and the Hamilton concept provide consistent approaches to derive the governing equations. We study the system behavior analyzing the forces that cause motion with the corresponding evolution of velocity and displacement. Using the displacement (position) vector $\vec{\mathbf{r}}$, the Newton's equation in the vector form is

$$\sum\vec{F}(t,\vec{\mathbf{r}})=m\vec{a},\tag{2.1}$$

where $\sum\vec{F}(t,\vec{\mathbf{r}})$ is the vector sum of all forces (*net* force) acted on the object; $\vec{a}$ is the vector of acceleration of the body with respect to an inertial reference frame; m is the mass of the body.

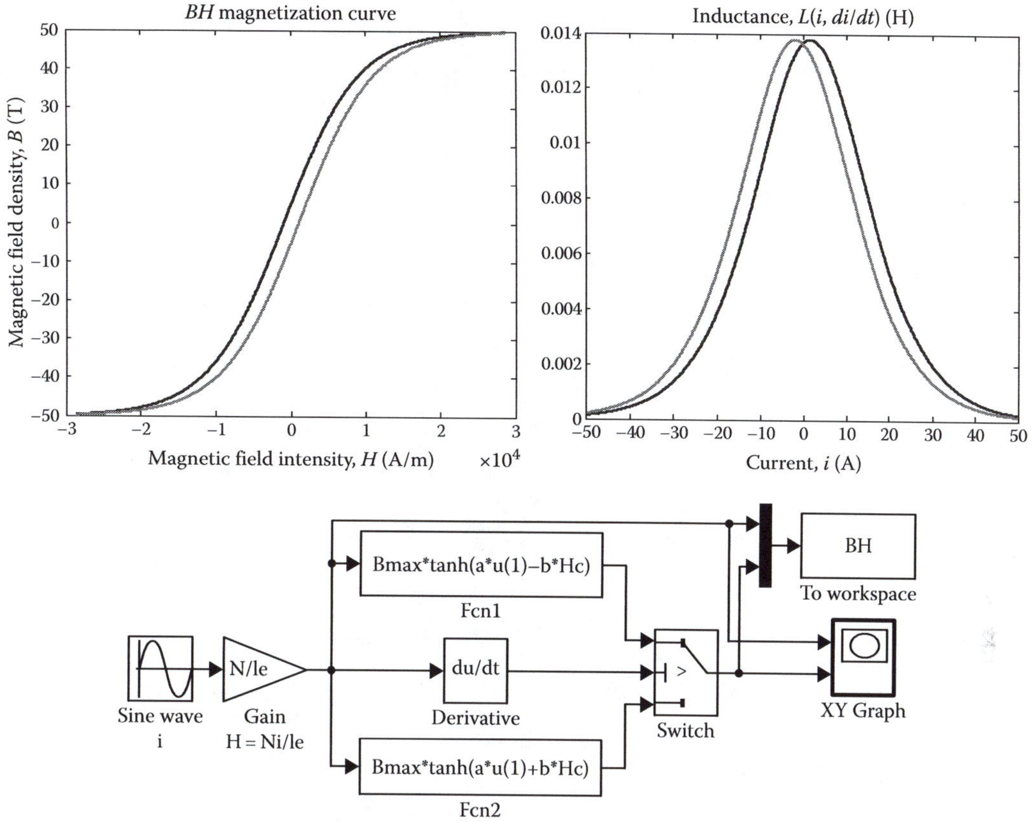

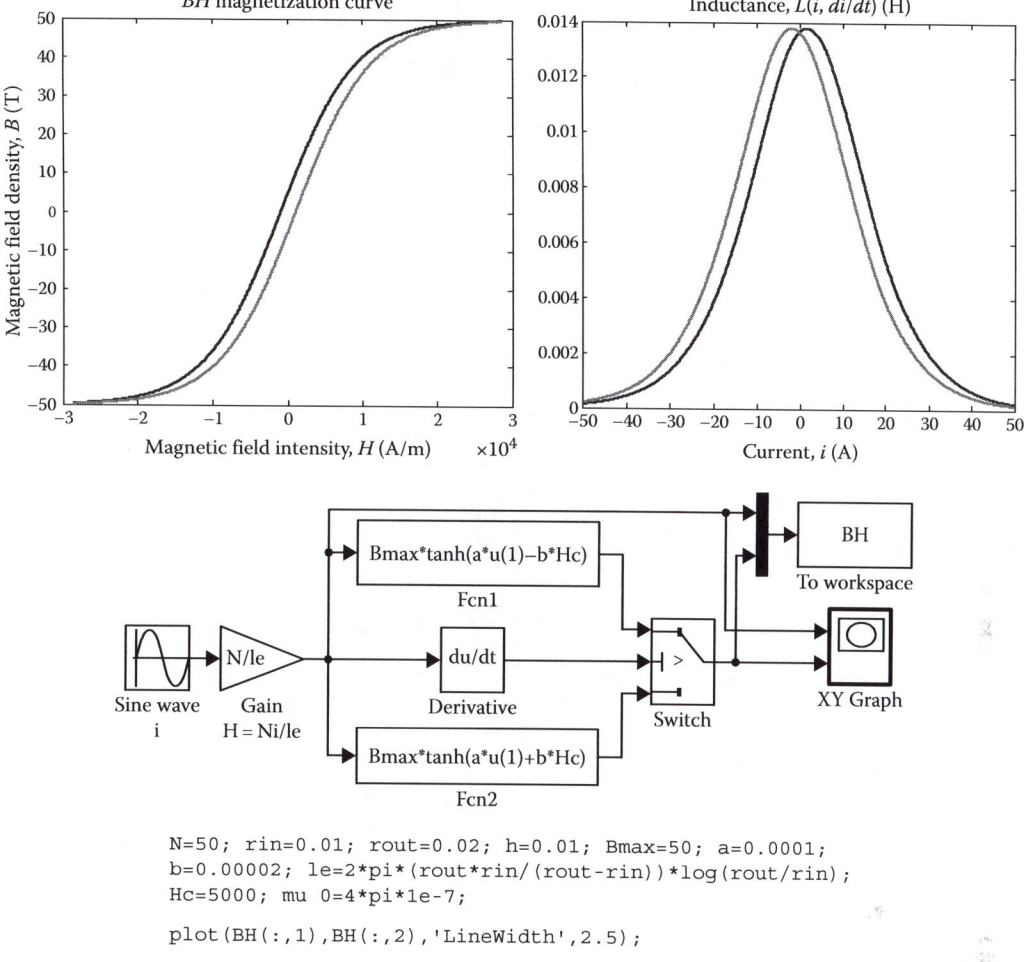

```
N=50; rin=0.01; rout=0.02; h=0.01; Bmax=50; a=0.0001;
b=0.00002; le=2*pi*(rout*rin/(rout-rin))*log(rout/rin);
Hc=5000; mu 0=4*pi*1e-7;

plot(BH(:,1),BH(:,2),'LineWidth',2.5);
```

FIGURE 2.6 Nonlinear magnetization and inductance plots. The Simulink® model to calculate and plot the *B–H* curve.

In (2.1), $m\vec{a}$ is not a force. A body is at equilibrium (the object is at rest or is moving with constant speed and the acceleration is zero ($a = 0$) if $\sum \vec{F} = 0$. In the Cartesian system, the mechanical equations of motion in the *xyz* coordinates are

$$\sum \vec{F}(t, \vec{\mathbf{r}}) = m\vec{a} = m\begin{bmatrix} \vec{a}_x \\ \vec{a}_y \\ \vec{a}_z \end{bmatrix} = m\frac{d^2\vec{\mathbf{r}}}{dt^2} = m\begin{bmatrix} \dfrac{d^2\vec{x}}{dt^2} \\ \dfrac{d^2\vec{y}}{dt^2} \\ \dfrac{d^2\vec{z}}{dt^2} \end{bmatrix}, \quad \sum \vec{F}_x = m\vec{a}_x, \quad \sum \vec{F}_y = m\vec{a}_y, \quad \sum \vec{F}_z = m\vec{a}_z.$$

One obtains the second-order ordinary differential equations, which are the rigid-body mechanical equations of motion. The forces to control the motion are developed by actuators. Newton's second law in terms of the linear momentum $\vec{p} = m\vec{v}$ is

$$\sum \vec{F} = \frac{d\vec{p}}{dt} = \frac{d(m\vec{v})}{dt},$$

where $\vec{v}$ is the velocity vector. The force is equal to the rate of change of the momentum. The object or particle moves uniformly if

$$\frac{d\vec{p}}{dt} = 0, \quad \vec{p} = \text{const.}$$

Using the potential energy function $\Pi(\vec{\mathbf{r}})$, for the conservative mechanical system we have $\sum \vec{F}(\vec{\mathbf{r}}) = -\nabla\Pi(\vec{\mathbf{r}})$.

The work done per unit time is

$$\frac{dW}{dt} = \sum \vec{F}(\vec{\mathbf{r}})\frac{d\vec{\mathbf{r}}}{dt} = -\nabla\Pi(\vec{\mathbf{r}})\frac{d\vec{\mathbf{r}}}{dt} = -\frac{d\Pi(\vec{\mathbf{r}})}{dt}.$$

From the Newton second law (2.1), one obtains $m\vec{a} - \sum \vec{F}(\vec{\mathbf{r}}) = 0$.

For a conservative system, using the kinetic and potential energies, we have

$$m\frac{d^2\vec{\mathbf{r}}}{dt^2} + \nabla\Pi(\vec{\mathbf{r}}) = 0.$$

The total kinetic energy is $\Gamma = \frac{1}{2}mv^2$. In the Largange equations, the *generalized* coordinates $(q_1,\ldots,q_n)$ and *generalized* velocities $\left(\frac{dq_1}{dt},\ldots,\frac{dq_n}{dt}\right)$ are used. The Lagrangian is $L = \Gamma - \Pi$, where Γ is the total kinetic energy of the system and Π is the potential energy. The total kinetic $\Gamma\left(q_1,\ldots,q_n,\frac{dq_1}{dt},\ldots,\frac{dq_n}{dt}\right)$ and potential $\Pi(q_1,\ldots,q_n)$ energies are found using q_i and dq_i/dt. From Newton's second law of motion (2.1), given using Γ and Π, we have $\frac{d}{dt}\left(\frac{\partial\Gamma}{\partial\dot{q}_i}\right) + \frac{\partial\Pi}{\partial q_i} = 0.$

Example 2.4

For one-dimensional linear motion, the Newton and Lagrange equations of motion are

$$m\frac{d^2x}{dt^2} + \frac{\partial\Pi}{\partial x} = 0 \quad \text{and} \quad \frac{d}{dt}\left(\frac{\partial\Gamma}{\partial\dot{q}_i}\right) + \frac{\partial\Pi}{\partial q_i} = 0, \quad q = x.$$ ■

Example 2.5

Consider a positioning table actuated by an actuator. The work required to accelerate a 20 g payload ($m = 0.02$ kg) from $v_0 = 0$ m/sec to $v_f = 1$ m/sec is

$$W = \frac{1}{2}\left(mv_f^2 - mv_0^2\right) = \frac{1}{2}20\times 10^{-3}\times 1^2 = 0.01 \text{ J.}$$

The work-energy theorem is applied. The work done by the *net* force on a particle equals the change in the object's kinetic energy, $W_{total} = \Gamma_2 - \Gamma_1 = \Delta\Gamma$. For a varying force, one finds the total work done by the *net* force as

$$W = \int_{x_1}^{x_2} F dx.$$

Using

$$a = \frac{dv}{dt} = \frac{dv}{dx}\frac{dx}{dt} = v\frac{dv}{dx},$$

we have

$$W = \int_{x_1}^{x_2} F dx = \int_{x_1}^{x_2} madx = \int_{x_1}^{x_2} mv\frac{dv}{dx}dx = \int_{v_1}^{v_2} mvdv. \qquad \blacksquare$$

Example 2.6

Consider a body with mass m in the XY coordinate system. The free-body diagram is shown in Figure 2.7a. The force $\vec{F}_a$ is applied in the x direction. Let $\vec{F}_a(t,x) = x\sin 10te^{-t}$. Assume that the *Coulomb* and static frictions are negligible and that the viscous friction force is $F_{friction} = B_v v$, where B_v is the viscous friction coefficient. We find the governing equations.

The sum of the forces, acting in the y direction, is $\sum \vec{F}_Y = \vec{F}_N - \vec{F}_g$, where $\vec{F}_g = mg$ is the gravitational force acting on the mass m; $\vec{F}_N$ is the normal force that is equal and opposite to the gravitational force. From (2.1), the equation of motion in the y direction is $\vec{F}_N - \vec{F}_g = ma_y = m\frac{d^2y}{dt^2}$. From $\vec{F}_N = \vec{F}_g$, the resulting equation is $\frac{d^2y}{dt^2} = 0$.

The sum of the forces acting in the x direction is found using the applied force $\vec{F}_a$ and the friction force $\vec{F}_{friction}$. We have $\sum \vec{F}_X = \vec{F}_a - \vec{F}_{friction}$. The applied force can be time invariant $\vec{F}_a = \text{const}$ or time varying $\vec{F}_a(t) = f(t,x,y,z)$. Using (2.1), the equation motion in the x direction is

$$\vec{F}_a - \vec{F}_{friction} = ma_x = m\frac{dv}{dt} = m\frac{d^2x}{dt^2}.$$

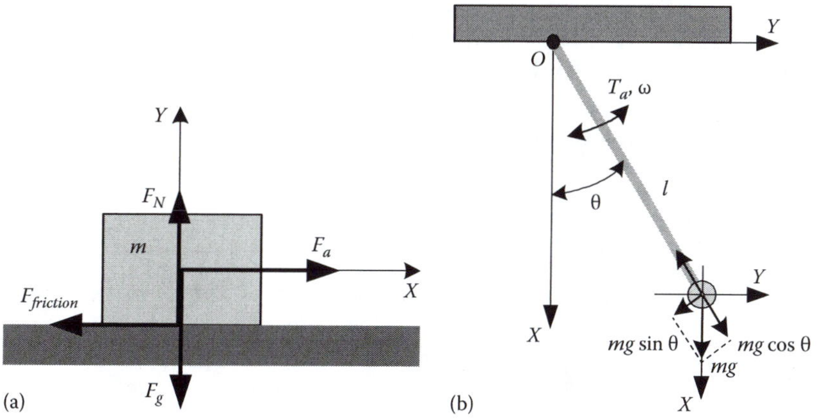

FIGURE 2.7 (a) Free-body diagram; (b) A simple pendulum (Example 2.10).

One obtains the second-order nonlinear differential equation, which describes the rigid-body dynamics in the x direction

$$\frac{d^2x}{dt^2} = \frac{1}{m}\left(F_a - B_v\frac{dx}{dt}\right) = \frac{1}{m}\left[x\sin 10te^{-t} - B_v\frac{dx}{dt}\right].$$

A set of two first-order linear differential equations is

$$\frac{dx}{dt} = v, \quad \frac{dv}{dt} = \frac{1}{m}\left[x\sin 10te^{-t} - B_v v\right], \quad t \geq 0. \quad \blacksquare$$

2.4.1.2 *Rotational Motion*

For rotational devices, the angular displacement and acceleration are used. The rotational Newton's second law for a rigid body is

$$\sum \vec{T}\left(t,\vec{\theta}\right) = J\vec{\alpha}, \tag{2.2}$$

where $\sum \vec{T}$ is the *net* torque (N-m); J is the moment of inertia (kg-m^2); $\vec{\alpha}$ is the angular acceleration vector (rad/sec^2), $\vec{\alpha} = \frac{d}{dt}\frac{d\vec{\theta}}{dt} = \frac{d^2\vec{\theta}}{dt^2} = \frac{d\vec{\omega}}{dt}$; $\vec{\theta}$ is the angular displacement; $\vec{\omega}$ is the angular velocity. Using the angular momentum

$$\vec{L}_M = \vec{R} \times \vec{p} = \vec{R} \times m\vec{v}, \quad \sum \vec{T} = \frac{d\vec{L}_M}{dt} = \vec{R} \times \vec{F},$$

where $\vec{R}$ is the position vector with respect to the origin. For the rigid body, rotating around the axis of symmetry, we have $\vec{L}_M = J\vec{\omega}$. For one-dimensional systems, using the *net* moment, we have $\sum M = J\alpha$.

Example 2.7

Let the actuator accelerate, and the angular velocity of the rotor be $\omega_r = 10t^3$, $t \geq 0$. One can find the angular momentum and the developed electromagnetic torque. Assume that the load and friction torques are zero. If the equivalent moment of inertia is $J = 1$ kg-m^2, the angular momentum is $L_M = J\omega_r = 10t^3$. The electromagnetic torque is $T_e = \frac{dL_M}{dt} = 30t^2$ N-m. $\blacksquare$

The analysis of motion was performed using the energy and momentum quantities, which are conserved. The principle of conservation of energy states that energy can be converted only from one form to another. Kinetic energy is associated with motion, while potential energy is a function of position. The sum of the kinetic Γ, potential Π, and dissipation D energies yields the total energy of the system Σ_t, which is conserved. The total energy remains constant, and, $\Sigma_t = \Gamma + \Pi + D =$ const.

Example 2.8

Consider the translational motion of a body attached to an ideal spring that exerts the spring force described by an ideal Hooke's law. The translational kinetic energy is $\Gamma = \frac{1}{2}mv^2$. The elastic potential energy of the spring is

$$\Pi = \frac{1}{2}k_s x^2,$$

where k_s is the spring force constant. Neglecting friction, the total energy is

$$\Sigma_t = \Gamma + \Pi = \frac{1}{2}(mv^2 + k_s x^2) = \text{const.}$$

For rotational motion and torsional spring, we have

$$\Sigma_t = \Gamma + \Pi = \frac{1}{2}(J\omega^2 + k_s \theta^2) = \text{const},$$

where the rotational kinetic energy and the elastic potential energy are

$$\Gamma = \frac{1}{2}J\omega^2 \quad \text{and} \quad \Pi = \frac{1}{2}k_s\theta^2.$$

■

If a rigid body exhibits translational and rotational motion, the kinetic energy is

$$\Gamma = \frac{1}{2}(mv^2 + J\omega^2).$$

The motion of the rigid body is represented as a combination of translational motion of the center of mass and rotational motion about the axis through the center of mass. The moment of inertia J depends on how the mass is distributed with respect to the axis. This J is different for different axes of rotation. If the body is uniform in density, J can be calculated for regularly shaped bodies. For a rigid cylinder of mass m (which is uniformly distributed), radius R, and length l, one has the horizontal and vertical moments of inertia

$$J_{horizontal} = \frac{1}{2}mR^2 \quad \text{and} \quad J_{vertical} = \frac{1}{4}mR^2 + \frac{1}{12}ml^2.$$

The *radius of gyration* can be found for irregularly shaped objects, yielding J.
Assuming that the body is rigid and the moment of inertia is constant, one has

$$\vec{T}\, d\vec{\theta} = J\vec{\alpha}\, d\vec{\theta} = J\frac{d\vec{\omega}}{dt}d\vec{\theta} = J\frac{d\vec{\theta}}{dt}d\vec{\omega} = J\vec{\omega}d\vec{\omega}.$$

The total work

$$W = \int_{\theta_0}^{\theta_{final}} \vec{T}d\vec{\theta} = \int_{\omega_0}^{\omega_{final}} J\vec{\omega}\, d\vec{\omega} = \frac{1}{2}\left(J\omega_{final}^2 - J\omega_0^2\right),$$

represents the change of the kinetic energy. Furthermore,

$$\frac{dW}{dt} = \vec{T}\frac{d\vec{\theta}}{dt} = \vec{T}\times\vec{\omega},$$

and the power is $P = \vec{T}\times\vec{\omega}$. This equation is an analog of $P = \vec{F}\times\vec{v}$, which is applied for translational motion.

Example 2.9

Assume that the rated power and angular velocity of a motor are 1 W and 1000 rad/sec. The electromagnetic torque is $T_e = \frac{P}{\omega_r} = \frac{1}{1000} = 1\times 10^{-3}$ N-m. ■

Example 2.10

Suppose a point mass m suspended by a massless unstretchable rod or string of length l, as shown in Figure 2.7b. For a simple pendulum, the restoring force $-mg\sin\theta$ is the tangential component of the *net* force. The sum of the moments about the pivot point O is

$$\sum M = -mgl\sin\theta + T_a,$$

where T_a is the applied torque; l is the length of the pendulum measured from the point of rotation.

Using (2.2), one obtains the equation of motion

$$J\alpha = J\frac{d^2\theta}{dt^2} = -mgl\sin\theta + T_a, \quad \frac{d^2\theta}{dt^2} = \frac{1}{J}\left(-mgl\sin\theta + T_a\right),$$

where J is the moment of inertial of the mass about the point O.

A set of two first-order differential equations is

$$\frac{d\omega}{dt} = \frac{1}{J}\left(-mgl\sin\theta + T_a\right), \quad \frac{d\theta}{dt} = \omega.$$

The moment of inertia is $J = ml^2$. The following differential equations result

$$\frac{d\omega}{dt} = -\frac{g}{l}\sin\theta + \frac{1}{ml^2}T_a, \quad \frac{d\theta}{dt} = \omega. \quad \blacksquare$$

2.4.2 Lagrange Equations of Motion

Electromechanical systems comprise mechanical, electromagnetic, and electronic components. One may apply Newtonian's dynamics. Then, using the coenergy concept, one derives the expression for the electromagnetic (or electrostatic) force or torque, which are functions of current, voltage, electromagnetic field quantities, displacement, etc. In addition to Newtonian dynamics, Kirchhoff's laws are used.

The Lagrange and Hamilton concepts are based on the energy analysis of an entire system. Using the Lagrange equations, one combines the mechanical and circuitry-electromagnetic dynamics. Correspondingly, the Lagrange and Hamilton concepts are more general. Using the system variables, one finds the total kinetic

$$\Gamma\left(t, q_1, \ldots, q_n, \frac{dq_1}{dt}, \ldots, \frac{dq_n}{dt}\right),$$

dissipation

$$D\left(t, q_1, \ldots, q_n, \frac{dq_1}{dt}, \ldots, \frac{dq_n}{dt}\right),$$

and potential $\Pi(t, q_1, \ldots, q_n)$ energies.

Using the total energies, the Lagrange equations of motion are

$$\frac{d}{dt}\left(\frac{\partial \Gamma}{\partial \dot{q}_i}\right) - \frac{\partial \Gamma}{\partial q_i} + \frac{\partial D}{\partial \dot{q}_i} + \frac{\partial \Pi}{\partial q_i} = Q_i, \tag{2.3}$$

where q_i and Q_i are the generalized coordinates and the generalized forces (applied forces and disturbances).

Using the displacement and charges as the generalized coordinates q_i, one finds energies Γ, D, and Π. For conservative (losseless) systems $D = 0$. From (2.3), one obtains

$$\frac{d}{dt}\left(\frac{\partial\Gamma}{\partial\dot{q}_i}\right)-\frac{\partial\Gamma}{\partial q_i}+\frac{\partial\Pi}{\partial q_i}=Q_i.$$

Example 2.11: Simple Pendulum

We derive the equations of motion for a simple pendulum as depicted in Figure 2.7b. The equations of motion were derived in Example 2.10 using the Newtonian mechanics. For the studied conservative (lossless) system with no friction, $D = 0$.

The kinetic energy of the pendulum is $\Gamma=\frac{1}{2}m(l\dot{\theta})^2$. The potential energy is $\Pi=mgl(1-\cos\theta)$.

The angular displacement is the generalized coordinate, $q_1 = \theta$. The kinetic and potential energies are $\Gamma=\frac{1}{2}m(l\dot{q}_1)^2$ and $\Pi = mgl(1 - \cos q_1)$. The generalized force is the applied torque T_a, $Q_1 = T_a$. Hence,

$$\frac{d}{dt}\left(\frac{\partial\Gamma}{\partial\dot{q}_1}\right)-\frac{\partial\Gamma}{\partial q_1}+\frac{\partial\Pi}{\partial q_1}=Q_1,$$

where

$$\frac{\partial\Gamma}{\partial\dot{q}_1}=\frac{\partial\Gamma}{\partial\dot{\theta}}=ml^2\dot{\theta};$$

$$\frac{d}{dt}\left(\frac{\partial\Gamma}{\partial\dot{\theta}}\right)=ml^2\frac{d^2\theta}{dt^2}+2ml\frac{dl}{dt}\frac{d\theta}{dt};$$

$$\frac{\partial\Gamma}{\partial q_1}=\frac{\partial\Gamma}{\partial\theta}=0;$$

$$\frac{\partial\Pi}{\partial q_1}=\frac{\partial\Pi}{\partial\theta}=mgl\sin\theta.$$

Assuming that the rod is unstretchable, we have $dl/dt = 0$. If this assumption is not valid, one should use the appropriate expression for dl/dt. For $dl/dt = 0$, we have

$$ml^2\frac{d^2\theta}{dt^2}+mgl\sin\theta=T_a.$$

One obtains

$$\frac{d^2\theta}{dt^2}=\frac{1}{ml^2}\left(-mgl\sin\theta+T_a\right).$$

The equation of motion, derived by using Newtonian mechanics, is

$$\frac{d^2\theta}{dt^2}=\frac{1}{J}\left(-mgl\sin\theta+T_a\right),$$

where $J = ml^2$.

One concludes that the results are the same, and

$$\frac{d\omega}{dt} = -\frac{g}{l}\sin\theta + \frac{1}{ml^2}T_a, \quad \frac{d\theta}{dt} = \omega.$$

The Lagrange equations of motion provide general results. The coordinate-dependent system parameters can be accounted for. For example, l can be a function of θ. ■

Example 2.12: Double Pendulum

Consider a two-degree-of-freedom double pendulum, shown in Figure 2.8.

The angular displacements θ_1 and θ_2 are the independent generalized coordinates q_1 and q_2. In the xy plane, let (x_1, y_1) and (x_2, y_2) be the rectangular coordinates of point masses m_1 and m_2. We obtain

$$x_1 = l_1\cos\theta_1, \quad x_2 = l_1\cos\theta_1 + l_2\cos\theta_2, \quad y_1 = l_1\sin\theta_1, \quad y_2 = l_1\sin\theta_1 + l_2\sin\theta_2.$$

The Lagrange equations of motion are

$$\frac{d}{dt}\left(\frac{\partial\Gamma}{\partial\dot{\theta}_1}\right) - \frac{\partial\Gamma}{\partial\theta_1} + \frac{\partial\Pi}{\partial\theta_1} = 0, \quad \frac{d}{dt}\left(\frac{\partial\Gamma}{\partial\dot{\theta}_2}\right) - \frac{\partial\Gamma}{\partial\theta_2} + \frac{\partial\Pi}{\partial\theta_2} = 0.$$

The total kinetic energy Γ is a nonlinear function of the displacements. We have

$$\Gamma = \frac{1}{2}m_1\left(\dot{x}_1^2 + \dot{y}_1^2\right) + \frac{1}{2}m_2\left(\dot{x}_2^2 + \dot{y}_2^2\right) = \frac{1}{2}(m_1 + m_2)l_1^2\dot{\theta}_1^2 + m_2l_1l_2\dot{\theta}_1\dot{\theta}_2\cos(\theta_2 - \theta_1) + \frac{1}{2}m_2l_2^2\dot{\theta}_2^2.$$

One obtains

$$\frac{\partial\Gamma}{\partial\theta_1} = m_2l_1l_2\sin(\theta_2 - \theta_1)\dot{\theta}_1\dot{\theta}_2, \quad \frac{\partial\Gamma}{\partial\dot{\theta}_1} = (m_1 + m_2)l_1^2\dot{\theta}_1 + m_2l_1l_2\cos(\theta_2 - \theta_1)\dot{\theta}_2,$$

$$\frac{\partial\Gamma}{\partial\theta_2} = -m_2l_1l_2\sin(\theta_1 - \theta_2)\dot{\theta}_1\dot{\theta}_2, \quad \frac{\partial\Gamma}{\partial\dot{\theta}_2} = m_2l_1l_2\cos(\theta_2 - \theta_1)\dot{\theta}_1 + m_2l_1^2\dot{\theta}_2.$$

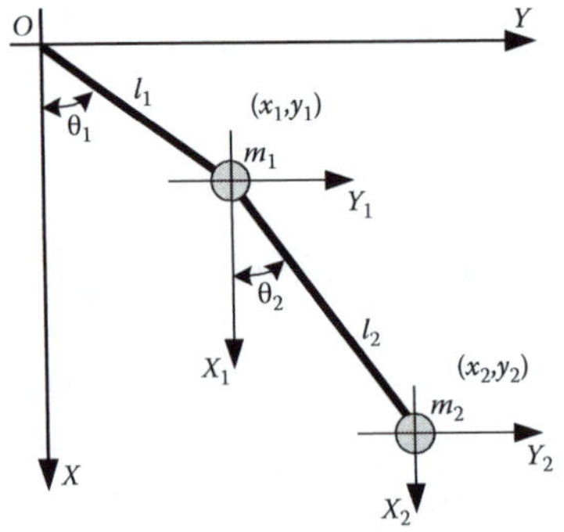

FIGURE 2.8 Double pendulum.

The total potential energy is

$$\Pi = m_1 g x_1 + m_2 g x_2 = (m_1 + m_2) g l_1 \cos\theta_1 + m_2 g l_2 \cos\theta_2,$$

yielding $\dfrac{\partial \Pi}{\partial \theta_1} = -(m_1 + m_2) g l_1 \sin\theta_1$ and $\dfrac{\partial \Pi}{\partial \theta_2} = -m_2 g l_2 \sin\theta_2$.

The differential equations are

$$l_1\Big[(m_1+m_2)l_1\ddot{\theta}_1 + m_2 l_2\cos(\theta_2-\theta_1)\ddot{\theta}_2 - m_2 l_2\sin(\theta_2-\theta_1)\dot{\theta}_2^2 - m_2 l_2\sin(\theta_2-\theta_1)\dot{\theta}_1\dot{\theta}_2 - (m_1+m_2)g\sin\theta_1\Big]=0,$$

$$m_2 l_2\Big[l_2\ddot{\theta}_2 + l_1\cos(\theta_2-\theta_1)\ddot{\theta}_1 + l_1\sin(\theta_2-\theta_1)\dot{\theta}_1^2 + l_1\sin(\theta_2-\theta_1)\dot{\theta}_1\dot{\theta}_2 - g\sin\theta_2\Big]=0.$$

If the torques T_1 and T_2 are applied to the first and second joints (two-degree-of-freedom robot), the following equations of motions result

$$l_1\Big[(m_1+m_2)l_1\ddot{\theta}_1 + m_2 l_2\cos(\theta_2-\theta_1)\ddot{\theta}_2 - m_2 l_2\sin(\theta_2-\theta_1)\dot{\theta}_2^2 - m_2 l_2\sin(\theta_2-\theta_1)\dot{\theta}_1\dot{\theta}_2 - (m_1+m_2)g\sin\theta_1\Big]=T_1,$$

$$m_2 l_2\Big[l_2\ddot{\theta}_2 + l_1\cos(\theta_2-\theta_1)\ddot{\theta}_1 + l_1\sin(\theta_2-\theta_1)\dot{\theta}_1^2 + l_1\sin(\theta_2-\theta_1)\dot{\theta}_1\dot{\theta}_2 - g\sin\theta_2\Big]=T_2. \quad ■$$

Example 2.13: Electric Circuits

Consider the electric circuits shown in Figures 2.9.

We use the electric charges as the generalized coordinates. The electric charges in the first and the second loops q_1 and q_2 are the independent generalized coordinates, as shown in Figures 2.9. These generalized coordinates are related to the currents $q_1 = \int i_1\,dt$, $i_1 = \dot{q}_1$ and $q_2 = \int i_2\,dt$, $i_2 = \dot{q}_2$. The Lagrange equations of motion are

$$\frac{d}{dt}\left(\frac{\partial \Gamma}{\partial \dot{q}_1}\right) - \frac{\partial \Gamma}{\partial q_1} + \frac{\partial D}{\partial \dot{q}_1} + \frac{\partial \Pi}{\partial q_1} = Q_1, \quad \frac{d}{dt}\left(\frac{\partial \Gamma}{\partial \dot{q}_2}\right) - \frac{\partial \Gamma}{\partial q_2} + \frac{\partial D}{\partial \dot{q}_2} + \frac{\partial \Pi}{\partial q_2} = 0.$$

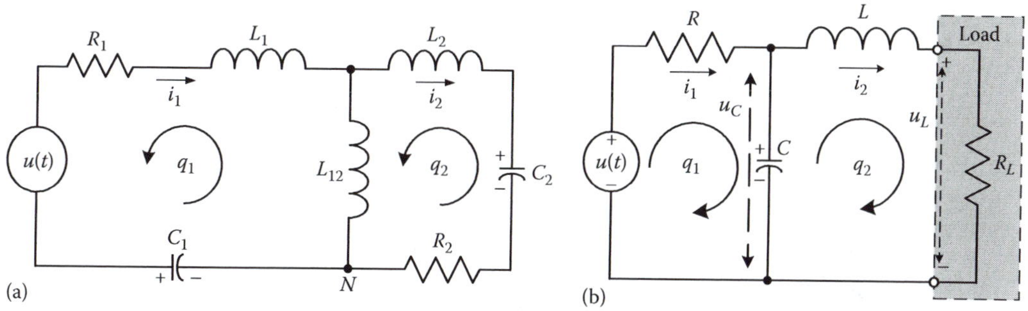

FIGURE 2.9 Electric circuits: (a) *RLC* network; (b) *RLC* circuit with a load.

For the *RLC* network illustrated in Figure 2.9a, the supplied voltage u is the generalized force Q_1 applied, $u(t) = Q_1$. The total magnetic energy (kinetic energy) is

$$\Gamma = \frac{1}{2}L_1\dot{q}_1^2 + \frac{1}{2}L_{12}\left(\dot{q}_1 - \dot{q}_2\right)^2 + \frac{1}{2}L_2\dot{q}_2^2.$$

Hence,

$$\frac{\partial \Gamma}{\partial q_1} = 0, \quad \frac{\partial \Gamma}{\partial \dot{q}_1} = \left(L_1 + L_{12}\right)\dot{q}_1 - L_{12}\dot{q}_2, \quad \frac{\partial \Gamma}{\partial q_2} = 0, \quad \frac{\partial \Gamma}{\partial \dot{q}_2} = -L_{12}\dot{q}_1 + \left(L_2 + L_{12}\right)\dot{q}_2.$$

The total electric energy (potential energy) is

$$\Pi = \frac{1}{2}\frac{q_1^2}{C_1} + \frac{1}{2}\frac{q_2^2}{C_2}, \quad \text{and} \quad \frac{\partial \Pi}{\partial q_1} = \frac{q_1}{C_1}, \quad \frac{\partial \Pi}{\partial q_2} = \frac{q_2}{C_2}.$$

The total heat energy dissipated is

$$D = \frac{1}{2}R_1\dot{q}_1^2 + \frac{1}{2}R_2\dot{q}_2^2,$$

yielding

$$\frac{\partial D}{\partial \dot{q}_1} = R_1\dot{q}_1, \quad \frac{\partial D}{\partial \dot{q}_2} = R_2\dot{q}_2.$$

The differential equations are

$$\left(L_1 + L_{12}\right)\ddot{q}_1 - L_{12}\ddot{q}_2 + R_1\dot{q}_1 + \frac{q_1}{C_1} = u, \quad -L_{12}\ddot{q}_1 + \left(L_2 + L_{12}\right)\ddot{q}_2 + R_2\dot{q}_2 + \frac{q_2}{C_2} = 0.$$

The resulting differential equations are

$$\ddot{q}_1 = \frac{1}{L_1 + L_{12}}\left(-\frac{q_1}{C_1} - R_1\dot{q}_1 + L_{12}\ddot{q}_2 + u\right), \quad \ddot{q}_2 = \frac{1}{L_2 + L_{12}}\left(L_{12}\ddot{q}_1 - \frac{q_2}{C_2} - R_2\dot{q}_2\right).$$

For an electric circuit with a load, as shown in Figure 2.9b, the total kinetic energy is $\Gamma = \frac{1}{2}L\dot{q}_2^2$. Therefore,

$$\frac{\partial \Gamma}{\partial q_1} = 0, \quad \frac{\partial \Gamma}{\partial \dot{q}_1} = 0, \quad \frac{d}{dt}\left(\frac{\partial \Gamma}{\partial \dot{q}_1}\right) = 0, \quad \frac{\partial \Gamma}{\partial q_2} = 0, \quad \frac{\partial \Gamma}{\partial \dot{q}_2} = L\dot{q}_2 \quad \text{and} \quad \frac{d}{dt}\left(\frac{\partial \Gamma}{\partial \dot{q}_2}\right) = L\ddot{q}_2.$$

The total potential energy is

$$\Pi = \frac{1}{2}\frac{\left(q_1 - q_2\right)^2}{C},$$

yielding

$$\frac{\partial \Pi}{\partial q_1} = \frac{q_1 - q_2}{C}, \quad \frac{\partial \Pi}{\partial q_2} = \frac{-q_1 + q_2}{C}.$$

The total dissipated energy is

$$D = \frac{1}{2}R\dot{q}_1^2 + \frac{1}{2}R_L\dot{q}_2^2.$$

Therefore,

$$\frac{\partial D}{\partial \dot{q}_1} = R\dot{q}_1, \; \frac{\partial D}{\partial \dot{q}_2} = R_L\dot{q}_2.$$

One obtains two differential equations

$$R\dot{q}_1 + \frac{q_1 - q_2}{C} = u, \quad L\ddot{q}_2 + R_L\dot{q}_2 + \frac{-q_1 + q_2}{C} = 0.$$

We found a set of differential equations

$$\dot{q}_1 = \frac{1}{R}\left(\frac{-q_1 + q_2}{C} + u\right), \quad \ddot{q}_2 = \frac{1}{L}\left(-R_L\dot{q}_2 + \frac{q_1 - q_2}{C}\right).$$

The equations of motion derived using the Lagrange concept are equivalent to the model developed using Kirchhoff's law:

$$\frac{du_C}{dt} = \frac{1}{C}\left(-\frac{u_C}{R} - i_2 + \frac{u}{R}\right), \quad \frac{di_2}{dt} = \frac{1}{L}\left(u_C - R_L i_2\right).$$

From $i_1 = \dot{q}_1$ and $i_2 = \dot{q}_2$, using $C\frac{du_C}{dt} = i_1 - i_2$, we obtain $u_C = \frac{q_1 - q_2}{C}$. ■

Example 2.14: Electromechanical Actuator

Electromechanical devices can be modeled using the Lagrange equations of motion. Newton's laws can be used only to model the rigid-body dynamics unless the *electromechanical analogies* are applied. Consider an electromechanical motion device that actuates the load (robotic arm, pointer, etc.). The actuator has two independently excited stator and rotor windings as shown in Figure 2.10. The magnetic coupling between the stator and rotor windings L_{sr} results in an electromagnetic torque T_e. The developed T_e is countered by the torsional spring. The load torque is T_L.

The following notations are used: i_s and i_r are the currents in the stator and rotor windings; u_s and u_r are the applied voltages to the stator and rotor windings; ω_r and θ_r are the rotor angular velocity and displacement; T_e and T_L are the electromagnetic and load torques; r_s and r_r are the resistances of the stator and rotor windings; L_s and L_r are the self-inductances of the stator and rotor windings; L_{sr} is the mutual inductance of the stator and rotor windings; $\Re_m$ is the reluctance of the magnetizing path; N_s and N_r are the number of turns in the stator and rotor windings; J is the equivalent moment of inertia of the rotor and attached load; B_m is the viscous friction coefficient; and k_s is the spring constant.

The independent generalized coordinates are q_1, q_2, and q_3, where q_1 and q_2 are the electric charges in the stator and rotor windings; q_3 is the rotor angular displacement. The generalized forces applied to a system are Q_1, Q_2, and Q_3, where Q_1 and Q_2 are the applied voltages to the stator and rotor windings; Q_3 is the load torque. Hence, $q_1 = \int i_s dt$, $q_2 = \int i_r dt$, $q_3 = \theta_r$, $\dot{q}_1 = i_s$, $\dot{q}_2 = i_r$, $\dot{q}_3 = \omega_r$, $Q_1 = u_s$, $Q_2 = u_r$, and $Q_3 = -T_L$.

The Lagrange equations are expressed in terms of each independent coordinate, yielding

$$\frac{d}{dt}\left(\frac{\partial \Gamma}{\partial \dot{q}_1}\right) - \frac{\partial \Gamma}{\partial q_1} + \frac{\partial D}{\partial \dot{q}_1} + \frac{\partial \Pi}{\partial q_1} = Q_1, \quad \frac{d}{dt}\left(\frac{\partial \Gamma}{\partial \dot{q}_2}\right) - \frac{\partial \Gamma}{\partial q_2} + \frac{\partial D}{\partial \dot{q}_2} + \frac{\partial \Pi}{\partial q_2} = Q_2,$$

$$\frac{d}{dt}\left(\frac{\partial \Gamma}{\partial \dot{q}_3}\right) - \frac{\partial \Gamma}{\partial q_3} + \frac{\partial D}{\partial \dot{q}_3} + \frac{\partial \Pi}{\partial q_3} = Q_3.$$

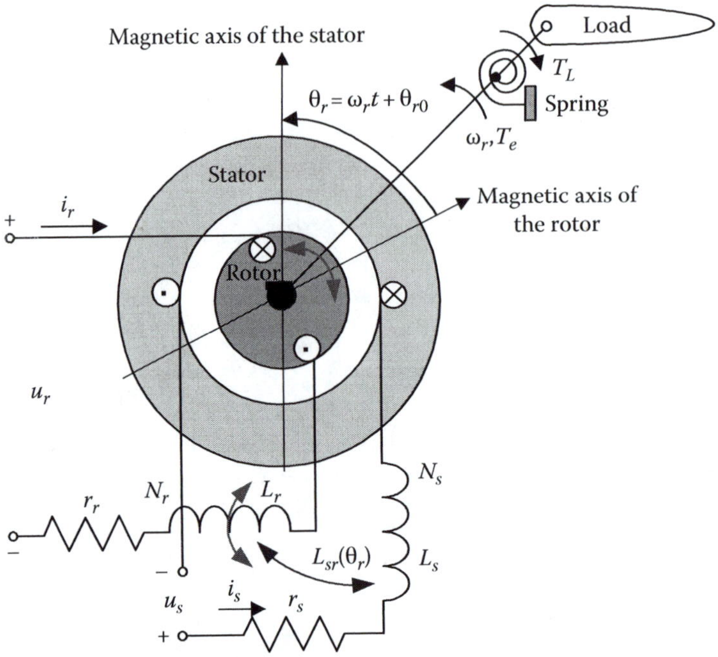

FIGURE 2.10 Actuator with stator and rotor windings.

The total kinetic energy of electrical and mechanical systems is found as a sum of the total electromagnetic (magnetic, electrical, etc.) Γ_E and mechanical Γ_M energies. The total kinetic energy of the stator and rotor circuitry is $\Gamma_E = \frac{1}{2}L_s\dot{q}_1^2 + L_{sr}\dot{q}_1\dot{q}_2 + \frac{1}{2}L_r\dot{q}_2^2$. The total kinetic energy of the mechanical system is $\Gamma_M = \frac{1}{2}J\dot{q}_3^2$. Therefore, $\Gamma = \Gamma_E + \Gamma_M = \frac{1}{2}L_s\dot{q}_1^2 + L_{sr}\dot{q}_1\dot{q}_2 + \frac{1}{2}L_r\dot{q}_2^2 + \frac{1}{2}J\dot{q}_3^2$.

The mutual inductance depends on the displacement of the rotor winding with respect to the stator winding. If the windings are orthogonal, we have $L_{sr} = 0$. Rotor rotates, and $L_{sr}(\theta_r)$ is a periodic function with a period 2π of the angular displacement, $L_{sr\,\min} \le L_{sr}(\theta_r) \le L_{sr\,\max}$. Let the mutual inductance L_{sr} be $L_{sr}(\theta_r) = L_M\cos\theta_r = L_M\cos q_3$, where L_M is the amplitude of the mutual inductance variations.

The total kinetic energy is

$$\Gamma = \frac{1}{2}L_s\dot{q}_1^2 + L_M\dot{q}_1\dot{q}_2\cos q_3 + \frac{1}{2}L_r\dot{q}_2^2 + \frac{1}{2}J\dot{q}_3^2.$$

Hence

$$\frac{\partial\Gamma}{\partial q_1} = 0,\ \frac{\partial\Gamma}{\partial\dot{q}_1} = L_s\dot{q}_1 + L_M\dot{q}_2\cos q_3,\ \frac{\partial\Gamma}{\partial q_2} = 0,\ \frac{\partial\Gamma}{\partial\dot{q}_2} = L_M\dot{q}_1\cos q_3 + L_r\dot{q}_2,\ \frac{\partial\Gamma}{\partial q_3} = -L_M\dot{q}_1\dot{q}_2\sin q_3,\ \frac{\partial\Gamma}{\partial\dot{q}_3} = J\dot{q}_3.$$

The potential energy of the spring with constant k_s is $\Pi = \frac{1}{2}k_s q_3^2$. Thus,

$$\frac{\partial\Pi}{\partial q_1} = 0,\quad \frac{\partial\Pi}{\partial q_2} = 0,\quad \frac{\partial\Pi}{\partial q_3} = k_s q_3.$$

The total heat energy dissipated is $D = D_E + D_M$, where D_E is the heat energy dissipated in the stator and rotor windings,

$$D_E = \frac{1}{2}r_s\dot{q}_1^2 + \frac{1}{2}r_r\dot{q}_2^2;$$

D_M is the heat energy dissipated by the mechanical system,

$$D_M = \frac{1}{2} B_m \dot{q}_3^2.$$

From

$$D = \frac{1}{2} r_s \dot{q}_1^2 + \frac{1}{2} r_r \dot{q}_2^2 + \frac{1}{2} B_m \dot{q}_3^2, \quad \frac{\partial D}{\partial \dot{q}_1} = r_s \dot{q}_1, \quad \frac{\partial D}{\partial \dot{q}_2} = r_r \dot{q}_2, \quad \text{and} \quad \frac{\partial D}{\partial \dot{q}_3} = B_m \dot{q}_3.$$

The relationships between the generalized coordinates and state variables are

$$q_1 = \int i_s \, dt, q_2 = \int i_r \, dt, q_3 = \theta_r, \dot{q}_1 = i_s, \dot{q}_2 = i_r, \dot{q}_3 = \omega_r, Q_1 = u_s, Q_2 = u_r, \quad \text{and} \quad Q_3 = -T_L.$$

We have three differential equations

$$L_s \frac{di_s}{dt} + L_M \cos\theta_r \frac{di_r}{dt} - L_M i_r \sin\theta_r \frac{d\theta_r}{dt} + r_s i_s = u_s,$$

$$L_r \frac{di_r}{dt} + L_M \cos\theta_r \frac{di_s}{dt} - L_M i_s \sin\theta_r \frac{d\theta_r}{dt} + r_r i_r = u_r,$$

$$J \frac{d^2\theta_r}{dt^2} + L_M i_s i_r \sin\theta_r + B_m \frac{d\theta_r}{dt} + k_s \theta_r = -T_L.$$

The nonlinear differential equations in Cauchy's form are

$$\frac{di_s}{dt} = \frac{-r_s L_r i_s - \frac{1}{2} L_M^2 i_s \omega_r \sin 2\theta_r + r_r L_M i_r \cos\theta_r + L_r L_M i_r \omega_r \sin\theta_r + L_r u_s - L_M \cos\theta_r u_r}{L_s L_r - L_M^2 \cos^2\theta_r},$$

$$\frac{di_r}{dt} = \frac{r_s L_M i_s \cos\theta_r + L_s L_M i_s \omega_r \sin\theta_r - r_r L_s i_r - \frac{1}{2} L_M^2 i_r \omega_r \sin 2\theta_r - L_M \cos\theta_r u_s + L_s u_r}{L_s L_r - L_M^2 \cos^2\theta_r},$$

$$\frac{d\omega_r}{dt} = \frac{1}{J}\left(-L_M i_s i_r \sin\theta_r - B_m \omega_r - k_s \theta_r - T_L\right),$$

$$\frac{d\theta_r}{dt} = \omega_r.$$

These nonlinear differential equations cannot be linearized. ■

Example 2.15: Beam Equations of Motion

Consider an elastic beam of length l with constant cross-sectional area A, uniform weight per unit volume (density) ρ, Young's modulus of elasticity E, and moment of inertia of the cross section about its neural axis I. As illustrated in Figure 2.11, the vertical displacement at the free end of the beam is q.

One finds the kinetic and potential energies. If the load F applied at the free end, or, load F_x is uniformly distributed, the beam deflection equations are

$$y(x) = \frac{1}{6EI} F\left(3lx^2 - x^3\right)$$

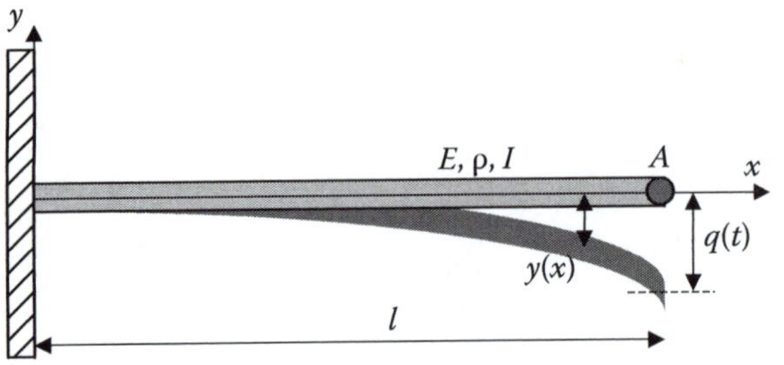

FIGURE 2.11 Beam in the *xy* plane.

and

$$y(x)=\frac{1}{24EI}F_x\left(x^4+6l^2x^2-4lx^3\right).$$

For the concentrated, uniformly varying, and other loads, the expressions for $y(x)$ can be found by solving the governing partial differential equations. Consider the third-order polynomial for $y(x)$, $y(x)=\frac{1}{2}\left(3\frac{x^2}{l^2}-\frac{x^3}{l^3}\right)q$. Using $q(t)$, we have $y(t,x)=\frac{1}{2}\left(3\frac{x^2}{l^2}-\frac{x^3}{l^3}\right)q(t)$.

The kinetic energy is

$$\Gamma(q)=\frac{1}{2}\int_0^l \dot{y}^2\,dm=\frac{1}{2}A\rho\int_0^l \frac{1}{4}\left(3\frac{x^2}{l^2}-\frac{x^3}{l^3}\right)^2\dot{q}^2\,dx=\frac{33}{280}A\rho l\dot{q}^2.$$

The potential energy of elastic deformation is

$$\Pi(q)=\frac{1}{2}\int_0^l EI\left(\frac{\partial^2 y}{\partial x^2}\right)^2 dx=\frac{1}{2}EI\int_0^l \frac{3}{2l^3}\left(1-\frac{x}{l}\right)^2 d\left(\frac{x}{l}\right)=\frac{3}{2}\frac{EI}{l^3}q^2$$

From the Lagrange equation

$$\frac{d}{dt}\left(\frac{\partial\Gamma}{\partial\dot{q}}\right)-\frac{\partial\Gamma}{\partial q}+\frac{\partial\Pi}{\partial q}=Q,$$

we have

$$\frac{d^2q}{dt^2}=-12.7\frac{EI}{A\rho l^4}q+F_q(t,x).$$

The potential energy of elastic beam is

$$\Pi=\frac{1}{2}\int_{\mathbf{r}}\sigma_{ij}\varepsilon_{ij}d\mathbf{r}+\int_{\mathbf{r}}T(\mathbf{r})w(\mathbf{r})d\mathbf{r}+\int_{\mathbf{r}}F(\mathbf{r})w(\mathbf{r})d\mathbf{r},$$

where $T(\mathbf{r})$ and $F(\mathbf{r})$ are the beam surface traction and force.

The term $\frac{1}{2}\sigma_{ij}\varepsilon_{ij}$ gives the strain energy stored. Explicit equations of motions can be derived. For laterally distributed load $T(x)$, the equation for the beam bending is $a_b\frac{d^4w}{dx^4}=T(x)$, $a_b = EI$. One may solve $\xi EI(x)\frac{\partial^5 y(t,x)}{\partial x^4\partial t}+EI(x)\frac{\partial^4 y(t,x)}{\partial x^4}+m_0(x)\frac{\partial^2 y(t,x)}{\partial t^2}+m(x)\frac{d^2\varphi}{dt^2}=F(t,x)$, where $F(t, x)$ is the distributed force through the beam. ■

2.4.3 Hamilton Equations of Motion

Applying the Hamilton concept, the differential equations are found using the generalized momenta p_i, $p_i=\partial L/\partial\dot{q}_i$. The Lagrangian function for the conservative systems is the difference between the total kinetic and potential energies $L\left(t,q_1,\ldots,q_n,\frac{dq_1}{dt},\ldots,\frac{dq_n}{dt}\right)=\Gamma\left(t,q_1,\ldots,q_n,\frac{dq_1}{dt},\ldots,\frac{dq_n}{dt}\right)-\Pi\left(t,q_1,\ldots,q_n\right)$. One concludes that the Lagrangian L is the function of $2n$ independent variables.

The Hamiltonian function is

$$H\left(t,q_1,\ldots,q_n,p_1,\ldots,p_n\right)=-L\left(t,q_1,\ldots,q_n,\frac{dq_1}{dt},\ldots,\frac{dq_n}{dt}\right)+\sum_{i=1}^{n}p_i\dot{q}_i,$$

and

$$H\left(t,q_1,\ldots,q_n,\frac{dq_1}{dt},\ldots,\frac{dq_n}{dt}\right)=\Gamma\left(t,q_1,\ldots,q_n,\frac{dq_1}{dt},\ldots,\frac{dq_n}{dt}\right)+\Pi\left(t,q_1,\ldots,q_n\right),$$

or

$$H\left(t,q_1,\ldots,q_n,p_1,\ldots,p_n\right)=\Gamma\left(t,q_1,\ldots,q_n,p_1,\ldots,p_n\right)+\Pi\left(t,q_1,\ldots,q_n\right).$$

The Hamiltonian represents the total energy, and H is a function of the generalized coordinates and generalized momenta. The Hamiltonian equations of motion are

$$\dot{p}_i=-\frac{\partial H}{\partial q_i},\quad \dot{q}_i=\frac{\partial H}{\partial p_i}. \tag{2.4}$$

One obtains the system of $2n$ first-order differential equations to describe the system dynamics. Using the Lagrange equations of motion, the system of n second-order differential equations results. However, the derived differential equations are equivalent.

Example 2.16

Consider the harmonic oscillator formed by the sliding mass m attached to the spring assuming that there is no friction. The total energy is given as the sum of the kinetic and potential energies, e.g.,

$$\Sigma_t=\Gamma+\Pi=\frac{1}{2}(mv^2+k_sx^2).$$

Using $q=x$, one finds Lagrangian

$$L\left(x,\frac{dx}{dt}\right)=\Gamma-\Pi=\frac{1}{2}(mv^2-k_sx^2)=\frac{1}{2}(m\dot{x}^2-k_sx^2).$$

From the Lagrange equation

$$\frac{d}{dt}\frac{\partial L}{\partial \dot{x}} - \frac{\partial L}{\partial x} = 0,$$

we find the second-order differential equation

$$m\frac{d^2x}{dt^2} + k_s x = 0.$$

The Newton second law yields the second-order differential equation

$$m\frac{d^2x}{dt^2} + k_s x = 0.$$

The Hamiltonian function is

$$H(x,p) = \Gamma + \Pi = \frac{1}{2}(mv^2 + k_s x^2) = \frac{1}{2}\left(\frac{1}{m}p^2 + k_s x^2\right).$$

From the Hamiltonian equations of motion (2.4), one obtains the following differential equations:

$$\dot{p} = -\frac{\partial H}{\partial x} = -k_s x, \quad \dot{x} = \dot{q} = \frac{\partial H}{\partial p} = \frac{p}{m}.$$

The equivalence of the resulting equations of motion is obvious. ■

2.5 FRICTION IN MOTION DEVICES

Friction is a very complex nonlinear phenomenon, which should be examined. The friction is exhibited by bearing, gear, and brushes–commutator, as shown in Figure 2.12. The *Coulomb* friction is a retarding frictional force or torque that changes its sign with the reversal of the direction of motion, and the amplitude of the frictional force or torque remains the same. For translational and rotational motions, the *Coulomb* friction force and torque are

$$F_{Coulomb} = k_{Fc}\,\mathrm{sgn}(v) = k_{Fc}\,\mathrm{sgn}\left(\frac{dx}{dt}\right) \quad \text{and} \quad T_{Coulomb} = k_{Tc}\,\mathrm{sgn}(\omega) = k_{Tc}\,\mathrm{sgn}\left(\frac{d\theta}{dt}\right),$$

where k_{Fc} and k_{Tc} are the *Coulomb* friction coefficients.

Figure 2.13a illustrates the *Coulomb* friction.

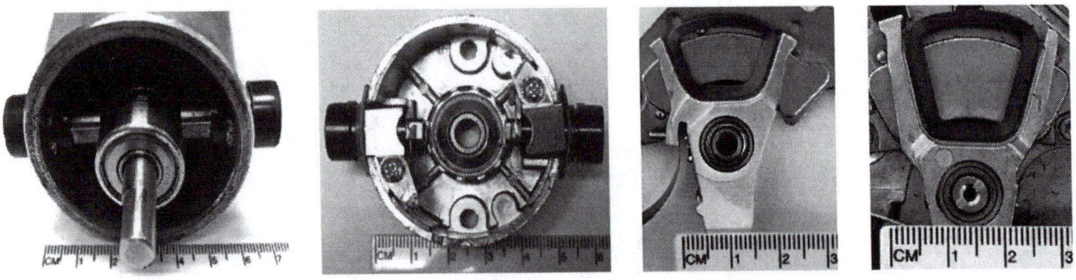

FIGURE 2.12 *Coulomb*, viscous, and static frictions are inherent phenomena in electromechanical motion devices.

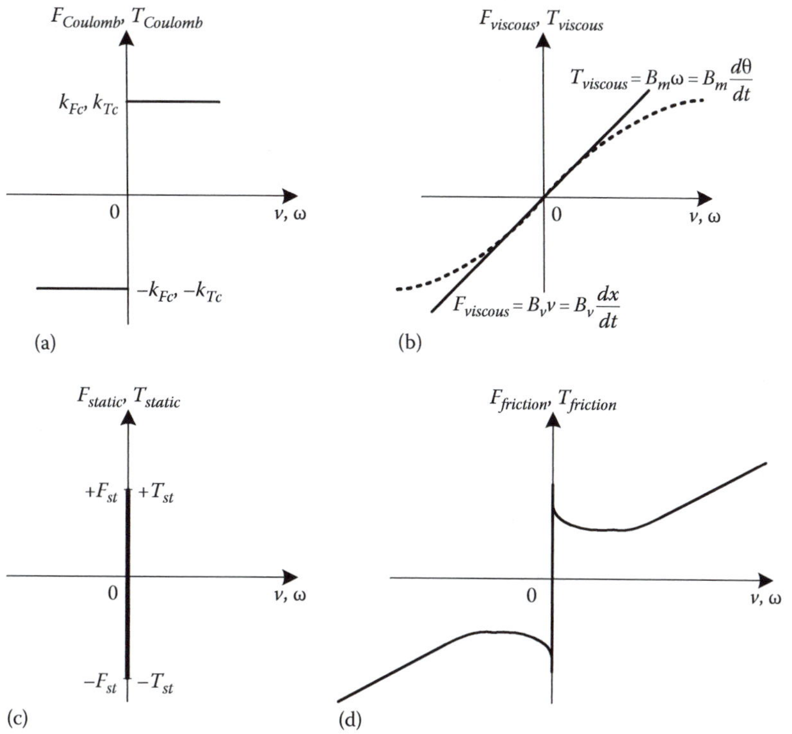

FIGURE 2.13 (a) *Coulomb* friction; (b) Viscous friction; (c) Static friction; (d) Friction force and torque.

Viscous friction is a retarding force or torque, which is a nonlinear function of linear or angular velocity. The representative characteristics of viscous friction force and torque versus velocities are shown in Figure 2.13b. The linear expressions to describe the viscous friction for translational and rotational motions are $F_{viscous} = B_v v$ and $T_{viscous} = B_m\omega$, where B_v and B_m are the viscous friction coefficients. The viscous friction is described by

$$T_{viscous} = p_1 \tanh(p_2\omega) + p_3\omega, \quad T_{viscous} = \frac{p_1 + p_2 e^{-p_3\omega}}{1 + p_4\omega}\omega,$$

and by other equations, where p_i are the unknown coefficients.

The static friction exists only when the body is stationary and vanishes as motion begins. The static friction is a force F_{static} or torque T_{static}, which may be expressed as

$$F_{static} = \pm F_{st}\Big|_{v=\frac{dx}{dt}=0}$$

and

$$T_{static} = \pm T_{st}\Big|_{\omega=\frac{d\theta}{dt}=0}.$$

The static friction is a retarding force or torque that tends to prevent the initial translational or rotational motion at the beginning, as shown in Figure 2.13c.

The friction force and torque are nonlinear functions that are modeled using frictional memory, presliding conditions, etc. The equations commonly used are

$$F_{friction} = \left(k_1 - k_2 e^{-k_4|v|} + k_3|v|\right)\mathrm{sgn}(v)$$

and

$$T_{friction} = \left(k_1 - k_2 e^{-k_4|\omega|} + k_3|\omega|\right)\mathrm{sgn}(\omega).$$

The typifying plots for $F_{friction}$ and $T_{friction}$ are shown in Figure 2.13d.

In electric drives, the viscous friction may be approximated as $T_{friction} = B_m\omega_r$. To design high-performance drives and servos, $T_{friction}$ must be accurately evaluated and measured. The steady-state and differential equations are found using the fluid lubrication hydrodynamics, dynamic viscosity, solid and fluid frictions on geometrical surfaces, surfaces interaction, and other phenomena. For the viscous and *Coulomb* frictions, the possible approximations are

$$T_{friction} = B_m\omega_r + b\,\mathrm{sgn}(\omega_r)\left(1 - e^{-c|\omega_r|}\right) \quad \text{and} \quad T_{friction} = B_m\omega_r + b\,\mathrm{sgn}(\omega_r)\left(1 - e^{-c\omega_r^2}\right),$$

where b and c are the coefficients, $b > 0$ and $c > 0$.

The inherent load-dependent time-varying asymmetry, eccentricity, lubrication, surface nonuniformity, wearing, temperature, surface roughness, and other phenomena significantly affect friction. The consistent expression is

$$T_{friction} = \sum_i B_{mi}\,\mathrm{sgn}(\omega_r)|\omega_r|^{\frac{i}{1+2\mu_1}} + \sum_j B_{mj}\omega_r^{j+2\mu_2}$$

$$+\,\mathrm{sgn}(\omega_r)\left[\sum_i b_i\left(1 - e^{-c_i\,\mathrm{sgn}(\omega_r)|\omega_r|^{i/(1+2\gamma_1)}}\right) + \sum_j b_j\left(1 - e^{-c_j|\omega_r|^{j+2\gamma_2}}\right)\right], \tag{2.5}$$

where $\mu_i = 0, 1, \ldots; \gamma_i = 0, 1, \ldots$.

Using the experimental data, the proposed (2.5) must be parametrized. The nonlinear mixed polynomial-exponential interpolation problem must be solved. The viscous friction depends on the angular velocity, loading, temperature, and other variables. Hence, $T_{friction}$ may be a multivariate function, and $T_{friction} = f(\omega_r, T_{Load}, T_{Temperature})$.

Example 2.17: Experimental Studies of Viscous Friction in Drives

For an electric motor, illustrated in Figure 2.12, we experimentally measure $T_{friction}$ in the velocity envelope $\omega_r \in [90 \quad 385]$ rad/sec. For $\omega_r \in [90 \quad 187 \quad 284 \quad 384]$ rad/sec, we have $T_{friction} = [3.28\times10^{-2} \quad 3.9\times10^{-2} \quad 4.37\times10^{-2} \quad 4.68\times10^{-2}]$ N-m. The viscous friction is a nonlinear function of ω_r as given by (2.5). Consider

$$T_{viscous} = B_m(\omega_r)\omega_r, \quad B_m(\omega_r) = T_{viscous}/\omega_r.$$

The experimentally measured data in Figure 2.14a is indicated by dots. Various exponential, transcendental, and other functions can be used to model $T_{viscous}(\omega_r)$. From (2.15), one may find the physics-consistent models

$$T_{viscous} = B_m(\omega_r)\omega_r = p_1\tanh(p_2\omega_r) + p_3\omega_r$$

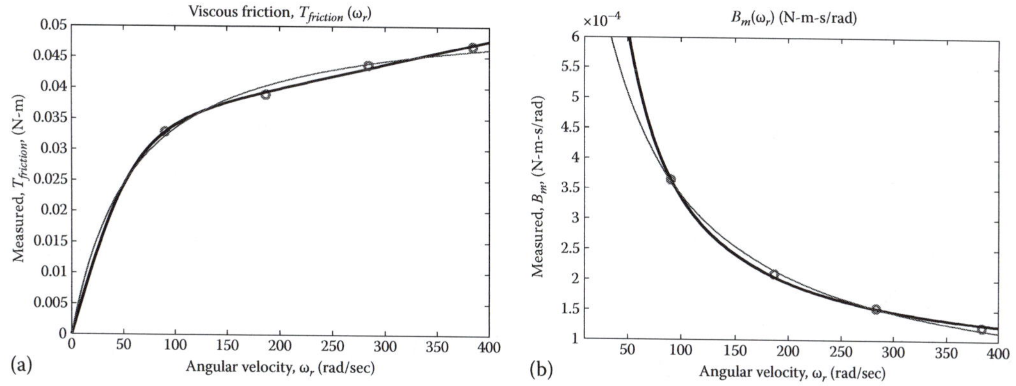

FIGURE 2.14 (a) Measured viscous friction data (dots) and the viscous friction torque by $T_{viscous} = 3.22 \times 10^{-2}\tanh(1.7 \times 10^{-2}\omega_r) + 3.88 \times 10^{-5}\omega_r$ and $T_{viscous} = \frac{9.05\times10^{-4} - 1.59e^{-41\omega_r}}{1+1.71\times10^{-2}\omega_r}\omega_r$; (b) Variations of $B_m(\omega_r)$: Using the measured four values for B_m (dots), one finds $B_m = \frac{2.82\times10^{-2}\tanh(3.1\omega_r)+5.31\times10^{-5}\omega_r}{\omega_r}$ and $B_m(\omega_r) = \frac{0.001+e^{-\omega_r}}{1+0.02\omega_r}$.

and

$$T_{viscous} = B_m(\omega_r)\omega_r = \frac{p_1 + p_2 e^{-p_3\omega_r}}{1+p_4\omega_r}\omega_r.$$

The unknown coefficients p_i are found solving the nonlinear interpolation problem in MATLAB. We find

$$T_{viscous} = 3.22\times10^{-2}\tanh(1.7\times10^{-2}\omega_r) + 3.88\times10^{-5}\omega_r$$

and

$$T_{viscous} = \frac{9.05\times10^{-4} - 1.59e^{-41\omega_r}}{1+1.71\times10^{-2}\omega_r}\omega_r.$$

The measured values for B_m at $\omega_r = \begin{bmatrix} 90 & 187 & 284 & 384 \end{bmatrix}$ rad/sec are as follows: $B_m = \begin{bmatrix} 3.64\times10^{-4} & 2.09\times10^{-4} & 1.54\times10^{-4} & 1.22\times10^{-4} \end{bmatrix}$ N-m-sec/rad. These $B_m(\omega_r)$ are shown in Figure 2.14b by four dots. From the $T_{viscous}(\omega_r)$ models,

$$B_m(\omega_r) = \frac{p_1\tanh(p_2\omega_r) + p_3\omega_r}{\omega_r}$$

and

$$B_m(\omega_r) = \frac{p_1 + p_2 e^{-p_3\omega_r}}{1+p_4\omega_r}.$$

The unknown coefficients p_i are found by using nonlinear interpolation supported by MATLAB. Using the **nlinfit** command, we have

```
format short e; % Deriving the Approximations for the Viscous Friction Torque
wr=[90 187 284 384]; Tfriction=[3.28e-02 3.9e-02 4.37e-02 4.68e-02]; Bm=Tfriction./wr;
plot(wr,Tfriction,'o','linewidth',3);
xlabel('Angular Velocity, {\it\omega_r} [rad/sec]','FontSize',18);
ylabel('Measured {\itT}_f_r_i_c_t_i_o_n, [N-m]','FontSize',18);
title('Viscous Friction {\itT}_f_r_i_c_t_i_o_n({\it\omega_r})' ,'FontSize',18);
ModelFunctionT1=@(p,x) p(1).*tanh(p(2).*x)+p(3).*x; StartingValuesT1=[1e-2 1e-2 1e-5];
CoefficientsT1=nlinfit(wr,Tfriction,ModelFunctionT1,StartingValuesT1); xgrid=linspace(0,400,100);
line(xgrid,ModelFunctionT1(CoefficientsT1,xgrid),'Color','k','linewidth',3); pause
ModelFunctionT2=@(p,x) x.*(p(1)+p(2).*exp(-p(3).*x))./(1+p(4).*x);
StartingValuesT2=[1e-2 1e-3 1e-2 0];
CoefficientsT2=nlinfit(wr, Tfriction,ModelFunctionT2,StartingValuesT2)
line(xgrid,ModelFunctionT2(CoefficientsT2,xgrid),'Color','b','linewidth',1); pause;
  % Deriving the Approximations for the Friction Mapping Bm
plot(wr,Bm,'o','linewidth',3); axis([10 400 1e-4 6e-4]);
xlabel('Angular Velocity, {\it\omega_r} [rad/sec]','FontSize',18);
ylabel('Measured {\itB_m}, [N-m-s/rad]','FontSize',18);
title('{\itB_m}({\it\omega_r}), [N-m-s/rad]' ,'FontSize',18);
ModelFunctionB1=@(p,x) (p(1).*tanh(p(2).*x)+p(3).*x)./x;StartingValuesB1=[1e-2 1e-2 1e-5];
CoefficientsB1=nlinfit(wr,Bm,ModelFunctionB1,StartingValuesB1); xgrid=linspace(0,400,100);
line(xgrid,ModelFunctionB1(CoefficientsB1,xgrid),'Color','k','linewidth',3); pause;
ModelFunctionB2=@(p,x) (p(1)+p(2).*exp(-p(3).*x))./(1+p(4).*x);
StartingValuesB2=[1e-2 1 1 1e-2];
CoefficientsB2=nlinfit(wr,Bm,ModelFunctionB2,StartingValuesB2); xgrid=linspace(0,400,100);
line(xgrid,ModelFunctionB2(CoefficientsB2,xgrid),'Color','b','linewidth',1);
```

For

$$B_m(\omega_r) = \frac{p_1 \tanh(p_2\omega_r) + p_3\omega_r}{\omega_r}$$

and

$$B_m(\omega_r) = \frac{p_1 + p_2 e^{-p_3\omega_r}}{1 + p_4\omega_r},$$

the unknown p_i are found to be

```
CoefficientsB1 =
2.8172e-02  3.0958e+00  5.3127e-05
CoefficientsB2 =
1.0170e-03  1.0000e+00  1.0000e+00  2.0005e-02
```

Hence,

$$B_m = \frac{2.82\times10^{-2}\tanh(3.1\omega_r) + 5.31\times10^{-5}\omega_r}{\omega_r} \quad \text{and} \quad B_m(\omega_r) = \frac{0.001 + e^{-\omega_r}}{1 + 0.02\omega_r}.$$

The plots for $B_m(\omega_r)$ are shown in Figure 2.14b. The interpolation error, convergence, and stability depend on algorithms, numerics, etc. Interpolation and estimates are found in an expanded operating envelope $\omega_r \in [\omega_{r\min} \quad \omega_{r\max}]$ obtaining the physics-consistent models for $T_{viscous}(\omega_r)$.

***Note*:** The polynomial and spline interpolants can be applied. We use physics-consistent rational, trigonometric, and exponential functions. The nonlinear trigonometric polynomials and mixed trigonometric-exponential interpolations are physics-consistent.

Frequently, one solves the bicubic, trilinear (three variable), and multivariate interpolation problems. For example, the viscous friction may be a multivariate function, $T_{friction} = f(\omega_r, T_{Load}, T_{Temperature})$. ■

Example 2.18: Experimental Studies of Friction in Servos

For the servo illustrated in Figure 2.12, the experimentally measured $T_{friction}$ in the angular velocity envelope $\omega_r \in [-25 \quad 25]$ rad/sec is shown in Figures 2.15. Using (2.5), we found

$$T_{friction} = B_{m10}\omega_r + B_{m11}\,\mathrm{sgn}(\omega_r)|\omega_r|^{\frac{1}{3}} + b\,\mathrm{sgn}(\omega_r)\left(1 - e^{-c\omega_r^2}\right), \quad B_{m10} = 6.1\times10^{-5}, \quad B_{m11} = 1.92\times10^{-4},$$

$$b = 5.8\times10^{-4}, \quad c = 19.5.$$

The experimentally measured $T_{friction}$ (dots) and corresponding plots for the derived $T_{friction}(\omega_r)$ are shown in Figures 2.15.

Note: Nonlinear interpolation is examined in Examples 2.17 and 2.28 using the **`nlinfit`** and **`lsqnonlin`** solvers. ■

2.6 *APPLICATION OF ELECTROMAGNETICS AND MECHANICS TO ELECTROMECHANICAL SYSTEMS*

The laws of electromagnetics and mechanics are applied to describe physical phenomena and examine electromechanical motion devices. Electromechanical devices can be described and modeled using different concepts. Forces and torques can be found using Maxwell's stress tensor and energies. Nonlinear partial differential equations and tensor calculus may ensure high-fidelity modeling. Due to complexity, the lumped-parameter models are widely used, which ensure accuracy, practicality, and consistency. Consider a rotor (bar magnet, current loop, and solenoid) in a uniform magnetic field, as illustrated in Figure 2.16.

The torque tends to align the magnetic moment $\vec{m}$ with $\vec{B}$, and $\vec{T} = \vec{m} \times \vec{B}$.

Consider a magnetic bar with the length of dipole l and the magnetic pole strength Q_m. The magnetic dipole moment is $m = Q_m l$, while the force is $F = Q_m B$. The north pole (+) produces a force $Q_m B$ to the right, while the south pole (−) exhibits a force $Q_m B$ to the left. As illustrated in Figure 2.16, the electromagnetic torque is $T = 2F\frac{1}{2}l\sin\alpha = Q_m lB\sin\alpha = mB\sin\alpha$. The torque tends to align the bar with the magnetic field $\vec{B}$, and it rotates the rotor clockwise.

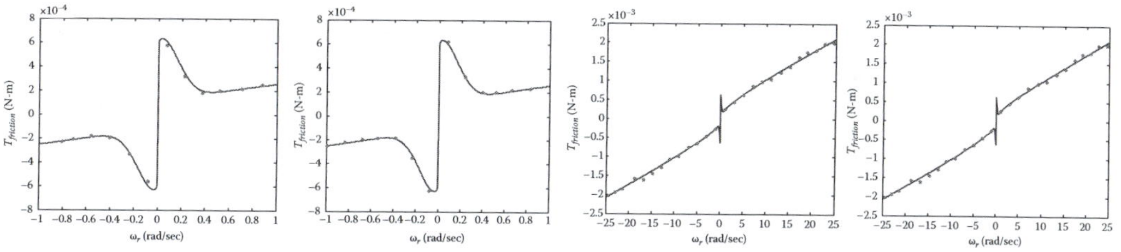

FIGURE 2.15 Experimental data for the measured $T_{friction}$ (dots) and the *Coulomb*, static, and viscous friction model $T_{friction} = B_{m10}\omega_r + B_{m11}\omega_r^{\frac{1}{3}} + b\,\mathrm{sgn}(\omega_r)(1 - e^{-c\omega_r^2})$ (solid line) in the full angular velocity ω_r envelope.

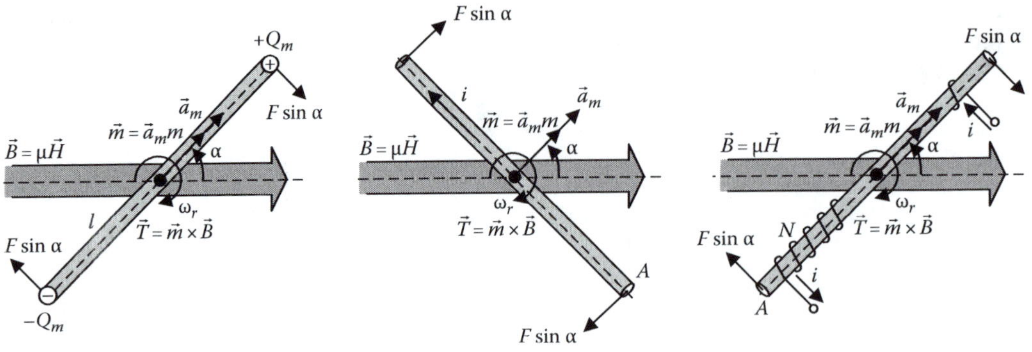

FIGURE 2.16 Clockwise rotation of magnetic bar, current loop, and solenoid.

Using the vector notations, one obtains $\vec{T} = \vec{m} \times \vec{B} = m\vec{a}_m \times \vec{B} = Q_m l\vec{a}_m \times \vec{B}$, where $\vec{a}_m$ is the unit vector in the magnetic moment direction.

For a current loop with the cross-sectional loop area A, $\vec{T} = \vec{m} \times \vec{B} = m\vec{a}_m \times \vec{B} = iA\vec{a}_m \times \vec{B}$.

For a solenoid with N turns, shown in Figure 2.16, one obtains $\vec{T} = \vec{m} \times \vec{B} = \vec{a}_m m \times \vec{B} = iAN\vec{a}_m \times \vec{B}$.

The expression for the electromagnetic torque is used in the *torsional–mechanical* dynamics

$$J\frac{d\omega_r}{dt} = \sum \vec{T}_\Sigma.$$

The total magnetic flux through the surface is $\Phi = \int \vec{B} \cdot d\vec{s}$. The Ampere circuital law is $\oint_l \vec{B} \cdot d\vec{l} = \mu_0 \int_s \vec{J} \cdot d\vec{s}$. For the filamentary current, Ampere's law relates the magnetic flux with the algebraic sum of the enclosed (linked) currents (*net current*) i_n, $\oint_l \vec{B} \cdot d\vec{l} = \mu i_n$. The time-varying magnetic field produces the electromotive force (*emf*) $\mathscr{E}$, which induces the current in the closed circuit. Faraday's law relates the *emf* (induced voltage due to conductor's motion in the magnetic field) to the rate of change of the magnetic flux Φ penetrating the loop. Lenz's law is used to find the direction of *emf* and the current induced. The *emf* is in a direction to produce a current whose flux, if added to the original flux, would reduce the magnitude of the *emf*. According to Faraday's law of induction, one has

$$\mathscr{E} = \oint_s \vec{E}(t) \cdot d\vec{l} = -\int_s \frac{\partial \vec{B}(t)}{\partial t} \cdot d\vec{s} = -N\frac{d\Phi}{dt} = -\frac{d\psi}{dt}.$$

The current flows in an opposite direction to the flux linkages ψ. The *emf* is measured in volts. It represents a magnitude of the potential difference V in a circuit carrying current. We have

$$V = -ir + \mathscr{E} = -ir - \frac{d\psi}{dt}.$$

The Kirchhoff voltage law states that around a closed path in an electric circuit, the algebraic sum of the *emf* is equal to the algebraic sum of the voltage drop across the resistance. The algebraic sum of the voltages around any closed path is zero. The Kirchhoff current law states that the algebraic sum of the currents at any node is zero.

The magnetomotive force (*mmf*) is the line integral of the time-varying magnetic field intensity $\vec{H}(t)$,

$$mmf = \oint_l \vec{H}(t) \cdot d\vec{l}.$$

The unit for the *mmf* is amperes or ampere-turns. The inductance is the ratio of the total flux linkages to the current which they link,

$$L = \frac{N\Phi}{i}.$$

The reluctance is the ratio of the *mmf* to the total flux, $\Re = \frac{mmf}{\Phi}$. Hence,

$$\Re = \frac{\oint_l \vec{H} \cdot d\vec{l}}{\int_s \vec{B} \cdot d\vec{s}}.$$

The *emf* and *mmf* are used to find inductance and reluctance. The equation $L = \psi/i$ yields

$$\mathscr{E} = -\frac{d\psi}{dt} = -\frac{d(Li)}{dt} = -L\frac{di}{dt} - i\frac{dL}{dt}.$$

If $L = \text{const}$, $\mathscr{E} = -L\frac{di}{dt}$. The self-inductance is the magnitude of the self-induced *emf* per unit rate of change of current.

The force–energy and torque–energy relations in electromagnetic and electrostatic actuators are examined. The energy stored in the capacitor is $\frac{1}{2}CV^2$, while the energy stored in the inductor is $\frac{1}{2}Li^2$. The energy in the capacitor is stored in the electric field between plates, while the energy in the inductor is stored in the magnetic field within the coils. In the variable-reluctance relays, solenoids, and magnetic levitation systems, the reluctance and inductance vary. One finds the electromagnetic force and torque using the coenergy. In many electromagnetic motion devices, the magnetic coupling is between windings that are carrying currents and the stationary magnetic field developed by permanent magnets or electromagnets. In separately exited DC, conventional synchronous generators and induction machines, there is a magnetic coupling between windings due to their mutual inductances. To derive the electromagnetic force or torque, one applies equations

$$F_e = -i\oint_l \vec{B}(t) \cdot d\vec{l} \quad \text{or} \quad \vec{T} = \vec{m} \times \vec{B}.$$

Using the coenergy $W_c[i, L(x)]$ or $W_c[i, L(\theta)]$, we have

$$F_e(i,x) = -\nabla W_c = -\frac{\partial W_c[i, L(x)]}{\partial x}$$

and

$$T_e(i,x) = -\nabla W_c = -\frac{\partial W_c[i, L(\theta)]}{\partial \theta}.$$

***Note*:** The magnetic flux crossing surface is $\Phi = \oint_s \vec{B} \cdot d\vec{s}$. The expression $\Phi = BA$ must be used with a great caution. In general, $\oint_s \vec{B} \cdot d\vec{s} \neq BA$. ■

Example 2.19

Solenoids have movable and stationary members made from high-permeability ferromagnetic materials. These electromechanical devices, convert electrical energy to mechanical energy. Solenoids and relays operate due to the varying reluctance $\Re$. The force is produced due to the changes in the magnetizing inductance $L = 1/\Re$. Performance of solenoids is strongly affected by the magnetic system, materials, geometry, relative permeability, etc. Let $L(x) = e^{-2x}$. From $W_c(x) = \frac{1}{2}e^{-2x}i^2$, the electromagnetic force is

$$F_e(x) = -\frac{\partial W_c(x)}{\partial x} = e^{-2x}i^2.$$

The stored magnetic energy is $W_m = \frac{1}{2}Li^2$. ■

Example 2.20

Two coils have a mutual inductance $L_{12} = 0.5$ H. The current in the first coil is $i_1 = \sqrt{\sin 4t}$, $\sin 4t > 0$. One can find the induced *emf* in the second coil as $\mathscr{E}_2 = L_{12}\frac{di_1}{dt}$. By using the power rule, for the time-varying current in the first coil $i_1 = \sqrt{\sin 4t}$, we have $\frac{di_1}{dt} = \frac{2\cos 4t}{\sqrt{\sin 4t}}$. Hence, $\xi_2 = \frac{\cos 4t}{\sqrt{\sin 4t}}$, $\sin 4t > 0$. ■

2.7 SIMULATION OF SYSTEMS USING MATLAB®

MATLAB (MATrix LABoratory, http://www.mathworks.com) is a high-performance interacting software environment for high-efficiency engineering and scientific numerical calculations [9,10]. This environment can be applied to perform heterogeneous simulations and data-intensive analysis of dynamic systems. One can solve a wide spectrum of control, optimization, identification, data acquisition, and other problems using this software. Excellent interactive capabilities, flexibility, and versatility are attained. In addition, MATLAB allows compiling features with high-level programming languages and possesses consistent graphical and interface capabilities. A family of application-specific toolboxes, with a collection and libraries of m-files for solving problems, guarantees effectiveness. The graphical mouse-driven interactive Simulink environment enables MATLAB. A great number of outstanding books and MathWorks user manuals in Simulink and MATLAB and its toolboxes are available. The MathWorks Inc. educational URL is http://www.mathworks.com/academia.

This section introduces the MATLAB environment. The MATLAB version 8.6 is used. MATLAB documentation and user manuals (thousands of pages each) are available in portable document format. This book focuses on MATLAB applications to electromechanical systems, educating one how to solve practical problems. This section is not aimed to substitute hundreds of excellent stand-alone books on MATLAB. To start MATLAB, double-click the following icon. The MATLAB Command Window with Launch Pad and Command History will appear on the screen, as shown in Figure 2.17. Typing `ver` or `demo`, the available toolboxes are listed, as shown in Figure 2.17.

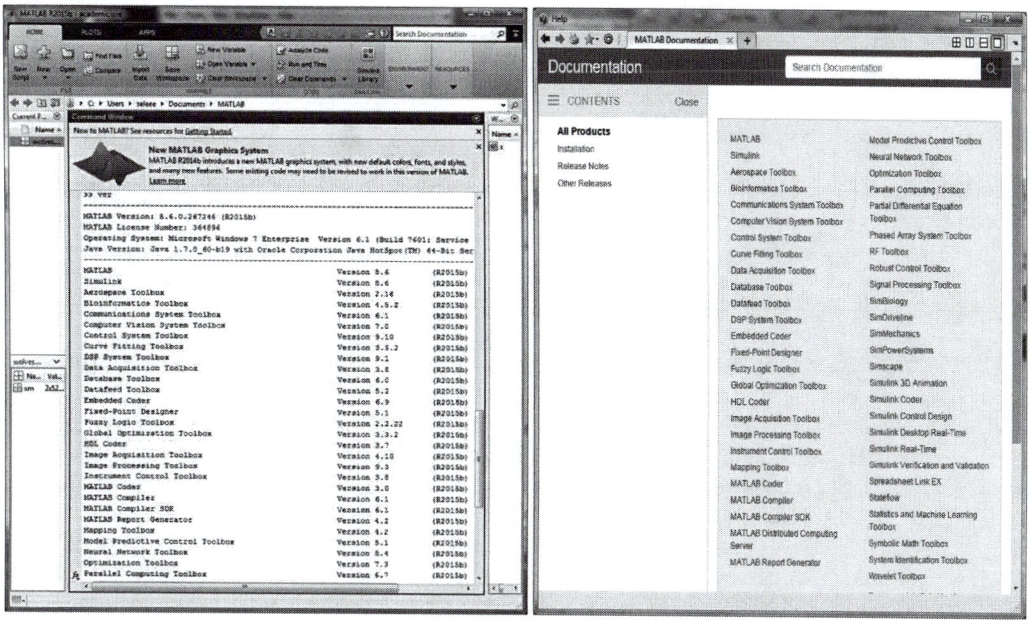

FIGURE 2.17 MATLAB® Command Window and MATLAB toolboxes.

The line

```
>>
```

is the MATLAB prompt. Typing

```
>> a=1+2+3*4
```

and pressing the Enter key, we have `a = 15`

Example 2.21

Calculate and plot functions $y = \sin(2x)$, $y = \sin(2x)\cos(2x)$, and $y = \sin(2x)e^{-x/2}$ if x varies from 0 to 3π. Let the increment be 0.025π. The MATLAB statement typed in the Command Window is

```
x=0:0.025*pi:3*pi; y1=sin(2*x); y2=sin(2*x).*cos(2*x); y3=sin(2*x).*exp(-x./2);
plot(x,y1,'-',x,y1,'o','Linewidth',2); axis([0 3*pi -1.05  1.05]);
title('Function y=sin(2{\itx})','FontSize',18); pause;
plot(x,y2,'-',x,y2,'o','Linewidth',2); axis([0 3*pi -0.6  0.6]);
title('Function y=sin(2{\itx})cos(2{\itx})','FontSize',18); pause;
plot(x,y3,'-',x,y3,'o','Linewidth',2); axis([0 3*pi -0.35  0.75]);
title('Function y=sin(2{\itx}){\ite}^{-0.5\itx}','FontSize',18);
```

The calculations are made, and the figures appear as shown in Figures 2.18. To capture these plots, one clicks the Edit icon and selects the Copy Figure option. One may refine the plotting statement. The help is displayed if one types >> `help plot` and presses the Enter key. ■

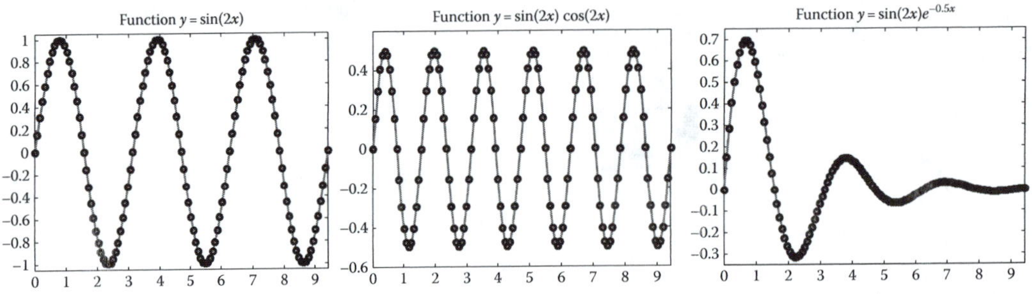

FIGURE 2.18 Plots for $y = \sin(2x)$, $y = \sin(2x)\cos(2x)$, and $y = \sin(2x)e^{-x/2}$.

The nonlinear differential equations that describe the dynamics of electromechanical systems usually cannot be solved analytically. For simple equations, analytic solution can be found using MATLAB. For translational and rotational rigid-body, one-dimensional mechanical systems, we found the second-order differential equations

$$m\frac{d^2x}{dt^2} + B_v\frac{dx}{dt} + k_s x = F_a(t)$$

and

$$J\frac{d^2\theta}{dt^2} + B_m\frac{d\theta}{dt} + k_s\theta = T_a(t),$$

where $F_a(t)$ and $T_a(t)$ are the time-varying applied force and torque.

For parallel and series *RLC* circuits, illustrated in Figures 2.19a, one obtains

$$C\frac{d^2u}{dt^2} + \frac{1}{R}\frac{du}{dt} + \frac{1}{L}u = \frac{di_a}{dt} \quad \text{or} \quad \frac{d^2u}{dt^2} + \frac{1}{RC}\frac{du}{dt} + \frac{1}{LC}u = \frac{1}{C}\frac{di_a}{dt},$$

and

$$L\frac{d^2i}{dt^2} + R\frac{di}{dt} + \frac{1}{C}i = \frac{du_a}{dt} \quad \text{or} \quad \frac{d^2i}{dt^2} + \frac{R}{L}\frac{di}{dt} + \frac{1}{LC}i = \frac{1}{L}\frac{du_a}{dt}.$$

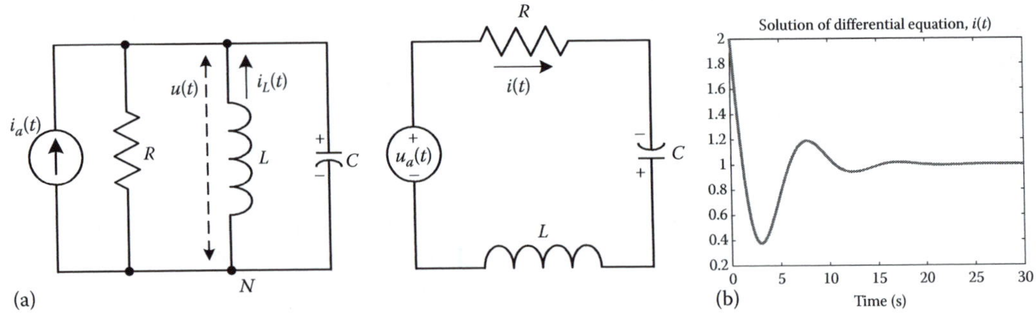

FIGURE 2.19 (a) Parallel and series *RLC* circuits; (b) Dynamics due to the unit step and initial conditions.

The analytic solutions of linear differential equations with constant coefficients are known. The general solution of the second-order linear differential equation is found by using the characteristic roots (eigenvalues) of the characteristic equation. The damping coefficient ξ and the resonant frequency ω_0 for the *RLC* circuits (parallel and series) and translational motion are $\xi=\frac{1}{2RC}$, $\omega_0=\frac{1}{\sqrt{LC}}$, $\xi=\frac{R}{2L}$, $\omega_0=\frac{1}{\sqrt{LC}}$, and $\xi=\frac{B_v}{2\sqrt{k_s m}}$, $\omega_0=\sqrt{\frac{k_s}{m}}$, respectively.

For a linear second-order differential equation

$$\frac{d^2x}{dt^2}+2\xi\frac{dx}{dt}+\omega_0 x=f(t),$$

using the Laplace operator $s = d/dt$, $s^2 = d^2/dt^2$, the characteristic equation is

$$\left|s^2+2\xi s+\omega_0\right|=0,\quad s^2+2\xi s+\omega_0^2=(s-s_1)(s-s_2)=0.$$

The characteristic roots (eigenvalues) are

$$s_{1,2}=-\xi\pm\sqrt{\xi^2-\omega_0^2}.$$

Case 1: If $\xi^2>\omega_0^2$, the real distinct characteristic roots s_1 and s_2 result. The general solution is $x(t)=ae^{s_1t}+be^{s_2t}+c_f$, where coefficients a and b are found using the initial conditions; c_f is the solution due to the *forcing* function *f*. For the *RLC* circuits, *f* is $i_a(t)$ or $u_a(t)$.

Case 2: For $\xi^2=\omega_0^2$, the characteristic roots are real and identical, e.g., $s_1 = s_2 = -\xi$. The solution of the second-order differential equation is $x(t) = (a + b)e^{-\xi t} + c_f$.

Case 3: If $\xi^2<\omega_0^2$, the complex distinct characteristic roots are $s_{1,2}=-\xi\pm j\sqrt{\omega_0^2-\xi^2}$. The general solution is

$$x(t)=e^{-\xi t}\left[a\cos\left(\sqrt{\omega_0^2-\xi^2 t}\right)+b\sin\left(\sqrt{\omega_0^2-\xi^2 t}\right)\right]+c_f=e^{-\xi t}\sqrt{a^2+b^2}\cos\left[\left(\sqrt{\omega_0^2-\xi^2 t}\right)+\tan^{-1}\left(\frac{-b}{a}\right)\right]+c_f.$$

Example 2.22

Consider the series *RLC* circuit illustrated in Figure 2.19a. We find and plot the transient response due to the unit step input. Let $R = 0.5$ ohm, $L = 1$ H, $C = 2$ F, $a = 1$, and $b = -1$.

The series *RLC* circuit is described by the differential equation

$$\frac{d^2i}{dt^2}+\frac{R}{L}\frac{di}{dt}+\frac{1}{LC}i=\frac{1}{L}\frac{du_a}{dt},$$

which yields the following characteristic equation:

$$s^2+\frac{R}{L}s+\frac{1}{LC}=0.$$

The characteristic eigenvalues are

$$s_1=-\frac{R}{2L}-\sqrt{\left(\frac{R}{2L}\right)^2-\frac{1}{LC}}\quad\text{and}\quad s_2=-\frac{R}{2L}+\sqrt{\left(\frac{R}{2L}\right)^2-\frac{1}{LC}}.$$

If $\left(\frac{R}{2L}\right)^2 > \frac{1}{LC}$, the characteristic eigenvalues are real and distinct. For

$$\left(\frac{R}{2L}\right)^2 = \frac{1}{LC},$$

the eigenvalues are real and identical. If

$$\left(\frac{R}{2L}\right)^2 < \frac{1}{LC},$$

the eigenvalues are complex.

For the assigned values for R, L, and C, the characteristic eigenvalues are complex. The dynamics is underdamped with a solution

$$i(t) = e^{-\xi t}\left[a\cos\left(\sqrt{\omega_0^2 - \xi^2 t}\right) + b\sin\left(\sqrt{\omega_0^2 - \xi^2 t}\right)\right] + c_f, \quad \xi = \frac{R}{2L} = 0.25, \quad \text{and} \quad \omega_0 = \frac{1}{\sqrt{LC}} = 0.71.$$

The MATLAB statements to calculate $i(t)$ and perform plotting are

```
R=0.5; L=1; C=2; a=1; b=-1; cf=1; e=R/2*L; w0=1/sqrt(L*C);
t=0:.01:30; x=exp(-e*t).*(a*cos(sqrt(w0^2-e^2)*t)+b*sin(sqrt(w0^2-e^2)*t))+cf;
plot(t,x ,'Linewidth',3); xlabel('Time [seconds]','FontSize',18);
title('Solution of Differential Equation {\iti}({\itt})','FontSize',18);
```

The resulting dynamics for $i(t)$ is documented in Figure 2.19b. ■

Example 2.23: Application of MATLAB to Analytically Solve Differential Equations

Using the Symbolic Toolbox, we analytically solve the third-order differential equation

$$\frac{d^3x}{dt^3} + 2\frac{dx}{dt} + 3x = 10f.$$

Using the analytic differential equation solver command `dsolve`, we have

```
x=dsolve('D3x+2*Dx+3*x=10*f')
```

The resulting solution is displayed

```
x =
(10*f)/3 + C3*exp(-t) + C1*exp(t/2)*cos((11^(1/2)*t)/2) + C2*exp(t/2)*sin((11^(1/2)*t)/2)
```

Using the `pretty` command, we find

```
>> pretty(x)
10 f                                        / sqrt(11) t \                        / sqrt(11) t \
---- + C3 exp(-t) + C1 exp(t/2) cos| ---------- | + C2 exp(t/2) sin| ---------- |
  3                                         \     2      /                        \     2      /
```

Hence, $x(t) = \frac{10}{3}f + c_3e^{-t} + c_1e^{0.5t}\cos\left(\frac{1}{2}\sqrt{11}t\right) + c_2e^{0.5t}\sin\left(\frac{1}{2}\sqrt{11}t\right)$.

Assigning the initial conditions, the unknown constants are found. Let $\left(\frac{d^2x}{dt^2}\right)_0 = 5$, $\left(\frac{dx}{dt}\right)_0 = 15$ and $x_0 = -20$. Hence,

```
>> x=dsolve('D3x+2*Dx+3*x=10*f','D2x(0)=5','Dx(0)=15','x(0)=-20'); pretty(x)
```

The resulting solution with the derived c_1, c_2, and c_3 is

```
10 f                                      / sqrt(11) t \ / 4 f     \
---- - exp(-t) (2 f + 14) - exp(t/2) cos| ---------- || --- + 6  |
  3                                       \     2      / \  3      /
                      / sqrt(11) t \
     sqrt(11) exp(t/2) sin| ---------- | (f - 3) 8
                          \     2      /
   - -----------------------------------------------
                         33
```

Thus, $x(t) = \frac{10}{3} f - e^{-t}(2f+14) - e^{0.5t}\cos\left(\frac{1}{2}\sqrt{11}t\right)\left(\frac{4}{3}f+6\right) - \frac{\sqrt{11}}{33}e^{0.5t}\sin\left(\frac{1}{2}\sqrt{11}t\right)(f-3)8.$

For time-varying the forcing functions, the analytic solution of $\frac{d^3x}{dt^3} + 2\frac{dx}{dt} + 3x = f(t)$ is found by using the statement `x=dsolve('D3x+2*Dx+3*x=f(t)'); pretty(x)`. Letting $f(t) = 50\cos(10t)$ and by using

```
x=dsolve('D3x+2*Dx+3*x=50*cos(5*t)','D2x(0)=5','Dx(0)=15','x(0)=-20'); pretty(x)
```

we found a solution for $x(t)$ as follows:

```
  2875               75               187
- ---- sin(5 t) + ---- cos(5 t)  -  --- exp(-t)
  6617              6617              13
       5702                            1/2       1/2
     + ---- exp(1/2 t) sin(1/2 11    t) 11
       5599
       2864                          1/2
     - ---- exp(1/2 t) cos(1/2 11  t)
        509
```

Hence,

$$x(t) = -\frac{2875}{6617}\sin 5t + \frac{75}{6617}\cos 5t - \frac{187}{13}e^{-t} + \frac{5702}{5599}e^{0.5t}\sin\left(\frac{1}{2}\sqrt{11}t\right)\sqrt{11} - \frac{2864}{509}e^{0.5t}\cos\left(\frac{1}{2}\sqrt{11}t\right).$$

■

For many nonlinear differential equations, an analytic solution cannot be derived. Therefore, numerical solutions must be found. The following example illustrates the application of MATLAB to numerically solve nonlinear differential equations.

Example 2.24

Using the MATLAB `ode45` command (ordinary differential equations solver), we numerically solve a system of highly nonlinear differential equations

$$\frac{dx_1(t)}{dt} = -20x_1 + |x_2x_3| + 10x_1x_2x_3, \quad x_1(t_0) = x_{10},$$

$$\frac{dx_2(t)}{dt} = -5x_1x_2 - 10\cos x_1 - \sqrt{|x_3|}, \quad x_2(t_0) = x_{20},$$

$$\frac{dx_3(t)}{dt} = -5x_1x_2 + 50x_2\cos x_1 - 25x_3, \quad x_3(t_0) = x_{30}.$$

The initial conditions are $x_0 = \begin{bmatrix} x_{10} \\ x_{20} \\ x_{30} \end{bmatrix} = \begin{bmatrix} 2 \\ 1 \\ -2 \end{bmatrix}$. Two m-files (`ch2_1.m` and `ch2_2.m`) are developed to numerically simulate a set of nonlinear differential equations. The dynamics of the state variables $x_1(t)$, $x_2(t)$, and $x_3(t)$ are plotted using the `plot` command. Comments, which are not executed, appear after the % symbol. These comments explain sequential steps. The MATLAB file (`ch2_1.m`) with the `ode45` solver, two- and three-dimensional plotting using the `plot` and `plot3` commands, is as follows:

```
echo on; clear all
t0=0; tfinal=1; tspan=[t0 tfinal];               % initial and final time
y0=[2 1 -2]';                                     % initial conditions for state variables
[t,y]=ode45('ch2_2',tspan,y0);                    %ode45 MATLAB solver using ode45 solver
% Plot of the transient dynamics by solving differential equations
% These differential equations are given in file ch2_2.m
plot(t,y(:,1),'-',t,y(:,2),'--',t,y(:,3),':','Linewidth',2.5);
% plot the transient dynamics
xlabel('Time (seconds)','FontSize',16); ylabel('State Variables','FontSize',16);
title('Solution of Differential Equations: {\itx}_1(t), {\itx}_2(t), {\itx}_3(t)','FontSize',16);
pause
% 3-D plot using x1, x2 and x3
plot3(y(:,1),y(:,2),y(:,3),'Linewidth',2.5); xlabel('{\itx}_1','FontSize',16);
ylabel('{\itx}_2','FontSize',16); zlabel('{\itx}_3','FontSize',16);
title('Three-Dimensional States Evolution:{\itx}_1({\itt}),{\itx}_2({\itt}),{\itx}_3({\itt})','FontSize',16);
text(0,-2.5,2,'Origin','FontSize',14);
```

The second MATLAB file (`ch2_2.m`), with a set of differential equations to be numerically solved, is

```
% Simulation of the third-order differential equations
function yprime = difer(t,y);
% Differential equations parameters
a11=-20; a12=1; a13=10; a21=-5; a22=-10; a31=-5; a32=50; a33=-25;
% Three differential equations:
yprime=[a11*y(1,:)+a12*abs(y(2,:)*y(3,:))+a13*y(1,:)*y(2,:)*y(3,:);...
a21*y(1,:)*y(2,:)+a22*cos(y(1,:))+sqrt(abs(y(3,:)));...
a31*y(1,:)*y(2,:)+a32*cos(y(1,:))*y(2,:)+a33*y(3,:)];
```

To calculate the transient dynamics and plot the transient dynamics, one types in the Command window `ch2_1` and presses the Enter key. The resulting transient behavior is documented in Figure 2.20a. The three-dimensional evolution of the state variables is illustrated in Figure 2.20b.

The resulting data for `x`, which is displayed in the Command Window typing `x`, is reported below. We have four columns for time t, as well as for three variables $x_1(t)$, $x_2(t)$, and $x_3(t)$. That is $x = [t, x_1, x_2, x_3]$.

```
x =
      0     2.0000    1.0000   -2.0000
 0.0013     1.9025    0.9941   -1.9724
.............................................
 1.0000     0.8543   -1.2543   -1.4336
```

One can perform plotting, data mining, filtering, and other advanced numerics using MATLAB. ■

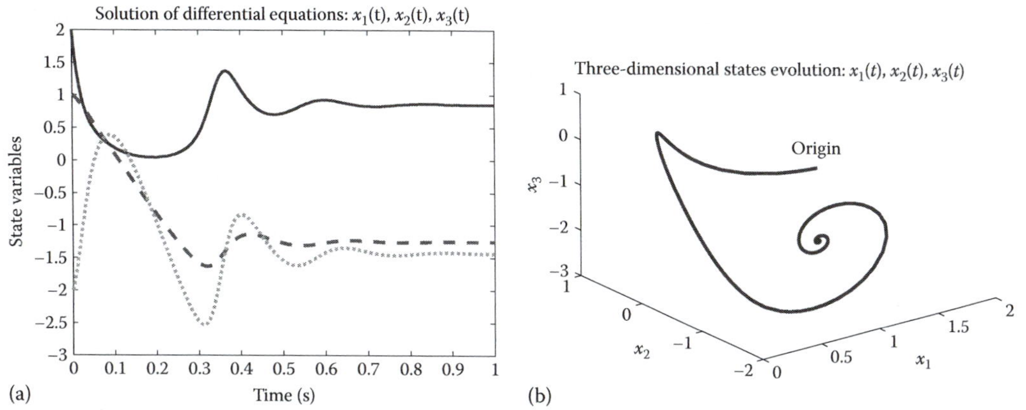

FIGURE 2.20 (a) Dynamics and evolution of the state variables (x_1, x_2, and x_3 are plotted using the solid, dashed, and dotted lines); (b) Three-dimensional plot for (x_1, x_2, x_3).

Simulink is an interactive computing environment for the simulation of dynamic systems. Simulink allows one to numerically simulate and analyze systems by developing a model's block diagrams. Simulink is applied to linear, nonlinear, continuous-time, discrete-time, multivariable, multirate, and hybrid systems. Blocksets are built-in blocks in Simulink that provide a block library for different system components. A C-code from block diagrams is generated using the Real-Time Workshop Toolbox. Using a mouse-driven block-diagram interface, the Simulink models (block-diagrams) can be built. Simulink provides a graphical user interface (GUI) for building models.

A library of sinks, sources, linear and nonlinear components (blocks), connectors, and customized blocks (S-functions) provide flexibility, interactability, and efficiency. Complex systems can be built using high- and low-level blocks. Systems can be numerically simulated by solving differential equations using a number of MATLAB commands. Different methods and various MATLAB algorithms are supported and used by Simulink. The easy-to-use Simulink menu suits interactive simulations, analyses, and visualizations. To start Simulink, one may type in the Command Window `Simulink` and press the Enter key. Clicking on the Simulink library icon brings up the Library Browser window, as shown in Figure 2.21. To run various Simulink demonstration programs, type `demo Simulink`. The interactive Simulink demo window is documented in Figure 2.21.

One can learn and explore Simulink using the Simulink and MATLAB Demos. Different MATLAB and Simulink versions exist. The Simulink manuals are available in the Portable Document Format (pdf). This section is not aimed to rewrite the excellent user manuals. With the ultimate goal of providing supplementary coverage and educate the reader on how to solve practical problems, we introduce Simulink with step-by-step instructions and practical examples. Figure 2.22 report the Simulink demo features with various simple, medium complexity and advanced examples which are ready to be used. For example, the friction model was covered in Section 2.5. The Index and Search icon can be used. As illustrated in Figure 2.22, MATLAB offers the model of friction. Various examples, from aerospace to automotive applications, from electronics to mechanical systems, are available.

Note: The designer must consistently assess the fitness, practicality, applicability, and validity of MATLAB, Simulink, and other toolboxes for specific problems and analysis objectives. The presumably ready-to-use files, blocks, diagrams, and other tools may not be adequate or require significant refinements.

Using the nonlinear, differential, difference, or other equations, one builds the Simulink model using Simulink blocks, some of which are reported in Figure 2.23. Various nonlinear deterministic and stochastic systems can be simulated as reported in the following examples.

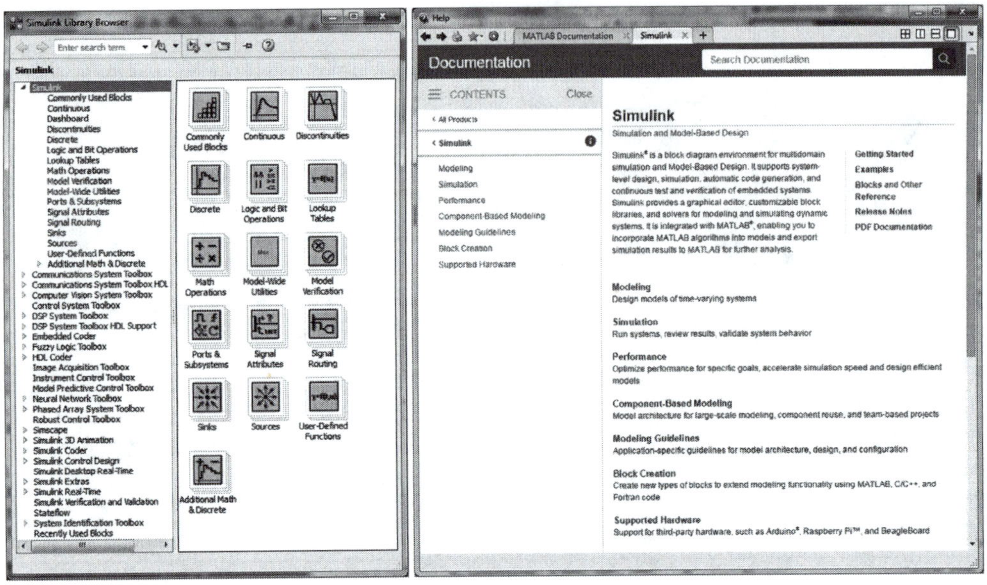

FIGURE 2.21 Simulink® Library Browser and Simulink demo window.

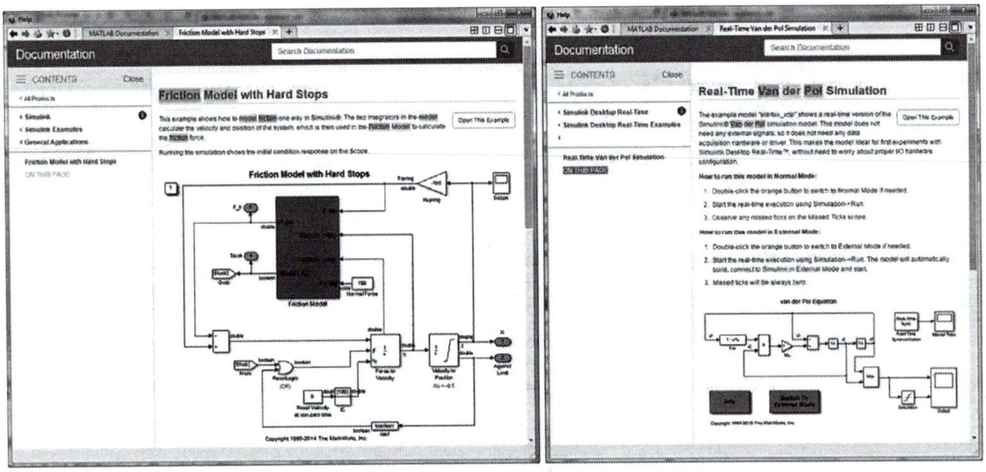

FIGURE 2.22 Simulink® demo features.

Example 2.25: Van der Pol Differential Equations: Simulations Using Simulink

The van der Pol oscillator is described by the second-order nonlinear differential equation

$$\frac{d^2x}{dt^2} - k\left(1 - x^2\right)\frac{dx}{dt} + x = d(t),$$

where $d(t)$ is the forcing function.

The differential equation is rewritten as a system of two first-order differential equations

$$\frac{dx_1(t)}{dt} = x_2, \quad x_1(t_0) = x_{10},$$

$$\frac{dx_2(t)}{dt} = -x_1 + kx_2 - kx_1^2x_2 + d(t), \quad x_2(t_0) = x_{20}.$$

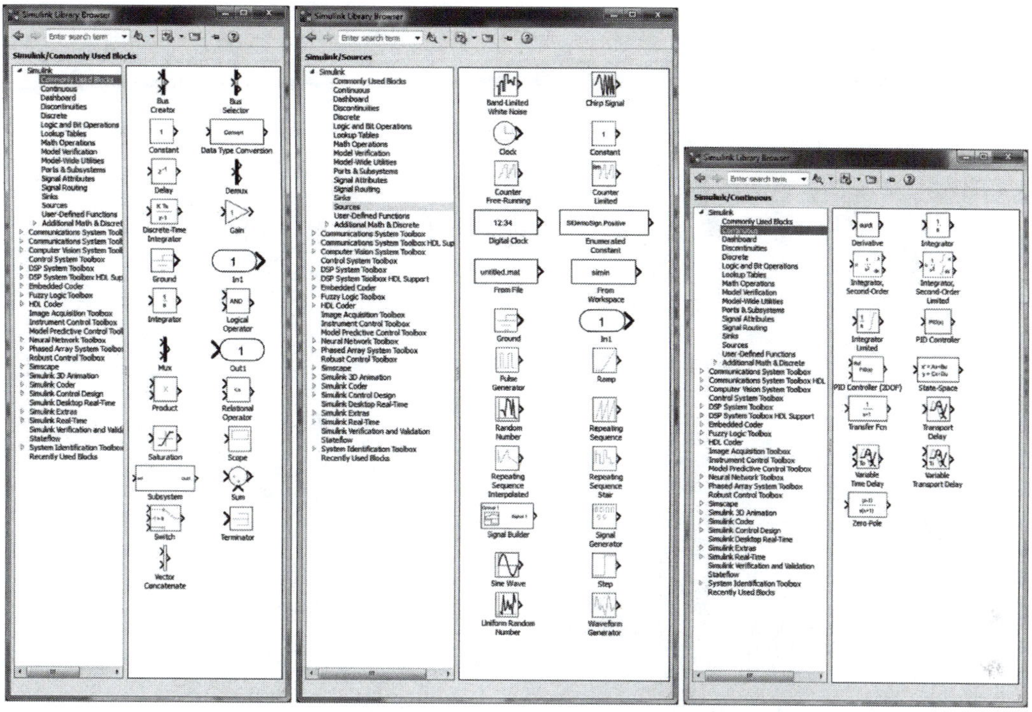

FIGURE 2.23 Commonly Used Blocks, Sources, and Continuous blocks from the Simulink® library.

In the literature, the differential equations for the van der Pol oscillator are given as

$$\frac{dx_1(t)}{dt} = x_2, \quad \frac{dx_2(t)}{dt} = -x_1 + \mu(1 - x_1^2)x_2.$$

The Simulink model can be built using the Function, Integrator, Mux, Signal Generator, Scope, and other blocks from the Simulink Block Library, see Figure 2.23. The Simulink model is developed as depicted in Figure 2.24. Simulations of the transient dynamics are performed for $k = 1$, $d = 0$, and initial conditions

$$x_0 = \begin{bmatrix} x_{10} \\ x_{20} \end{bmatrix} = \begin{bmatrix} 1 \\ -1 \end{bmatrix}.$$

The model coefficients must be uploaded. One can type in the Command Window $k = 1$. By double-clicking the Signal Generator block, one may select the different waveforms specifying the amplitude and frequency. The Band-Limited White Noise, Constant, and other blocks can be used for the specific d. The initial conditions are set by double-clicking the Integrator blocks and setting x_{10} and x_{20}. After specifying the simulation time to be 25 sec, the Simulink model is run by clicking the ▸ icon. The results are illustrated in Figure 2.24 as displayed in the Scopes. Saving the Scope data in the array format as a variable x, the plot command is used. The resulting plots are illustrated in Figure 2.24.

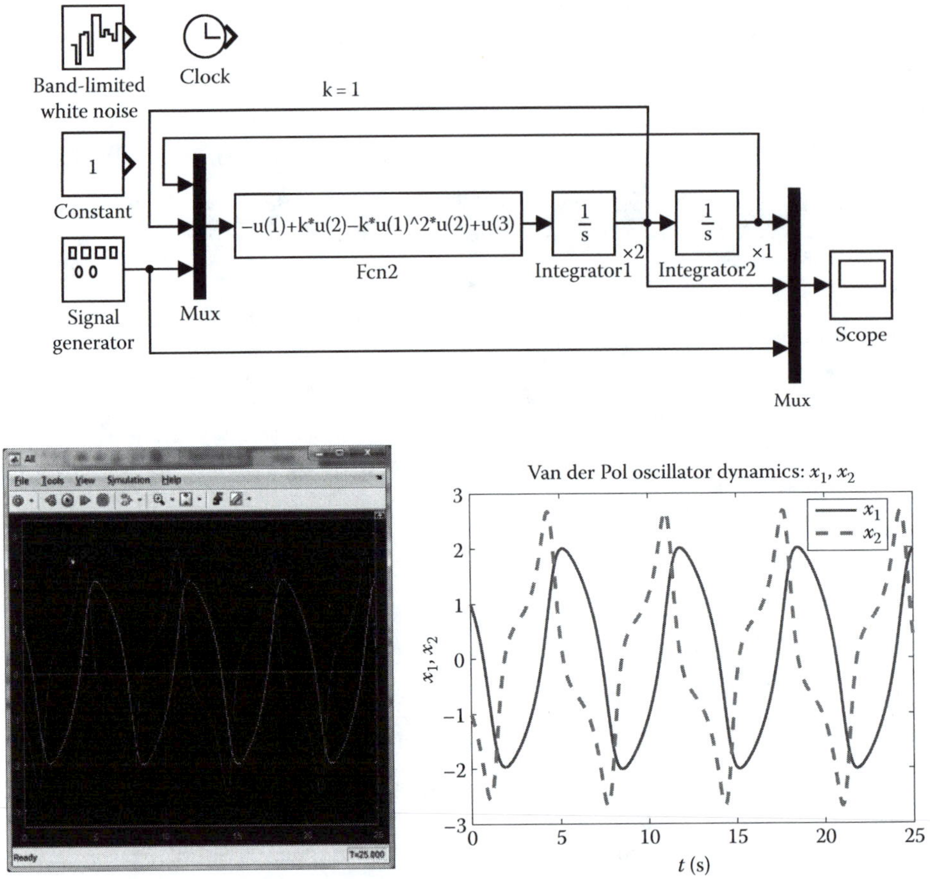

FIGURE 2.24 Simulink® block diagram (VanDerPolV01), transient dynamics displayed in the Scope, and dynamics of the state variables $x_1(t)$ and $x_2(t)$ plotted using the plot command.

```
plot(x(:,1),x(:,2),'-',x(:,1),x(:,3),'--','LineWidth',2.5);
legend('{\itx}_1','{\itx}_2')
xlabel('{\itt}, [sec]','FontSize',18); ylabel('{\itx}_1, {\itx}_2','Fontsize',18);
title('Van der Pol Oscillator Dynamics: {\itx}_1, {\itx}_2','Fontsize',18);
```
■

Example 2.26: Lotka–Volterra Differential Equations

The deterministic and stochastic differential equations are examined. Consider the deterministic Lotka–Volterra equations

$$\frac{dx_1}{dt} = a_{11}x_1 - a_{12}x_1x_2, \quad x_1(t_0) = x_{10}, \quad \forall a_{ii} > 0,$$

$$\frac{dx_2}{dt} = -a_{21}x_2 + a_{22}x_1x_2, \quad x_2(t_0) = x_{20}.$$

The governing dynamics of a system with states x_1 and x_2 is studied for given equilibriums and bifurcations with critical points, saddle points, orbits, and limit cycles. The Simulink block diagram is documented in Figure 2.25. The system evolution is affected by the positive-definite coefficients, $a_{ii} > 0$. For $a_{11} = 1$, $a_{12} = 1$, $a_{21} = 10$, and

$a_{22} = 2$, Figure 2.25 reports the transient dynamics and orbits for $x_1(t)$ and $x_2(t)$ with $x_{10} = x_{20} = 5$. The plotting statements are

```
plot(x(:,1),x(:,2),'-',x(:,1),x(:,3),'--','LineWidth',2.5);
legend('{\itx}_1','{\itx}_2')
xlabel('{\itt}, [sec]','FontSize',18); ylabel('{\itx}_1, {\itx}_2','Fontsize',18);
title('Lotka-Volterra Dynamics: {\itx}_1, {\itx}_2','Fontsize',18); pause
plot(x(:,2),x(:,3),'-','LineWidth',2.5);
xlabel('{\itx}_1','FontSize',18); ylabel('{\itx}_2','Fontsize',18);
```

The stochastic disturbances affect dynamic systems. Consider the stochastic equations to simulate the perturbed dynamics under disturbances. We study the second-order system

$$\frac{dx_1}{dt} = a_{11}x_1 - a_{12}x_1x_2 + b_{11}\xi_1, \quad x_1(t_0) = x_{10}, \quad \forall a_{ii} > 0,$$

$$\frac{dx_2}{dt} = -a_{21}x_2 + a_{22}x_1x_2 + b_{22}\xi_2, \quad x_2(t_0) = x_{20}.$$

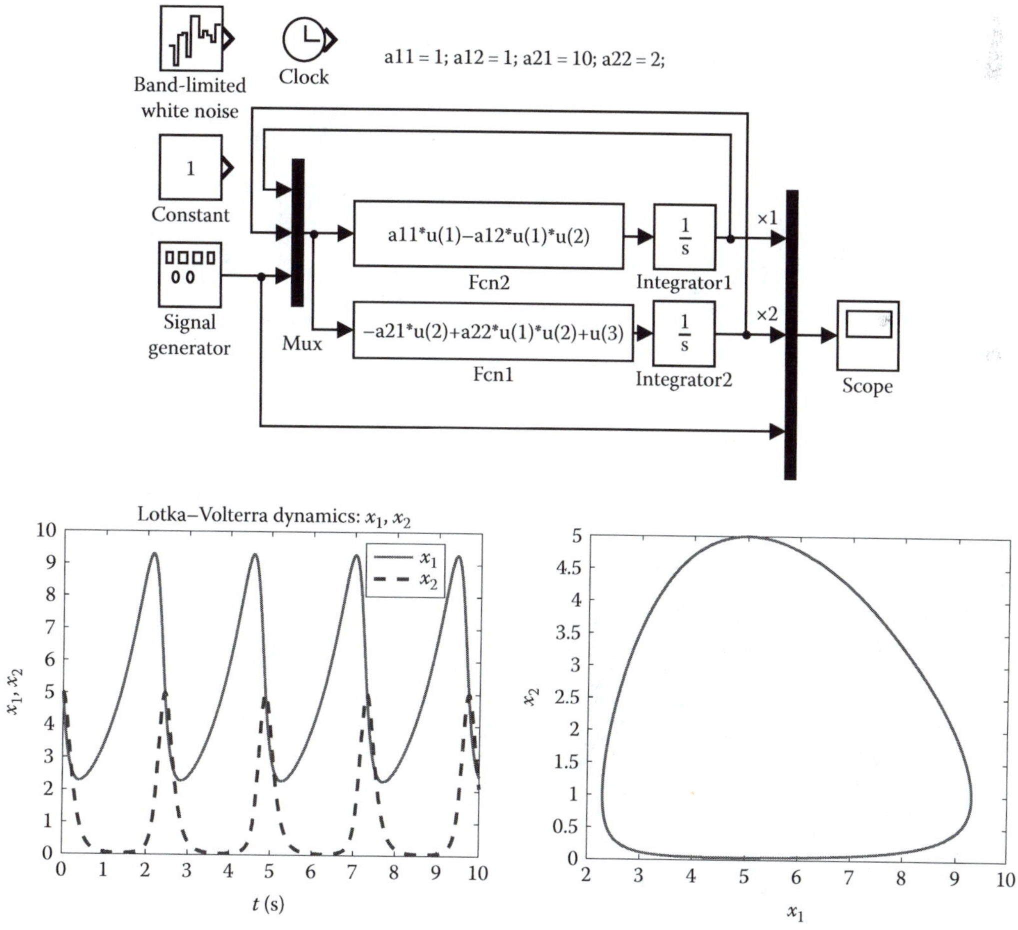

FIGURE 2.25 Simulink® block diagram (`LotkaVolterraV01`), dynamics of $x_1(t)$ (solid line) and $x_2(t)$ (dashed line), and periodic solution with elliptical orbits of states x_1 and x_2.

Here, ξ_1 and ξ_2 are the white noise. The real-valued ξ_1 and ξ_2 are characterized by the finite variances σ_1^2 and σ_2^2, and covariance $E[\xi_1(t_1)\cdot\xi_1(t_2)] = E[\xi_2(t_1)\cdot\xi_2(t_2)] = 0$, $\forall t_1 \neq t_2$. A white noise is represented by a set of independent and identically distributed random variables. The samples are statistically uncorrelated and identically distributed with σ^2. The covariance matrix is

$$C_{xx} = \begin{bmatrix} \sigma^2 & \cdots & 0 \\ \vdots & \ddots & \vdots \\ 0 & \cdots & \sigma^2 \end{bmatrix} = \sigma^2 I.$$

The probability density function is

$$f_X(x) = \frac{1}{\sigma\sqrt{2\pi}} e^{-\frac{1}{2}\sigma^{-2}x^2}.$$

Figure 2.26 reports the Simulink block diagram. The stochastic dynamics of $x_1(t)$ and $x_2(t)$ are documented in Figures 2.26 if $\sigma_1^2 = 0.1$ and $\sigma_2^2 = 0.1$. These results demonstrate the effect of disturbances ξ_1 and ξ_2. ■

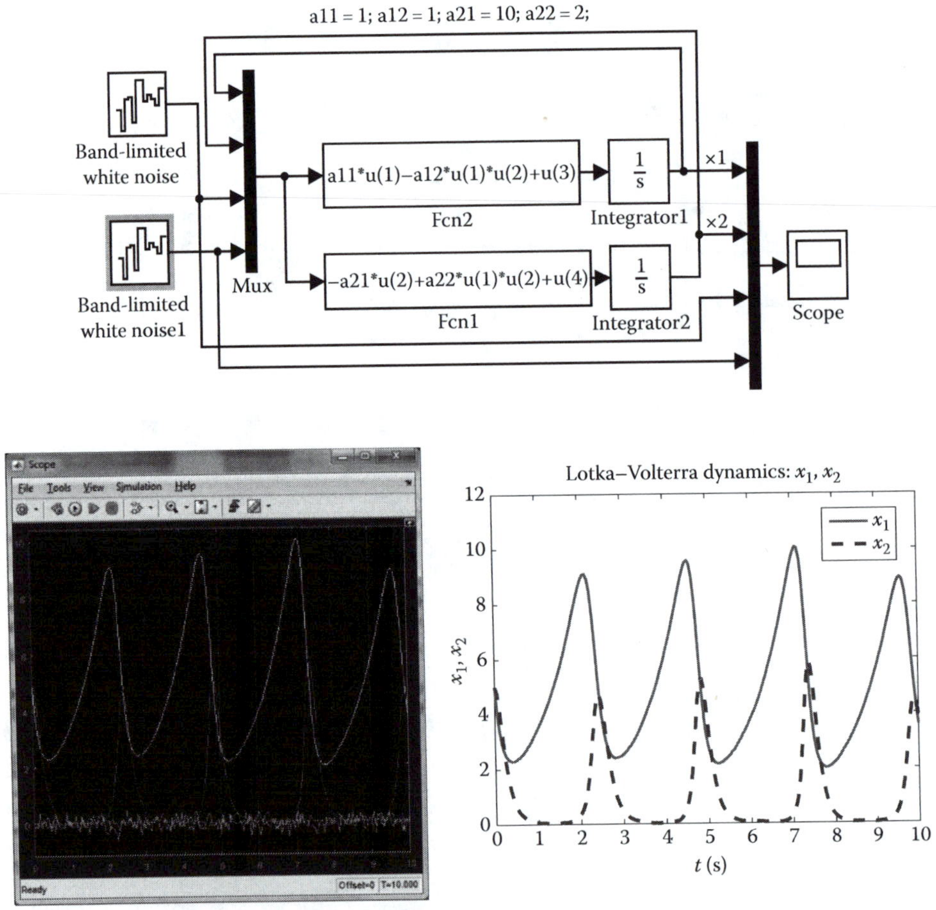

FIGURE 2.26 Simulink® block diagram and perturbed stochastic dynamics of $x_1(t)$ and $x_2(t)$.

Example 2.27

The Lorenz differential equations are

$$\frac{dx_1}{dt} = a_{11}\left(-x_1 + x_2\right) + b_{11}d(t),$$

$$\frac{dx_2}{dt} = a_{21}x_1 + a_{22}x_2 - a_{23}x_1x_3,$$

$$\frac{dx_3}{dt} = a_{31}x_3 + x_1x_2.$$

The Simulink model is reported in Figure 2.27. The positive-definite constants $a_{ii} > 0$ yield the specified equilibriums, bifurcations with quantified orbits, or limit cycles. Let $a_{11} = 5$, $a_{21} = 12.5$, $a_{22} = 2$, $a_{23} = 250$, $a_{31} = 2$, $b_{11} = 1$, and $d = 0$. The initial conditions are $x_{10} = 0.5$, $x_{20} = 0$, and $x_{30} = 0$. The systems dynamics and three-dimensional evolution (x_1, x_2, x_3) are given in Figures 2.27. The plotting statements are

```
plot(x(:,1),x(:,2),'-',x(:,1),x(:,3),'--',x(:,1),x(:,4),':','LineWidth',2.5);
axis([0 15 -0.75 1]); xlabel('{\itt} [sec]','FontSize',18);
title('Lorenz Equations Dynamics: {\itx}_1, {\itx}_2, {\itx}_3','FontSize',18); pause;
plot3(x(:,2),x(:,3),x(:,4),'-','LineWidth',1.5);
axis([-0.7 0.75 -0.85 1.05 0 0.125])
xlabel('{\itx}_1','FontSize',18); ylabel('{\itx}_2','FontSize',18); zlabel('{\itx}_3','FontSize',18);
```
■

Example 2.28: Nonlinear Interpolation

The MATLAB lsqnonlin solver solves the nonlinear least-squares interpolation (data fitting) problem. Let $y = f(x) = 2e^{-0.5x}\cos 5x$ ($p_1 = 2$, $p_2 = 0.5$, $p_3 = 5$) be measured with a superimposed zero-mean high-variance σ^2 noise n_ξ. The measured data set is $y_{data} = 2e^{-0.5x}\cos 5x + n_\zeta$ with the specified sampling.

For $y = f(x) = 2e^{-0.5x}\cos 5x$, using $y_{data} = 2e^{-0.5x}\cos 5x + n_\zeta$, one needs to find the unknown parameters p_1, p_2, and p_3 of $f_p(p,x) = p_1e^{-p_2x}\cos p_3x$. The sampled measured $y_{data} = 2e^{-0.5x}\cos 5x + n_\zeta$ is shown in Figures 2.28 by dots. The unknown parameter tuple (p_1, p_2, p_3) is found as illustrated below. The initial unknown parameter values are $p_{10} = 1$, $p_{20} = 1$, and $p_{30} = 10$. Due to the high-variance perturbations on measurements (normally distributed pseudorandom numbers n_ξ, modeled using the `randn` command), the p_1, p_2, and p_3 vary for different data sets y_{data}. However, the convergence is guaranteed and the error is minimized despite high variance. The resulting plots for a function $f(x)$ (dotted blue line), y_{data} used for interpolation (dots), and interpolated $f_p(p, x)$ (solid black line) are depicted in Figures 2.28. For a given $f(x)$, accurate interpolation $f_p(p, x)$ is found, error is minimized, and the convergence is guaranteed.

The MATLAB file is

```
x=linspace(0,10,50); yfx=2.*exp(-0.5*x) .*cos(5*x); ydata=yfx+0.5*randn(size(x));
fun=@(p) p(1).*exp(-p(2)*x).*cos(p(3)*x)-ydata;
p0=[1 1 10]; p=lsqnonlin(fun,p0), yint=p(1).*exp(-p(2)*x) .*cos(p(3)*x);
plot(x,yfx,'b:',x,ydata,'ko',x,yint,'k-','LineWidth',3);
title('Function {\ity}={\itf}({\itx}), {\ity}_d_a_t_a, Interpolation {\ity_p}={\itf}({\itp},{\itx})',
   'Fontsize',18);
xlabel('{\itx}','Fontsize',18);
legend('Function {\ity}={\itf}({\itx})','Data {\ity}_d_a_t_a ','Interpolation {\ity_p}={\itf}({\itp},{\itx})')
```
■

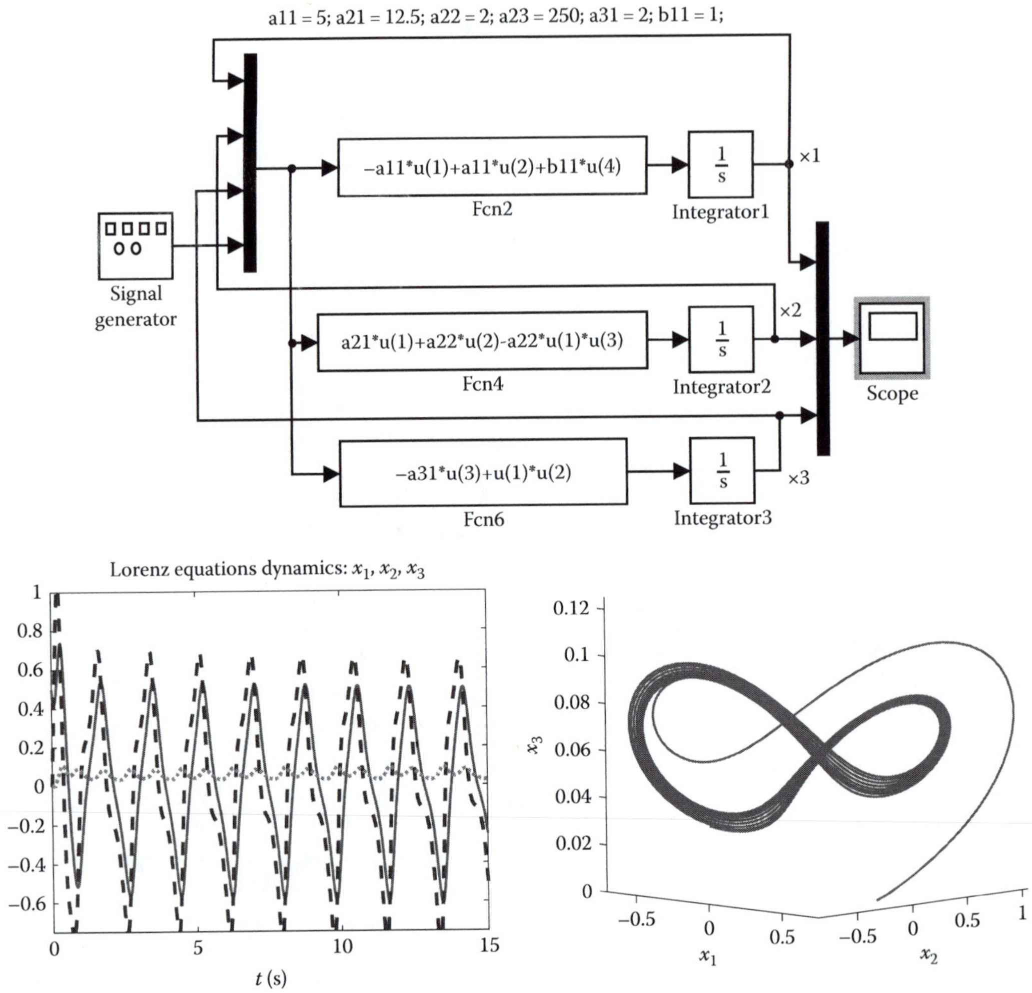

FIGURE 2.27 Simulink® model, dynamics of $x_1(t)$, $x_2(t)$, and $x_3(t)$ (solid, dashed, and dotted lines), and three-dimensional evolution (x_1, x_2, x_3).

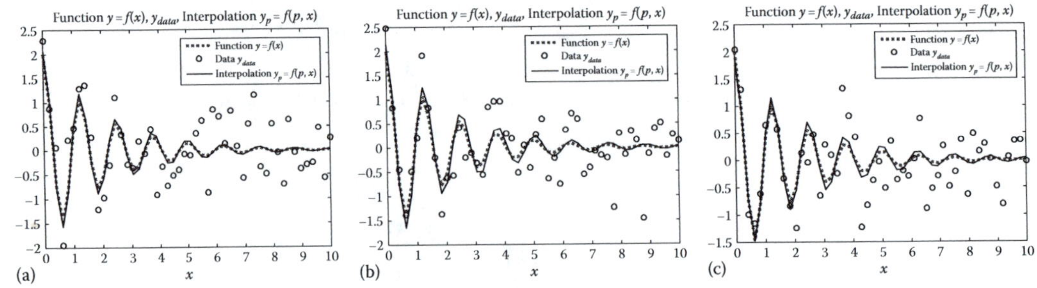

FIGURE 2.28 Plots for $y = f(x) = 2e^{-0.5x}\cos 5x$, $y_{data} = 2e^{-0.5x}\cos 5x + n_\zeta$, and $f_p(p,x) = p_1 e^{-p_2 x}\cos p_3 x$. The unknown parameters of $f_p(p,x) = p_1 e^{-p_2 x}\cos p_3 x$ are as follows: (a) $p_1 = 2.0744$, $p_1 = 0.4518$, $p_1 = 5.01$; (b) $p_1 = 2.1436$, $p_1 = 0.42$, $p_1 = 4.9469$; (c) $p_1 = 1.8714$, $p_1 = 0.3887$, $p_1 = 5.0271$.

HOMEWORK PROBLEMS

2.1 Figure 2.29 illustrates a motion devise with windings on rotor. The electromagnetic torque is developed due to the interaction of the 20-turn rectangular coil (l = 15 cm and w = 5 cm) in the yz plane and stationary magnetic field (established by magnets on stator windings).

a. Determine the magnetic moment. Find the electromagnetic torque acting on the coil. Let the current in the coil be i = 10 A. The magnetic field density is $\vec{B} = 0.02(\vec{a}_x + 2\vec{a}_y)$ T.

b. Derive at which angle ϕ, $T_e = 0$? At what angle ϕ, T_e is maximum? Determine the value of $T_{e\max}$. Derive the solution from: (i) the electromagnetics viewpoint and (ii) the mathematical (mini-max problem) viewpoint.

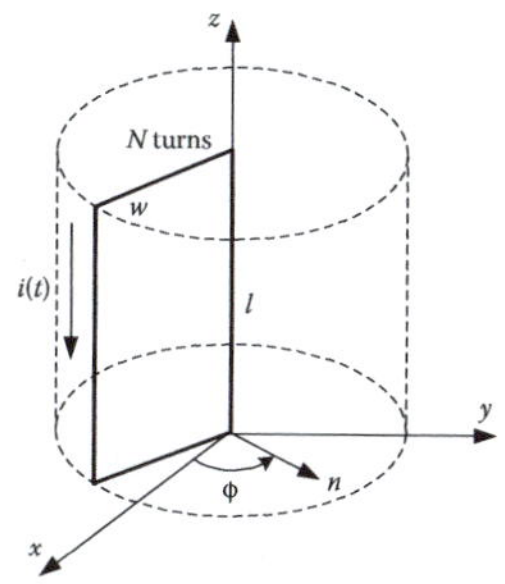

FIGURE 2.29 Rotor's winding in the magnetic field.

2.2 The differential equations were derived for a circuit reported in Figure 2.13a (Example 2.15). In particular,

$$\ddot{q}_1 = \frac{1}{(L_1 + L_{12})}\left(-\frac{q_1}{C_1} - R_1\dot{q}_1 + L_{12}\ddot{q}_2 + u_a\right), \quad \ddot{q}_2 = \frac{1}{(L_2 + L_{12})}\left(L_{12}\ddot{q}_1 - \frac{q_2}{C_2} - R_2\dot{q}_2\right).$$

Develop the Simulink model and perform simulation if L_1 = 0.01 H, L_2 = 0.005 H, L_{12} = 0.0025 H, C_1 = 0.02 F, C_2 = 0.1 F, R_1 = 10 ohm, R_2 = 5 ohm, and u_a = 100sin(200t) V. Report plots for the generalized coordinates and currents.

2.3 The differential equations were derived for a circuit reported in Figure 2.13a (Example 2.15). In particular,

$$\ddot{q}_1 = \frac{1}{(L_1 + L_{12})}\left(-\frac{q_1}{C_1} - R_1\dot{q}_1 + L_{12}\ddot{q}_2 + u_a\right), \quad \ddot{q}_2 = \frac{1}{(L_2 + L_{12})}\left(L_{12}\ddot{q}_1 - \frac{q_2}{C_2} - R_2\dot{q}_2\right).$$

Develop the Simulink model and perform simulation if L_1 = 0.01 H, L_2 = 0.005 H, L_{12} = 0.0025 H, C_1 = 0.02 F, C_2 = 0.1 F, R_1 = 10 ohm, R_2 = 5 ohm, and u_a = 100sin(200t) V. Report plots for the generalized coordinates and currents.

2.4 Consider these nonlinear differential equations (see Example 2.27)

$$\frac{dx_1}{dt} = a_{11}(-x_1 + x_2) + \sin(x_1x_2) + b_{11}d(t),$$

$$\frac{dx_2}{dt} = a_{21}x_1 + a_{22}x_2 + \cos(x_1x_3) - a_{23}x_1x_3,$$

$$\frac{dx_3}{dt} = a_{31}x_3 + x_1x_2.$$

Develop the Simulink model and perform simulations if $a_{11} = 5$, $a_{21} = 12.5$, $a_{22} = 2$, $a_{23} = 250$, $a_{31} = 2$, $b_{11} = 1$, and $d = e^{-\sin t}$. The initial conditions are $x_{10} = 0.5$, $x_{20} = 0$, and $x_{30} = 0$.

Note: The Simulink clock block may be used to generate the time t to implement the forcing function $d = e^{-\sin t}$.

REFERENCES

1. S. J. Chapman, *Electric Machinery Fundamentals*, McGraw-Hill, New York, 2011.
2. A. E. Fitzgerald, C. Kingsley, and S. D. Umans, *Electric Machinery*, McGraw-Hill, New York, 2003.
3. P. C. Krause and O. Wasynczuk, *Electromechanical Motion Devices*, McGraw-Hill, New York, 1989.
4. P. C. Krause, O. Wasynczuk, S. D. Sudhoff, and S. Pekarek, *Analysis of Electric Machinery*, Wiley-IEEE Press, New York, 2013.
5. W. Leonhard, *Control of Electrical Drives*, Springer, Berlin, Germany, 2001.
6. S. E. Lyshevski, *Electromechanical Systems, Electric Machines, and Applied Mechatronics*, CRC Press, Boca Raton, FL, 1999.
7. G. R. Slemon, *Electric Machines and Drives*, Addison-Wesley Publishing Company, Reading, MA, 1992.
8. White D. C. and Woodson H. H., *Electromechanical Energy Conversion*, Wiley, New York, 1959.
9. S. E. Lyshevski, *Engineering and Scientific Computations Using MATLAB®*, Wiley, Hoboken, NJ, 2003.
10. MATLAB, R2015b, MathWorks, Inc., Natick, MA, 2015.
11. D. R. Lide, *Handbook of Chemistry and Physics*, 83rd edn., CRC Press, Boca Raton, FL, 2002.
12. R. C. Dorf, *Handbook of Engineering Tables*, CRC Press, Boca Raton, FL, 2003.
13. S. E. Lyshevski, *Nano- and Micro-Electromechanical Systems: Fundamentals of Nano- and Microengineering*, CRC Press, Boca Raton, FL, 2004.

3 Electrostatic and Electromagnetic Motion Devices

3.1 INTRODUCTION AND DISCUSSIONS

In this chapter, we consider widely commercialized electrostatic and electromagnetic transducers. In sensing and low-power actuator applications, the electrostatic devices may ensure desired performance and capabilities. The mini- and microscale electrostatic transducers, actuators, and sensors are fabricated using *bulk* and *surface* micromachining, which complies with microelectronic technologies. The commercialized microelectromechanical systems (MEMS) technology guarantees affordability, high yield, robustness, etc. The electromagnetic devices ensure high force, torque and power densities. There are variable-reluctance (solenoids, relays, electromagnets, levitation systems, etc.) [1–5], permanent-magnet, and other electromagnetic transducers. The stored electric and magnetic volume energy densities ρ_{We} and ρ_{Wm} for electrostatic and electromagnetic transducers are

$$\rho_{We} = \frac{1}{2}\varepsilon E^2 \quad \text{and} \quad \rho_{Wm} = \frac{1}{2}\mu^{-1}B^2 = \frac{1}{2}\mu H^2,$$

where ε is the permittivity, $\varepsilon = \varepsilon_0\varepsilon_r$; ε_0 and ε_r are the permittivity of free space and relative permittivity, $\varepsilon_0 = 8.85 \times 10^{-12}$ F/m; E is the electric field intensity; μ is the permeability, $\mu = \mu_0\mu_r$; μ_0 and μ_r are the permeability of free space and relative permeability, $\mu_0 = 4\pi \times 10^{-7}$ T-m/A; B and H are the magnetic field density and intensity.

The maximum energy density of electrostatic actuators is limited by the maximum field (voltage), which can be applied before an electrostatic breakdown occurs. In mini- and microstructures, the maximum electric field cannot exceed E_{max} resulting in the maximum energy density $\rho_{We\max} = \frac{1}{2}\varepsilon_0\varepsilon_r E_{\max}^2$. In ~100 × 100 μm to millimeter size structures with a few micrometers air gap, E_{max} may reach ~3×10^6 V/m. Depending on the device, the relative permittivity ε_r may vary from 1 to ~10. One estimates $\rho_{We\max}$ to be less than 100 J/m^3. For electromagnetic actuators, the maximum energy density $\rho_{Wm\max}$ is limited by the saturation flux density B_{sat} (which is ~1 T), material permeability μ_r, $(BH)_{\max}$, etc. The resulting magnetic energy density $\rho_{Wm\max}$ may reach ~100,000 J/m^3. Hence, $\rho_{Wm}/\rho_{We} \gg 1$. However, in many applications, electrostatic MEMS are an effective, affordable, and consistent solution. To actuate any mechanism, the electrostatic force or torque must be greater than the load force or torque. The load forces and disturbances can be small. For example, ~1,000,000 electrostatic micromirrors (each ~10 × 10 μm) are repositioned in the Texas Instruments digital light processing (DLP) module, as shown in Figure 3.1a. In the Texas Instruments DLP5500, there are 1024 × 768 aluminum ~10.8 × 10.8 μm mirrors that are actuated by a small electrostatic force. These DLPs are used in high-definition displays and projection systems. Electrostatic microactuators are individually controlled ensuring ~8000 Hz bandwidth. Images of the ADMP401 MEMS electrostatic microphone (100–15,000 Hz) are illustrated in Figure 3.1b. Using the MEMS technology, various electrostatic actuators and sensors can be fabricated, diced, bonded, and packaged as illustrated in Figure 3.1c.

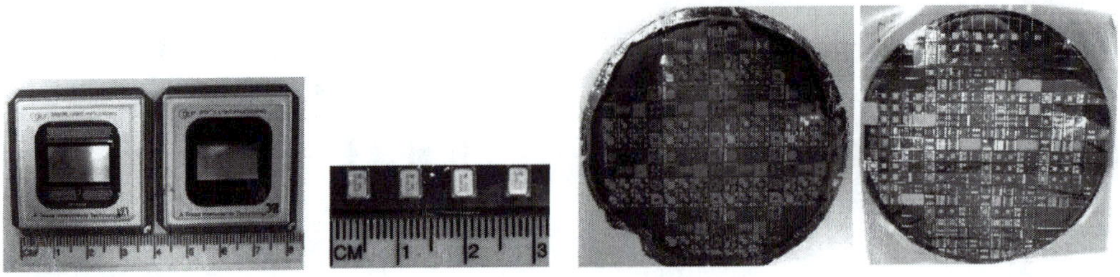

(a) (b) (c)

FIGURE 3.1 (a) Texas Instrument DLP with ~1,000,000 electrostatic torsional micromirrors that are individually controlled; (b) ADMP401 MEMS microphones with −42 dBV sensitivity, 250 μA current consumption, 100 Hz to 15 kHz frequency response; (c) Rochester Institute of Technology MEMS transducers and devices on the silicon wafer.

Using the *bulk* and *surface* micromachining, the fabricated diced structures, sensors, electrostatic and electromagnetic actuators are reported in Figures 3.2. Figures 3.2 document images of various electromagnetic, variable-capacitance electrostatic transducers, and multifunctional sensors. All MEMS must be interfaced with electronics and packaged. For example, 15 μm gold wire–bonded packaged devices are illustrated in Figure 3.2b. These MEMS devices are interconnected with microelectronics to perform sensing and actuation. The Texas Instruments MEMS photodiode with on-chip transimpedance amplifier OPT101 and visible light sensors OPT3001DNPR are illustrated in Figure 3.2c. The physical quantities are measured. For example, for the micromachined actuators, documented in Figure 3.2e, the deflection of the diaphragm, see Figures 3.2d and e is measured by using the variations of resistances of four polysilicon resistors that form the Wheatstone bridge. By measuring the varying capacitance $C(x)$, the displacement can also be measured. The energy stored in the electric field between two surfaces in capacitors is $W_e = \frac{1}{2}QV = \frac{1}{2}CV^2$. The energy stored in the inductor is $\frac{1}{2}Li^2$. The electrostatic energy variations, due to the varying capacitance $C(x)$ or $C(\theta_r)$, is the foundation of the device physics. The MEMS transducers are controlled by applying the voltage as depicted in Figure 3.2f. The electrostatic force and torque are

$$F_e = \frac{\partial W_e}{\partial x} = \frac{\partial}{\partial x}\frac{1}{2}C(x)V^2 = \frac{1}{2}\frac{\partial C(x)}{\partial x}V^2 \quad \text{and} \quad T_e = \frac{\partial W_e}{\partial \theta_r} = \frac{\partial}{\partial \theta_r}\frac{1}{2}C(\theta_r)V^2 = \frac{1}{2}\frac{\partial C(\theta_r)}{\partial \theta_r}V^2.$$

The physics of electromagnetic devices is based on the following principles:

- *Variable-reluctance electromagnetics*: The force (torque) is produced to minimize or align the reluctance of the electromagnetic system (electromagnets, solenoids, relays, magnetic levitation systems, reluctance motors, etc.);
- *Induction electromagnetics*: The phase voltages are induced in the rotor windings due to the time-varying stator magnetic field and motion of the rotor with respect to the stator. The electromagnetic torque (force) results from the interaction between time-varying electromagnetic fields;
- *Synchronous electromagnetics*: The torque (force) results from the interaction between the time-varying magnetic field established by the stator windings and the stationary magnetic field established by the permanent magnets or electromagnets on the rotor.

Permanent-magnet electromechanical motion devices usually surpass induction and variable-reluctance devices. However, there are fundamental, technological, and market limits and constraints on permanent-magnet motion devices. Various variable-reluctance devices are used in many systems. In this section, we cover different radial and axial topologies for translational and rotational electrostatic variable-capacitance actuators and electromagnetic variable-reluctance transducers.

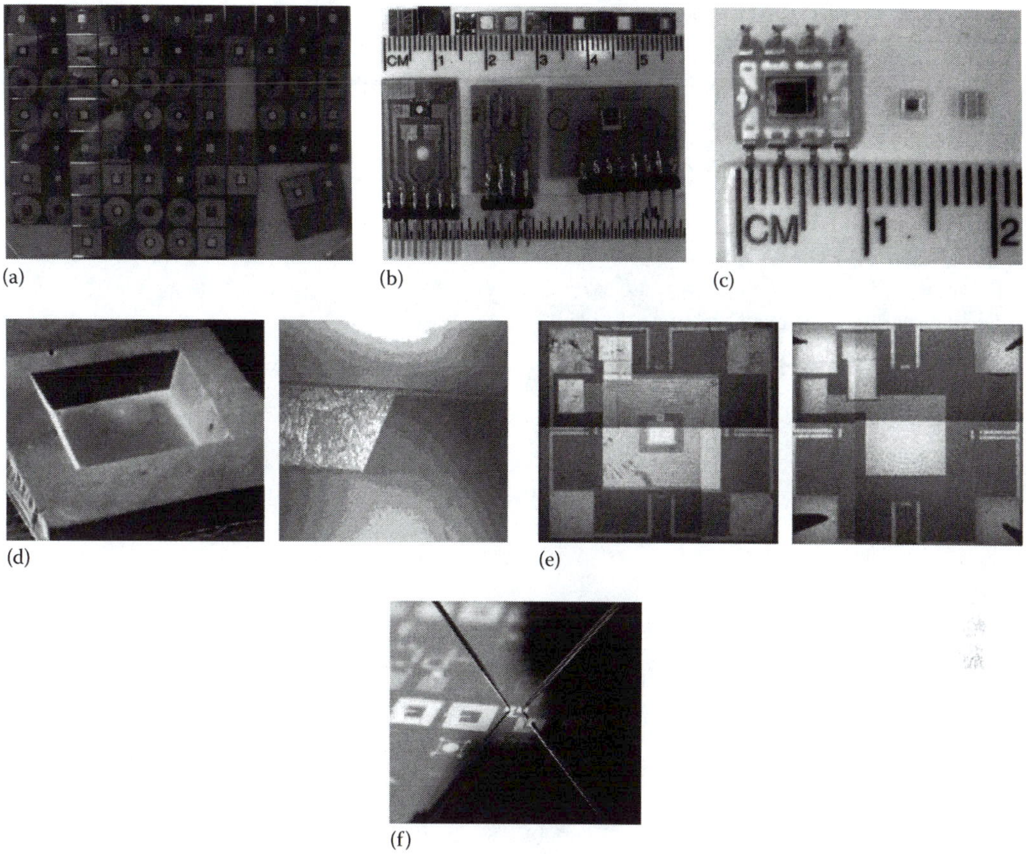

FIGURE 3.2 (a) Diced MEMS with coils and structures deposited on ~30 μm silicon diaphragm; (b) Wire-bonded packaged MEMS devices on evaluation boards for testing and characterization; (c) The Texas Instruments OPT101 monolithic photodiode with on-chip transimpedance amplifier and OPT3001DNPR visible light sensor; (d) Etched silicon structure and cross section of a ~35 μm silicon diaphragm; (e) Micromachined ~3 × 3 mm MEMS actuators with displacement sensors. The displacement is measured using the changes in four polysilicon resistors that form the Wheatstone bridge; (f) Micromachined electrostatic actuator with the suspended movable plate: voltage is applied to displace the top plate by the electrostatic force $F_e(x)$.

The following sequential steps ensure a consistent design flow:

1. For a given application, consistently define limits, constraints, specifications and requirements;
2. Evaluate and advance existing technologies and electromechanical motion devices by examining device physics, operating principles, topologies, electromagnetic systems, etc.;
3. Perform electromagnetic, energy conversion, mechanical, thermal, and sizing-dimensional estimates;
4. Conduct data-intensive analysis to select or design electromechanical devices;
5. Perform electromagnetic, mechanical, vibroacoustic, and thermodynamic analyses;
6. Define materials, processes, and technologies to fabricate or select structures (stator and rotor with windings, bearing, etc.), and assemble and package them as applicable;
7. Define matching power electronics and control solutions (this task is partitioned to many subtasks and problems related to power converter topologies, filters and control designs, controller implementation, actuator–sensor–ICs integration, etc.);

8. Integrate components, devices, and modules;
9. Test, characterize, evaluate, justify, and substantiate devices, modules, and system;
10. Optimize and redesign systems, ensuring best performance and *achievable* capabilities.

3.2 ELECTROSTATIC ACTUATORS

Consider the translational and rotational electrostatic actuators. These affordable actuators are fabricated using cost-effective and high-yield micromachining, thin film, electroplating, and other technologies. The images of electrostatic MEMS are documented in Figures 3.1 and 3.2. We perform electrostatic and electromechanical analyses.

3.2.1 PARALLEL-PLATE ELECTROSTATIC ACTUATORS

Consider the parallel-plate capacitor charged to a voltage V. The separation between two plates is x. The dielectric permittivity is ε. We neglect the fringing effect at the edges and assume that the electric field is uniform, such that $E = V/x$. The stored electrostatic energy is

$$W_e = \frac{1}{2}\int_v \varepsilon\left|\vec{E}\right|^2 dv = \frac{1}{2}\int_v \varepsilon\left(\frac{V}{x}\right)^2 dv = \frac{1}{2}\varepsilon\frac{V^2}{x^2}Ax = \frac{1}{2}\varepsilon\frac{A}{x}V^2 = \frac{1}{2}C(x)V^2,$$

where $C(x) = \varepsilon(A/x)$; A is the effective area.

Using $C(x) = \varepsilon(A/x)$, one finds the electrostatic force as a nonlinear function of the voltage applied V and displacement x

$$F_e = \frac{\partial W_e}{\partial x} = \frac{1}{2}\frac{\partial C(x)}{\partial x}V^2 = -\frac{1}{2}\varepsilon A\frac{1}{x^2}V^2.$$

The resulting equations of motion are

$$\frac{dv}{dt} = \frac{1}{m}\left(F_e - F_{air} - F_{elastic} - F_L\right),$$

$$\frac{dx}{dt} = v,$$

where the air friction, elastic, and load (perturbation) forces F_{air}, $F_{elastic}$, and F_L are device- and application-specific. For example, in vacuum, there is no air friction force on the movable suspended plate, cantilever, diaphragm, or membrane, and $F_{air} = 0$.

Consider a movable 250 × 250 μm silicon plate with the thickness ~30 μm. Deposited aluminum thin films with the thickness ~0.5 μm on the movable plate and substrate form a parallel plate capacitor. The voltage $0 \le u \le u_{max}$ is applied to actuate a micromirror. The image of the micromachined electrostatic actuator is reported in Figure 3.3a. The repositioning from the equilibrium ($x_e = 0$) to the final position must be accomplished with high bandwidth and minimal settling time. The top plate is suspended, and the separation between the capacitor surfaces is $x_0 = 10$ μm. From $C(x) = \varepsilon\frac{A}{x + x_0}$, the electrostatic force is $F_e = \frac{1}{2}\varepsilon A\frac{1}{\left(x + x_0\right)^2}u^2$. The experimental open-loop dynamics is reported in Figure 3.3b when the voltage pulses ~9 and 13.5 V are applied to ensure ~3 μm and 5 μm repositioning. The actuator repositions to the equilibrium $x_e = 0$ as a result of the restoration of the elastic force when the applied voltage is $u = 0$. The electrostatic actuator ensures only one-directional active control capabilities. The settling time is ~0.002 sec.

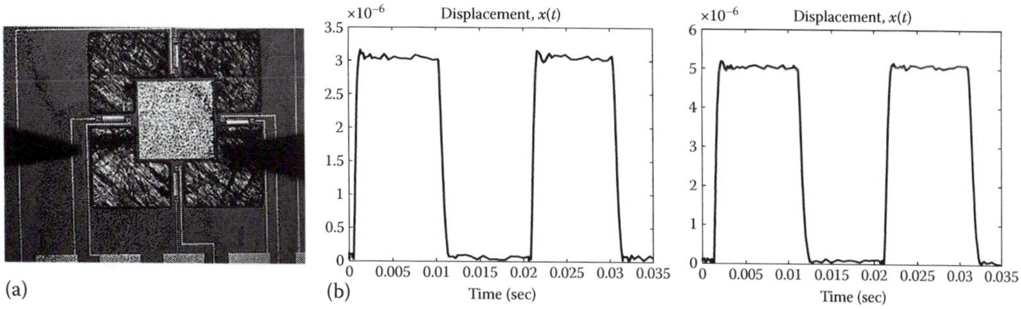

FIGURE 3.3 (a) Electrostatic microactuator; (b) Open-loop system response: Dynamics of $x(t)$ if the voltage pulses 9 and 13.5 V are applied to reposition actuator for ~3 and 5 μm.

The elastic and air friction force are modeled as $F_{air} = a_1 v + a_2 v|v|$ and $F_{elastic} = b_1 x + b_2 x|x|$. Using the Newtonian mechanics $m\mathbf{a} = \Sigma\mathbf{F}$ and $F_e = \frac{1}{2}\varepsilon A \frac{1}{(x+x_0)^2}u^2$, we have

$$\frac{dv}{dt} = \frac{1}{m}\left(F_e - F_{air} - F_{elastic}\right) = \underbrace{1.5\times10^{-10}\frac{u^2}{(x+x_0)^2}}_{F_e} - \underbrace{\left(7\times10^3 v + 1\times10^6 v|v|\right)}_{F_{air}} - \underbrace{\left(2\times10^7 x + 4\times10^{11} x|x|\right)}_{F_{elastic}},$$

$$\frac{dx}{dt} = v.$$

The Simulink® model and simulation results are reported in Figures 3.4. One concludes that the simulated dynamics corresponds to the experimental results.

3.2.2 Rotational Electrostatic Actuators

We study rotational electrostatic transducers. These transducers are used only as the limited displacement angle devices. One must interconnect two conducting surfaces, including the rotating rotor. In actuators, the electromagnetic torque can be developed only in one direction. Figure 3.5 shows ~500 μm diameter limited-angle electrostatic micromachined actuators.

As the voltage V is applied to the parallel conducting rotor and stator surfaces, the charge is $Q = CV$, where C is the capacitance, $C = \varepsilon(A/g) = \varepsilon(WL/g)$; A is the overlapping area of the plates, $A = WL$; W and L are the width and length of the plates; g is the air gap between the plates.

The energy associated with the electric potential is $W_e = \frac{1}{2}CV^2$. The electrostatic force at each overlapping plate segment $F_{el} = \frac{\partial W_e}{\partial g} = -\frac{1}{2}\frac{\varepsilon WL}{g^2}V^2$ is balanced by the opposite segment. We assume an ideal fabrication for which W, L, and g are the same for all conducting surfaces.

The electrostatic tangential force due to misalignment is $F_t = \frac{\partial W_e}{\partial y} = \frac{1}{2}\frac{\varepsilon}{g}\frac{\partial (WL)}{\partial y}V^2$, where y is the direction in which misalignment potentially may occur.

The capacitance of a cylindrical capacitor is found to derive the electrostatic torque. The voltage between the cylinders are obtained by integrating the electric field. The electric field at a distance r from a conducting cylinder has only a radial component, $E_r = \rho/2\pi\varepsilon r$, where ρ is the linear charge density, and $Q = \rho L$. The potential difference is $\Delta V = V_a - V_b = \int_a^b \vec{E}\cdot d\vec{l} = \int_a^b E_r \cdot dr = \frac{\rho}{2\pi\varepsilon}\int_{r_1}^{r_2}\frac{1}{r}dr = \frac{\rho}{2\pi\varepsilon}\ln\frac{r_2}{r_1}$, where r_1 and r_2 are the radii of the rotor and stator where the conducting plates are positioned. Thus, $C = \frac{Q}{\Delta V} = \frac{2\pi\varepsilon L}{\ln\frac{r_2}{r_1}}$. The capacitance per unit length is $\frac{C}{L} = \frac{\rho}{\Delta V} = \frac{2\pi\varepsilon}{\ln\frac{r_2}{r_1}}$.

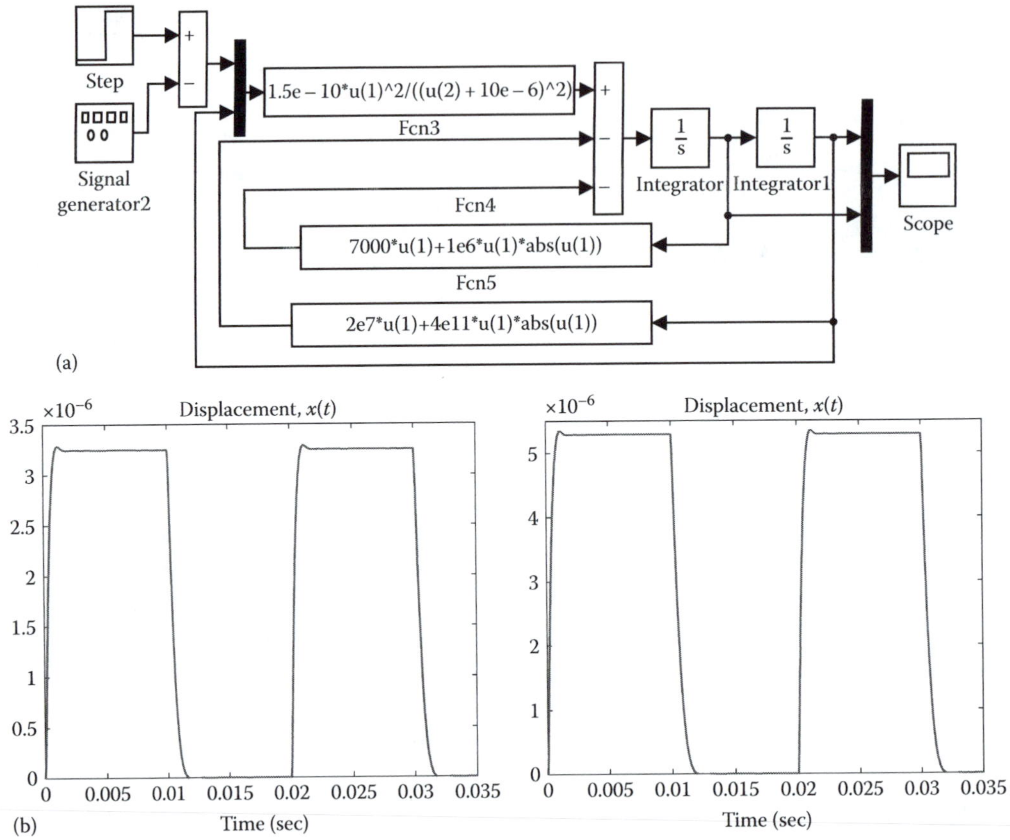

FIGURE 3.4 (a) Simulink® model; (b) Open-loop system response $x(t)$ if the voltage pulses 9 and 13.5 V are applied to reposition actuator.

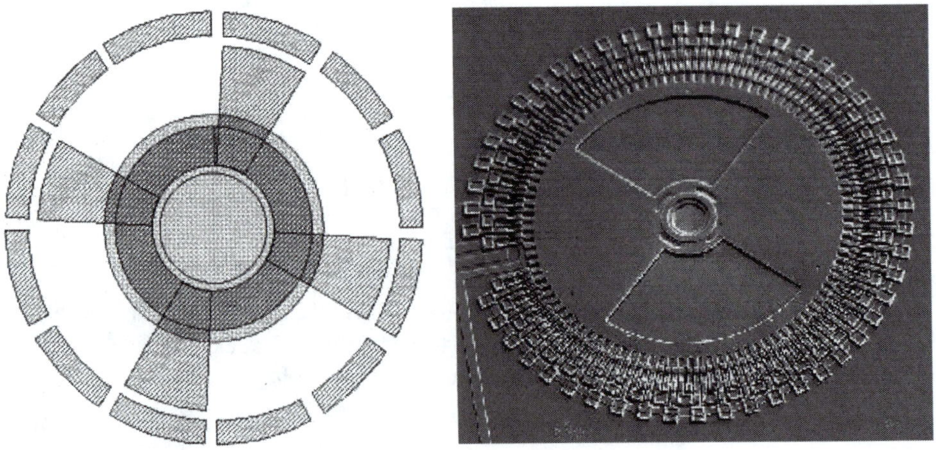

FIGURE 3.5 Electrostatic actuators.

Using the stator–rotor conducting surfaces overlap, the capacitance is a function of the angular displacement $C(\theta_r) = N\dfrac{2\pi\varepsilon}{\ln\dfrac{r_2}{r_1}}\theta_r$, where N is the number of overlapping stator–rotor surfaces. The electrostatic torque is $T_e = \dfrac{1}{2}\dfrac{\partial C(\theta_r)}{\partial \theta_r}V^2 = N\dfrac{\pi\varepsilon}{\ln\dfrac{r_2}{r_1}}V^2 = N\dfrac{\pi\varepsilon}{\ln r_2 - \ln r_1}V^2$. Other expressions for $C(\theta_r)$ can be found. Assuming that viscous friction is $T_{friction} = B_m\omega_r$, the *torsional–mechanical* equations of motion are

$$\frac{d\omega_r}{dt} = \frac{1}{J}\left(T_e - T_{friction} - T_L\right) = \frac{1}{J}\left(N\frac{\pi\varepsilon}{\ln r_2 - \ln r_1}V^2 - B_m\omega_r - T_L\right),$$

$$\frac{d\theta_r}{dt} = \omega_r.$$

The actuator rotates in the specified direction if $T_e > (T_L + T_{friction})$. The motor must develop the electrostatic torque $T_e = N\dfrac{\pi\varepsilon}{\ln r_2 - \ln r_1}V^2$ higher than the rated or peak load torque. Estimating the load torque $T_{L\max}$ and assigning the desired acceleration capabilities yields T_e. One evaluates the effect of N, r_1, r_2, and J. The motor sizing estimates can be found and the fabrication technologies (processes and materials) asserted. The applied voltage V is bounded. The fabrication technologies and processes affect the motor dimensions and parameters. For example, one may attempt to minimize the air gap to attain the minimal value of $(r_2 - r_1)$. By minimizing $\ln(r_2/r_1)$, one maximizes T_e. The moment of inertia J can be minimized by reducing the rotor mass using cavities, polymers, etc. There are physical limits on the maximum V and E. The technologies affect and define materials, tolerance, r_2/r_1 ratio, etc. Another critical issue is the need for a contact with the conducting rotor surfaces. This fact limits the application of rotational electrostatic actuators as limited-angle actuators.

3.3 VARIABLE-RELUCTANCE ELECTROMAGNETIC ACTUATORS

Variable-reluctance electromagnetic devices are widely used. We consider the translational (solenoids, relays, electromagnets, and magnetic levitation systems) and rotational (variable-reluctance synchronous motors) electromagnetic devices. The varying reluctance results in the electromagnetic force and torque.

3.3.1 SOLENOIDS, RELAYS, AND MAGNETIC LEVITATION SYSTEMS

Solenoid and relay usually consist of a movable member (called plunger or rotor) and a stationary member [1–6]. High-permeability ferromagnetic materials are used. The windings are wound in a helical pattern. The electromagnetic force is developed due to the varying reluctance. The performance of variable-reluctance devices is defined by the magnetic system, materials, relative permeability, friction, etc. Solenoids and relay (electromagnet) with a movable member are shown in Figures 3.6a and b. Magnetic levitation systems with the suspended ferromagnetic ball are depicted in Figure 3.6c. The electromagnetic system is formed by stationary and movable members. When the voltage is applied to the winding, current flows in the winding, magnetic flux is produced within the flux linkages path, and the electromagnetic force is developed. The movable members (plunder and ball) move to minimize the reluctance. When the applied voltage becomes zero, the plunger resumes its equilibrium position due to the returning spring. The undesirable phenomena such as residual magnetism and friction must be minimized. Different materials for the central guide (nonmagnetic sleeve) and plunger coating (plating) are used to minimize friction and wear. Glass-filled nylon and brass (for the guide), copper, aluminum, tungsten, platinum, or other low-friction plunger coatings are used. Depending on the surface roughness and material composition, the friction coefficients of lubricated (solid film and oil) and unlubricated materials are tungsten on tungsten 0.04–0.1 and ~0.3; copper on copper ~0.04 to 0.1 and ~1.2; aluminum on aluminum 0.04–0.12 and ~1; platinum on platinum 0.04–0.25 and ~1.2; titanium on titanium 0.04–0.1 and ~0.6.

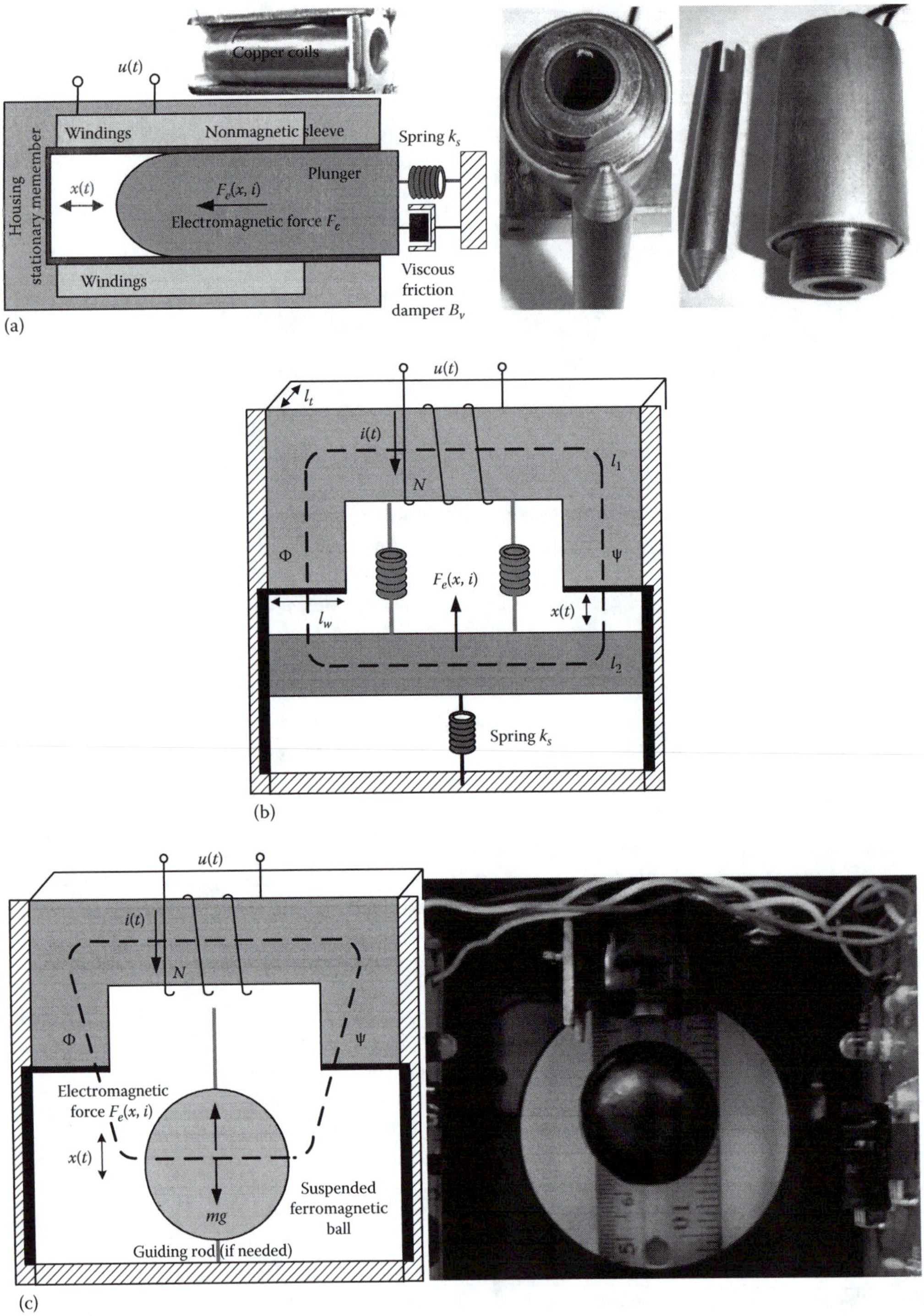

FIGURE 3.6 (a) Schematic and images of solenoids; (b) Schematic of a relay or electromagnet. The spring placement depends on device applications. *Note*: Helical springs are designed for compression and tension. The tension spring is designed to operate with a tension load, and the spring stretches as the force is applied. The compression spring operates with a compression load, and the spring contracts as the force is applied. In variable-reluctance devices, both springs usually have a helical coil, *flat*, or V-spring designs ensuring robustness, strength, and elasticity; (c) Schematics and image of a magnetic levitation system with the suspended ferromagnetic ball.

To analyze variable-reluctance devices, we apply laws of electromagnetics and mechanics. Consider the electromagnet reported in Figure 3.6b. The current i in N coils produces the flux Φ. Assume that the flux is constant. The displacement (the virtual displacement is dx) changes the magnetic energy stored in two air gaps. From $W_m = \frac{1}{2}\int_v \mu\left|\vec{H}\right|^2 dv = \frac{1}{2}\int_v \mu^{-1}\left|\vec{B}\right|^2 dv$, we have $dW_m = dW_{m\,airgap} = 2\frac{B^2}{2\mu_0}A\,dx = \frac{\Phi^2}{\mu_0 A}dx$, where A is the cross-sectional area, $A = l_w l_t$.

The flux Φ is constant if i = const. Hence, the increase in the air gap dx leads to increase in the stored magnetic energy. Using $F_e = -\nabla W_m = -\frac{\partial W_m}{\partial x}$, one finds the electromagnetic force as $\vec{F}_e = -\vec{a}_x\frac{\Phi^2}{\mu_0 A}$. The force tends to reduce the air gap length. That is, the reluctance is minimized. The movable member, for which the gravitational force is mg, is attached to the restoring spring.

The reluctances of the ferromagnetic materials of stationary and movable members are $\Re_1 = \frac{l_1}{\mu_0\mu_{r1}A}$ and $\Re_2 = \frac{l_2}{\mu_0\mu_{r2}A}$. The total air gap reluctance with two air gaps in series is $\Re_g = \frac{2x}{\mu_0 A}$.

Using the *fringing* effect, the air gap reluctance is $\Re_g = \frac{2x}{\mu_0\left(k_{g1}l_w l_t + k_{g2}x^2\right)}$, where k_{g1} and k_{g2} are the nonlinear functions of the ferromagnetic material, l_t/l_w ratio, B–H curve, electromagnetic load, etc.

The magnetizing inductance is $L(x) = \frac{N^2}{\Re_{total}(x)} = \frac{N^2}{\Re_g(x)+\Re_1+\Re_2}$.

The electromagnetic force is $F_e = -\nabla W_m = -\frac{\partial W_m}{\partial x} = -\frac{1}{2}i^2\frac{\partial}{\partial x}L(x) = -\frac{1}{2}i^2\frac{\partial}{\partial x}\left(\frac{N^2}{\Re_g(x)+\Re_1+\Re_2}\right)$.

Using $\Re_{total}(x)$ and $L(x)$, one finds F_e and *emf*. The governing differential equations result.

Example 3.1

Figure 3.7a documents a cross-sectional view of a variable-reluctance actuator (solenoid or relay) with N turns. The equivalent magnetic circuit with the reluctances is illustrated in Figure 3.7b. The source of the flux linkages in the ferromagnetic members and air gaps is the *magnetomotive force* (*mmf*), $mmf = Ni = \Sigma_j H_j l_j = H_1 l_1 + H_2 l_2 + 2H_x x$, $\oint_l \vec{H}\cdot d\vec{l} = \int_s \vec{J}\cdot d\vec{s}$.

The air gap (separation) between the stationary and movable members is $x(t)$. The movable member evolves in $x \in \left[x_{\min} \quad x_{\max}\right]$, $x_{\min} \le x \le x_{\max}$, $x_{\min} \ge 0$. The stroke is $\Delta x = (x_{\max} - x_{\min})$. The mean lengths of the stationary and movable members are l_1 and l_2, and the cross-sectional area is A. One can find the force exerted on the movable member as a function of the current $i(t)$ and displacement $x(t)$. The permeabilities of stationary and movable members are μ_{r1} and μ_{r2}.

The electromagnetic force is $F_e = -\nabla W_m = -\frac{\partial W_m}{\partial x}$, where $W_m = \frac{1}{2}Li^2$.

The magnetizing inductance is $L = \frac{N\Phi}{i(t)} = \frac{\psi}{i(t)}$, where the magnetic flux is $\Phi = \frac{Ni(t)}{\Re_1+\Re_x+\Re_x+\Re_2}$. The reluctances of the ferromagnetic stationary and movable members $\Re_1$ and $\Re_2$ and the air gap reluctance $\Re_x$ are $\Re_1 = \frac{l_1}{\mu_0\mu_{r1}A}$, $\Re_2 = \frac{l_2}{\mu_0\mu_{r2}A}$, and $\Re_x = \frac{x(t)}{\mu_0 A}$.

Using reluctances, we find $\psi = N\Phi = \frac{N^2 i(t)}{\frac{l_1}{\mu_0\mu_{r1}A} + \frac{2x(t)}{\mu_0 A} + \frac{l_2}{\mu_0\mu_{r2}A}}$.

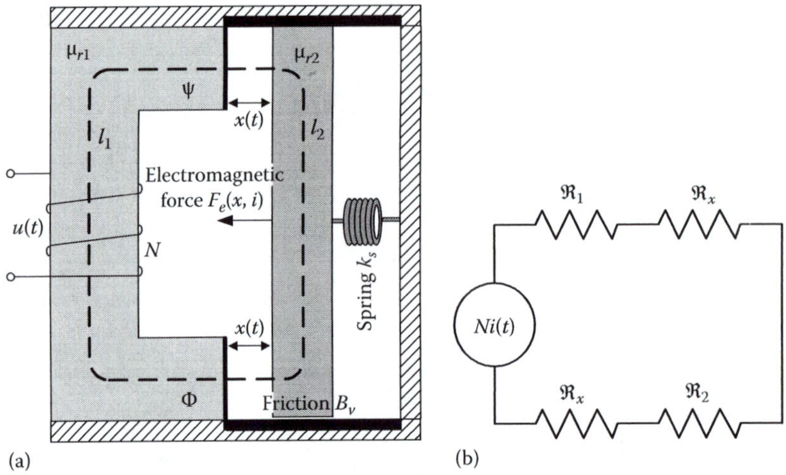

FIGURE 3.7 (a) Schematic of a variable-reluctance actuator (relay or solenoid). The plunger displaces in $x \in [x_{\min} \quad x_{\max}]$, $x_{\min} \le x \le x_{\max}$, $x_{\min} \ge 0$. The stroke is $\Delta x = (x_{\max} - x_{\min})$; (b) Equivalent magnetic circuit.

The magnetizing inductance is a nonlinear function of the displacement, and

$$L(x) = \frac{N^2}{\dfrac{l_1}{\mu_0\mu_{r1}A} + \dfrac{2x(t)}{\mu_0 A} + \dfrac{l_2}{\mu_0\mu_{r2}A}} = \frac{N^2\mu_0\mu_{r1}\mu_{r2}A}{\mu_{r2}l_1 + 2\mu_{r1}\mu_{r2}x(t) + \mu_{r1}l_2}.$$

From $F_e = -\nabla W_m = -\dfrac{\partial W_m}{\partial x} = -\dfrac{1}{2}\dfrac{\partial}{\partial x}L\big(x(t)\big)i^2(t)$, the one-directional electromagnetic force in the x direction is $F_e = \dfrac{N^2\mu_0\mu_{r1}^2\mu_{r2}^2A}{(\mu_{r2}l_1 + 2\mu_{r1}\mu_{r2}x + \mu_{r1}l_2)^2}i^2$.

The voltage $u(t)$ is applied changing $i(t)$. We use the Kirchhoff voltage law $u = ri + \dfrac{d\psi}{dt}$, where the flux linkage is $\psi = L(x)i$. From $u = ri + L(x)\dfrac{di}{dt} + i\dfrac{dL(x)}{dx}\dfrac{dx}{dt}$, one finds

$$\frac{di}{dt} = \frac{1}{L(x)}\left[-ri - \frac{2N^2\mu_0\mu_{r1}^2\mu_{r2}^2A}{(\mu_{r2}l_1 + 2\mu_{r1}\mu_{r2}x + \mu_{r1}l_2)^2}iv + u\right].$$

The force, acceleration, velocity, and displacement are vectors. The electromagnetic force F_e, which is a vector, is developed to minimize the reluctance. If voltage u is applied, F_e is developed to minimize the air gap x, and $x_{\min} \ge 0$. The tension spring exhibits the force when it stretches, and, using an ideal Hooke's law we have $F_{spring} = k_s x_{spring}$. At the steady state, $F_e = F_{spring}$. In differential equations, the movable member displacement is used as a variable x.

Consider a tension spring that exhibits zero force at *zero-length* when spring is relaxed. Using the plunger displacement $x(t)$, $x_{\min} \le x \le x_{\max}$ and the *zero-length* $(x_0 - x_{\max})$, we have $F_{spring} = k_s(x_0 - x)$, $x_0 = x_{\max}$. One has $F_{spring\,\max} = k_s(x_0 - x_{\min})$, $x_{\min} \ge 0$. At the *zero-length* $(x_0 - x_{\max})$, the spring is relaxed and exhibits zero force $F_{spring\,\min} = k_s(x_0 - x_{\max}) = 0$.

The motional *emf* opposes the voltage applied. We have a set of three differential equations

$$\frac{di}{dt} = \frac{\mu_{r2}l_1 + 2\mu_{r1}\mu_{r2}x + \mu_{r1}l_2}{N^2\mu_0\mu_{r1}\mu_{r2}A}\left[-ri - \frac{2N^2\mu_0\mu_{r1}^2\mu_{r2}^2A}{(\mu_{r2}l_1 + 2\mu_{r1}\mu_{r2}x + \mu_{r1}l_2)^2}iv + u\right],$$

$$\frac{dv}{dt} = \frac{1}{m}\left[\frac{N^2\mu_0\mu_{r1}^2\mu_{r2}^2A}{(\mu_{r2}l_1 + \mu_{r1}\mu_{r2}2x(t) + \mu_{r1}l_2)^2}i^2 - B_v v - k_s(x_0 - x)\right],$$

$$\frac{dx}{dt} = v, \quad x_{\min} \le x \le x_{\max}, \quad x_{\min} \ge 0.$$

There are limits on the plunger displacement $x_{\min} \le x \le x_{\max}$ and $x \in [x_{\min} \;\; x_{\max}]$, $x_{\min} \ge 0$. As the voltage u is applied, the plunger moves to the left minimizing the air gap until x becomes $x_{\min}$, which is a mechanical limit. For $u = 0$ V, the return spring restores the plunger to the equilibrium position $x_{\max}$ at which $F_{spring} = k_s(x_0 - x) = 0$. ■

Example 3.2

Figure 3.8a illustrates a solenoid with a stationary member and a movable plunger.

The magnetizing inductance is $L(x) = \dfrac{N^2}{\Re_f + \Re_x} = \dfrac{N^2\mu_0\mu_r A_f A_x}{A_x l_f + A_f\mu_r(x + 2d)}$, where $\Re_f$ is the total reluctances of the ferromagnetic movable members (stator and plunger are made using the same electric steel grade with μ_r); $\Re_x$ is reluctance of the *effective* air gap, which is $(x + 2d)$ due to thickness d of the nonmagnetic sleeve; A_f and A_x are the cross section areas; l_f is the equivalent lengths of the magnetic path in the ferromagnetic stationary member and plunger; d is the nonmagnetic sleeve thickness, which is usually ~1 μm. In general, $l_{f\,plunger}$ varies if x changes. However, $l_{f\,plunger}(x)$ variations do not affect the results due to high μ_r. Hence, we let l_f = const.

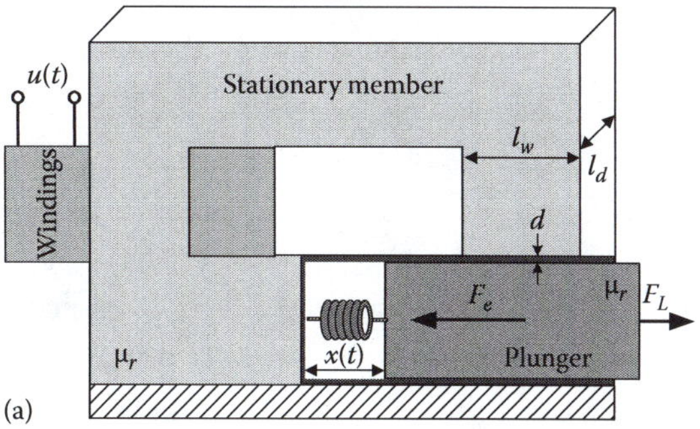

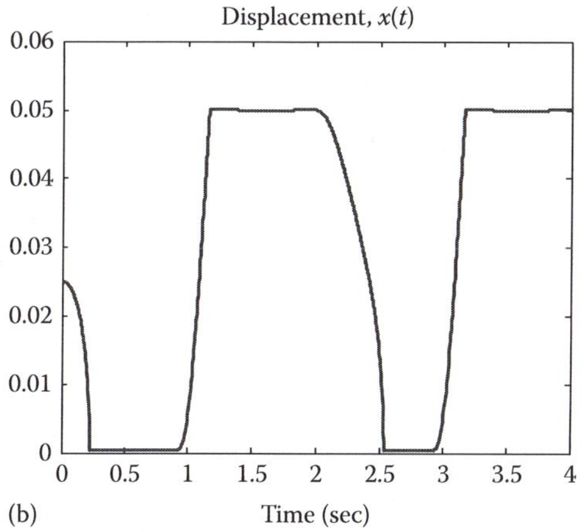

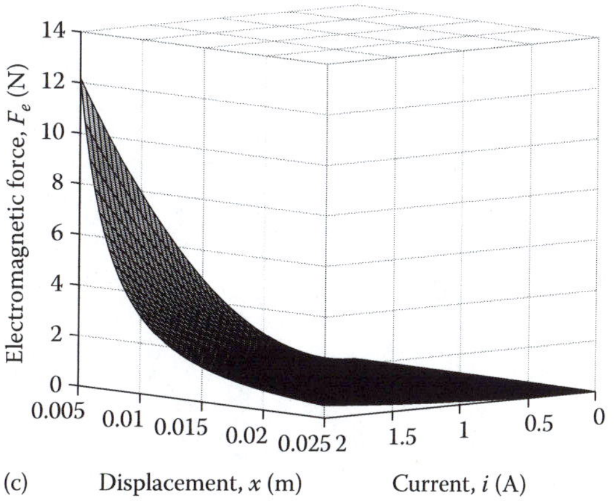

FIGURE 3.8 (a) Solenoid schematics; (b) Dynamics of $x(t)$; (c) Plot for $F_e(x, i)$.

Assuming that the magnetic system is linear, the coenergy is $W_c(i,x)=\frac{1}{2}L(x)i^2$. The electromagnetic force is $F_e(i,x)=-\frac{\partial W_c(i,x)}{\partial x}=-\frac{1}{2}i^2\frac{dL(x)}{dx}$, where $\frac{dL}{dx}=-\frac{N^2\mu_0\mu_r^2A_f{}^2A_x}{[A_xl_f+A_f\mu_r(x+2d)]^2}$.

Using Kirchhoff's law $u=ri+\frac{d\psi}{dt}$, $\psi=L(x)i$, one has $u=ri+L(x)\frac{di}{dt}+i\frac{dL(x)}{dx}\frac{dx}{dt}$.

We apply Newton's second law for translational motion $m\frac{d^2x}{dt^2}=F_e-B_v\frac{dx}{dt}-k_s(x_0-x)-F_L$.

The resulting nonlinear differential equations are

$$\frac{di}{dt}=-\frac{A_xl_f+A_f\mu_r(x+2d)}{N^2\mu_0\mu_rA_fA_x}ri-\frac{\mu_rA_f}{A_xl_f+A_f\mu_r(x+2d)}iv+\frac{A_xl_f+A_f\mu_r(x+2d)}{N^2\mu_0\mu_rA_fA_x}u,$$

$$\frac{dv}{dt}=\frac{1}{m}\left[\frac{N^2\mu_0\mu_r^2A_f{}^2A_x}{2[A_xl_f+A_f\mu_r(x+2d)]^2}i^2-B_vv-k_s(x_0-x)-F_L\right],$$

$$\frac{dx}{dt}=v,\quad x_{\min}\le x\le x_{\max}.$$

The *back emf* opposes the applied voltage. The friction and spring forces act against the electromagnetic force. ■

Example 3.3

Consider a solenoid depicted in Figure 3.8a with the stroke 5 cm. Let the relative permeabilities of the stationary member and plunger be different, $\mu_{rs}\neq\mu_{rp}$. We obtain the mathematical model and perform simulations with the following parameters: r = 8.5 ohm, L_l = 0.001 H, N = 700, m = 0.095 kg, μ_{rs} = 4500, μ_{rp} = 5000, l_{fp} = 0.055 m, l_{fs} = 0.095 m, $A_f=A_x$ = 0.00025 m^2, B_v = 0.06 N-sec/m, k_s = 10 N/m, and x_0 = 0.05 m. The subscripts p and s stand for the plunger and stationary member.

Variations of $l_{fp}(x)$ result in very minor overall changes while complicating the equations and obscuring the expression for F_e. The assumptions l_{fp} = const and d = 0 result in error less than 0.1%. The magnetizing inductance is

$$L(x)=\frac{N^2}{\Re_{fs}+\Re_{fp}+\Re_x}=\frac{N^2}{\frac{l_{fs}}{\mu_0\mu_{rs}A}+\frac{l_{fp}}{\mu_0\mu_{rp}A}+\frac{x}{\mu_0A}}=\frac{N^2\mu_0\mu_{rs}\mu_{rp}A}{\mu_{rp}l_{fs}+\mu_{rs}l_{fp}+\mu_{rs}\mu_{rp}x}.$$

The electromagnetic force is $F_e=-\frac{\partial}{\partial x}\frac{1}{2}L(x)i^2=\frac{N^2\mu_0\mu_{rs}^2\mu_{rp}^2A}{2\left(\mu_{rp}l_{fs}+\mu_{rs}l_{fp}+\mu_{rs}\mu_{rp}x\right)^2}i^2$.

The electromagnetic force is developed to minimize the air gap. Using the leakage inductance L_l, and applying an ideal Hooke's law, one finds the following nonlinear differential equations

$$\frac{di}{dt}=\frac{1}{L(x)+L_l}\left[-ri-\frac{N^2\mu_0\mu_{rs}^2\mu_{rp}^2A}{2\left(\mu_{rp}l_{fs}+\mu_{rs}l_{fp}+\mu_{rs}\mu_{rp}x\right)^2}iv+u\right],\quad L(x)=\frac{N^2\mu_0\mu_{rs}\mu_{rp}A}{\mu_{rp}l_{fs}+\mu_{rs}l_{fp}+\mu_{rs}\mu_{rp}x},$$

$$\frac{dv}{dt}=\frac{1}{m}\left[\frac{N^2\mu_0\mu_{rs}^2\mu_{rp}^2A}{2\left(\mu_{rp}l_{fs}+\mu_{rs}l_{fp}+\mu_{rs}\mu_{rp}x\right)^2}i^2-B_vv-k_s(x_0-x)-F_L\right],$$

$$\frac{dx}{dt}=v,\quad x_{\min}\le x\le x_{\max}.$$

The displacement of the plunger $x(t)$ is constrained by hard mechanical limits, and $x \in [0.0005 \;\; 0.05]$ m. As the voltage u is applied, the plunger moves to the left, minimizing the air gap until x becomes 0.0005 m. The simulations are performed. The limits on $x(t)$ can be set in the integrator, or using the Saturation block. For $u = 0$ V, the return spring restores the plunger to the left equilibrium position 0.05 m to the *zero-length* spring position at which $F_{spring} = k_s(x_0 - x) = 0$. Figures 3.8b and c illustrate the evolution of $x(t)$ as well as the three-dimensional plot $F_e(x, i)$ for $x \in [0.005 \;\; 0.025]$ m and $i \in [0 \;\; 2]$ A. In catalogs, the plots for $F_e(x)$ are reported for different currents or voltages. The statement is

```
N=700; mu0=4*pi*1e-7; mus=4500; mup=5000; ls=0.095; lp=0.0275; A=2.5e-4;
x=linspace(0.005,0.025,50); i=linspace(0,2,50); [X,Y]=meshgrid(x,i);
Fe=(0.5*N^2*mu0*mup^2*mus^2*A*Y.*Y)./(lp*mus+ls*mup+mup*mus*X).^2 ; surf(x,i,Fe);
xlabel('Displacement, {\itx} [m]','FontSize',18);
ylabel('Current, {\iti} [A]','FontSize',18);
zlabel('Electromagnetic Force, {\itF_e} [N]','FontSize',18);
```

■

The equations for physical quantities and governing differential equations were found based on assumptions. Nonlinear magnetization, nonlinear magnetic system, secondary effects, cross section area variations, nonuniformity, and other effects were not considered or were simplified. Using the experimental data, one obtains consistent and high-fidelity models using the results reported. For example, the experimental $L(x)$ and $F_e(x)$ can be measured.

Example 3.4

The magnetizing inductance $L(x)$ and electromagnetic force F_e can be experimentally measured. Inductance $L(x)$ can be measured using the solenoid or relay as a transformer adding additional coils on the plunger. The electromagnetic force F_e is measured by loading a device. The use of reluctances and coenergy result in the following expressions

$$L(x) = \frac{a_1}{b+cx}, \quad F_e = -\frac{\partial}{\partial x}\left(\frac{1}{2}L(x)i^2\right) = \frac{a_1 c}{(b+cx)^2}i^2 = \frac{a}{(b+cx)^2}i^2, \quad a_1 > 0, b > 0, \text{and } c > 0.$$

The unknown coefficients a_1, b, and c were found using the solenoid parameters such as lengths, area, μ_r, N, etc. For a solenoid, the experimental data for the measured $F_e(x)$ if $i = 1$ A is depicted in Figure 3.9a. The maximum plunger displacement is ~5 cm. We find the unknown constants a, b, and c for the following approximations of the electromagnetic force

$$F_e = \frac{a}{(b+cx)^2}i^2, \quad F_e = ae^{-bx}i^2, \quad F_e = ae^{-b\sin cx}i^2, \quad \text{and} \quad F_e = \frac{ae^{-dx}}{(b+cx)^2}i^2.$$

The aforementioned expressions for $F_e(x)$ correspond to the device physics recalling that $F_e = \frac{\partial}{\partial x}\frac{1}{2}L(x)i^2 = \frac{a}{(b+cx)^2}i^2$.

(a) Let $F_e = \dfrac{a}{(b+cx)^2}i^2$. The unknowns a, b, and c are found by using nonlinear interpolation. The MATLAB® file is developed. The command **nlinfit** is used. Let $a_0 = 100$, $b_0 = 0.1$, and $c_0 = 100$.

```
% Measured data
x=[0.005 0.01 0.015 0.021 0.025 0.03 0.035 0.04]; Fe1=[35 20 12 7 4.5 2.5 1.5 1];
plot(x,Fe1,'ko','linewidth',3); xlabel('Displacement, {\itx} [m]','FontSize',15);
title('Electromagnetic Force, {\itF_e}({\itx}), [N]','FontSize',15);
modelFun = @(p,x) p(1)./((p(2)+p(3).*x).^2); startingValues=[100 0.1 100];
CoefEsts = nlinfit(x, Fe1, modelFun, startingValues)
xgrid=linspace(0,0.05,100);
line(xgrid, modelFun(CoefEsts,xgrid), 'Color','r','linewidth',3);
```

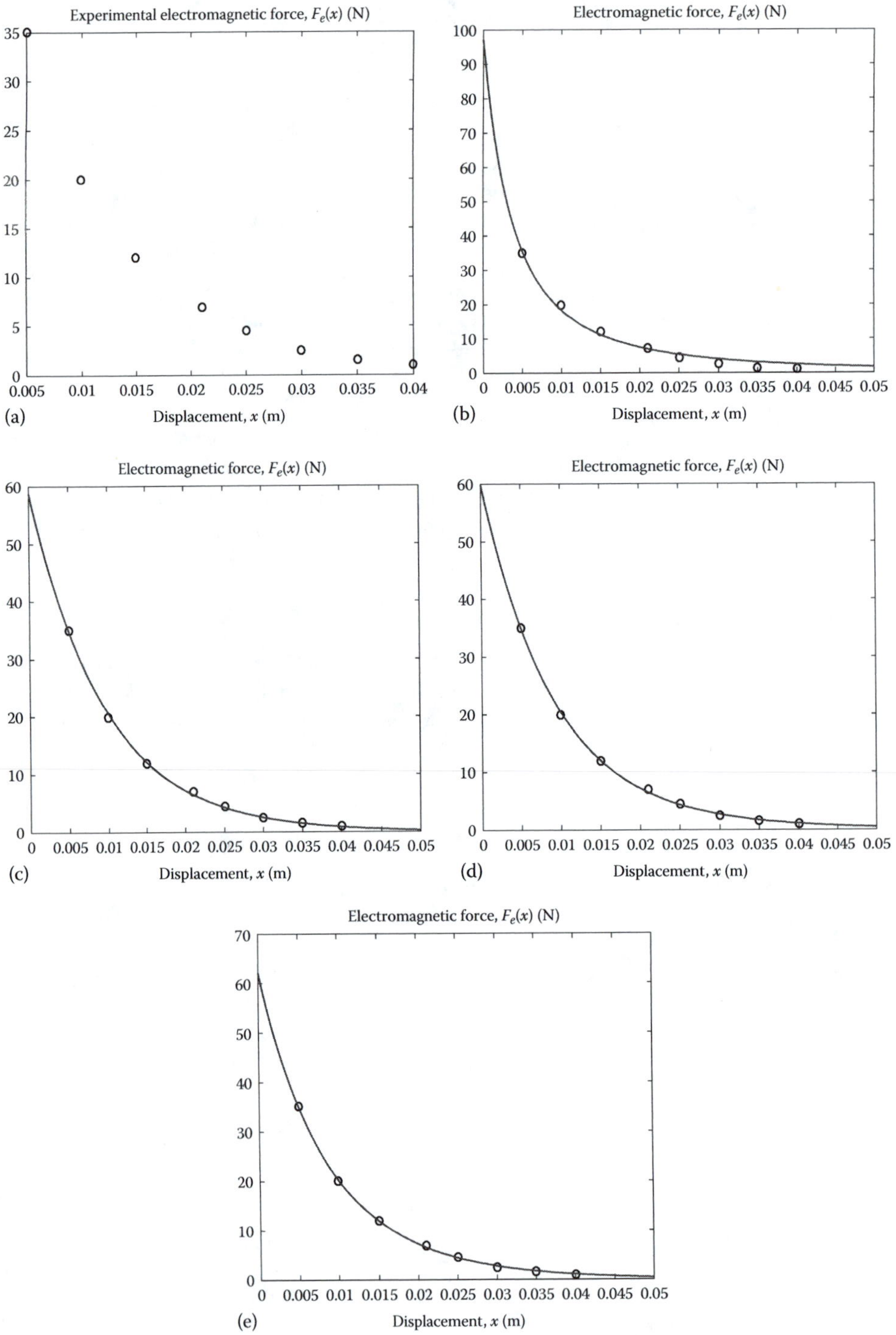

FIGURE 3.9 (a) Experimental data for the measured $F_e(x)$, i = 1 A; (b) Plots of the measured $F_e(x)$ (dots) and $F_e = \frac{a}{(b+cx)^2}i^2$, $a = 52.77$, $b = 0.737$, and $c = 96.1$ (solid line); (c) Plots of the measured $F_e(x)$ (dots) and $F_e = ae^{-bx}i^2$, $a = 58.7$, and $b = 105$ (solid line); (d) Plots of the measured $F_e(x)$ (dots) and $F_e = ae^{-b\text{sinc}x}i^2$, $a = 59.8$, $b = 6.31$, and $c = 17.2$; (e) Plots of the measured $F_e(x)$ (dots) and approximation $F_e = \frac{ae^{-dx}}{(b+cx)^2}i^2$, $a = 62.5$, $b = 1$, $c = 27.1$, and $d = 64.3$.

The unknown coefficients are found as displayed in the Command Window

```
CoefEsts =
   5.2773e+01    7.3694e-01    9.6106e+01
```

Hence, we have $a = 52.77$, $b = 0.737$ and $c = 96.1$. Therefore, $F_e = \frac{a}{(b+cx)^2} i^2 = \frac{52.77}{(0.737+96.1x)^2} i^2$. The resulting plots of the experimental $F_e(x)$ and derived approximation are documented in Figure 3.9b.

(b) For $F_e = ae^{-bx}i^2$, the MATLAB statements are modified to find the unknown a and b coefficients. In particular, the function is assigned as

```
modelFun=@(p,x) (p(1).* exp(-p(2).*x));startingValues=[10 10];
```

The initial values of coefficient are $a_0 = 10$, $b_0 = 10$. One obtains $a = 58.7$ and $b = 105$. Hence, $F_e = ae^{-bx}i^2 = 58.7e^{-105x}i^2$. The plots are depicted in Figure 3.9c.

(c) Using $F_e = ae^{-b\sin cx}i^2$, the unknowns a, b, and c are found. Let $a_0 = 10$, $b_0 = 10$, and $c_0 = 10$. One has

```
modelFun=@(p,x) p(1).* exp(-p(2).*(sin(p(3).*x))); startingValues=[10 10 10];
```

We find $a = 59.8$, $b = 6.31$ and $c = 17.2$. Therefore, $F_e = ae^{-b\sin cx}i^2 = 59.8e^{-6.31\sin(17.2x)}i^2$. Figure 3.9d reports the resulting plots.

(d) For $F_e = \frac{ae^{-dx}}{(b+cx)^2} i^2$, the unknown coefficients are found to be $a = 62.5$, $b = 1$, $c = 27.1$, and $d = 64.3$ using

```
modelFun=@(p,x) (p(1).*exp(-p(4).*x))./((p(2)+p(3).*x).^2); startingValues=[100 0.1 100 1].
```

The plot for $F_e = \frac{62.5e^{-64.3x}}{(1+27.1x)^2} i^2$ is depicted in Figure 3.9e.

Note: Nonlinear interpolation is examined in Examples 2.17 and 2.28 using the **nlinfit** and **lsqnonlin** solvers. ■

Example 3.5

Consider a solenoid if the measured electromagnetic force is $F_e(x) = ae^{-bx}i^2$, $0 \le x \le 0.05$, $a > 0$ and $b > 0$. Our goal is to derive the resulting equations of motion.

From $F_e(i,x) = -\frac{\partial W_c(i,x)}{\partial x} = -\frac{1}{2} i^2 \frac{dL(x)}{dx}$, we conclude $\frac{dL}{dx} = -2ae^{-bx}$.

The integration gives $-\int 2ae^{-bx}\, dx = 2\frac{a}{b} e^{-bx}$. The positive-definite magnetizing inductance is $L(x) = 2\frac{a}{b} e^{-bx}$. Using Kirchhoff's law, we have $u = ri + \frac{d\psi}{dt} = ri + L(x)\frac{di}{dt} + i\frac{dL(x)}{dx}\frac{dx}{dt}$. Applying Newtonian mechanics, one obtains

$$\frac{di}{dt} = \frac{b}{2ae^{-bx}}\left[-ri - 2ae^{-bx}iv + u\right],$$

$$\frac{dv}{dt} = \frac{1}{m}\left[ae^{-bx}i^2 - B_v v - k_s(x_0 - x) - F_L\right],$$

$$\frac{dx}{dt} = v, \quad x_{\min} \le x \le x_{\max}.$$

We perform simulations using the parameters found in Example 3.4. In particular, $a = 58.7$, $b = 105$, $r = 15$ ohm, $m = 0.1$ kg, $B_v = 0.06$ N-sec/m, and $k_s = 10$ N/m. The Simulink model is developed as depicted in Figure 3.10a. The following parameters are uploaded

```
>> a=58.7; b=105; r=15; m=0.1; x0=0.05; Bv=0.06; ks=10;
```

The plunger displacement is constrained, $x_{\min} \le x \le x_{\max}$, $0_{\min\ airgap} \le x \le 4\ \text{cm}_{\max\ stroke}$. The voltage u is applied as steps $u = \begin{cases} 3\ \text{V} \\ 2\ \text{V} \end{cases}$, see Figure 3.10b. The displacement is limited, $0 \le x \le 0.04$ m.

The dynamics of the current $i(t)$, velocity $v(t)$, and position $x(t)$ are reported in Figures 3.10c and d for different initial plunger displacement $x_{t=0}$. Consider the plunger at $x_{t=0} = 4$ cm and $x_{t=0} = 0$. As the voltage is applied, the F_e is developed. As illustrated in Figures 3.10b and c, for the specified voltages $u = \begin{cases} 3\ \text{V} \\ 2\ \text{V} \end{cases}$, the plunger is repositioned to the equilibrium positions x_e at which $F_e = F_{spring}$. ■

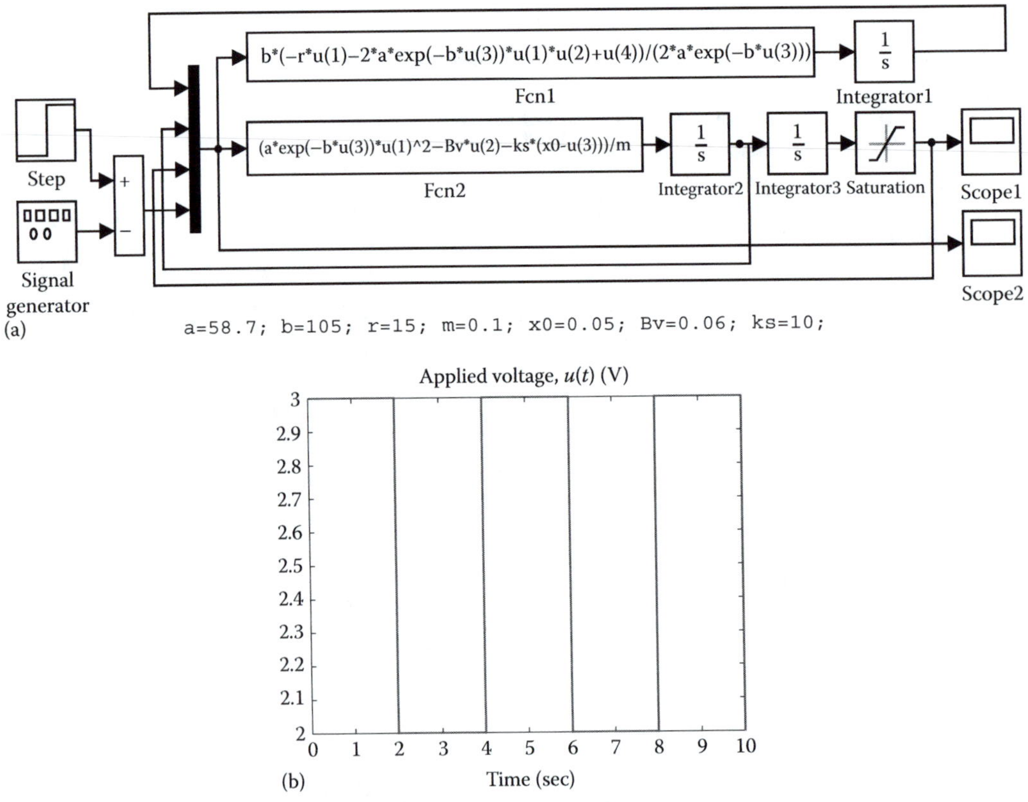

FIGURE 3.10 (a) Simulink® model; (b) Applied voltage pulses $u = \begin{cases} 3\ \text{V} \\ 2\ \text{V} \end{cases}$. *(Continued)*

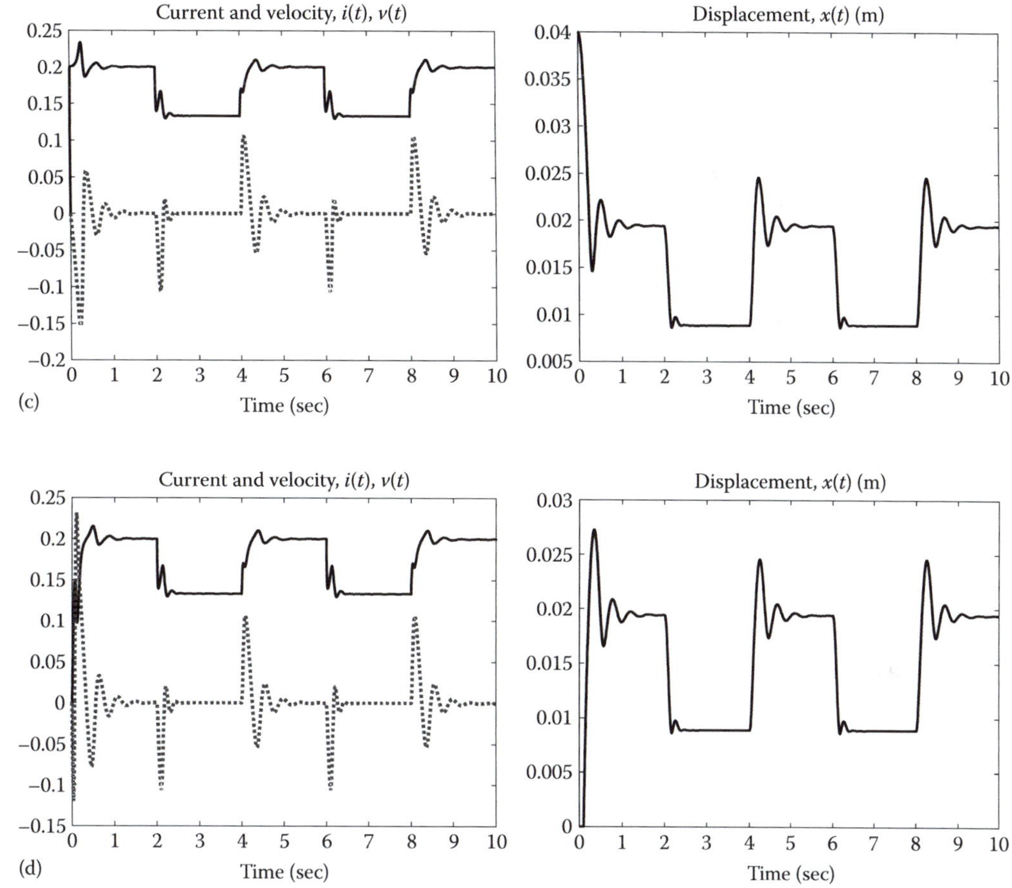

FIGURE 3.10 (*Continued*) (c) Open-loop system dynamics for the current $i(t)$, velocity $v(t)$, and $x(t)$ if $x_{t=0}=0.04$ m, $0 \le x \le 0.04$ m; (d) Open-loop system dynamics for the current $i(t)$, velocity $v(t)$, and displacement $x(t)$ if $x_{t=0}=0$.

Example 3.6

We derive the equations of motion for a solenoid with the magnetizing inductance $L(x)=\dfrac{ae^{-bx}}{(c+dx)}$, where a, b, c, $d>0$.

One applies $F_e(i,x)=-\dfrac{\partial W_c(i,x)}{\partial x}=-\dfrac{1}{2}i^2\dfrac{dL(x)}{dx}$. Using the quotient rule, for a function f/g, the derivative is $(f'g-g'f)/g^2$. Hence, $F_e=\dfrac{1}{2}ae^{-bx}\dfrac{b(c+dx)+d}{(c+dx)^2}i^2$.

The maximum electromagnetic force at $x=0$ is

$$F_{e\max}=F_{e\,x=0}=\frac{1}{2}ae^{-bx}\frac{b(c+dx)+d}{(c+dx)^2}i^2\bigg|_{x=0}=\frac{1}{2}a\frac{bc+d}{c^2}i^2.$$

Using the Faraday law of inductance, the total *emf* is

$$-\frac{d\psi}{dt}=-\frac{d}{dt}\big(L(x)i\big)=-\frac{di}{dt}\frac{ae^{-bx}}{(c+dx)}+ae^{-bx}\frac{b(c+dx)+d}{(c+dx)^2}iv,$$

where the motional *emf* is $ae^{-bx}\dfrac{b(c+dx)+d}{(c+dx)^2}iv$.

The resulting nonlinear differential equations are

$$\frac{di}{dt}=\frac{c+dx}{ae^{-bx}}\left[-ri-ae^{-bx}\frac{b(c+dx)+d}{(c+dx)^2}iv+u\right]$$

$$\frac{dv}{dt}=\frac{1}{m}\left[\frac{1}{2}ae^{-bx}\frac{b(c+dx)+d}{(c+dx)^2}i^2-B_v v-k_s\left(x_0-x\right)-F_L\right],$$

$$\frac{dx}{dt}=v,\quad x_{\min}\le x\le x_{\max}.$$ ■

3.3.2 Experimental Analysis and Control of a Solenoid

Consider a solenoid illustrated in Figures 3.6a and 3.11a. The equivalent magnetic circuit is reported in Figure 3.11b. One finds the reluctances of the stationary member $\Re_{fs}=\dfrac{l_{fs}}{\mu_0\mu_r A_1}$, the stationary member which faces the plunger $\Re_{fsp}=\dfrac{l_{fsp}}{\mu_0\mu_r A_2}$, the air gap $\Re_x=\dfrac{x}{\mu_0 A_2}$, and the plunger $\Re_{fp}=\dfrac{l_{fp}}{\mu_0\mu_r A_2}$.

The equivalent magnetic circuit, as depicted in Figure 3.11b, yields

$$\frac{1}{2}Ni=\Re_{fs}\Phi_1+(\Re_{fsp}+\Re_x+\Re_{fp})\Phi_3\quad\text{and}\quad\frac{1}{2}Ni=\Re_{fs}\Phi_2+(\Re_{fsp}+\Re_x+\Re_{fp})\Phi_3.$$

From $Ni=\Re_{fs}(\Phi_1+\Phi_2)+2(\Re_{fsp}+\Re_x+\Re_{fp})\Phi_3$, using $\Phi_1+\Phi_2=\Phi_3$, we obtain $Ni=(\Re_{fs}+2\Re_{fsp}+2\Re_x+2\Re_{fp})\Phi_3$. The magnetic flux Φ_3 and flux linkages $\psi=N\Phi_3$ are

$$\Phi_3=\frac{Ni}{\Re_{fs}+2\Re_{fsp}+2\Re_x+2\Re_{fp}}\quad\text{and}\quad\psi=\frac{N^2 i}{\Re_{fs}+2\Re_{fsp}+2\Re_x+2\Re_{fp}}.$$

The magnetizing inductance is $L(x)=\dfrac{N^2}{\Re_{fs}+2\Re_{fsp}+2\Re_x+2\Re_{fp}}=\dfrac{N^2\mu_0\mu_r A_1A_2}{l_{fs}A_2+2l_{fsp}A_1+2A_1\mu_r x+2l_{fp}A_1}$.

Using the coenergy, the electromagnetic force is $F_e=-\dfrac{\partial}{\partial x}\dfrac{1}{2}L(x)i^2=\dfrac{N^2\mu_0\mu_r^2A_1^2A_2}{\left(l_{fs}A_2+2l_{fsp}A_1+2A_1\mu_r x+2l_{fp}A_1\right)^2}i^2$.

The Kirchhoff voltage law and Newton law yield

$$\frac{di}{dt}=\frac{1}{L(x)+L_l}\left[-ri-\frac{2N^2\mu_0\mu_r^2A_1^2A_2}{\left(l_{fs}A_2+2l_{fsp}A_1+2A_1\mu_r x+2l_{fp}A_1\right)^2}iv+u\right],$$

$$\frac{dv}{dt}=\frac{1}{m}\left[\frac{N^2\mu_0\mu_r^2A_1^2A_2}{\left(l_{fs}A_2+2l_{fsp}A_1+2A_1\mu_r x+2l_{fp}A_1\right)^2}i^2-B_v v-k_s\left(x_0-x\right)-F_L\right],$$

$$\frac{dx}{dt}=v,\quad x_{\min}\le x\le x_{\max}.$$

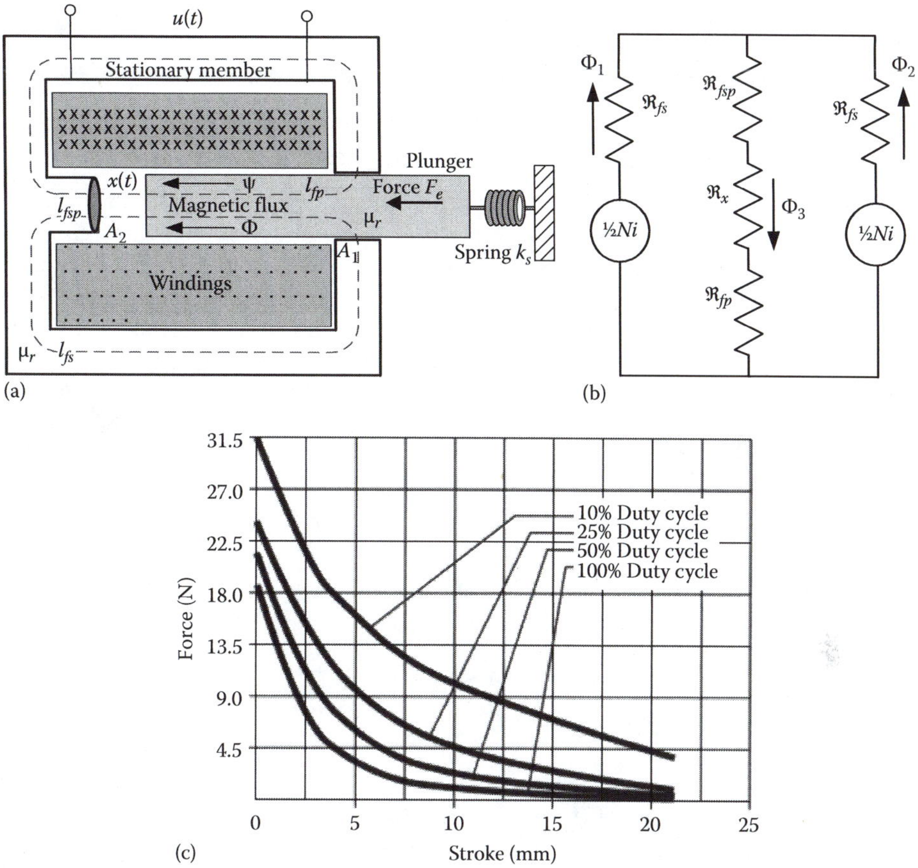

FIGURE 3.11 (a) Solenoid schematics; (b) Equivalent magnetic circuit for a solenoid; (c) The force-displacement characteristics of the Ledex B11M-254 solenoid when u_{max} is 12, 17, 24, and 38 V for 100%, 50%, 25%, and 10% duty cycle operations, $r = 17.3$ ohm, and $0 \le x \le 2.2$ cm.

Numerical analysis can be performed using the model developed and by simulating the solenoid. The experimental studies are of great interest. We examine a Ledex B11M-254 solenoid, a driving circuit, and a proportional–integral control law. For a "pull" operation solenoid, the parameters are as follows: $u_{max} = 12$, 17, 24 and 38 V (at 100%, 50%, 25%, and 10% duty cycle operation), 15.5 N holding force, A-class coil insulation, 105°C maximum temperature, the plunger weight is 17 g. The $F_e(x)$ characteristics are depicted in Figure 3.11c. The maximum stroke is 2.2 cm. The solenoid parameters are $r = 17.3$ ohm, $L_{self} = 0.0064$ H, $L_l = 0.001$ H, $N = 1780$, $m = 0.017$ kg, $\mu_{rs} = \mu_{rp} = 5500$, $l_{fp} = 0.02$ m, $l_{fs} = 0.048$ m, $l_{fsp} = 0.08$ m, $A_f = A_x = 2 \times 10^{-4}$ m^2, $B_v = 0.25$ N-sec/m, and $k_s = 5$ N/m.

A closed-loop system includes solenoid, a position sensor, a one-quadrant PWM regulator, filters, and a proportional–integral controller. The solenoid displacement $x(t)$ is measured and compared with the reference displacement $r(t)$ to obtain the tracking error $e(t) = r(t) - x(t)$. Using pulse width modulation (PWM), we control the duty cycle of the MOSFET transistor, thereby changing the *average* voltage $u(t)$ applied to a winding. The control voltage $u_c(t)$ is compared with a periodic (triangular or sinusoidal) signal u_t. The electronics schematics and images of the system are depicted in Figures 3.12. Notations, definitions, subcircuitry, components, and signals are labeled and accentuated correspondingly.

The first component of the circuit is the error circuit to obtain $e(t)$. The error circuit adds the reference voltage signal $r(t)$ with the inverted linear potentiometers output signal which corresponds to the measured plunger displacement $x(t)$. The error circuit is implemented using a unit-gain instrumentation amplifier INA128. The tracking error $e(t) = r(t) - x(t)$ is fed into the controller circuit, which implements a proportional-integral control law $u_c = k_p e + k_i \int e dt$, where k_p and k_i are the proportional and integral feedback gains. The proportional term $k_p e$ is implemented by using an operational amplifier TLC277 in an inverting configuration. The proportional gain k_p is realized by using the input and feedback resistors, $k_p = -R_{P2}/R_{P1}$. To implement the integral feedback $k_i \int e dt$, an integrator is implemented using an inverted operational amplifier with an input resistor R_I and a feedback capacitor C_I. The integral feedback gain is $k_i = -1/R_I C_I$. Due to the inverting configuration for the proportional and integral terms, the outputs are inverted and summed to yield the control voltage u_c. To perform the inversion and summation, instrumentation amplifiers are used due to their robustness, tolerance, low noise, linearity, etc. By fixing the positive input of the instrumentation amplifier to ground and feeding the respective signal to the negative input, a simple inverter is implemented. By feeding the output of an inverting amplifier to the reference node of the following circuitry, a summing circuit is implemented.

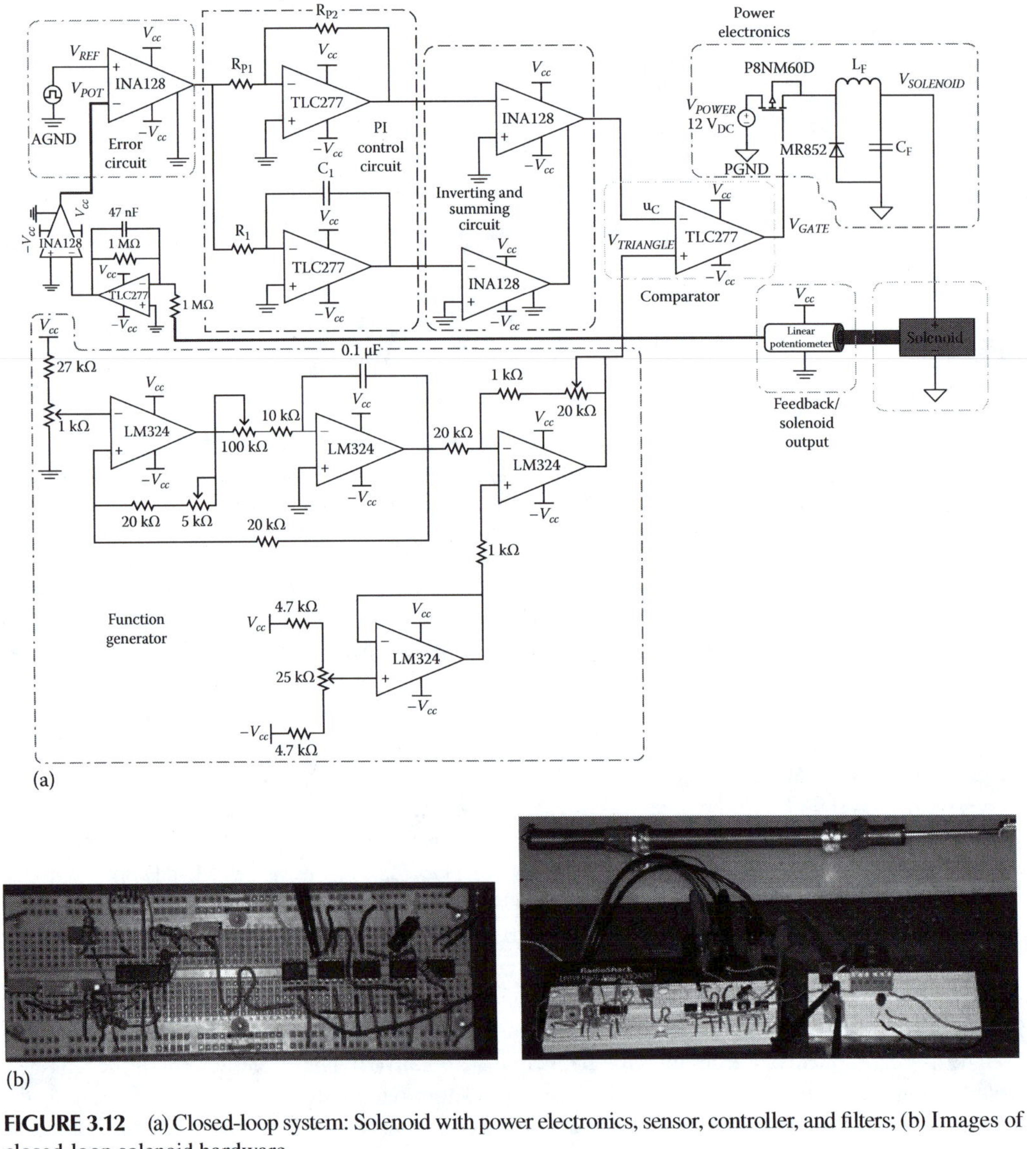

FIGURE 3.12 (a) Closed-loop system: Solenoid with power electronics, sensor, controller, and filters; (b) Images of closed-loop solenoid hardware.

The control signal u_c is supplied to a comparator that compares two inputs to produce an output, see Figure 3.12a. The comparator is implemented using an operational amplifier with the control signal u_c supplied to its inverting terminal and a periodic near-triangular waveform u_t applied to its noninverting terminal. The comparator outputs a positive rail voltage V_{cc} for the duration of time when its positive input is greater than its negative input. The comparator will output $-V_{cc}$ when its negative input is greater than its positive input. The comparator develops the PWM waveform changing the duty cycle of a square wave, whose amplitude is equal to the comparators rail, with respect to control signal $u_c = k_p e + k_i \int e dt$. The periodic triangular waveform is established by a function generator.

As depicted in Figure 3.12a, the first LM324 operational amplifier in the function generator circuit is the comparator. The positive input of the comparator is the output of the second operational amplifier in the circuit, which is an integrator. The oscillation of the comparator produces a square waveform, which is integrated by the second operational amplifier with a capacitor in the negative feedback path to produce a near-triangular waveform u_t. The frequency of u_t is determined by the time constant defined by the input resistance and feedback capacitance of the integrator. The remaining two LM324 operational amplifiers in the function generator circuit are the buffer/attenuator amplifier for u_t and an adjustable DC offset. For the reported circuitry, the maximum frequency of robust operation is ~6.8 kHz.

Various waveform generating and switching circuits can be used. The oscillating frequency of the n-stage ring pulse-oscillator is $f = \frac{1}{n(\tau_1 + \tau_2)}$, where τ_1 and τ_2 are the intrinsic propagation delays; n is the number of inverters (n is odd). The oscillators, wave generation and shaping circuit design using a Schmitt trigger, the Wien bridge oscillator, the Colpitts oscillator, the Hartley oscillator, the RC phase-shift concept, and other solutions are well known and can be used.

The PWM waveform, produced by comparing the control signal u_c and near-triangular waveform u_t, allows one to control the *average* voltage u supplied to the solenoid. The TLC277 operational amplifier that produces the PWM waveform can output a maximum current and voltage of ~30 mA and 18 V. For the studied solenoid, the rated current and voltage are ~1 A and up to ~40 V. The TLC277 comparator cannot ensure the necessary current. We use a one-quadrant power electronics stage, as reported in Figure 3.12a. A power MOSFET switch is controlled by applying the voltage to a gate. The comparator output is connected to the MOSFET gate to control the switching activity. A low pass filter is implemented using an inductor L_F and capacitor C_F. The filter corner frequency $f_c = \frac{1}{2\pi\sqrt{L_F C_F}}$. Let f_c = 3.4 kHz, with L_F = 500 μH, and C_F = 4.7 μF. The plunger of the solenoid is connected to a linear potentiometer to measure displacement $x(t)$. One finds $e(t)$. A second-order low pass filter is implemented to attenuate the noise. The low pass filter is documented in Figure 3.12a between the potentiometer and error circuit input. Our goal is to ensure a high signal-to-noise ratio within the operating solenoid bandwidth. For an unity-gain filter, the cutoff frequency is $f_c = \frac{1}{2\pi R_f C_f}$. To ensure f_c = 3 Hz, R_f = 1 Mohm and C_f = 47 nF.

The control voltage u_c (output of the proportional–integral controller) and the near-triangular signal u_t are illustrated in Figure 3.13a. These signals are compared by the comparator, and a PWN signal drives a MOSFET. The MOSFET gate voltage and the voltage applied to the solenoid are shown in Figure 3.13b.

The transient dynamics for the closed-loop system are reported in Figure 3.14. The comparison of the reference $r(t)$ and plunger displacements $x(t)$ provide an evidence that: (1) fast nonoscillatory repositioning is accomplished; (2) stability is guaranteed; (3) the steady-state tracking error is zero within the positioning sensor accuracy; (4) the disturbances are attenuated (F_L is applied when x reaches the steady-state value at the second bottom plot for x); (5) robustness to parameter variations is accomplished. One-directional control of the displacement $x(t)$ is accomplished. The returning spring restores the plunger position at the equilibrium if $u = 0$. Consistent analysis and design are accomplished and verified using fundamental results, circuits design, and experimental studies.

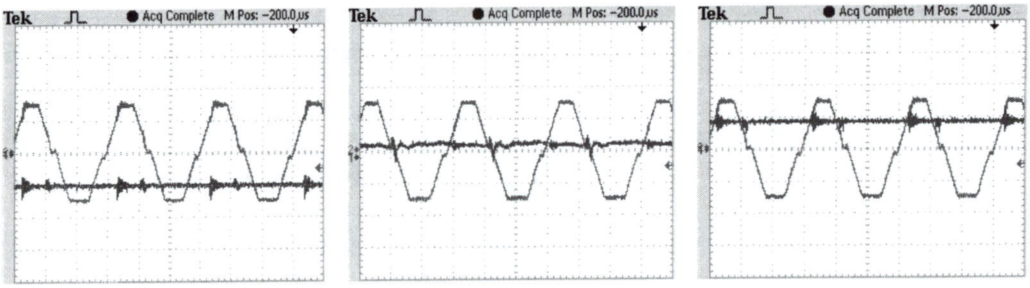

(a)

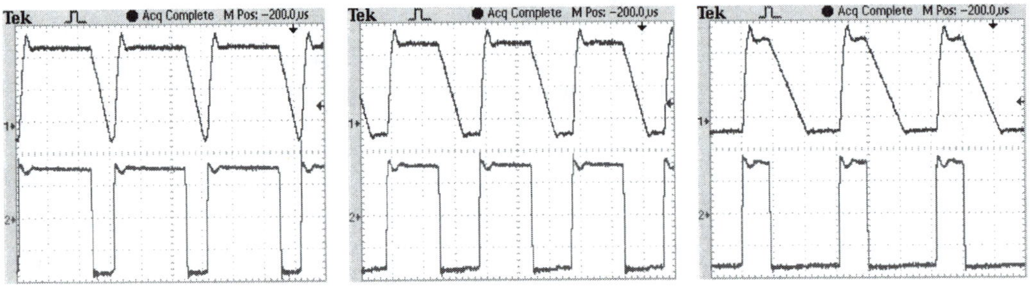

(b)

FIGURE 3.13 (a) Near-triangular waveform u_t and control voltage u_c for three reference r plunger positions: $r = 0.5$, $r = 1$, and $r = 1.5$ cm; (b) MOSFET gate voltage and voltage applied to a solenoid winding u when $r = 0.5$ cm (3.2 and 2.6 V), $r = 1$ cm (8.4 and 6.8 V), and $r = 1.5$ cm (12.2 and 9.8 V).

3.3.3 Synchronous Variable-Reluctance Rotational Actuators

We examine the radial topology limited-angle reluctance actuators. The synchronous reluctance motors are examined in detail in Section 6.2. Our goal is to study and analyze these transducers, which are used in many applications, such as limited-angle rotational actuators, relays, etc. The considered problem is relevant to variable-reluctance stepper motors. A single-phase four-pole synchronous reluctance actuator (rotational relay) is illustrated in Figure 3.15a. The path for ψ_{as} is illustrated. If rotor rotates, the positive-definite reluctance $\Re(\theta_r) > 0$ varies with the period $\pi/2$, $\Re_{\min} \le \Re(\theta_r) \le \Re_{\max}$. The magnetizing inductance $L(\theta_r) > 0$ is a periodic function with the period $\pi/2$, and $L_{\min} \le L(\theta_r) \le L_{\max}$. The variations of inductance $L_m(\theta_r)$ is studied in Section 6.2 using the *quadrature* and *direct* magnetic axes. Having found the *quadrature* (q) and *direct* (d) axes reluctances, $\Re_{mq}$ and $\Re_{md}$, $\Re_{mq} > \Re_{md}$, the inductances are $L_{mq} = 1/\Re_{mq}$ and $L_{md} = 1/\Re_{md}$. For the *direct* axis, reported in Figure 3.15a, the reluctance $\Re_{md}$ is minimum because the air gap is minimum; The reluctance in the *quadrature* axis $\Re_{mq}$ is maximum due to the maximum air gap. Using the magnetizing inductances L_{mq} and L_{md}, $\bar{L}_m = \frac{1}{2}\left(L_{mq} + L_{md}\right)$ and $L_{\Delta m} = \frac{1}{2}\left(L_{md} - L_{mq}\right)$.

Let the stator and rotor magnetic system and geometry be designed such that $L_m(\theta_r) = \bar{L}_m - L_{\Delta m}\cos^3\left(4\theta_r - \frac{\pi}{16}\right)$, where $\bar{L}_m$ is the average inductance; $L_{\Delta m}$ is the magnitude of inductance variations.

Using the coenergy $W_c(i_{as},\theta_r) = \frac{1}{2}\left(\bar{L}_m - L_{\Delta m}\cos^3(4\theta_r - \frac{\pi}{16})\right)i_{as}^2$, the electromagnetic torque T_e, developed by a single-phase reluctance actuator, is

$$T_e = \frac{\partial W_c(i_{as},\theta_r)}{\partial \theta_r} = \frac{\partial}{\partial \theta_r}\frac{1}{2}\left(\bar{L}_m - L_{\Delta m}\cos^3\left(4\theta_r - \frac{\pi}{16}\right)\right)i_{as}^2 = 6L_{\Delta m}\sin\left(4\theta_r - \frac{\pi}{16}\right)\cos^2\left(4\theta_r - \frac{\pi}{16}\right)i_{as}^2.$$

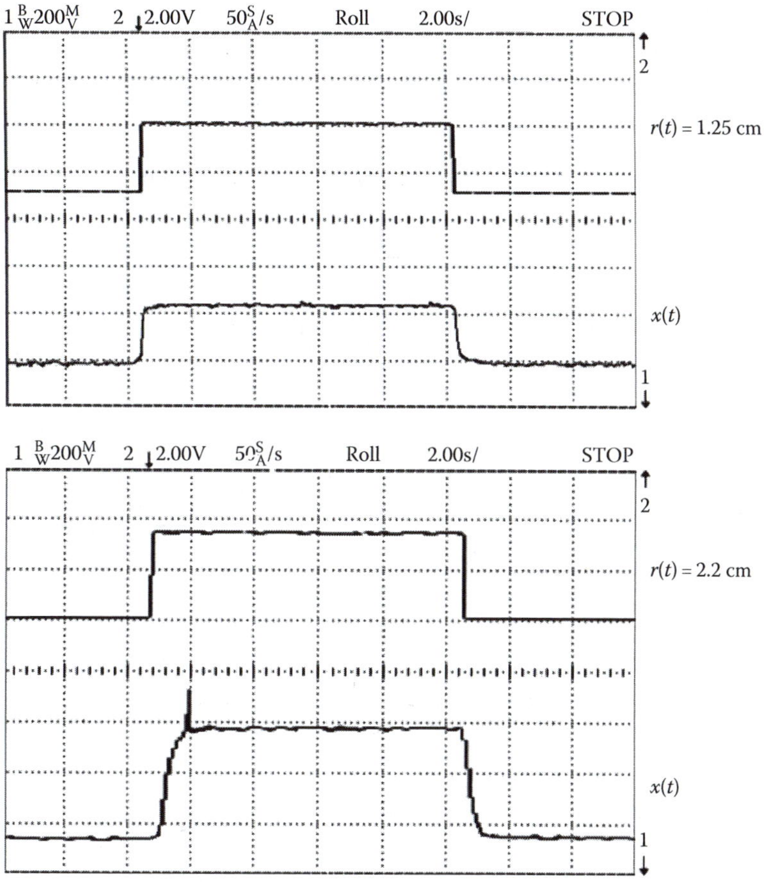

FIGURE 3.14 Reference inputs r (top plots, $r = 1.25$ and 2.2 cm) and the plunger displacement x (bottom plots) with the plunger at its initial state ($x = 2.2$ cm, $u = 0$ V). The bottom figure reports the maximum stroke repositioning.

By applying a DC voltage u_{as}, one has a DC current i_{as}. Within a limited angle $\theta_{r\,\min} \le \theta_r \le \theta_{r\,\max}$, the electromagnetic torque is developed. The variable-reluctance actuator (relay) rotates to minimize the reluctance path. The reluctance $\Re(\theta_r)$ is minimum when the rotor assumes a position when the stator–rotor air gap is minimum. We apply the Kirchhoff law $u_{as} = r_{as}i_{as} + \dfrac{d\psi_{as}}{dt}$, where the total *emf* is

$$\frac{d\psi_{as}}{dt} = \frac{d}{dt}L_m(\theta_r)i_{as} = \frac{d}{dt}\left(\bar{L}_m - L_{\Delta m}\cos^3\left(4\theta_r - \frac{\pi}{16}\right)\right)i_{as}$$

$$= \left(\bar{L}_m - L_{\Delta m}\cos^3\left(4\theta_r - \frac{\pi}{16}\right)\right)\frac{di_{as}}{dt} - 6L_{\Delta m}\sin\left(4\theta_r - \frac{\pi}{16}\right)\cos^2\left(4\theta_r - \frac{\pi}{16}\right)i_{as}\omega_r.$$

The Newton law for rotational motion is $\dfrac{d\omega_r}{dt} = \dfrac{1}{J}\left(T_e - T_{friction} - T_{spring} - T_L\right)$, $\dfrac{d\theta_r}{dt} = \omega_r$.

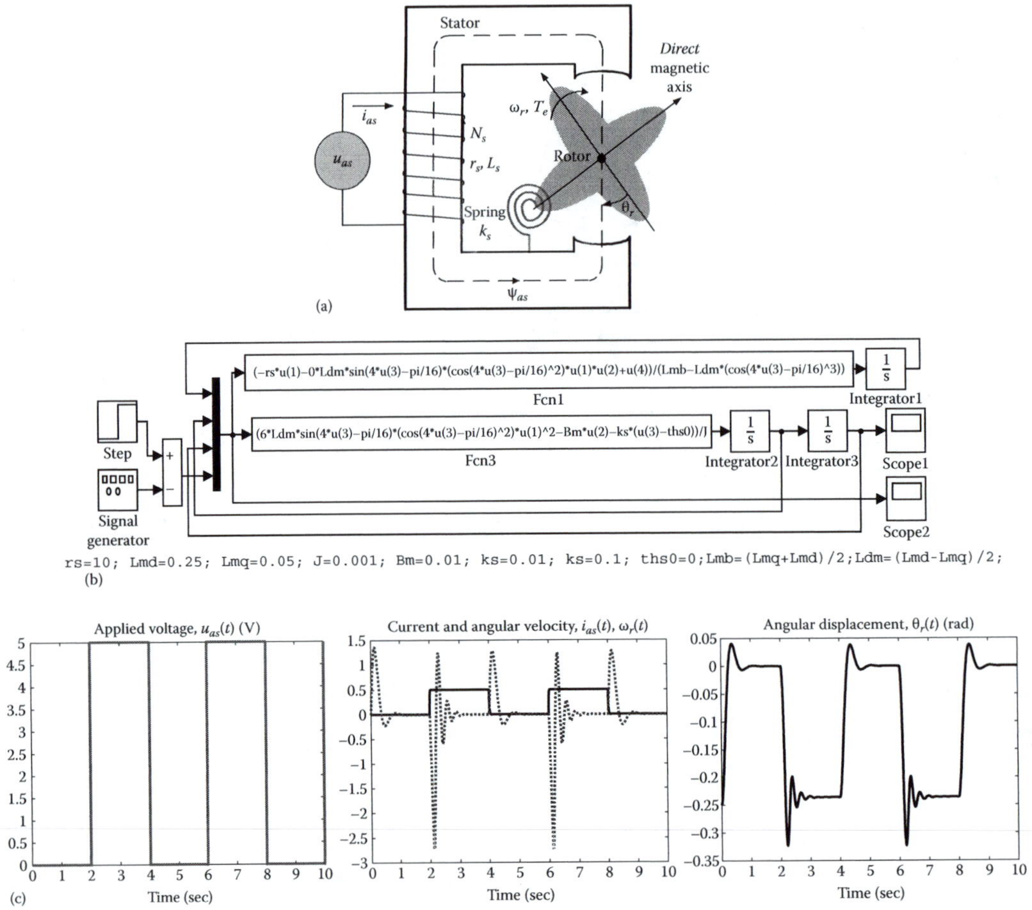

FIGURE 3.15 (a) Radial topology limited-angle reluctance actuator with a torsional spring; (b) Simulink® model to simulate a limited-angle reluctance actuator; (c) Open-loop dynamics for $i(t)$, $\omega_r(t)$ and $\theta_r(t)$ if $u_{as} = \begin{cases} 0 \\ 5\text{ V} \end{cases}$, $f = 0.25$ Hz, $\theta_{r0} = -0.25$ rad.

One obtains a set of three first-order nonlinear differential equations

$$\frac{di_{as}}{dt} = \frac{1}{\bar{L}_m - L_{\Delta m}\cos^3\left(4\theta_r - \frac{\pi}{16}\right)}\left[-r_s i_{as} - 6L_{\Delta m}\sin\left(4\theta_r - \frac{\pi}{16}\right)\cos^2\left(4\theta_r - \frac{\pi}{16}\right) i_{as}\omega_r + u_{as}\right],$$

$$\frac{d\omega_r}{dt} = \frac{1}{J}\left[6L_{\Delta m}\sin\left(4\theta_r - \frac{\pi}{16}\right)\cos^2\left(4\theta_r - \frac{\pi}{16}\right) i_{as}^2 - B_m\omega_r - k_s(\theta_r - \theta_{r0}) - T_L\right], \tag{3.1}$$

$$\frac{d\theta_r}{dt} = \omega_r, \quad \theta_{r\min} \le \theta \le \theta_{r\max}.$$

In this chapter, electrostatic and variable-reluctance electromechanical motion devices were covered. We examined basic physics and functionality of electromechanical transducers. Modeling, simulation, and analysis were accomplished. The use of MATLAB and experimental data in evaluation of performance and capabilities were reported.

PRACTICE PROBLEMS

3.1 Consider a single-phase, limited-angle actuator shown in Figure 3.15a. The parameters are $r_s = 10$ ohm, $L_{md} = 0.25$ H, $L_{mq} = 0.05$ H, $J = 0.001$ kg-m^2, $B_m = 0.01$ N-m-sec/rad, and $k_s = 0.1$ N-m/rad. The Simulink model, developed using differential equations (3.1), is documented in Figure 3.15b. One uploads the parameters

```
rs=10; Lmd=0.25; Lmq=0.05; J=0.001; Bm=0.01; ks=0.1; ths0=0;
Lmb=(Lmq+Lmd)/2; Ldm=(Lmd-Lmq)/2;
```

To operate the limited-angle actuator, one supplies the DC voltage. The simulations are performed for the DC voltage pulses $u_{as} = \begin{cases} 0 \\ u \end{cases}$ at $f = 0.25$ Hz. Let the initial rotor displacement be $\theta_{r0} = -0.25$ rad. The dynamics of the current $i(t)$, angular velocity $\omega_r(t)$, and angular displacement $\theta_r(t)$ are reported in Figure 3.15c for $u_{as} = \begin{cases} 0 \\ 5\text{ V} \end{cases}$. The electromagnetic torque T_e is developed to minimize the air gap. As u_{as} is applied, the rotor rotates to assume a position corresponding to the minimal *direct* axis reluctance $\Re_{md}$. If the adequate voltage is applied, $\theta_{r\,final} = 0$. With the torsional spring, at steady state, $T_e = T_{spring}$.

3.2 Derive the electromagnetic force and equations of motion for a magnetic levitation system if $L(x) = \dfrac{a}{(b + cx + dx^2)}$, $a, b, c, d > 0$. Analyze the electromagnetic force.

For the electromagnetic force, one has $F_e(i,x) = -\dfrac{\partial W_c(i,x)}{\partial x} = -\dfrac{1}{2} i^2 \dfrac{dL(x)}{dx}$.

Using the chain rule $\dfrac{df(u)}{dx} = \dfrac{df}{du}\dfrac{du}{dx}$, denote $u = b + cx + dx^2$. Hence, $\dfrac{d}{du} u^{-1} = -\dfrac{1}{u^2}$.

From $\dfrac{d}{dx}(b + cx + dx^2) = c + 2dx$, one finds $\dfrac{d}{dx} L(x) = a\left(-\dfrac{1}{u^2}\right)(c + 2dx) = -\dfrac{a(c+2dx)}{(b+cx+dx^2)^2}$.

Therefore,

$$F_e = \frac{a(c+2dx)}{2(b+cx+dx^2)^2} i^2.$$

The maximum electromagnetic force at $x = 0$ is $F_{e\max} = F_e(i,x)\big|_{x=0} = \left.\dfrac{a(c+2dx)}{2(b+cx+dx^2)^2} i^2\right|_{x=0} = \dfrac{ac}{2b^2} i^2$.

Using the Faraday law of inductance, the total *emf* is

$$-\frac{d\psi}{dt} = -\frac{d}{dt}\big(L(x)i\big) = -\frac{di}{dt}\frac{a}{(b+cx+dx^2)} + \frac{a(c+2dx)}{(b+cx+dx^2)^2} iv.$$

The resulting nonlinear differential equations are

$$\frac{di}{dt} = \frac{b+cx+dx^2}{a}\left[-ri - \frac{a(c+2dx)}{(b+cx+dx^2)^2} iv + u\right],$$

$$\frac{dv}{dt} = \frac{1}{m}\left[\frac{a(c+2dx)}{2(b+cx+dx^2)^2} i^2 - B_v v - mg - F_L\right],$$

$$\frac{dx}{dt} = v, \quad x_{\min} \le x \le x_{\max}.$$

3.3 Consider a solenoid with a magnetizing inductance $L(x) = a\ \text{csch}(bx)$, $a > 0$ and $b > 0$ (for example, $a = 0.01$ and $b = 100$). The domain for this $L(x)$ is $\{x \in \Re\text{: } b \neq 0 \text{ and } x \neq 0\}$.

The electromagnetic force is $F_e(i,x) = -\dfrac{\partial W_c(i,x)}{\partial x} = -\dfrac{1}{2}i^2\dfrac{dL(x)}{dx} = \tfrac{1}{2}ab\coth(bx)\text{csch}(bx)i^2$.

Using the Kirchhoff and Faraday laws, we have

$$u = ri + \frac{d\psi}{dt} = ri + L(x)\frac{di}{dt} + i\frac{dL(x)}{dx}\frac{dx}{dt} = ri + a\ \text{csch}(bx)\frac{di}{dt} - ab\coth(bx)\text{csch}(bx)iv.$$

Applying the laws of electromagnetics and Newtonian mechanics, the governing equations are

$$\frac{di}{dt} = \frac{1}{a\ \text{csch}(bx)}\left[-ri - ab\coth(bx)\text{csch}(bx)iv + u\right],$$

$$\frac{dv}{dt} = \frac{1}{m}\left[\frac{1}{2}ab\coth(bx)\text{csch}(bx)i^2 - B_v v - k_s\left(x_0 - x\right) - F_L\right],$$

$$\frac{dx}{dt} = v, \quad x_{\min} \le x \le x_{\max}, \quad x_{\min} > 0, \quad x_{\min} \neq 0.$$

HOMEWORK PROBLEMS

3.1 A force $\vec{F} = 3\vec{i} + 2\vec{j} + 4\vec{k}$ acts through the point with a position vector $\vec{R} = 2\vec{i} + \vec{j} + 3\vec{k}$. Derive a torque $\vec{T} = \vec{R} \times \vec{F}$.

3.2 A spherical electrostatic actuator, as documented in Figure 3.16a, is designed using spherical conducting shells separated by the flexible material (for example, parylene, teflon, and polyethylene have relative permittivity ~3). The inner shell has a total charge $+q_i$ and a diameter r_i. The charge of the outer shell $q_0(t)$ is seminegative and time-varying. The diameter of the outer shell is denoted by r_0.

a. Derive the expression for the capacitance $C(r)$. Calculate the numerical value for capacitance if $r_i = 1$ cm, $r_0 = 1.5$ cm, $q_i = 1$ C, and $q_0(t) = [\sin(t) + 1]$ C.

b. Derive the expression for the electrostatic force using the coenergy $W_c[u,C(r)] = \dfrac{1}{2}C(r)u^2$. Recall that $F_e(u,r) = \dfrac{\partial W_c[u,C(r)]}{\partial r}$. Calculate the electrostatic force between the inner and outer shells assigning time-varying applied voltage u, $u_{\max} < 1000$ V.

c. For a flexible material (parylene, teflon, or polyethylene), find the resulting displacements. Use the expression for the elastic force. The approximation $F_s = k_s r$ can be applied, where $k_s = 1$ N/m.

d. Develop the differential equations that describe the spherical actuator dynamics. Examine the actuator performance and capabilities. Perform simulation in MATLAB.

3.3 Consider a magnetic levitation system with a suspended ball, as illustrated in Figure 3.16b.

a. Derive a mathematical model. That is, find the differential equations that describe the system dynamics. Find the expression for the electromagnetic force.

b. Assign magnetic levitation system dimensions and derive the parameters. For example, let the total length of the magnetic path be ~0.1 m. Assuming that the diameter of the copper wire is 1 mm, one layer winding can include ~10 turns. You may have multilayered winding. The geometry (shape) and diameter of the moving mass (ball) and electric steel permeability result in the value for m, A, μ_r, etc. (See image in Figure 3.6c.)

c. In MATLAB, perform numerical simulations of the magnetic levitation system. Analyze the dynamics and assess the performance.

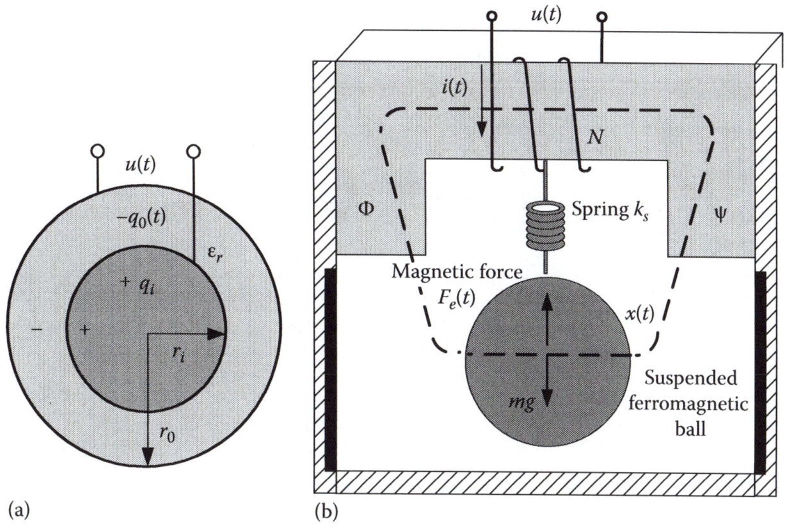

FIGURE 3.16 (a) Electrostatic actuator; (b) Magnetic levitation system.

3.4 The solenoid's magnetizing inductance is $L(x) = ae^{-bx}$, $x_{\min} \le x \le x_{\max}$, $x_{\min} = 0$, $x_{\max} = 1$ cm, $x \ge 0$, $a > 0$ and $b > 0$.

a. Derive and report an explicit equation for the electromagnetic force F_e.

b. Find and report an explicit equation (expression) for the $F_{e\,\max}$.

c. Let $a = 0.1$, $b = 500$. For $I = 1$ A, calculate the numeric value for $F_{e\,\max}$.

d. The solenoid stroke is 1 cm, $(x_{\max} - x_{\min}) = 0.01$ m. The spring force is given by an ideal Hooke's law. The spring can be stretched by 1 cm. Calculate the spring constant k_s.

Note: A tension spring exhibits zero force at *zero-length* when the spring is relaxed. Using the plunger displacement $x(t)$, $x_{\min} \le x \le x_{\max}$ and the *zero-length* $(x_0 - x_{\max})$, we have $F_{spring} = k_s(x_0 - x)$, $x_0 = x_{\max}$. One has $F_{spring\,\max} = k_s(x_0 - x_{\min})$, $x_{\min} \ge 0$. At the *zero-length* $(x_0 - x_{\max})$, the spring is relaxed and exhibits zero force $F_{spring\,\min} = k_s(x_0 - x_{\max}) = 0$.

3.5 Let solenoid's magnetizing inductance be $L(x) = cb^{ax}$, $x > 0$, $a = 1, 2, 3, \ldots$, $b = 1/n$, $n = 2, 3, 4, \ldots$, $c > 0$.

a. Derive and report an explicit equation (expression) for the electromagnetic force F_e.

b. Find and report the explicit equations (expressions) for the *total* and *motional emfs.*

c. Derive and report a mathematical model (three differential equations) for the considered solenoid.

3.6 For the solenoid, the magnetizing inductance is $L_m(x) = e^{-ax}\sec(x)$. Recall that $\sec(x) = 1/\cos(x)$.

a. Derive an explicit (complete) equation (expression) for the electromagnetic force F_e.

b. Derive an explicit (complete) equation (expression) for the motional *emf.*

REFERENCES

1. A. E. Fitzgerald, C. Kingsley, and S. D. Umans, *Electric Machinery*, McGraw-Hill, New York, 2003.
2. P. C. Krause and O. Wasynczuk, *Electromechanical Motion Devices*, McGraw-Hill, New York, 1989.
3. P. C. Krause, O. Wasynczuk, S. D. Sudhoff, and S. Pekarek, *Analysis of Electric Machinery*, Wiley-IEEE Press, New York, 2013.
4. S. E. Lyshevski, *Electromechanical Systems, Electric Machines, and Applied Mechatronics*, CRC Press, Boca Raton, FL, 1999.
5. S. E. Lyshevski, *Electromechanical Systems and Devices*, CRC Press, Boca Raton, FL, 2007.
6. White D. C. and Woodson H. H., *Electromechanical Energy Conversion*, Wiley, New York, 1959.

4 Permanent-Magnet Direct-Current Motion Devices and Actuators

4.1 PERMANENT-MAGNET MOTION DEVICES AND ELECTRIC MACHINES: INTRODUCTION

The principle of energy conversion and electromagnetism in electromechanical motion devices was examined and demonstrated by Michael Faraday in 1821. The first commutator DC electric motors were designed, tested, and demonstrated by Anyos Jedlik in 1828 and William Sturgeon in 1832. Alternating current machines (synchronous and induction) were invented and demonstrated by Nicola Tesla in the 1880s. By 1882, Tesla pioneered, developed, and experimentally substantiated the principles of time-varying AC electromagnetic field for two-phase induction motors. These pioneering developments are the cornerstones of wireless communication, AC electric machines, radio detection and ranging (RADAR) technology, etc. He developed, demonstrated, patented, and commercialized the induction motor, AC power transmission technologies, and communication devices during the 1880s and 1900s. The first three-phase, squirrel-cage induction motor was designed and demonstrated by Mikhail Dolivo-Dobrovolosky in 1891.

In this chapter, we study various high-performance translational and rotational direct current (DC) electromechanical motion devices and actuators, which operate due to the electromagnetic interactions between windings and permanent magnets. As covered in Section 2.5, the torque tends to align the magnetic moment $\vec{m}$ with $\vec{B}$, and $\vec{T} = \vec{m} \times \vec{B}$. One recalls that the torque is $\vec{T} = \vec{R} \times \vec{F}$, where the expression for the electromagnetic force is $\vec{F} = -i\oint_l \vec{B} \times d\vec{l}$. For a uniform magnetic flux density, $\vec{F} = -i\vec{B} \times \oint_l d\vec{l}$. We examined electrostatic and electromagnetic variable-reluctance actuators in Chapter 3.

This chapter examines high-power, high-force, and high-torque-density transducers (actuators, generators, motors, and sensors) with permanent magnets that establish a stationary magnetic field. The windings are placed on the rotating rotor. The voltage is applied to the windings using brushes and commutator. The permanent magnets are placed on the stator.

The device physics is based on the electromagnetic force $\vec{F} = -i\vec{B} \times \oint_l d\vec{l}$ or torque $\vec{T} = \vec{m} \times \vec{B}$ developed between the windings on the moving (or stationary) member and the permanent magnets on the stationary (or moving) member. Images of permanent-magnet DC motors, a limited-angle axial-topology actuator, and a speaker are documented in Figure 4.1. Superior performance, excellent capabilities, and affordability are ensured by permanent-magnet DC and AC electromechanical motion devices. These devices are the preferred choice and widely applied. The rated power of permanent-magnet machines reaches ~100 kW with the overloading capability (for a short time depending on a duty cycle) of ~10.

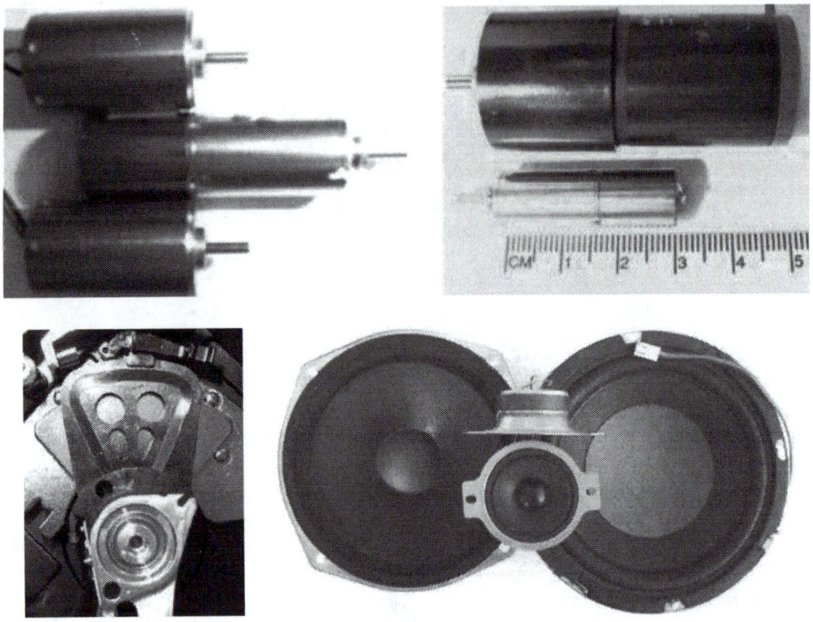

FIGURE 4.1 Permanent-magnet DC motors, servo-motors with gear heads, limited-angle axial-topology actuator, and loudspeakers (translational permanent-magnet DC actuators).

4.2 RADIAL TOPOLOGY PERMANENT-MAGNET DIRECT-CURRENT ELECTRIC MACHINES AND POWER ELECTRONIC SOLUTIONS

4.2.1 ELECTRIC MACHINES

The principle of energy conversion by electromechanical motion devices were demonstrated by Michael Faraday in 1821 using the homopolar motor. The practical commutator-type DC electric motor was invented and substantiated by William Sturgeon in 1832. The DC generator-motor system was studied by Zenobe Gramme in 1873.

In this chapter, we focus on high-performance electromechanical motion devices with permanent magnets. These electric machines guarantee high power and torque densities, efficiency, affordability, reliability, ruggedness, overloading capabilities, and other advantages [1–8]. The power range of permanent-magnet DC electric machines (motors and generators) is from μW to ~100 kW and the dimensions are from ~1 mm to ~1 m in diameter and length. A permanent-magnet electric machine can be used as a motor or generator. Only permanent-magnet synchronous machines, which do not have brushes and commutator, surpass permanent-magnet DC machines. Permanent-magnet DC and synchronous electric machines and actuators are used in aerospace, automotive, power, naval, robotics, and other applications. To rotate a computer hard drive, household fan, small pump, manufacturing robot, and drive a 60 ton tank (track), respectively, ~1 W, ~10 W, ~100 W, ~1 kW, and ~100 kW machines are needed. The ~1 μW to ~100 kW power range covers the major consumer and industrial systems. In submegawatt and megawatt range applications (ships, locomotives, high-power energy systems, etc.), DC machines with electromagnets as well as induction and synchronous machines are used.

Note: A typical Navy destroyer needs two propulsion motors, each rated at 36.5 MW, 120 rpm. These DC motors have been built using conventional technology and transformative superconducting technology, developed by the AMSC and Northrop Grumman in the 2000s. There are two 44 MW DC motors on the *Queen Elizabeth* cruise ships. The electromagnets are used to produce high stationary fields. If a magnetic field should be higher than the ferromagnetic limit ~2 T, superconducting electromagnets with windings cooled in liquid helium are used, and, the field strength reaches 20 T.

Permanent-magnet DC electric machines are rotating, energy-transforming electromechanical motion devices that convert energies. Motors (actuators) convert electrical energy to mechanical energy. Generators convert mechanical energy to electrical energy. A permanent-magnet electric machine can operate as a motor (if voltage is applied, the electromagnetic torque is produced) or generator. In generators, if the torque is applied to rotate the machine, the voltage is induced. Electric machines have stationary (stator) and rotating (rotor) members, separated by an air gap. The armature windings are placed in the rotor slots and connected to a rotating commutator, see Figure 4.2. One supplies the armature voltage u_a to the rotor windings using brushes and commutator which are connected to the armature windings. The armature winding consists of identical uniformly distributed coils. The excitation magnetic field is produced by permanent magnets. The images of a permanent-magnet DC electric machine with these components are shown in Figure 4.2.

Using the commutator (circular conducting copper segments on the rotor as depicted in Figure 4.2), the voltage is supplied to the armature windings on rotor. The armature windings and permanent magnets produce stationary *magnetomotive force (mmfs)*, which are displaced by 90 electrical degrees. The armature magnetic force is along the *quadrature* (rotor) magnetic axis, while the *direct* axis stands for a permanent-magnet magnetic axis. The electromagnetic torque is produced due to the interaction of the magnetic dipole moment and stationary magnetic field. A consistent electromagnetic design results in a near-optimal symmetric electromagnetic system characterized by symmetry, homogeneity, magnetic field uniformity (as viewed from the energized rotor winding), etc. From $\vec{F} = -i\oint_l \vec{B}\times d\vec{l}$, we have $F_e = Nl_{ef}B_{ef}i_a$, where N is the number of turns; l_{ef} is the effective winding length; B_{ef} is the effective field that depends on magnets, air gap, winding design, etc.; i_a is the armature current. Hence,

$$T_e = R_\perp F_e = k_a i_a,$$

where k_a is the torque constant; $R_\perp$ is the perpendicular radius.

The electromagnetic torque can be found by using the coenergy

$$T_e(i,\theta) = \frac{\partial W_c(i,\theta)}{\partial\theta}, \quad W_c = \int_i \psi\, di.$$

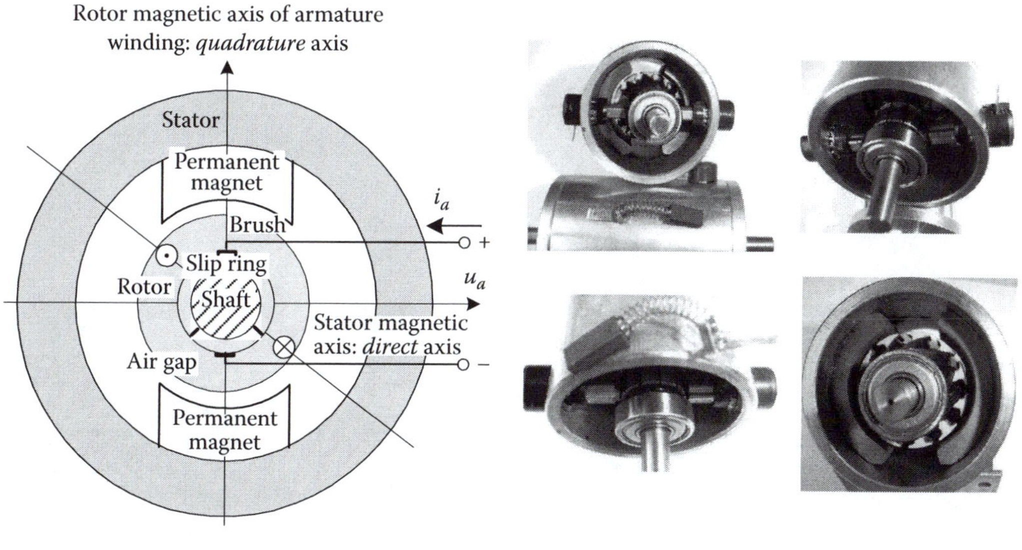

FIGURE 4.2 Radial topology permanent-magnet DC electric machine schematics and images.

The magnetic flux crossing a surface is $\Phi = \oint_s \vec{B} \cdot d\vec{s}$. The expression for T_e agrees with the equation for a torque experienced by a current loop in the magnetic field $\vec{T} = \vec{m} \times \vec{B} = \vec{a}_m m \times \vec{B} = iA\vec{a}_m \times \vec{B}$. In permanent-magnet electric machines, the stationary near-uniform magnetic field $\vec{B}$ is produced by permanent magnets, and B and A are assumed to be constant. Hence, $T_e = k_a i_a$.

For motors, Kirchhoff's law yields the steady-state equation for the armature voltage u_a

$$u_a - E_a = r_a i_a,$$

where r_a is the armature resistance; E_a is the *back emf*, which is also denoted as $\mathscr{E}$; i_a is the current in the armature winding.

The difference between the applied voltage and the *emf* is the voltage drop across the armature resistance r_a. The motor rotates at an angular velocity ω_r at which the *emf* E_a, induced in the armature winding, *balances* the armature voltage u_a supplied. If an electric machine operates as a motor, the induced *emf* is less than the voltage applied to the windings. If a machine operates as a generator, the generated (induced) *emf* is greater than the terminal voltage. For generators, the armature current i_a is in the same direction as the induced *emf*, and the terminal voltage is $(E_a - r_a i_a)$. A schematic diagram of a permanent-magnet DC motor is depicted in Figure 4.3a.

One recalls that the *emf* and *mmf* are

$$emf = \oint_l \vec{E} \cdot d\vec{l} = -\oint_s \frac{\partial \vec{B}}{\partial t} \cdot d\vec{s} \quad \text{or} \quad emf = \oint_l (\vec{v} \times \vec{B}) \cdot d\vec{l} \quad \text{(Faraday law of induction)},$$

and

$$mmf = \oint_l \vec{H} \cdot d\vec{l} = \oint_s \vec{J} \cdot d\vec{s} + \oint_s \frac{\partial \vec{D}}{\partial t} \cdot d\vec{s}.$$

Applying the expressions for the *emf*, under adequate assumptions (magnetic field is stationary and uniform, magnetic system is linear, magnetic *susceptibility* is constant, etc.), we have the following expression for the *back emf*

$$E_a = k_a \omega_r,$$

where k_a is the *back emf* constant.

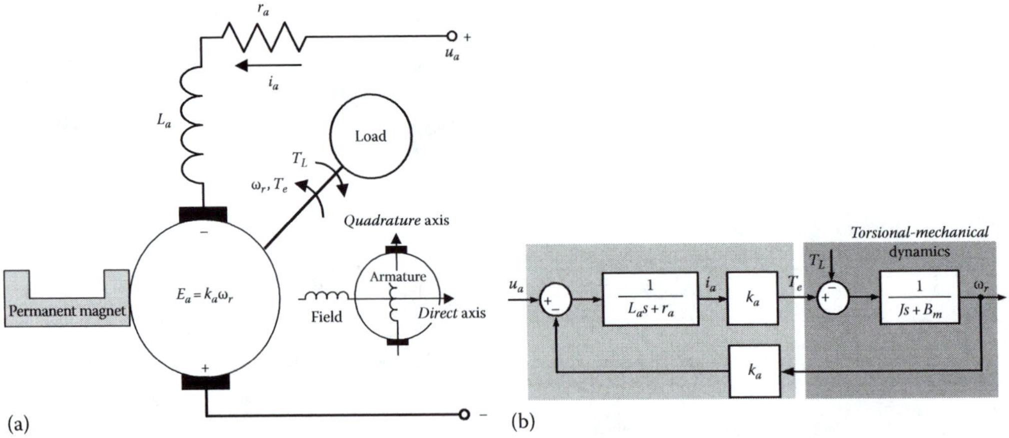

FIGURE 4.3 (a) Schematic diagram of a permanent-magnet DC electric machine (current direction corresponds to the motor operation); (b) *s*-domain diagram of permanent-magnet DC motors.

Using Kirchhoff's voltage law $u_a = r_a i_a + \frac{d\psi}{dt}$ and Newton's second law of motion

$$\frac{d\omega_r}{dt} = \frac{1}{J}\left(T_e - B_m\omega_r - T_L\right),$$

the differential equations for permanent-magnet DC motors are derived. Assume that the *susceptibility* of permanent magnets is constant and that the field established by the magnetic field is uniform. While the *susceptibility* varies as a function of temperature and uniformity may not be ensured, the linear differential equations may be adequate to describe permanent-magnet electric machines. We have

$$\frac{di_a}{dt} = -\frac{r_a}{L_a}i_a - \frac{k_a}{L_a}\omega_r + \frac{1}{L_a}u_a,$$
$$\frac{d\omega_r}{dt} = \frac{k_a}{J}i_a - \frac{B_m}{J}\omega_r - \frac{1}{J}T_L. \tag{4.1}$$

In the state-space form we have

$$\frac{d\mathbf{x}}{dt} = A\mathbf{x} + B\mathbf{u},\ \mathbf{x} = \begin{bmatrix} i_a \\ \omega_r \end{bmatrix},\quad \mathbf{u} = u_a,\quad A \in \mathbb{R}^{2\times 2} \text{ and } B \in \mathbb{R}^{2\times 1}.$$

From (4.1)

$$\begin{bmatrix} \frac{di_a}{dt} \\ \frac{d\omega_r}{dt} \end{bmatrix} = \begin{bmatrix} -\frac{r_a}{L_a} & -\frac{k_a}{L_a} \\ \frac{k_a}{J} & -\frac{B_m}{J} \end{bmatrix}\begin{bmatrix} i_a \\ \omega_r \end{bmatrix} + \begin{bmatrix} \frac{1}{L_a} \\ 0 \end{bmatrix} u_a - \begin{bmatrix} 0 \\ \frac{1}{J} \end{bmatrix} T_L. \tag{4.2}$$

An s-domain diagram of permanent-magnet DC motors is illustrated in Figure 4.3b.

From the differential equation $\frac{di_a}{dt} = \frac{1}{L_a}\left(-r_a i_a - k_a\omega_r + u_a\right)$, for the steady-state operation $0 = -r_a i_a - k_a\omega_r + u_a$. Hence, $\omega_r = \frac{u_a - r_a i_a}{k_a}$. The electromagnetic torque is $T_e = k_a i_a$. In the steady-state $T_e = T_{friction} + T_L$, or $T_e = T_L$ if $T_{friction} = 0$. The torque-speed characteristics are given by

$$\omega_r = \frac{u_a - r_a i_a}{k_a} = \frac{u_a}{k_a} - \frac{r_a}{k_a^2}T. \tag{4.3}$$

One changes the applied armature voltage u_a to vary the angular velocity. If the load is applied, the angular velocity reduces. The slope of the torque-speed characteristic is $-r_a/k_a^2$. The torque-speed characteristics are illustrated in Figure 4.4a for different u_a, and $|u_a| \le u_{a\,\max}$, where $u_{a\,\max}$ is the maximum (rated) voltage. To reduce the angular velocity, one decreases u_a. The angular velocity at which motor rotates is found as the intersection of the torque-speed characteristic and the load characteristic. For example, if u_{a2} is applied, the angular velocity is ω_{r2}. From Newton's second law, neglecting friction, $\frac{d\omega_r}{dt} = \frac{1}{J}(T_e - T_L)$. At $T_e = T_L$, the motor rotates at the constant angular velocity. At no load, from (4.3) one finds that the angular velocity is $\omega_r = u_a/k_a$. The angular velocity can be reversed if the polarity of the applied voltage is changed.

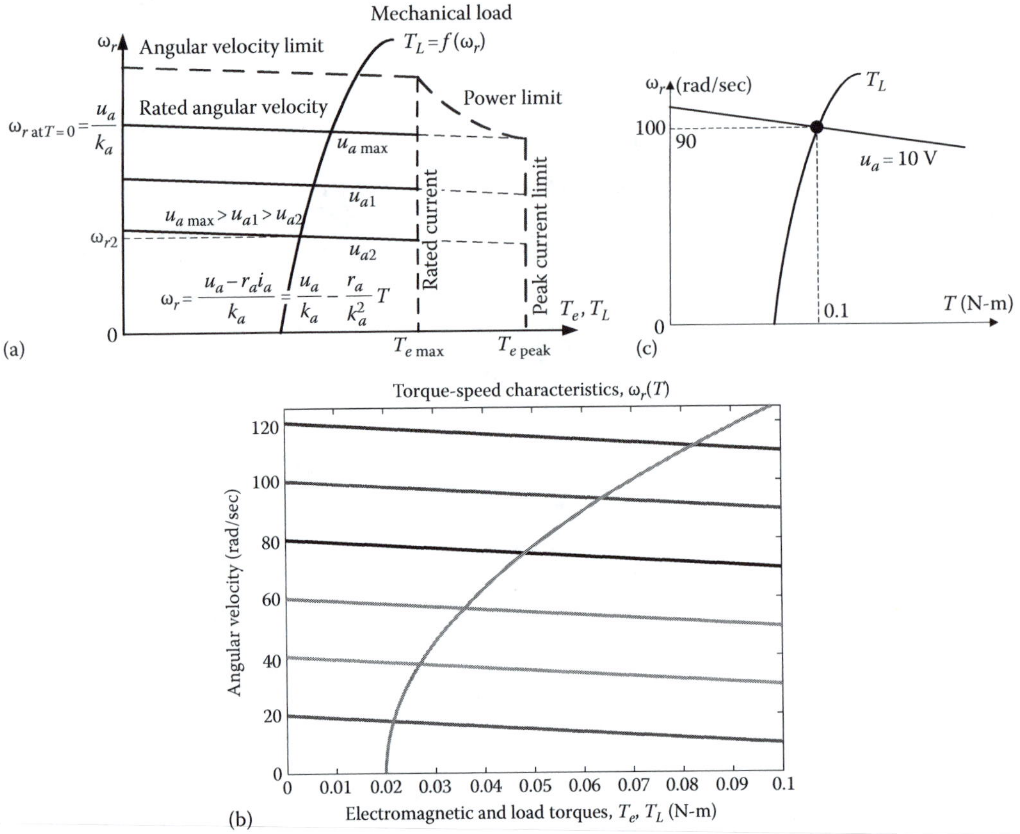

FIGURE 4.4 (a) Torque-speed characteristics of permanent-magnet DC motors. The overloading capabilities are specified by the manufacturer. For a short time, $T_{e\,peak}/T_{e\,max}$ could reach ~10; (b) Torque-speed and load characteristics for a permanent-magnet DC motor (Example 4.1); (c) Torque-speed and load characteristics (Example 4.2).

If the permanent-magnet electric machine is used as a generator with a resistive load R_L, the induced *motional emf* is $E_a = k_a\omega_r$. In steady-state, the induced *emf* is proportional to the angular velocity. A generator is rotated by a prime mover that applies the torque T_{PM} of the aerodynamic (wind power), hydrodynamic (hydro power), thermal or other origin. The differential equations for a DC generator are

$$\frac{di_a}{dt} = \frac{1}{L_a}\left[-\left(r_a + R_L\right)i_a + k_a\omega_r\right],$$
$$\frac{d\omega_r}{dt} = \frac{1}{J}\left(-k_a i_a - B_m\omega_r + T_{PM}\right). \tag{4.4}$$

Example 4.1

Calculate and plot the torque–speed characteristics for a 10 V (rated) permanent-magnet DC motor with the following parameters: $r_a = 1$ ohm and $k_a = 0.1$ V-sec/rad or N-m/A. The load is a nonlinear function of the angular velocity, $T_L = f(\omega_r) = 0.02 + 0.000005\omega_r^2$ N-m.

The torque-speed characteristics are governed by (4.3). Using different values for the armature voltage u_a, the steady-state characteristics are calculated and plotted as depicted in Figure 4.4b. The load $T_L = f(\omega_r)$ is also illustrated. The following MATLAB® program is used to perform calculations and plotting:

```
% parameters of a permanent-magnet DC motor
ra=1; ka=0.1; Te=0:0.001:0.1; % torque in N-m
for ua=2:2:12;  % applied voltage
wr=ua/ka-(ra/ka^2)*Te; % angular velocity for different voltages
wrl=0:1:225; Tl=0.02+5e-6*wrl.^2; % load torque at different velocities
plot(Te,wr,'-',Tl,wrl,'-','LineWidth',3);
title('Torque-Speed Characteristics, \omega_r(T)','FontSize',18);
xlabel('Electromagnetic and Load Torques, T_e, T_L [N-m]','FontSize',15);
ylabel('Angular Velocity [rad/sec]','FontSize',18); hold on;
axis([0, 0.1, 0, 125]); end;
```

■

Example 4.2

Let $u_a = 10$ V. A motor rotates at an angular velocity of 100 rad/sec if $T_L = 0$ N-m, and, 90 rad/sec for $T_L = 0.1$ N-m, see Figure 4.4c. Assume that friction may be neglected. One needs to find k_a, r_a, i_a, and efficiency η.

From (4.3) $\omega_r = \dfrac{u_a - r_a i_a}{k_a} = \dfrac{u_a}{k_a} - \dfrac{r_a}{k_a^2}T$, we have $k_a = \dfrac{u_a - r_a i_a}{\omega_r}$. This equation does not result in k_a because r_a and i_a are not given. At $T_L = 0$, neglecting friction, $k_a = \dfrac{u_a}{\omega_r} = \dfrac{10}{100} = 0.1$ V-sec/rad.

The armature resistance is found using the slope $-\dfrac{r_a}{k_a^2}T$. One finds $r_a = 1$ ohm.

Neglecting friction, at $T_L = 0.1$ N-m, $T_e = k_a i_a = T_L$. Hence, $i_a = 1$ A.

The motor efficiency is $\eta = \dfrac{P_{output}}{P_{input}} = \dfrac{T_L \Omega_r}{U_a I_a} \times 100\% = \dfrac{9}{10} \times 100\% = 90\%$. ■

Efficiency and losses analyses: We study the losses and efficiency. There are mechanical, magnetic, electrical (copper winding) and other losses. The mechanical losses P_M are due to the friction between the bearings and the shaft, friction between the brushes and commutator, air drag losses, etc. The induced (*back*) *emf* in the armature winding varies, which results in the magnetic losses P_m, such as the hysteresis iron losses and eddy current losses.

The electrical losses P_E depend on r_a and current i_a, $P_E = r_a i_a^2$.

Considering the essential electrical and mechanical losses, and assuming that the magnetic, air drag, and other losses are negligible, one may define the losses as

$$P_{losses} = P_E + P_M = r_a i_a^2 + B_m(\cdot)\omega_r^2,$$

where $B_m(\cdot)$ is the real-valued nonlinear friction function, which depends on the electromagnetic and mechanical loads, angular velocity, temperature, etc.

The efficiency η is found for the steady-state operation of electric machines. At equilibrium, for motors, using the input and output powers P_{input} and P_{output}, we have

$$\eta = \frac{P_{output}}{P_{input}} = \frac{T_L \Omega_r}{U_a I_a} \quad \text{or} \quad \eta = \frac{P_{output}}{P_{input}} \times 100\% = \frac{T_L \Omega_r}{U_a I_a} \times 100\%.$$

The load torque may not be directly measured. One may find the efficiency recalling that $P_{output} = (P_{input} - P_{losses})$. We have $\eta = \frac{P_{output}}{P_{input}} = \frac{P_{input} - P_{losses}}{P_{input}}$. It is difficult to find explicitly P_{losses}. If T_L is not measured, using the main losses, one may estimate η as

$$\eta = \frac{P_{output}}{P_{input}} = \frac{P_{input} - P_{losses}}{P_{input}} = \frac{U_a I_a - \left[r_a I_a^2 + B_m(\cdot)\Omega_r^2\right]}{U_a I_a}.$$

For a generator, $\eta = \frac{P_{output}}{P_{input}} = \frac{U_a I_a}{T_L \Omega_r}$ or $\eta = \frac{P_{output}}{P_{input}} \times 100\% = \frac{U_a I_a}{T_L \Omega_r} \times 100\%$.

Example 4.3

For $P_{losses} = r_a i_a^2 + B_m(\omega_r)\omega_r^2$, we have $P_{output} = P_{input} - P_{losses} = U_a I_a - \left[r_a I_a^2 + B_m(\Omega_r)\Omega_r^2\right]$.

Hence, $\eta = \frac{P_{output}}{P_{input}} = \frac{P_{input} - P_{losses}}{P_{input}} = \frac{U_a I_a - \left[r_a I_a^2 + B_m(\Omega_r)\Omega_r^2\right]}{U_a I_a}$.

Using the experimental results, the physics-consistent $T_{viscous}(\omega_r)$ and $B_m(\omega_r)$ are found in Example 2.17. One has

$T_{viscous} = B_m(\omega_r)\omega_r = p_1 \tanh(p_2\omega_r) + p_3\omega_r$ and $B_m(\omega_r) = \frac{p_1 \tanh(p_2\omega_r) + p_3\omega_r}{\omega_r}$.

One may apply $P_{losses} = r_a i_a^2 + B_m(\omega_r)\omega_r^2$ to estimate η. ■

4.2.2 Simulation and Experimental Studies of Permanent-Magnet DC Machines

The majority of electromechanical motion devices are described by nonlinear differential equations. Permanent-magnet DC electric machines are among a very limited class of motion devices, which can be described by linear differential equations. The linear theory may not always be applied to permanent-magnet DC machines because there are constraints on the applied voltage $|u_a| \le u_{a\,\max}$, nonlinear magnetic system, nonlinear friction, etc.

The state-space model of permanent-magnet DC motors was found as given by (4.2). For a 23 NEMA-size Torquemaster 2620 (E winding and G tachogenerator winding), 250 W, 70 V (rated) permanent-magnet DC motor, the experimentally found parameters are as follows: $r_a = 3.15$ ohm, $k_a = 0.16$ V-sec/rad (N-m/A), $L_a = 0.0066$ H, $B_m = 0.0001$ N-m-sec/rad, and $J = 0.0002$ kg-m^2.

In the Command Window, we enter these parameters as

```
ra=3.15; La=0.0066; ka=0.156; J=0.0002; Bm=0.0001;
```

The state-space model (4.2) is $\frac{d\mathbf{x}}{dt} = A\mathbf{x} + B\mathbf{u}$, $\begin{bmatrix} \frac{di_a}{dt} \\ \frac{d\omega_r}{dt} \end{bmatrix} = \begin{bmatrix} -\frac{r_a}{L_a} & -\frac{k_a}{L_a} \\ \frac{k_a}{J} & -\frac{B_m}{J} \end{bmatrix} \begin{bmatrix} i_a \\ \omega_r \end{bmatrix} + \begin{bmatrix} \frac{1}{L_a} \\ 0 \end{bmatrix} u_a - \begin{bmatrix} 0 \\ \frac{1}{J} \end{bmatrix} T_L$.

The state vector is $\mathbf{x} = \begin{bmatrix} x_1 & x_2 \end{bmatrix}^T = [i_a \quad \omega_r]^T$. The output $y = \omega_r$ results in the output equation $\mathbf{y} = H\mathbf{x} + D\mathbf{u}$ with $H = \begin{bmatrix} 0 & 1 \end{bmatrix}$ and $D = [0]$. The matrices $A \in \mathbb{R}^{2\times 2}$, $B \in \mathbb{R}^{2\times 1}$, $H \in \mathbb{R}^{1\times 2}$, and $D \in \mathbb{R}^{1\times 1}$ are found as $A = \begin{bmatrix} -477.3 & -23.6 \\ 780 & -0.5 \end{bmatrix}$, $B = \begin{bmatrix} 151.5 \\ 0 \end{bmatrix}$, $H = \begin{bmatrix} 0 & 1 \end{bmatrix}$, and $D = [0]$. The MATLAB file solves the simulation problem using the `lsim` command:

```
A=[-ra/La -ka/La; ka/J -Bm/J]; B=[1/La; 0]; H=[0 1]; D=[0];
t0=0.04; tf=0.25; x10=0; x20=0; t=t0:0.001:tf; x0=[x10 x20];
Ua=60; u=Ua*ones(size(t)); [y,x]=lsim(A,B,H,D,u,t,x0);
plot(t,10*x(:,1),'k-',t,x(:,2),'b-','LineWidth',3); xlabel('Time (seconds)','FontSize',15);
title('Angular Velocity {\it\omega_r} [rad/sec] and Current {\iti_a} [A]','FontSize',15);
```

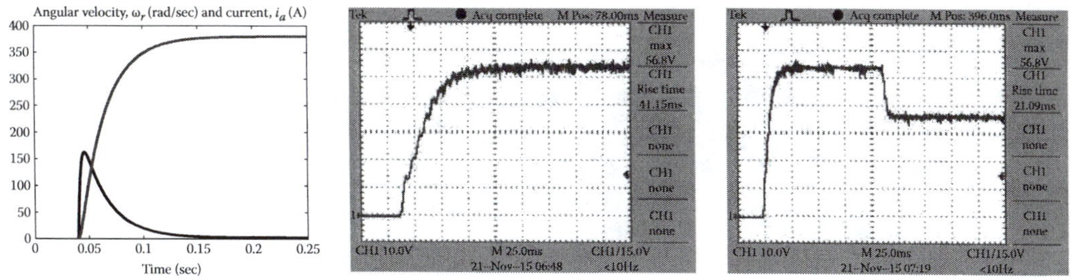

FIGURE 4.5 Dynamics of the current $i_a(t)$ (multiplied by 10) and $\omega_r(t)$. Experimental results: Acceleration of the motor if $u_a = 60$ V (the tachogenerator induced voltage is proportional to the angular velocity, and $\omega_{r\,measured} = k_{tach}u_{tach}$, $k_{tach} = 6.91$ V-rad/sec).

For the specified voltage applied $u_a = 60$ V, the motor current and angular velocity are plotted in Figure 4.5. To visualize the dynamics for $i_a(t)$, the current is multiplied by a factor of 10. The simulation results illustrate that the motor reaches the final angular velocity ~380 rad/sec within 0.1 sec. For $T_L = 0$, neglecting friction, the steady-state angular velocity should be $\omega_r = u_a/k_a = 384.6$ rad/sec, see (4.3). However, ω_r is 380 rad/sec due to the viscous friction torque, $T_{friction} = B_m\omega_r = 0.038$ N-m. At $T_L = 0$, $T_{e\ steady\text{-}state} = T_{friction} = B_m\omega_r = k_a i_a$. The simulation results agree with the experimental results reported in Figure 4.5. The last oscilloscope data corresponds to the motor loading with $T_L = 1$ N-m.

The experimental studies can be performed using the motor–generator system with two identical (or different) permanent-magnet DC machines. One machine operates as a motor, while other is the loading generator. The experimental setup with two identical 23 NEMA-size Torquemaster 2620 permanent-magnet DC machines, coupled using a flexible coupler, is illustrated in Figure 4.6a. The prime mover (permanent-magnet DC motor) drives a generator, and the circuit is shown in Figure 4.6b.

Using the subscript g for a generator and notations reported in Figure 4.6b, one finds the following equations of motion for a motor–generator system

$$\frac{di_a}{dt} = \frac{1}{L_a}\left(-r_a i_a - k_a\omega_r + u_a\right),$$

$$\frac{di_{ag}}{dt} = \frac{1}{L_{ag}}\left(-(r_{ag} + R_L)i_{ag} + k_{ag}\omega_r\right),$$

$$\frac{d\omega_r}{dt} = \frac{1}{J + J_g}\left(k_a i_a - (B_m + B_{mg})\omega_r - k_{ag}i_{ag}\right).$$

The simulation and analysis of permanent-magnet DC electric machines can be effectively performed in Simulink®. The Simulink model is depicted in Figures 4.7. The Signal Generator block is used to specify the applied voltage to be ±60, for example, $u_a = \pm 60\ \text{rect}(2t)$ V. The load torque T_L is specified to be ±1 N-m, e.g., ±1 rect(2t) N-m. The transient responses for the current $i_a(t)$ and velocity $\omega_r(t)$, as well as the evolution of η, are illustrated in Figures 4.8a and b. The steady-state value of η must be used, and $\eta = 64\%$. To plot the motor dynamics, one may use the `plot` command saving the data in the Scopes in an arrays format. The resulting plot is reported in Figure 4.8c using the plotting statement

```
plot(x(:,1),x(:,2),'k-',x(:,1),x(:,3),'b-','LineWidth',3);
xlabel('Time (seconds)','FontSize',15);
title('Angular Velocity {\it\omega_r} [rad/sec] and Current {\iti_a} [A]','FontSize',15);
```

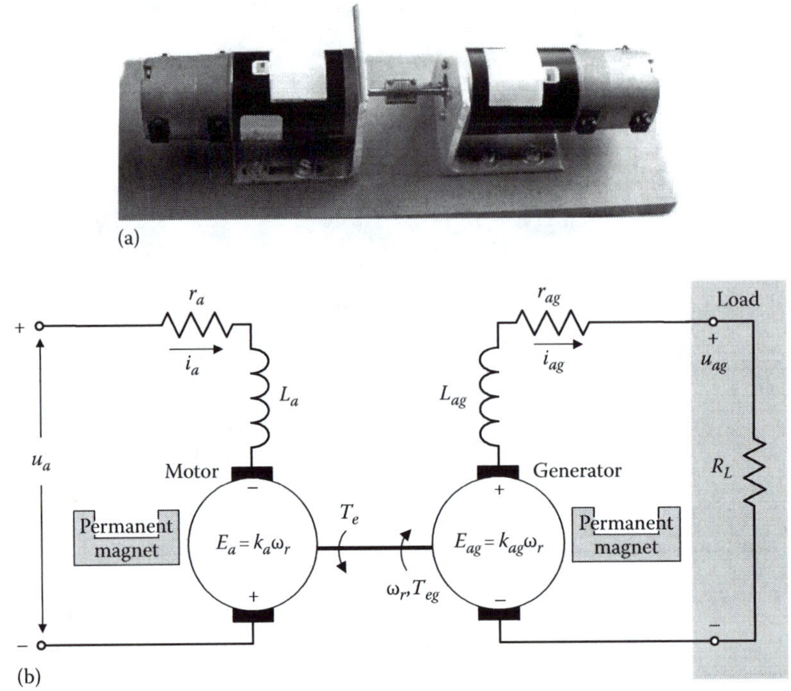

FIGURE 4.6 (a) A motor-generator system with two permanent-magnet DC machines. The angular velocity is measured by a tachogenerator (tachometer), which is mounted on the shaft of each machine (voltage induced by a tachogenerator is proportional to the angular velocity); (b) Permanent-magnet DC motor-generator with a resistive load R_L and the terminal voltage u_{ag}.

One analyzes the steady-state and dynamics responses, settling time, stability, efficiency, losses, etc. Using the input and output powers, the efficiency is $\eta = \frac{P_{output}}{P_{input}} = \frac{T_L\Omega_r}{U_a I_a}$.

If T_L is not measured $\eta = \frac{P_{output}}{P_{input}} = \frac{P_{input} - P_{losses}}{P_{input}} = \frac{U_a I_a - \left[r_a I_a^2 + B_m \Omega_r^2\right]}{U_a I_a}$.

One finds $\eta = 64\%$ at load $T_L = 1$ N-m as plotted in Figure 4.8b. The steady-state and dynamic analysis of efficiency, power, and losses can be accomplished at varying inputs, loads, and disturbances. One may modify the Simulink models. In Example 2.17, for the studied motor we found $B_m(\omega_r) = \frac{p_1 \tanh(p_2\omega_r) + p_3\omega_r}{\omega_r}$. Therefore, if T_L is not measured,

$$\eta = \frac{P_{output}}{P_{input}} = \frac{P_{input} - P_{losses}}{P_{input}} = \frac{U_a I_a - \left[r_a I_a^2 + B_m(\Omega_r)\Omega_r^2\right]}{U_a I_a} = \frac{U_a I_a - \left[r_a I_a^2 + \left[p_1 \tanh(p_2\Omega_r) + p_3\Omega_r\right]\Omega_r\right]}{U_a I_a}.$$

4.2.3 Electromechanical Devices with Power Electronics

The applied voltage to permanent-magnet DC motors is supplied by the high-frequency switching DC–DC converters. We briefly cover high-performance single- and four-quadrant PWM amplifiers, which control electromagnetic actuators. Various converters are reported in Chapter 7. We consider the schematics of a high-frequency *step-down* switching converter with a permanent-magnet motor as illustrated Figure 4.9.

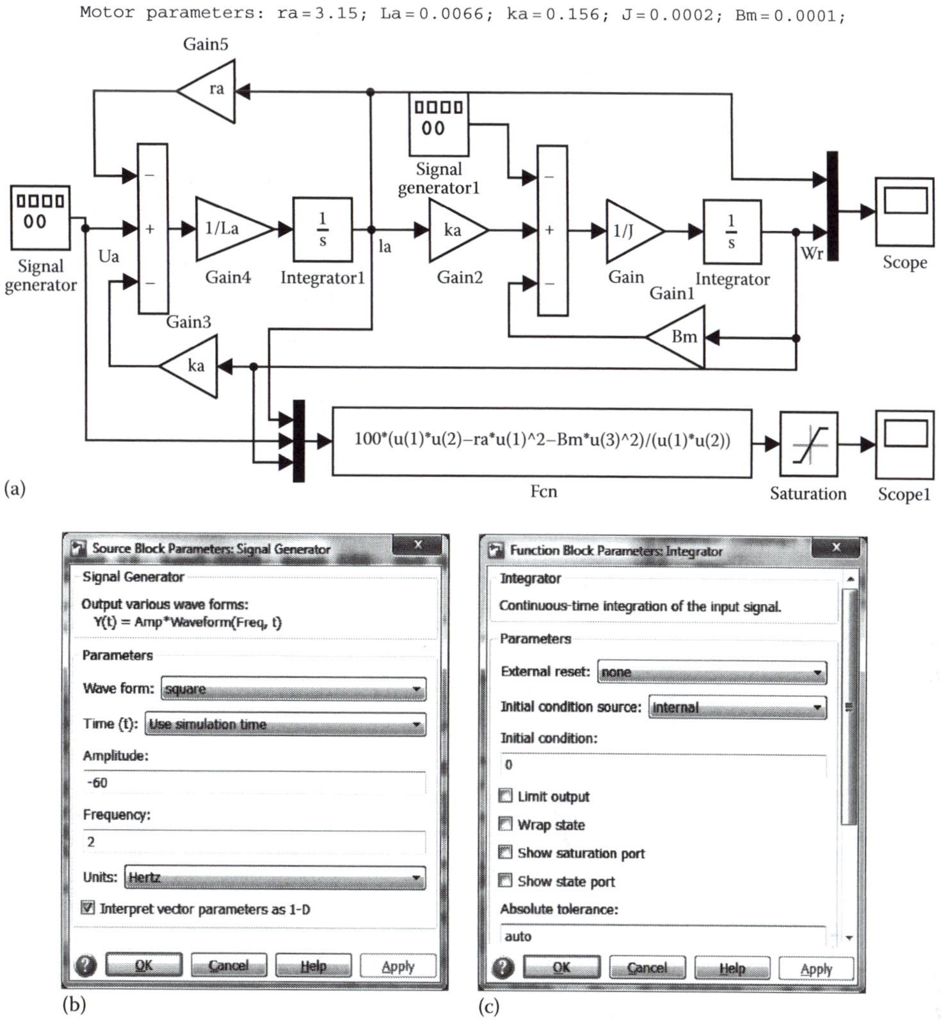

FIGURE 4.7 (a) Simulink® model to simulate permanent-magnet motors; (b) Signal generators; (c) Integrator blocks.

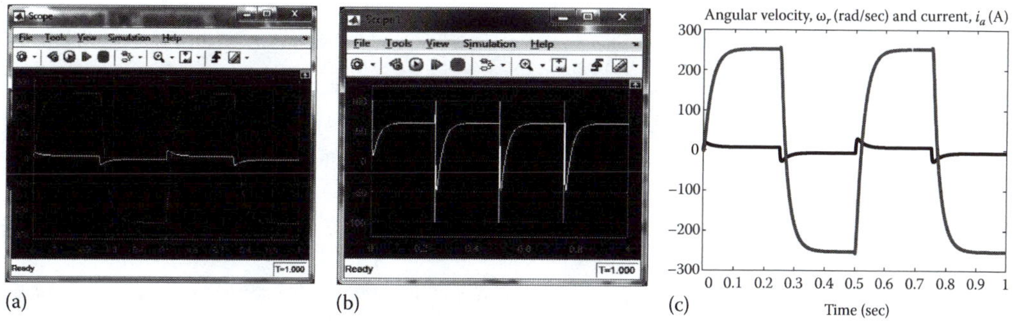

FIGURE 4.8 Permanent-magnet motor dynamics: (a) scope for the current $i_a(t)$ and angular velocity $\omega_r(t)$; (b) Scope data for the evolution of η; (c) Plots of current $i_a(t)$ and angular velocity $\omega_r(t)$.

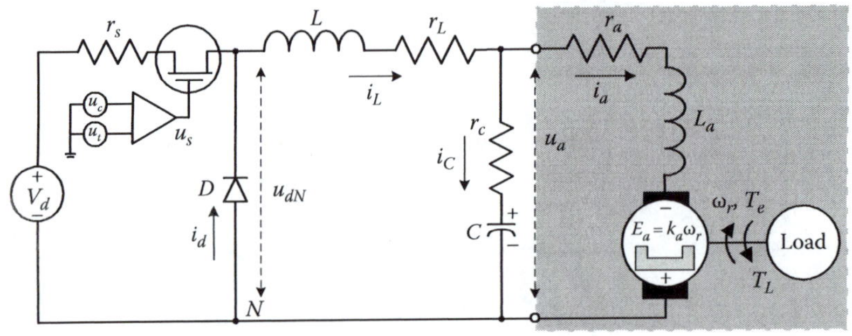

FIGURE 4.9 Permanent-magnet DC motor with *step-down* switching converter and filter.

The voltage applied to the motor winding u_a is regulated by controlling the switching *on* and *off* durations t_{on} and t_{off}. One changes the duty ratio d_D, $d_D = \dfrac{t_{on}}{t_{on}+t_{off}}$. The converter with a filter is examined in Chapter 7, and, the differential equations are

$$\frac{du_C}{dt} = \frac{1}{C}\left(i_L - i_a\right), \quad \frac{di_L}{dt} = \frac{1}{L}\left(-u_C - \left(r_L + r_c\right)i_L + r_c i_a - r_s i_L d_D + V_d d_D\right),$$
$$\frac{di_a}{dt} = \frac{1}{L_a}\left(u_C + r_c i_L - \left(r_a + r_c\right)i_a - E_a\right).$$

Using (4.1), we have a set of four first-order differential equations

$$\begin{aligned}
\frac{du_C}{dt} &= \frac{1}{C}\left(i_L - i_a\right),\\
\frac{di_L}{dt} &= \frac{1}{L}\left(-u_C - \left(r_L + r_c\right)i_L + r_c i_a - r_s i_L d_D + V_d d_D\right),\\
\frac{di_a}{dt} &= \frac{1}{L_a}\left(u_C + r_c i_L - \left(r_a + r_c\right)i_a - k_a\omega_r\right),\\
\frac{d\omega_r}{dt} &= \frac{1}{J}\left(k_a i_a - B_m\omega_r - T_L\right).
\end{aligned} \tag{4.5}$$

The duty ratio is regulated by changing the signal-level control voltage u_c, which is bounded as $u_{t\,\min} \le u_c \le u_{t\,\max}$. For $u_{t\,\min} = 0$, we have $d_D = \dfrac{u_c}{u_{t\,\max}} \in [0 \quad 1]$, $u_c \in [0 \quad u_{c\max}]$, $u_{c\max} = u_{t\max}$. The signal-level control voltage u_c is a control input u. The second equation in (4.5) becomes

$$\frac{di_L}{dt} = \frac{1}{L}\left(-u_C - \left(r_L + r_c\right)i_L + r_c i_a - \frac{r_s}{u_{t\max}} i_L u_c + \frac{V_d}{u_{t\max}} u_c\right).$$

The mathematical model of permanent-magnet DC motors with the *buck* converter is nonlinear. The control bounds are $0 \le u_c \le u_{c\,\max}$, $u_c \in [0 \quad u_{c\max}]$, $d_D \in [0 \quad 1]$. The armature voltage applied to the motor windings can be regulated by *boost* (*step-up*), Ćuk, and other DC–DC high-frequency switching converters.

Example 4.4: Permanent-Magnet DC Motor with a Step-Down Converter

Consider a permanent-magnet DC motor with a *step-down* converter shown in Figure 4.9. The converter parameters are as follows: $r_s = 0.025$ ohm, $r_L = 0.02$ ohm, $r_c = 0.15$ ohm, $C = 0.003$ F, and $L = 0.0007$ H. A low-pass filter ensures ~5% voltage ripple. Let $d_D = 0.5$ and $V_d = 50$ V. For motor, let $r_a = 2$ ohm, $k_a = 0.05$ V-sec/rad (N-m/A), $L_a = 0.005$ H, $B_m = 0.0001$ N-m-sec/rad, and $J = 0.0001$ kg-m^2.

Using differential equations (4.5), two m-files are developed to perform the simulation assuming that the motor accelerates from the stall and that $T_L = 0$. The MATLAB file **ch4_1.m** is

```
t0=0; tfinal=0.4; tspan=[t0 tfinal]; y0=[0 0 0 0]';
[t,y]=ode45('ch4_2',tspan,y0);
subplot(2,2,1); plot(t,y(:,1),'-');
xlabel('Time (seconds)','FontSize',12); title('Voltage u_C, [V]','FontSize',12);
subplot(2,2,2); plot(t,y(:,2),'-');
xlabel('Time (seconds)','FontSize',12); title('Current i_L, [A]','FontSize',12);
subplot(2,2,3); plot(t,y(:,3),'-');
xlabel('Time (seconds)','FontSize',12); title('Current i_a, [A]','FontSize',12);
subplot(2,2,4); plot(t,y(:,4),'-');
xlabel('Time (seconds)','FontSize',12);
title('Angular Velocity \omega_r, [rad/sec]','FontSize',12);
```

while the second file ch4_2.m is

```
% Dynamics of the PM DC motor with buck converter
function yprime=difer(t,y);
% parameters
Vd=50; D=0.5; rs=0.025; rl=0.02; rc=0.15; C=0.003; L=0.0007;
ra=2; ka=0.05; La=0.005; Bm=0.00001; J=0.0001; Tl=0;
% differential equations for PM DC Motor - Buck Converters
yprime=[(y(2,:)-y(3,:))/C;...
(-y(1,:)-(rl+rc)*y(2,:)+rc*y(3,:)-rs*y(2,:)*D+Vd*D)/L;...
(y(1,:)+rc*y(2,:)-(rc+ra)*y(3,:)-ka*y(4,:))/La;...
(ka*y(3,:)-Bm*y(4,:)-Tl)/J];
```

The transient dynamics for the state variables $u_C(t)$, $i_L(t)$, $i_a(t)$, and $\omega_r(t)$ are illustrated in Figure 4.10. The settling time is ~0.4 sec, and the motor reaches the steady-state angular velocity 496 rad/sec. ■

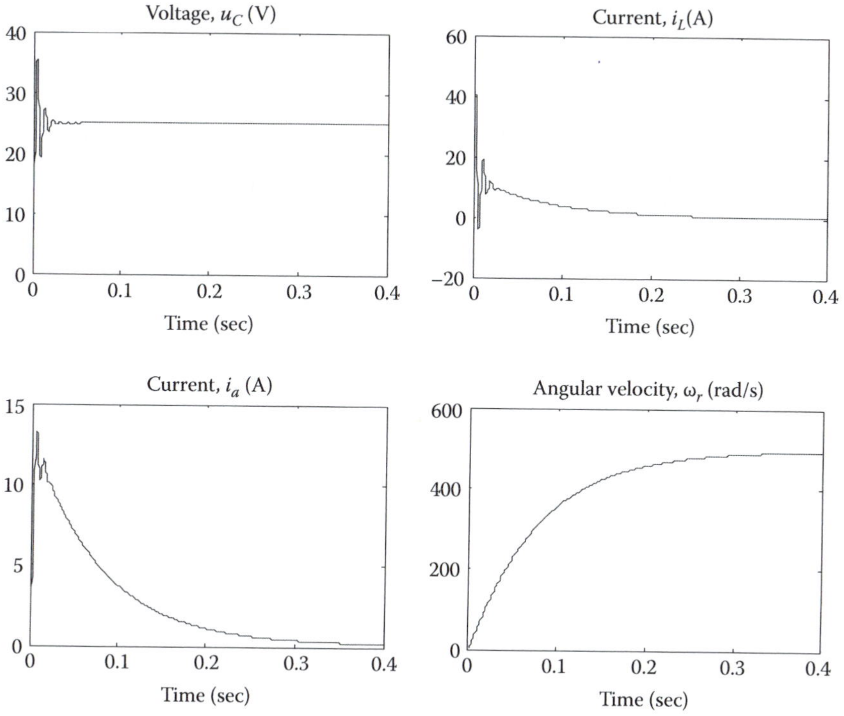

FIGURE 4.10 Transient dynamics of a permanent-magnet DC motor with a *buck* converter.

To regulate the angular velocity, one changes u_a. The power electronics and PWM amplifiers are covered in Chapter 7. The four-quadrant H-configured power stages guarantee high performance, efficiency, and other desired capabilities. To rotate motor clockwise and counterclockwise, the bipolar voltage u_a should be applied. Permanent-magnet DC electric machines are made in different sizes. The schematics of a four-quadrant 25A PWM servo amplifier (20–80 V, ±12.5 A continuous, ±25 A peak, 22 kHz, 129 × 76 × 25 mm dimensions) for a ~500 W permanent-magnet DC motor is shown in Figure 4.11. The motor armature winding is connected to P2-1 and P2-2. To control the angular velocity, one specifies the reference voltage to P1-4. The reference is compared with the tachogenerator voltage, proportional to the motor angular velocity, which is supplied to P1-6. This amplifier can be used in the servo systems. The voltage proportional to the displacement should be specified at P1-6. The proportional-integral analog controller is implemented. One can change the proportional and integral feedback gains adjusting the corresponding potentiometers.

The size of permanent-magnet DC electric machine can be less that the size of operational amplifier. High-performance 2 and 4 mm diameter permanent-magnet DC motors are manufactured and commercialized. One must estimate the load torque of such motors. To guarantee the rotation, $T_e > T_L$ must be met. The acceleration capability is $(T_e - T_L)/J$. Using T_e and ω_r, the power is $P = T_e\omega_r$. The sizing and volumetric features can be estimated using the power density estimate ~1 W/cm^3. The power density is significantly affected by the design, electromagnetic system, dimensions, magnets used, angular velocity, etc.

Small ~10 W permanent-magnet DC motors can be driven by dual operational amplifiers. The schematics is depicted in Figure 4.12a. The TCA0372 dual power operational amplifiers (40 V and 1 A) can be used to ensure bidirectional rotation of motors. The transient dynamics are reported in Figure 4.12a. The motor angular velocity is controlled by changing u_a.

Monolithic PWM amplifiers are available. The MC33030 DC servo motor controller/driver (36 V and 1 A) integrates operational amplifiers, comparators, driving logics, PWM four-quadrant H-bridge, etc. The built-in proportional controller changes the armature voltage u_a using the difference between the reference and actual angular velocity or displacement. The schematics of the MC33030 are documented in Figure 4.12b. The VNH2SP30, VNH3SP30, and other fully integrated H-bridge permanent-magnet DC motor drivers can be used, as reported in Chapter 7.

4.3 AXIAL TOPOLOGY PERMANENT-MAGNET DIRECT-CURRENT ELECTRIC MACHINES

4.3.1 Fundamentals of Axial Topology Permanent-Magnet Machines

Radial topology permanent-magnet DC electric machines were covered. We now examine the axial topology electric machines commonly used as *disc* motors, hard disk drive actuators, *limited angle* motors, etc. The planar segmented permanent magnets are placed on the stator, while the planar windings are on the rotor. Brushes and commutator are used to supply the armature voltage to the windings. The planar windings significantly simplify fabrication. The advantages of the axial topology are:

1. Affordability and simplicity to fabricate and assemble machines because permanent magnets and windings are planar. It is easy to fabricate single- and multilayered planar windings on the planar (flat) rotor;
2. There are relaxed shape-geometry and sizing requirements imposed on the magnets and windings (however, device performance is affected by the topology, geometry, spacing, etc.);
3. There is no rotor back ferromagnetic material required (silicon, polymers, or other materials can be used);
4. The air gap and magnet-winding separation can be adjusted.

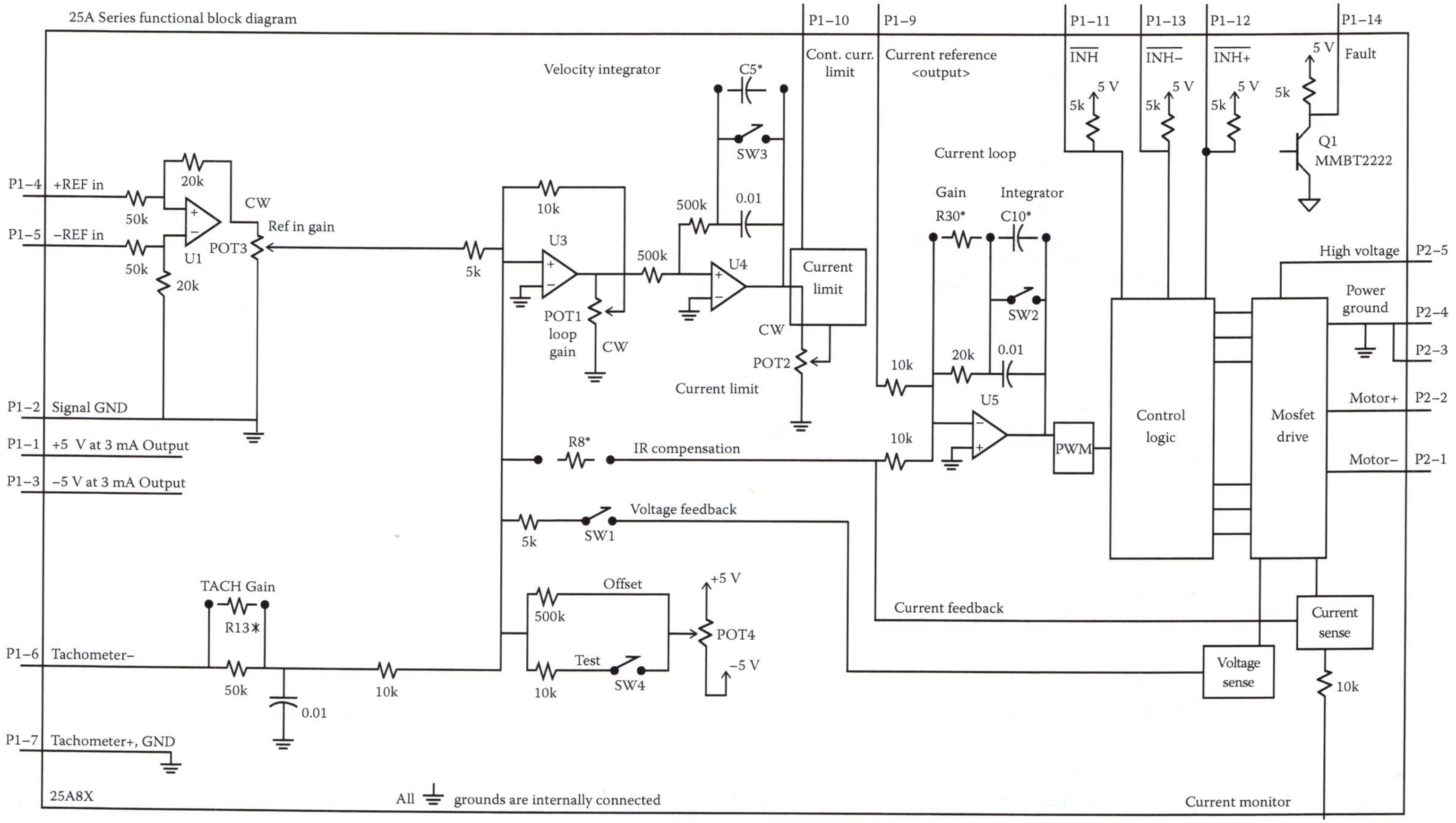

FIGURE 4.11 25A PWM servo amplifier. (From Lyshevski, S.E., *Electromechanical Systems, Electric Machines, and Applied Mechatronics*, CRC Press, Boca Raton, FL, 1999; Courtesy of Advanced Motion Controls, Camarillo, CA, www.a-m-c.com.)

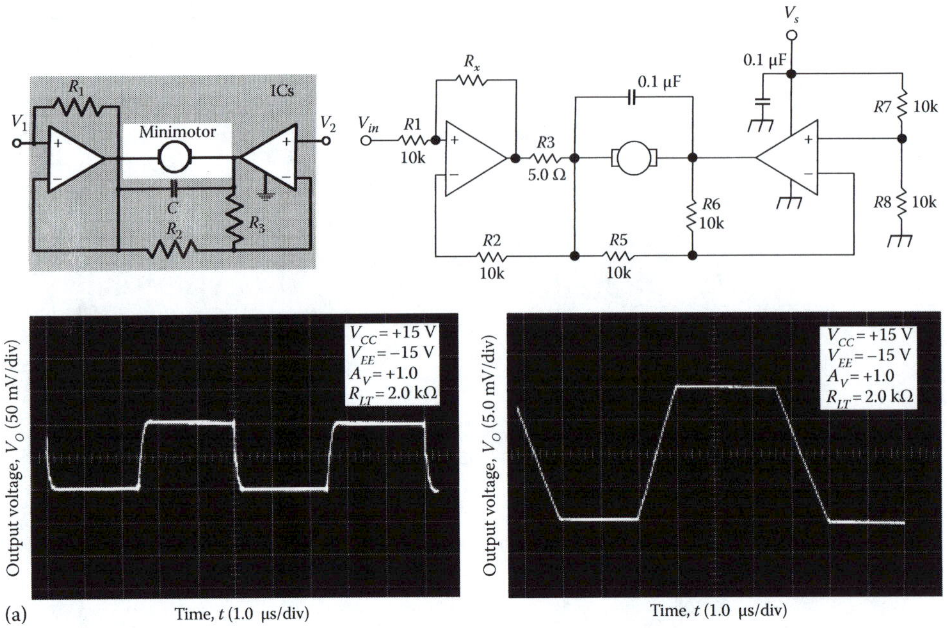

(a)

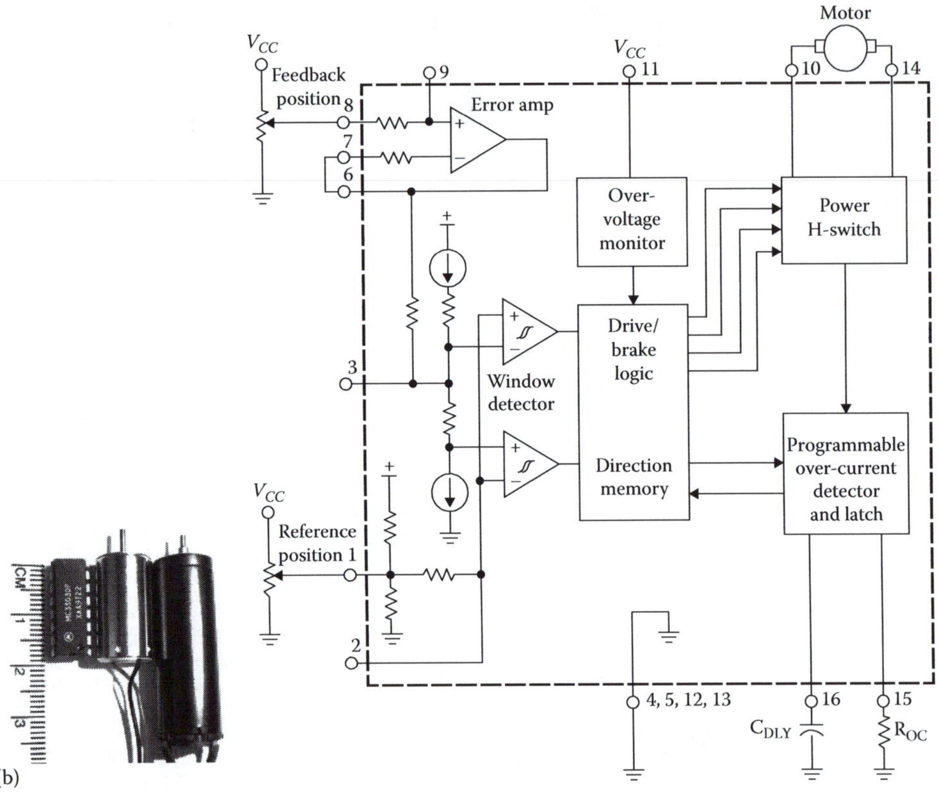

(b)

FIGURE 4.12 (a) Application of dual operational amplifier to control permanent-magnet DC minimotors: Bidirectional speed control of minimotors using TCA0372 dual power operational amplifier (40 V and 1 A). The TCA0372 transient responses; (b) Image of the MC33030 DC servo-motor controller/driver and 1 and 3 W permanent-magnet DC minimotor. The MC33030 DC servo-motor controller/driver schematics. (From Lyshevski, S.E., *Electromechanical Systems, Electric Machines, and Applied Mechatronics*, CRC Press, Boca Raton, FL, 1999; Copyright of Motorola, used with permission.)

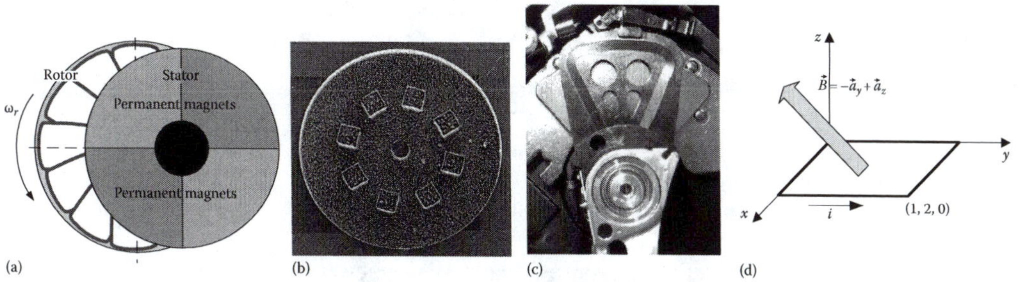

FIGURE 4.13 (a) Axial topology permanent-magnet electric machine schematics; (b) Rotor with magnets; (c) Limited-angle axial topology hard disk drive actuator; (d) Rectangular planar current loop in a uniform magnetic field (Example 4.5).

The axial topology permanent-magnet DC electric machine is schematically depicted in Figure 4.13a. Mini- and microelectromechanical motion devices can be fabricated. The image of the fabricated microstructure (stator or rotor) with segmented magnets is illustrated in Figure 4.13b. The magnets are in the plane of the coils. An image of the limited-angle axial topology hard disk drive actuator is depicted in Figure 4.13c. The force on the left and right side filaments should be developed in the same direction to ensure clockwise or counterclockwise motion. The top and bottom winding sides experience low magnetic field from the left and right magnets. These two forces are in opposite directions and may be countered.

Consider a current loop in the magnetic field, produced by magnets, as seen in the hard disk drive in Figure 4.13c and representative drawing in Figure 4.13d. The torque on a planar current loop of any size and shape in the uniform magnetic field is $\vec{T} = i\vec{s} \times \vec{B} = \vec{m} \times \vec{B}$, where i is the current in the loop (winding); $\vec{m}$ is the magnetic dipole moment (A-m^2).

The torque is $\vec{T} = \vec{R} \times \vec{F}$, where $\vec{F} = -i\oint_l \vec{B} \times d\vec{l}$. In the differential form, $d\vec{F} = id\vec{l} \times \vec{B}$.

For a uniform magnetic field, $\vec{F} = -i\vec{B} \times \oint_l d\vec{l}$.

The torque on the current loop tends to turn the loop to align the magnetic field produced by the loop, which causes the electromagnetic torque. The interaction between the magnetic dipole moment $\vec{m} = i\vec{s}$ and the magnets $\vec{B}$ results in the electromagnetic force and torque. Magnets are magnetized in the specified direction to ensure design consistency.

Example 4.5

Consider a 1 × 2 cm current loop in a uniform magnetic field $\vec{B} = -\vec{a}_y + \vec{a}_z$ as illustrated in Figure 4.13c. The torque is $\vec{T} = i\vec{s} \times \vec{B} = \vec{m} \times \vec{B} = i\begin{bmatrix} \vec{a}_x & \vec{a}_y & \vec{a}_z \\ 0.01 & 0.02 & 0 \\ 0 & -1 & 1 \end{bmatrix} = i\left(0.02\vec{a}_x - 0.01\vec{a}_y - 0.01\vec{a}_z\right)$N-m.

The magnetic dipole moment of the coils (winding) is $\vec{m} = Ni\vec{s}$, where i is the current; N is the number of turns. The cross product of $\vec{m} = \begin{bmatrix} m_i \\ m_j \\ m_k \end{bmatrix}$ and $\vec{B} = \begin{bmatrix} B_i \\ B_j \\ B_k \end{bmatrix}$ is $\vec{T} = \vec{m} \times \vec{B} = \begin{vmatrix} \vec{i} & \vec{j} & \vec{k} \\ m_i & m_j & m_k \\ B_i & B_j & B_k \end{vmatrix} = \vec{i}\begin{vmatrix} m_J & m_k \\ B_j & B_k \end{vmatrix} - \vec{j}\begin{vmatrix} m_i & m_k \\ B_i & B_k \end{vmatrix} + \vec{k}\begin{vmatrix} m_i & m_j \\ B_i & B_j \end{vmatrix}$. The cross product of two vectors is a vector that is orthogonal to both the vectors.

The solution is verified by finding $\vec{m} \cdot (\vec{m} \times \vec{B}) = 0$ and $\vec{B} \cdot (\vec{m} \times \vec{B}) = 0$. ■

Consider a consistently designed axial topology actuator. A limited-angle axial topology hard disk drive actuator is illustrated in Figure 4.13c. The planar coils are above the segmented NdFeB nickel-plated magnets. The expressions $\vec{T} = \vec{R} \times \vec{F}$, $\vec{F} = i\vec{l} \times \vec{B}$ are simplified to a one-dimensional case for optimal structural and electromagnetic designs. The electromagnetic forces F_{eL} and F_{eR} act on the left and right filaments. These F_{eL} and F_{eR} are perpendicular to filaments, and produce the electromagnetic torque $T_e = T_{eL} + T_{eR}$ to rotate the rotor around a point O.

Consider the *effective* flux density $B(\theta_r)$, which varies as a function of the angular displacement θ_r due to the angular displacement of rotor with windings relative to the stator with magnets that produce the stationary field. Depending on the magnets magnetization, geometry, and shape, as well as coil–magnet placement, one applies distinct expressions for $B(\theta_r)$. For motors, the flux density $B(\theta_r)$, as viewed from the windings, is a periodic function of θ_r. Let each magnet produce a uniform magnetic field, and, there is a number of magnets N_m ($N_m = 2\,m$, m is the integer), which are oppositely magnetized through the thickness. If there is no spacing between magnets in a segmented magnet array (which can be uniformly magnetized), or if there is a spacing (or magnets shape or magnetization vary), one may use

$$B(\theta_r) = B_{\max}\,\mathrm{sgn}\left(\sin\left(\frac{1}{2}N_m\theta_r\right)\right), \quad B(\theta_r) = \sum_{n=1}^{\infty} B_{\max n} \sin^{2n-1}\left(\frac{1}{2}N_m\theta_r\right),$$

where $B_{\max}$ is the effective flux density produced by the magnets as viewed from the winding ($B_{\max}$ depends on the magnets used, magnet-winding separation, temperature, etc.); N_m is the number of magnets; n is the integer, which is a function of the magnet magnetization, geometry, shape, width, thickness, separation, etc.

Example 4.6

Consider $B(\theta_r)$ as given by

$$B(\theta_r) = B_{\max} \sin\left(\frac{1}{2}N_m\theta_r\right), \quad B(\theta_r) = B_{\max}\,\mathrm{sgn}\left(\sin\left(\frac{1}{2}N_m\theta_r\right)\right), \quad B(\theta_r) = B_{\max} \sin^5\left(\frac{1}{2}N_m\theta_r\right)$$

with $B_{\max} = 0.9$ T and $N_m = 4$. We calculate and plot $B(\theta_r)$ using the following statements:

```
th=0:0.01:2*pi; Nm=4; Bmax=0.9; B=Bmax*sin(Nm*th/2); plot(th,B,'k-','LineWidth',3);
xlabel('Rotor Displacement, {\it\theta_r} [rad]','FontSize',18);
ylabel('{\itB}({\it\theta_r}) [T]','FontSize',18);
title('Field as a Function on Displacement,{\itB(\theta_r)}','FontSize',18);
```

and

```
th=0:0.01:2*pi; Nm=4; Bmax=0.9; B=Bmax*sign(sin(Nm*th/2)); plot(th,B,'k-','LineWidth',3);
xlabel('Rotor Displacement, {\it\theta_r} [rad]','FontSize',18);
ylabel('{\itB}({\it\theta_r}) [T]','FontSize',18);
title('Field as a Function on Displacement,{\itB(\theta_r)}','FontSize',18);
```

and

```
th=0:0.01:2*pi; Nm=4; Bmax=0.9; B=Bmax*sin(Nm*th/2).^5; plot(th,B,'k-','LineWidth',3);
xlabel('Rotor Displacement, {\it\theta_r} [rad]','FontSize',18);
ylabel('{\itB}({\it\theta_r}) [T]','FontSize',18);
title('Field as a Function on Displacement,{\itB(\theta_r)}','FontSize',18);
```

The resulting plots for $B(\theta_r)$ are reported in Figures 4.14.

As illustrated, $B(\theta_r)$ can be given by

$$B(\theta_r) = B_{\max}\,\mathrm{sgn}\left(\sin\left(\frac{1}{2}N_m\theta_r\right)\right), \quad B(\theta_r) = \sum_{n=1}^{\infty} B_{\max n} \sin^{2n-1}\left(\frac{1}{2}N_m\theta_r\right)$$

or other periodic continuous or discontinuous functions.

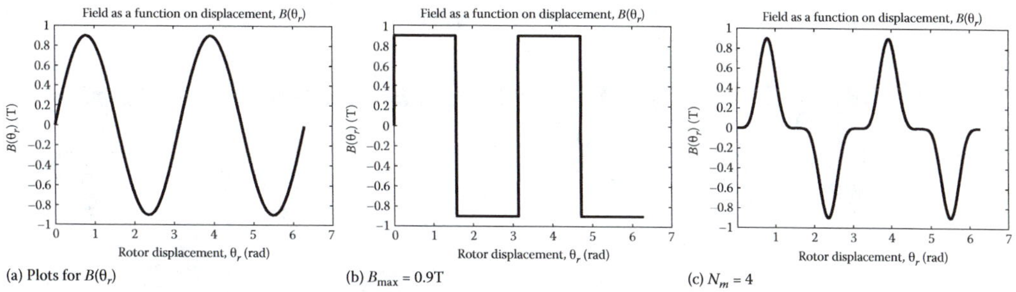

FIGURE 4.14 (a) Plot if $B(\theta_r) = B_{\max}\sin\left(\frac{1}{2}N_m\theta_r\right)$; (b) Plot for $B(\theta_r) = B_{\max}\,\text{sgn}\left(\sin\left(\frac{1}{2}N_m\theta_r\right)\right)$; (c) Plot for $B(\theta_r) = B_{\max}\sin^5\left(\frac{1}{2}N_m\theta_r\right)$, $B_{\max} = 0.9$ T, $N_m = 4$.

The analytic expressions for $B(\theta_r)$ are used to derive *emf* and T_e. The expression $\vec{T} = \vec{R}\times\vec{F}$, $\vec{F} = i\vec{l}\times\vec{B}$ is simplified to

$$T_e = R_\perp N l_{eq} B(\theta_r) i_a,$$

where $R_\perp$ is the perpendicular radius (lever arm); N is the number of turns; l_{eq} is the *effective* coil filament length.

Due to the use of brushes and commutator, the electromagnetic torque, acting on the left and right coil filaments, is maximized by properly commutating coils to which u_a is applied. Depending on designs, for many axial and radial topology permanent-magnet DC motors, the expression for the electromagnetic torque is simplified to $T_e = k_a i_a$. For example, for $B(\theta_r) = B_{\max}\,\text{sgn}\left(\sin\left(\frac{1}{2}N_m\theta_r\right)\right)$, one has $k_a = R_\perp N l_{eq} B_{\max}$.

Using Kirchhoff's voltage law $u_a = r_a i_a + \frac{d\psi}{dt}$ and Newton's second law of motion $\frac{d\omega_r}{dt} = \frac{1}{J}\left(T_e - B_m\omega_r - T_L\right)$, the differential equations for the axial topology permanent-magnet DC motors are

$$\frac{di_a}{dt} = \frac{1}{L_a}\left(-r_a i_a - k_a\omega_r + u_a\right), \quad \frac{d\omega_r}{dt} = \frac{1}{J}\left(k_a i_a - B_m\omega_r - T_L\right).$$

Remark. The topology, geometry, permanent magnets, windings, dimensionality, and other factors significantly affect the resulting equations. We performed the quantitative analysis. The equations were derived assuming optimal or near-optimal design of electromechanical motion devices. There is an indefinite number of possible solutions. One can use the three-dimensional Maxwell's equations and tensor calculus to perform high-fidelity modeling and data-intensive analyses. Analytic and numerical results must be validated by performing experiments, testing, characterization, and evaluation. ■

4.3.2 Axial Topology Hard Disk Drive Actuators

Consider axial topology hard disk drive actuators with two nickel-plated NdFeB segmented magnets, see Figure 4.15. To rotate rotor in the *xy* plane clockwise or counterclockwise the polarity of the applied voltage u_a is changed. The force on the left F_{eL} and right F_{eR} side current loop filaments should be developed in the same direction to ensure clockwise or counterclockwise motion. The top and bottom winding sides experience low magnetic field from the left and right magnets. These two forces

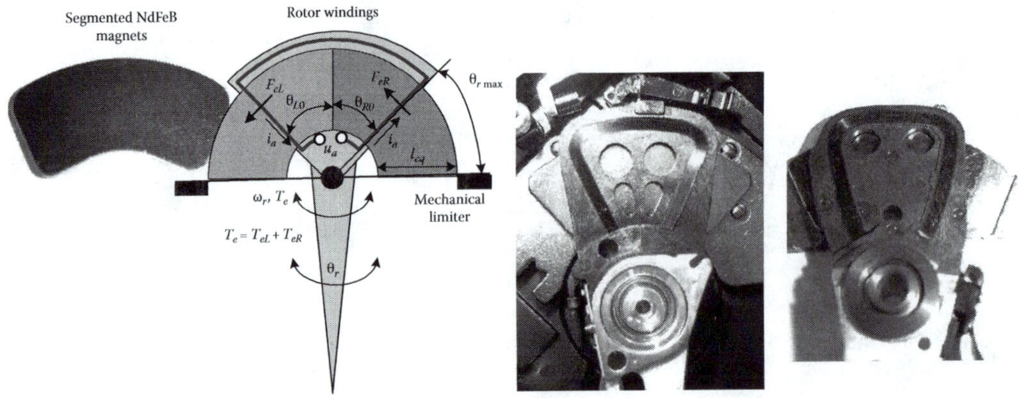

FIGURE 4.15 Limited-angle axial topology hard disk drive actuators: A planar coil is above the nickel-plated NdFeB magnets.

are in opposite directions and may be countered. One applies the right-hand rule. Due to opposite magnetization of the left and right magnets, the electromagnetic torque (force) is developed in the same direction, and, $T_e = T_{eL} + T_{eR}$. The mechanical limiters restrict the rotor angular displacement to $-\theta_{r\,max} \le \theta_r \le \theta_{r\,max}$. For the computer and camera hard drives illustrated in Figure 4.15, the mechanical limits on displacement are usually $-10° \le \theta_r \le 10°$. The interaction between the stationary magnetic field $B_L(\theta_r)$ by the left magnet and coil current i_a results in F_{eL} and T_{eL}. Assume that magnetic field $B_R(\theta_r)$ by the right magnet does not affect F_{eL} and T_{eL} on the left coil filament. The same analysis is true for the right filament. The images of two different actuators are reported in Figure 4.15. For a limited angle actuator, the commutator is not required. The voltage is supplied by using a flexible cord. For 360-degree-rotation of motors, the commutator is used to supply voltage to coils.

Example 4.7

Examine the variations of $B(\theta_r)$ as viewed from the left and right coil filaments. With different magnetization of magnets, coil design, and orientation, the effective $B(\theta_r)$ will be different. This $B(\theta_r)$, which can be measured, significantly affects the overall performance. Consider typifying cases when two magnets are magnetized and a planar winding is positioned such that the effective $B(\theta_r)$ is:

1. $B(\theta_r) = kB_{max}\theta_r$, $k > 0$ $(k = 5)$;
2. $B(\theta_r) = B_{max}\tanh(a\theta_r)$ and $B(\theta_r) = B_{max}\tanh^{11}(a\theta_r)$, $a \gg 1$, $(a = 100)$;
3. $B(\theta_r) = B_{max}\sin(a\theta_r)$, $a > 0$ $(a = 8)$;
4. $B(\theta_r) = 0.5B_{max}\tan^{-1}(a\theta_r)$, $a > 0$ $(a = 100)$;
5. $B(\theta_r) = B_{max}\,\mathrm{sgn}(\theta_r)$ and $B(\theta_r) = B_{max}\dfrac{\theta_r}{\sqrt{a+\theta_r^2}}$, $a > 0$, $a \ll 1$ $(a = 0.0001)$.

Let $B_{max} = 1$ T and $-0.2 \le \theta_r \le 0.2$ rad. We have

```
th=-0.2:.0001:0.2; Bmax=1; k=5; B1=k*Bmax*th;
a2=100; B21=Bmax*tanh(a2*th); B22=Bmax*(tanh(a2*th)).^11; a3=8; B3=Bmax*sin(a3*th);
a4=100; B4=0.5*Bmax*atan(a4*th); B51=Bmax*sign(th);
a5=1e-4; B52=Bmax*th./sqrt(a5+th.^2);
plot(th,B1,'r--',th,B21,'k-',th,B22,'b:','LineWidth',3);
% plot(th,B3,'r-',th,B4,'k--',th,B51,'k-',th,B52,'b:','LineWidth',3);
axis([-0.2 0.2 -1.05 1.05]);
xlabel('Displacement, {\it\theta_r} [rad]','FontSize',18);
ylabel('{\itB(\theta_r)} [T]','FontSize',18);
title('Field as a Function of Displacement, {\itB(\theta_r)}','FontSize',18);
```

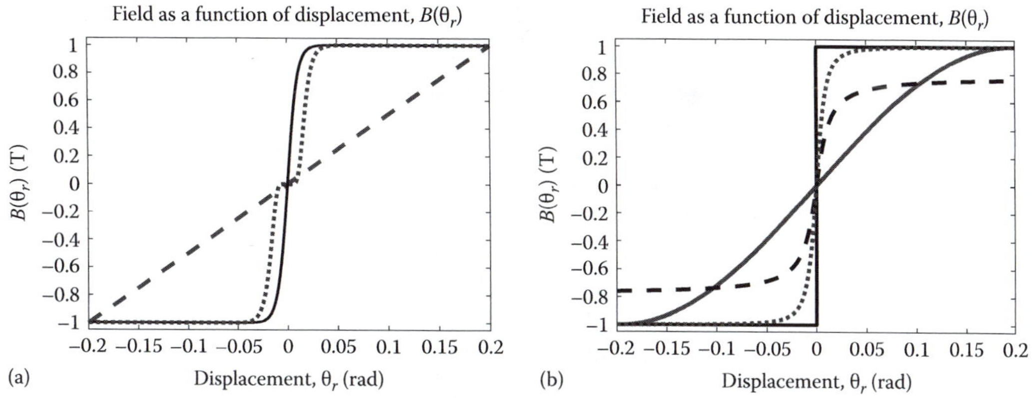

FIGURE 4.16 (a) Plots for $B(\theta_r)=kB_{\max}\theta_r$, $k=5$ (dashed line), $B(\theta_r)=B_{\max}\tanh(a\theta_r)$, and $B(\theta_r)=B_{\max}\tanh^{11}(a\theta_r)$, $a = 100$ (solid dotted lines); (b) Plots for $B(\theta_r) = B_{\max}\sin(a\theta_r)$, $a = 8$ (solid line), $B(\theta_r) = 0.5\, B_{\max}\tan^{-1}(a\theta_r)$, $a = 100$ (dashed line), $B(\theta_r) = B_{\max}\,\mathrm{sgn}(\theta_r)$, and $B(\theta_r) = B_{\max}\dfrac{\theta_r}{\sqrt{a+\theta_r^2}}$, $a = 0.0001$ (solid dotted lines).

The resulting plots for $B(\theta_r)$ are depicted in Figures 4.16. Performing measurements, one usually finds that the accurate approximation for $B(\theta_r)$ is $B(\theta_r) = B_{\max}\tanh^{2q+1}(a\theta_r)$, $a \gg 1$, $q = 1, 2, 3, \ldots$

Kirchhoff's voltage law is $u_a = r_a i_a + \dfrac{d\psi}{dt}$. The motional *emf* is $\mathscr{E} = -\oint_s \dfrac{\partial \vec{B}}{\partial t}\cdot d\vec{s}$.

Therefore, $\dfrac{di_a}{dt} = \dfrac{1}{L_a}\left(-r_a i_a - \dfrac{d}{dt}\oint_s B(\theta_r)\cdot d\vec{s} + u_a\right)$.

Remark. The magnetic flux is $\Phi = \oint_s \vec{B}\cdot d\vec{s}$ and $\oint_s \vec{B}\cdot d\vec{s} \neq BA$. ■

The torque is found using the force and the position vector, $\vec{T} = \vec{R}\times\vec{F}$. For adequate symmetric designs, when the coil filaments point to the center of rotation, using the perpendicular radius (*lever arm*) $R_\perp$, one has the electromagnetic torque $T_e = T_{eL} + T_{eR}$, $T_{eL} = R_\perp F_{eL}$, $T_{eL} = R_\perp F_{eL}$. Using the magnetic fields on the left and right filaments, $\vec{T} = \vec{R}\times\left(\vec{F}_{eL} + \vec{F}_{eR}\right) = -i\vec{R}\times\left(\oint_l \vec{B}_L\times d\vec{l} + \oint_l \vec{B}_R\times d\vec{l}\right)$.

Denote the number of turns N and equivalent filament length l_{eq}. Using the magnetic fields and the relative angular filament displacements $\theta_L(t)$ and $\theta_R(t)$, $T_{eL} = R_\perp N l_{eq} B_L(\theta_r) i_a$ and $T_{eR} = R_\perp N l_{eq} B_R(\theta_r) i_a$. As documented in Figure 4.15, the mechanical limit is $-\theta_{r\max} \le \theta_r \le \theta_{r\max}$, $|\theta_{r\min}| = |\theta_{r\max}|$. The left and right coil filaments take the angular positions $\theta_L(t)$ and $\theta_R(t)$ with respect to two stationary magnetic fields produced by magnets. Assuming $\theta_{r0} = 0$ (pointer at the center), $\theta_{L0} = |\theta_{r\max}|$ rad, and $\theta_{R0} = |\theta_{r\max}|$. For a symmetric kinematics, $\theta_L(t) = \theta_{L0} - \theta_r(t)$, $\theta_R(t) = \theta_{R0} + \theta_r(t)$ and $-\theta_{r\max} \le \theta_r \le \theta_{r\max}$. Newton's second law of motion is

$$\frac{d\omega_r}{dt} = \frac{1}{J}\left[T_e - T_{friction} - T_L\right] = \frac{1}{J}\left[R_\perp N l_{ef}\left(B_L(\theta_r) + B_R(\theta_r)\right)i_a - T_{friction} - T_L\right],\quad T_e = T_{eR} + T_{eL},$$

$$\frac{d\theta_r}{dt} = \omega_r,\quad -\theta_{r\max} \le \theta_r \le \theta_{r\max}.$$

The analysis is performed using the device-specific $B(\theta_r)$, which yields $\dfrac{d}{dt}\oint_s \vec{B}(t)\cdot d\vec{s}$ and the expressions for $B_L(\theta_r)$ and $B_R(\theta_r)$.

Example 4.8: Electromagnetic Theory and Limited Angle Axial Topology Actuators

Explicit expressions for the electromagnetic torque and *emf* are derived for common magnets magnetizations. In many actuators, $B(\theta_r) = B_{\max}\tanh(a\theta_r)$. The induced *emf* is

$$\mathscr{E} = -N\frac{d}{dt}\int_{r_{in}}^{r_{out}}\int_{\theta_R}^{\theta_L} B_{\max}\tanh\left(a\theta_r\right) r\,dr\,d\theta.$$

For a continuous real-valued function *f*, defined in a closed interval $[a, b]$, let F be an antiderivative of *f*. For a definite integral, we have $\int_a^b f(x)dx = F(b) - F(a)$. Thus,

$$\mathscr{E} = -\frac{r_{out}^2 - r_{in}^2}{2} NB_{\max}\left(\tanh a\theta_L - \tanh a\theta_R\right)\omega_r$$

$$= -\frac{r_{out}^2 - r_{in}^2}{2} NB_{\max}\left[\tanh a\left(\theta_{L0} - \theta_r\right) - \tanh a\left(\theta_{R0} + \theta_r\right)\right]\omega_r.$$

For $B(\theta_r) = B_{\max}\tanh(a\theta_r)$, the electromagnetic torque is $T_e = R_\perp N l_{eq} B_{\max}(\tanh a\theta_L + \tanh a\theta_R)i_a$. Hence,

$$T_e = R_\perp N l_{eq} B_{\max}\left[\tanh a(\theta_{L0} - \theta_r) + \tanh a(\theta_{R0} + \theta_r)\right]i_a.$$

Kirchhoff's and Newton's laws yield

$$\frac{di_a}{dt} = \frac{1}{L_a}\left[-r_a i_a - \frac{r_{\text{out}}^2 - r_{\text{in}}^2}{2} NB_{\max}\left[\tanh a(\theta_{L0} - \theta_r) - \tanh a(\theta_{R0} + \theta_r)\right]\omega_r + u_a\right],$$

$$\frac{d\omega_r}{dt} = \frac{1}{J}\left[R_\perp N l_{eq} B_{\max}\left[\tanh a(\theta_{L0} - \theta_r) + \tanh a(\theta_{R0} + \theta_r)\right]i_a - T_{\text{friction}} - T_L\right], \tag{4.6}$$

$$\frac{d\theta_r}{dt} = \omega_r, \quad -\theta_{r\max} \le \theta_r \le \theta_{r\max}.$$

Here, $\theta_{L0} = \theta_{R0}$. ■

Example 4.9

Using the required repositioning rate, displacement, bandwidth, and other specifications, one performs data-intensive electromagnetic design and analysis. The actuator parameters can be measured. Rudimentary calculations and estimates can be performed. For example, for a copper winding with $N = 300$, coil diameter 0.1 mm, and the total length $(l_L + l_{top} + l_{bottom} + l_R)$ 6 cm, we have $r_a = \frac{N(l_L + l_{top} + l_{bottom} + l_R)\sigma}{A} = \frac{300\times 0.06\times 1.72\times 10^{-8}}{\pi(0.00005)^2} = 39$ ohm. The circular loop self-inductance (loop radius is R_l and coil diameter is d) is estimated as $L_a = N^2 R_l\mu_0\mu_r\left(\ln\frac{16R_l}{d} - 2\right) = 300^2\times 0.0075\times 4\pi\times 10^{-7}\left(\ln\frac{8\times 0.0075}{0.00005} - 2\right) = 0.0037$ H. The equation for the moment of inertia of thin disk is $J = mR_{disk}^2$, which yields an overestimated J. ■

Example 4.10

In high-performance actuators $a \gg 1$. Hence, $\tanh(a\theta_i) \approx \pm 1$, $\theta_i \neq 0$.

Therefore, $T_e = T_{eL} + T_{eR} = 2R_\perp N l_{eq} B_{\max} i_a$, and, $\mathscr{E} = -\left(r_{out}^2 - r_{in}^2\right) N B_{\max} \omega_r$.

For $a \gg 1$, assuming that the torque exhibited by the flexible cable is $k_s\theta_r$, and letting $T_{friction} = B_m\omega_r$, one finds the following differential equations

$$\frac{di_a}{dt} = \frac{1}{L_a}\left[-r_a i_a - \left(r_{out}^2 - r_{in}^2\right) N B_{\max}\omega_r + u_a\right],$$

$$\frac{d\omega_r}{dt} = \frac{1}{J}\left[2R_\perp l_{eq} N B_{\max} i_a - B_m\omega_r - k_s\theta_r\right],$$

$$\frac{d\theta_r}{dt} = \omega_r, \quad -\theta_{r\max} \le \theta_r \le \theta_{r\max}.$$

■

Example 4.11

Consider an axial topology limited angle actuator if $B(\theta_r) = B_{\max}\tanh(a\theta_r)$. Using (4.6), for $T_{friction} = B_m\omega_r$ and $T_{elastic} = k_s\theta_r$, we have the following differential equations

$$\frac{di_a}{dt} = \frac{1}{L_a}\left[-r_a i_a - \frac{r_{out}^2 - r_{in}^2}{2} N B_{\max}\left[\tanh a(\theta_{L0} - \theta_r) - \tanh a(\theta_{R0} + \theta_r)\right]\omega_r + u_a\right], \quad A_{eq} = \frac{1}{2}\left(r_{out}^2 - r_{in}^2\right),$$

$$\frac{d\omega_r}{dt} = \frac{1}{J}\left[R_\perp N l_{eq} B_{\max}\left[\tanh a(\theta_{L0} - \theta_r) + \tanh a(\theta_{R0} + \theta_r)\right] i_a - B_m\omega_r - k_s\theta_r\right],$$

$$\frac{d\theta_r}{dt} = \omega_r, \quad -\theta_{r\max} \le \theta_r \le \theta_{r\max}.$$

For hard disk drive actuators, $-10° \le \theta_r \le 10°$, $-0.175 \le \theta_r \le 0.175$ rad. Assuming $\theta_{r0} = 0$ (pointer at the center), $\theta_{L0} = \theta_{R0} = 0.175$ rad. The parameters are $B_{\max} = 1$, $a = 100$, $r_a = 35$ ohm, $L_a = 4.1 \times 10^{-3}$ H, $R_\perp = 0.02$ m, $N = 100$, $l_{eq} = 0.0125$ m, $B_m = 5 \times 10^{-4}$, $k_s = 0.05$, $J = 1.5 \times 10^{-6}$ kg-m^2, $r_{in} = 0.015$ m, and $r_{out} = 0.025$ m.

We upload the parameters and constants as

```
Bmax=1; a=100; ra=35; La=4.1e-3; Rp=0.02; N=100; leq=1.25e-2; Bm=5e-4; ks=0.05; J=1.5e-6;
TR0=0.175; Rin=0.015; Rou=0.025; Aeq=(Rou^2-Rin^2)/2;
```

For the found equations of motion, the corresponding Simulink model is reported in Figure 4.17a. The transient dynamics for $\theta_r(t)$ is depicted in Figure 4.17b if u_a are the steps ± 5 V with $f = 10$ Hz.

■

4.4 *TRANSLATIONAL PERMANENT-MAGNET ELECTROMECHANICAL MOTION DEVICES*

Various rotational permanent-magnet devices and DC electric machines were covered. The translational (linear) devices are used in many applications. For example, loudspeakers and microphones are actuators (motors) and generators, respectively. Alexander Graham Bell received a patent on the electromagnetic loudspeaker in 1876. The moving coil speaker was proposed and demonstrated by Oliver Lodge in 1898. In early designs, the stationary field was established by electromagnets. The performance of loudspeakers and microphones was significantly enhanced by using permanent magnets, which establish a stationary magnetic field. Some ironless loudspeakers with radially magnetized magnets are illustrated in Figure 4.18a.

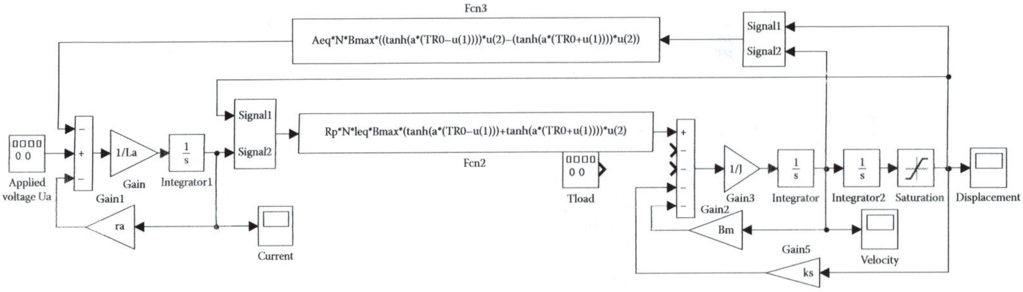


```
Bmax=1; a=100; ra=35; La=4.1e-3; Rp=0.02; N=100; leq=1.25e-2; Bm=5e-4; ks=0.05; J=1.5e-6;
TR0=0.175; Rin=0.015; Rou=0.025; Aeq=(Rou^2-Rin^2)/2;
```

(a)

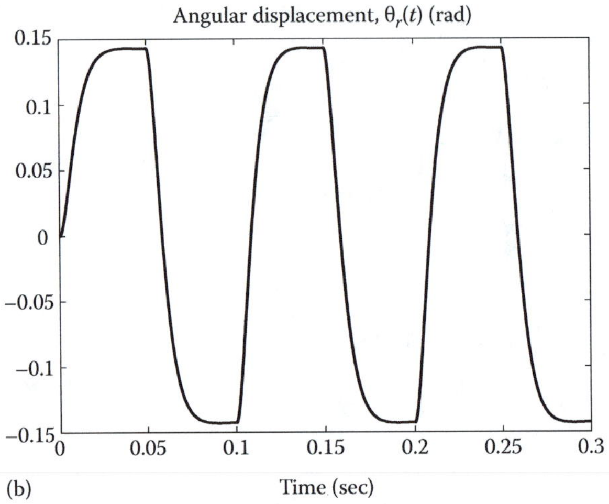

(b)

FIGURE 4.17 (a) Simulink® model for an axial topology limited-angle actuator when $B(\theta_r) = B_{max}\tanh(a\theta_r)$; (b) Transient dynamics for the angular displacement $\theta_r(t)$.

(a)

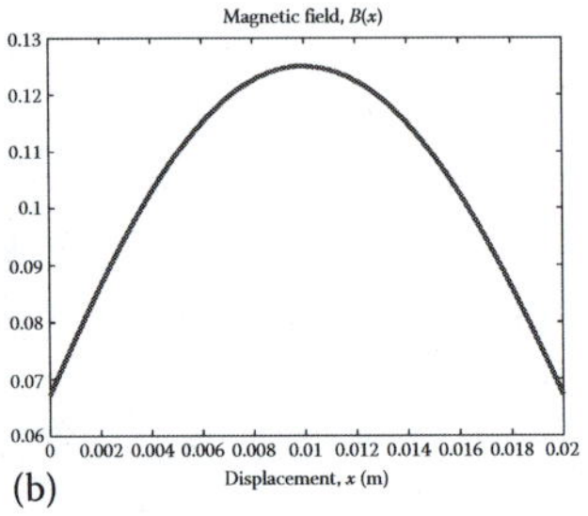

(b)

FIGURE 4.18 (a) Ironless loudspeakers with translational limited-displacement permanent-magnet DC actuators; (b) Plot of $B(x) = B_{max}\left[1 - \dfrac{1}{1+e^{-a\left(x-\frac{1}{2}(l_{max}-l_{min})\right)^2}}\right]$, B_{max} = 0.25 T, $l_{min} \le x \le l_{max}$, $0 \le x \le 0.02$ m, $x_0 = 0.01$ m, $a = 10000$.

A lightweight cone- or dome-shape diaphragm is connected to a rigid frame using a flexible suspension. A variety of different materials are used, such as coated paper, plastic, and composite materials. An N-turn winding (*voice coil*) is under the stationary magnetic field established by radially magnetized permanent magnets, as illustrated in Figure 4.18a. To displace a diaphragm, one applies the voltage to the winding. The electromagnetic force $\vec{F} = \oint_l i d\vec{l} \times \vec{B} = -i\oint_l \vec{B} \times d\vec{l}$ is produced. The suspension system maintains the coil centered within the gap and provides a restoring elastic force to make the cone return to an equilibrium position if voltage is not applied. Insulated copper and silver wire is used to fabricate a *voice coil* within a circular, rectangular, or hexagonal cross section. The coil is oriented coaxially inside the gap. Ceramic, ferrite, alnico, and rear-earth magnets are used.

The analysis, design, and optimization tasks can be performed. The electromagnetic force and *emf* are $\vec{F}=\oint_l i d\vec{l}\times\vec{B}=-i\oint_l \vec{B}\times d\vec{l}$ and $emf=\oint_l \vec{E}\cdot d\vec{l}=-\oint_s \frac{\partial \vec{B}}{\partial t}\cdot d\vec{s}$. Applying Kirchhoff's voltage law and Newtonian mechanics, one obtains the resulting equations. These loudspeaker and microphone equations are design dependent. There are various magnetic systems, geometry, and kinematics that are found to optimize the performance of electromagnetic motion devices. Piezoelectric and electrostatic speakers are also used in low-performance applications when the low cost is a factor.

Example 4.12

Consider a loudspeaker with a radially magnetized ring magnet as depicted in Figure 4.18a. The distribution of field *B*, as viewed from the coil, significantly affects the overall performance. The magnet is magnetized to ensure the desired field orientation. One approximates *B* using various real-valued continuous differentiable functions. For one-dimensional fields, trigonometric, exponential, sigmoid

$$B=B_{\max}\frac{1}{1+e^{-ax}},\quad B=B_{\max}\frac{1}{1+e^{-a|x|}},\quad B=B_{\max}\frac{1}{1+e^{-ax^2}},\quad B=B_{\max}\frac{1}{1+e^{-ax^3}},\quad a>0$$

and other functions are used depending on the magnet's magnetization, coil displacement with respect to the magnet, magnet and coil geometry and orientation, air gap, etc.

In equilibrium x_0, $\Sigma F = 0$. Within the constrained one-dimensional cone displacement $x_{\min}\le x\le x_{\max}$, the field $B(x)$ is used to find $F(x)=-i\oint_l B(x)dl$. Under the electromagnetic force, the cone displaces with respect to x_0. The field $B(x)$ may be measured in $x_{\min}\le x\le x_{\max}$, $l_{\min}\le l_{w\,coil}\le l_{\max}$.

There are different loudspeaker electromagnetic system designs. The widths of the coils and magnets $l_{w\,coil}$ and $l_{w\,magnet}$ can be $l_{w\,coil}>l_{w\,magnet}$, $l_{w\,coil}=l_{w\,magnet}$, or $l_{w\,coil}<l_{w\,magnet}$.

As an example, the plot for the axial field $B(x)$ on $x \in [0\quad 0.02]$, $B(x)=B_{\max}\left[1-\frac{1}{1+e^{-a\left(x-\frac{1}{2}(l_{\max}-l_{\min})\right)2}}\right]$, $B_{\max}=0.25$ T, $a=10{,}000$, and $0\le x\le 0.02$ m is depicted in Figure 4.18b. The MATLAB statement to perform calculations and plotting is

```
lmin=0; lmax=0.02; x=0:(lmax-lmin)/1000:lmax; Bmax=0.25; a=10000;
B=Bmax.*(1-1./(1+exp(-a*(x-(lmax-lmin)/2).^2))); plot(x,B,'LineWidth',3);
xlabel('Displacement {\itx} [m]','FontSize',18);
title('Magnetic Field {\itB}({\itx})','FontSize',18);
```

■

Using the Kirchhoff voltage law and Newtonian dynamics, for the one-dimensional case, one obtains

$$\frac{di_a}{dt}=\frac{1}{L_a}\left(-r_a i_a-\frac{d}{dt}\oint_s B(x)\cdot d\vec{s}+u_a\right),$$

$$\frac{dv}{dt}=\frac{1}{m}\left(i_a\oint_l B(x)dl-F_{air}-F_{elastic}-F_\xi\right),$$

$$\frac{dx}{dt}=v,\quad x_{\min}\le x\le x_{\max},$$

where F_{air} is the air friction force; $F_{elastic}$ is the elastic restoring force; F_ξ is the disturbance force.

For a preliminary design, one may apply the following approximations: $F_{elastic}=k_{elastic}x$ and $F_{air}=k_{air1}v+k_{air2}v|v|$, where $k_{elastic}$, k_{air1} and k_{air2} are the constants.

PRACTICE PROBLEMS

4.1 Consider a PM DC machine with $r_a = 1$ ohm. For $u_a = 50$ V, the angular velocity is 500 rad/sec. At no-load conditions, the current i_a is 0.1 A. Derive the value for the viscous friction coefficient B_m.

The torque-speed characteristics are given by $\omega_r = \dfrac{u_a - r_a i_a}{k_a} = \dfrac{u_a}{k_a} - \dfrac{r_a}{k_a^2}T$.

One finds the unknown k_a by using the expression $k_a = \dfrac{u_a - r_a i_a}{\omega_r}$.

We have k_a =0.0998 V-sec/rad. By estimating k_a assuming that the friction is negligible, we have $k_a = u_a/\omega_r = 50/500 = 0.1$ V-sec/rad.
At the steady state if $T_L = 0$, $T_e = T_{friction}$. Hence, $k_a i_a = B_m \omega_r$.
One finds $B_m = k_a i_a/\omega_r = 0.0998 \times 0.1/500 = 1.996 \times 10^{-5}$ N-m-sec/rad.

4.2 Model an axial topology limited angle actuator if $B(\theta_r) = B_{\max} k\theta_r$. The angular displacement is bounded, $-0.175 \le \theta_r \le 0.175$ rad, and, $\theta_{L0} = \theta_{R0} = 0.175$ rad. The parameters are $B_{\max} = 1$, $k = 5$, $r_a = 35$, $L_a = 4.1 \times 10^{-3}$, $R_\perp = 0.02$ m, $N = 100$, $l_{eq} = 0.0125$ m, $B_m = 5 \times 10^{-4}$, $k_s = 0.05$, $J = 1.5 \times 10^{-6}$ kg-m², $\theta_{R0} = 0.175$ rad, $r_{in} = 0.015$ m, $r_{out} = 0.025$ m, $A_{eq} = \frac{1}{2}\left(r_{out}^2 - r_{in}^2\right)$.

For $B(\theta_r) = B_{\max} k\theta_r$, one finds

$$\mathscr{E} = -N\frac{d}{dt}\int_{r_{in}}^{r_{out}}\int_{\theta_R}^{\theta_L} B_{\max}k\theta_r r\, dr\, d\theta_i = -\frac{r_{out}^2 - r_{in}^2}{2}NB_{\max}k\left(\theta_L - \theta_R\right)\omega_r,$$

$$\theta_L(t) = \theta_{L0} - \theta_r(t) \quad \text{and} \quad \theta_R(t) = \theta_{R0} + \theta_r(t).$$

Thus, $\mathscr{E} = -\dfrac{r_{out}^2 - r_{in}^2}{2}NB_{\max}k\left[\left(\theta_{L0} - \theta_r\right) - \left(\theta_{R0} + \theta_r\right)\right]\omega_r = \left(r_{out}^2 - r_{in}^2\right)NB_{\max}k\theta_r\omega_r,\ \theta_{L0} = \theta_{R0}$.

The electromagnetic torque is $T_e = R_\perp N l_{eq} B_{\max} k(\theta_L + \theta_R) i_a = 2R_\perp N l_{eq} B_{\max} k\theta_{R0} i_a$.

From the Kirchhoff voltage law and Newton's second law, one has

$$\frac{di_a}{dt} = \frac{1}{L_a}\left[-r_a i_a - \left(r_{out}^2 - r_{in}^2\right)NB_{\max}k\theta_r\omega_r + u_a\right],$$

$$\frac{d\omega_r}{dt} = \frac{1}{J}\left[2R_\perp N l_{eq} B_{\max} k\theta_{R0} i_a - T_{friction} - T_L\right], \quad T_{friction} = B_m\omega_r, \quad T_{elastic} = k_s\theta_r,$$

$$\frac{d\theta_r}{dt} = \omega_r, \quad -\theta_{r\max} \le \theta_r \le \theta_{r\max}.$$

We upload the parameters and constants as

```
Bmax=1; k=5; ra=35; La=4.1e-3; Rp=0.02; N=100; leq=1.25e-2; Bm=5e-4; ks=0.05; J=1.5e-6;
TR0=0.175; Rin=0.015; Rou=0.025; Aeq=(Rou^2-Rin^2)/2;
```

The Simulink model is reported in Figure 4.19a. The transient dynamics for $i_a(t)$ and $\theta_r(t)$ is depicted in Figure 4.19b if u_a is applied as steps ±5 V with $f = 10$ Hz.

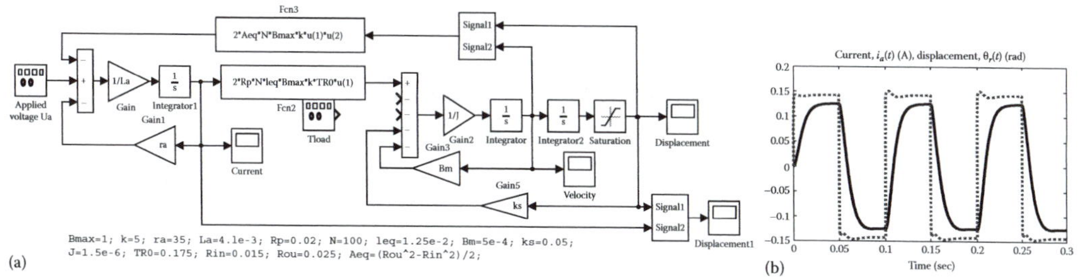

FIGURE 4.19 (a) Simulink® model to simulate the axial topology limited angle actuator, $B(\theta_r) = B_{max}k\theta_r$; (b) Transient dynamics for $i_a(t)$ and $\theta_r(t)$ (dotted and solid lines, respectively).

HOMEWORK PROBLEMS

4.1 Consider a PM DC generator that is rotated by a prime mover with $\omega_r = 100$ rad/sec, and, $T_{PM} = 1$ N-m.
The armature voltage at the generator terminal $u_{a\,terminal}$ is 10 V, while the current i_a is 8 A.
The armature resistance r_a is 0.1 ohm.
Find the efficiency. Calculate k_a.

4.2 In Section 2.5 (Examples 2.17 and 2.18), adequate friction models were reported with the experimental substantiations. The physics-consistent $T_{viscous}(\omega_r)$ and $B_m(\omega_r)$ may be found in an operating envelope. Let in $\omega_r \in [\omega_{r\min} \quad \omega_{r\max}]$, $B_m(\omega_r) = ae^{-b\omega_r}$, $a > 0$, $b > 0$. Report the mechanical and *total* losses. If T_L is not measured, document the equation for η.

4.3 Perform simulations in MATLAB of a 30 V, 10 A, 300 rad/sec permanent-magnet DC motor. The motor parameters are as follows: $r_a = 1$ ohm, $k_a = 0.1$ V-sec/rad, $L_a = 0.005$ H, $B_m = 0.0001$ N-m-sec/rad, and $J = 0.0001$ kg-m^2. Analyze losses for the unloaded and loaded motor, examine efficiency, and study the dynamics.

4.4 Study an axial topology limited angle actuator. Let one measure $B(\theta_r) = B_{\max}\dfrac{\theta_r}{\sqrt{a+\theta_r^2}}$, $a > 0$, $a \ll 1$. This $B(\theta_r)$ was studied in Example 4.7, and, plotted in Figure 4.16b. A continuous $B(\theta_r) = B_{\max}\dfrac{\theta_r}{\sqrt{a+\theta_r^2}}$ is relevant to the piece-wise continuous sign function $B(\theta_r) = B_{\max}\operatorname{sgn}(\theta_r)$, which cannot be ensured. Derive a mathematical model.

REFERENCES

1. S. J. Chapman, *Electric Machinery Fundamentals*, McGraw-Hill, New York, 2011.
2. A. E. Fitzgerald, C. Kingsley, and S. D. Umans, *Electric Machinery*, McGraw-Hill, New York, 2003.
3. P. C. Krause and O. Wasynczuk, *Electromechanical Motion Devices*, McGraw-Hill, New York, 1989.
4. P. C. Krause, O. Wasynczuk, S. D. Sudhoff, and S. Pekarek, *Analysis of Electric Machinery*, Wiley-IEEE Press, New York, 2013.
5. S. E. Lyshevski, *Electromechanical Systems, Electric Machines, and Applied Mechatronics*, CRC Press, Boca Raton, FL, 1999.
6. S. E. Lyshevski, *Electromechanical Systems and Devices*, CRC Press, Boca Raton, FL, 2008.
7. G. R. Slemon, *Electric Machines and Drives*, Addison-Wesley Publishing Company, Reading, MA, 1992.
8. D. C. White and H. H. Woodson, *Electromechanical Energy Conversion*, Wiley, New York, 1959.

5 Induction Motors

5.1 INTRODUCTION AND FUNDAMENTALS

In high-performance drives and servos, permanent-magnet electric machines are the preferable choice. Robust low-cost fractional horsepower single- and three-phase induction machines are widely used [1–6]. In industrial applications, three-phase induction motors are very effectively used in medium- and high-power drives up to thousands kilowatts. Compared with permanent-magnet machines with the rated power limit of ~100 kW, induction motors have lower torque and power densities and may not be effective as generators. However, induction motors guarantee high power and require simple power electronics. Induction motors were invented and demonstrated by Nicola Tesla in the 1880s. He made indispensable contributions to science, engineering, and technology by inventing AC electric machines (synchronous and induction machines), transmission lines, transformers, radars, wireless communication, etc. Tesla pioneered, developed, and commercialized the theory and technology of time-varying AC electromagnetics. He controlled induction motors by polyphase AC voltage systems, demonstrating the speed control capabilities in the late 1880s. Nicola Tesla designed and demonstrated two-phase induction motors in 1883. Three-phase squirrel-cage induction motors were demonstrated by Mickail Dolivo-Dobrovolsky in the 1890s. In induction motors, the electromagnetic torque results due to the interaction of the time-varying electromagnetic fields. The images of an induction and permanent-magnet machines are illustrated in Figures 5.1.

The phase voltages are supplied to the stator windings. In squirrel-cage induction motors, the voltages in the short-circuited rotor windings are induced due to time-varying stator magnetic field and motion of the rotor with respect to the stator. To design electric machines, three-dimensional electromagnetic, mechanical, thermal, vibroacoustic, and structural designs are performed. This chapter focuses on consistent analysis, which includes modeling, simulation, performance evaluation, capabilities assessment, and control. The analyses in the *machine* (*abc* phase) variables and in the *quadrature-direct* quantities are covered. Though the *quadrature-direct* premise can be used in modeling, this concept may not offer advantages because AC machines are controlled supplying the *abc* voltages. One varies the phase stator and rotor voltages u_{as}, u_{bs}, u_{cs}, and u_{ar}, u_{br}, u_{cr}. In squirrel-cage induction motors, one changes u_{as}, u_{bs}, u_{cs} because the rotor windings are short-circuited. The highest acceleration capabilities and minimal settling time are achieved using the frequency control. To reduce the losses, the voltage–frequency control is applied. These control concepts are implemented using power electronic hardware.

5.2 TORQUE–SPEED CHARACTERISTICS AND CONTROL OF INDUCTION MOTORS

5.2.1 Torque–Speed Characteristics

The angular velocity of induction motors must be controlled. We study the torque–speed characteristics $\omega_r = \Omega_T(T_e)$. The electromagnetic torque is a function of the stator and rotor currents. Induction motors are controlled by changing the frequency f and magnitude u_M of the voltages supplied to the phase windings. The magnitude of the voltages applied to the stator windings cannot exceed the rated voltage $u_{M\max}$, and $u_{M\min} \le u_M \le u_{M\max}$. The angular frequency of the applied phase voltages is $\omega_f = 2\pi f$, $f_{\min} \le f \le f_{\max}$, $f_{\min} > 0$.

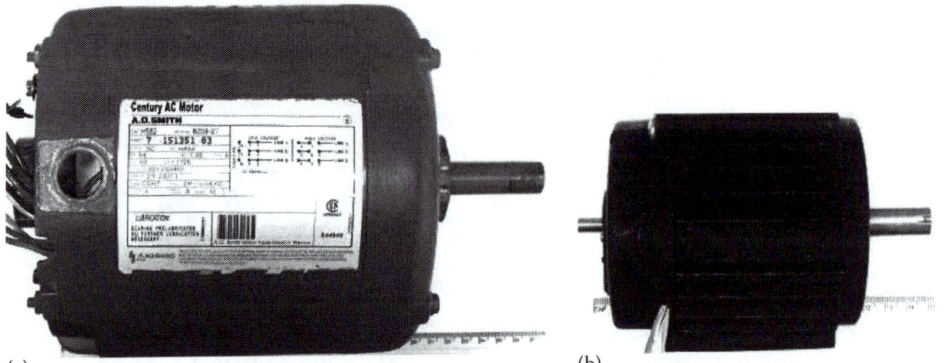

(a) (b)

FIGURE 5.1 (a) A 3/4 horsepower (0.56 kW) induction motor; (b) A 0.5 kW permanent-magnet synchronous electric machines (NEMA 56 frame size induction motors and a NEMA 34 frame size synchronous machine, NEMA 56 size means 5.6 in. or 142 mm diameter).

The synchronous angular velocity ω_e of induction machines is a function of f. Using the number of poles P, $\omega_e = 4\pi f/P$. The electrical angular velocity of induction motors ω_r is less or equal (at no load and no friction) to ω_e, $\omega_r \leq \omega_e$. In contrast, synchronous motors rotate at ω_e, $\omega_r = \omega_e$. The steady-state torque–speed characteristics $\omega_r = \Omega_T(T_e)$ are found by plotting the angular velocity versus the electromagnetic torque. The National Electric Manufacturers Association (NEMA) in the USA and the International Electromechanical Commission (IEC) in Europe define four basic classes A, B, C, and D of induction motors. Typical steady-state torque–speed characteristics are depicted in Figure 5.2a, where the *slip* is

$$slip = \frac{\omega_e - \omega_r}{\omega_e}, \quad \omega_e = \frac{4\pi f}{P}, \quad \text{and} \quad \omega_r = \left(1 - slip\right)\omega_e.$$

The torque–speed characteristics can be found using the experimental data by measuring the torque T_e and angular velocity ω_r. The steady-state $\omega_r = \Omega_T(T_e)$ can be derived by *averaging* the measured experimental dynamic characteristics $\omega_r(t) = \Omega_T[T_e(t)]$. If $T_e(t)$ is not directly measurable, one measures or observes the phase currents and angular displacement obtaining $T_e(t)$.

The motor angular velocity is found at the intersection of the torque–speed $\omega_r = \Omega_T(T_e)$ and load $T_L(\omega_r)$ characteristics as illustrated in Figure 5.2b. From the Newton second law, neglecting the friction, we have

$$\frac{d\omega_r}{dt} = \frac{1}{J}(T_e - T_L).$$

Hence, $\omega_r = \text{const}$ if $T_e = T_L$. If $T_e > T_L$ and $T_{e\,start} > T_{L0}$, the motor accelerates reaching ω_r at which $T_e = T_L$. The critical angular velocity $\omega_{r\,critical}$ is shown in Figure 5.2b. The maximum torque is $T_{e\max} = T_{e\,critical\,\max}$. The motor operating envelope is

$$\omega_r \in \left[\omega_{r\,critical}(f, u_M) \quad \omega_e(f)\right], \quad T_e \in \left[0 \quad T_{e\max}\left(f, u_M\right)\right], \quad u_M \in \left[u_{M\min} \quad u_{M\max}\right],$$

$$f \in \left[f_{\min} \quad f_{\max}\right], \quad T_e\left(f, u_M\right) \geq T_L, \quad \forall T_L.$$

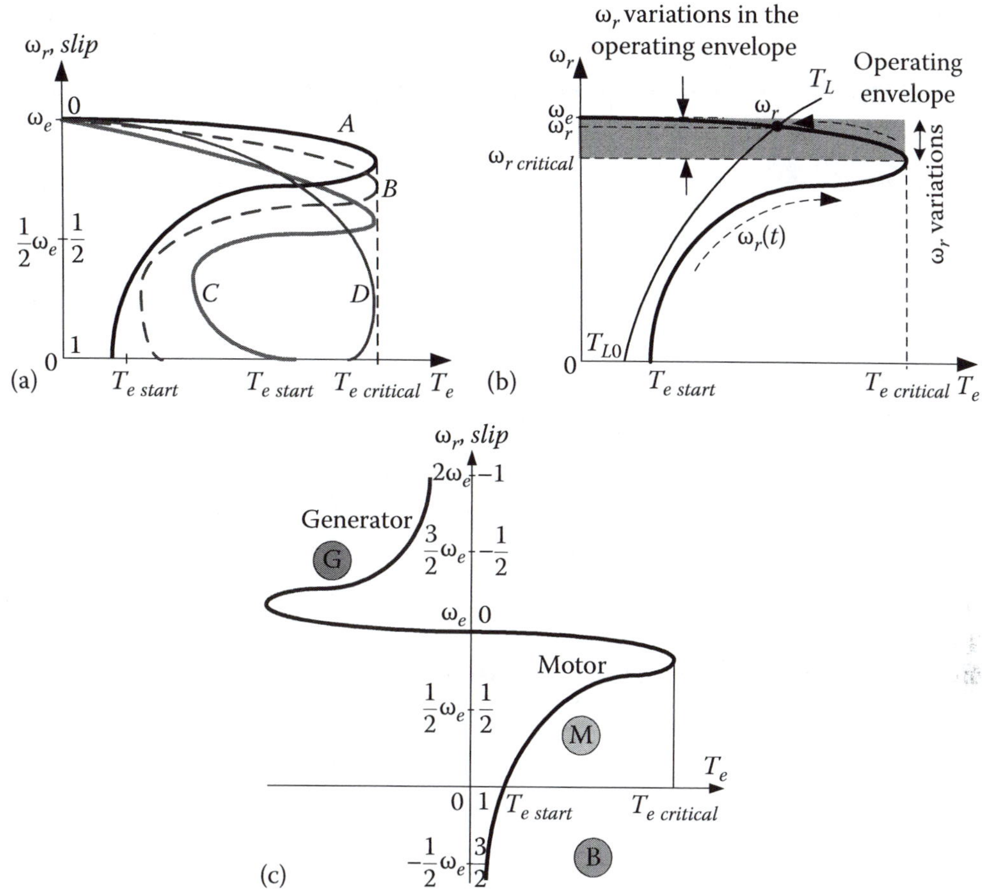

FIGURE 5.2 (a) Typical torque–speed characteristics of the A, B, C, and D class induction motors; (b) Torque–speed and load curves: Motor accelerates to rotate at the angular velocity ω_r, $\omega_r \le \omega_e$; (c) Torque–speed characteristics in the motor (M), generator (G), and breaking (B) regions.

The industrial induction motors are usually A or B class machines. These motors have normal starting torque and low *slip*, which is ~0.05. The C class induction motors have higher starting torque due to double-rotor design, and *slip* > 0.05. The D class induction motors have high rotor resistance and exhibit high starting torque. Typically, the *slip* of the D class induction motors is from 0.5 to 0.9. The E and F class induction motors have very low starting torque, and the rotor bars are deeply buried, resulting in high leakage inductances. Figure 5.2c depicts the torque–speed characteristic of A and B class induction machines in the motor, generator, and braking regions.

Experimentally, one measures $\omega_r(t)$ and $T_e(t)$, yielding $\omega_r(t) = \Omega_T[T_e(t)]$. Under assumptions and simplifications, the steady-state torque–speed characteristics can be obtained by using the equivalent circuits. Assigning different values for magnitude u_M and frequency f of the phase voltage, one applies

$$T_e = \frac{3\left(u_M \dfrac{X_M}{X_s + X_M}\right)^2 \dfrac{r_r'}{slip}}{\omega_e \left[\left(r_s\left(\dfrac{X_M}{X_s + X_M}\right)^2 + \dfrac{r_r'}{slip}\right)^2 + \left(X_s + X_r'\right)^2\right]}, \quad slip = \frac{\omega_e - \omega_r}{\omega_e}, \quad \omega_e = \frac{4\pi f}{P},$$

where X_s and X_r' are the stator and rotor reactances; X_M is the magnetizing reactance.

Note: The complex impedance is $\mathbf{Z} = |\mathbf{Z}|\angle\phi = R + jX$, where the *resistance* is $R = \text{Re}\mathbf{Z} = |\mathbf{Z}|\cos\phi$ and the reactance is $X = \text{Im}\mathbf{Z} = |\mathbf{Z}|\sin\phi$. The impedance is inductive if $X > 0$ and capacitive if $X < 0$. In general, $X = \omega L - \frac{1}{\omega C}$. Using the magnetic inductance L, the magnetic reactance is $X = \omega L$.

Example 5.1

Calculate and plot the torque–speed characteristics for four-pole induction motors. The parameters are: (1) $r_s = 24.5$ ohm, $r_r' = 23$ ohm, $X_s = 10$ ohm, $X_r' = 40$ ohm, and $X_M = 25$ ohm; (2) $r_s = 24.5$ ohm, $r_r' = 23$ ohm, $X_s = 50$ ohm, $X_r' = 200$ ohm, and $X_M = 125$ ohm. The rated voltage is $u_{M\,\max} = 110$ V. The maximum frequency of the supplied phase voltages is $f_{\max} = 60$ Hz.

Let the phase voltages are supplied with frequencies 20, 40, and 60 Hz. For each f, the synchronous angular velocity is $\omega_e = \frac{4\pi f}{P}$. The torque–speed characteristics are found assigning different values for f, and MATLAB® is used to calculate and plot $\omega_r = \Omega_T(T_e)$

```
clear all
P=4; rs=24.5; rr=23; Xs=50; Xr=200; Xm=125; % parameters of an induction motor
uM=80; f=60; we=4*pi*f/P;
% calculation of a torque-speed characteristic
for wr=[1:0.25:4*pi*f/P];
% angular velocity slip=(we-wr)/we;
% slipTe=3*(uM*Xm/(Xs+Xm))^2*(rr/slip)/(we*((rs*(Xm/(Xs+Xm))^2+rr/slip)^2+(Xs+Xr)^2));
plot(Te,wr,'o'); title('Torque-Speed Characteristics','FontSize',18);
xlabel('Electromagnetic Torque {\itT_e}, N-m','FontSize',18);
ylabel('Angular Velocity {\it\omega_r}, rad/sec','FontSize',18);
hold on; end;
```

The resulting torque–speed characteristics are documented in Figures 5.3a and b by changing the frequency f to be 20, 40, and 60 Hz. Figure 5.3c depicts the torque–speed characteristics if u_M is 50, 80, and 110 V if $f = 60$ Hz. One observes the frequency and voltage control principles. Using $\omega_r = \Omega_T(T_e)$, one assesses the acceleration capabilities, control, etc. The frequency control is a key principle in high-performance electric drives with induction motors. The starting $T_{e\,start}$ is maximum at the minimum frequency $f_{\min}$, and $T_{e\,start}$ is found. By changing the frequency f, one controls ω_r and T_e. ■

5.2.2 Control of Induction Motors

To control induction motors, one must vary the angular velocity ω_r and electromagnetic torque T_e. To vary ω_r and T_e, one changes the frequency and magnitude of phase voltages applied to the windings. The electromagnetic torque T_e in two- and three-phase induction motors is given by (5.8) or (5.26)

$$T_e = -\frac{P}{2} L_{ms}\left[\left(i_{as}i'_{ar} + i_{bs}i'_{br}\right)\sin\theta_r + \left(i_{as}i'_{br} - i_{bs}i'_{ar}\right)\cos\theta_r\right],$$

$$T_e = -\frac{P}{2} L_{ms}\left[\left(i_{as}i'_{ar} + i_{bs}i'_{br} + i_{cs}i'_{cr}\right)\sin\theta_r + \left(i_{as}i'_{cr} + i_{bs}i'_{ar} + i_{cs}i'_{br}\right)\sin\left(\theta_r - \frac{2}{3}\pi\right)\right.$$
$$\left. + \left(i_{as}i'_{br} + i_{bs}i'_{cr} + i_{cs}i'_{ar}\right)\sin\left(\theta_r + \frac{2}{3}\pi\right)\right].$$

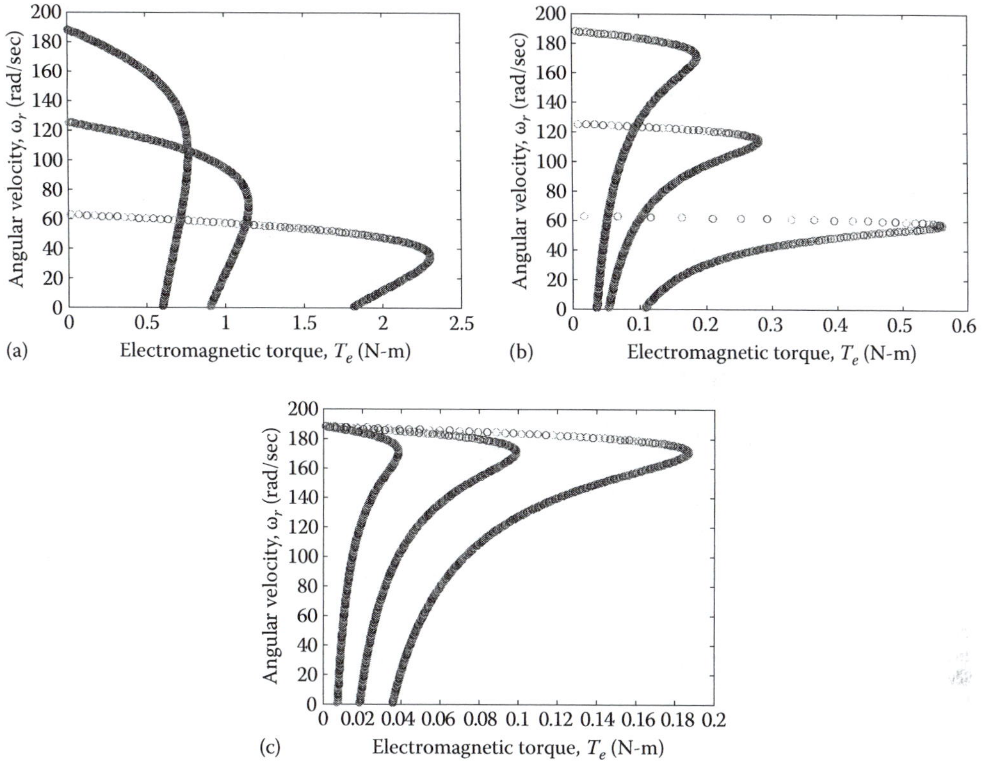

FIGURE 5.3 Torque–speed characteristics of three-phase induction motors:
(a) f is 20, 40, and 60 Hz ($r_s = 24.5$ ohm, $r_r' = 23$ ohm, $X_s = 10$ ohm, $X_r' = 40$ ohm, $X_M = 25$ ohm);
(b) f is 20, 40, and 60 Hz ($r_s = 24.5$ ohm, $r_r' = 23$ ohm, $X_s = 50$ ohm, $X_r' = 200$ ohm, $X_M = 125$ ohm);
(c) u_M is 50, 80, and 110 V, $f = 60$ Hz ($r_s = 24.5$ ohm, $r_r' = 23$ ohm, $X_s = 50$ ohm, $X_r' = 200$ ohm, $X_M = 125$ ohm).

To guarantee the balanced operation of two-phase induction motors, one applies the AC phase voltages with frequency f to the stator windings

$$u_{as}(t) = \sqrt{2}u_M \cos(\omega_f t), \quad u_{bs}(t) = \sqrt{2}u_M \sin(\omega_f t), \quad \omega_f = 2\pi f, \quad u_{M\min} \le u_M \le u_{M\max}.$$

The sinusoidal steady-state phase currents are

$$i_{as}(t) = \sqrt{2}i_M \cos(\omega_f t - \varphi_i), \quad i_{bs}(t) = \sqrt{2}i_M \sin(\omega_f t - \varphi_i), \quad \omega_f = 2\pi f, \quad i_{M\min} \le i_M \le i_{M\max}.$$

Here, u_M and i_M are the magnitudes of the *as* and *bs* stator voltages and currents; φ_i is the phase difference.

For three-phase induction motors,

$$u_{as}(t) = \sqrt{2}u_M \cos(\omega_f t), \quad u_{bs}(t) = \sqrt{2}u_M \cos\left(\omega_f t - \frac{2}{3}\pi\right), \quad u_{cs}(t) = \sqrt{2}u_M \cos\left(\omega_f t + \frac{2}{3}\pi\right),$$
$$\omega_f = 2\pi f, \quad u_{M\min} \le u_M \le u_{M\max}.$$

The applied voltage to the motor windings cannot exceed the rated voltage, $u_{M\min} \le u_M \le u_{M\max}$. The synchronous angular velocity is $\omega_e = 4\pi f/P$, where $f_{\min} \le f \le f_{\max}$ due to the power electronics limits on f_{min} and the mechanical limits on the maximum angular velocity $\omega_{r\,max}$, which yields f_{max}. To vary ω_r, one can change both the magnitude of the applied voltages u_M and frequency f. The control concepts are application specific and depend on the induction motor class and power electronics.

Voltage control: By changing the magnitude u_M of the applied phase voltages to the stator windings, the angular velocity is regulated in the stable operating region, see Figure 5.4a. It was emphasized that $u_{M\,min} \le u_M \le u_{M\,max}$. By reducing u_M, one reduces $T_{e\,start}$ and $T_{e\,critical}$. The operating envelope for the angular velocity is $\omega_r \in \left[\omega_{rcritical} \quad \omega_e\right]$. For A, B, and C class induction motors, one is not able to effectively regulate the angular velocity by changing u_M. The voltage control is applicable for low-efficiency D class induction motors, frequently used in household items to avoid the use of power converters.

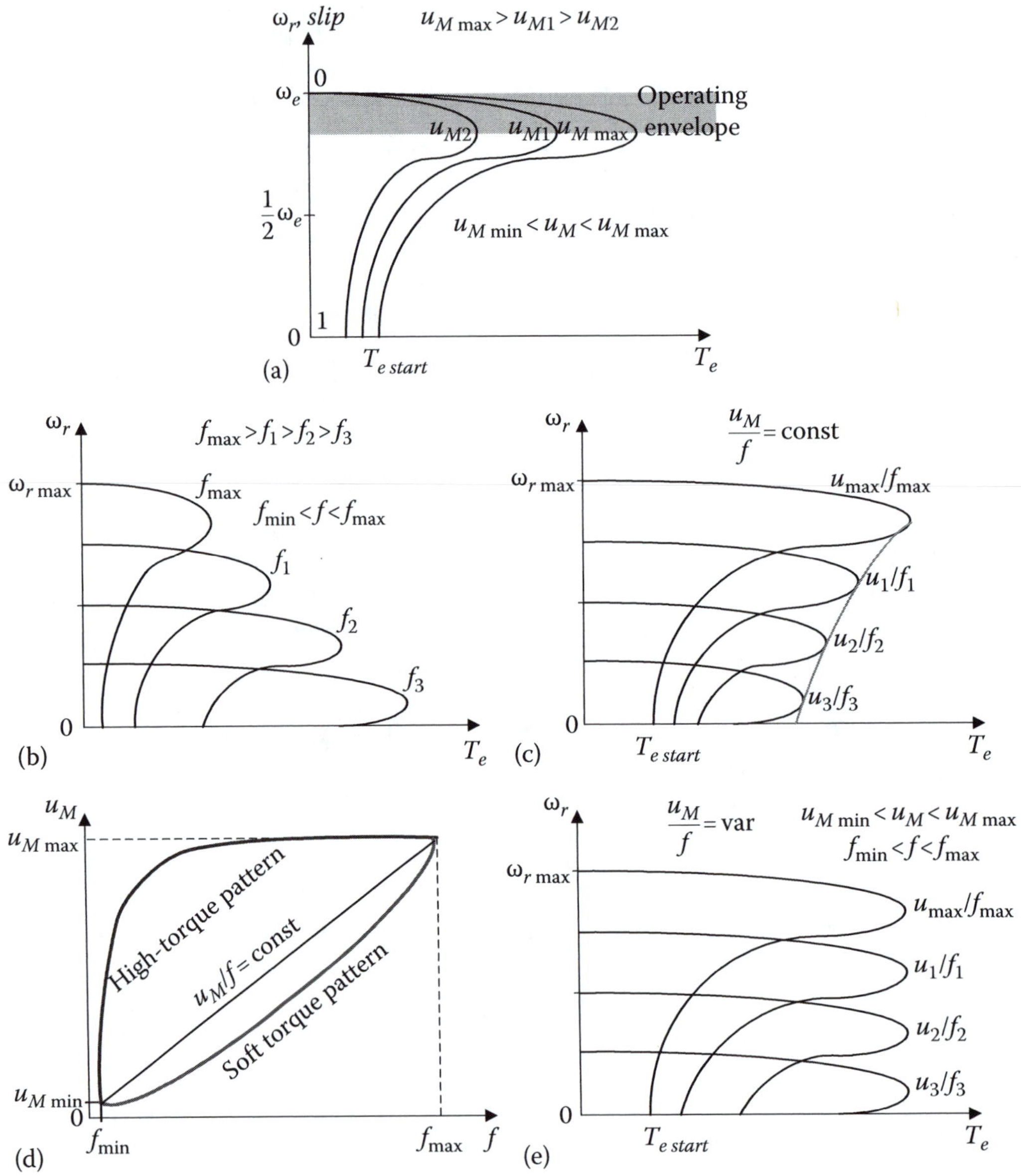

FIGURE 5.4 Torque–speed characteristics $\omega_r = \Omega_T(T_e)$: (a) Voltage control; (b) Frequency control; (c) Voltage–frequency control: *Constant-volts-per-hertz* control; (d) Voltage–frequency patterns, $u_M = \phi(f)$; (e) Varying voltage–frequency control.

Frequency control: The magnitude of the supplied phase voltages is constant u_M = const, and the angular velocity is regulated by varying the voltage frequency f, $f_{\min} \le f \le f_{\max}$. Power electronics may ensure $f_{\min}$ ~2 Hz. By varying f, one changes the synchronous angular velocity $\omega_e = 4\pi f/P$. The torque–speed characteristics for different values of f are shown in Figure 5.4b.

Voltage–frequency control: To minimize losses, the voltage magnitude u_M and frequency f are regulated. The *constant-volts-per-hertz* control is ensured if u_M is proportional to f, e.g., u_M/f = const. The resulting torque–speed characteristics are documented in Figure 5.4c.

Defining the voltage–frequency patterns, one may apply $\sqrt{u_M/f}$ = const to change u_M depending on f. To attain the required acceleration, settling time, and other specifications, the general-purpose *constant-volts-per-hertz* control, soft-starting, and high-starting torque patterns are illustrated in Figure 5.4d. Assigning $u_M = \phi(f)$ with domain $f_{\min} \le f \le f_{\max}$ and range $u_{M\min} \le u_M \le u_{M\max}$, one has u_M/f = var. The desired torque–speed characteristics, documented in Figure 5.4e, can be ensured using

$$\frac{u_M}{f} = \text{const}, \quad \sqrt{\frac{u_M}{f}} = \text{const}, \quad \text{or} \quad \frac{u_M}{f} = \text{var}.$$

Alternative control schemes—Control of induction motors using arbitrary reference frame quantities and vector control: Induction motors can be analyzed and controlled using the *quadrature, direct,* and *zero* ($qd0$) axis components of voltages, currents, and flux linkages. This concept was developed in the 1930s to reduce the mathematical complexity in analysis, and allow analytic solutions of the resulting differential equations. For motor control, practicality of $qd0$ quantities is debatable. Induction, synchronous, and DC motors are open-loop stable. One does not experience any difficulties in controlling these motors. Induction and synchronous machines can be analyzed in the *arbitrary* reference frame, which rotates with the angular velocity ω. The stationary, rotor, and synchronous frames are used. The synchronous reference frame "rotates" with ω_e. The $qd0$-axis components of stator and rotor AC voltages, currents, and flux linkages become DC quantities in the synchronous reference frames. In synchronous reference frames, applying the Park transformation, one finds $u_{qs}^e = u_M$, $u_{ds}^e = 0$, and $u_{0s}^e = 0$. These $qd0$ quantities mathematically correspond to the physical *machine* variables. One does not directly measure or observe the $qd0$-axis components. The directly measured currents and voltages are i_{as}, i_{bs}, i_{cs} and u_{as}, u_{bs}, u_{cs}. The phase voltages u_{as}, u_{bs}, and u_{cs} must be applied to the stator windings.

One cannot apply the mathematical operators u_{qs}^e, u_{ds}^e, and u_{0s}^e to the phase windings. If one attempts to use the $qd0$ quantities $\mathbf{u}_{qd0s}^e$, the Park transformation is applied to compute $\mathbf{u}_{abcs} = (u_{as}, u_{bs}, u_{cs})$ in real-time

$$\mathbf{u}_{abcs} = (\mathbf{K}_s^e)^{-1}\mathbf{u}_{qd0s}^e, \quad \mathbf{K}_s^e = \frac{2}{3}\begin{bmatrix} \cos\theta_e & \cos\left(\theta_e - \frac{2}{3}\pi\right) & \cos\left(\theta_e + \frac{2}{3}\pi\right) \\ \sin\theta_e & \sin\left(\theta_e - \frac{2}{3}\pi\right) & \sin\left(\theta_e + \frac{2}{3}\pi\right) \\ \frac{1}{2} & \frac{1}{2} & \frac{1}{2} \end{bmatrix}.$$

The specialized digital signal processors and other hardware are needed. The frequency f is varied by the power converters. While the *vector* control of induction motors can be applied, a limited practical benefit may be expected. The variable voltage–frequency control

$$\frac{u_{Mi}}{f_i} = \text{var}, \quad u_{M\min} \le u_M \le u_{M\max}, \quad f_{\min} \le f \le f_{\max},$$

guarantees the high-torque patterns, which surpass the capabilities of the *vector* control or other $qd0$-centric concepts [6]. The highest $T_{e\,start}$ and $T_{e\,critical}$ are developed using the well-established frequency control. The $T_{e\,start\,\max}$ corresponds to $f_{\min}$. This $f_{\min}$ is defined by the converter topology, solid-state devices, efficiency, etc. Control in the *machine* variables and practical concepts are prioritized.

5.3 TWO-PHASE INDUCTION MOTORS

5.3.1 Modeling of Two-Phase Induction Motors

We study a two-phase induction motor as illustrated in Figure 5.5. The stator (*as* and *bs*) and rotor (*ar* and *br*) windings, and, the stator–rotor magnetic coupling, are depicted. To rotate squirrel-cage induction motors and control the angular velocity, one may vary the frequency as well as the magnitude of the phase voltages u_{as} and u_{bs} supplied to the stator windings. For the wound-rotor induction motors, one also may vary the voltages u_{ar} and u_{br} supplied to the rotor windings. In induction machines, the *motional emfs* (voltages) are induced in the rotor windings due to the time-varying stator magnetic field and motion of the rotor with respect to the stator. The electromagnetic torque results due to the interaction of time-varying electromagnetic fields.

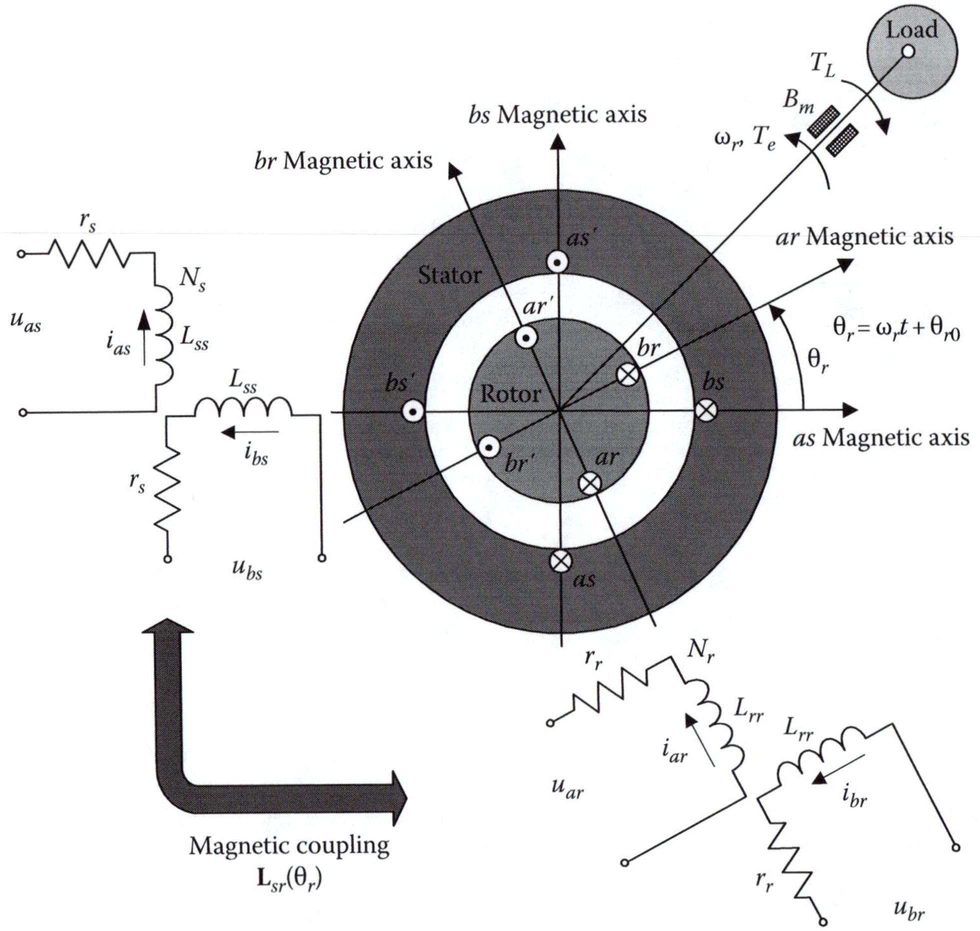

FIGURE 5.5 Two-phase symmetric induction motor.

To derive the governing equations, we examine the stator–rotor circuitry-electromagnetic and *torsional–mechanical* dynamics. The Kirchhoff voltage law relates the *as*, *bs*, *ar*, and *br* voltages, currents, and flux linkages. We have

$$u_{as} = r_s i_{as} + \frac{d\psi_{as}}{dt}, \quad u_{bs} = r_s i_{bs} + \frac{d\psi_{bs}}{dt},$$
$$u_{ar} = r_r i_{ar} + \frac{d\psi_{ar}}{dt}, \quad u_{br} = r_r i_{br} + \frac{d\psi_{br}}{dt}, \tag{5.1}$$

where u_{as} and u_{bs} are the phase voltages supplied to the *as* and *bs* stator windings; u_{ar} and u_{br} are the phase voltages supplied to the *ar* and *br* rotor windings in wound induction motors (in squirrel-cage motors, the rotor windings are short-circuited, u_{ar}, u_{br} are not supplied, however, the motional *emfs* are induced in the *ar* and *br* windings); i_{as} and i_{bs} are the phase currents in the stator windings; i_{ar} and i_{br} are the phase currents in the rotor windings; ψ_{as} and ψ_{bs} are the stator flux linkages; ψ_{ar} and ψ_{br} are the rotor flux linkages; r_s and r_r are the resistances of the stator and rotor winding.

Using the vector notations, differential equations (5.1) are expressed as

$$\mathbf{u}_{abs} = \mathbf{r}_s \mathbf{i}_{abs} + \frac{d\boldsymbol{\psi}_{abs}}{dt}, \quad \mathbf{u}_{abr} = \mathbf{r}_r \mathbf{i}_{abr} + \frac{d\boldsymbol{\psi}_{abr}}{dt}, \tag{5.2}$$

where $\mathbf{u}_{abs} = \begin{bmatrix} u_{as} \\ u_{bs} \end{bmatrix}$, $\mathbf{u}_{abr} = \begin{bmatrix} u_{ar} \\ u_{br} \end{bmatrix}$, $\mathbf{i}_{abs} = \begin{bmatrix} i_{as} \\ i_{bs} \end{bmatrix}$, $\mathbf{i}_{abr} = \begin{bmatrix} i_{ar} \\ i_{br} \end{bmatrix}$, $\boldsymbol{\psi}_{abs} = \begin{bmatrix} \psi_{as} \\ \psi_{bs} \end{bmatrix}$, and $\boldsymbol{\psi}_{abr} = \begin{bmatrix} \psi_{ar} \\ \psi_{br} \end{bmatrix}$ are the vectors of phase voltages, currents, and flux linkages; $\mathbf{r}_s = \begin{bmatrix} r_s & 0 \\ 0 & r_s \end{bmatrix}$ and $\mathbf{r}_r = \begin{bmatrix} r_r & 0 \\ 0 & r_r \end{bmatrix}$ are the matrices of the stator and rotor resistances.

The flux linkages are expressed using the self- and mutual inductances as

$$\psi_{as} = L_{asas} i_{as} + L_{asbs} i_{bs} + L_{asar} i_{ar} + L_{asbr} i_{br}, \quad \psi_{bs} = L_{bsas} i_{as} + L_{bsbs} i_{bs} + L_{bsar} i_{ar} + L_{bsbr} i_{br},$$

$$\psi_{ar} = L_{aras} i_{as} + L_{arbs} i_{bs} + L_{arar} i_{ar} + L_{arbr} i_{br}, \quad \psi_{br} = L_{bras} i_{as} + L_{brbs} i_{bs} + L_{brar} i_{ar} + L_{brbr} i_{br},$$

where L_{asas}, L_{bsbs}, L_{arar}, and L_{brbr} are the self-inductances of the stator and rotor windings; L_{asbs}, L_{asar}, L_{asbr}, L_{bsar}, L_{bsbr}, and L_{arbr} are the mutual inductances between the corresponding stator and rotor windings which are denoted using the corresponding subscripts, $L_{asbs} = L_{bsas}$, $L_{asar} = L_{aras}$, etc.

Assume that the magnetic system is linear. The stator and rotor self-inductances are denoted as L_{ss} and L_{rr}. Hence, $L_{ss} = L_{asas} = L_{bsbs}$ and $L_{rr} = L_{arar} = L_{brbr}$. The stator (*as* and *bs*) and rotor (*ar* and *br*) windings are orthogonal. There are no magnetic coupling between the *as* and *bs*, as well as between *ar* and *br* windings. Hence, for the mutual inductances we have $L_{asbs} = L_{bsas} = 0$ and $L_{arbr} = L_{brar} = 0$.

We examine the magnetic coupling between stator and rotor windings. The mutual inductances are periodic functions of the electrical angular displacement of rotor θ_r. The stator–rotor mutual inductances have minimum and maximum values. As illustrated in Figure 5.5, the stator is the stationary member, while the rotor rotates with the electrical angular velocity ω_r. The magnetic coupling is periodic with a period 2π. The maximum coupling occurs when the *ar* rotor winding is below the *as* stator winding. Induction motors are designed to ensure the sinusoidal mutual inductances between stator and rotor windings. The mutual inductances depend on the initial position of rotor with respect to stator. However, the resulting equations are same. Let the initial position of *as* with respect to *ar* be ensuring

$L_{asar} = L_{aras} = L_{sr}\cos\theta_r$, $L_{asbr} = L_{bras} = -L_{sr}\sin\theta_r$ and $L_{bsar} = L_{arbs} = L_{sr}\sin\theta_r$, $L_{bsbr} = L_{brbs} = L_{sr}\cos\theta_r$.

For the magnetically coupled windings, the expressions for the flux linkages are

$$\psi_{as} = L_{ss}i_{as} + L_{sr}\cos\theta_r i_{ar} - L_{sr}\sin\theta_r i_{br}, \quad \psi_{bs} = L_{ss}i_{bs} + L_{sr}\sin\theta_r i_{ar} + L_{sr}\cos\theta_r i_{br},$$

$$\psi_{ar} = L_{sr}\cos\theta_r i_{as} + L_{sr}\sin\theta_r i_{bs} + L_{rr}i_{ar}, \quad \psi_{br} = -L_{sr}\sin\theta_r i_{as} + L_{sr}\cos\theta_r i_{bs} + L_{rr}i_{br}.$$

The following expression for the flux linkage vectors results

$$\begin{bmatrix}\psi_{abs}\\ \psi_{abr}\end{bmatrix} = \begin{bmatrix}\mathbf{L}_s & \mathbf{L}_{sr}(\theta_r)\\ \mathbf{L}_{sr}^T(\theta_r) & \mathbf{L}_r\end{bmatrix}\begin{bmatrix}\mathbf{i}_{abs}\\ \mathbf{i}_{abr}\end{bmatrix}, \quad \mathbf{L}_s = \begin{bmatrix}L_{ss} & 0\\ 0 & L_{ss}\end{bmatrix}, \quad \mathbf{L}_r = \begin{bmatrix}L_{rr} & 0\\ 0 & L_{rr}\end{bmatrix},$$

$$\mathbf{L}_{sr}(\theta_r) = \begin{bmatrix}L_{sr}\cos\theta_r & -L_{sr}\sin\theta_r\\ L_{sr}\sin\theta_r & L_{sr}\cos\theta_r\end{bmatrix},$$

where $\mathbf{L}_s$ is the matrix of the stator self-inductances, $\mathbf{L}_s = \begin{bmatrix}L_{ss} & 0\\ 0 & L_{ss}\end{bmatrix}$, $L_{ss} = L_{ls} + L_{ms}$, $L_{ms} = \dfrac{N_s^2}{\Re_m}$; $\mathbf{L}_r$ is the matrix of the rotor self-inductances, $\mathbf{L}_r = \begin{bmatrix}L_{rr} & 0\\ 0 & L_{rr}\end{bmatrix}$, $L_{rr} = L_{lr} + L_{mr}$, $L_{mr} = \dfrac{N_r^2}{\Re_m}$; $\mathbf{L}_{sr}(\theta_r)$ is the stator–rotor mutual inductance mapping, $\mathbf{L}_{sr}(\theta_r) = \begin{bmatrix}L_{sr}\cos\theta_r & -L_{sr}\sin\theta_r\\ L_{sr}\sin\theta_r & L_{sr}\cos\theta_r\end{bmatrix}$, $L_{sr} = \dfrac{N_sN_r}{\Re_m}$; L_{ms} and L_{mr} are the stator and rotor magnetizing inductances; L_{ls} and L_{lr} are the stator and rotor leakage inductances; N_s and N_r are the number of turns of the stator and rotor windings; $\Re_m$ is the magnetizing reluctance.

Using the number of turns of the stator and rotor windings, we have

$$\mathbf{i}'_{abr} = \frac{N_r}{N_s}\mathbf{i}_{abr}, \quad \mathbf{u}'_{abr} = \frac{N_s}{N_r}\mathbf{u}_{abr}, \quad \text{and} \quad \psi'_{abr} = \frac{N_s}{N_r}\psi_{abr}.$$

Applying the turn ratio, the flux linkages are

$$\begin{bmatrix}\psi_{abs}\\ \psi'_{abr}\end{bmatrix} = \begin{bmatrix}\mathbf{L}_s & \mathbf{L}'_{sr}(\theta_r)\\ \mathbf{L}'^T_{sr}(\theta_r) & \mathbf{L}'_r\end{bmatrix}\begin{bmatrix}\mathbf{i}_{abs}\\ \mathbf{i}'_{abr}\end{bmatrix}, \quad \mathbf{L}'_r = \left(\frac{N_s}{N_r}\right)^2\mathbf{L}_r = \begin{bmatrix}L'_{rr} & 0\\ 0 & L'_{rr}\end{bmatrix},$$

$$\mathbf{L}'_{sr}(\theta_r) = \left(\frac{N_s}{N_r}\right)\mathbf{L}_{sr}(\theta_r) = L_{ms}\begin{bmatrix}\cos\theta_r & -\sin\theta_r\\ \sin\theta_r & \cos\theta_r\end{bmatrix},$$

where $L'_{rr} = L'_{lr} + L'_{mr}$, $L_{ms} = \dfrac{N_s}{N_r}L_{sr}$, $L'_{mr} = \left(\dfrac{N_s}{N_r}\right)^2 L_{mr}$, $L'_{mr} = L_{ms} = \dfrac{N_s}{N_r}L_{sr}$ and $L'_{rr} = L'_{lr} + L_{ms}$.

From the reported $\mathbf{L}_s$, $\mathbf{L}'_r$, and $\mathbf{L}'_{sr}(\theta_r)$, one obtains

$$\begin{bmatrix}\psi_{as}\\ \psi_{bs}\\ \psi'_{ar}\\ \psi'_{br}\end{bmatrix} = \begin{bmatrix}L_{ss} & 0 & L_{ms}\cos\theta_r & -L_{ms}\sin\theta_r\\ 0 & L_{ss} & L_{ms}\sin\theta_r & L_{ms}\cos\theta_r\\ L_{ms}\cos\theta_r & L_{ms}\sin\theta_r & L'_{rr} & 0\\ -L_{ms}\sin\theta_r & L_{ms}\cos\theta_r & 0 & L'_{rr}\end{bmatrix}\begin{bmatrix}i_{as}\\ i_{bs}\\ i'_{ar}\\ i'_{br}\end{bmatrix}. \tag{5.3}$$

The differential equations (5.2) are rewritten as

$$\mathbf{u}_{abs}=\mathbf{r}_s\mathbf{i}_{abs}+\frac{d\boldsymbol{\psi}_{abs}}{dt},\quad \mathbf{u}'_{abr}=\mathbf{r}'_r\mathbf{i}'_{abr}+\frac{d\boldsymbol{\psi}'_{abr}}{dt},\quad \mathbf{r}'_r=\frac{N_s^2}{N_r^2}\mathbf{r}_r=\frac{N_s^2}{N_r^2}\begin{bmatrix} r'_r & 0\\ 0 & r'_r\end{bmatrix}. \tag{5.4}$$

The self-inductances L_{ss} and L'_{rr} are time invariant. Furthermore, L_{ms} is constant. From (5.4), using (5.3), one obtains a set of four nonlinear differential equations

$$\begin{aligned}
&L_{ss}\frac{di_{as}}{dt}+L_{ms}\frac{d\left(i'_{ar}\cos\theta_r\right)}{dt}-L_{ms}\frac{d\left(i'_{br}\sin\theta_r\right)}{dt}=-r_si_{as}+u_{as},\\
&L_{ss}\frac{di_{bs}}{dt}+L_{ms}\frac{d\left(i'_{ar}\sin\theta_r\right)}{dt}+L_{ms}\frac{d\left(i'_{br}\cos\theta_r\right)}{dt}=-r_si_{bs}+u_{bs},\\
&L_{ms}\frac{d\left(i_{as}\cos\theta_r\right)}{dt}+L_{ms}\frac{d\left(i_{bs}\sin\theta_r\right)}{dt}+L'_{rr}\frac{di'_{ar}}{dt}=-r'_ri'_{ar}+u'_{ar},\\
&-L_{ms}\frac{d\left(i_{as}\sin\theta_r\right)}{dt}+L_{ms}\frac{d\left(i_{bs}\cos\theta_r\right)}{dt}+L'_{rr}\frac{di'_{br}}{dt}=-r'_ri'_{br}+u'_{br}.
\end{aligned} \tag{5.5}$$

The *emf* is $emf=\oint_l \vec{E}\cdot d\vec{l}=\oint_l\left(\vec{v}\times\vec{B}\right)\cdot d\vec{l}=-\int_s\frac{\partial\vec{B}}{\partial t}\cdot d\vec{s}$. The Faraday law of induction is $\mathscr{E}=\oint_l \vec{E}(t)\cdot d\vec{l}=-N\frac{d\Phi}{dt}=-\frac{d\psi}{dt}$. From (5.5), the *total emfs* in the rotor windings are

$$emf_{ar}=-L_{ms}\frac{d\left(i_{as}\cos\theta_r\right)}{dt}-L_{ms}\frac{d\left(i_{bs}\sin\theta_r\right)}{dt}-L'_{rr}\frac{di'_{ar}}{dt},$$

$$emf_{br}=L_{ms}\frac{d\left(i_{as}\sin\theta_r\right)}{dt}-L_{ms}\frac{d\left(i_{bs}\cos\theta_r\right)}{dt}-L'_{rr}\frac{di'_{br}}{dt}.$$

The motional *emfs* (voltages) are induced in the rotor windings. For the steady-state operation, the motional *emfs* are

$$emf_{ar\omega}=L_{ms}\left(i_{as}\sin\theta_r-i_{bs}\cos\theta_r\right)\omega_r \quad\text{and}\quad emf_{br\omega}=L_{ms}\left(i_{as}\cos\theta_r+i_{bs}\sin\theta_r\right)\omega_r.$$

From (5.5), Cauchy's form of differential equations are found as given by the first four equations in (5.10). The *torsional–mechanical* equation of motion is derived using Newton's second law

$$\begin{aligned}
&\frac{d\omega_{rm}}{dt}=\frac{1}{J}(T_e-B_m\omega_{rm}-T_L),\\
&\frac{d\theta_{rm}}{dt}=\omega_{rm}.
\end{aligned} \tag{5.6}$$

The mechanical angular velocity of the rotor ω_{rm} is expressed by using the electrical angular velocity ω_r and the number of poles *P*. In particular, $\omega_{rm}=\frac{2}{P}\omega_r$.

The mechanical angular displacement θ_{rm} is $\theta_{rm}=\frac{2}{P}\theta_r$.

It is convenient to derive the equations of motion using the electrical angular velocity ω_r and displacement θ_r. From (5.6), one finds

$$\frac{d\omega_r}{dt} = \frac{1}{J}\left(\frac{P}{2}T_e - B_m\omega_r - \frac{P}{2}T_L\right),$$
$$\frac{d\theta_r}{dt} = \omega_r. \tag{5.7}$$

The electromagnetic torque, developed by induction motors, is found using the coenergy. We have

$$T_e = \frac{P}{2}\frac{\partial W_c\left(\mathbf{i}_{abs}, \mathbf{i}'_{abr}, \theta_r\right)}{\partial\theta_r}.$$

Assuming that the magnetic system is linear, the coenergy is

$$W_c = \frac{1}{2}\mathbf{i}_{abs}^T\left(\mathbf{L}_s - L_{ls}\mathbf{I}\right)\mathbf{i}_{abs} + \mathbf{i}_{abs}^T\mathbf{L}'_{sr}(\theta_r)\mathbf{i}'_{abr} + \frac{1}{2}\mathbf{i}'^T_{abr}\left(\mathbf{L}'_r - L'_{lr}\mathbf{I}\right)\mathbf{i}'_{abr}.$$

The self-inductances L_{ss} and L'_{rr}, as well as the leakage inductances L_{ls} and L'_{lr}, are not functions of the angular displacement θ_r. In (5.3), assuming the sinusoidal stator–rotor mutual inductances, we have

$$\mathbf{L}'_{sr}(\theta_r) = L_{ms}\begin{bmatrix}\cos\theta_r & -\sin\theta_r\\ \sin\theta_r & \cos\theta_r\end{bmatrix}.$$

For P-pole two-phase induction motors, the electromagnetic torque is

$$T_e = \frac{P}{2}\frac{\partial W_c\left(\mathbf{i}_{abs}, \mathbf{i}'_{abr}, \theta_r\right)}{\partial\theta_r} = \frac{P}{2}\mathbf{i}_{abs}^T\frac{\partial\mathbf{L}'_{sr}(\theta_r)}{\partial\theta_r}\mathbf{i}'_{abr} = \frac{P}{2}L_{ms}\begin{bmatrix}i_{as} & i_{bs}\end{bmatrix}\begin{bmatrix}-\sin\theta_r & -\cos\theta_r\\ \cos\theta_r & -\sin\theta_r\end{bmatrix}\begin{bmatrix}i'_{ar}\\ i'_{br}\end{bmatrix}$$
$$= -\frac{P}{2}L_{ms}\left[\left(i_{as}i'_{ar} + i_{bs}i'_{br}\right)\sin\theta_r + \left(i_{as}i'_{br} - i_{bs}i'_{ar}\right)\cos\theta_r\right]. \tag{5.8}$$

The electromagnetic torque T_e is a vector that defines the clockwise or counterclockwise rotor rotation. Using (5.7) and (5.8), the *torsional–mechanical* equations are

$$\frac{d\omega_r}{dt} = \frac{1}{J}\left[-\frac{P^2}{4}L_{ms}\left[\left(i_{as}i'_{ar} + i_{bs}i'_{br}\right)\sin\theta_r + \left(i_{as}i'_{br} - i_{bs}i'_{ar}\right)\cos\theta_r\right] - B_m\omega_r - \frac{P}{2}T_L\right],$$
$$\frac{d\theta_r}{dt} = \omega_r. \tag{5.9}$$

Applying the phase voltages u_{as} and u_{bs}, one rotates the motor clockwise or counterclockwise. The electromagnetic torque T_e counteracts the load and friction torques. The friction torque acts against the electromagnetic torque, while the load torques may be bidirectional. To control the direction of rotation, one changes the sign for T_e. Using the circuitry-electromagnetic and *torsional–mechanical* equations (5.5) and (5.9), one obtains nonlinear differential equations

$$\begin{aligned}
\frac{di_{as}}{dt} &= -\frac{L'_{rr}r_s}{L_\Sigma}i_{as} + \frac{L_{ms}^2}{L_\Sigma}i_{bs}\omega_r + \frac{L_{ms}L'_{rr}}{L_\Sigma}i'_{ar}\left(\omega_r\sin\theta_r + \frac{r'_r}{L'_{rr}}\cos\theta_r\right)\\
&\quad+\frac{L_{ms}L'_{rr}}{L_\Sigma}i'_{br}\left(\omega_r\cos\theta_r - \frac{r'_r}{L'_{rr}}\sin\theta_r\right) + \frac{L'_{rr}}{L_\Sigma}u_{as} - \frac{L_{ms}}{L_\Sigma}\cos\theta_r u'_{ar} + \frac{L_{ms}}{L_\Sigma}\sin\theta_r u'_{br},\\
\frac{di_{bs}}{dt} &= -\frac{L'_{rr}r_s}{L_\Sigma}i_{bs} - \frac{L_{ms}^2}{L_\Sigma}i_{as}\omega_r - \frac{L_{ms}L'_{rr}}{L_\Sigma}i'_{ar}\left(\omega_r\cos\theta_r - \frac{r'_r}{L'_{rr}}\sin\theta_r\right)\\
&\quad+\frac{L_{ms}L'_{rr}}{L_\Sigma}i'_{br}\left(\omega_r\sin\theta_r + \frac{r'_r}{L'_{rr}}\cos\theta_r\right) + \frac{L'_{rr}}{L_\Sigma}u_{bs} - \frac{L_{ms}}{L_\Sigma}\sin\theta_r u'_{ar} - \frac{L_{ms}}{L_\Sigma}\cos\theta_r u'_{br},\\
\frac{di'_{ar}}{dt} &= -\frac{L_{ss}r'_r}{L_\Sigma}i'_{ar} + \frac{L_{ms}L_{ss}}{L_\Sigma}i_{as}\left(\omega_r\sin\theta_r + \frac{r_s}{L_{ss}}\cos\theta_r\right) - \frac{L_{ms}L_{ss}}{L_\Sigma}i_{bs}\left(\omega_r\cos\theta_r - \frac{r_s}{L_{ss}}\sin\theta_r\right)\\
&\quad-\frac{L_{ms}^2}{L_\Sigma}i'_{br}\omega_r - \frac{L_{ms}}{L_\Sigma}\cos\theta_r u_{as} - \frac{L_{ms}}{L_\Sigma}\sin\theta_r u_{bs} + \frac{L_{ss}}{L_\Sigma}u'_{ar},\\
\frac{di'_{br}}{dt} &= -\frac{L_{ss}r'_r}{L_\Sigma}i'_{br} + \frac{L_{ms}L_{ss}}{L_\Sigma}i_{as}\left(\omega_r\cos\theta_r - \frac{r_s}{L_{ss}}\sin\theta_r\right) + \frac{L_{ms}L_{ss}}{L_\Sigma}i_{bs}\left(\omega_r\sin\theta_r + \frac{r_s}{L_{ss}}\cos\theta_r\right)\\
&\quad+\frac{L_{ms}^2}{L_\Sigma}i'_{ar}\omega_r + \frac{L_{ms}}{L_\Sigma}\sin\theta_r u_{as} - \frac{L_{ms}}{L_\Sigma}\cos\theta_r u_{bs} + \frac{L_{ss}}{L_\Sigma}u'_{br},\\
\frac{d\omega_r}{dt} &= -\frac{P^2}{4J}L_{ms}\left[\left(i_{as}i'_{ar} + i_{bs}i'_{br}\right)\sin\theta_r + \left(i_{as}i'_{br} - i_{bs}i'_{ar}\right)\cos\theta_r\right] - \frac{B_m}{J}\omega_r - \frac{P}{2J}T_L\\
\frac{d\theta_r}{dt} &= \omega_r,
\end{aligned} \tag{5.10}$$

where $L_\Sigma = L_{ss}L'_{rr} - L_{ms}^2$.

5.3.2 Lagrange Equations of Motion

The mathematical models can be derived using the Lagrange equations of motions

$$\frac{d}{dt}\left(\frac{\partial\Gamma}{\partial\dot q_i}\right) - \frac{\partial\Gamma}{\partial q_i} + \frac{\partial D}{\partial\dot q_i} + \frac{\partial\Pi}{\partial q_i} = Q_i, \tag{5.11}$$

where Γ, D, and Π are the total kinetic, dissipation, and potential energies; q_i and Q_i are the *generalized* independent coordinates and forces.

The *generalized* independent coordinates q_i are the charges and rotor angular displacement. That is, $q_1 = \int i_{as}\,dt, q_2 = \int i_{bs}\,dt, q_3 = \int i'_{ar}\,dt, q_4 = \int i'_{br}\,dt, q_5 = \theta_r$. The *generalized* forces Q_i are the voltages and load torque, $Q_1 = u_{as}$, $Q_2 = u_{bs}$, $Q_3 = u'_{ar}$, $Q_4 = u'_{br}$, $Q_5 = -T_L$. Using (5.11), five Lagrange equations are

$$\begin{aligned}
&\frac{d}{dt}\left(\frac{\partial\Gamma}{\partial\dot q_1}\right) - \frac{\partial\Gamma}{\partial q_1} + \frac{\partial D}{\partial\dot q_1} + \frac{\partial\Pi}{\partial q_1} = Q_1, \quad \frac{d}{dt}\left(\frac{\partial\Gamma}{\partial\dot q_2}\right) - \frac{\partial\Gamma}{\partial q_2} + \frac{\partial D}{\partial\dot q_2} + \frac{\partial\Pi}{\partial q_2} = Q_2,\\
&\frac{d}{dt}\left(\frac{\partial\Gamma}{\partial\dot q_3}\right) - \frac{\partial\Gamma}{\partial q_3} + \frac{\partial D}{\partial\dot q_3} + \frac{\partial\Pi}{\partial q_3} = Q_3, \quad \frac{d}{dt}\left(\frac{\partial\Gamma}{\partial\dot q_4}\right) - \frac{\partial\Gamma}{\partial q_4} + \frac{\partial D}{\partial\dot q_4} + \frac{\partial\Pi}{\partial q_4} = Q_4,\\
&\frac{d}{dt}\left(\frac{\partial\Gamma}{\partial\dot q_5}\right) - \frac{\partial\Gamma}{\partial q_5} + \frac{\partial D}{\partial\dot q_5} + \frac{\partial\Pi}{\partial q_5} = Q_5.
\end{aligned} \tag{5.12}$$

The total kinetic, potential, and dissipation energies, used in (5.12), are

$$\Gamma=\frac{1}{2}L_{ss}\dot{q}_1^2+L_{ms}\dot{q}_1\dot{q}_3\cos q_5-L_{ms}\dot{q}_1\dot{q}_4\sin q_5+\frac{1}{2}L_{ss}\dot{q}_2^2+L_{ms}\dot{q}_2\dot{q}_3\sin q_5+L_{ms}\dot{q}_2\dot{q}_4\cos q_5$$
$$+\frac{1}{2}L'_{rr}\dot{q}_3^2+\frac{1}{2}L'_{rr}\dot{q}_4^2+\frac{1}{2}J\dot{q}_5^2,$$

$$\Pi=0,$$

and

$$D=\frac{1}{2}\left(r_s\dot{q}_1^2+r_s\dot{q}_2^2+r'_r\dot{q}_3^2+r'_r\dot{q}_4^2+B_m\dot{q}_5^2\right).$$

The derivative terms of (5.12) are

$$\frac{\partial\Gamma}{\partial q_1}=0,\quad \frac{\partial\Gamma}{\partial\dot{q}_1}=L_{ss}\dot{q}_1+L_{ms}\dot{q}_3\cos q_5-L_{ms}\dot{q}_4\sin q_5,\quad \frac{\partial\Gamma}{\partial q_2}=0,\quad \frac{\partial\Gamma}{\partial\dot{q}_2}=L_{ss}\dot{q}_2+L_{ms}\dot{q}_3\sin q_5+L_{ms}\dot{q}_4\cos q_5,$$
$$\frac{\partial\Gamma}{\partial q_3}=0,\quad \frac{\partial\Gamma}{\partial\dot{q}_3}=L'_{rr}\dot{q}_3+L_{ms}\dot{q}_1\cos q_5+L_{ms}\dot{q}_2\sin q_5,\quad \frac{\partial\Gamma}{\partial q_4}=0,\quad \frac{\partial\Gamma}{\partial\dot{q}_4}=L'_{rr}\dot{q}_4-L_{ms}\dot{q}_1\sin q_5+L_{ms}\dot{q}_2\cos q_5,$$
$$\frac{\partial\Gamma}{\partial q_5}=-L_{ms}\dot{q}_1\dot{q}_3\sin q_5-L_{ms}\dot{q}_1\dot{q}_4\cos q_5+L_{ms}\dot{q}_2\dot{q}_3\cos q_5-L_{ms}\dot{q}_2\dot{q}_4\sin q_5$$
$$=-L_{ms}\left[\left(\dot{q}_1\dot{q}_3+\dot{q}_2\dot{q}_4\right)\sin q_5+\left(\dot{q}_1\dot{q}_4-\dot{q}_2\dot{q}_3\right)\cos q_5\right],$$
$$\frac{\partial\Gamma}{\partial\dot{q}_5}=J\dot{q}_5,$$
$$\frac{\partial\Pi}{\partial q_1}=0,\quad \frac{\partial\Pi}{\partial q_2}=0,\quad \frac{\partial\Pi}{\partial q_3}=0,\quad \frac{\partial\Pi}{\partial q_4}=0,\quad \frac{\partial\Pi}{\partial q_5}=0,$$
$$\frac{\partial D}{\partial\dot{q}_1}=r_s\dot{q}_1,\quad \frac{\partial D}{\partial\dot{q}_2}=r_s\dot{q}_2,\quad \frac{\partial D}{\partial\dot{q}_3}=r'_r\dot{q}_3,\quad \frac{\partial D}{\partial\dot{q}_4}=r'_r\dot{q}_4,\quad \frac{\partial D}{\partial\dot{q}_5}=B_m\dot{q}_5. \tag{5.13}$$

The *generalized* coordinates and forces are expressed by using the machine variables as follows

$\dot{q}_1=i_{as}$, $\dot{q}_2=i_{bs}$, $\dot{q}_3=i'_{ar}$, $\dot{q}_4=i'_{br}$, $\dot{q}_5=\omega_r$, and $Q_1=u_{as}$, $Q_2=u_{bs}$, $Q_3=u'_{ar}$, $Q_4=u'_{br}$, $Q_5=-T_L$.

From (5.12) and (5.13), one obtains the resulting differential equations

$$\begin{aligned}
&L_{ss}\frac{di_{as}}{dt}+L_{ms}\frac{d\left(i'_{ar}\cos\theta_r\right)}{dt}-L_{ms}\frac{d\left(i'_{br}\sin\theta_r\right)}{dt}+r_si_{as}=u_{as},\\
&L_{ss}\frac{di_{bs}}{dt}+L_{ms}\frac{d\left(i'_{ar}\sin\theta_r\right)}{dt}+L_{ms}\frac{d\left(i'_{br}\cos\theta_r\right)}{dt}+r_si_{bs}=u_{bs},\\
&L_{ms}\frac{d\left(i_{as}\cos\theta_r\right)}{dt}+L_{ms}\frac{d\left(i_{bs}\sin\theta_r\right)}{dt}+L'_{rr}\frac{di'_{ar}}{dt}+r'_ri'_{ar}=u'_{ar},\\
&-L_{ms}\frac{d\left(i_{as}\sin\theta_r\right)}{dt}+L_{ms}\frac{d\left(i_{bs}\cos\theta_r\right)}{dt}+L'_{rr}\frac{di'_{br}}{dt}+r'_ri'_{br}=u'_{br},\\
&J\frac{d^2\theta_r}{dt^2}+L_{ms}\left[\left(i_{as}i'_{ar}+i_{bs}i'_{br}\right)\sin\theta_r+\left(i_{as}i'_{br}-i_{bs}i'_{ar}\right)\cos\theta_r\right]+B_m\frac{d\theta_r}{dt}=-T_L.
\end{aligned} \tag{5.14}$$

From (5.14), for P-pole induction motors, by using $\frac{d\theta_r}{dt}=\omega_r$, six differential equations (5.5) and (5.9) result. The advantage of the Lagrange concept is that no coenergy and no Kirchhoff's, Newton's,

Faraday's, or Lorenz laws are not used to derive the resulting models. The Lagrange equations of motion provide a general and consistent procedure. One finds the *emfs*, electromagnetic torque, etc. For example, from

$$\frac{d}{dt}\left(\frac{\partial \Gamma}{\partial \dot{q}_5}\right)-\frac{\partial \Gamma}{\partial q_5}+\frac{\partial D}{\partial \dot{q}_5}+\frac{\partial \Pi}{\partial q_5}=Q_5,$$

one concludes that the electromagnetic torque is

$$T_e=\frac{\partial \Gamma}{\partial q_5}=\frac{\partial \Gamma}{\partial \theta_r}=-L_{ms}\left[\left(\dot{q}_1\dot{q}_3+\dot{q}_2\dot{q}_4\right)\sin q_5+\left(\dot{q}_1\dot{q}_4-\dot{q}_2\dot{q}_3\right)\cos q_5\right]. \tag{5.15}$$

Expression (5.15) results in

$$T_e=-L_{ms}\left[\left(i_{as}i'_{br}+i_{bs}i'_{br}\right)\sin\theta_r+\left(i_{as}i'_{br}-i_{bs}i'_{ar}\right)\cos\theta_r\right],$$

which is the same as (5.8). Equation for T_e (5.8) was derived by applying the coenergy.

Example 5.2

We simulate two two-phase 115 V (*rms*), 60 Hz, four-pole (P = 4) induction motors. The dynamics is described by differential equations (5.10). The objective is to examine the performance of induction motors by analyzing the acceleration capabilities, settling time, etc. The torque–speed characteristics will be studied by solving the differential equations to compare the dynamic $\omega_r(t) = \Omega_T[T_e(t)]$ and steady-state $\omega_r = \Omega_T(T_e)$. The parameters of the A and D class motors are as follows:

1. A Class: r_s = 1.2 ohm, r'_r =1.5 ohm, L_{ms} = 0.16 H, L_{ls} = 0.02 H, $L_{ss} = L_{ls} + L_{ms}$, L'_{lr} = 0.02 H, $L'_{rr} = L'_{lr} + L_{ms}$, $B_m = 1 \times 10^{-6}$ N-m-sec/rad, and J = 0.005 kg-m^2;
2. D Class: r_s = 24.5 ohm, r'_r = 23 ohm, L_{ms} = 0.27 H, L_{ls} = 0.027 H, $L_{ss} = L_{ls} + L_{ms}$, L'_{lr} = 0.027 H, $L'_{rr} = L'_{lr} + L_{ms}$, $B_m = 1 \times 10^{-6}$ N-m-sec/rad, and J = 0.001 kg-m^2.

To guarantee the balanced operation, the supplied phase voltages are $u_{as}(t)=\sqrt{2}u_M\cos\left(\omega_f t\right)$ and $u_{bs}(t)=\sqrt{2}u_M\sin\left(\omega_f t\right)$. For the rated voltage, $u_{as}(t)=\sqrt{2}115\cos\left(377t\right)$ and $u_{bs}(t)=\sqrt{2}115\sin\left(377t\right)$.

No load and loaded conditions are examined assigning the load torque to be 0 and 0.5 N-m. The simulations are performed, and transient dynamics of the stator and rotor currents $i_{as}(t), i_{bs}(t), i'_{ar}(t)$, and $i'_{br}(t)$, as well as the mechanical angular velocity $\omega_{rm}(t)$, are potted in Figures 5.6 and 5.7. Figures 5.6 illustrate the transient dynamics of an A class motor. The motor accelerates from stall, e.g. ω_{rm0} = 0 rad/sec. Figures 5.6a and b depict the motor dynamics if T_L = 0, and if T_L = 0.5 N-m is applied at t = 0 sec. The dynamics and acceleration capabilities of the D class motor are shown in Figures 5.7.

The A class induction motor reaches the steady-state angular velocity within 0.8 sec (with no load). The settling time is 1.4 sec if motor operates under T_L = 0.5 N-m. The acceleration capabilities are studied for the D class motor. The settling time is 0.5 and 0.8 sec for no load and loaded motor. The D class induction motors may develop higher starting electromagnetic torque, see Figure 5.2a. However, the D class motors do not possess high $T_{e\,start}$ and $T_{e\,critical}$ as compared to the A class motors, which are controlled using the frequency or voltage–frequency control. As illustrated in Figures 5.6 and 5.7, the A class induction motors have higher $T_{e\,start}$ and $T_{e\,critical}$ and higher $T_{e\,critical}/T_{e\,start}$ ratio. For the A and D class induction motors, the $T_{e\,critical}$ are ~3 and 1.1 N-m, respectively. The efficiency of D class induction motors is low due to high r'_r. Figures 5.6 and 5.7 illustrate that the electromagnetic torque for A and D class motors reaches 4.1 and 1.8 N-m, respectively. Hence, A class induction motors ensure better performance, exhibit high efficiency, and have low losses.

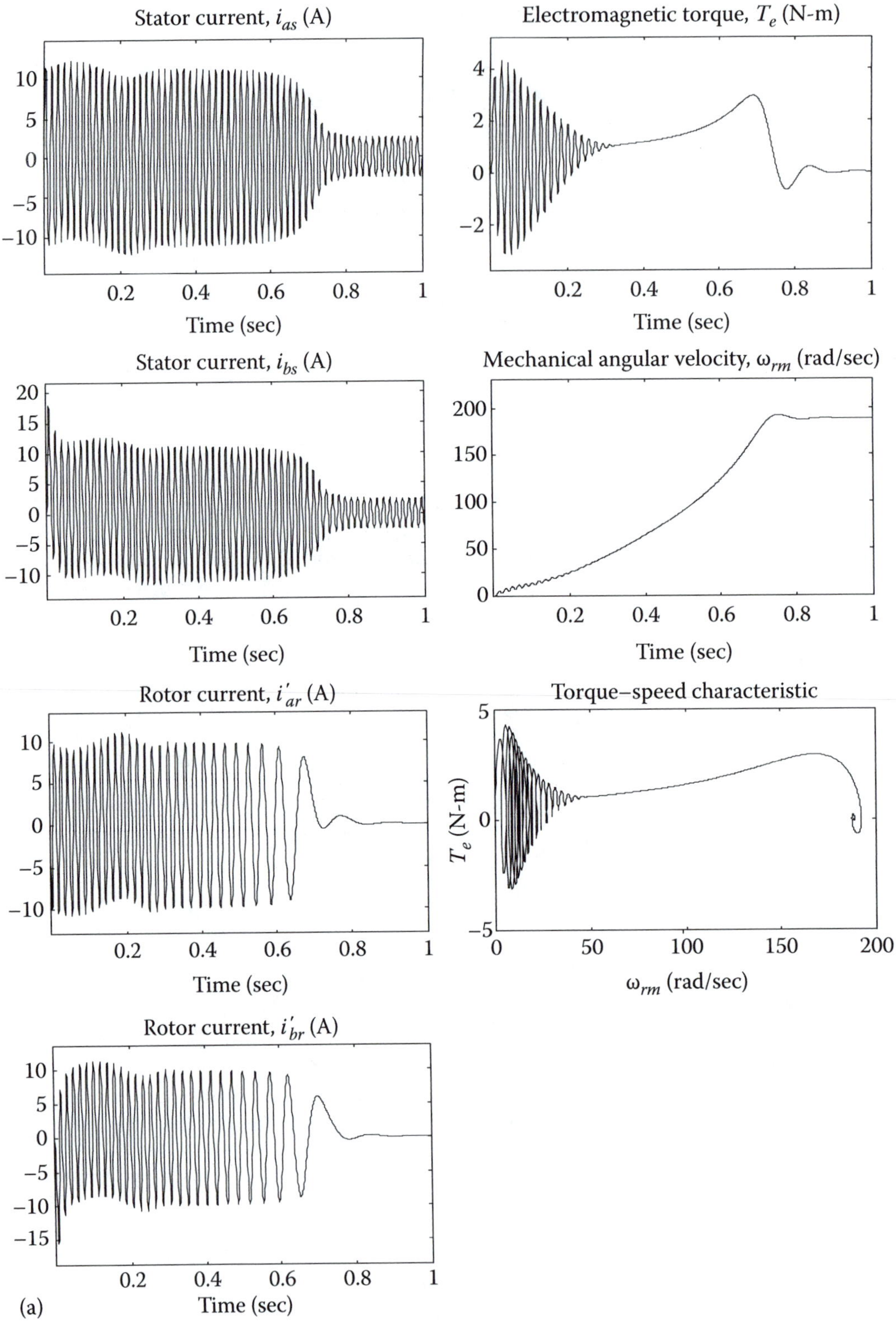

FIGURE 5.6 Dynamics and torque–speed characteristic of an A class induction motor: (a) $T_L = 0$ N-m.

(Continued)

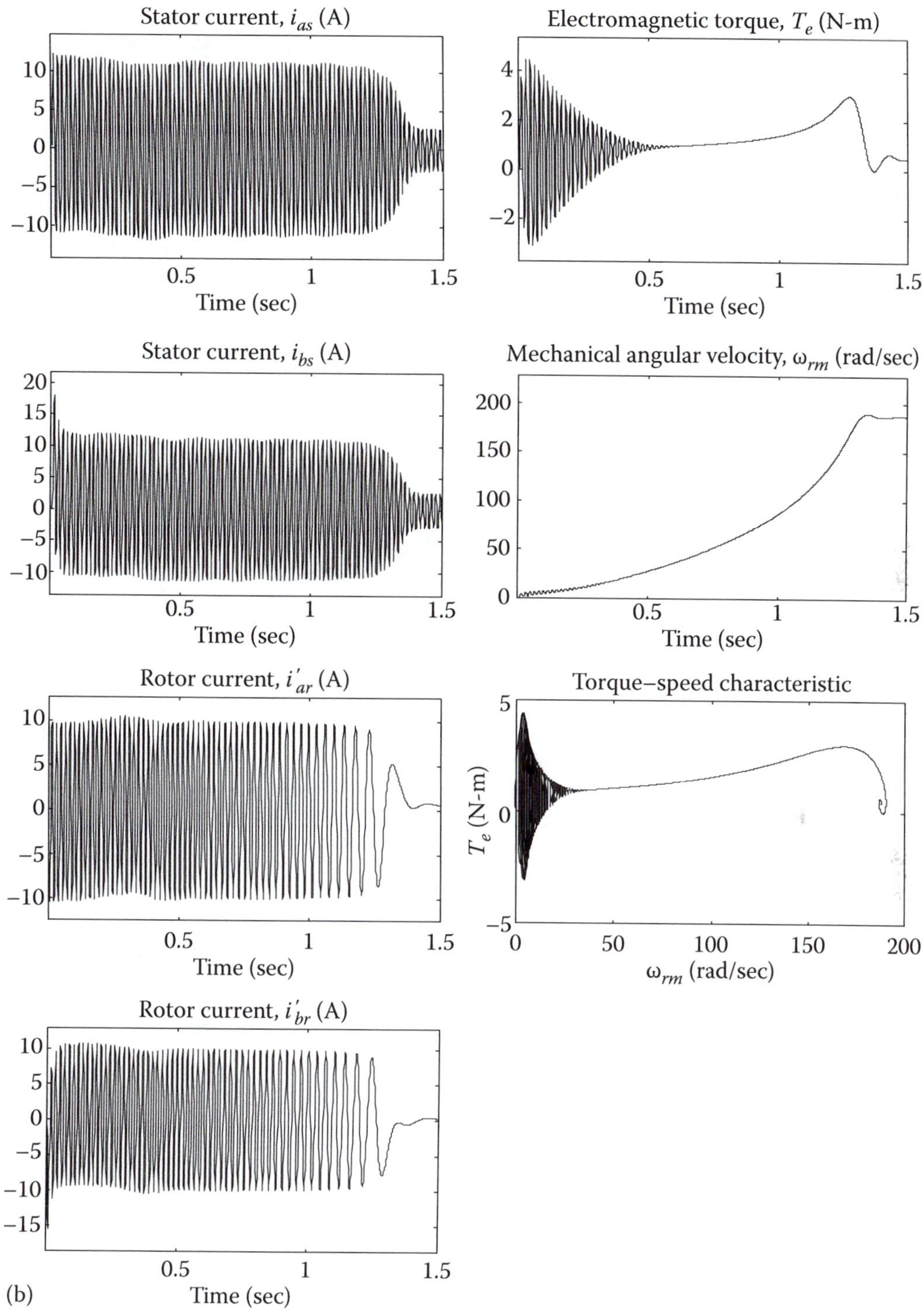

FIGURE 5.6 (*Continued*) Dynamics and torque–speed characteristic of an A class induction motor: (b) $T_L = 0.5$ N-m.

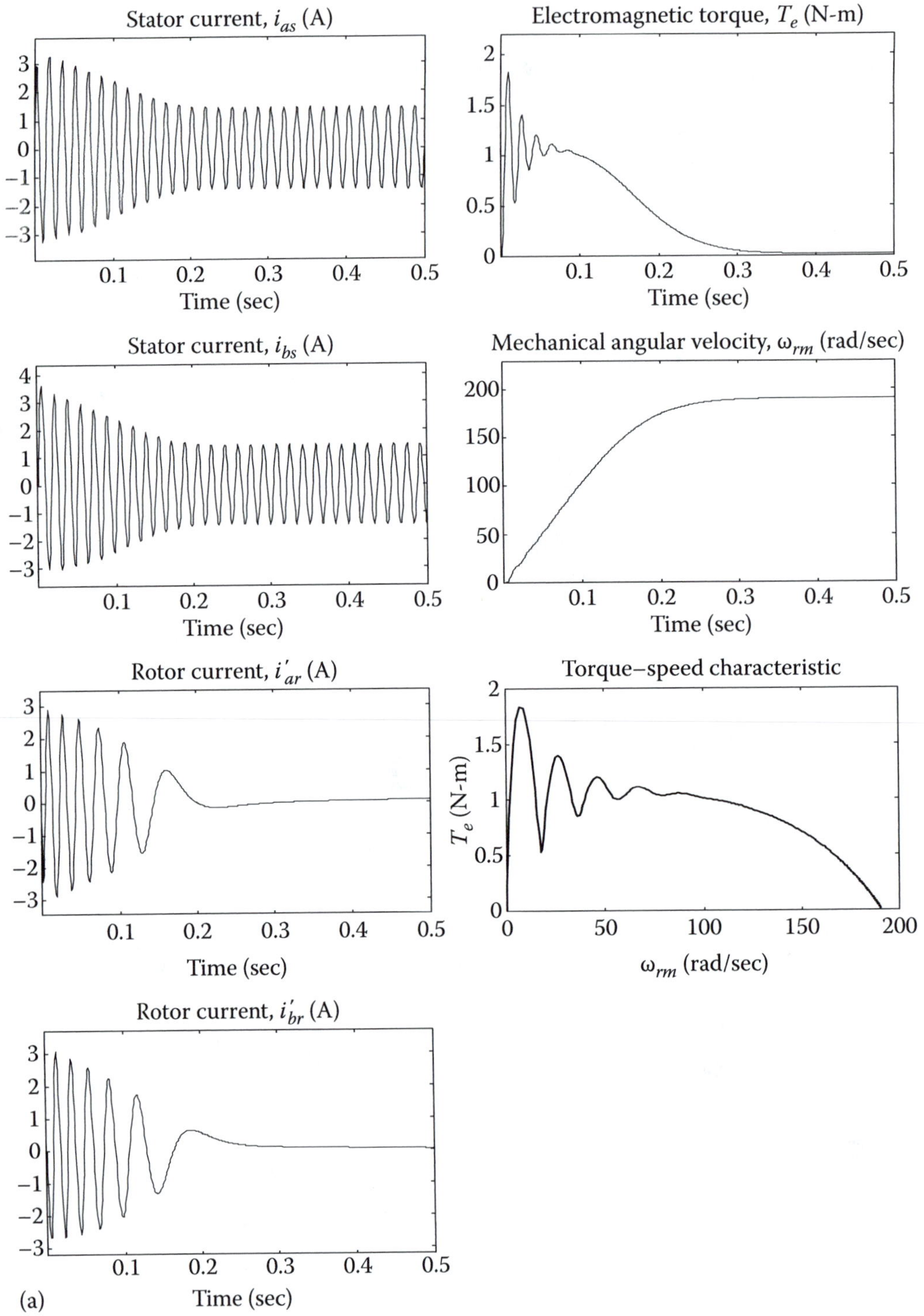

FIGURE 5.7 Dynamics and torque–speed characteristics of the D class induction motor: (a) $T_L = 0$ N-m.

(Continued)

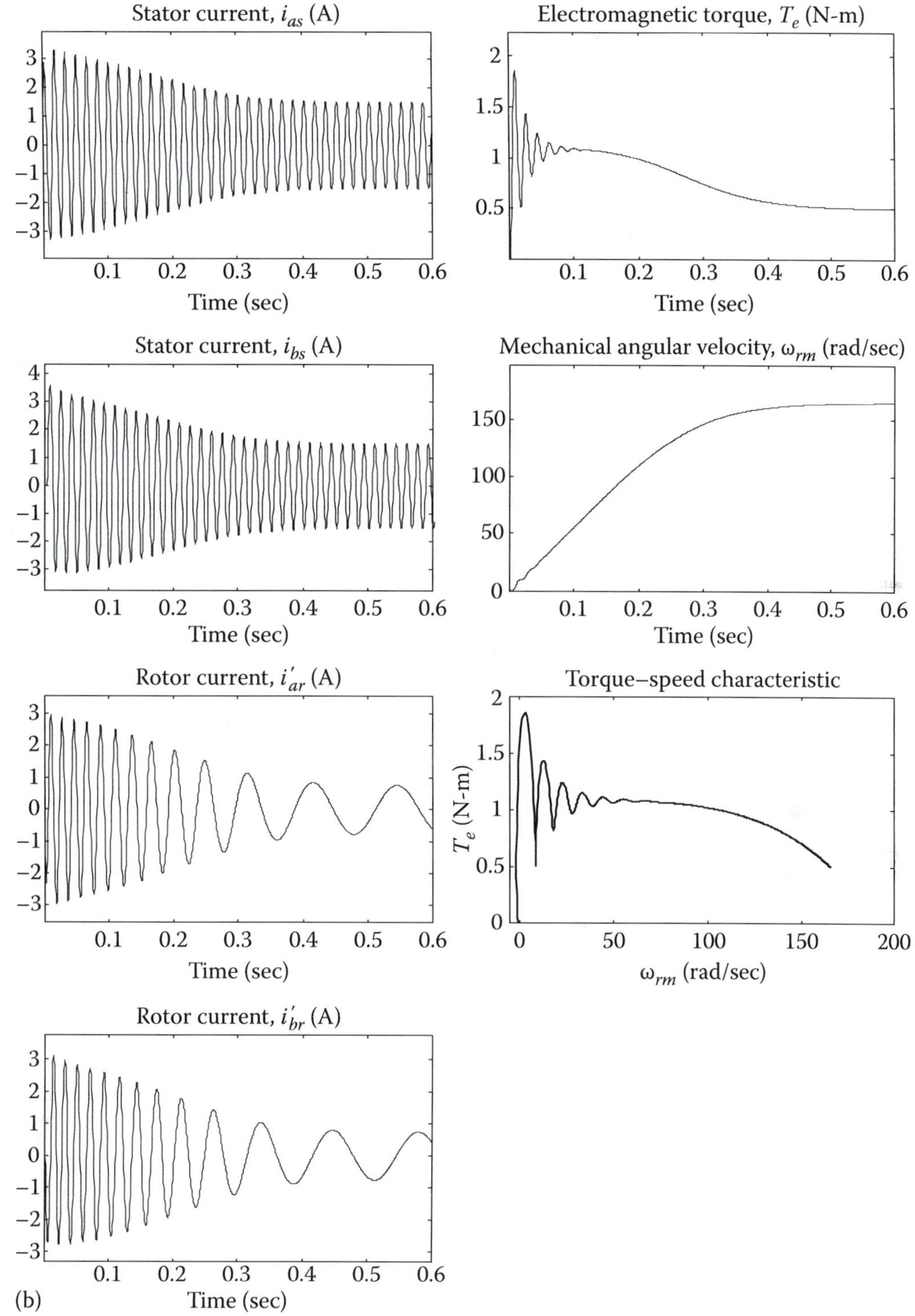

FIGURE 5.7 (*Continued*) Dynamics and torque–speed characteristics of the D class induction motor: (b) $T_L = 0.5$ N-m.

The dynamics of $\omega_{rm}(t)$ and evolutions of $T_e(t)$ are shown in Figures 5.6 and 5.7. The dynamic characteristics $\omega_{rm}(t) = \Omega_T[T_e(t)]$ are obtained by plotting the mechanical angular velocity versus the electromagnetic torque. Figures 5.6 show the torque–speed characteristics of an A class motor, while Figures 5.7 illustrate $\omega_{rm}(t) = \Omega_T[T_e(t)]$ for the D class motor. One finds the steady-state torque–speed characteristics $\omega_{rm} = \Omega_T(T_e)$ by *averaging* $\omega_{rm}(t) = \Omega_T[T_e(t)]$. An analysis of motor transients enables the assessment of acceleration, disacceleration, loading, and other capabilities. ■

5.3.3 Advanced Topics in the Analysis of Induction Machines

The analysis was performed assuming a consistent optimal electromagnetic system design. The designer can achieve near-optimal design in the specified operating envelope. The undesired effects can degrade performance and capabilities. The magnetic coupling between stator and rotor windings may not be ideally sinusoidal. Hence, an assumption that

$$L_{asar} = L_{aras} = L_{sr}\cos\theta_r, \quad L_{asbr} = L_{bras} = -L_{sr}\sin\theta_r, \quad L_{bsar} = L_{arbs} = L_{sr}\sin\theta_r, \quad \text{and}$$

$$L_{bsbr} = L_{brbs} = L_{sr}\cos\theta_r$$

may not be ensured in the full operating envelope particularly under peak loads. A nonideal sinusoidal winding distribution leads to the torque ripple, current chattering, overheating, vibration, noise, etc. Depending on the induction machine's overall design, electric steel *B*–*H* curve, in the operating envelope, one may have

$$L_{asar} = L_{aras} = \sum_{n=1}^{\infty} L_{sr\,n}\cos^{2n-1}\theta_r, \quad L_{asbr} = L_{bras} = -\sum_{n=1}^{\infty} L_{sr\,n}\sin^{2n-1}\theta_r,$$
$$L_{bsar} = L_{arbs} = \sum_{n=1}^{\infty} L_{sr\,n}\sin^{2n-1}\theta_r, \quad L_{bsbr} = L_{brbs} = \sum_{n=1}^{\infty} L_{sr\,n}\cos^{2n-1}\theta_r. \tag{5.16}$$

Discussions on high-fidelity analysis: The stator–rotor mutual inductances can be experimentally found. Complex magnetic coupling is observed. Using the results reported earlier and in Section 6.5, one may have

$$L_{asar} = L_{aras} = \sum_{n=1}^{\infty} L_{sr\,n}\cos^{2n-1}\theta_r + \operatorname{sgn}(\cos\theta_r)\sum_{k,l=1}^{\infty} L_{sr\,l,k}\left|\cos^{\frac{2l-1}{2k-1}}\theta_r\right|.$$

In the full operating envelope, using the element-by-element product, for three-phase induction motors

$$L_{asar} = L_{aras} = \left[L_1 + \sum_{n=1}^{\infty} L_{sr\,n}\cos^{2n-1}\theta_r + \operatorname{sgn}(\cos\theta_r)\sum_{k,l=1}^{\infty} L_{sr\,l,k}\left|\cos^{\frac{2l-1}{2k-1}}\theta_r\right|\right] \circ \left[L_2 + L_{sr\,p}\, e^{\sum_{p=1}^{\infty}\left(a_p \sin^p \frac{1}{c_p}\theta_r + b_p \cos^p \frac{1}{c_p}\theta_r\right)}\right].$$

The illustrative results and examples are studied in Section 6.5. The Practice and Engineering Problems 5.4 and 5.5 are formulated and solved in the "Practice and Engineering Problems" section.

From $L_{asar}(\theta_r)$, using the electrical winding displacement angle for two- and three-phase induction motors, one yields the following stator–rotor mutual inductances

$$L_{asbr}=L_{bras},\quad L_{ascr}=L_{cras},\quad L_{bsar}=L_{arbs},\quad L_{bsbr}=L_{brbs},\quad L_{bscr}=L_{crbs},\quad L_{csar}=L_{arcs},$$

$$L_{csbr}=L_{brcs},\quad \text{and}\quad L_{cscr}=L_{crcs}.$$

One finds

$$L_{asar}=L_{aras}=\left[L_1+\sum_{n=1}^{\infty}L_{sr\,n}\cos^{2n-1}\theta_r+\operatorname{sgn}(\cos\theta_r)\sum_{k,l=1}^{\infty}L_{sr\,l,k}\left|\cos^{\frac{2l-1}{2k-1}}\theta_r\right|\right]$$
$$\circ\left[L_2+L_{sr\,p}e^{\sum_{p=1}^{\infty}\left(a_p\sin^p\frac{1}{cp}\theta_r+b_p\cos^p\frac{1}{cp}\theta_r\right)}\right],$$

$$L_{asbr}=\left[L_1+\sum_{n=1}^{\infty}L_{sr\,n}\cos^{2n-1}\left(\theta_r-\frac{2}{3}\pi\right)+\operatorname{sgn}\left(\cos\left(\theta_r-\frac{2}{3}\pi\right)\right)\sum_{k,l=1}^{\infty}L_{sr\,l,k}\left|\cos^{\frac{2l-1}{2k-1}}\left(\theta_r-\frac{2}{3}\pi\right)\right|\right]$$
$$\circ\left[L_2+L_{sr\,p}e^{\sum_{p=1}^{\infty}\left(a_p\sin^p\frac{1}{cp}\left(\theta_r-\frac{2}{3}\pi\right)+b_p\cos^p\frac{1}{cp}\left(\theta_r-\frac{2}{3}\pi\right)\right)}\right],$$

$$L_{ascr}=\left[L_1+\sum_{n=1}^{\infty}L_{sr\,n}\cos^{2n-1}\left(\theta_r+\frac{2}{3}\pi\right)+\operatorname{sgn}\left(\cos\left(\theta_r+\frac{2}{3}\pi\right)\right)\sum_{k,l=1}^{\infty}L_{sr\,l,k}\left|\cos^{\frac{2l-1}{2k-1}}\left(\theta_r+\frac{2}{3}\pi\right)\right|\right]$$
$$\circ\left[L_2+L_{sr\,p}e^{\sum_{p=1}^{\infty}\left(a_p\sin^p\frac{1}{cp}\left(\theta_r+\frac{2}{3}\pi\right)+b_p\cos^p\frac{1}{cp}\left(\theta_r+\frac{2}{3}\pi\right)\right)}\right].$$

The high-fidelity analysis can be accomplished using the motor design and experimental data.

The motor parameters, induced *emf*, torque, magnetic field distribution, and other quantities can be experimentally obtained in the full operating envelope. The characterization can be accomplished using the analytic and experimental data yielding (5.16) or other motor-consistent mutual inductances. One finds the inductance mapping $\mathbf{L}'_{sr}(\theta_r)$. Using (5.4) and (5.16), the circuitry-electromagnetic equations are

$$\mathbf{u}_{abs}=\mathbf{r}_s\mathbf{i}_{abs}+\frac{d\psi_{abs}}{dt},\quad \mathbf{u}'_{abr}=\mathbf{r}'_r\mathbf{i}'_{abr}+\frac{d\psi'_{abr}}{dt},\quad \begin{bmatrix}\psi_{abs}\\ \psi_{abr}\end{bmatrix}=\begin{bmatrix}\mathbf{L}_s & \mathbf{L}'_{sr}(\theta_r)\\ \mathbf{L}'^{T}_{sr}(\theta_r) & \mathbf{L}'_r\end{bmatrix}\begin{bmatrix}\mathbf{i}_{abs}\\ \mathbf{i}'_{abr}\end{bmatrix},\tag{5.17}$$

$$\mathbf{L}_s=\begin{bmatrix}L_{ss} & 0\\ 0 & L_{ss}\end{bmatrix},\quad \mathbf{L}'_r=\begin{bmatrix}L'_{rr} & 0\\ 0 & L'_{rr}\end{bmatrix},\quad \mathbf{L}'_{sr}(\theta_r)=\begin{bmatrix}\sum_{n=1}^{\infty}L_{sr\,n}\cos^{2n-1}\theta_r & -\sum_{n=1}^{\infty}L_{sr\,n}\sin^{2n-1}\theta_r\\ \sum_{n=1}^{\infty}L_{sr\,n}\sin^{2n-1}\theta_r & \sum_{n=1}^{\infty}L_{sr\,n}\cos^{2n-1}\theta_r\end{bmatrix}.$$

The electromagnetic torque is

$$T_e = \frac{P}{2}\frac{\partial W_c\left(\mathbf{i}_{abs}, \mathbf{i}'_{abr}, \theta_r\right)}{\partial \theta_r} = \frac{P}{2}\mathbf{i}_{abs}^T \frac{\partial \mathbf{L}'_{sr}(\theta_r)}{\partial \theta_r}\mathbf{i}'_{abr}. \tag{5.18}$$

The *torsional–mechanical* dynamics is given by (5.7).

Example 5.3

Using (5.16), assume that

$$L_{asar} = L_{aras} = L_{sr1}\cos\theta_r + L_{sr2}\cos^3\theta_r, \quad L_{asbr} = L_{bras} = -L_{sr1}\sin\theta_r - L_{sr2}\sin^3\theta_r,$$

$$L_{bsar} = L_{arbs} = L_{sr1}\sin\theta_r + L_{sr2}\sin^3\theta_r \quad \text{and} \quad L_{bsbr} = L_{brbs} = L_{sr1}\cos\theta_r + L_{sr2}\cos^3\theta_r.$$

Applying the turn ratio, the flux linkages are

$$\begin{bmatrix}\psi_{abs}\\ \psi'_{abr}\end{bmatrix} = \begin{bmatrix}\mathbf{L}_s & \mathbf{L}'_{sr}(\theta_r)\\ \mathbf{L}'^{T}_{sr}(\theta_r) & \mathbf{L}'_r\end{bmatrix}\begin{bmatrix}\mathbf{i}_{abs}\\ \mathbf{i}'_{abr}\end{bmatrix},$$

$$\mathbf{L}'_{sr}(\theta_r) = \left(\frac{N_s}{N_r}\right)\mathbf{L}_{sr}(\theta_r) = \begin{bmatrix}L_{ms1}\cos\theta_r + L_{ms2}\cos^3\theta_r & -L_{ms1}\sin\theta - L_{ms2}\sin^3\theta_r\\ L_{ms1}\sin\theta_r + L_{ms2}\sin^3\theta_r & L_{ms1}\cos\theta_r + L_{ms2}\cos^3\theta_r\end{bmatrix}.$$

Hence,

$$\begin{bmatrix}\psi_{as}\\ \psi_{bs}\\ \psi'_{ar}\\ \psi'_{br}\end{bmatrix} = \begin{bmatrix}L_{ss} & 0 & L_{ms1}\cos\theta_r + L_{ms2}\cos^3\theta_r & -L_{ms1}\sin\theta_r - L_{ms2}\sin^3\theta_r\\ 0 & L_{ss} & L_{ms1}\sin\theta_r + L_{ms2}\sin^3\theta_r & L_{ms1}\cos\theta_r + L_{ms2}\cos^3\theta_r\\ L_{ms1}\cos\theta_r + L_{ms2}\cos^3\theta_r & L_{ms1}\sin\theta_r + L_{ms2}\sin^3\theta_r & L'_{rr} & 0\\ -L_{ms1}\sin\theta_r - L_{ms2}\sin^3\theta_r & L_{ms1}\cos\theta_r + L_{ms2}\cos^3\theta_r & 0 & L'_{rr}\end{bmatrix}\begin{bmatrix}i_{as}\\ i_{bs}\\ i'_{ar}\\ i'_{br}\end{bmatrix}$$

From (5.17), one finds

$$L_{ss}\frac{di_{as}}{dt} + L_{ms1}\frac{d\left(i'_{ar}\cos\theta_r\right)}{dt} + L_{ms2}\frac{d\left(i'_{ar}\cos^3\theta_r\right)}{dt} - L_{ms1}\frac{d\left(i'_{br}\sin\theta_r\right)}{dt} - L_{ms2}\frac{d\left(i'_{br}\sin^3\theta_r\right)}{dt} = -r_s i_{as} + u_{as},$$

$$L_{ss}\frac{di_{bs}}{dt} + L_{ms1}\frac{d\left(i'_{ar}\sin\theta_r\right)}{dt} + L_{ms2}\frac{d\left(i'_{ar}\sin^3\theta_r\right)}{dt} + L_{ms1}\frac{d\left(i'_{br}\cos\theta_r\right)}{dt} + L_{ms2}\frac{d\left(i'_{br}\cos^3\theta_r\right)}{dt} = -r_s i_{bs} + u_{bs},$$

$$L_{ms1}\frac{d\left(i_{as}\cos\theta_r\right)}{dt} + L_{ms2}\frac{d\left(i_{as}\cos^3\theta_r\right)}{dt} + L_{ms1}\frac{d\left(i_{bs}\sin\theta_r\right)}{dt} + L_{ms2}\frac{d\left(i_{bs}\sin^3\theta_r\right)}{dt} + L'_{rr}\frac{di'_{ar}}{dt} = -r'_r i'_{ar} + u'_{ar},$$

$$-L_{ms1}\frac{d\left(i_{as}\sin\theta_r\right)}{dt} - L_{ms2}\frac{d\left(i_{as}\sin^3\theta_r\right)}{dt} + L_{ms1}\frac{d\left(i_{bs}\cos\theta_r\right)}{dt} + L_{ms2}\frac{d\left(i_{bs}\cos^3\theta_r\right)}{dt} + L'_{rr}\frac{di'_{br}}{dt} = -r'_r i'_{br} + u'_{br}.$$

In the *ar* and *br* rotor phases, the induced motional *emfs* in the steady-state operation are

$$emf_{ar\omega} = \left(L_{ms1}i_{as}\sin\theta_r + 3L_{ms2}i_{as}\sin\theta_r\cos^2\theta_r - L_{ms1}i_{bs}\cos\theta_r - 3L_{ms2}i_{bs}\cos\theta_r\sin^2\theta_r\right)\omega_r$$

and

$$emf_{br\omega} = \left(L_{ms1} i_{as} \cos\theta_r + 3L_{ms2} i_{as} \cos\theta_r \sin^2\theta_r + L_{ms1} i_{bs} \sin\theta_r + 3L_{ms2} i_{bs} \sin\theta_r \cos^2\theta_r\right)\omega_r.$$

The electromagnetic torque is

$$T_e = \frac{P}{2}\frac{\partial W_c\left(\mathbf{i}_{abs}, \mathbf{i}'_{abr}, \theta_r\right)}{\partial\theta_r} = \frac{P}{2}\mathbf{i}_{abs}^T \frac{\partial \mathbf{L}'_{sr}(\theta_r)}{\partial\theta_r}\mathbf{i}'_{abr}$$

$$= \frac{P}{2}\begin{bmatrix} i_{as} & i_{bs}\end{bmatrix}\begin{bmatrix} -L_{ms1}\sin\theta_r - 3L_{ms2}\sin\theta_r\cos^2\theta_r & -L_{ms1}\cos\theta_r - 3L_{ms2}\cos\theta_r\sin^2\theta_r \\ L_{ms1}\cos\theta_r + 3L_{ms2}\cos\theta_r\sin^2\theta_r & -L_{ms1}\sin\theta_r - 3L_{ms2}\sin\theta_r\cos^2\theta_r \end{bmatrix}\begin{bmatrix} i'_{ar} \\ i'_{br}\end{bmatrix}$$

$$= -\frac{P}{2}\Big\{L_{ms1}\left[\left(i_{as}i'_{ar} + i_{bs}i'_{br}\right)\sin\theta_r + \left(i_{as}i'_{br} - i_{bs}i'_{ar}\right)\cos\theta_r\right]$$

$$+ 3L_{ms2}\left[\left(i_{as}i'_{ar} + i_{bs}i'_{br}\right)\sin\theta_r\cos^2\theta_r + \left(i_{as}i'_{br} - i_{bs}i'_{ar}\right)\cos\theta_r\sin^2\theta_r\right]\Big\}.$$

Using (5.7), the *torsional–mechanical* equations are

$$\frac{d\omega_r}{dt} = -\frac{P^2}{4J}\Big\{L_{ms1}\left[\left(i_{as}i'_{ar} + i_{bs}i'_{br}\right)\sin\theta_r + \left(i_{as}i'_{br} - i_{bs}i'_{ar}\right)\cos\theta_r\right]$$

$$+ 3L_{ms2}\left[\left(i_{as}i'_{ar} + i_{bs}i'_{br}\right)\sin\theta_r\cos^2\theta_r + \left(i_{as}i'_{br} - i_{bs}i'_{ar}\right)\cos\theta_r\sin^2\theta_r\right]\Big\} - \frac{B_m}{J}\omega_r - \frac{P}{2J}T_L,$$

$$\frac{d\theta_r}{dt} = \omega_r.$$

■

Example 5.4

We simulate an A class two-phase, 115 V (*rms*), 60 Hz, four-pole ($P = 4$) induction motor if:

1. $L_{asar} = L_{aras} = L_{sr}\cos\theta_r$, $L_{asbr} = L_{bras} = -L_{sr}\sin\theta_r$, $L_{bsar} = L_{arbs} = L_{sr}\sin\theta_r$ and $L_{bsbr} = L_{brbs} = L_{sr}\cos\theta_r$;
2. $L_{asar} = L_{aras} = L_{sr1}\cos\theta_r + L_{sr2}\cos^3\theta_r$, $L_{asbr} = L_{bras} = -L_{sr1}\sin\theta_r - L_{sr2}\sin^3\theta_r$, $L_{bsar} = L_{arbs} = L_{sr1}\sin\theta_r + L_{sr2}\sin^3\theta_r$ and $L_{bsbr} = L_{brbs} = L_{sr1}\cos\theta_r + L_{sr2}\cos^3\theta_r$.

The motor parameters are as in Example 5.2. We have $r_s = 1.2$ ohm, $r_r' = 1.5$ ohm, $L_{ms1} = 0.145$ H, $L_{ms2} = 0.005$ H, $L_{ls} = 0.02$ H, $L_{ss} = L_{ls} + L_{ms1} + 3L_{ms2}$, $L'_{lr} = 0.02$ H, $L'_{rr} = L'_{lr} + L_{ms1} + 3L_{ms2}$, $B_m = 1 \times 10^{-6}$ N-m-sec/rad and $J = 0.005$ kg-m^2.

The supplied phase voltages are $u_{as}(t) = \sqrt{2}115\cos(377t)$ and $u_{bs}(t) = \sqrt{2}115\sin(377t)$.

The transient dynamic and torque–speed characteristics for

$$\mathbf{L}'_{sr}(\theta_r) = \begin{bmatrix} L_{sr}\cos\theta_r & -L_{sr}\sin\theta_r \\ L_{sr}\sin\theta_r & L_{sr}\cos\theta_r \end{bmatrix}$$

are reported in Figure 5.8a.

For

$$\mathbf{L}'_{sr}(\theta_r) = \begin{bmatrix} L_{sr1}\cos\theta_r + L_{sr2}\cos^3\theta_r & -L_{sr1}\sin\theta_r - L_{sr2}\sin^3\theta_r \\ L_{sr1}\sin\theta_r + L_{sr2}\sin^3\theta_r & L_{sr1}\cos\theta_r + L_{sr2}\cos^3\theta_r \end{bmatrix},$$

the results are documented in Figures 5.8b and c for unloaded and loaded ($T_L = 0.5$ N-m at $t = 1.5$ sec) motors.

The motor accelerates from the stall, and one can assess: (1) Acceleration capabilities; (2) Transient dynamics of the state variables $i_{as}(t)$, $i_{bs}(t)$, $i'_{ar}(t)$, $i'_{br}(t)$, $\omega_r(t)$, and $\theta_r(t)$; (3) Evolution of the electromagnetic torque T_e; (4) Dynamic torque–speed characteristics $\omega_r(t) = \Omega_T[T_e(t)]$; (5) Efficiency and losses; (6) Motional *emfs* induced in the rotor windings; etc. The results indicate that even small deviations from the ideal sinusoidal stator–rotor magnetic coupling result in the degradation of the motor performance and capabilities. We found degraded acceleration capabilities, and torque ripple (which results in vibration, noise, mechanical wearing, etc.), higher losses. The analysis performed supports the need for consistent design and analysis with minimum level of simplifications and assumptions. ■

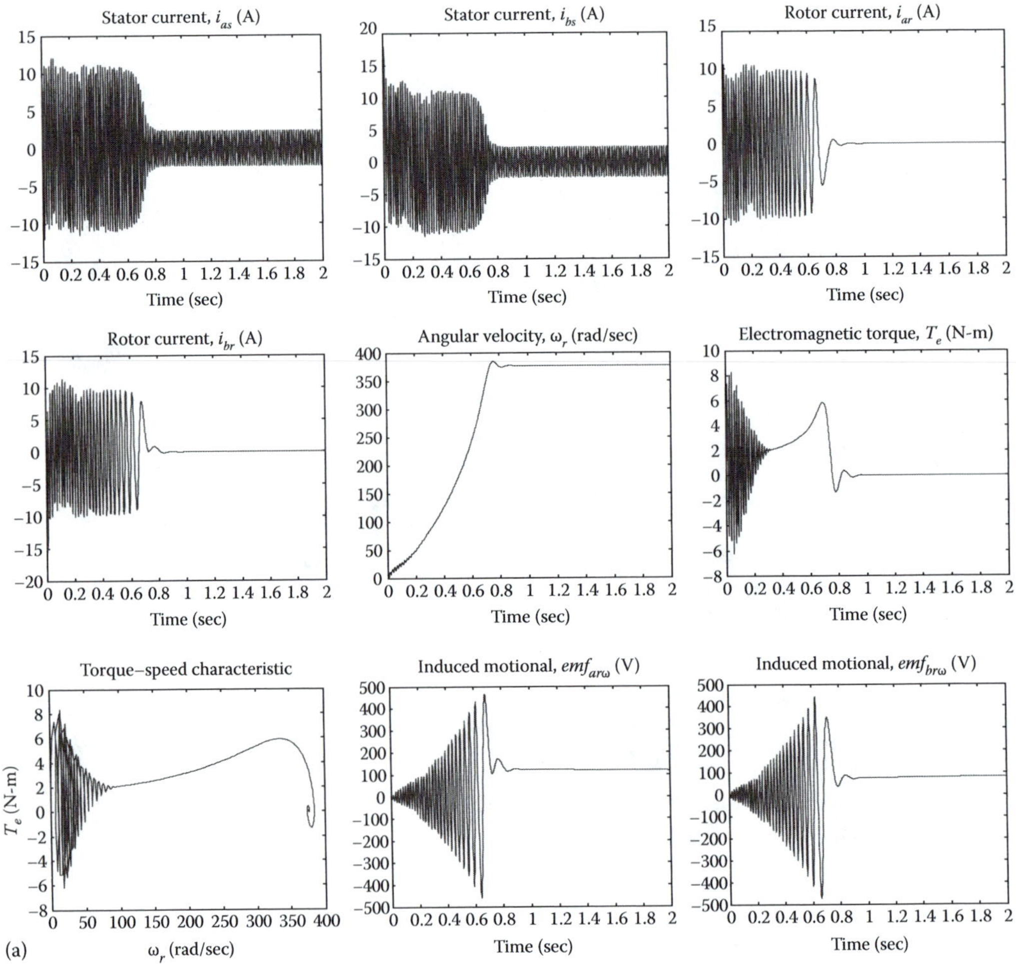

FIGURE 5.8 Dynamics of an A class induction motor: (a) $T_L = 0$ N-m, $\mathbf{L}'_{sr}(\theta_r) = \begin{bmatrix} L_{sr}\cos\theta_r & -L_{sr}\sin\theta_r \\ L_{sr}\sin\theta_r & L_{sr}\cos\theta_r \end{bmatrix}$;

(Continued)

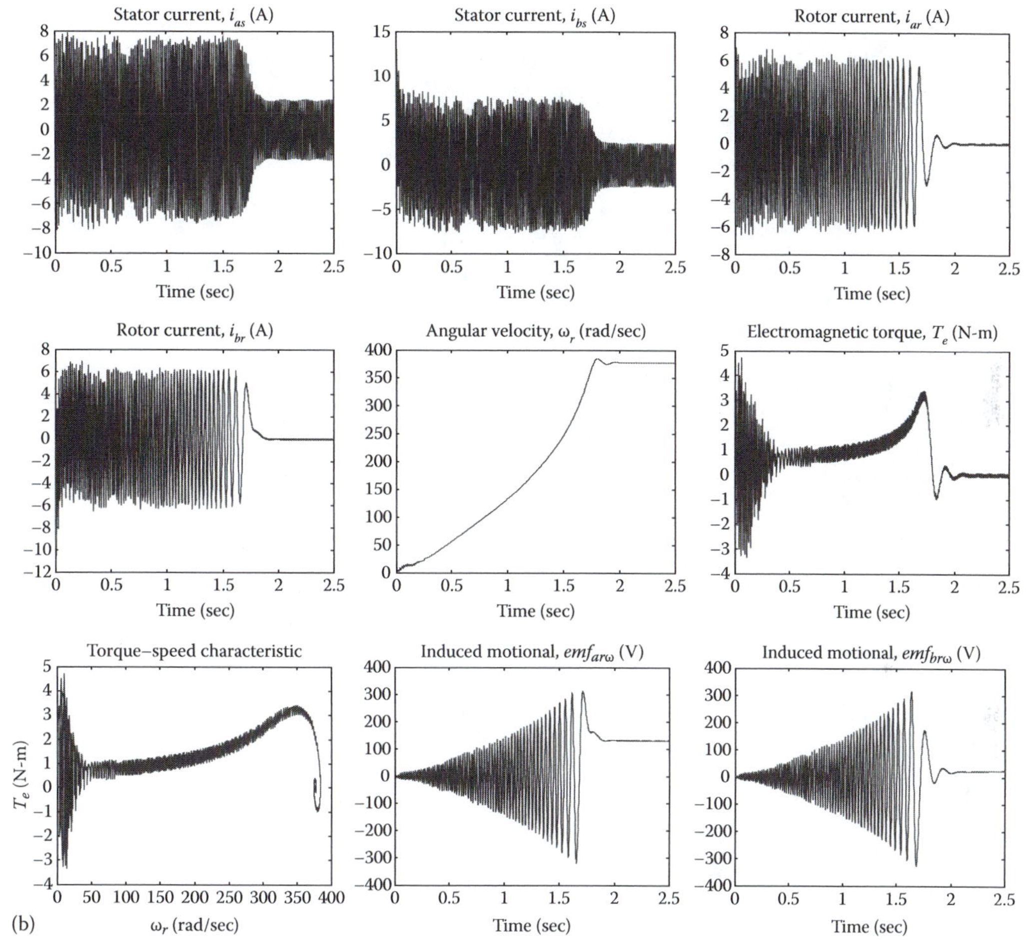

FIGURE 5.8 (*Continued*) Dynamics of an A class induction motor:

(b) $T_L = 0$ N-m, $\mathbf{L}'_{sr}(\theta_r) = \begin{bmatrix} L_{sr1}\cos\theta_r + L_{sr2}\cos^3\theta_r & -L_{sr1}\sin\theta_r - L_{sr2}\sin^3\theta_r \\ L_{sr1}\sin\theta_r + L_{sr2}\sin^3\theta_r & L_{sr1}\cos\theta_r + L_{sr2}\cos^3\theta_r \end{bmatrix}$; (*Continued*)

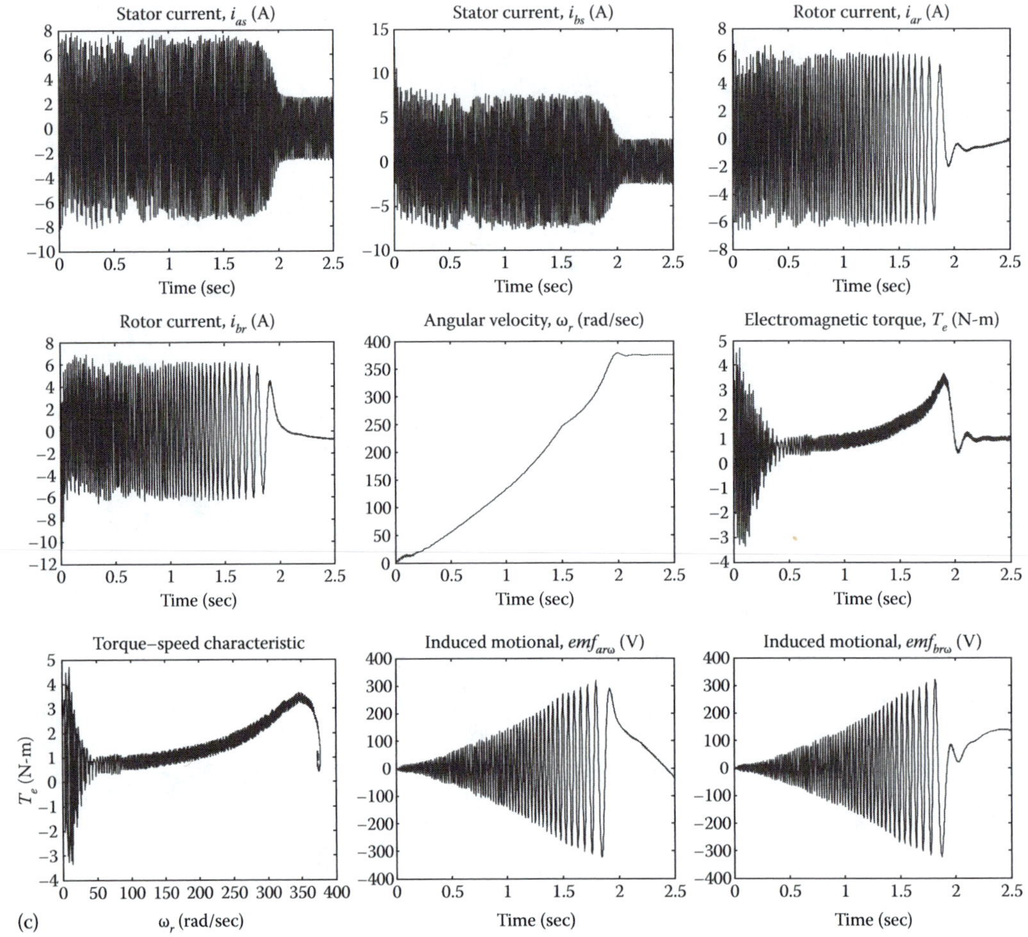

FIGURE 5.8 (*Continued*) Dynamics of an A class induction motor:

(c) $T_L = 0.5$ N-m (at $t = 1.5$ sec), $\mathbf{L}'_{sr}(\theta_r) = \begin{bmatrix} L_{sr1}\cos\theta_r + L_{sr2}\cos^3\theta_r & -L_{sr1}\sin\theta_r - L_{sr2}\sin^3\theta_r \\ L_{sr1}\sin\theta_r + L_{sr2}\sin^3\theta_r & L_{sr1}\cos\theta_r + L_{sr2}\cos^3\theta_r \end{bmatrix}$.

5.4 THREE-PHASE INDUCTION MOTORS IN THE MACHINE VARIABLES

The majority of industrial induction machines are three-phase motors. Our goal is to accomplish various analysis and design tasks for three-phase induction motors shown in Figure 5.9.

Kirchhoff's voltage law for the *abc* stator and rotor voltages, currents, and flux linkages yields

$$\begin{aligned} u_{as} &= r_s i_{as} + \frac{d\psi_{as}}{dt}, \quad u_{bs} = r_s i_{bs} + \frac{d\psi_{bs}}{dt}, \quad u_{cs} = r_s i_{cs} + \frac{d\psi_{cs}}{dt} \\ u_{ar} &= r_r i_{ar} + \frac{d\psi_{ar}}{dt}, \quad u_{br} = r_r i_{br} + \frac{d\psi_{br}}{dt}, \quad u_{cr} = r_r i_{cr} + \frac{d\psi_{cr}}{dt}, \end{aligned} \tag{5.19}$$

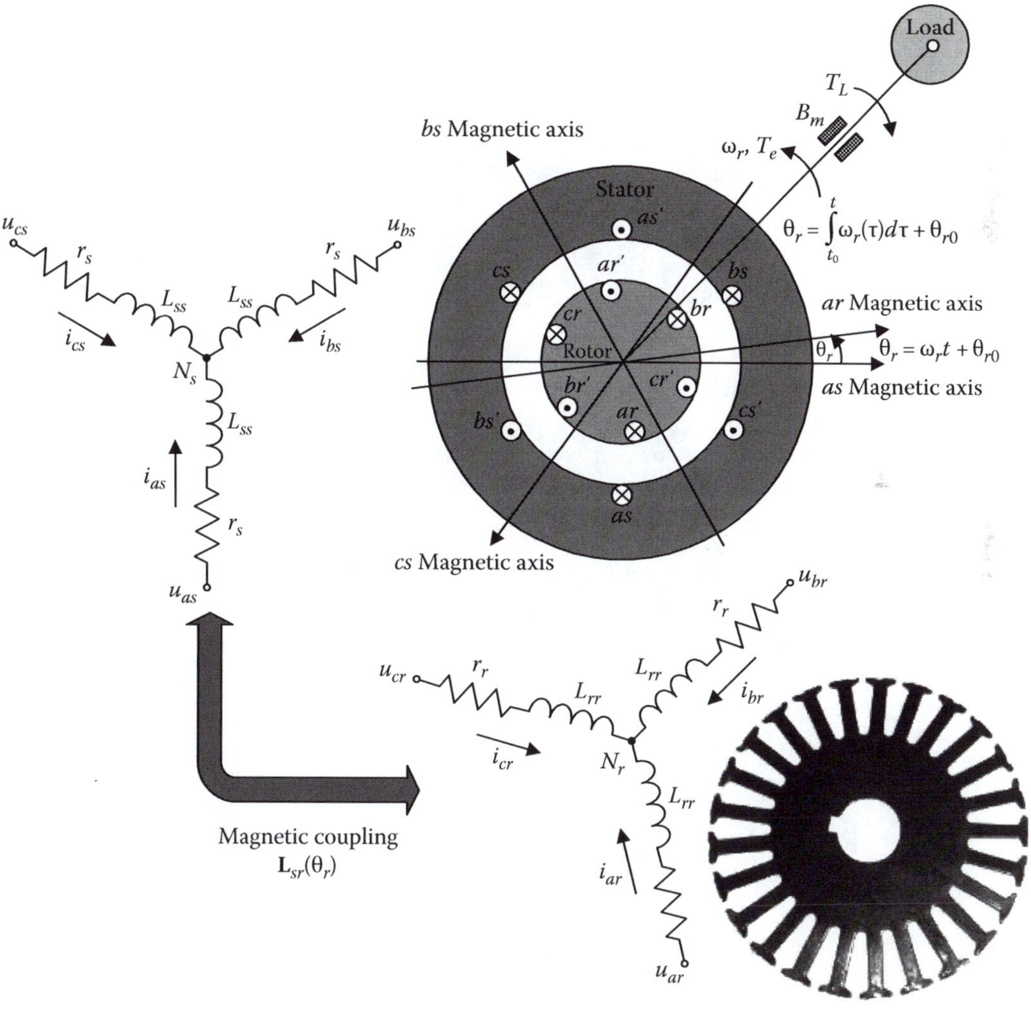

FIGURE 5.9 Three-phase symmetric induction motor: Rotor windings are placed in the slots in the laminated rotor made from electric steel.

where u_{as}, u_{bs}, and u_{cs} are the phase voltages supplied to the *as*, *bs*, and *cs* stator windings; u_{ar}, u_{br}, and u_{cr} are the phase voltages supplied to the *ar*, *br*, and *cr* rotor windings in wound induction motors (in squirrel-cage motors, the rotor windings are short-circuited, and, u_{ar}, u_{br}, u_{cr} are not supplied, and the motional *emfs* are induced in the *ar*, *br*, and *cr* windings); i_{as}, i_{bs}, and i_{cs} are the phase currents in the stator windings; i_{ar}, i_{br}, and i_{cr} are the phase currents in the rotor windings; ψ_{as}, ψ_{bs}, and ψ_{cs} are the stator flux linkages; ψ_{ar}, ψ_{br}, and ψ_{cr} are the rotor flux linkages.

From (5.19), using the vector notations for the *abc* voltages, currents, and flux linkages, one has

$$\mathbf{u}_{abcs} = \mathbf{r}_s \mathbf{i}_{abcs} + \frac{d\boldsymbol{\psi}_{abcs}}{dt}, \quad \mathbf{u}_{abcs} = \begin{bmatrix} u_{as} \\ u_{bs} \\ u_{cs} \end{bmatrix}, \quad \mathbf{i}_{abcs} = \begin{bmatrix} i_{as} \\ i_{bs} \\ i_{cs} \end{bmatrix}, \quad \boldsymbol{\psi}_{abcs} = \begin{bmatrix} \psi_{as} \\ \psi_{bs} \\ \psi_{cs} \end{bmatrix},$$
$$\mathbf{u}_{abcr} = \mathbf{r}_r \mathbf{i}_{abcr} + \frac{d\boldsymbol{\psi}_{abcr}}{dt}, \quad \mathbf{u}_{abcr} = \begin{bmatrix} u_{ar} \\ u_{br} \\ u_{cr} \end{bmatrix}, \quad \mathbf{i}_{abcr} = \begin{bmatrix} i_{ar} \\ i_{br} \\ i_{cr} \end{bmatrix}, \quad \boldsymbol{\psi}_{abcr} = \begin{bmatrix} \psi_{ar} \\ \psi_{br} \\ \psi_{cr} \end{bmatrix}, \tag{5.20}$$

where the diagonal stator and rotor resistances matrices are

$$\mathbf{r}_s = \begin{bmatrix} r_s & 0 & 0 \\ 0 & r_s & 0 \\ 0 & 0 & r_s \end{bmatrix} \text{ and } \mathbf{r}_r = \begin{bmatrix} r_r & 0 & 0 \\ 0 & r_r & 0 \\ 0 & 0 & r_r \end{bmatrix}.$$

Using the self- and mutual inductances, the flux linkages are found as functions of the corresponding currents in the stator and rotor windings. The analysis of the stator and rotor magnetically coupled system, as represented in Figure 5.9, yields

$$\psi_{as} = L_{asas} i_{as} + L_{asbs} i_{bs} + L_{ascs} i_{cs} + L_{asar} i_{ar} + L_{asbr} i_{br} + L_{ascr} i_{cr},$$

$$\psi_{bs} = L_{bsas} i_{as} + L_{bsbs} i_{bs} + L_{bscs} i_{cs} + L_{bsar} i_{ar} + L_{bsbr} i_{br} + L_{bscr} i_{cr},$$

$$\psi_{cs} = L_{csas} i_{as} + L_{csbs} i_{bs} + L_{cscs} i_{cs} + L_{csar} i_{ar} + L_{csbr} i_{br} + L_{cscr} i_{cr},$$

$$\psi_{ar} = L_{aras} i_{as} + L_{arbs} i_{bs} + L_{arcs} i_{cs} + L_{arar} i_{ar} + L_{arbr} i_{br} + L_{arcr} i_{cr},$$

$$\psi_{br} = L_{bras} i_{as} + L_{brbs} i_{bs} + L_{brcs} i_{cs} + L_{brar} i_{ar} + L_{brbr} i_{br} + L_{brcr} i_{cr},$$

$$\psi_{cr} = L_{cras} i_{as} + L_{crbs} i_{bs} + L_{crcs} i_{cs} + L_{crar} i_{ar} + L_{crbr} i_{br} + L_{crcr} i_{cr},$$

where L_{asas}, L_{bsbs}, L_{cscs}, L_{arar}, L_{brbr}, and L_{crcr} are the stator and rotor self-inductances; L_{asbs}, L_{ascs}, L_{asar}, L_{asbr}, L_{ascr}, …, L_{cras}, L_{crbs}, L_{crcs}, L_{crar}, and L_{crbr} are the mutual inductances between stator–stator, stator–rotor, and rotor–rotor windings.

The stator and rotor windings are identical and displaced magnetically by $2\pi/3$. There exists a coupling between the *abc* stator and rotor windings. The mutual inductances between the stator windings are equal, and

$$L_{asbs} = L_{ascs} = L_{bscs} = L_{ms}\cos\left(\frac{2}{3}\pi\right) = -\frac{1}{2}L_{ms},\ L_{ms} = \frac{N_s^2}{\Re_m}.$$

The rotor windings are also displaced by 120 electrical degrees. The mutual inductances between the rotor windings are

$$L_{arbr} = L_{arcr} = L_{brcr} = L_{mr}\cos\left(\frac{2}{3}\pi\right) = -\frac{1}{2}L_{mr},\quad L_{mr} = \frac{N_r^2}{\Re_m} = \frac{N_r^2}{N_s^2}L_{ms}.$$

The stator and rotor self-inductances are $L_{ss} = L_{ls} + L_{ms}$ and $L_{rr} = L_{lr} + L_{mr}$.

The matrices of self- and mutual inductances $\mathbf{L}_s$ and $\mathbf{L}_r$ are

$$\mathbf{L}_s = \begin{bmatrix} L_{ls}+L_{ms} & -\frac{1}{2}L_{ms} & -\frac{1}{2}L_{ms} \\ -\frac{1}{2}L_{ms} & L_{ls}+L_{ms} & -\frac{1}{2}L_{ms} \\ -\frac{1}{2}L_{ms} & -\frac{1}{2}L_{ms} & L_{ls}+L_{ms} \end{bmatrix} \text{ and } \mathbf{L}_r = \begin{bmatrix} L_{lr}+L_{mr} & -\frac{1}{2}L_{mr} & -\frac{1}{2}L_{mr} \\ -\frac{1}{2}L_{mr} & L_{lr}+L_{mr} & -\frac{1}{2}L_{mr} \\ -\frac{1}{2}L_{mr} & -\frac{1}{2}L_{mr} & L_{lr}+L_{mr} \end{bmatrix}.$$

The mutual inductances between the stator and rotor windings are periodic functions of the electrical angular displacement θ_r. The period of variations is 2π. Assume that the mutual inductances are sinusoidal functions and the rotor and stator windings are initially positioned such as

$$L_{asar} = L_{aras} = L_{sr}\cos\theta_r,\quad L_{asbr} = L_{bras} = L_{sr}\cos\left(\theta_r + \frac{2}{3}\pi\right),\quad L_{ascr} = L_{cras} = L_{sr}\cos\left(\theta_r - \frac{2}{3}\pi\right),$$

$$L_{bsar} = L_{arbs} = L_{sr}\cos\left(\theta_r - \frac{2}{3}\pi\right),\quad L_{bsbr} = L_{brbs} = L_{sr}\cos\theta_r,\quad L_{bscr} = L_{crbs} = L_{sr}\cos\left(\theta_r + \frac{2}{3}\pi\right),$$

$$L_{csar} = L_{arcs} = L_{sr}\cos\left(\theta_r + \frac{2}{3}\pi\right),\quad L_{csbr} = L_{brcs} = L_{sr}\cos\left(\theta_r - \frac{2}{3}\pi\right),\quad L_{cscr} = L_{crcs} = L_{sr}\cos\theta_r,$$

where $L_{sr} = N_sN_r/\Re_m$.

The stator–rotor mutual inductance mapping is

$$\mathbf{L}_{sr}(\theta_r) = L_{sr}\begin{bmatrix} \cos\theta_r & \cos\left(\theta_r + \frac{2}{3}\pi\right) & \cos\left(\theta_r - \frac{2}{3}\pi\right) \\ \cos\left(\theta_r - \frac{2}{3}\pi\right) & \cos\theta_r & \cos\left(\theta_r + \frac{2}{3}\pi\right) \\ \cos\left(\theta_r + \frac{2}{3}\pi\right) & \cos\left(\theta_r - \frac{2}{3}\pi\right) & \cos\theta_r \end{bmatrix}.$$

One obtains

$$\begin{bmatrix} \psi_{abcs} \\ \psi_{abcr} \end{bmatrix} = \begin{bmatrix} \mathbf{L}_s & \mathbf{L}_{sr}(\theta_r) \\ \mathbf{L}_{sr}^T(\theta_r) & \mathbf{L}_r \end{bmatrix}\begin{bmatrix} \mathbf{i}_{abcs} \\ \mathbf{i}_{abcr} \end{bmatrix},\quad \psi_{abcs} = \mathbf{L}_s\mathbf{i}_{abcs} + \mathbf{L}_{sr}(\theta_r)\mathbf{i}_{abcr},\quad \psi_{abcr} = \mathbf{L}_{sr}^T(\theta_r)\mathbf{i}_{abcs} + \mathbf{L}_r\mathbf{i}_{abcr}. \tag{5.21}$$

Using the number of turns N_s and N_r,

$$\mathbf{u}'_{abcr} = \frac{N_s}{N_r}\mathbf{u}_{abcr}, \quad \mathbf{i}'_{abcr} = \frac{N_r}{N_s}\mathbf{i}_{abcr}, \quad \text{and} \quad \boldsymbol{\psi}'_{abcr} = \frac{N_s}{N_r}\boldsymbol{\psi}_{abcr}.$$

The inductances are

$$L_{ms} = \frac{N_s}{N_r}L_{sr}, \quad L_{sr} = \frac{N_s N_r}{\Re_m}, \quad \text{and} \quad L_{ms} = \frac{N_s^2}{\Re_m}.$$

One finds

$$\mathbf{L}'_{sr}(\theta_r) = \frac{N_s}{N_r}\mathbf{L}_{sr}(\theta_r) = L_{ms}\begin{bmatrix} \cos\theta_r & \cos\left(\theta_r + \frac{2}{3}\pi\right) & \cos\left(\theta_r - \frac{2}{3}\pi\right) \\ \cos\left(\theta_r - \frac{2}{3}\pi\right) & \cos\theta_r & \cos\left(\theta_r + \frac{2}{3}\pi\right) \\ \cos\left(\theta_r + \frac{2}{3}\pi\right) & \cos\left(\theta_r - \frac{2}{3}\pi\right) & \cos\theta_r \end{bmatrix},$$

and

$$\mathbf{L}'_r = \frac{N_s^2}{N_r^2}\mathbf{L}_r = \begin{bmatrix} L'_{lr} + L_{ms} & -\frac{1}{2}L_{ms} & -\frac{1}{2}L_{ms} \\ -\frac{1}{2}L_{ms} & L'_{lr} + L_{ms} & -\frac{1}{2}L_{ms} \\ -\frac{1}{2}L_{ms} & -\frac{1}{2}L_{ms} & L'_{lr} + L_{ms} \end{bmatrix}, \quad L'_{lr} = \frac{N_s^2}{N_r^2}L_{lr}.$$

Equations (5.21) for the flux linkages are rewritten as

$$\begin{bmatrix} \boldsymbol{\psi}_{abcs} \\ \boldsymbol{\psi}'_{abcr} \end{bmatrix} = \begin{bmatrix} \mathbf{L}_s & \mathbf{L}'_{sr}(\theta_r) \\ \mathbf{L}'^{T}_{sr}(\theta_r) & \mathbf{L}'_r \end{bmatrix}\begin{bmatrix} \mathbf{i}_{abcs} \\ \mathbf{i}'_{abcr} \end{bmatrix}, \tag{5.22}$$

$$\begin{bmatrix} \psi_{as} \\ \psi_{bs} \\ \psi_{cs} \\ \psi'_{ar} \\ \psi'_{br} \\ \psi'_{cr} \end{bmatrix} = \begin{bmatrix} L_{ls} + L_{ms} & -\frac{1}{2}L_{ms} & -\frac{1}{2}L_{ms} & L_{ms}\cos\theta_r & L_{ms}\cos\left(\theta_r + \frac{2}{3}\pi\right) & L_{ms}\cos\left(\theta_r - \frac{2}{3}\pi\right) \\ -\frac{1}{2}L_{ms} & L_{ls} + L_{ms} & -\frac{1}{2}L_{ms} & L_{ms}\cos\left(\theta_r - \frac{2}{3}\pi\right) & L_{ms}\cos\theta_r & L_{ms}\cos\left(\theta_r + \frac{2}{3}\pi\right) \\ -\frac{1}{2}L_{ms} & -\frac{1}{2}L_{ms} & L_{ls} + L_{ms} & L_{ms}\cos\left(\theta_r + \frac{2}{3}\pi\right) & L_{ms}\cos\left(\theta_r - \frac{2}{3}\pi\right) & L_{ms}\cos\theta_r \\ L_{ms}\cos\theta_r & L_{ms}\cos\left(\theta_r - \frac{2}{3}\pi\right) & L_{ms}\cos\left(\theta_r + \frac{2}{3}\pi\right) & L'_{lr} + L_{ms} & -\frac{1}{2}L_{ms} & -\frac{1}{2}L_{ms} \\ L_{ms}\cos\left(\theta_r + \frac{2}{3}\pi\right) & L_{ms}\cos\theta_r & L_{ms}\cos\left(\theta_r - \frac{2}{3}\pi\right) & -\frac{1}{2}L_{ms} & L'_{lr} + L_{ms} & -\frac{1}{2}L_{ms} \\ L_{ms}\cos\left(\theta_r - \frac{2}{3}\pi\right) & L_{ms}\cos\left(\theta_r + \frac{2}{3}\pi\right) & L_{ms}\cos\theta_r & -\frac{1}{2}L_{ms} & -\frac{1}{2}L_{ms} & L'_{lr} + L_{ms} \end{bmatrix}\begin{bmatrix} i_{as} \\ i_{bs} \\ i_{cs} \\ i'_{ar} \\ i'_{br} \\ i'_{cr} \end{bmatrix}.$$

Using (5.20) and (5.22), we obtain

$$\begin{aligned} \mathbf{u}_{abcs} &= \mathbf{r}_s\mathbf{i}_{abcs} + \frac{d\boldsymbol{\psi}_{abcs}}{dt} = \mathbf{r}_s\mathbf{i}_{abcs} + \mathbf{L}_s\frac{d\mathbf{i}_{abcs}}{dt} + \frac{d(\mathbf{L}'_{sr}(\theta_r)\mathbf{i}'_{abcr})}{dt}, \\ \mathbf{u}'_{abcr} &= \mathbf{r}'_r\mathbf{i}'_{abcr} + \frac{d\boldsymbol{\psi}_{abcr}}{dt} = \mathbf{r}'_r\mathbf{i}'_{abcr} + \mathbf{L}'_r\frac{d\mathbf{i}'_{abcr}}{dt} + \frac{d(\mathbf{L}'^{T}_{sr}(\theta_r)\mathbf{i}_{abcs})}{dt}, \end{aligned} \tag{5.23}$$

where $\mathbf{r}'_r = \frac{N_s^2}{N_r^2}\mathbf{r}_r$.

The total derivatives of the flux linkages $\frac{d\psi_{abcs}}{dt}$ and $\frac{d\psi'_{abcr}}{dt}$ yield the expressions for the *emfs*. Equations (5.23) are written in expanded form as follows

$$u_{as} = r_s i_{as} + (L_{ls} + L_{ms})\frac{di_{as}}{dt} - \frac{1}{2}L_{ms}\frac{di_{bs}}{dt} - \frac{1}{2}L_{ms}\frac{di_{cs}}{dt} + L_{ms}\frac{d(i'_{ar}\cos\theta_r)}{dt} + L_{ms}\frac{d\left(i'_{br}\cos\left(\theta_r + \frac{2\pi}{3}\right)\right)}{dt}$$

$$+ L_{ms}\frac{d\left(i'_{cr}\cos\left(\theta_r - \frac{2\pi}{3}\right)\right)}{dt},$$

$$u_{bs} = r_s i_{bs} - \frac{1}{2}L_{ms}\frac{di_{as}}{dt} + (L_{ls} + L_{ms})\frac{di_{bs}}{dt} - \frac{1}{2}L_{ms}\frac{di_{cs}}{dt} + L_{ms}\frac{d\left(i'_{ar}\cos\left(\theta_r - \frac{2\pi}{3}\right)\right)}{dt} + L_{ms}\frac{d(i'_{br}\cos\theta_r)}{dt}$$

$$+ L_{ms}\frac{d\left(i'_{cr}\cos\left(\theta_r + \frac{2\pi}{3}\right)\right)}{dt},$$

$$u_{cs} = r_s i_{cs} - \frac{1}{2}L_{ms}\frac{di_{as}}{dt} - \frac{1}{2}L_{ms}\frac{di_{bs}}{dt} + (L_{ls} + L_{ms})\frac{di_{cs}}{dt} + L_{ms}\frac{d\left(i'_{ar}\cos\left(\theta_r + \frac{2\pi}{3}\right)\right)}{dt} + L_{ms}\frac{d\left(i'_{br}\cos\left(\theta_r - \frac{2\pi}{3}\right)\right)}{dt}$$

$$+ L_{ms}\frac{d(i'_{cr}\cos\theta_r)}{dt},$$

$$u'_{ar} = r'_r i'_{ar} + L_{ms}\frac{d(i_{as}\cos\theta_r)}{dt} + L_{ms}\frac{d\left(i_{bs}\cos\left(\theta_r - \frac{2\pi}{3}\right)\right)}{dt} + L_{ms}\frac{d\left(i_{cs}\cos\left(\theta_r + \frac{2\pi}{3}\right)\right)}{dt}$$

$$+ (L'_{lr} + L_{ms})\frac{di'_{ar}}{dt} - \frac{1}{2}L_{ms}\frac{di'_{br}}{dt} - \frac{1}{2}L_{ms}\frac{di'_{cr}}{dt},$$

$$u'_{br} = r'_r i'_{br} + L_{ms}\frac{d\left(i_{as}\cos\left(\theta_r + \frac{2\pi}{3}\right)\right)}{dt} + L_{ms}\frac{d(i_{bs}\cos\theta_r)}{dt} + L_{ms}\frac{d\left(i_{cs}\cos\left(\theta_r - \frac{2\pi}{3}\right)\right)}{dt} - \frac{1}{2}L_{ms}\frac{di'_{ar}}{dt}$$

$$+ (L'_{lr} + L_{ms})\frac{di'_{br}}{dt} - \frac{1}{2}L_{ms}\frac{di'_{cr}}{dt},$$

$$u'_{cr} = r'_r i'_{cr} + L_{ms}\frac{d\left(i_{as}\cos\left(\theta_r - \frac{2\pi}{3}\right)\right)}{dt} + L_{ms}\frac{d\left(i_{bs}\cos\left(\theta_r + \frac{2\pi}{3}\right)\right)}{dt} + L_{ms}\frac{d(i_{cs}\cos\theta_r)}{dt} - \frac{1}{2}L_{ms}\frac{di'_{ar}}{dt}$$

$$- \frac{1}{2}L_{ms}\frac{di'_{br}}{dt} + (L'_{lr} + L_{ms})\frac{di'_{cr}}{dt}.$$

We obtain equations that describe the circuitry-electromagnetic dynamics of three-phase induction motors

$$u_{as} = r_s i_{as} + (L_{ls} + L_{ms})\frac{di_{as}}{dt} - \frac{1}{2}L_{ms}\frac{di_{bs}}{dt} - \frac{1}{2}L_{ms}\frac{di_{cs}}{dt} + L_{ms}\cos\theta_r \frac{di'_{ar}}{dt} + L_{ms}\cos\left(\theta_r + \frac{2\pi}{3}\right)\frac{di'_{br}}{dt}$$

$$+ L_{ms}\cos\left(\theta_r - \frac{2\pi}{3}\right)\frac{di'_{cr}}{dt} - L_{ms}\left[i'_{ar}\sin\theta_r + i'_{br}\sin\left(\theta_r + \frac{2\pi}{3}\right) + i'_{cr}\sin\left(\theta_r - \frac{2\pi}{3}\right)\right]\omega_r,$$

$$u_{bs} = r_s i_{bs} - \frac{1}{2}L_{ms}\frac{di_{as}}{dt} + (L_{ls} + L_{ms})\frac{di_{bs}}{dt} - \frac{1}{2}L_{ms}\frac{di_{cs}}{dt} + L_{ms}\cos\left(\theta_r - \frac{2\pi}{3}\right)\frac{di'_{ar}}{dt} + L_{ms}\cos\theta_r \frac{di'_{br}}{dt}$$

$$+ L_{ms}\cos\left(\theta_r + \frac{2\pi}{3}\right)\frac{di'_{cr}}{dt} - L_{ms}\left[i'_{ar}\sin\left(\theta_r - \frac{2\pi}{3}\right) + i'_{br}\sin\theta_r + i'_{cr}\sin\left(\theta_r + \frac{2\pi}{3}\right)\right]\omega_r,$$

$$u_{cs} = r_s i_{cs} - \frac{1}{2}L_{ms}\frac{di_{as}}{dt} - \frac{1}{2}L_{ms}\frac{di_{bs}}{dt} + (L_{ls} + L_{ms})\frac{di_{cs}}{dt} + L_{ms}\cos\left(\theta_r + \frac{2\pi}{3}\right)\frac{di'_{ar}}{dt} + L_{ms}\cos\left(\theta_r - \frac{2\pi}{3}\right)\frac{di'_{br}}{dt}$$

$$+ L_{ms}\cos\theta_r \frac{di'_{cr}}{dt} - L_{ms}\left[i'_{ar}\sin\left(\theta_r + \frac{2\pi}{3}\right) + i'_{br}\sin\left(\theta_r - \frac{2\pi}{3}\right) + i'_{cr}\sin\theta_r\right]\omega_r,$$

$$u'_{ar} = r'_r i'_{ar} + L_{ms}\cos\theta_r \frac{di_{as}}{dt} + L_{ms}\cos\left(\theta_r - \frac{2\pi}{3}\right)\frac{di_{bs}}{dt} + L_{ms}\cos\left(\theta_r + \frac{2\pi}{3}\right)\frac{di_{cs}}{dt} + (L'_{lr} + L_{ms})\frac{di'_{ar}}{dt}$$

$$- \frac{1}{2}L_{ms}\frac{di'_{br}}{dt} - \frac{1}{2}L_{ms}\frac{di'_{cr}}{dt} - L_{ms}\left[i_{as}\sin\theta_r + i_{bs}\sin\left(\theta_r - \frac{2\pi}{3}\right) + i_{cs}\sin\left(\theta_r + \frac{2\pi}{3}\right)\right]\omega_r,$$

$$u'_{br} = r'_r i'_{br} + L_{ms}\cos\left(\theta_r + \frac{2\pi}{3}\right)\frac{di_{as}}{dt} + L_{ms}\cos\theta_r \frac{di_{bs}}{dt} + L_{ms}\cos\left(\theta_r - \frac{2\pi}{3}\right)\frac{di_{cs}}{dt} - \frac{1}{2}L_{ms}\frac{di'_{ar}}{dt}$$

$$+ (L'_{lr} + L_{ms})\frac{di'_{br}}{dt} - \frac{1}{2}L_{ms}\frac{di'_{cr}}{dt} - L_{ms}\left[i_{as}\sin\left(\theta_r + \frac{2\pi}{3}\right) + i_{bs}\sin\theta_r + i_{cs}\sin\left(\theta_r - \frac{2\pi}{3}\right)\right]\omega_r,$$

$$u'_{cr} = r'_r i'_{cr} + L_{ms}\cos\left(\theta_r - \frac{2\pi}{3}\right)\frac{di_{as}}{dt} + L_{ms}\cos\left(\theta_r + \frac{2\pi}{3}\right)\frac{di_{bs}}{dt} + L_{ms}\cos\theta_r \frac{di_{cs}}{dt} - \frac{1}{2}L_{ms}\frac{di'_{ar}}{dt}$$

$$- \frac{1}{2}L_{ms}\frac{di'_{br}}{dt} + (L'_{lr} + L_{ms})\frac{di'_{cr}}{dt} - L_{ms}\left[i_{as}\sin\left(\theta_r - \frac{2\pi}{3}\right) + i_{bs}\sin\left(\theta_r + \frac{2\pi}{3}\right) + i_{cs}\sin\theta_r\right]\omega_r. \quad (5.24)$$

Differential Equations (5.24) yield the differential equations in Cauchy's form as

$$
\begin{bmatrix} \frac{di_{as}}{dt} \\ \frac{di_{bs}}{dt} \\ \frac{di_{cs}}{dt} \\ \frac{di'_{ar}}{dt} \\ \frac{di'_{br}}{dt} \\ \frac{di'_{cr}}{dt} \end{bmatrix}
= \frac{1}{L_{\Sigma L}}
\begin{bmatrix}
-r_s L_{\Sigma m} & -\frac{1}{2} r_s L_{ms} & -\frac{1}{2} r_s L_{ms} & 0 & 0 & 0 \\
-\frac{1}{2} r_s L_{ms} & -r_s L_{\Sigma m} & -\frac{1}{2} r_s L_{ms} & 0 & 0 & 0 \\
-\frac{1}{2} r_s L_{ms} & -\frac{1}{2} r_s L_{ms} & -r_s L_{\Sigma m} & 0 & 0 & 0 \\
0 & 0 & 0 & -r_r L_{\Sigma m} & -\frac{1}{2} r_r L_{ms} & -\frac{1}{2} r_r L_{ms} \\
0 & 0 & 0 & -\frac{1}{2} r_r L_{ms} & -r_r L_{\Sigma m} & -\frac{1}{2} r_r L_{ms} \\
0 & 0 & 0 & -\frac{1}{2} r_r L_{ms} & -\frac{1}{2} r_r L_{ms} & -r_r L_{\Sigma m}
\end{bmatrix}
\begin{bmatrix} i_{as} \\ i_{bs} \\ i_{cs} \\ i'_{ar} \\ i'_{br} \\ i'_{cr} \end{bmatrix}
$$

$$
+ \frac{1}{L_{\Sigma L}}
\begin{bmatrix}
0 & 0 & 0 & r_r L_{ms} \cos\theta_r & r_r L_{ms} \cos\left(\theta_r + \frac{2}{3}\pi\right) & r_r L_{ms} \cos\left(\theta_r - \frac{2}{3}\pi\right) \\
0 & 0 & 0 & r_r L_{ms} \cos\left(\theta_r - \frac{2}{3}\pi\right) & r_r L_{ms} \cos\theta_r & r_r L_{ms} \cos\left(\theta_r + \frac{2}{3}\pi\right) \\
0 & 0 & 0 & r_r L_{ms} \cos\left(\theta_r + \frac{2}{3}\pi\right) & r_r L_{ms} \cos\left(\theta_r - \frac{2}{3}\pi\right) & r_r L_{ms} \cos\theta_r \\
r_s L_{ms} \cos\theta_r & r_s L_{ms} \cos\left(\theta_r - \frac{2}{3}\pi\right) & r_s L_{ms} \cos\left(\theta_r + \frac{2}{3}\pi\right) & 0 & 0 & 0 \\
r_s L_{ms} \cos\left(\theta_r + \frac{2}{3}\pi\right) & r_s L_{ms} \cos\theta_r & r_s L_{ms} \cos\left(\theta_r - \frac{2}{3}\pi\right) & 0 & 0 & 0 \\
r_s L_{ms} \cos\left(\theta_r - \frac{2}{3}\pi\right) & r_s L_{ms} \cos\left(\theta_r + \frac{2}{3}\pi\right) & r_s L_{ms} \cos\theta_r & 0 & 0 & 0
\end{bmatrix}
\begin{bmatrix} i_{as} \\ i_{bs} \\ i_{cs} \\ i'_{ar} \\ i'_{br} \\ i'_{cr} \end{bmatrix}
$$

$$
+ \frac{1}{L_{\Sigma L}}
\begin{bmatrix}
0 & 1.299 L_{ms}^2 \omega_r & -1.299 L_{ms}^2 \omega_r & L_{\Sigma ms}\omega_r \sin\theta_r & L_{\Sigma ms}\omega_r \sin\left(\theta_r + \frac{2}{3}\pi\right) & L_{\Sigma ms}\omega_r \sin\left(\theta_r - \frac{2}{3}\pi\right) \\
-1.299 L_{ms}^2 \omega_r & 0 & 1.299 L_{ms}^2 \omega_r & L_{\Sigma ms}\omega_r \sin\left(\theta_r - \frac{2}{3}\pi\right) & L_{\Sigma ms}\omega_r \sin\theta_r & L_{\Sigma ms}\omega_r \sin\left(\theta_r + \frac{2}{3}\pi\right) \\
1.299 L_{ms}^2 \omega_r & -1.299 L_{ms}^2 \omega_r & 0 & L_{\Sigma ms}\omega_r \sin\left(\theta_r + \frac{2}{3}\pi\right) & L_{\Sigma ms}\omega_r \sin\left(\theta_r - \frac{2}{3}\pi\right) & L_{\Sigma ms}\omega_r \sin\theta_r \\
L_{\Sigma ms}\omega_r \sin\theta_r & L_{\Sigma ms}\omega_r \sin\left(\theta_r - \frac{2}{3}\pi\right) & L_{\Sigma ms}\omega_r \sin\left(\theta_r + \frac{2}{3}\pi\right) & 0 & -1.299 L_{ms}^2 \omega_r & 1.299 L_{ms}^2 \omega_r \\
L_{\Sigma ms}\omega_r \sin\left(\theta_r + \frac{2}{3}\pi\right) & L_{\Sigma ms}\omega_r \sin\theta_r & L_{\Sigma ms}\omega_r \sin\left(\theta_r - \frac{2}{3}\pi\right) & 1.299 L_{ms}^2 \omega_r & 0 & -1.299 L_{ms}^2 \omega_r \\
L_{\Sigma ms}\omega_r \sin\left(\theta_r - \frac{2}{3}\pi\right) & L_{\Sigma ms}\omega_r \sin\left(\theta_r + \frac{2}{3}\pi\right) & L_{\Sigma ms}\omega_r \sin\theta_r & -1.299 L_{ms}^2 \omega_r & 1.299 L_{ms}^2 \omega_r & 0
\end{bmatrix}
\begin{bmatrix} i_{as} \\ i_{bs} \\ i_{cs} \\ i'_{ar} \\ i'_{br} \\ i'_{cr} \end{bmatrix}
$$

$$
+ \frac{1}{L_{\Sigma L}}
\begin{bmatrix}
2L_{ms} + L'_{lr} & \frac{1}{2} L_{ms} & \frac{1}{2} L_{ms} & -L_{ms}\cos\theta_r & -L_{ms}\cos\left(\theta_r + \frac{2}{3}\pi\right) & -L_{ms}\cos\left(\theta_r - \frac{2}{3}\pi\right) \\
\frac{1}{2} L_{ms} & 2L_{ms} + L'_{lr} & \frac{1}{2} L_{ms} & -L_{ms}\cos\left(\theta_r - \frac{2}{3}\pi\right) & -L_{ms}\cos\theta_r & -L_{ms}\cos\left(\theta_r + \frac{2}{3}\pi\right) \\
\frac{1}{2} L_{ms} & \frac{1}{2} L_{ms} & 2L_{ms} + L'_{lr} & -L_{ms}\cos\left(\theta_r + \frac{2}{3}\pi\right) & -L_{ms}\cos\left(\theta_r - \frac{2}{3}\pi\right) & -L_{ms}\cos\theta_r \\
-L_{ms}\cos\theta_r & -L_{ms}\cos\left(\theta_r - \frac{2}{3}\pi\right) & -L_{ms}\cos\left(\theta_r + \frac{2}{3}\pi\right) & 2L_{ms} + L'_{lr} & \frac{1}{2} L_{ms} & \frac{1}{2} L_{ms} \\
-L_{ms}\cos\left(\theta_r + \frac{2}{3}\pi\right) & -L_{ms}\cos\theta_r & -L_{ms}\cos\left(\theta_r - \frac{2}{3}\pi\right) & \frac{1}{2} L_{ms} & 2L_{ms} + L'_{lr} & \frac{1}{2} L_{ms} \\
-L_{ms}\cos\left(\theta_r - \frac{2}{3}\pi\right) & -L_{ms}\cos\left(\theta_r + \frac{2}{3}\pi\right) & -L_{ms}\cos\theta_r & \frac{1}{2} L_{ms} & \frac{1}{2} L_{ms} & 2L_{ms} + L'_{lr}
\end{bmatrix}
\begin{bmatrix} u_{as} \\ u_{bs} \\ u_{cs} \\ u'_{ar} \\ u'_{br} \\ u'_{cr} \end{bmatrix}.
\tag{5.25}
$$

Here, $L_{\Sigma L} = \left(3L_{ms} + L'_{lr}\right)L'_{lr}$, $L_{\Sigma m} = 2L_{ms} + LL'_{lr}$, and $L_{\Sigma ms} = \frac{3}{2}L_{ms}^2 + L_{ms}L'_{lr}$.

The expression for the electromagnetic torque is obtained using the coenergy $W_c\left(\mathbf{i}_{abcs}, \mathbf{i}'_{abcr}, \theta_r\right)$. For P-pole three-phase induction machines

$$T_e = \frac{P}{2}\frac{\partial W_c\left(\mathbf{i}_{abcs}, \mathbf{i}'_{abcr}, \theta_r\right)}{\partial \theta_r}.$$

The coenergy is

$$W_c = \frac{1}{2}\mathbf{i}_{abcs}^T\left(\mathbf{L}_s - L_{ls}\mathbf{I}\right)\mathbf{i}_{abcs} + \mathbf{i}_{abcs}^T\mathbf{L}'_{sr}(\theta_r)\mathbf{i}'_{abcr} + \frac{1}{2}\mathbf{i}'^T_{abcr}\left(\mathbf{L}'_r - L'_{lr}\mathbf{I}\right)\mathbf{i}'_{abcr}.$$

The matrices $\mathbf{L}_s$, $L_{ls}\mathbf{I}$, $\mathbf{L}'_r$, and $L'_{lr}\mathbf{I}$ are not functions of θ_r. Using the inductance mapping $\mathbf{L}'_{sr}(\theta_r)$, the electromagnetic torque is

$$T_e = \frac{P}{2}\mathbf{i}_{abcs}^T\frac{\partial \mathbf{L}'_{sr}\left(\theta_r\right)}{\partial \theta_r}\mathbf{i}'_{abcr}$$

$$= -\frac{P}{2}L_{ms}\begin{bmatrix} i_{as} & i_{bs} & i_{cs}\end{bmatrix}\begin{bmatrix} \sin\theta_r & \sin\left(\theta_r + \frac{2}{3}\pi\right) & \sin\left(\theta_r - \frac{2}{3}\pi\right) \\ \sin\left(\theta_r - \frac{2}{3}\pi\right) & \sin\theta_r & \sin\left(\theta_r + \frac{2}{3}\pi\right) \\ \sin\left(\theta_r + \frac{2}{3}\pi\right) & \sin\left(\theta_r - \frac{2}{3}\pi\right) & \sin\theta_r \end{bmatrix}\begin{bmatrix} i'_{ar} \\ i'_{br} \\ i'_{cr}\end{bmatrix}$$

$$= -\frac{P}{2}L_{ms}\left[\left(i_{as}i'_{ar} + i_{bs}i'_{br} + i_{cs}i'_{cr}\right)\sin\theta_r + \left(i_{as}i'_{cr} + i_{bs}i'_{ar} + i_{cs}i'_{br}\right)\sin\left(\theta_r - \frac{2}{3}\pi\right)\right.$$

$$\left. + \left(i_{as}i'_{br} + i_{bs}i'_{cr} + i_{cs}i'_{ar}\right)\sin\left(\theta_r + \frac{2}{3}\pi\right)\right]. \quad (5.26)$$

From (5.26), one may find

$$T_e = -\frac{P}{2}L_{ms}\left\{\left[i_{as}\left(i'_{ar} - \frac{1}{2}i'_{br} - \frac{1}{2}i'_{cr}\right) + i_{bs}\left(i'_{br} - \frac{1}{2}i'_{ar} - \frac{1}{2}i'_{cr}\right) + i_{cs}\left(i'_{cr} - \frac{1}{2}i'_{br} - \frac{1}{2}i'_{ar}\right)\right]\sin\theta_r\right.$$

$$\left. + \frac{\sqrt{3}}{2}\left[i_{as}\left(i'_{br} - i'_{cr}\right) + i_{bs}\left(i'_{cr} - i'_{ar}\right) + i_{cs}\left(i'_{ar} - i'_{br}\right)\right]\cos\theta_r\right\}.$$

Having found T_e, given by (5.26), the *torsional–mechanical* equations are

$$\frac{d\omega_r}{dt} = \frac{1}{J}\left(\frac{P}{2}T_e - B_m\omega_r - \frac{P}{2}T_L\right)$$

$$= \frac{1}{J}\left[-\frac{P^2}{4}L_{ms}\left[\left(i_{as}i'_{ar} + i_{bs}i'_{br} + i_{cs}i'_{cr}\right)\sin\theta_r + \left(i_{as}i'_{cr} + i_{bs}i'_{ar} + i_{cs}i'_{br}\right)\sin\left(\theta_r - \frac{2}{3}\pi\right)\right.\right.$$

$$\left.\left. + \left(i_{as}i'_{br} + i_{bs}i'_{cr} + i_{cs}i'_{ar}\right)\sin\left(\theta_r + \frac{2}{3}\pi\right)\right] - B_m\omega_r - \frac{P}{2}T_L\right], \quad (5.27)$$

$$\frac{d\theta_r}{dt} = \omega_r.$$

Combining differential equations (5.25) and (5.27), one obtains the resulting model for three-phase induction motors in the *machine* variables. For induction motors, analysis can be performed using Cauchy's and non-Cauchy's forms of differential equations.

Example 5.5: Simulation of Three-Phase Induction Motors

One may use MATLAB and Simulink® to simulate three-phase induction motors modeled in Cauchy's form (5.25) and (5.27) as well as in non-Cauchy's form (5.24) [6, 7]. Using (5.25) and (5.27), two MATLAB files are developed. For a 220 V, 60 Hz, two-pole induction motor, the parameters are: $r_s = 0.8$ ohm, $r_r' = 1$ ohm, $L_{ms} = 0.1$ H, $L_{ls} = 0.01$ H, $L_{lr}' = 0.01$ H, $B_m = 4 \times 10^{-4}$ N-m-sec/rad, and $J = 0.002$ kg-m^2. The three-phase balanced voltage set is applied with $u_M = 220$ V. The load torque is $T_L = 40$ N-m at 0.7 sec.

The first MATLAB file (ch5 _ 03.m) is

```
% Simulation of Three-Phase Induction Motors in Machine Variables
function yprime = motor(t,y);
global mag freq P J Rs Rr L Lms Bm TL0
% The Load Torque is Applied at 0.5 sec
if t(1,:) < 0.7
          TL=0;
     else TL=TL0;
end
UAR=0; UBR=0; UCR=0; % Squirrel-Cage Induction Motor: Rotor Windings are Short-Circuited
% Balanced Voltage Set
UAS=sqrt(2)*mag*cos(freq*2*pi*t); UBS=sqrt(2)*mag*cos(freq*2*pi*t-2*pi/3);
UCS=sqrt(2)*mag*cos(freq*2*pi*t+2*pi/3);
theta=y(8,:); A=cos(theta); B=cos(theta+2*pi/3); C=cos(theta-2*pi/3);
S1=sin(theta); S2=sin(theta+2*pi/3); S3=sin(theta-2*pi/3);
IAS=y(1,:); IBS=y(2,:); ICS=y(3,:); IAR=y(4,:); IBR=y(5,:); ICR=y(6,:); W=y(7,:);
TE=-0.5*P*Lms.*((IAS.*IAR+IBS.*IBR+ICS.*ICR).*S1+(IAS.*IBR+IBS.*ICR+ICS.*IAR).*S2+...
(IAS.*ICR+IBS.*IAR+ICS.*IBR).*S3);
LS1=1/(L*(L+3*Lms));
% Differential Equations
yprime=[LS1*(-Rs*IAS*(2* Lms+L)-0.5*Rs*Lms*(IBS+ICS)+Rr*Lms*(A*IAR+B*IBR+C*ICR)+...
1.299*(Lms^2)*W*(IBS-ICS)+(L*Lms+1.5*Lms^2)*W*(S1*IAR+S2*IBR+S3*ICR)+...
(2*Lms+L)*UAS+0.5*Lms*(UBS+UCS)-Lms*(A*UAR+B*UBR+C*UCR));...
LS1*(-Rs*IBS*(2*Lms+L)-0.5*Rs*Lms*(IAS+ICS)+Rr*Lms*(C*IAR+A*IBR+B*ICR)+...
1.299*(Lms^2)*W*(ICS-IAS)+(L*Lms+1.5*Lms^2)*W*(S3*IAR+S1*IBR+S2*ICR)+...
(2*Lms+L)*UBS+0.5*Lms*(UAS+UCS)-Lms*(C*UAR+A*UBR+B*UCR));...
LS1*(-Rs*ICS*(2*Lms+L)-0.5*Rs*Lms*(IAS+IBS)+Rr*Lms*(B*IAR+C*IBR+A*ICR)+...
1.299*(Lms^2)*W*(IAS-IBS)+(L*Lms+1.5*Lms^2)*W*(S2*IAR+S3*IBR+S1*ICR)+...
(2*Lms+L)*UCS+0.5*Lms*(UAS+UBS)-Lms*(B*UAR+C*UBR+A*UCR));...
LS1*(-Rr*IAR*(2*Lms+L)-0.5*Rr*Lms*(ICR+IBR)+Rs*Lms*(A*IAS+C*IBS+B*ICS)...
+1.299*(Lms^2)*W*(ICR-IBR)+(L*Lms+1.5*Lms^2)*W*(S1*IAS+S3*IBS+S2*ICS)...
+(2*Lms+L)*UAR+0.5*Lms*(UBR+UCR)-Lms*(A*UAS+C*UBS+B*UCS));...
LS1*(-Rr*IBR*(2*Lms+L)-0.5*Rr*Lms*(IAR+ICR)+Rs*Lms*(B*IAS+A*IBS+C*ICS)...
+1.299*(Lms^2)*W*(IAR-ICR)+(L*Lms+1.5*Lms^2)*W*(S2*IAS+S1*IBS+S3*ICS)+...
(2*Lms+L)*UBR+0.5*Lms*(UAR+UCR)-Lms*(B*UAS+A*UBS+C*UCS));...
LS1*(-Rr*ICR*(2*Lms+L)-0.5*Rr*Lms*(IAR+IBR)+Rs*Lms*(C*IAS+B*IBS+A*ICS)+...
1.299*(Lms^2)*W*(IBR-IAR)+(L*Lms+1.5*Lms^2)*W*(S3*IAS+S2*IBS+S1*ICS)+...
(2*Lms+L)*UCR+0.5*Lms*(UBR+UAR)-Lms*(C*UAS+B*UBS+A*UCS));...
(P/(2*J)*(TE-TL))-Bm*W/J;...
W];
```

The second MATLAB file (ch5 _ 04.m) is

```
% Simulation of Three-Phase Induction Motors in Machine Variables
echo on; clc; clear all;
global mag freq P J Rs Rr L Lms Bm TL0
```

```
% Motor Parameters
P=2;                  % Number of Poles
Rs=0.8; Rr=1;         % Stator and Rotor Resistances
Lms=0.1;              % Mutual Stator-Rotor Inductance
L=0.01;               % Leakage Inductance
Bm=0.0004;            % Viscous Friction Coefficient
J=0.002;              % Moment of Inertia
TL0=40;               % Load Torque Applied
time=1;               % Final Time for Simulations
mag=220;              % Applied Voltage Magnitude to the abc Windings
freq=60;              % Frequency of the Applied Voltage
tspan=[0 time]; y0=[0 0 0 0 0 0 0 0];   % initial conditions
options=odeset('RelTol',1e-4,'AbsTol',[1e-4 1e-4 1e-4 1e-4 1e-4 1e-4 1e-4 1e-4]);
[t,y]=ode45('ch5_03',tspan,y0,options);
UAS=sqrt(2)*mag*cos(freq*2*pi*t); UBS=sqrt(2)*mag*cos(freq*2*pi*t-2*pi/3);
UCS=sqrt(2)*mag*cos(freq*2*pi*t+2*pi/3);
theta=y(:,8); S1=sin(theta); S2=sin(theta+2*pi/3); S3=sin(theta-2*pi/3);
IAS=y(:,1); IBS=y(:,2); ICS=y(:,3); IAR=y(:,4); IBR=y(:,5); ICR=y(:,6);
W=y(:,7);
TE=-0.5*P*Lms.*(S1.*(IAR.*IAS+ICS.*ICR+IBR.*IBS)+S2.*(ICS.*IAR+IBR.*IAS+IBS.*ICR)+...
S3.*(IAR.*IBS+ICS.*IBR+ICR.*IAS));
%  Plots
plot(t,UAS,t,UBS,t,UCS);
title('Stator Phase Voltages Applied, u_a_s, u_b_s and u_c_s [V]','FontSize',14);
axis([0 0.1 -sqrt(2)*225 sqrt(2)*225]);
xlabel('Time [seconds]','FontSize',14); ylabel('u_a_s, u_b_s, u_c_s','FontSize',14);
grid; pause;
plot(t,y(:,1)); title('Stator Current, i_a_s [A]','FontSize',14);
xlabel('Time [seconds]','FontSize',14); ylabel('i_a_s','FontSize',14); grid; pause;
plot(t,y(:,2)); title('Stator Current, i_b_s [A]','FontSize',14);
xlabel('Time [seconds]','FontSize',14); ylabel('i_b_s','FontSize',14); grid; pause;
plot(t,y(:,3)); title('Stator Current, i_c_s [A]','FontSize',14);
xlabel('Time [seconds]','FontSize',14); ylabel('i_c_s','FontSize',14); grid; pause;
plot(t,y(:,4)); title('Rotor Current, i_a_r [A]','FontSize',14);
xlabel('Time [seconds]','FontSize',14); ylabel('i_a_r','FontSize',14); grid; pause;
plot(t,y(:,5)); title('Rotor Current, i_b_r [A]','FontSize',14);
xlabel('Time [seconds]','FontSize',14); ylabel('i_b_r','FontSize',14); grid; pause;
plot(t,y(:,6)); title('Rotor Current, i_c_r [A]','FontSize',14);
xlabel('Time [seconds]','FontSize',14); ylabel('i_c_r','FontSize',14); grid; pause;
plot(t,y(:,7)); title('Angular Velocity, \omega_r [rad/sec]','FontSize',14);
xlabel('Time [seconds]','FontSize',14); ylabel('\omega_r','FontSize',14); grid; pause;
Te(:,1)=(-P*M/2)*((y(:,1).*(y(:,4)-0.5*y(:,5)-0.5*y(:,6))+y(:,2).*(y(:,5)-0.5*y(:,4)...
-0.5*y(:,6))+y(:,3).*(y(:,6)-0.5*y(:,5)-...
0.5*y(:,4))).*sin(y(:,8))+0.865*(y(:,1).*(y(:,5)-y(:,6))+y(:,2).*(y(:,6)-...
y(:,4))+y(:,3).*(y(:,5)-y(:,4))).*cos(y(:,8)));
plot(t,TE); title('Electromagnetic Torque, T_e [N-m]','FontSize',14);
xlabel('Time [seconds]','FontSize',14); ylabel('T_e [N-m]','FontSize',14); grid; pause;
plot(W,TE); title('Torque-Speed Characteristic','FontSize',14);
xlabel('Angular Velocity, \omega_r [rad/sec]','FontSize',14);
ylabel('Electromagnetic Torque, T_e [N-m]','FontSize',14);
```

Using the ode45 differential equations solver, we numerically solve a set of eight differential equations (5.25) and (5.27). The transients of the stator and rotor currents $i_{as}(t), i_{bs}(t), i_{cs}(t), i'_{ar}(t), i'_{br}(t), i'_{cr}(t)$, as well as the angular velocity $\omega_r(t)$, are plotted using the MATLAB statements. The evolution of the electromagnetic torque $T_e(t)$, as well as the torque–speed characteristic, are depicted in Figures 5.10. The load torque is applied at 0.7 sec and $T_L = 40$ N-m.

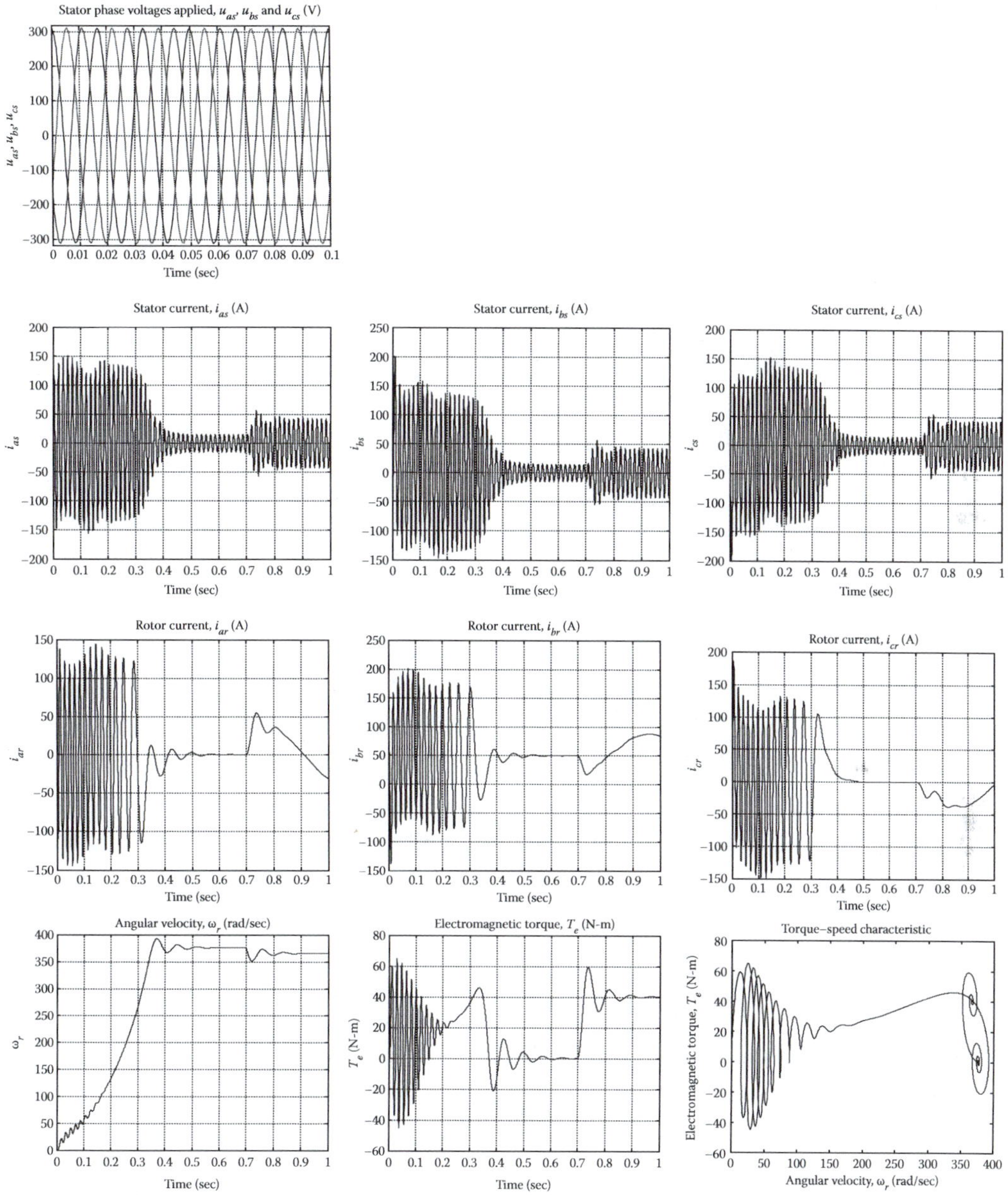

FIGURE 5.10 Evolution of balanced phase voltages applied, phase currents, angular velocity $\omega_r(t)$, $T_e(t)$, and dynamic torque–speed characteristic $\omega_r(t) = \Omega_T[T_e(t)]$.

One may use the circuitry-electromagnetic equations (5.24) and the *torsional–mechanical* equations of motion (5.27). One finds the following differential equations

$$\frac{di_{as}}{dt}=\frac{1}{L_{ls}+L_{ms}}\left[-r_s i_{as}+\frac{1}{2}L_{ms}\frac{di_{bs}}{dt}+\frac{1}{2}L_{ms}\frac{di_{cs}}{dt}-L_{ms}\frac{d\left(i'_{ar}\cos\theta_r\right)}{dt}-L_{ms}\frac{d\left(i'_{br}\cos\left(\theta_r+\frac{2\pi}{3}\right)\right)}{dt}-L_{ms}\frac{d\left(i'_{cr}\cos\left(\theta_r-\frac{2\pi}{3}\right)\right)}{dt}+u_{as}\right],$$

$$\frac{di_{bs}}{dt}=\frac{1}{L_{ls}+L_{ms}}\left[-r_s i_{bs}+\frac{1}{2}L_{ms}\frac{di_{as}}{dt}+\frac{1}{2}L_{ms}\frac{di_{cs}}{dt}-L_{ms}\frac{d\left(i'_{ar}\cos\left(\theta_r-\frac{2\pi}{3}\right)\right)}{dt}-L_{ms}\frac{d\left(i'_{br}\cos\theta_r\right)}{dt}-L_{ms}\frac{d\left(i'_{cr}\cos\left(\theta_r+\frac{2\pi}{3}\right)\right)}{dt}+u_{bs}\right],$$

$$\frac{di_{cs}}{dt}=\frac{1}{L_{ls}+L_{ms}}\left[-r_s i_{cs}+\frac{1}{2}L_{ms}\frac{di_{as}}{dt}+\frac{1}{2}L_{ms}\frac{di_{bs}}{dt}-L_{ms}\frac{d\left(i'_{ar}\cos\left(\theta_r+\frac{2\pi}{3}\right)\right)}{dt}-L_{ms}\frac{d\left(i'_{br}\cos\left(\theta_r-\frac{2\pi}{3}\right)\right)}{dt}-L_{ms}\frac{d\left(i'_{cr}\cos\theta_r\right)}{dt}+u_{cs}\right],$$

$$\frac{di'_{ar}}{dt}=\frac{1}{L'_{lr}+L_{ms}}\left[-r'_r i'_{ar}-L_{ms}\frac{d\left(i_{as}\cos\theta_r\right)}{dt}-L_{ms}\frac{d\left(i_{bs}\cos\left(\theta_r-\frac{2\pi}{3}\right)\right)}{dt}-L_{ms}\frac{d\left(i_{cs}\cos\left(\theta_r+\frac{2\pi}{3}\right)\right)}{dt}+\frac{1}{2}L_{ms}\frac{di'_{br}}{dt}+\frac{1}{2}L_{ms}\frac{di'_{cr}}{dt}+u'_{ar}\right],$$

$$\frac{di'_{br}}{dt}=\frac{1}{L'_{lr}+L_{ms}}\left[-r'_r i'_{br}-L_{ms}\frac{d\left(i_{as}\cos\left(\theta_r+\frac{2\pi}{3}\right)\right)}{dt}-L_{ms}\frac{d\left(i_{bs}\cos\theta_r\right)}{dt}-L_{ms}\frac{d\left(i_{cs}\cos\left(\theta_r-\frac{2\pi}{3}\right)\right)}{dt}+\frac{1}{2}L_{ms}\frac{di'_{ar}}{dt}+\frac{1}{2}L_{ms}\frac{di'_{cr}}{dt}+u'_{br}\right],$$

$$\frac{di'_{cr}}{dt}=\frac{1}{L'_{lr}+L_{ms}}\left[-r'_r i'_{cr}-L_{ms}\frac{d\left(i_{as}\cos\left(\theta_r-\frac{2\pi}{3}\right)\right)}{dt}-L_{ms}\frac{d\left(i_{bs}\cos\left(\theta_r+\frac{2\pi}{3}\right)\right)}{dt}-L_{ms}\frac{d\left(i_{cs}\cos\theta_r\right)}{dt}+\frac{1}{2}L_{ms}\frac{di'_{ar}}{dt}+\frac{1}{2}L_{ms}\frac{di'_{br}}{dt}+u'_{cr}\right]$$

$$\frac{d\omega_r}{dt}=\frac{1}{J}\left[-\frac{P^2}{4}L_{ms}\left[\left(i_{as}i'_{ar}+i_{bs}i'_{br}+i_{cs}i'_{cr}\right)\sin\theta_r+\left(i_{as}i'_{cr}+i_{bs}i'_{ar}+i_{cs}i'_{br}\right)\sin\left(\theta_r-\frac{2}{3}\pi\right)+\left(i_{as}i'_{br}+i_{bs}i'_{cr}+i_{cs}i'_{ar}\right)\sin\left(\theta_r+\frac{2}{3}\pi\right)\right]\right.$$
$$\left.-B_m\omega_r-\frac{P}{2}T_L\right],$$

$$\frac{d\theta_r}{dt}=\omega_r. \tag{5.28}$$

The simulations are performed. Figure 5.10 documents the balanced three-phase voltage set

$$u_{as}(t)=\sqrt{2}u_M\cos\left(\omega_f t\right),\quad u_{bs}(t)=\sqrt{2}u_M\cos\left(\omega_f t-\frac{2}{3}\pi\right),\quad u_{cs}(t)=\sqrt{2}u_M\cos\left(\omega_f t+\frac{2}{3}\pi\right),$$

$$u_M=220\text{ V}.$$

The frequency of the supplied voltage is 60 Hz. Hence, $\omega_f=4\pi f/P=377$ rad/sec. Figures 5.10 illustrate the transient dynamics for the phase currents $i_{as}(t), i_{bs}(t), i_{cs}(t), i'_{ar}(t), i'_{br}(t), i'_{cr}(t)$ and the angular velocity $\omega_r(t)$. The dynamic torque–speed characteristics $\omega_r(t)=\Omega_T[T_e(t)]$ are documented. The changes occur at 0.7 sec when the load $T_L=40$ N-m is applied. The settling time, acceleration, losses, and other performance characteristics are found. ■

5.5 ANALYSIS OF INDUCTION MOTORS USING QUADRATURE AND DIRECT QUANTITIES

5.5.1 Arbitrary, Stationary, Rotor, and Synchronous Reference Frames

The differential equations to examine induction machines were developed in the *machine* variables by using Kirchhoff's second law, Newton's law of motion, and the Lagrange equations. We used the real-valued *abc* stator and rotor voltages, currents, and flux linkages. The *quadrature*, *direct*, and *zero* (*qd*0) quantities can be applied to reduce the complexity of differential equations. For three-phase induction and synchronous machines, the transformations of the *machine* variables (stator and rotor voltages, currents and flux linkages) to the *quadrature-*, *direct-*, and *zero*-axis components are performed using the Park and other transformations. The general results are found by using the *arbitrary* reference frame. The reference frames can be fixed with the rotor and/or stator, and, the frame "rotates". In the *arbitrary* reference frame, the frame angular velocity ω is not specified. The stationary ($\omega = 0$), rotor ($\omega = \omega_r$), and synchronous ($\omega = \omega_e$) reference frames are used. Assigning the frame angular velocity to be $\omega = 0$, $\omega = \omega_r$, or $\omega = \omega_e$, the models in the three reference frames are found. The angular displacement θ of the reference frame is used. Table 5.1 documents the transformations of the *machine* variables to the *qd*0 quantities, and vice versa.

TABLE 5.1
Transformations of the *Machine* and *qd*0 Quantities

Stator, Rotor, *Quadrature*, and *Direct* Magnetic Axes

Arbitrary reference frame

The angular velocity ω is not specified

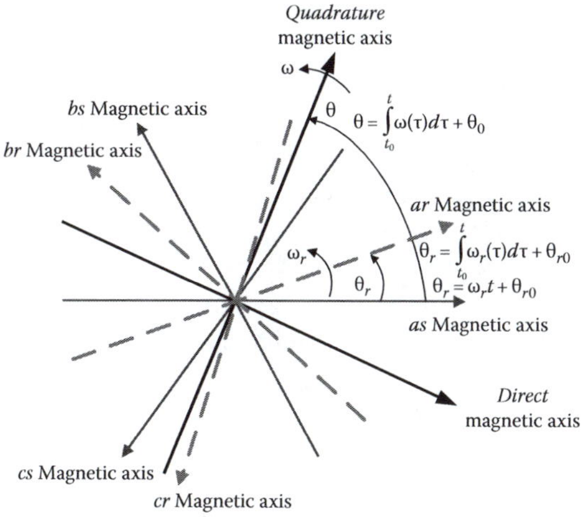

Transformation of Variables Using Transformation Matrices

Direct transformation (stator variables)

$$\mathbf{u}_{qd0s} = \mathbf{K}_s \mathbf{u}_{abcs},\ \mathbf{i}_{qd0s} = \mathbf{K}_s \mathbf{i}_{abcs},\ \boldsymbol{\psi}_{qd0s} = \mathbf{K}_s \boldsymbol{\psi}_{abcs}$$

Inverse transformation (stator variables)

$$\mathbf{u}_{abcs} = \mathbf{K}_s^{-1} \mathbf{u}_{qd0s},\ \mathbf{i}_{abcs} = \mathbf{K}_s^{-1} \mathbf{i}_{qd0s},\ \boldsymbol{\psi}_{abcs} = \mathbf{K}_s^{-1} \boldsymbol{\psi}_{qd0s}$$

Stator transformation matrices

$$\mathbf{K}_s = \frac{2}{3}\begin{bmatrix} \cos\theta & \cos\left(\theta - \frac{2}{3}\pi\right) & \cos\left(\theta + \frac{2}{3}\pi\right) \\ \sin\theta & \sin\left(\theta - \frac{2}{3}\pi\right) & \sin\left(\theta + \frac{2}{3}\pi\right) \\ \frac{1}{2} & \frac{1}{2} & \frac{1}{2} \end{bmatrix},$$

$$\mathbf{K}_s^{-1} = \begin{bmatrix} \cos\theta & \sin\theta & 1 \\ \cos\left(\theta - \frac{2}{3}\pi\right) & \sin\left(\theta - \frac{2}{3}\pi\right) & 1 \\ \cos\left(\theta + \frac{2}{3}\pi\right) & \sin\left(\theta + \frac{2}{3}\pi\right) & 1 \end{bmatrix}$$

Direct transformation (rotor variables)

$$\mathbf{u}_{qd0r} = \mathbf{K}_r \mathbf{u}_{abcr},\ \mathbf{i}_{qd0r} = \mathbf{K}_r \mathbf{i}_{abcr},\ \boldsymbol{\psi}_{qd0r} = \mathbf{K}_r \boldsymbol{\psi}_{abcr}$$

Inverse transformation (rotor variables)

$$\mathbf{u}_{abcr} = \mathbf{K}_r^{-1} \mathbf{u}_{qd0r},\ \mathbf{i}_{abcr} = \mathbf{K}_r^{-1} \mathbf{i}_{qd0r},\ \boldsymbol{\psi}_{abcr} = \mathbf{K}_r^{-1} \boldsymbol{\psi}_{qd0r}$$

(*Continued*)

TABLE 5.1 (*Continued*)
Transformations of the *Machine* and *qd*0 Quantities

Stator, Rotor, *Quadrature*, and *Direct* Magnetic Axes	Transformation of Variables Using Transformation Matrices
	Rotor transformation matrices $\mathbf{K}_r=\frac{2}{3}\begin{bmatrix}\cos(\theta-\theta_r) & \cos\left(\theta-\theta_r-\frac{2}{3}\pi\right) & \cos\left(\theta-\theta_r+\frac{2}{3}\pi\right)\\ \sin(\theta-\theta_r) & \sin\left(\theta-\theta_r-\frac{2}{3}\pi\right) & \sin\left(\theta-\theta_r+\frac{2}{3}\pi\right)\\ \frac{1}{2} & \frac{1}{2} & \frac{1}{2}\end{bmatrix}$, $\mathbf{K}_r^{-1}=\begin{bmatrix}\cos(\theta-\theta_r) & \sin(\theta-\theta_r) & 1\\ \cos\left(\theta-\theta_r-\frac{2}{3}\pi\right) & \sin\left(\theta-\theta_r-\frac{2}{3}\pi\right) & 1\\ \cos\left(\theta-\theta_r+\frac{2}{3}\pi\right) & \sin\left(\theta-\theta_r+\frac{2}{3}\pi\right) & 1\end{bmatrix}$
	For the stationary, rotor and synchronous reference frames, $\omega=0$, $\theta=0$ $(\theta_0=0)$, $\omega=\omega_r$, $\theta=\theta_r$ $(\theta_0=0)$, and, $\omega=\omega_e$, $\theta=\theta_e$ $(\theta_0=0)$.
Stationary reference frame $\omega=0$, $\theta=0$ $(\theta_0=0)$	*Direct* transformation (stator variables) $\mathbf{u}_{qd0s}^{s}=\mathbf{K}_s^s\mathbf{u}_{abcs}$, $\mathbf{i}_{qd0s}^{s}=\mathbf{K}_s^s\mathbf{i}_{abcs}$, $\boldsymbol{\psi}_{qd0s}^{s}=\mathbf{K}_s^s\boldsymbol{\psi}_{abcs}$
	Inverse transformation (stator variables) $\mathbf{u}_{abcs}=\mathbf{K}_s^{s^{-1}}\mathbf{u}_{qd0s}^{s}$, $\mathbf{i}_{abcs}=\mathbf{K}_s^{s^{-1}}\mathbf{i}_{qd0s}^{s}$, $\boldsymbol{\psi}_{abcs}=\mathbf{K}_s^{s^{-1}}\boldsymbol{\psi}_{qd0s}^{s}$
	Direct transformation (rotor variables) $\mathbf{u}_{qd0r}^{s}=\mathbf{K}_r^s\mathbf{u}_{abcr}$, $\mathbf{i}_{qd0r}^{s}=\mathbf{K}_r^s\mathbf{i}_{abcr}$, $\boldsymbol{\psi}_{qd0r}^{s}=\mathbf{K}_r^s\boldsymbol{\psi}_{abcr}$
	Inverse transformation (rotor variables) $\mathbf{u}_{abcr}=\mathbf{K}_r^{s^{-1}}\mathbf{u}_{qd0r}^{s}$, $\mathbf{i}_{abcr}=\mathbf{K}_r^{s^{-1}}\mathbf{i}_{qd0r}^{s}$, $\boldsymbol{\psi}_{abcr}=\mathbf{K}_r^{s^{-1}}\boldsymbol{\psi}_{qd0r}^{s}$
Rotor reference frame $\omega=\omega_r$, $\theta=\theta_r$ $(\theta_0=0)$	*Direct* transformation (stator variables) $\mathbf{u}_{qd0s}^{r}=\mathbf{K}_s^r\mathbf{u}_{abcs}$, $\mathbf{i}_{qd0s}^{r}=\mathbf{K}_s^r\mathbf{i}_{abcs}$, $\boldsymbol{\psi}_{qd0s}^{r}=\mathbf{K}_s^r\boldsymbol{\psi}_{abcs}$
	Inverse transformation (stator variables) $\mathbf{u}_{abcs}=\mathbf{K}_s^{r^{-1}}\mathbf{u}_{qd0s}^{r}$, $\mathbf{i}_{abcs}=\mathbf{K}_s^{r^{-1}}\mathbf{i}_{qd0s}^{r}$, $\boldsymbol{\psi}_{abcs}=\mathbf{K}_s^{r^{-1}}\boldsymbol{\psi}_{qd0s}^{r}$
	Direct transformation (rotor variables) $\mathbf{u}_{qd0r}^{r}=\mathbf{K}_r^r\mathbf{u}_{abcr}$, $\mathbf{i}_{qd0r}^{r}=\mathbf{K}_r^r\mathbf{i}_{abcr}$, $\boldsymbol{\psi}_{qd0r}^{r}=\mathbf{K}_r^r\boldsymbol{\psi}_{abcr}$
	Inverse transformation (rotor variables) $\mathbf{u}_{abcr}=\mathbf{K}_r^{r^{-1}}\mathbf{u}_{qd0r}^{r}$, $\mathbf{i}_{abcr}=\mathbf{K}_r^{r^{-1}}\mathbf{i}_{qd0r}^{r}$, $\boldsymbol{\psi}_{abcr}=\mathbf{K}_r^{r^{-1}}\boldsymbol{\psi}_{qd0r}^{r}$
Synchronous reference frame $\omega=\omega_e$, $\theta=\theta_e$ $(\theta_0=0)$	*Direct* transformation (stator variables) $\mathbf{u}_{qd0s}^{e}=\mathbf{K}_s^e\mathbf{u}_{abcs}$, $\mathbf{i}_{qd0s}^{e}=\mathbf{K}_s^e\mathbf{i}_{abcs}$, $\boldsymbol{\psi}_{qd0s}^{e}=\mathbf{K}_s^e\boldsymbol{\psi}_{abcs}$
	Inverse transformation (stator variables) $\mathbf{u}_{abcs}=\mathbf{K}_s^{e^{-1}}\mathbf{u}_{qd0s}^{e}$, $\mathbf{i}_{abcs}=\mathbf{K}_s^{e^{-1}}\mathbf{i}_{qd0s}^{e}$, $\boldsymbol{\psi}_{abcs}=\mathbf{K}_s^{e^{-1}}\boldsymbol{\psi}_{qd0s}^{e}$
	Direct transformation (rotor variables) $\mathbf{u}_{qd0r}^{e}=\mathbf{K}_r^e\mathbf{u}_{abcr}$, $\mathbf{i}_{qd0r}^{e}=\mathbf{K}_r^e\mathbf{i}_{abcr}$, $\boldsymbol{\psi}_{qd0r}^{e}=\mathbf{K}_r^e\boldsymbol{\psi}_{abcr}$
	Inverse transformation (rotor variables) $\mathbf{u}_{abcr}=\mathbf{K}_r^{e^{-1}}\mathbf{u}_{qd0r}^{e}$, $\mathbf{i}_{abcr}=\mathbf{K}_r^{e^{-1}}\mathbf{i}_{qd0r}^{e}$, $\boldsymbol{\psi}_{abcr}=\mathbf{K}_r^{e^{-1}}\boldsymbol{\psi}_{qd0r}^{e}$

Example 5.6

Consider the balanced three-phase voltage set (a set of equal-amplitude sinusoidal voltages displaced by 120°) with $f = 60$ Hz. The angular frequency of voltages is 377 rad/sec, and

$$u_{as}(t) = 100\cos(377t)\ \text{V},\quad u_{bs}(t) = 100\cos\left(377t - \frac{2}{3}\pi\right) \text{ and } u_{cs}(t) = 100\cos\left(377t + \frac{2}{3}\pi\right)\text{V}.$$

One finds the $qd0$-axis components of voltages in the *arbitrary* reference frame. As given in Table 5.1, one applies the transformation matrix $\mathbf{K}_s$ using the *direct* Park transformation

$$\mathbf{u}_{qd0s} = \mathbf{K}_s\mathbf{u}_{abcs},\quad \mathbf{K}_s = \frac{2}{3}\begin{bmatrix} \cos\theta & \cos\left(\theta - \frac{2}{3}\pi\right) & \cos\left(\theta + \frac{2}{3}\pi\right) \\ \sin\theta & \sin\left(\theta - \frac{2}{3}\pi\right) & \sin\left(\theta + \frac{2}{3}\pi\right) \\ \frac{1}{2} & \frac{1}{2} & \frac{1}{2} \end{bmatrix},$$

where θ is the angular displacement of the reference frame.

In the *arbitrary* reference frame, ω and θ are not specified. We have

$$\begin{bmatrix} u_{qs} \\ u_{ds} \\ u_{0s} \end{bmatrix} = \frac{2}{3}\begin{bmatrix} \cos\theta & \cos\left(\theta - \frac{2}{3}\pi\right) & \cos\left(\theta + \frac{2}{3}\pi\right) \\ \sin\theta & \sin\left(\theta - \frac{2}{3}\pi\right) & \sin\left(\theta + \frac{2}{3}\pi\right) \\ \frac{1}{2} & \frac{1}{2} & \frac{1}{2} \end{bmatrix}\begin{bmatrix} u_{as} \\ u_{bs} \\ u_{cs} \end{bmatrix}.$$

Hence,

$$u_{qs}(t) = \frac{2}{3}\left(\cos(\theta)u_{as}(t) + \cos\left(\theta - \frac{2}{3}\pi\right)u_{bs}(t) + \cos\left(\theta + \frac{2}{3}\pi\right)u_{cs}(t)\right),$$

$$u_{ds}(t) = \frac{2}{3}\left(\sin(\theta)u_{as}(t) + \sin\left(\theta - \frac{2}{3}\pi\right)u_{bs}(t) + \sin\left(\theta + \frac{2}{3}\pi\right)u_{cs}(t)\right),\ u_{0s}(t) = \frac{1}{3}\left(u_{as}(t) + u_{bs}(t) + u_{cs}(t)\right)$$

For given $u_{as}(t) = 100\cos(377t)$, $u_{bs}(t) = 100\cos\left(377t - \frac{2}{3}\pi\right)$, $u_{cs}(t) = 100\cos\left(377t + \frac{2}{3}\pi\right)$, we obtain the *quadrature-*, *direct-*, and *zero-*axis components of voltages in the *arbitrary* reference frame

$$u_{qs}(t) = \frac{200}{3}\left(\cos(\theta)\cos(377t) + \cos\left(\theta - \frac{2}{3}\pi\right)\cos\left(377t - \frac{2}{3}\pi\right) + \cos\left(\theta + \frac{2}{3}\pi\right)\cos\left(377t + \frac{2}{3}\pi\right)\right),$$

$$u_{ds}(t) = \frac{200}{3}\left(\sin(\theta)\cos(377t) + \sin\left(\theta - \frac{2}{3}\pi\right)\cos\left(377t - \frac{2}{3}\pi\right) + \sin\left(\theta + \frac{2}{3}\pi\right)\cos\left(377t + \frac{2}{3}\pi\right)\right),$$

$$u_{0s}(t) = \frac{200}{3}\left(\cos(377t) + \cos\left(377t - \frac{2}{3}\pi\right) + \cos\left(377t + \frac{2}{3}\pi\right)\right).$$

With $f = 60$ Hz, the synchronous reference frame is studied because the angular frequency of voltages is 377 rad/sec. Assuming $\theta_0 = 0$, we have $\theta_e = \omega_e t$. Using the trigonometric identities, one finds

$$u_{qs}^e(t) = \frac{200}{3}\left(\cos^2(377t) + \cos^2\left(377t - \frac{2}{3}\pi\right) + \cos^2\left(377t + \frac{2}{3}\pi\right)\right) = \frac{200}{3}\frac{3}{2} = 100 \text{ V},$$

$$u_{ds}^e(t) = \frac{200}{3}\left(\sin(377t)\cos(377t) + \sin\left(377t - \frac{2}{3}\pi\right)\cos\left(377t - \frac{2}{3}\pi\right) + \sin\left(377t + \frac{2}{3}\pi\right)\cos\left(377t + \frac{2}{3}\pi\right)\right)$$
$$= 0 \text{ V},$$

$$u_{0s}^e(t) = \frac{200}{3}\left(\cos(377t) + \cos\left(377t - \frac{2}{3}\pi\right) + \cos\left(377t + \frac{2}{3}\pi\right)\right) = 0 \text{ V}.$$

The resulting $qd0$ components $u_{qs}^e(t)$, $u_{ds}^e(t)$, and $u_{0s}^e(t)$ are the DC voltages. Furthermore $u_{ds}^e(t) = 0$ and $u_{0s}^e(t) = 0$.

In the stationary reference frame, $\omega = 0$ and $\theta = 0$. From

$$\mathbf{K}_s = \frac{2}{3}\begin{bmatrix} \cos\theta & \cos\left(\theta - \frac{2}{3}\pi\right) & \cos\left(\theta + \frac{2}{3}\pi\right) \\ \sin\theta & \sin\left(\theta - \frac{2}{3}\pi\right) & \sin\left(\theta + \frac{2}{3}\pi\right) \\ \frac{1}{2} & \frac{1}{2} & \frac{1}{2} \end{bmatrix},$$

the stationary reference frame Park transformation matrix is

$$\mathbf{K}_s^s = \frac{2}{3}\begin{bmatrix} \cos\theta & \cos\left(\theta - \frac{2}{3}\pi\right) & \cos\left(\theta + \frac{2}{3}\pi\right) \\ \sin\theta & \sin\left(\theta - \frac{2}{3}\pi\right) & \sin\left(\theta + \frac{2}{3}\pi\right) \\ \frac{1}{2} & \frac{1}{2} & \frac{1}{2} \end{bmatrix}\Bigg|_{\theta=0} = \frac{2}{3}\begin{bmatrix} 1 & -\frac{1}{2} & -\frac{1}{2} \\ 0 & -\frac{\sqrt{3}}{2} & \frac{\sqrt{3}}{2} \\ \frac{1}{2} & \frac{1}{2} & \frac{1}{2} \end{bmatrix} = \begin{bmatrix} \frac{2}{3} & -\frac{1}{3} & -\frac{1}{3} \\ 0 & -\frac{1}{\sqrt{3}} & \frac{1}{\sqrt{3}} \\ \frac{1}{3} & \frac{1}{3} & \frac{1}{3} \end{bmatrix}.$$

From

$$\mathbf{u}_{qd0s}^s = \mathbf{K}_s^s\mathbf{u}_{abcs}, \begin{bmatrix} u_{qs}^s \\ u_{ds}^s \\ u_{os}^s \end{bmatrix} = \begin{bmatrix} \frac{2}{3} & -\frac{1}{3} & -\frac{1}{3} \\ 0 & -\frac{1}{\sqrt{3}} & \frac{1}{\sqrt{3}} \\ \frac{1}{3} & \frac{1}{3} & \frac{1}{3} \end{bmatrix}\begin{bmatrix} u_{as} \\ u_{bs} \\ u_{cs} \end{bmatrix},$$

one finds

$$u_{qs}^s(t)=\frac{2}{3}u_{as}(t)-\frac{1}{3}u_{bs}(t)-\frac{1}{3}u_{cs}(t),\quad u_{ds}^s(t)=-\frac{1}{\sqrt{3}}u_{bs}(t)+\frac{1}{\sqrt{3}}u_{cs}(t),$$

$$u_{os}^s(t)=\frac{1}{3}u_{as}(t)+\frac{1}{3}u_{bs}(t)+\frac{1}{3}u_{cs}(t).$$

Therefore,

$$u_{qs}^s(t)=\frac{200}{3}\cos(377t)-\frac{100}{3}\cos\left(377t-\frac{2}{3}\pi\right)-\frac{100}{3}\cos\left(377t+\frac{2}{3}\pi\right)=100\cos(377t),$$

$$u_{ds}^s(t)=-\frac{100}{\sqrt{3}}\cos\left(377t-\frac{2}{3}\pi\right)+\frac{100}{\sqrt{3}}\cos\left(377t+\frac{2}{3}\pi\right)=-100\sin(377t),$$

$$u_{0s}^s(t)=\frac{100}{3}\cos(377t)+\frac{100}{3}\cos\left(377t-\frac{2}{3}\pi\right)+\frac{100}{3}\cos\left(377t+\frac{2}{3}\pi\right)=0.$$ ■

5.5.2 Induction Motors in the Arbitrary Reference Frame

We derive the equations of motion for three-phase induction machines in the *arbitrary* reference frame when the frame angular velocity ω is not specified. Assigning the frame angular velocities $\omega = 0$, $\omega = \omega_r$, or $\omega = \omega_e$, the models in the stationary ($\omega = 0$), rotor ($\omega = \omega_r$), and synchronous ($\omega = \omega_e$) reference frames result. Consider three-phase induction motors with *quadrature* and *direct* magnetic axes, as shown in Figure 5.11.

The *abc* stator and rotor variables are mapped by the *quadrature*, *direct*, and *zero* quantities. To transform the *machine* (*abc*) stator voltages, currents, and flux linkages to the *qd*0-axis components, the *direct* Park transformation is used. Using the transformations reported in Table 5.1, we have

$$\mathbf{u}_{qd0s}=\mathbf{K}_s\mathbf{u}_{abcs},\quad \mathbf{i}_{qd0s}=\mathbf{K}_s\mathbf{i}_{abcs},\quad \psi_{qd0s}=\mathbf{K}_s\psi_{abcs},\tag{5.29}$$

where the stator transformation matrix $\mathbf{K}_s$ is

$$\mathbf{K}_s=\frac{2}{3}\begin{bmatrix}\cos\theta & \cos\left(\theta-\frac{2}{3}\pi\right) & \cos\left(\theta+\frac{2}{3}\pi\right)\\ \sin\theta & \sin\left(\theta-\frac{2}{3}\pi\right) & \sin\left(\theta+\frac{2}{3}\pi\right)\\ \frac{1}{2} & \frac{1}{2} & \frac{1}{2}\end{bmatrix}.\tag{5.30}$$

The angular displacement of the reference frame is $\theta=\int_{t_0}^{t}\omega(\tau)d\tau+\theta_0$. Using the rotor transformations matrix $\mathbf{K}_r$, the *qd*0-axis components of rotor voltages, currents, and flux linkages are found by using the *abc* rotor voltages, currents, and flux linkages as

$$\mathbf{u}'_{qd0r}=\mathbf{K}_r u'_{abcr},\quad \mathbf{i}'_{qd0r}=\mathbf{K}_r\mathbf{i}'_{abcr},\quad \psi'_{qd0r}=\mathbf{K}_r\psi'_{abcr},\tag{5.31}$$

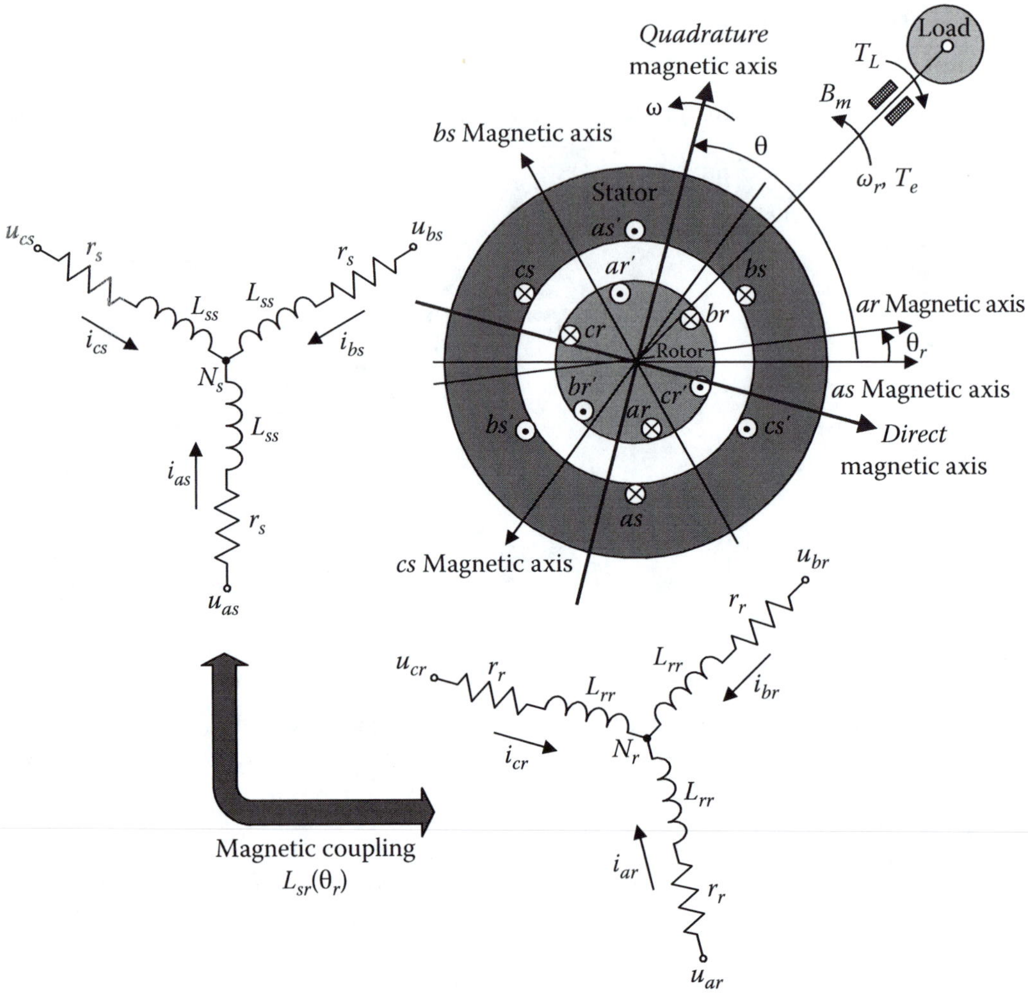

FIGURE 5.11 Three-phase symmetric induction motor with a rotating reference frame.

where the rotor transformation matrix $\mathbf{K}_r$ is

$$\mathbf{K}_r = \frac{2}{3}\begin{bmatrix} \cos(\theta-\theta_r) & \cos\left(\theta-\theta_r-\frac{2}{3}\pi\right) & \cos\left(\theta-\theta_r+\frac{2}{3}\pi\right) \\ \sin(\theta-\theta_r) & \sin\left(\theta-\theta_r-\frac{2}{3}\pi\right) & \sin\left(\theta-\theta_r+\frac{2}{3}\pi\right) \\ \frac{1}{2} & \frac{1}{2} & \frac{1}{2} \end{bmatrix}. \tag{5.32}$$

From the differential equations (5.23)

$$\mathbf{u}_{abcs} = \mathbf{r}_s\mathbf{i}_{abcs} + \frac{d\boldsymbol{\psi}_{abcs}}{dt}, \quad \mathbf{u}'_{abcr} = \mathbf{r}'_r i'_{abcr} + \frac{d\boldsymbol{\psi}'_{abcr}}{dt},$$

using the *inverse* Park transformation matrices $\mathbf{K}_s^{-1}$ and $\mathbf{K}_r^{-1}$, we have

$$\mathbf{K}_s^{-1}\mathbf{u}_{qd0s} = \mathbf{r}_s\mathbf{K}_s^{-1}\mathbf{i}_{qd0s} + \frac{d\left(\mathbf{K}_s^{-1}\boldsymbol{\psi}_{qd0s}\right)}{dt},$$
$$\mathbf{K}_r^{-1}u'_{qd0r} = \mathbf{r}'_r\mathbf{K}_r^{-1}\mathbf{i}'_{qd0r} + \frac{d\left(\mathbf{K}_r^{-1}\boldsymbol{\psi}'_{qd0r}\right)}{dt}. \tag{5.33}$$

Using (5.30) and (5.32), one finds inverse matrices $\mathbf{K}_s^{-1}$ and $\mathbf{K}_r^{-1}$ as

$$\mathbf{K}_s^{-1} = \begin{bmatrix} \cos\theta & \sin\theta & 1 \\ \cos\left(\theta - \frac{2}{3}\pi\right) & \sin\left(\theta - \frac{2}{3}\pi\right) & 1 \\ \cos\left(\theta + \frac{2}{3}\pi\right) & \sin\left(\theta + \frac{2}{3}\pi\right) & 1 \end{bmatrix} \text{ and } \mathbf{K}_r^{-1} = \begin{bmatrix} \cos(\theta-\theta_r) & \sin(\theta-\theta_r) & 1 \\ \cos\left(\theta-\theta_r - \frac{2}{3}\pi\right) & \sin\left(\theta-\theta_r - \frac{2}{3}\pi\right) & 1 \\ \cos\left(\theta-\theta_r + \frac{2}{3}\pi\right) & \sin\left(\theta-\theta_r + \frac{2}{3}\pi\right) & 1 \end{bmatrix}.$$

Multiplying the left- and right-hand sides of (5.33) by $\mathbf{K}_s$ and $\mathbf{K}_r$, yields

$$\mathbf{u}_{qd0s} = \mathbf{K}_s\mathbf{r}_s\mathbf{K}_s^{-1}\mathbf{i}_{qd0s} + \mathbf{K}_s\frac{d\mathbf{K}_s^{-1}}{dt}\boldsymbol{\psi}_{qd0s} + \mathbf{K}_s\mathbf{K}_s^{-1}\frac{d\boldsymbol{\psi}_{qd0s}}{dt},$$
$$\mathbf{u}'_{qd0r} = \mathbf{K}_r\mathbf{r}'_r\mathbf{K}_r^{-1}\mathbf{i}'_{qd0r} + \mathbf{K}_r\frac{d\mathbf{K}_r^{-1}}{dt}\boldsymbol{\psi}'_{qd0r} + \mathbf{K}_r\mathbf{K}_r^{-1}\frac{d\boldsymbol{\psi}'_{qd0r}}{dt}. \tag{5.34}$$

The matrices of the stator and rotor resistances $\mathbf{r}_s$ and $\mathbf{r}'_r$ are diagonal. Hence, $\mathbf{K}_s\mathbf{r}_s\mathbf{K}_s^{-1} = \mathbf{r}_s$ and $\mathbf{K}_r\mathbf{r}'_r\mathbf{K}_r^{-1} = \mathbf{r}'_r$. In (5.34),

$$\frac{d\mathbf{K}_s^{-1}}{dt} = \omega\begin{bmatrix} -\sin\theta & \cos\theta & 0 \\ -\sin\left(\theta - \frac{2}{3}\pi\right) & \cos\left(\theta - \frac{2}{3}\pi\right) & 0 \\ -\sin\left(\theta + \frac{2}{3}\pi\right) & \cos\left(\theta + \frac{2}{3}\pi\right) & 0 \end{bmatrix}$$

and

$$\frac{d\mathbf{K}_r^{-1}}{dt} = (\omega - \omega_r)\begin{bmatrix} -\sin(\theta-\theta_r) & \cos(\theta-\theta_r) & 0 \\ -\sin\left(\theta-\theta_r - \frac{2}{3}\pi\right) & \cos\left(\theta-\theta_r - \frac{2}{3}\pi\right) & 0 \\ -\sin\left(\theta-\theta_r + \frac{2}{3}\pi\right) & \cos\left(\theta-\theta_r + \frac{2}{3}\pi\right) & 0 \end{bmatrix}.$$

Therefore,

$$\mathbf{K}_s\frac{d\mathbf{K}_s^{-1}}{dt}=\omega\begin{bmatrix}0&1&0\\-1&0&0\\0&0&0\end{bmatrix}$$

and

$$\mathbf{K}_r\frac{d\mathbf{K}_r^{-1}}{dt}=(\omega-\omega_r)\begin{bmatrix}0&1&0\\-1&0&0\\0&0&0\end{bmatrix}.$$

From (5.34), one obtains the following equations in the *arbitrary* reference frame

$$\begin{aligned}\mathbf{u}_{qd0s}&=\mathbf{r}_s\mathbf{i}_{qd0s}+\begin{bmatrix}0&\omega&0\\-\omega&0&0\\0&0&0\end{bmatrix}\boldsymbol{\psi}_{qd0s}+\frac{d\boldsymbol{\psi}_{qd0s}}{dt},\\ \mathbf{u}'_{qd0r}&=\mathbf{r}'_r\mathbf{i}'_{qd0r}+\begin{bmatrix}0&\omega-\omega_r&0\\-\omega+\omega_r&0&0\\0&0&0\end{bmatrix}\boldsymbol{\psi}'_{qd0r}+\frac{d\boldsymbol{\psi}'_{qd0r}}{dt}.\end{aligned}\tag{5.35}$$

Using (5.35), six differential equations are

$$\begin{aligned}&u_{qs}=r_si_{qs}+\omega\psi_{ds}+\frac{d\psi_{qs}}{dt},\quad u_{ds}=r_si_{ds}-\omega\psi_{qs}+\frac{d\psi_{ds}}{dt},\quad u_{0s}=r_si_{0s}+\frac{d\psi_{0s}}{dt},\\ &u'_{qr}=r'_ri'_{qr}+(\omega-\omega_r)\psi'_{dr}+\frac{d\psi'_{qr}}{dt},\quad u'_{dr}=r'_ri'_{dr}-(\omega-\omega_r)\psi'_{qr}+\frac{d\psi'_{dr}}{dt},\quad u'_{0r}=r'_ri'_{0r}+\frac{d\psi'_{0r}}{dt}.\end{aligned}\tag{5.36}$$

From

$$\begin{bmatrix}\boldsymbol{\psi}_{abcs}\\ \boldsymbol{\psi}'_{abcr}\end{bmatrix}=\begin{bmatrix}\mathbf{L}_s&\mathbf{L}'_{sr}(\theta_r)\\ \mathbf{L}'^{T}_{sr}(\theta_r)&\mathbf{L}'_r\end{bmatrix}\begin{bmatrix}\mathbf{i}_{abcs}\\ \mathbf{i}'_{abcr}\end{bmatrix},$$

we have $\boldsymbol{\psi}_{abcs}=\mathbf{L}_s\mathbf{i}_{abcs}+\mathbf{L}'_{sr}(\theta_r)\mathbf{i}'_{abcr}$ and $\boldsymbol{\psi}'_{abcr}=\mathbf{L}'^{T}_{sr}(\theta_r)\mathbf{i}_{abcs}+\mathbf{L}'_r\mathbf{i}'_{abcr}$.

For the *abc* flux linkages, the *qd*0 flux linkages components are

$$\mathbf{K}_s^{-1}\boldsymbol{\psi}_{qd0s}=\mathbf{L}_s\mathbf{K}_s^{-1}\mathbf{i}_{qd0s}+\mathbf{L}'_{sr}(\theta_r)\mathbf{K}_r^{-1}\mathbf{i}'_{qd0r},\quad \mathbf{K}_r^{-1}\boldsymbol{\psi}'_{qd0r}=\mathbf{L}'^{T}_{sr}(\theta_r)\mathbf{K}_s^{-1}\mathbf{i}_{qd0s}+\mathbf{L}'_r\mathbf{K}_r^{-1}\mathbf{i}'_{qd0r}.$$

Multiplying these equations by $\mathbf{K}_s$ and $\mathbf{K}_r$, respectively, yields

$$\begin{aligned}\boldsymbol{\psi}_{qd0s}&=\mathbf{K}_s\mathbf{L}_s\mathbf{K}_s^{-1}\mathbf{i}_{qd0s}+\mathbf{K}_s\mathbf{L}'_{sr}(\theta_r)\mathbf{K}_r^{-1}\mathbf{i}'_{qd0r},\\ \boldsymbol{\psi}'_{qd0r}&=\mathbf{K}_r\mathbf{L}'^{T}_{sr}(\theta_r)\mathbf{K}_s^{-1}\mathbf{i}_{qd0s}+\mathbf{K}_r\mathbf{L}'_r\mathbf{K}_r^{-1}\mathbf{i}'_{qd0r}.\end{aligned}\tag{5.37}$$

Multiplying the transformation matrices by

$$\mathbf{L}_s = \begin{bmatrix} L_{ls}+L_{ms} & -\frac{1}{2}L_{ms} & -\frac{1}{2}L_{ms} \\ -\frac{1}{2}L_{ms} & L_{ls}+L_{ms} & -\frac{1}{2}L_{ms} \\ -\frac{1}{2}L_{ms} & -\frac{1}{2}L_{ms} & L_{ls}+L_{ms} \end{bmatrix},$$

$$\mathbf{L}'_{sr}(\theta_r) = L_{ms}\begin{bmatrix} \sin\theta_r & \sin\left(\theta_r+\frac{2}{3}\pi\right) & \sin\left(\theta_r-\frac{2}{3}\pi\right) \\ \sin\left(\theta_r-\frac{2}{3}\pi\right) & \sin\theta_r & \sin\left(\theta_r+\frac{2}{3}\pi\right) \\ \sin\left(\theta_r+\frac{2}{3}\pi\right) & \sin\left(\theta_r-\frac{2}{3}\pi\right) & \sin\theta_r \end{bmatrix}, \text{ and } \mathbf{L}'_r = \begin{bmatrix} L'_{lr}+L_{ms} & -\frac{1}{2}L_{ms} & -\frac{1}{2}L_{ms} \\ -\frac{1}{2}L_{ms} & L'_{lr}+L_{ms} & -\frac{1}{2}L_{ms} \\ -\frac{1}{2}L_{ms} & -\frac{1}{2}L_{ms} & L'_{lr}+L_{ms} \end{bmatrix}$$

gives

$$\mathbf{K}_s\mathbf{L}_s\mathbf{K}_s^{-1} = \begin{bmatrix} L_{ls}+M & 0 & 0 \\ 0 & L_{ls}+M & 0 \\ 0 & 0 & L_{ls} \end{bmatrix}, \quad \mathbf{K}_s\mathbf{L}'_{sr}(\theta_r)\mathbf{K}_r^{-1} = \mathbf{K}_r\mathbf{L}'^{T}_{sr}(\theta_r)\mathbf{K}_s^{-1} = \begin{bmatrix} M & 0 & 0 \\ 0 & M & 0 \\ 0 & 0 & 0 \end{bmatrix}$$

and

$$\mathbf{K}_r\mathbf{L}'_r\mathbf{K}_r^{-1} = \begin{bmatrix} L'_{lr}+M & 0 & 0 \\ 0 & L'_{lr}+M & 0 \\ 0 & 0 & L'_{lr} \end{bmatrix},$$

where $M = \frac{3}{2}L_{ms}$.

From (5.37), the flux linkage equations are

$$\psi_{qd0s} = \begin{bmatrix} L_{ls}+M & 0 & 0 \\ 0 & L_{ls}+M & 0 \\ 0 & 0 & L_{ls} \end{bmatrix}\mathbf{i}_{qd0s} + \begin{bmatrix} M & 0 & 0 \\ 0 & M & 0 \\ 0 & 0 & 0 \end{bmatrix}\mathbf{i}'_{qd0r},$$

$$\psi_{qd0r} = \begin{bmatrix} M & 0 & 0 \\ 0 & M & 0 \\ 0 & 0 & 0 \end{bmatrix}\mathbf{i}_{qd0s} + \begin{bmatrix} L'_{lr}+M & 0 & 0 \\ 0 & L'_{lr}+M & 0 \\ 0 & 0 & L'_{lr} \end{bmatrix}\mathbf{i}'_{qd0r} \tag{5.38}$$

$$\psi_{qs} = (L_{ls}+M)i_{qs} + Mi'_{qr}, \quad \psi_{ds} = (L_{ls}+M)i_{ds} + Mi'_{dr}, \quad \psi_{0s} = L_{ls}i_{0s},$$

$$\psi'_{qr} = Mi_{qs} + (L'_{lr}+M)i'_{qr}, \quad \psi'_{dr} = Mi_{ds} + (L'_{lr}+M)i'_{dr}, \quad \psi'_{0r} = L'_{lr}i'_{0r}.$$

Substituting (5.38) in (5.36), the following differential equations result

$$u_{qs} = r_s i_{qs} + \omega\left(L_{ls}i_{ds} + Mi_{ds} + Mi'_{dr}\right) + \frac{d\left(L_{ls}i_{qs} + Mi_{qs} + Mi'_{qr}\right)}{dt},$$

$$u_{ds} = r_s i_{ds} - \omega\left(L_{ls}i_{qs} + Mi_{qs} + Mi'_{qr}\right) + \frac{d\left(L_{ls}i_{ds} + Mi_{ds} + Mi'_{dr}\right)}{dt},$$

$$u_{0s} = r_s i_{0s} + \frac{d\left(L_{ls} i_{0s}\right)}{dt},$$

$$u'_{qr} = r'_r i'_{qr} + \left(\omega - \omega_r\right)\left(L'_{lr} i'_{dr} + M i_{ds} + M i'_{dr}\right) + \frac{d\left(M i_{qs} + L'_{lr} i'_{qr} + M i'_{qr}\right)}{dt},$$

$$u'_{dr} = r'_r i'_{dr} - \left(\omega - \omega_r\right)\left(L'_{lr} i'_{qr} + M i_{qs} + M i'_{qr}\right) + \frac{d\left(M i_{ds} + L'_{lr} i'_{dr} + M i'_{dr}\right)}{dt},$$

$$u'_{0r} = r'_r i'_{0r} + \frac{d\left(L'_{lr} i'_{0r}\right)}{dt}.$$

Cauchy's forms of the differential equations are

$$\frac{di_{qs}}{dt} = \frac{1}{L_{SM} L_{RM} - M^2}\left[-L_{RM} r_s i_{qs} - \left(L_{SM} L_{RM} - M^2\right)\omega i_{ds} + M r'_r i'_{qr} - M\left(M i_{ds} + L_{RM} i'_{dr}\right)\omega_r + L_{RM} u_{qs} - M u'_{qr}\right],$$

$$\frac{di_{ds}}{dt} = \frac{1}{L_{SM} L_{RM} - M^2}\left[\left(L_{SM} L_{RM} - M^2\right)\omega i_{qs} - L_{RM} r_s i_{ds} + M r'_r i'_{dr} + M\left(M i_{qs} + L_{RM} i'_{qr}\right)\omega_r + L_{RM} u_{ds} - M u'_{dr}\right],$$

$$\frac{di_{0s}}{dt} = \frac{1}{L_{ls}}\left(-r_s i_{0s} + u_{0s}\right),$$

$$\frac{di'_{qr}}{dt} = \frac{1}{L_{SM} L_{RM} - M^2}\left[M r_s i_{qs} - L_{SM} r'_r i'_{qr} - \left(L_{SM} L_{RM} - M^2\right)\omega i'_{dr} + L_{SM}\left(M i_{ds} + L_{RM} i'_{dr}\right)\omega_r - M u_{qs} + L_{SM} u'_{qr}\right],$$

$$\frac{di'_{dr}}{dt} = \frac{1}{L_{SM} L_{RM} - M^2}\left[M r_s i_{ds} + \left(L_{SM} L_{RM} - M^2\right)\omega i'_{qr} - L_{SM} r'_r i'_{dr} - L_{SM}\left(M i_{qs} + L_{RM} i'_{qr}\right)\omega_r - M u_{ds} + L_{SM} u'_{dr}\right],$$

$$\frac{di'_{0r}}{dt} = \frac{1}{L'_{lr}}\left(-r'_r i'_{0r} + u'_{0r}\right). \tag{5.39}$$

where

$$L_{SM} = L_{ls} + M = L_{ls} + \frac{3}{2} L_{ms},$$

$$L_{RM} = L'_{lr} + M = L'_{lr} + \frac{3}{2} L_{ms}$$

The *torsional–mechanical* equations are

$$J\frac{d\omega_{rm}}{dt} = T_e - B_m \omega_{rm} - T_L,$$

$$\frac{d\theta_{rm}}{dt} = \omega_{rm}. \tag{5.40}$$

The expression for T_e should be obtained in terms of the $qd0$-axis components of stator and rotor currents. Using

$$W_c = \frac{1}{2}\mathbf{i}_{abcs}^T\left(\mathbf{L}_s - L_{ls}\mathbf{I}\right)\mathbf{i}_{abcs} + \mathbf{i}_{abcs}^T\mathbf{L}'_{sr}\left(\theta_r\right)\mathbf{i}'_{abcr} + \frac{1}{2}\mathbf{i}_{abcr}^T\left(\mathbf{L}'_r - L'_{lr}\mathbf{I}\right)\mathbf{i}'_{abcr},$$

one finds

$$T_e = \frac{P}{2}\frac{\partial W_c\left(\mathbf{i}_{abcs}, \mathbf{i}'_{abcr}, \theta_r\right)}{\partial \theta_r} = \frac{P}{2}\mathbf{i}_{abcs}^T\frac{\partial \mathbf{L}'_{sr}(\theta_r)}{\partial \theta_r}\mathbf{i}'_{abcr},$$

$$\mathbf{L}'_{sr}\left(\theta_r\right) = L_{ms}\begin{bmatrix} \sin\theta_r & \sin\left(\theta_r + \frac{2}{3}\pi\right) & \sin\left(\theta_r - \frac{2}{3}\pi\right) \\ \sin\left(\theta_r - \frac{2}{3}\pi\right) & \sin\theta_r & \sin\left(\theta_r + \frac{2}{3}\pi\right) \\ \sin\left(\theta_r + \frac{2}{3}\pi\right) & \sin\left(\theta_r - \frac{2}{3}\pi\right) & \sin\theta_r \end{bmatrix}.$$

Therefore,

$$T_e = \frac{P}{2}\left(\mathbf{K}_s^{-1}\mathbf{i}_{qd0s}\right)^T\frac{\partial \mathbf{L}'_{sr}\left(\theta_r\right)}{\partial \theta_r}\mathbf{K}_r^{-1}\mathbf{i}'_{qd0r}$$

$$= \frac{P}{2}\mathbf{i}_{qdos}^T\mathbf{K}_s^{-1T}\left(\frac{\partial}{\partial \theta_r}L_{ms}\begin{bmatrix} \sin\theta_r & \sin\left(\theta_r + \frac{2}{3}\pi\right) & \sin\left(\theta_r - \frac{2}{3}\pi\right) \\ \sin\left(\theta_r - \frac{2}{3}\pi\right) & \sin\theta_r & \sin\left(\theta_r + \frac{2}{3}\pi\right) \\ \sin\left(\theta_r + \frac{2}{3}\pi\right) & \sin\left(\theta_r - \frac{2}{3}\pi\right) & \sin\theta_r \end{bmatrix}\right)\mathbf{K}_r^{-1}\mathbf{i}'_{qd0r}.$$

The matrix multiplication yields

$$T_e = \frac{3P}{4}M\left(i_{qs}i'_{dr} - i_{ds}i'_{qr}\right). \tag{5.41}$$

From (5.40) and (5.41), one has

$$\begin{aligned} \frac{d\omega_r}{dt} &= \frac{3P^2}{8J}M\left(i_{qs}i'_{dr} - i_{ds}i'_{qr}\right) - \frac{B_m}{J}\omega_r - \frac{P}{2J}T_L, \\ \frac{d\theta_r}{dt} &= \omega_r. \end{aligned} \tag{5.42}$$

Combining the circuitry-electromagnetic and *torsional–mechanical* dynamics, as given by (5.39) and (5.42), the model for three-phase induction motors in the *arbitrary* reference frame is given as a set of eight nonlinear differential equations (5.39) and (5.42). Omitting the differential equation $\frac{d\theta_r}{dt} = \omega_r$, the following state-space model results

$$\begin{bmatrix} \frac{di_{qs}}{dt} \\ \frac{di_{ds}}{dt} \\ \frac{di_{0s}}{dt} \\ \frac{di'_{qr}}{dt} \\ \frac{di'_{dr}}{dt} \\ \frac{di'_{0r}}{dt} \\ \frac{d\omega_r}{dt} \end{bmatrix} = \begin{bmatrix} -\frac{L_{RM} r_s}{L_{SM}L_{RM}-M^2} & -\omega & 0 & \frac{Mr'_r}{L_{SM}L_{RM}-M^2} & 0 & 0 & 0 \\ \omega & -\frac{L_{RM} r_s}{L_{SM}L_{RM}-M^2} & 0 & 0 & \frac{Mr'_r}{L_{SM}L_{RM}-M^2} & 0 & 0 \\ 0 & 0 & -\frac{r_s}{L_{ls}} & 0 & 0 & 0 & 0 \\ \frac{Mr_s}{L_{SM}L_{RM}-M^2} & 0 & 0 & -\frac{L_{SM} r'_r}{L_{SM}L_{RM}-M^2} & -\omega & 0 & 0 \\ 0 & \frac{Mr_s}{L_{SM}L_{RM}-M^2} & 0 & \omega & -\frac{L_{SM} r'_r}{L_{SM}L_{RM}-M^2} & 0 & 0 \\ 0 & 0 & 0 & 0 & 0 & -\frac{r'_r}{L'_{lr}} & 0 \\ 0 & 0 & 0 & 0 & 0 & 0 & -\frac{B_m}{J} \end{bmatrix} \begin{bmatrix} i_{qs} \\ i_{ds} \\ i_{0s} \\ i'_{qr} \\ i'_{dr} \\ i'_{0r} \\ \omega_r \end{bmatrix}$$

$$+ \begin{bmatrix} -\frac{M\left(Mi_{ds}+L_{RM}i'_{dr}\right)\omega_r}{L_{SM}L_{RM}-M^2} \\ \frac{M\left(Mi_{qs}+L_{RM}i'_{qr}\right)\omega_r}{L_{SM}L_{RM}-M^2} \\ 0 \\ \frac{L_{SM}\left(Mi_{ds}+L_{RM}i'_{dr}\right)\omega_r}{L_{SM}L_{RM}-M^2} \\ -\frac{L_{SM}\left(Mi_{qs}+L_{RM}i'_{qr}\right)\omega_r}{L_{SM}L_{RM}-M^2} \\ 0 \\ \frac{3P^2}{8J} M\left(i_{qs}i'_{dr}-i_{ds}i'_{qr}\right) \end{bmatrix} + \begin{bmatrix} \frac{L_{RM}}{L_{SM}L_{RM}-M^2} & 0 & 0 & -\frac{M}{L_{SM}L_{RM}-M^2} & 0 & 0 \\ 0 & \frac{L_{RM}}{L_{SM}L_{RM}-M^2} & 0 & 0 & -\frac{M}{L_{SM}L_{RM}-M^2} & 0 \\ 0 & 0 & \frac{1}{L_{ls}} & 0 & 0 & 0 \\ -\frac{M}{L_{SM}L_{RM}-M^2} & 0 & 0 & \frac{L_{SM}}{L_{SM}L_{RM}-M^2} & 0 & 0 \\ 0 & -\frac{M}{L_{SM}L_{RM}-M^2} & 0 & 0 & \frac{L_{SM}}{L_{SM}L_{RM}-M^2} & 0 \\ 0 & 0 & 0 & 0 & 0 & \frac{1}{L'_{lr}} \\ 0 & 0 & 0 & 0 & 0 & 0 \end{bmatrix} \begin{bmatrix} u_{qs} \\ u_{ds} \\ u_{0s} \\ u'_{qr} \\ u'_{dr} \\ u'_{0r} \end{bmatrix} - \begin{bmatrix} 0 \\ 0 \\ 0 \\ 0 \\ 0 \\ 0 \\ \frac{P}{2J} \end{bmatrix} T_L. \tag{5.43}$$

In squirrel-cage motors, the rotor windings are short-circuited. The three-phase balanced voltage set is

$$u_{as}(t) = \sqrt{2}u_M \cos\left(\omega_f t\right), \quad u_{bs}(t) = \sqrt{2}u_M \cos\left(\omega_f t - \frac{2}{3}\pi\right), \quad \text{and} \quad u_{cs}(t) = \sqrt{2}u_M \cos\left(\omega_f t + \frac{2}{3}\pi\right).$$

The *quadrature-*, *direct-*, and *zero*-axis components of stator voltages are found by using the stator Park transformation,

$$\mathbf{u}_{qd0s} = \mathbf{K}_s \mathbf{u}_{abcs}, \quad \mathbf{K}_s = \frac{2}{3}\begin{bmatrix} \cos\theta & \cos\left(\theta - \frac{2}{3}\pi\right) & \cos\left(\theta + \frac{2}{3}\pi\right) \\ \sin\theta & \sin\left(\theta - \frac{2}{3}\pi\right) & \sin\left(\theta + \frac{2}{3}\pi\right) \\ \frac{1}{2} & \frac{1}{2} & \frac{1}{2} \end{bmatrix}.$$

The stationary, rotor, and synchronous reference frames are commonly used. For the aforementioned reference frames, in (5.39) and (5.43), $\omega = 0$, $\omega = \omega_r$, and $\omega = \omega_e$. This results in the corresponding angular displacement. Letting $\theta_0 = 0$, for the stationary, rotor, and synchronous reference frames, $\theta = 0$, $\theta = \theta_r$, and $\theta = \theta_e$.

5.5.3 Induction Motors in the Synchronous Reference Frame

The synchronous reference frame is most commonly used to study induction and synchronous machines. The differential equations, obtained for the *arbitrary* reference frame, as given by (5.39), are modified using the specified $\omega = \omega_e$. From (5.39) and (5.43), we have

$$\frac{di_{qs}^e}{dt} = \frac{1}{L_{SM}L_{RM} - M^2}\left[-L_{RM}r_s i_{qs}^e - \left(L_{SM}L_{RM} - M^2\right)\omega_e i_{ds}^e + Mr_r' i_{qr}'^e - M\left(Mi_{ds}^e + L_{RM}i_{dr}'^e\right)\omega_r + L_{RM}u_{qs}^e - Mu_{qr}'^e\right],$$

$$\frac{di_{ds}^e}{dt} = \frac{1}{L_{SM}L_{RM} - M^2}\left[\left(L_{SM}L_{RM} - M^2\right)\omega_e i_{qs}^e - L_{RM}r_s i_{ds}^e + Mr_r' i_{dr}'^e + M\left(Mi_{qs}^e + L_{RM}i_{qr}'^e\right)\omega_r + L_{RM}u_{ds}^e - Mu_{dr}'^e\right],$$

$$\frac{di_{0s}^e}{dt} = \frac{1}{L_{ls}}\left(-r_s i_{0s}^e + u_{0s}^e\right),$$

$$\frac{di_{qr}'^e}{dt} = \frac{1}{L_{SM}L_{RM} - M^2}\left[Mr_s i_{qs}^e - L_{SM}r_r' i_{qr}'^e - \left(L_{SM}L_{RM} - M^2\right)\omega_e i_{dr}'^e + L_{SM}\left(Mi_{ds}^e + L_{RM}i_{dr}'^e\right)\omega_r - Mu_{qs}^e + L_{SM}u_{qr}'^e\right],$$

$$\frac{di_{dr}'^e}{dt} = \frac{1}{L_{SM}L_{RM} - M^2}\left[Mr_s i_{ds}^e + \left(L_{SM}L_{RM} - M^2\right)\omega_e i_{qr}'^e - L_{SM}r_r' i_{dr}'^e - L_{SM}\left(Mi_{qs}^e + L_{RM}i_{qr}'^e\right)\omega_r - Mu_{ds}^e + L_{SM}u_{dr}'^e\right],$$

$$\frac{di_{0r}'^e}{dt} = \frac{1}{L_{lr}'}\left(-r_r' i_{0r}'^e + u_{0r}'^e\right),$$

$$\frac{d\omega_r}{dt} = \frac{3P^2}{8J}M\left(i_{qs}^e i_{dr}'^e - i_{ds}^e i_{qr}'^e\right) - \frac{B_m}{J}\omega_r - \frac{P}{2J}T_L,$$

$$\frac{d\theta_r}{dt} = \omega_r. \tag{5.44}$$

The *quadrature*, *direct*, and *zero* components of stator voltages u_{qs}^e, u_{ds}^e, and u_{0s}^e are found by using $\mathbf{u}_{qd0s}^e = \mathbf{K}_s^e \mathbf{u}_{abcs}$. Using the Park transformation matrix

$$\mathbf{K}_s = \frac{2}{3}\begin{bmatrix} \cos\theta & \cos\left(\theta - \frac{2}{3}\pi\right) & \cos\left(\theta + \frac{2}{3}\pi\right) \\ \sin\theta & \sin\left(\theta - \frac{2}{3}\pi\right) & \sin\left(\theta + \frac{2}{3}\pi\right) \\ \frac{1}{2} & \frac{1}{2} & \frac{1}{2} \end{bmatrix} \quad \text{with } \theta = \theta_e,$$

one has

$$\mathbf{K}_s^e = \frac{2}{3}\begin{bmatrix} \cos\theta_e & \cos\left(\theta_e - \frac{2}{3}\pi\right) & \cos\left(\theta_e + \frac{2}{3}\pi\right) \\ \sin\theta_e & \sin\left(\theta_e - \frac{2}{3}\pi\right) & \sin\left(\theta_e + \frac{2}{3}\pi\right) \\ \frac{1}{2} & \frac{1}{2} & \frac{1}{2} \end{bmatrix}.$$

From

$$\begin{bmatrix} u_{qs}^e \\ u_{ds}^e \\ u_{0s}^e \end{bmatrix} = \frac{2}{3}\begin{bmatrix} \cos\theta_e & \cos\left(\theta_e - \frac{2}{3}\pi\right) & \cos\left(\theta_e + \frac{2}{3}\pi\right) \\ \sin\theta_e & \sin\left(\theta_e - \frac{2}{3}\pi\right) & \sin\left(\theta_e + \frac{2}{3}\pi\right) \\ \frac{1}{2} & \frac{1}{2} & \frac{1}{2} \end{bmatrix}\begin{bmatrix} u_{as} \\ u_{bs} \\ u_{cs} \end{bmatrix},$$

one obtains

$$u_{qs}^e(t) = \frac{2}{3}\left[u_{as}\cos\theta_e + u_{bs}\cos\left(\theta_e - \frac{2}{3}\pi\right) + u_{cs}\cos\left(\theta_e + \frac{2}{3}\pi\right)\right],$$

$$u_{ds}^e(t) = \frac{2}{3}\left[u_{as}\sin\theta_e + u_{bs}\sin\left(\theta_e - \frac{2}{3}\pi\right) + u_{cs}\sin\left(\theta_e + \frac{2}{3}\pi\right)\right], \quad \text{and} \quad u_{0s}^e(t) = \frac{1}{3}\left(u_{as} + u_{bs} + u_{cs}\right).$$

The three-phase balanced voltage set is

$$u_{as}(t) = \sqrt{2}u_M\cos\left(\omega_f t\right), \quad u_{bs}(t) = \sqrt{2}u_M\cos\left(\omega_f t - \frac{2}{3}\pi\right), \quad \text{and} \quad u_{cs}(t) = \sqrt{2}u_M\cos\left(\omega_f t + \frac{2}{3}\pi\right).$$

Assume that the initial displacement of the *quadrature* magnetic axis is zero, $\theta_0 = 0$. From $\theta_e = \omega_f t$, the $qd0$ components of stator voltages are

$$u_{qs}^e(t) = \sqrt{2}u_M, \quad u_{ds}^e(t) = 0, \quad u_{0s}^e(t) = 0. \tag{5.45}$$

The deviations and corresponding trigonometric identities were reported in Example 5.6.

The advantage of the stationary, rotor, and synchronous reference frames is the mathematical simplicity of equations of motion. Analysis tools and simulation software allow one to examine the most sophisticated induction machine models in the *machine* variable. There are disadvantages of the *arbitrary* (stationary, rotor, and synchronous) reference frame from the implementation standpoints as discussed in Section 5.2.2. The AC *machine* phase voltages u_{as}, u_{bs}, and u_{cs} are applied to the stator windings. One cannot apply $u_{qs}^e(t)$, $u_{ds}^e(t)$, and $u_{0s}^e(t)$. If control in the $qd0$ frames is searched, using $\mathbf{u}_{qd0s}^e$, one should calculate $\mathbf{u}_{abcs}$ by using the *inverse* Park transformation $\mathbf{u}_{abcs} = (\mathbf{K}_s^e)^{-1}\mathbf{u}_{qd0s}^e$. Though the *vector* control of induction motors can be used, the voltage–frequency control $\frac{u_{Mi}}{f_i} = \text{var}$, $u_{M\min} \le u_M \le u_{M\max}, f_{\min} \le f \le f_{\max}$, guarantees the high-torque patterns, which surpass the $qd0$-centric concepts. The highest $T_{e\,start}$ and $T_{e\,critical}$ are developed applying the frequency control. The $T_{e\,start\,\max}$ corresponds to $f_{\min}, f_{\min} \le f \le f_{\max}$. The $f_{\min}$ is defined by the converter topologies, MOSFETs limits, switching strategies, etc.

Example 5.7

We simulate a three-phase squirrel-cage motor in the synchronous reference frame. Consider a two-pole ($P = 2$), 220 V (*rms*), 60 Hz squirrel-cage induction motor with $r_s = 9.4$ ohm, $r_r' = 7.1$ ohm, $L_{ms} = 0.37$ H, $L_{ls} = 0.032$ H, $L_{lr}' = 0.032$ H, $B_m = 0.0001$ N-m-rad/sec, and $J = 0.001$ kg-m^2. The Simulink model is developed. The $qd0$ voltage components are $u_{qs}^e(t) = \sqrt{2}220$, $u_{ds}^e(t) = 0$, and $u_{os}^e(t) = 0$. Figures 5.12 illustrates the motor dynamics and the torque–speed characteristics if $T_L = 0$ and $T_L = 2.5$ N-m. The $qd0$-axis components of stator and rotor currents and voltages are DC quantities. ■

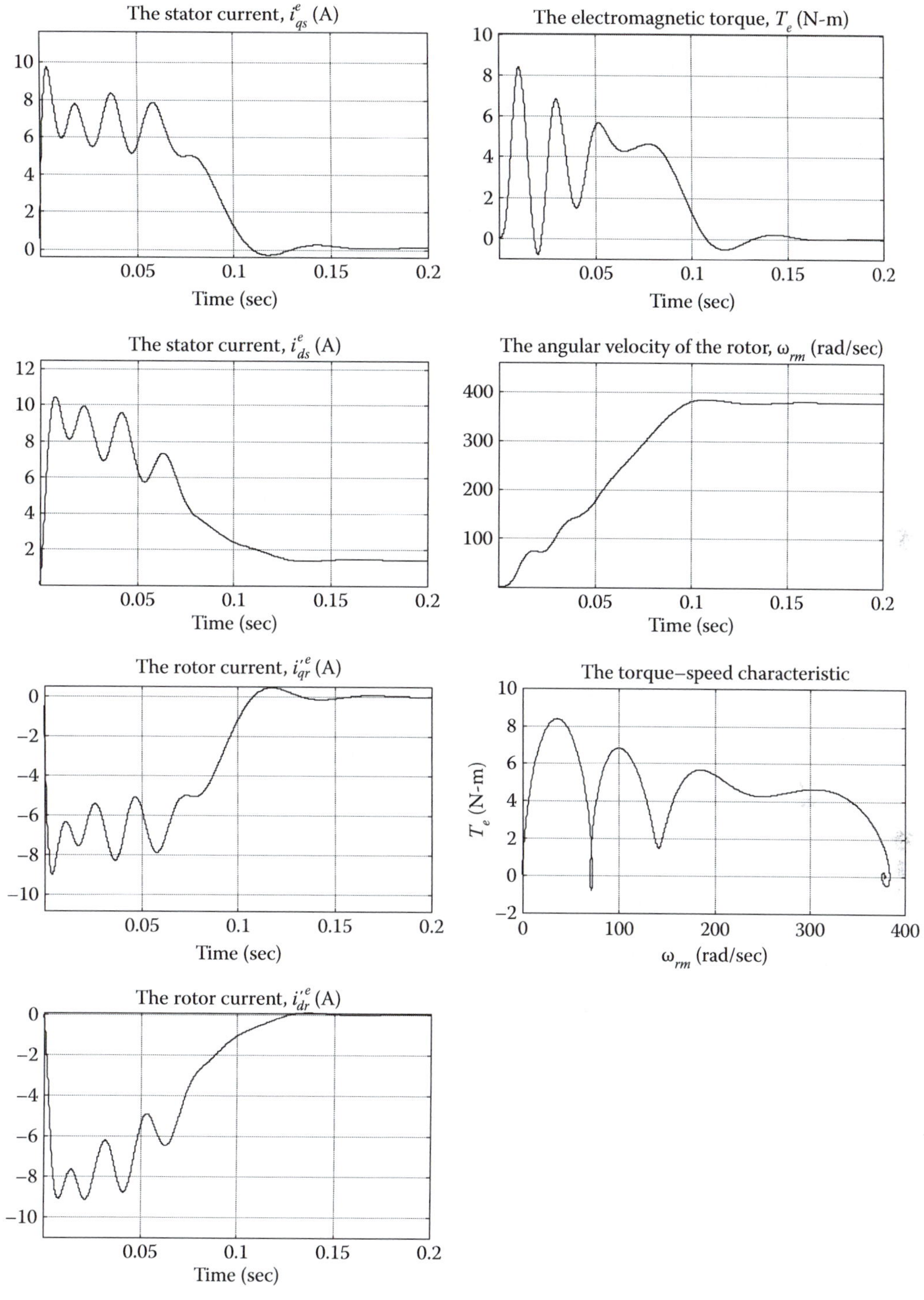

FIGURE 5.12 Dynamics and torque–speed characteristics of a three-phase induction motor for $T_L = 0$ N-m and $T_L = 2.5$ N-m, $u^e_{qs}(t) = \sqrt{2}220$, $u^e_{ds}(t) = 0$, $u^e_{os}(t) = 0$. *(Continued)*

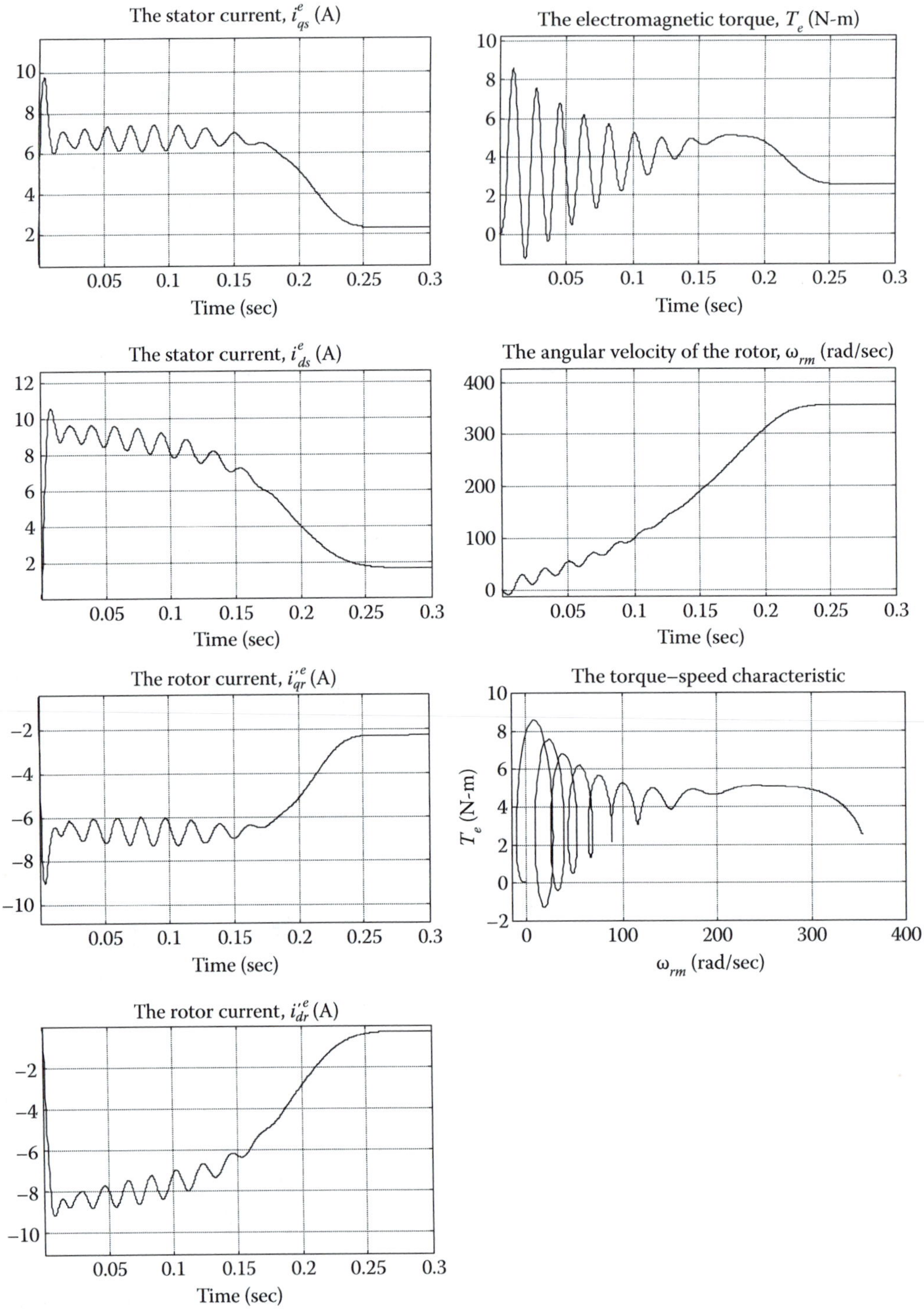

FIGURE 5.12 (*Continued*) Dynamics and torque–speed characteristics of a three-phase induction motor for $T_L = 0$ N-m and $T_L = 2.5$ N-m, $u_{qs}^e(t) = \sqrt{2}220$, $u_{ds}^e(t) = 0$, $u_{os}^e(t) = 0$.

5.5.4 Three-Phase Induction Motors in the Rotor Reference Frame

The mathematical model of induction motors in the rotor reference frame is found using the angular velocity of the frame $\omega = \omega_r$. From the derived differential equations (5.39) and (5.42), we have

$$\frac{di_{qs}^r}{dt} = \frac{1}{L_{SM}L_{RM} - M^2}\left[-L_{RM}r_s i_{qs}^r + Mr_r' i_{qr}'^r - L_{RM}\left(L_{SM}i_{ds}^r + Mi_{dr}'^r\right)\omega_r + L_{RM}u_{qs}^r - Mu_{qr}'^r\right],$$

$$\frac{di_{ds}^r}{dt} = \frac{1}{L_{SM}L_{RM} - M^2}\left[-L_{RM}r_s i_{ds}^r + Mr_r' i_{dr}'^r + L_{RM}\left(L_{SM}i_{qs}^r + Mi_{qr}'^r\right)\omega_r + L_{RM}u_{ds}^r - Mu_{dr}'^r\right],$$

$$\frac{di_{0s}^r}{dt} = \frac{1}{L_{ls}}\left(-r_s i_{0s}^r + u_{0s}^r\right),$$

$$\frac{di_{qr}'^r}{dt} = \frac{1}{L_{SM}L_{RM} - M^2}\left[Mr_s i_{qs}^r - L_{SM}r_r' i_{qr}'^r + M\left(L_{SM}i_{ds}^r + Mi_{dr}'^r\right)\omega_r - Mu_{qs}^r + L_{SM}u_{qr}'^r\right],$$

$$\frac{di_{dr}'^r}{dt} = \frac{1}{L_{SM}L_{RM} - M^2}\left[Mr_s i_{ds}^r - L_{SM}r_r' i_{dr}'^r - M\left(L_{SM}i_{qs}^r + Mi_{qr}'^r\right)\omega_r - Mu_{ds}^r + L_{SM}u_{dr}'^r\right],$$

$$\frac{di_{0r}'^r}{dt} = \frac{1}{L_{lr}'}\left(-r_r' i_{0r}'^r + u_{0r}'^r\right),$$

$$\frac{d\omega_r}{dt} = \frac{3P^2}{8J}M\left(i_{qs}^r i_{dr}'^r - i_{ds}^r i_{qr}'^r\right) - \frac{B_m}{J}\omega_r - \frac{P}{2J}T_L,$$

$$\frac{d\theta_r}{dt} = \omega_r.$$

One may find the $qd0$ voltage components u_{qs}^r, u_{ds}^r, and u_{os}^r. The Park transformation matrix $\mathbf{K}_s^r$ is used. In

$$\mathbf{K}_s = \frac{2}{3}\begin{bmatrix} \cos\theta & \cos\left(\theta - \frac{2}{3}\pi\right) & \cos\left(\theta + \frac{2}{3}\pi\right) \\ \sin\theta & \sin\left(\theta - \frac{2}{3}\pi\right) & \sin\left(\theta + \frac{2}{3}\pi\right) \\ \frac{1}{2} & \frac{1}{2} & \frac{1}{2} \end{bmatrix},$$

the substitution $\theta = \theta_r$ yields

$$\mathbf{K}_s^r = \frac{2}{3}\begin{bmatrix} \cos\theta_r & \cos\left(\theta_r - \frac{2}{3}\pi\right) & \cos\left(\theta_r + \frac{2}{3}\pi\right) \\ \sin\theta_r & \sin\left(\theta_r - \frac{2}{3}\pi\right) & \sin\left(\theta_r + \frac{2}{3}\pi\right) \\ \frac{1}{2} & \frac{1}{2} & \frac{1}{2} \end{bmatrix}.$$

Hence, $\mathbf{u}_{qdos}^{r} = \mathbf{K}_s^r \mathbf{u}_{abcs}$, or $\begin{bmatrix} u_{qs}^r \\ u_{ds}^r \\ u_{0s}^r \end{bmatrix} = \frac{2}{3}\begin{bmatrix} \cos\theta_r & \cos\left(\theta_r - \frac{2}{3}\pi\right) & \cos\left(\theta_r + \frac{2}{3}\pi\right) \\ \sin\theta_r & \sin\left(\theta_r - \frac{2}{3}\pi\right) & \sin\left(\theta_r + \frac{2}{3}\pi\right) \\ \frac{1}{2} & \frac{1}{2} & \frac{1}{2} \end{bmatrix}\begin{bmatrix} u_{as} \\ u_{bs} \\ u_{cs} \end{bmatrix}$.

Therefore, $u_{qs}^r(t) = \frac{2}{3}\left[u_{as}\cos\theta_r + u_{bs}\cos\left(\theta_r - \frac{2}{3}\pi\right) + u_{cs}\cos\left(\theta_r + \frac{2}{3}\pi\right)\right]$,

$u_{ds}^r(t) = \frac{2}{3}\left[u_{as}\sin\theta_r + u_{bs}\sin\left(\theta_r - \frac{2}{3}\pi\right) + u_{cs}\sin\left(\theta_r + \frac{2}{3}\pi\right)\right]$, and $u_{0s}^r(t) = \frac{1}{3}\left(u_{as} + u_{bs} + u_{cs}\right)$.

From $u_{as}(t) = \sqrt{2}u_M\cos\left(\omega_f t\right)$, $u_{bs}(t) = \sqrt{2}u_M\cos\left(\omega_f t - \frac{2}{3}\pi\right)$, $u_{cs}(t) = \sqrt{2}u_M\cos\left(\omega_f t + \frac{2}{3}\pi\right)$, one has $u_{qs}^r(t) = \sqrt{2}u_M\cos\left(\omega_f t - \theta_r\right)$, $u_{ds}^r(t) = -\sqrt{2}u_M\sin\left(\omega_f t - \theta_r\right)$, $u_{0s}^r(t) = 0$.

Example 5.8

Examine the dynamics of a squirrel-cage induction motor in the rotor reference frame. The parameters are reported in Example 5.7. The *qd*0 voltage set is $u_{qs}^r(t) = \sqrt{2}220\cos\left(377t - \theta_r\right)$, $u_{ds}^r(t) = -\sqrt{2}220\sin\left(377t - \theta_r\right)$, and $u_{0s}^r(t) = 0$. The evolution of the state variables and torque–speed characteristics are plotted in Figures 5.13 for different loading conditions. ■

5.5.5 Three-Phase Induction Motors in the Stationary Reference Frame

In the stationary reference frame, ω = 0 and θ = 0. Using the mathematical model of induction motors in the *arbitrary* reference frame, given by (5.39) and (5.42), one finds

$$\frac{di_{qs}^s}{dt} = \frac{1}{L_{SM}L_{RM} - M^2}\left[-L_{RM}r_s i_{qs}^s + Mr_r' i_{qr}'^s - M\left(Mi_{ds}^s + L_{RM}i_{dr}'^s\right)\omega_r + L_{RM}u_{qs}^s - Mu_{qr}'^s\right],$$

$$\frac{di_{ds}^s}{dt} = \frac{1}{L_{SM}L_{RM} - M^2}\left[-L_{RM}r_s i_{ds}^s + Mr_r' i_{dr}'^s + M\left(Mi_{qs}^s + L_{RM}i_{qr}'^s\right)\omega_r + L_{RM}u_{ds}^s - Mu_{dr}'^s\right],$$

$$\frac{di_{0s}^s}{dt} = \frac{1}{L_{ls}}\left(-r_s i_{0s}^s + u_{0s}^s\right),$$

$$\frac{di_{qr}'^s}{dt} = \frac{1}{L_{SM}L_{RM} - M^2}\left[Mr_s i_{qs}^s - L_{SM}r_r' i_{qr}'^s + L_{SM}\left(Mi_{ds}^s + L_{RM}i_{dr}'^s\right)\omega_r - Mu_{qs}^s + L_{SM}u_{qr}'^s\right],$$

$$\frac{di_{dr}'^s}{dt} = \frac{1}{L_{SM}L_{RM} - M^2}\left[Mr_s i_{ds}^s - L_{SM}r_r' i_{dr}'^s - L_{SM}\left(Mi_{qs}^s + L_{RM}i_{qr}'^s\right)\omega_r - Mu_{ds}^s + L_{SM}u_{dr}'^s\right],$$

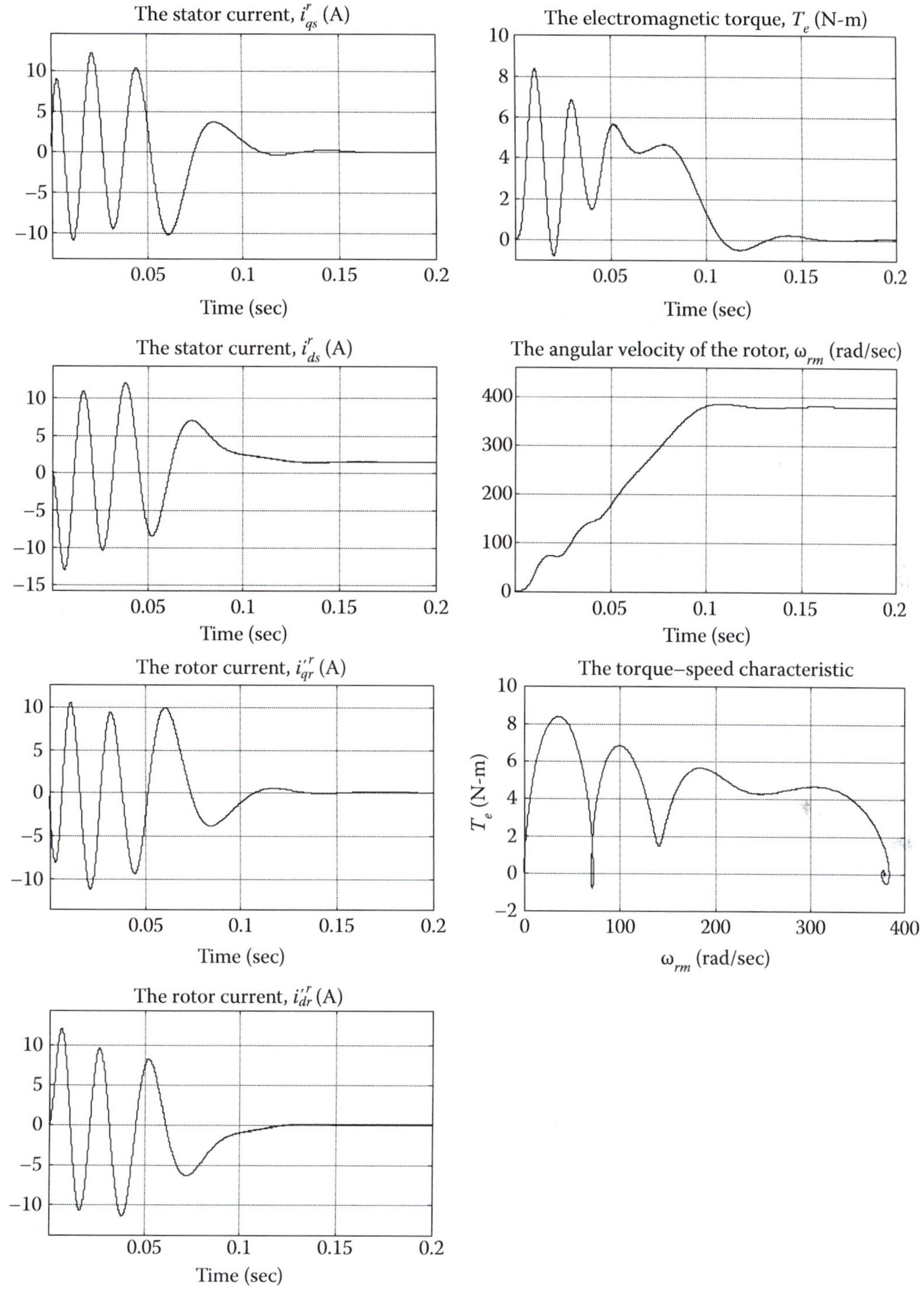

FIGURE 5.13 Dynamics and torque–speed characteristics of three-phase induction motor for $T_L = 0$ N-m and $T_L = 2.5$ N-m, $u_{qs}^r(t) = \sqrt{2}220\cos(377t - \theta_r)$, $u_{ds}^r(t) = -\sqrt{2}220\sin(377t - \theta_r)$, $u_{0s}^r(t) = 0$. *(Continued)*

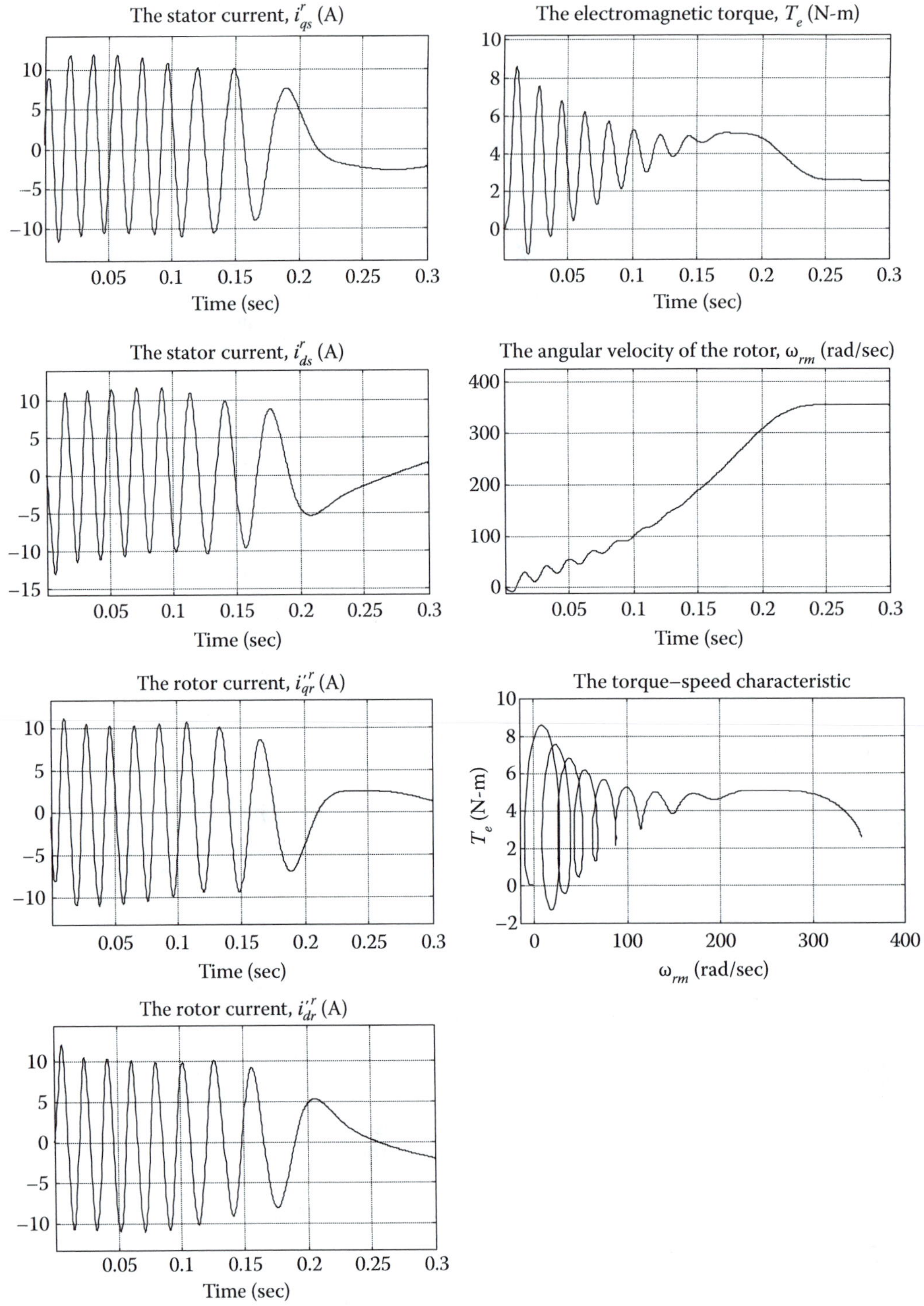

FIGURE 5.13 (*Continued*) Dynamics and torque–speed characteristics of three-phase induction motor for $T_L = 0$ N-m and $T_L = 2.5$ N-m, $u^r_{qs}(t) = \sqrt{2}220\cos(377t - \theta_r)$, $u^r_{ds}(t) = -\sqrt{2}220\sin(377t - \theta_r)$, $u^r_{0s}(t) = 0$.

$$\frac{di_{0r}'^{s}}{dt} = \frac{1}{L_{lr}'}\left(-r_r' i_{0r}'^{s} + u_{0r}'^{s}\right),$$

$$\frac{d\omega_r}{dt} = \frac{3P^2}{8J} M\left(i_{qs} i_{dr}'^{s} - i_{ds} i_{qr}'^{s}\right) - \frac{B_m}{J}\omega_r - \frac{P}{2J}T_L,$$

$$\frac{d\theta_r}{dt} = \omega_r.$$

The *quadrature*-, *direct*-, and *zero*-axis voltage components u_{qs}^s, u_{ds}^s, and u_{os}^s are obtained by using the stator Park transformation matrices. In

$$\mathbf{K}_s = \frac{2}{3}\begin{bmatrix} \cos\theta & \cos\left(\theta - \frac{2}{3}\pi\right) & \cos\left(\theta + \frac{2}{3}\pi\right) \\ \sin\theta & \sin\left(\theta - \frac{2}{3}\pi\right) & \sin\left(\theta + \frac{2}{3}\pi\right) \\ \frac{1}{2} & \frac{1}{2} & \frac{1}{2} \end{bmatrix}, \quad \theta = 0.$$

Hence,

$$\mathbf{u}_{qdos}^s = \mathbf{K}_s^s \mathbf{u}_{abcs} = \begin{bmatrix} \frac{2}{3} & -\frac{1}{3} & -\frac{1}{3} \\ 0 & -\frac{1}{\sqrt{3}} & \frac{1}{\sqrt{3}} \\ \frac{1}{3} & \frac{1}{3} & \frac{1}{3} \end{bmatrix}\mathbf{u}_{abcs} \quad \text{and} \quad \begin{bmatrix} u_{qs}^s \\ u_{ds}^s \\ u_{os}^s \end{bmatrix} = \begin{bmatrix} \frac{2}{3} & -\frac{1}{3} & -\frac{1}{3} \\ 0 & -\frac{1}{\sqrt{3}} & \frac{1}{\sqrt{3}} \\ \frac{1}{3} & \frac{1}{3} & \frac{1}{3} \end{bmatrix}\begin{bmatrix} u_{as} \\ u_{bs} \\ u_{cs} \end{bmatrix}.$$

Therefore,

$$u_{qs}^s(t) = \frac{2}{3}u_{as}(t) - \frac{1}{3}u_{bs}(t) - \frac{1}{3}u_{cs}(t),$$

$$u_{ds}^s(t) = -\frac{1}{\sqrt{3}}u_{bs}(t) + \frac{1}{\sqrt{3}}u_{cs}(t),$$

and

$$u_{0s}^s(t) = \frac{1}{3}u_{as}(t) + \frac{1}{3}u_{bs}(t) + \frac{1}{3}u_{cs}(t).$$

Using a balanced three-phase voltage set, one obtains

$$u_{qs}^s(t) = \sqrt{2}u_M \cos\left(\omega_f t\right), \quad u_{ds}^s(t) = -\sqrt{2}u_M \sin\left(\omega_f t\right), \quad u_{0s}^s(t) = 0.$$

5.6 POWER CONVERTERS

The angular velocity of squirrel-cage induction motors is regulated by changing the magnitude and frequency of the phase voltages applied to the stator windings using power converters as illustrated in Figure 5.14. The basic components of variable-frequency converters are rectifier, filter, and inverter. The simplest rectifiers are the single-phase half- and full-wave rectifiers.

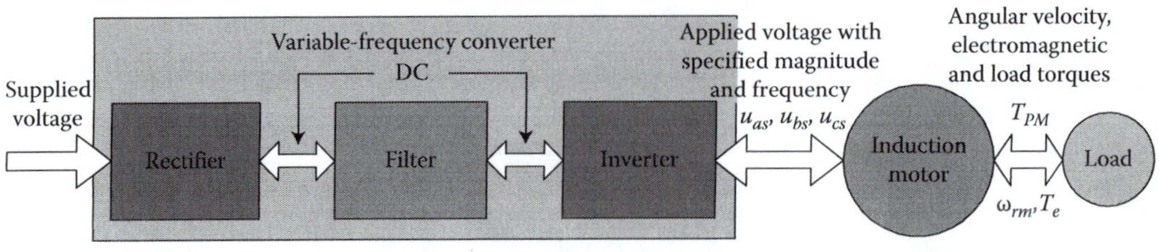

FIGURE 5.14 Variable-frequency power converter to drive induction motors.

To control medium- and high-power induction motors, *polyphase* rectifiers are used. *Polyphase* rectifiers contain several AC sources, and the rectified voltage is summated at the output. The rectified voltage is filtered to reduce the harmonic content of the rectifier output voltage. Passive and active harmonic reduction, harmonic elimination, and harmonic cancellation can be achieved by using passive and active filters. To change and control the voltage frequency *f*, inverters are used. Voltage- and current-fed inverters convert the DC voltage or current, respectively. Pulse width modulation (PWM) ensures controllability and high efficiency and reduces the total harmonic distortion. The PWM concept uses control and driving electronics to control the switching activity of switches (usually power MOSFETs) with a high frequency. The filtering is accomplished using the *LC* filter topologies.

Power converters produce sinusoidal voltages, which are applied to the induction motor windings. The DC voltage is obtained by rectifying and filtering the line voltage. The magnitude of the voltage can be controlled. The sinusoidal AC voltage with the regulated frequency is obtained by using DC-to-AC inverters. The design of high-performance electric drives depends on power electronics and power semiconductor devices. High switching frequencies ~200 V and 200 A power transistors with used in light- and medium drives. The specialized three-phase drivers are available. For high-power drives, ~3000 V, 1000 A insulated-gate bipolar transistors are integrated with diodes in the same package. The switching frequency ~200 kVA soft switching resonant-link inverters may reach ~100 kHz. The development of the gate turn-off (GTO) thyristor was the key that helped extend the power rating of electric drives with induction machines to the megawatt range. Power converters with GTO thyristors have found widespread applications in traction drives (electric drivetrains in ships and locomotives). Gate turn-off thyristors are current-controlled devices that require large gate current to enable turn-off the anode current. Large snubbers are needed to ensure turn-off without failures. Various high-performance topologies exist and are available from various manufacturers. There are two basic types of inverters. The voltage source inverters ensure variable frequency phase voltages. The variable frequency phase currents are fed to the induction motor windings by current source inverters. Figures 5.15 illustrate high-level diagrams of power converters that include a PWM voltage source inverter with an unregulated rectifier, a squire-wave voltage source inverter with a regulated rectifier, and current source inverter with a regulated rectifier [5, 6, 8].

Typical PWM power converter configurations consist of three legs, one for each phase, to control the frequency and the magnitude of the phase voltages as shown in Figures 5.16. The inverter converts the DC bus voltage into a *polyphase* AC voltage at the desired frequency. The drawbacks of hard-switching inverters are switching stresses, losses, high electromagnetic interference, etc. Soft-switching by using resonant-linked converters allows one to attain zero voltage across (current through) the switching device. The semiconductor device is switched when the voltage across it, or the current through it, is zero. Hence, low losses, high efficiency, and high switching frequency are achieved using soft PWM in the soft-switching inverters shown in Figure 5.16b.

For hard- and soft-switching inverters, shown in Figures 5.16, the phase voltage waveforms supplied to the motor windings are illustrated in Figures 5.17.

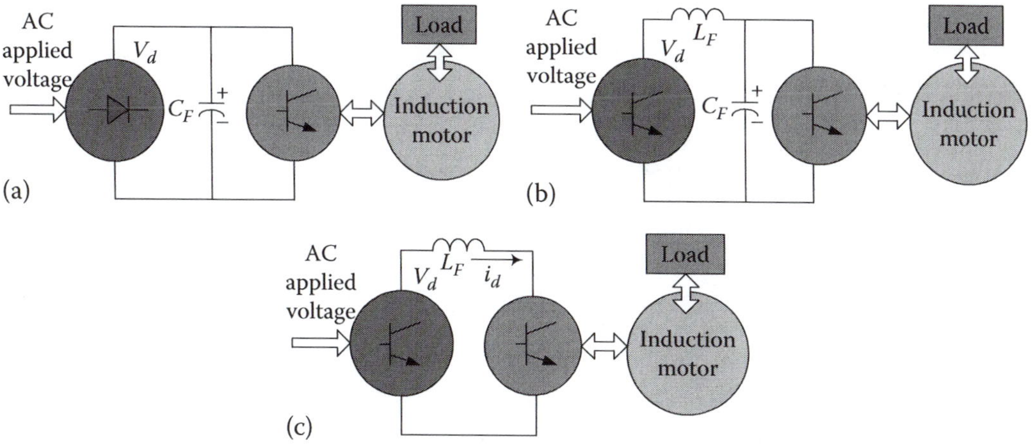

FIGURE 5.15 Variable-frequency power converters: (a) PWM voltage source inverter with an unregulated rectifier; (b) Squire-wave voltage source inverter with a regulated rectifier; (c) Current source inverter with a regulated rectifier.

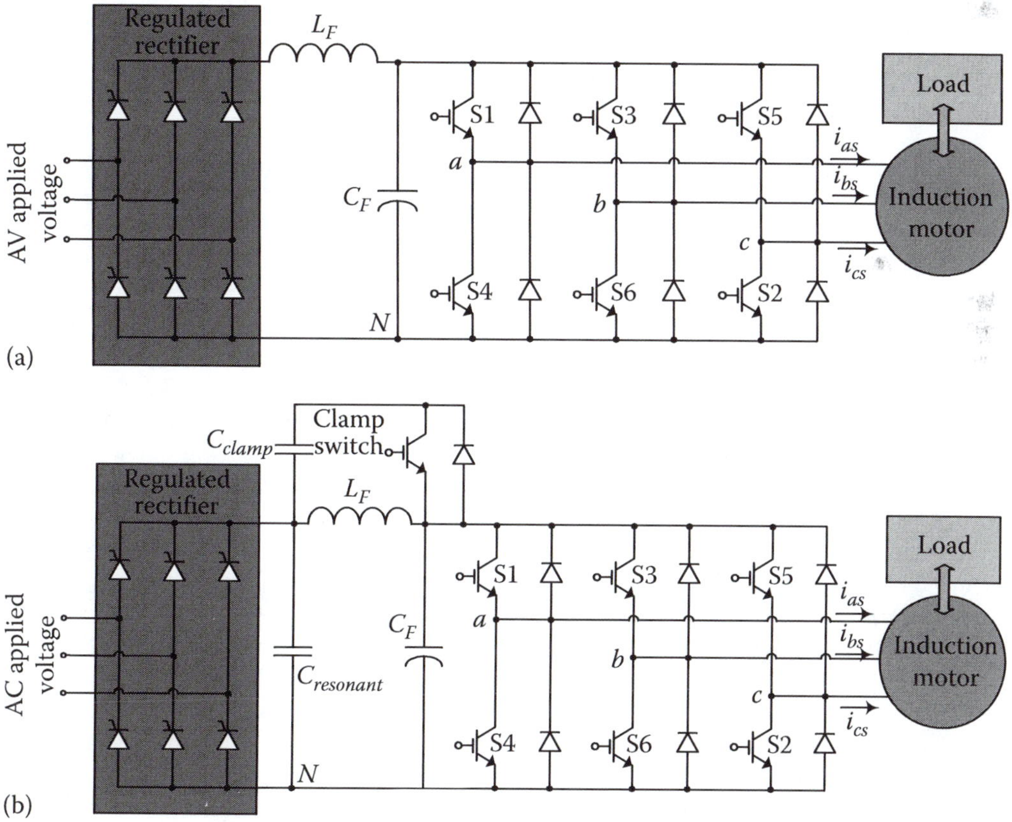

FIGURE 5.16 (a) Power converter with a three-phase hard-switching inverter; (b) Controlled power converter with a regulated three-phase soft-switching inverter.

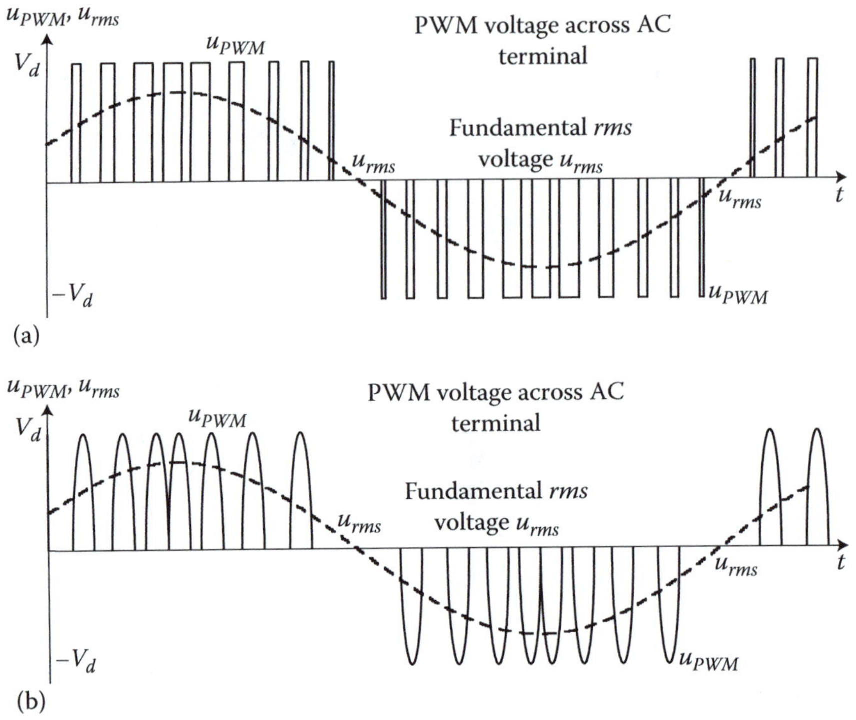

FIGURE 5.17 (a) Phase voltage waveforms in the hard-switching PWM inverter; (b) Phase voltage waveforms in the soft-switching PWM inverter.

High-power transistors and thyristors are current-controlled solid-state devices. Transistors require continuous drive signals, while thyristors need a momentary gate current to turn-on and turn-off. For example, a base current must be regulated to maintain the bipolar junction transistor in the conducting state. The turn-on and turn-off times depend on how rapidly the charge needed to be supplied (to turn-on) or removed (to turn-off) can be delivered to the base region. The turn-off switching speed is decreased by initially applying a spike of base current, and then reducing the current to the magnitude needed. In contrast, the turn-off switching speed is decreased by initially applying a spike of negative base current. Consistent microelectronics drive high-frequency transistors to vary the magnitude u_M and frequency f of the phase voltages. The closed-loop systems are designed for induction motors with power converters.

Consider the hard-switching inverter with three switch pairs shown in Figure 5.16a. To obtain three-phase balanced output voltages using a PWM concept, a triangular signal-level voltage is compared with three sinusoidal control signals shifted by 120° as shown in Figure 5.18a. High-frequency switches S1 and S4, S3 and S6, S5 and S2 close and open opposite each other. That is, switches in each pair are turned on and off simultaneously. If S1 and S4 are closed at the same instant, the circuit is short-circuited across the source. The instantaneous voltages u_{aN}, u_{bN}, and u_{cN} are either equal to V_d or 0. The signal level voltages u_{ac}, u_{bc}, and u_{cc} are compared with the triangular signal u_t. If $u_{ac} > u_t$, then S1 is closed, whereas S4 is open. If the signal-level voltage $u_{ac} < u_t$, then S4 is closed whereas S1 is open. The resulting waveform for the phase voltage u_{aN} is shown in Figure 5.18b. In the similar manner, the phase voltages u_{bN} and u_{cN} are defined by comparing the signal-level voltages u_{bc} and u_{cc} with u_t to open or close switches S3-S6 and S5-S2. The resulting voltages u_{bN} and u_{cN} possess the same pattern as the *aN* voltage, except that u_{bN} and u_{cN} are shifted by 120° and 240°, as illustrated in Figures 5.18c and d. The voltages u_{aN}, u_{bN}, and u_{cN} are measured with respect to the negative DC bus. These DC components are canceled as one uses the line-to-line voltage, which is plotted in Figure 5.18e. The line-to-line voltage u_{ab} is found by subtracting voltage u_{bN} from u_{aN}. One can analyze the waveforms of the instantaneous and *rms* voltages, shown in Figures 5.17 and 5.18e.

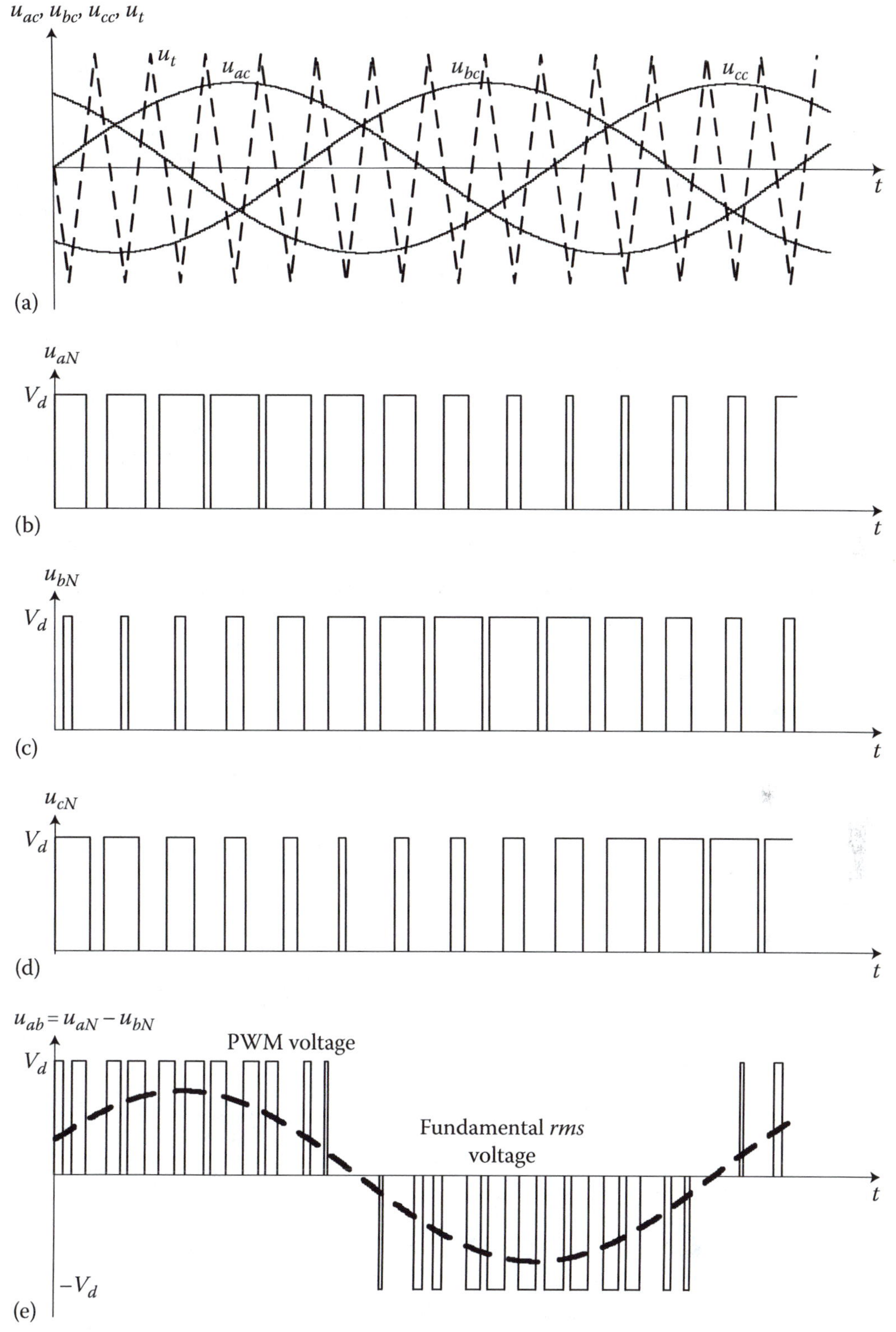

FIGURE 5.18 Voltage waveforms in three-phase hard-switching inverters.

The square-wave voltage source inverters, known as six-step inverters, are commonly used. The three-phase square-wave voltage source inverter bridge is shown in Figures 5.15. The rectifier rectifies the three-phase AC applied voltage, and a large electrolytic capacitor C_F maintains a near-constant DC voltage as well as providing a path for the rapidly changing currents drawn by the inverter. The inductor L_F attenuates current spikes. Assume that the inverter consists of six ideal switches. We consider the basic operation of the square-wave voltage inverters. Each switch is closed for 180° and is opened for the remaining 180° in a cyclic pattern. Furthermore, S3 is closed 120° after S1, S5 is closed 120° after S3, S4 is closed 180° after S1, S6 is closed 180° after S3, and S2 is closed 180° after S5, as shown in Figure 5.19. The result of this switching operation is that a combination of three switches are closed simultaneously for every 60° duration, as shown in Figure 5.19. That is, in three-phase six-step inverters, the switching appears at every 60° interval, e.g., withing $\frac{1}{6}T$ time interval.

To determine the voltage waveforms applied to the *abc* windings, consider the six-step inverter and motor circuitry as illustrated in Figure 5.20. During the interval from 0° to 60°, where switches S5, S6, and S1 are closed, the phase *a* is in parallel with *c*, and they are connected to the phase *b* in series, which is connected to the source via S6. The voltage waveforms as shown in Figure 5.20 result. In particular, $u_{aN} = u_{cN} = V_d$ and $u_{bN} = 0$. Hence, $u_{ab} = V_d$, $u_{bc} = -V_d$, and $u_{ca} = 0$. Because phases *a* and *c* are connected in parallel, the apparent impedance, seen from the neutral of the motor (depicted in Figure 5.20 as a point N'), is halved. Hence, the voltage drop across the phases *as* and *cs* is $\frac{2}{3}V_d$, whereas voltage drop across the phase *bs* is $\frac{1}{3}V_d$. That is, $u_{as} = \frac{1}{3}V_d$, $u_{bs} = -\frac{2}{3}V_d$, and $u_{cs} = \frac{1}{3}V_d$. Hence the voltage drop across the phase is always $\frac{1}{3}V_d$ or $\frac{2}{3}V_d$ depending on the connection of the phases (series or parallel). The waveforms for u_{as}, u_{bs}, and u_{cs} are shown in Figure 5.20.

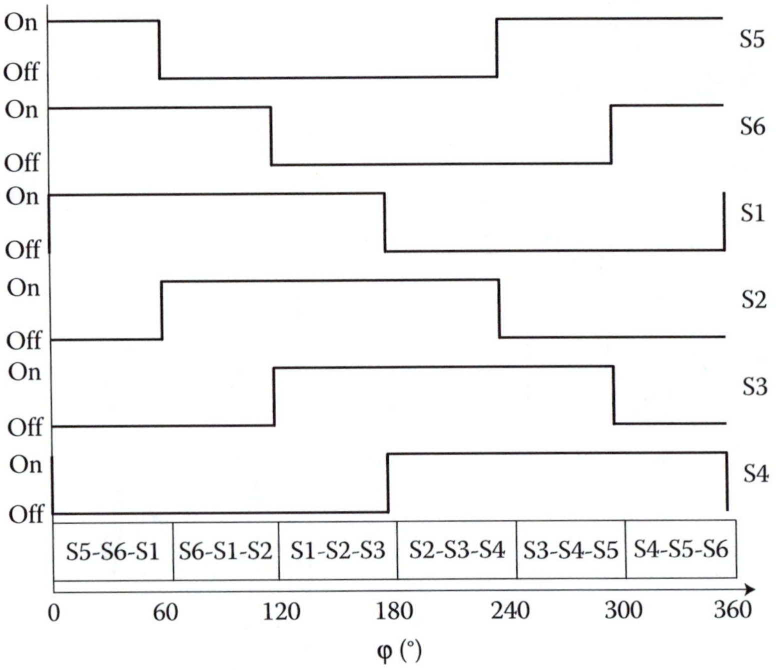

FIGURE 5.19 Switching pattern for three-phase six-step inverters.

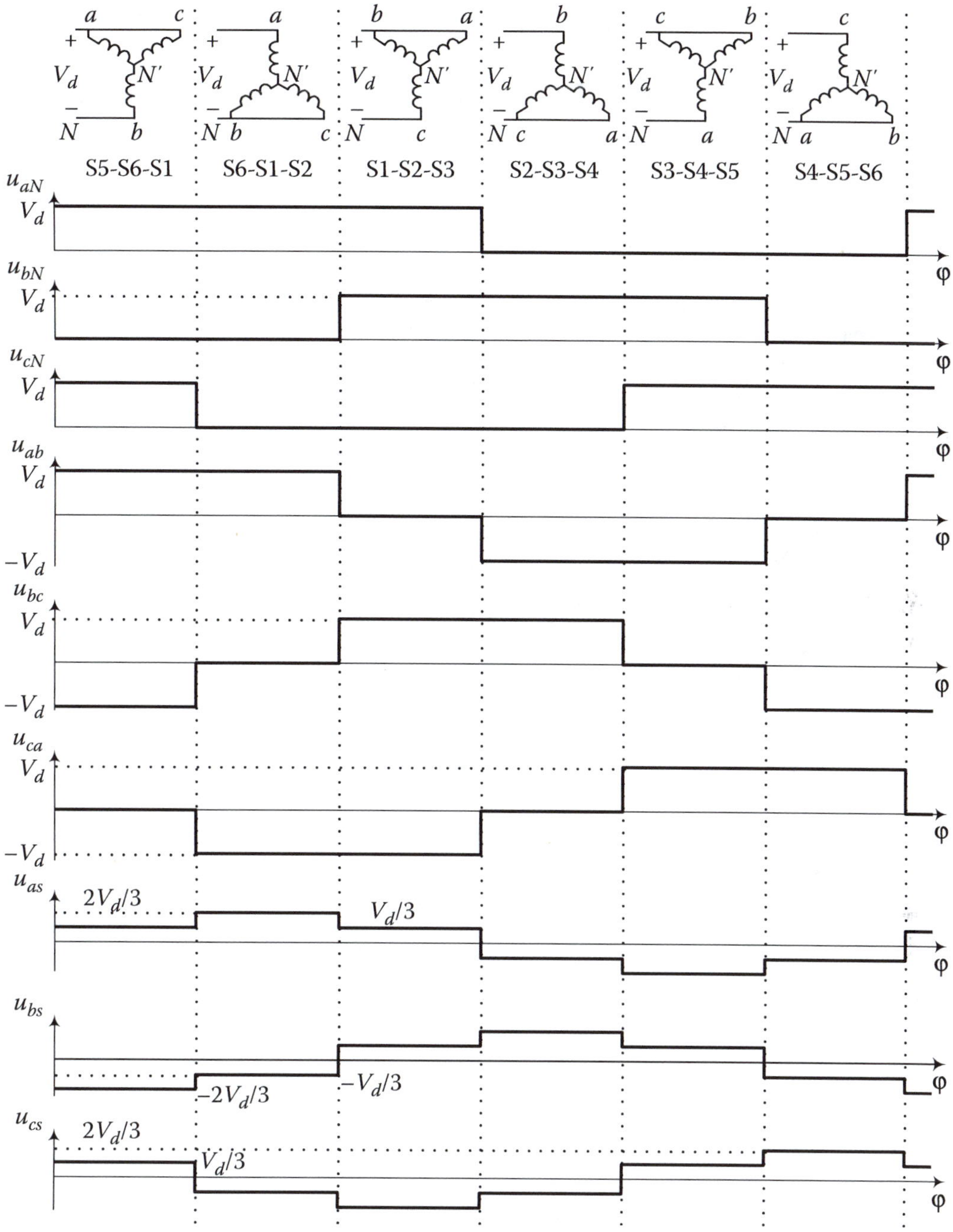

FIGURE 5.20 Voltage waveforms at the terminals *aN*, *bN*, and *cN*, line-to-line, and line-to-neutral voltages applied to the induction motor windings.

Voltage source inverters are different from current source inverters. The current source inverter is fed from a constant current source, which is generated by a controlled rectifier with a large DC link inductor L to smooth the current. The schematics of a current source inverter is shown in Figure 5.21.

At any time, only two thyristors conduct. In particular, one of the thyristors is connected to the positive dc link and the other is connected to the negative dc link. The current is switched sequentially into one of the phases of a three-phase induction motor by the top half of the inverter and returns from another phase to the dc link by the bottom half of the inverter. Since the current is constant, there will be a constant voltage drop across the stator winding of the motor and zero voltage drop across the self-inductance of the winding. Hence, the motor terminal voltage is not set by the inverter but by the resistance of the stator winding. The motor has sinusoidally distributed windings, and the voltages at the phase terminals are sinusoidal. The phase current waveforms are shown in Figure 5.22.

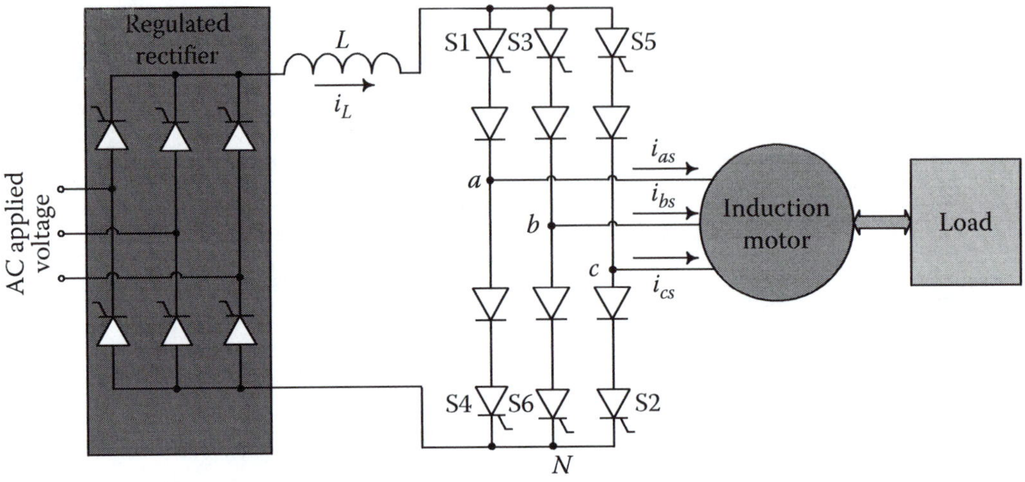

FIGURE 5.21 Power converter with a thyristor current source inverter.

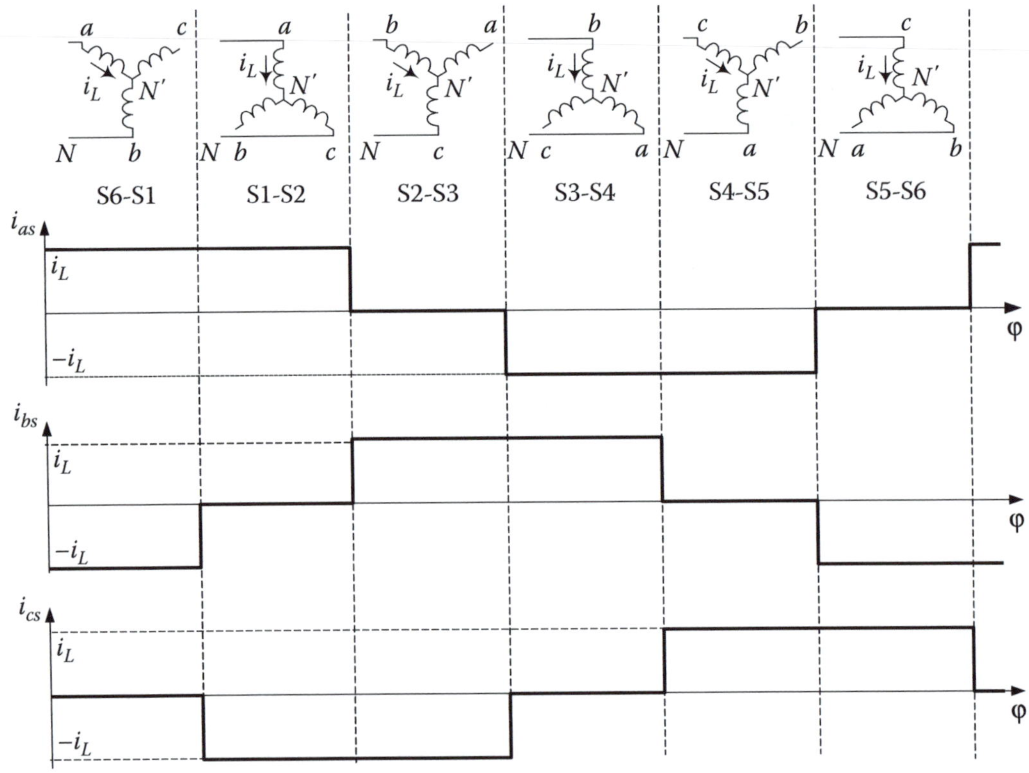

FIGURE 5.22 Phase currents for the induction motors fed by a current source inverter.

PRACTICE AND ENGINEERING PROBLEMS

5.1 Consider an A class two-pole induction motor. The synchronous angular velocity ω_e for the frequency of the applied phase voltages $f = 60$ Hz is

$$\omega_e = 4\pi f/P = 4 \times 3.14 \times 60/2 = 120\pi = 377 \text{ rad/sec.}$$

If at the steady state the *slip* is specified, one calculates the electrical angular velocity ω_r. Recall that

$$slip = \frac{\omega_e - \omega_r}{\omega_e}, \quad \omega_e = \frac{4\pi f}{P}.$$

If $slip = 0.01$, we have $\omega_r = 377 - 3.77 = 373.23$ rad/sec.

5.2 The torque–speed characteristics for the frequency control of induction motors are reported in Figures 5.3 and 5.4. The torque–speed equation is

$$T_e = \frac{3\left(u_M \dfrac{X_M}{X_s + X_M}\right)^2 \dfrac{r_r'}{slip}}{\omega_e\left[\left(r_s\left(\dfrac{X_M}{X_s+X_M}\right)^2 + \dfrac{r_r'}{slip}\right)^2 + \left(X_s+X_r'\right)^2\right]}, \quad slip = \frac{\omega_e - \omega_r}{\omega_e}.$$

The electromagnetic torque T_e depends on the frequency f of the applied phase voltages. Furthermore, $T_e \to \infty$ as $f \to 0$, and, $T_e \to 0$ as $f \to \infty$.

Recall that $\omega_e = \dfrac{4\pi f}{P}$ and $slip = \dfrac{\omega_e - \omega_r}{\omega_e}$.

At $\omega_r = \omega_e = 0$, $slip = 1$. Thus,

$$T_e = \frac{3\left(u_M \dfrac{X_M}{X_s + X_M}\right)^2 r_r'}{\dfrac{4\pi f}{P}\left[\left(r_s\left(\dfrac{X_M}{X_s + X_M}\right)^2 + r_r'\right)^2 + \left(X_s + X_r'\right)^2\right]} \cong k\frac{3u_M^2 P}{4\pi f},$$

where k is the constant, $k > 0$.

For the frequency control, u_M = const. Thus, T_e is proportional to $1/f$.

For low f, T_e is high, and, $T_e \to \infty$ if $f \to 0$. For high f, T_e is low. Furthermore, $T_e \to 0$ if $f \to \infty$.

5.3 We will examine the relationships between the stator and rotor currents assuming an ideal stator–rotor magnetic coupling. Consider a two-phase A class induction motor with an inductance mapping

$$\mathbf{L}_{sr}'(\theta_r) = L_{ms}\begin{bmatrix}\cos\theta_r & -\sin\theta_r \\ \sin\theta_r & \cos\theta_r\end{bmatrix}.$$

For P-pole two-phase induction motors, the electromagnetic torque T_e is

$$T_e = \frac{P}{2}\frac{\partial W_c\left(\mathbf{i}_{abs}, \mathbf{i}_{abr}', \theta_r\right)}{\partial\theta_r} = \frac{P}{2}\mathbf{i}_{abs}^T\frac{\partial \mathbf{L}_{sr}'(\theta_r)}{\partial\theta_r}\mathbf{i}_{abr}' = \frac{P}{2}L_{ms}\begin{bmatrix} i_{as} & i_{bs}\end{bmatrix}\begin{bmatrix}-\sin\theta_r & -\cos\theta_r \\ \cos\theta_r & -\sin\theta_r\end{bmatrix}\begin{bmatrix} i_{ar}' \\ i_{br}'\end{bmatrix}$$

$$= -\frac{P}{2}L_{ms}\left[\left(i_{as}i_{ar}' + i_{bs}i_{br}'\right)\sin\theta_r + \left(i_{as}i_{br}' - i_{bs}i_{ar}'\right)\cos\theta_r\right].$$

At the steady state, the following relationships (equalities) are guaranteed, ensuring the balanced operation of the induction motors:

$$\left(i_{as}i'_{ar}+i_{bs}i'_{br}\right)\equiv\sin\theta_r \quad \text{and} \quad \left(i_{as}i'_{br}-i_{bs}i'_{ar}\right)\equiv\cos\theta_r.$$

5.4 Consider a two-phase induction motor with $L_{asar}=L_{sr}\cos^3\theta_r e^{a\sin^4\theta_r}$ and $L_{asbr}=-L_{sr}\sin^3\theta_r e^{a\cos^4\theta_r}$, $a>0$. One calculates and plots the L_{asar} and L_{asbr} if $a=0.2$ and $a=2$. The MATLAB statement is

```
t=0*pi:1e-3:14; a=2; a=0.2; y1=(cos(t).^3).*exp(a*sin(t).^4); y2=-(sin(t).^3).*exp(a*cos(t).^4);
plot(t,y1,'k-',t,y2,'b:','linewidth',2.5);
title('Stator-Rotor Magnetic Inductances {\itL_a_s_a_r} and {\itL_a_s_b_r}','FontSize',14);
xlabel('Angular Displacement {\it\theta_r}, rad','FontSize',16);
```

The resulting plots are shown in Figures 5.23a and b.

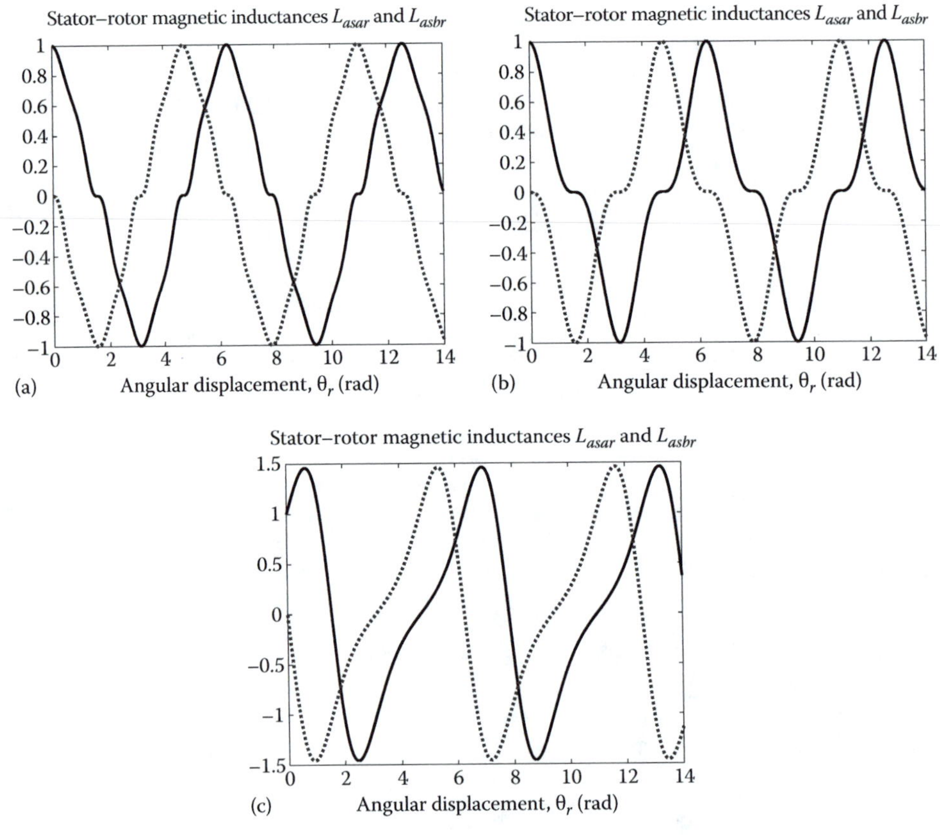

FIGURE 5.23 (a) Plots for $L_{asar}=L_{sr}\cos^3\theta_r e^{a\sin^4\theta_r}$ and $L_{asbr}=-L_{sr}\sin^3\theta_r e^{a\cos^4\theta_r}$ if $a=0.2$; (b) Plots for $L_{asar}=L_{sr}\cos^3\theta_r e^{a\sin^4\theta_r}$ and $L_{asbr}=-L_{sr}\sin^3\theta_r e^{a\cos^4\theta_r}$ if $a=2$; (c) Plots for $L_{asar}=e^{\sin\theta_r}\cos\theta_r$ and $L_{asbr}=-e^{\cos\theta_r}\sin\theta_r$. (Practice and Engineering Problem 5.5.)

The inductance mapping is

$$\mathbf{L}_{sr}(\theta_r)=\begin{bmatrix} L_{asar} & L_{asbr} \\ L_{bsar} & L_{bsbr} \end{bmatrix}=\begin{bmatrix} L_{sr}\cos^3\theta_r e^{a\sin^4\theta_r} & -L_{sr}\sin^3\theta_r e^{a\cos^4\theta_r} \\ L_{sr}\sin^3\theta_r e^{a\cos^4\theta_r} & L_{sr}\cos^3\theta_r e^{a\sin^4\theta_r} \end{bmatrix}.$$

The electromagnetic torque is

$$T_e=\frac{P}{2}\frac{\partial W_c\left(\mathbf{i}_{abs},\mathbf{i}'_{abr},\theta_r\right)}{\partial\theta_r}=\frac{P}{2}\begin{bmatrix} i_{as} & i_{bs} \end{bmatrix}\frac{\partial\mathbf{L}_{sr}(\theta_r)}{\partial\theta_r}\begin{bmatrix} i'_{ar} \\ i'_{br} \end{bmatrix}$$

$$=\frac{P}{2}\begin{bmatrix} i_{as} & i_{bs} \end{bmatrix}\frac{\partial}{\partial\theta_r}\begin{bmatrix} L_{sr}\cos^3\theta_r e^{a\sin^4\theta_r} & -L_{sr}\sin^3\theta_r e^{a\cos^4\theta_r} \\ L_{sr}\sin^3\theta_r e^{a\cos^4\theta_r} & L_{sr}\cos^3\theta_r e^{a\sin^4\theta_r} \end{bmatrix}\begin{bmatrix} i'_{ar} \\ i'_{br} \end{bmatrix}.$$

Hence,

$$T_e=\frac{P}{2}L_{sr}\left(3-4a\sin^2\theta_r\cos^2\theta_r\right)\Big[-i_{as}\cos^2\theta_r\sin\theta_r e^{a\sin^4\theta_r}+i_{bs}\sin^2\theta_r\cos\theta_r e^{a\cos^4\theta_r}$$

$$-i_{as}\sin^2\theta_r\cos\theta_r e^{a\cos^4\theta_r}-i_{bs}\cos^2\theta_r\sin\theta_r e^{a\sin^4\theta_r}\Big]\begin{bmatrix} i'_{ar} \\ i'_{br} \end{bmatrix}$$

$$=\frac{P}{2}L_{sr}\left(3-4a\sin^2\theta_r\cos^2\theta_r\right)\Big\{i'_{ar}\left[-i_{as}\cos^2\theta_r\sin\theta_r e^{a\sin^4\theta_r}+i_{bs}\sin^2\theta_r\cos\theta_r e^{a\cos^4\theta_r}\right]$$

$$-i'_{br}\left[i_{as}\sin^2\theta_r\cos\theta_r e^{a\cos^4\theta_r}+i_{bs}\cos^2\theta_r\sin\theta_r e^{a\sin^4\theta_r}\right]\Big\}.$$

5.5 For a two-phase induction motor, let the inductances be $L_{asar}=e^{\sin\theta_r}\cos\theta_r$ and $L_{asbr}=-e^{\cos\theta_r}\sin\theta_r$. The plots are depicted in Figure 5.23c. The MATLAB statement is

```
t=0*pi:1e-3:14; a=1; y1=cos(t).*exp(a*sin(t)); y2=-sin(t).*exp(a*cos(t));
plot(t,y1,'k-',t,y2,'b:','linewidth',2.5);
title('Stator-Rotor Magnetic Inducyances {\itL_a_s_a_r} and {\itL_a_s_b_r}','FontSize',14);
xlabel('Angular Displacement {\it\theta_r}, rad','FontSize',16);
```

Using the inductance mapping $\mathbf{L}_{sr}(\theta_r)=\begin{bmatrix} L_{asar} & L_{asar} \\ L_{bsar} & L_{bsar} \end{bmatrix}=\begin{bmatrix} e^{\sin\theta_r}\cos\theta_r & -e^{\cos\theta_r}\sin\theta_r \\ e^{\cos\theta_r}\sin\theta_r & e^{\sin\theta_r}\cos\theta_r \end{bmatrix}$, the electromagnetic torque is obtained as

$$T_e=\frac{P}{2}\frac{\partial W_c\left(\mathbf{i}_{abs},\mathbf{i}'_{abr},\theta_r\right)}{\partial\theta_r}=\frac{P}{2}\begin{bmatrix} i_{as} & i_{bs} \end{bmatrix}\frac{\partial\mathbf{L}_{sr}(\theta_r)}{\partial\theta_r}\begin{bmatrix} i'_{ar} \\ i'_{br} \end{bmatrix}$$

$$=\frac{P}{2}\begin{bmatrix} i_{as} & i_{bs} \end{bmatrix}\frac{\partial}{\partial\theta_r}\begin{bmatrix} e^{\sin\theta_r}\cos\theta_r & -e^{\cos\theta_r}\sin\theta_r \\ e^{\cos\theta_r}\sin\theta_r & e^{\sin\theta_r}\cos\theta_r \end{bmatrix}\begin{bmatrix} i'_{ar} \\ i'_{br} \end{bmatrix}.$$

Therefore,

$$T_e = \frac{P}{2}\Big[i_{as}e^{\sin\theta_r}\left(-\sin\theta_r + \cos^2\theta_r\right) + i_{bs}e^{\cos\theta_r}\left(\cos\theta_r - \sin^2\theta_r\right)$$

$$- i_{as}e^{\cos\theta_r}\left(\cos\theta_r - \sin^2\theta_r\right) + i_{bs}e^{\sin\theta_r}\left(-\sin\theta_r + \cos^2\theta_r\right)\Big]\begin{bmatrix} i'_{ar} \\ i'_{br}\end{bmatrix}$$

$$= \frac{P}{2}\Big\{i'_{ar}\Big[i_{as}e^{\sin\theta_r}\left(-\sin\theta_r + \cos^2\theta_r\right) + i_{bs}e^{\cos\theta_r}\left(\cos\theta_r - \sin^2\theta_r\right)\Big]$$

$$-i'_{br}\Big[i_{as}e^{\cos\theta_r}\left(\cos\theta_r - \sin^2\theta_r\right) + i_{bs}e^{\sin\theta_r}\left(-\sin\theta_r + \cos^2\theta_r\right)\Big]\Big\}.$$

5.6 Consider a symmetric six-phase, two-pole ($P = 2$) induction motor. Let the rotor–stator mutual magnetic inductance between the *ar* and *as* windings be $L_{aras} = L_{ms}\cos\theta_r$. We find the *total emf* in the *ar* phase.
For a six-phase induction motor

$$\psi_{ar} = L_{aras}i_{as} + L_{arbs}i_{bs} + L_{arcs}i_{cs} + L_{ards}i_{ds} + L_{ares}i_{es} + L_{arfs}i_{fs} + L_{arar}i_{ar}$$
$$+ L_{arbr}i_{br} + L_{arcr}i_{cr} + L_{ardr}i_{dr} + L_{arer}i_{er} + L_{arfr}i_{fr}.$$

From the given $L_{aras} = L_{sr}\cos\theta_r$, the *total emf* in the *ar* phase is

$$\frac{d\psi_{ar}}{dt} = L_{ms}\frac{d\left(i_{as}\cos\theta_r\right)}{dt} + L_{ms}\frac{d\left(i_{bs}\cos\left(\theta_r - \frac{1}{3}\pi\right)\right)}{dt} + L_{ms}\frac{d\left(i_{cs}\cos\left(\theta_r - \frac{2}{3}\pi\right)\right)}{dt}$$

$$+ L_{ms}\frac{d\left(i_{ds}\cos(\theta_r - \pi)\right)}{dt} + L_{ms}\frac{d\left(i_{es}\cos\left(\theta_r - \frac{4}{3}\pi\right)\right)}{dt} + L_{ms}\frac{d\left(i_{fs}\cos\left(\theta_r - \frac{5}{3}\pi\right)\right)}{dt}$$

$$+\left(L'_{lr} + L_{ms}\right)\frac{di'_{ar}}{dt} + \frac{1}{2}L_{ms}\frac{di'_{br}}{dt} - \frac{1}{2}L_{ms}\frac{di'_{cr}}{dt} - L_{ms}\frac{di'_{dr}}{dt} - \frac{1}{2}L_{ms}\frac{di'_{er}}{dt} + \frac{1}{2}L_{ms}\frac{di'_{fr}}{dt}.$$

The *motional emf*, induced in the *ar* winding, is

$$emf_{ar\omega} = L_{ms}\Big[i_{as}\sin\theta_r + i_{bs}\sin\left(\theta_r - \frac{1}{3}\pi\right) + i_{cs}\sin\left(\theta_r - \frac{2}{3}\pi\right) + i_{ds}\sin\left(\theta_r - \pi\right) + i_{es}\sin\left(\theta_r - \frac{4}{3}\pi\right)$$

$$+ i_{fs}\sin\left(\theta_r - \frac{5}{3}\pi\right)\Big]\omega_r.$$

HOMEWORK PROBLEMS

5.1 Consider a two-phase, four-pole induction motor.

a. Derive the expression for ψ_{ar} assuming that the stator–rotor mutual inductances have $\cos^3\theta_r$ distributions.

b. Derive the expression for the emf_{ar} induced in the rotor winding *ar*. Why is the studied AC motor called an induction motor?

c. The induction motor accelerates, and the motor's "instantaneous" angular velocity is denoted as $\omega_{r\ Inst}$ in Figure 5.24. What is the final angular velocity with which the motor will operate (report the numerical value of ω_r).

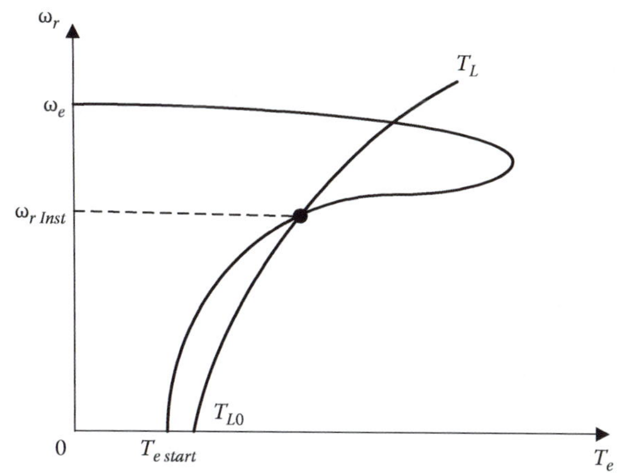

FIGURE 5.24 Torque–speed and load characteristics.

5.2 Consider an A class two-phase, 115 V (*rms*), 60 Hz, four-pole (P = 4) induction motor. Let $L_{asar} = L_{aras} = L_{ms1}\cos\theta_r + L_{ms3}\cos^5\theta_r$, $L_{asbr} = L_{bras} = -L_{ms1}\sin\theta_r - L_{ms3}\sin^5\theta_r$, $L_{bsar} = L_{arbs} = L_{ms1}\sin\theta_r + L_{ms3}\sin^5\theta_r$, and $L_{bsbr} = L_{brbs} = L_{ms1}\cos\theta_r + L_{ms3}\cos^5\theta_r$. The motor parameters are as follows: r_s = 1.2 ohm, r_r' = 1.5 ohm, L_{ms1} = 0.14 H, L_{ms2} = 0.004 H, L_{ls} = 0.02 H, $L_{ss} = L_{ls} + L_{ms1} + 5L_{ms3}$, L_{lr}' = 0.02 H, $L_{rr}' = L_{lr}' + L_{ms1} + 5L_{ms3}$, $B_m = 1 \times 10^{-6}$ N-m-sec/rad, and J = 0.005 kg-m². The phase voltages supplied are $u_{as}(t) = \sqrt{2}115\cos(377t)$ and $u_{bs}(t) = \sqrt{2}115\sin(377t)$.

a. In MATLAB, calculate, plot, and compare

$$L_{asar} = L_{ms}\cos\theta_r\ (L_{ms} = 0.16\text{ H})\quad\text{and}\quad L_{asar} = L_{ms1}\cos\theta_r + L_{ms3}\cos^5\theta_r;$$

b. Derive the circuitry-electromagnetic equations of motion;
c. Report emf_{as} and emf_{ar};
d. Find the expression for the electromagnetic torque;
e. Using Newton's second law, obtain the *torsional–mechanical* model;
f. Report the equations of motion in non-Cauchy's form;
g. Develop the Simulink mdl model or MATLAB file using ode45 differential equations solver to simulate induction motor dynamics;
h. Plot the transient dynamics for all state variables, e.g., report $i_{as}(t)$, $i_{bs}(t)$, $i_{ar}'(t)$, $i_{br}'(t)$, $\omega_r(t)$, and $\theta_r(t)$;
i. Plot the motional $emf_{ar}\omega$ and $emf_{br}\omega$ induced in the rotor windings;
j. Plot the torque–speed characteristics $\omega_r = \Omega_T(T_e)$ using the simulation results;
k. Analyze the induction motor's performance (acceleration, settling time, load attenuation, etc.).

REFERENCES

1. A. E. Fitzgerald, C. Kingsley, and S. D. Umans, *Electric Machinery*, McGraw-Hill, New York, 2003.
2. P. C. Krause and O. Wasynczuk, *Electromechanical Motion Devices*, McGraw-Hill, New York, 1989.
3. P. C. Krause, O. Wasynczuk, S. D. Sudhoff, and S. Pekarek, *Analysis of Electric Machinery*, Wiley-IEEE Press, New York, 2013.
4. W. Leonhard, *Control of Electrical Drives*, Springer, Berlin, Germany, 2001.
5. S. E. Lyshevski, *Electromechanical Systems, Electric Machines, and Applied Mechatronics*, CRC Press, Boca Raton, FL, 1999.
6. S. E. Lyshevski, *Electromechanical Systems and Devices*, CRC Press, Boca Raton, FL, 2008.
7. S. E. Lyshevski, *Engineering and Scientific Computations Using MATLAB®*, Wiley-Interscience, Hoboken, NJ, 2003.
8. N. Mohan, T. M. Undeland, and W. P. Robbins, *Power Electronics: Converters, Applications, and Design*, John Wiley & Sons, New York, 2002.

6 Synchronous Machines in Electromechanical and Energy Systems

6.1 SYNCHRONOUS MACHINES: INTRODUCTION

Direct-current exited, permanent-magnet, and variable-reluctance synchronous machines are widely used in high-performance electromechanical and energy systems [1–6]. Permanent-magnet synchronous machines guarantee superior performance and capabilities. These machines surpass other electric machines such as permanent-magnet DC and induction electric machines. In high-performance drives, servos, and power generation systems, up to ~100 kW rated and ~1000 kW peak, three-phase, permanent-magnet synchronous machines (motors and generators) are a preferable choice. There are translational (linear) and rotational synchronous machines. A three-phase radial-topology permanent-magnet machine is illustrated in Figure 6.1a. In motors, the electromagnetic torque results due to the interaction of time-varying magnetic field established by the phase windings and the magnetic field produced by magnets or field windings on the rotor [1–6]. As shown in Figure 6.1b, in generators, the voltages are induced in the stator windings if the machine is rotated by the prime mover. For example, the torque is applied to the generator shaft, resulting in rotation. In high-power power generation systems (~1 to 1000 MW), conventional three-phase synchronous generators are used.

In this section, we examine the energy conversion, torque production, control, and other aspects. The angular velocity of synchronous motors is fixed with the frequency of the phase voltages applied to the stator windings. The phase voltages are applied as functions of the rotor angular displacement θ_r. The steady-state torque-speed characteristics are the horizontal lines as depicted in Figure 6.1c. The electrical angular velocity ω_r is equal to the synchronous angular velocity $\omega_e = 4\pi f/P$. In the operating envelope, the peak electromagnetic torque must exceed maximum load torque, $T_{e\,peak} > T_{L\,\max}$. For a short period of time, one may overload electric machines. In permanent-magnet synchronous machines, the ratio $T_{e\,peak}/T_{e\,rated}$ could reach ~10. The limits on electromagnetic torque and power are due to the device physics and physical limits (nonlinear magnetic system, saturation, maximum current density, motor heating, maximum insulation temperature, load bearing, etc.) as well as constraints on the amplifier's peak voltages and currents. If $T_{e\,peak} > T_{L\,\max}$ is not guaranteed, the rotor magnetic field slips behind the stator field. Due to the loss of synchronization, the electromagnetic torque surges. The motors are controlled by the power converters, which are referred to as the pulse width modulation (PWM) amplifiers, controllers, or drivers. While the synchronous machines can be overloaded by the factor of ~10, the maximum current overloading of PWM drivers is usually up to ~2.

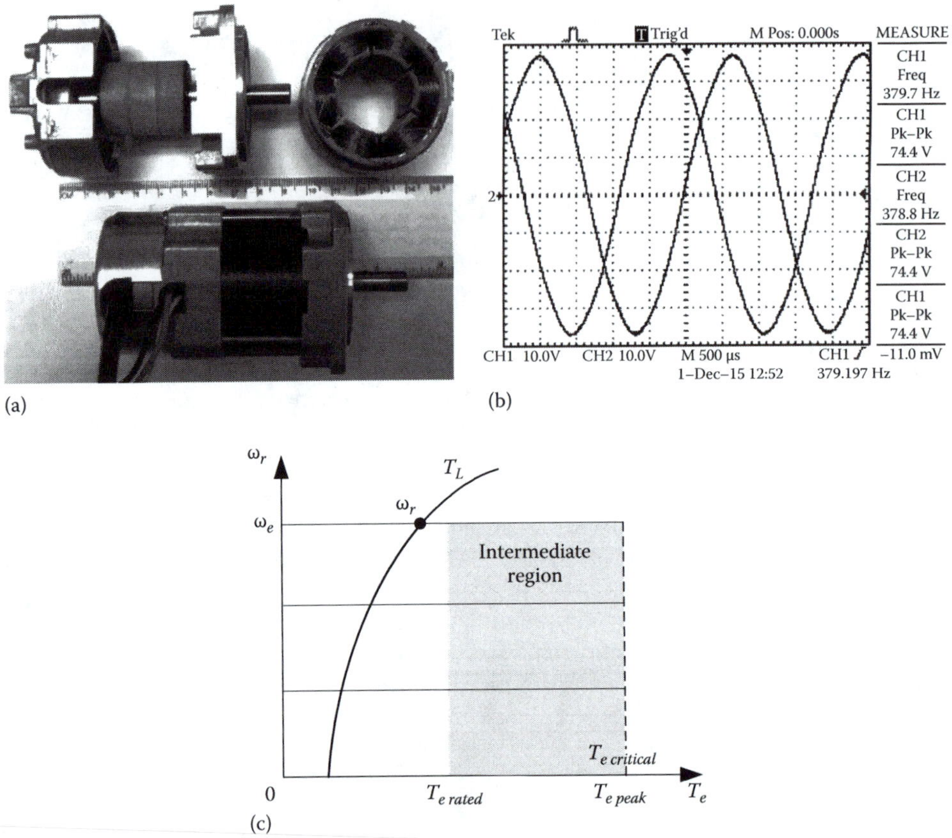

FIGURE 6.1 (a) NEMA 23 size, ~50 W rated, 500 W peak three-phase permanent-magnet synchronous machine, which can be used as a motor or generator. Three-phase windings are placed in the slots of laminated electric steel stator. The SmCo magnets are on the rotor;
(b) Induced phase voltages in a three-phase permanent-magnet synchronous generator, which is rotated by a prime mover;
(c) Torque-speed characteristics—synchronous motors rotate at synchronous angular velocity $\omega_r = \omega_e$.

Enabling technologies are used to advance electric machines performance and capabilities. Mini- and micromachines can be fabricated using micromachining technologies. The images of 2 and 4 mm diameter permanent-magnet synchronous machines are reported in Figure 6.2a. These synchronous machines are smaller than the controlling integrated circuits (ICs). The operating envelope (torque, force, load, load profile, angular velocity, etc.) defines T_e and ω_r, resulting in the motor dimensionality and characteristics. The acceleration capability, settling time, and repositioning rate depend on the ratio $(T_e - T_L)/J$. The torque and power densities, rated angular velocity, and other characteristics are defined by the machine design, dimensionality, materials, magnets, and other factors. For preliminary estimates, one may assume that the power density is ~1 W/cm^3. Figures 6.2b and c document the images of permanent-magnet synchronous motors for drives and servos with the PWM drivers. A stepper motor, which is a synchronous machine, with the PWM controller/driver is reported in Figure 6.2d. Consistent control strategies are needed to ensure best performance and *achievable* capabilities. Maximum efficiency, minimal losses, maximum torque and power densities, minimal vibrations and noise, and other improvements can be achieved. Figure 6.2e depicts magnetic field and temperature sensors used in enabled functionality systems.

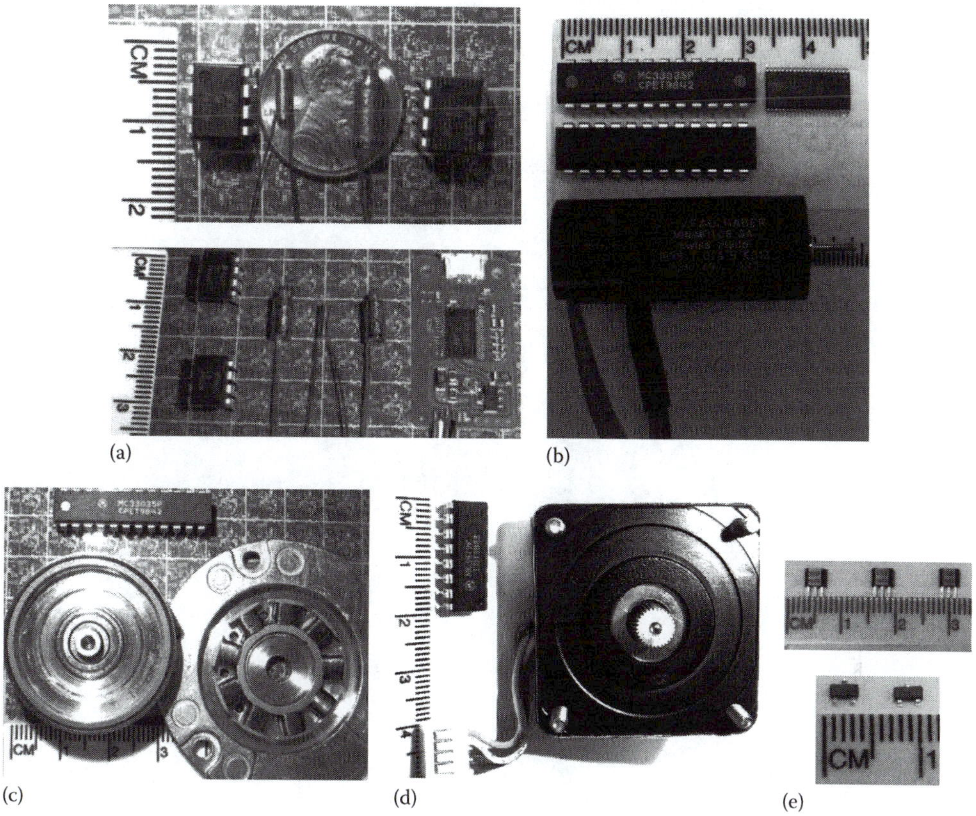

FIGURE 6.2 (a) Images of permanent-magnet synchronous machines and operational amplifiers on a silicon wafer; (b) The Motorola MC33035P PWM driver (~30 V and 1.5 A), the STMicroelectronics L6235N driver (~50 V, 2.8 A rated, 5.6 A peak, 100 kHz), and the Texas Instruments DRV8312DDWR driver (~50 V, 3 A rated, 6 A peak, 100–500 kHz, ~90%–97% efficiency). These drivers with analog controller can be used for a Faulhaber 1628 024B permanent-magnet synchronous (brushless DC) motor with SmCo magnets, $P=2$, 17 W, 24 V, 0.5 A rated, 1.5 A peak, 3000 rad/sec (7000 rad/sec maximum), 3.3 mN-m rated (11 mN-m peak), 15.2 ohm, 0.517 mH, 0.000000054 kg-m^2, up to 70% efficiency; (c) Permanent-magnet synchronous motor to drive a computer hard drive and a PWM controller/driver; (d) Stepper motor (permanent-magnet synchronous machine) and MC33035P PWM controller/driver; (e) Texas Instrument DRV5053 analog Hall-effect sensor (analog output voltage is linear to the magnetic flux density), and an LMT90 temperature sensor with voltage linearly proportional to temperature, 10 mV/°C.

6.2 SYNCHRONOUS RELUCTANCE MOTORS

6.2.1 Single-Phase Synchronous Reluctance Motors

The single-phase synchronous reluctance motor is illustrated in Figure 6.3a. We examine functionality, analyze torque production, and evaluate control concepts.

As shown in Figure 6.3a, the *quadrature* (corresponds to the maximum magnetizing reluctance $\Re_{mq}$) and *direct* (corresponds to the minimal magnetizing reluctance $\Re_{md}$) magnetic axes are fixed with the rotor. The rotor rotates with the angular velocity ω_r. The magnetic axes rotate with the angular velocity ω. Under normal operation, the angular velocity of synchronous machines is equal to the synchronous angular velocity ω_e. Hence, $\omega_r=\omega_e=\omega$. The angular displacements of the rotor θ_r and the *quadrature* magnetic axis θ are

$$\theta_r=\theta=\int_{t_0}^{t}\omega_r(\tau)d\tau=\int_{t_0}^{t}\omega(\tau)d\tau.$$

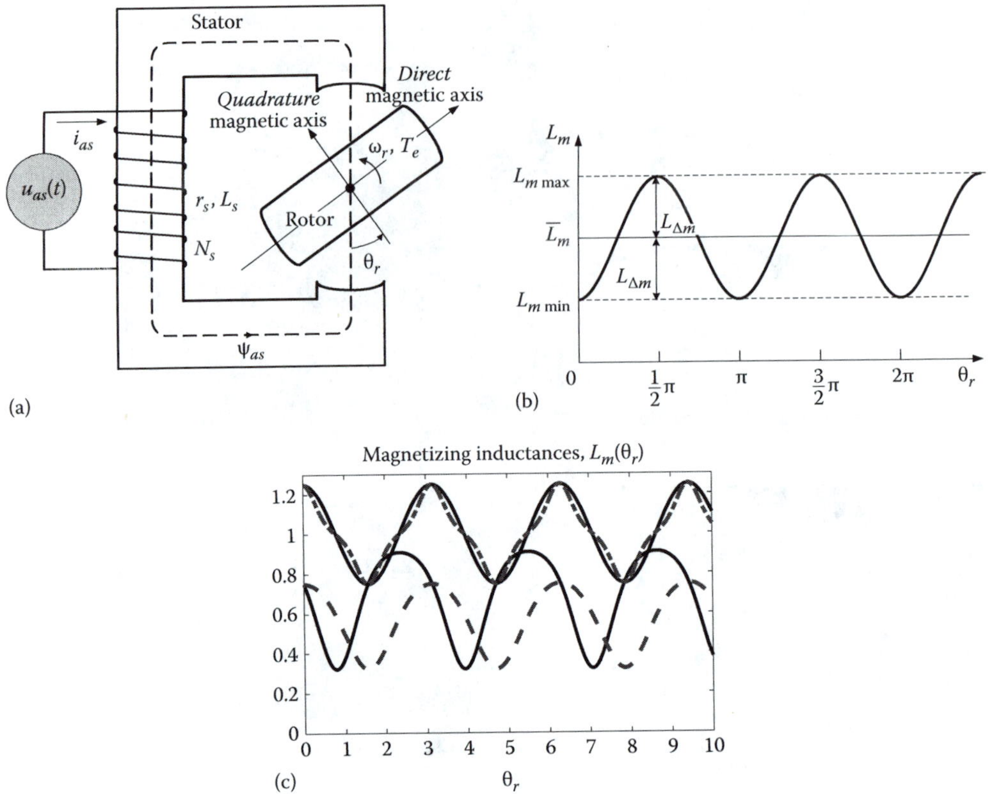

FIGURE 6.3 (a) A single-phase radial topology reluctance motor; (b) Sinusoidal magnetizing inductance $L_m(\theta_r)$, $L_m(\theta_r) = \bar{L}_m - L_{\Delta m}\cos 2\theta_r$; (c) Periodic magnetizing inductances (Example 6.2): $L_m(\theta_r) = \bar{L}_m - L_{\Delta m}e^{\sin 2\theta_r}$, $L_m(\theta_r) = \bar{L}_m - L_{\Delta m}e^{-\sin^2\theta_r}$ (two lower plots), and, $L_m(\theta_r) = \bar{L}_m + L_{\Delta m}\cos 2\theta_r$, $L_m(\theta_r) = \bar{L}_m + L_{\Delta m}\cos 2\theta_r e^{-\sin^2 2\theta_r}$ (two upper plots).

The magnetizing reluctance $\Re_m$ is a function of θ_r. Using the number of turns N_s, the magnetizing inductance is

$$L_m(\theta_r) = \frac{N_s^2}{\Re_m(\theta_r)}.$$

The $L_m(\theta_r)$ is a periodic function that varies twice per a rotor revolution,

$$L_{m\min} = \left.\frac{N_s^2}{\Re_{m\max}(\theta_r)}\right|_{\theta_r = 0, \pi, 2\pi, \ldots}$$

and

$$L_{m\max} = \left.\frac{N_s^2}{\Re_{m\min}(\theta_r)}\right|_{\theta_r = \frac{1}{2}\pi, \frac{3}{2}\pi, \frac{5}{2}\pi, \ldots}.$$

The magnetizing inductance varies as a periodic function with a period π. Letting $L_m(\theta_{r0}) = L_{m\min}$, assume

$$L_m(\theta_r) = \bar{L}_m - L_{\Delta m}\cos 2\theta_r,$$

where $\bar{L}_m$ is the average value of the magnetizing inductance; $L_{\Delta m}$ is the half of amplitude of the sinusoidal variation of the magnetizing inductance.

The plot for $L_m(\theta_r)$ is documented in Figure 6.3b. The *quadrature* (*q*) and *direct* (*d*) axes are depicted in Figure 6.3a. One finds the maximum and minimum reluctances $\Re_{mq}$ and $\Re_{md}$, $\Re_{mq} > \Re_{md}$. Using the magnetizing inductances L_{mq} and L_{md},

$$\bar{L}_m = \frac{1}{2}(L_{mq} + L_{md}) \quad \text{and} \quad L_{\Delta m} = \frac{1}{2}(L_{md} - L_{mq}).$$

The electromagnetic torque, developed by single-phase reluctance motors, is found using the coenergy $W_c(i_{as}, \theta_r)$. From

$$W_c(i_{as}, \theta_r) = \frac{1}{2}(L_{ls} + \bar{L}_m - L_{\Delta m}\cos 2\theta_r)i_{as}^2,$$

one finds

$$T_e = \frac{\partial W_c(i_{as}, \theta_r)}{\partial \theta_r} = \frac{\partial}{\partial \theta_r}\frac{1}{2}(L_{ls} + \bar{L}_m - L_{\Delta m}\cos 2\theta_r)i_{as}^2 = L_{\Delta m}\sin 2\theta_r i_{as}^2.$$

From $T_e = L_{\Delta m}\sin 2\theta_r i_{as}^2$, one may hypothesize that to maximize T_e and avoid the torque ripple, the phase current $i_{as} = i_M \dfrac{1}{\sqrt{\sin 2\theta_r}}$ should be fed. This i_{as} theoretically leads to $T_e = L_{\Delta m} i_M^2$. However, it is impossible to implement $i_{as} = i_M \dfrac{1}{\sqrt{\sin 2\theta_r}}$ because i_{as} cannot be complex (the denominator is complex if $\sin 2\theta_r < 0$), constraints $|i_{as}| \le i_{\max}$, singularity, etc. The real-valued phase current $i_{as}(\theta_r)$ must be found, ensuring electromagnetic and power electronics consistencies.

We have

$$T_{e\,average} \neq 0 \quad \text{if } i_{as} = \begin{cases} \dfrac{i_M}{\sqrt{\sin 2\theta_r}}, & \sin 2\theta_r > 0 \\ 0, & \sin 2\theta_r \le 0 \end{cases}, \quad \text{or,} \quad i_{as} = \begin{cases} i_M\sqrt{\sin 2\theta_r}, & \sin 2\theta_r > 0 \\ 0, & \sin 2\theta_r \le 0 \end{cases}, \; |i_{as}| \le i_{\max}.$$

For

$$i_{as} = i_M \operatorname{Re}\left(\sqrt{\sin(2\theta_r - \phi)}\right),$$

the torque is

$$T_e = L_{\Delta m}\sin 2\theta_r i_{as}^2 = L_{\Delta m} i_M^2 \sin 2\theta_r \left(\operatorname{Re}\sqrt{\sin(2\theta_r - \phi)}\right)^2.$$

The average torque is found using

$$T_{e\,average} = \frac{1}{\pi}\int_0^{\pi} L_{\Delta m}\sin 2\theta_r i_{as}^2(\theta_r)d\theta_r.$$

For example, for

$$i_{as} = \begin{cases} \dfrac{i_M}{\sqrt{\sin 2\theta_r}}, & \sin 2\theta_r > 0 \\ 0, & \sin 2\theta_r \le 0 \end{cases},$$

one finds $T_{e\,average}$ for given $|i_{as}| \le i_{max}$ recalling that

$$\int_0^{\frac{1}{2}\pi} \sin 2x\, dx = \frac{1}{\pi}.$$

The applied phase voltage u_{as}, as a function of θ_r, is found to ensure $T_{e\,average} \neq 0$. One may apply

$$u_{as} = \begin{cases} \dfrac{u_M}{\sqrt{\sin 2\theta_r}}, & \sin 2\theta_r > 0 \\ 0, & \sin 2\theta_r \le 0 \end{cases}, \quad |u_{as}| \le u_{max}.$$

The mathematical model for a single-phase reluctance motor is found by using the Kirchhoff law

$$u_{as} = r_s i_{as} + \frac{d\psi_{as}}{dt}, \quad \psi_{as} = \left(L_{ls} + \bar{L}_m - L_{\Delta m}\cos 2\theta_r\right) i_{as}$$

and the *torsional–mechanical* equation

$$J\frac{d^2\theta_r}{dt^2} = T_e - B_m\omega_r - T_L.$$

One obtains a set of three first-order nonlinear differential equations

$$\begin{aligned} \frac{di_{as}}{dt} &= \frac{1}{L_{ls} + \bar{L}_m - L_{\Delta m}\cos 2\theta_r}\left(-r_s i_{as} - 2L_{\Delta m}\sin 2\theta_r i_{as}\omega_r + u_{as}\right), \\ \frac{d\omega_r}{dt} &= \frac{1}{J}\left(L_{\Delta m}\sin 2\theta_r i_{as}^2 - B_m\omega_r - T_L\right), \\ \frac{d\theta_r}{dt} &= \omega_r. \end{aligned} \tag{6.1}$$

Example 6.1

Consider a single-phase synchronous reluctance motor as illustrated in Figure 6.3a. The parameters and variation of magnetizing inductance $L_m(\theta_r)$ can be directly measured and estimated. Let $L_m(\theta_r) = L_{ls} + \bar{L}_m - L_{\Delta m}\cos 2\theta_r$. The parameters are $r_s = 1$ ohm, $L_{md} = 0.25$ H, $L_{mq} = 0.05$ H, $L_{ls} = 0.01$ H, $J = 0.001$ kg-m^2, and $B_m = 0.0005$ N-m-sec/rad.

To guarantee rotation,

$$u_{as} = \begin{cases} \dfrac{u_M}{\sqrt{\sin(2\theta_r - \phi)}}, & \sin 2\theta_r > 0 \\ 0, & \sin 2\theta_r \le 0 \end{cases}, \quad u_{as} = \frac{u_M}{\mathrm{sgn}\left(\sqrt{\sin(2\theta_r - \phi)}\right)}$$

and other phase voltage u_{as} with bounds, $|u_{as}| \le u_{max}$ can be applied. The Simulink® model, developed using differential equations (6.1), is documented in Figure 6.4a. The applied voltage is $u_{as} = \dfrac{u_M}{\mathrm{sgn}\left(\sqrt{\sin(2\theta_r - \phi)}\right)}$, $u_M = 50$ V, $\phi = 1$.

The switch, complex-to-real, and other blocks, depicted in Figure 6.4a, can be used to implement adequate physics-consistent phase voltage $u_{as}(\theta_r)$.

The angular velocity $\omega_r(t)$ is plotted in Figure 6.4b. One observes the electromagnetic torque ripple and phase current chattering. These effects lead to low efficiency, heating, vibration, noise, mechanical wearing, and other undesired features. Single-phase synchronous motors are not used due to low performance and inadequate capabilities in high-speed applications. These electromechanical motion devices are widely used as limited-angle rotational relays and variable-reluctance stepper motors. ■

Example 6.2

Consider a single-phase synchronous reluctance motor. One may measure and calculate the magnetizing inductance $L_m(\theta_r)$, which is a periodic with a period π and has an average value $\bar{L}_m$. There could be asymmetric $L_m(\theta_r)$ due to varying cross-sectional areas on opposite rotor ends, unequal air gaps, nonuniformity, and other affects. For example,

$$L_m(\theta_r) = \bar{L}_m - L_{\Delta m} e^{\sin 2\theta_r} \quad \text{and} \quad L_m(\theta_r) = \bar{L}_m - L_{\Delta m} e^{-\sin^2 \theta_r}.$$

For a symmetric design, one may find

$$L_m(\theta_r) = \bar{L}_m + L_{\Delta m} \cos 2\theta_r \quad \text{and} \quad L_m(\theta_r) = \bar{L}_m + L_{\Delta m} \cos 2\theta_r e^{-\sin^2 2\theta_r}.$$

Let $\bar{L}_m = 1$ H and $L_{\Delta m} = 0.25$ H. The MATLAB® statements to calculate and plot symmetric and asymmetric $L_m(\theta_r)$ are

```
Lmb=1; LDm=0.25; th=0:0.01:10; Lm1=Lmb-LDm*exp(sin(2*th)); Lm2=Lmb-LDm*exp(sin(th).^2);
Lm3=Lmb+LDm*cos(2*th); Lm4=Lmb+LDm*cos(2*th).*exp(-sin(2*th).^2);
plot(th,Lm1,'k-',th,Lm2,'b--',th,Lm3,'k-',th,Lm4,'r-.','LineWidth',3);axis([0 10 0 1.3]);
xlabel('{\it\theta_r}','FontSize',30);
title('Magnetizing Inductances, {\itL_m}({\it\theta_r})','FontSize',22);
```

The resulting plots for $L_m(\theta_r)$ are depicted in Figure 6.3c.

The Kirchhoff law is $u_{as} = r_s i_{as} + \dfrac{d\psi_{as}}{dt}$. Consider $\psi_{as} = \left(\bar{L}_m - L_{\Delta m} e^{\sin 2\theta_r}\right) i_{as}$. The *total emf* is

$$\frac{d\psi_{as}}{dt} = \left(\bar{L}_m - L_{\Delta m} e^{\sin 2\theta_r}\right)\frac{di_{as}}{dt} - 2L_{\Delta m} \cos 2\theta_r e^{\sin 2\theta_r} i_{as}\omega_r.$$

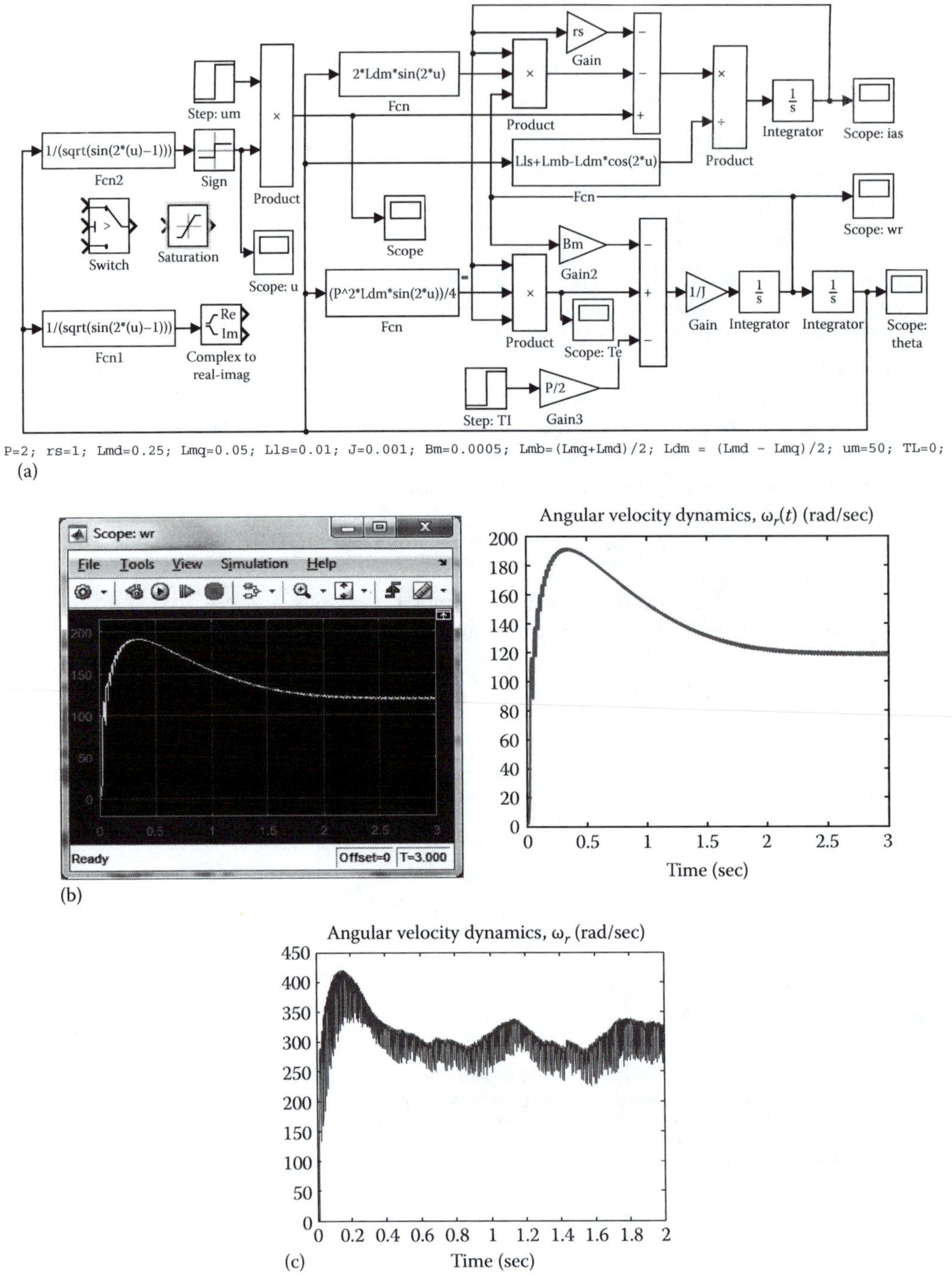

FIGURE 6.4 (a) Simulink® model to simulate a single-phase synchronous reluctance motor; (b) Dynamics of angular velocity $\omega_r(t)$; (c) Dynamics of angular velocity $\omega_r(t)$ for Example 6.4.

The electromagnetic torque is

$$T_e = -\frac{\partial W_c(i_{as},\theta_r)}{\partial \theta_r} = -\frac{\partial}{\partial \theta_r}\frac{1}{2}\left(\bar{L}_m - L_{\Delta m}e^{\sin 2\theta_r}\right)i_{as}^2 = L_{\Delta m}\cos 2\theta_r e^{\sin 2\theta_r} i_{as}^2.$$

One obtains a set of three first-order nonlinear differential equations

$$\frac{di_{as}}{dt} = \frac{1}{\bar{L}_m - L_{\Delta m}e^{\sin 2\theta_r}}\left(-r_s i_{as} + 2L_{\Delta m}\cos 2\theta_r e^{\sin 2\theta_r} i_{as}\omega_r + u_{as}\right),$$

$$\frac{d\omega_r}{dt} = \frac{1}{J}\left(L_{\Delta m}\cos 2\theta_r e^{\sin 2\theta_r} i_{as}^2 - B_m\omega_r - T_L\right),$$

$$\frac{d\theta_r}{dt} = \omega_r.$$

The phase current i_{as} or voltage u_{as} must be applied as a function of θ_r. In the equation $T_e = L_{\Delta m}\cos 2\theta_r e^{\sin 2\theta_r} i_{as}^2$, the term $e^{\sin 2\theta_r}$ is always positive. One may fed various phase currents, such as

$$i_{as} = \begin{cases} i_M \dfrac{1}{\sqrt{\cos 2\theta_r}} & \text{if } \cos 2\theta_r > 0 \\ 0 & \text{if } \cos 2\theta_r \le 0 \end{cases}, \; |i_{as}| \le i_{\max}, \quad \text{or, } i_{as} = \begin{cases} i_M \dfrac{1}{\sqrt{e^{\sin 2\theta_r}}\sqrt{\cos 2\theta_r}} & \text{if } \cos 2\theta_r > 0 \\ 0 & \text{if } \cos 2\theta_r \le 0 \end{cases}$$

with $T_{e\,average} = \frac{1}{2}L_{\Delta m}i_M^2$. ■

Example 6.3

Synchronous reluctance motors can be designed to guarantee near-triangular magnetizing inductance $L_m(\theta_r)$. Consider a single-phase synchronous reluctance motor, as shown in Figure 6.5a. Using the period T, for a triangular magnetizing inductance, one has

$$L_m(\theta_r) = \bar{L}_m - \frac{2L_{\Delta m}}{T}\arcsin\left(\sin\left(\frac{2\pi}{T}\theta_r\right)\right).$$

Correspondingly,

$$L_m(\theta_r) = \bar{L}_m - \frac{2L_{\Delta m}}{\pi}\arcsin\left(\sin 2\theta_r\right).$$

The plot for $L_m(\theta_r)$ is depicted in Figure 6.5b if $\bar{L}_m = 1$ H and $L_{\Delta m} = 0.25$ H. The MATLAB statements are

```
Lmb=1; LDm=0.25; th=0:0.01:10; Lm=Lmb-2*LDm*asin(sin(2*th))/pi;
plot(th, Lm,'k-','LineWidth',2.5); axis([0 10 0 1.3]); xlabel('{\it\theta_r}','FontSize',18);
title('Magnetizing Inductance, {\itL_m}({\it\theta_r})','FontSize',18);
```

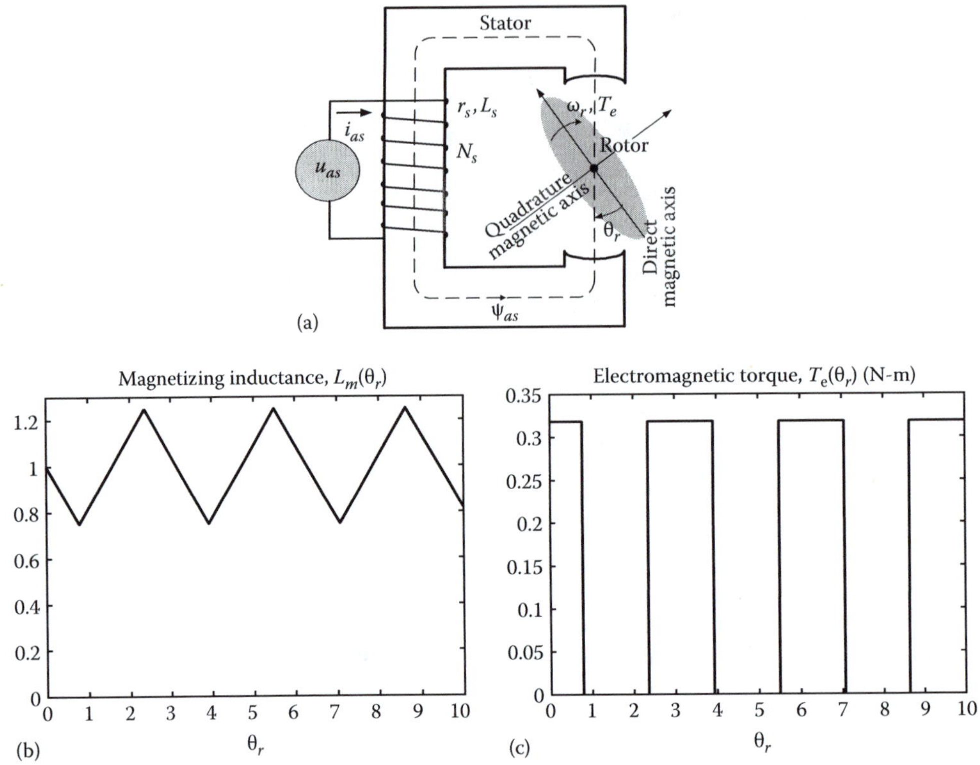

FIGURE 6.5 (a) A single-phase synchronous reluctance motor; (b) Magnetizing inductance $L_m(\theta_r) = \bar{L}_m - \frac{2L_{\Delta m}}{\pi}\arcsin(\sin 2\theta_r)$; (c) Electromagnetic torque $T_e(\theta_r) = \frac{L_{\Delta m}}{\pi}\frac{\cos(2\theta_r)}{|\cos(2\theta_r)|}$.

The electromagnetic torque is

$$T_e = -\frac{\partial W_c(i_{as}, \theta_r)}{\partial \theta_r} = -\frac{\partial}{\partial \theta_r}\frac{1}{2}\left(\bar{L}_m - \frac{2L_{\Delta m}}{\pi}\arcsin(2\theta_r)\right)i_{as}^2.$$

The derivative of a composite function $f(x) = \arcsin(\sin x)$ is found using the chain rule letting $u = \sin x$. One has

$$f'(x) = \frac{du}{dx}\frac{d}{du}\arcsin u = \cos x\frac{1}{\sqrt{1-u^2}} = \frac{\cos x}{1-\sin^2 x} = \frac{\cos x}{|\cos x|}.$$

Correspondingly, we have

$$T_e(\theta_r) = \frac{L_{\Delta m}}{\pi}\frac{\cos(2\theta_r)}{|\cos(2\theta_r)|}.$$

The plot for $T_e(\theta_r)$ is illustrated in Figure 6.5c. Although arcsin(sin$x$) is continuous for all x, its derivative is undefined at certain x. The derivative is undefined when $\cos(2\theta_r) = 0$, e.g., at $\theta_r = \frac{1}{4}\theta_r + \frac{1}{2}k\theta_r$, where k is an integer. The aforementioned issue does not cause problems in electromechanical devices. ■

Example 6.4

The magnetizing inductance depends on the electromagnetic and mechanical designs, such as the rotor and stator geometry, air gap, *B–H* curve, permeability, etc. One may have

$$L_m(\theta_r) = \bar{L}_m - \sum_{n=1}^{\infty} L_{\Delta m\,n} \cos^{2n-1} 2\theta_r.$$

Hence, $T_e = \dfrac{\partial W_c(i_{as}, \theta_r)}{\partial \theta_r} = i_{as}^2 \sum_{n=1}^{\infty} (2n-1) L_{\Delta mn} \sin 2\theta_r \cos^{2n-2} 2\theta_r.$

The resulting differential equations are found for the specific design, operating envelope, etc. For example, for $L_m(\theta_r) = \bar{L}_m - L_{\Delta m2} \cos^3 2\theta_r$, $L_{\Delta m2} \neq 0$ and $\forall L_{\Delta m\,n} = 0$, one finds

$$\frac{di_{as}}{dt} = \frac{1}{\bar{L}_m - L_{\Delta m2}\cos^3 2\theta_r}\left(-r_s i_{as} - 6L_{\Delta m2} \sin 2\theta_r \cos^2 2\theta_r i_{as}\omega_r + u_{as}\right),$$

$$\frac{d\omega_r}{dt} = \frac{1}{J}\left(3L_{\Delta m2} \sin 2\theta_r \cos^2 2\theta_r i_{as}^2 - B_m\omega_r - T_L\right),$$

$$\frac{d\theta_r}{dt} = \omega_r.$$

From $T_e = 3L_{\Delta m2} \sin 2\theta_r \cos^2 2\theta_r i_{as}^2$, one may find the following phase current and voltage

$$i_{as} = \begin{cases} \dfrac{i_M}{\sqrt{\sin 2\theta_r \cos 2\theta_r}}, & \sin 2\theta_r > 0 \\ 0, & \sin 2\theta_r \leq 0 \end{cases}, \quad |i_{as}| \leq i_{\max},$$

and,

$$u_{as} = \begin{cases} \dfrac{u_M}{\sqrt{\sin 2\theta_r \cos 2\theta_r}}, & \sin 2\theta_r > 0 \\ 0, & \sin 2\theta_r \leq 0 \end{cases}, \quad |u_{as}| \leq u_{\max}.$$

Simulations are performed using the motor parameters given in Example 6.1. The load torque $T_L = 0.0025$ N-m is applied at 1.25 sec. The transient dynamics for $\omega_r(t)$ is illustrated in Figure 6.4c if $u_M = 100$ V. The losses are examined using the root-mean-square values of the phase voltage, current, and angular velocity, e.g.,

$$u_{as_{rms}} = \sqrt{\frac{1}{T}\int_0^T u_{as}^2\, dt}, \quad i_{as_{rms}} = \sqrt{\frac{1}{T}\int_0^T i_{as}^2\, dt}, \quad \text{and} \quad \omega_{r_{rms}} = \sqrt{\frac{1}{T}\int_0^T \omega_r^2\, dt}.$$

The efficiency, which is a steady-state quantity, is $\eta = P_{output}/P_{input} = T_L\Omega_r/U_{as}I_{as}$, $P_{output} = P_{input} - P_{losses}$. Despite variations in $i_{as}(t)$, $\omega_r(t)$, and $u_{as}(t)$, one can estimate η by finding losses $P_{losses}(t) = r_s i_{as}^2(t) + B_m\omega_r^2(t)$. ■

6.2.2 Three-Phase Synchronous Reluctance Motors

6.2.2.1 Synchronous Reluctance Motors in the Machine Variables

Consider a three-phase synchronous reluctance motor depicted in Figure 6.6. The angular velocity is regulated by changing the phase voltages or phase currents. We study synchronous reluctance motors in the *machine abc* variables.

The machine parameters are the stator resistance r_s (the phase resistances are equal), the magnetizing inductances in the *quadrature* and *direct* axes L_{mq} and L_{md} ($L_{mq} \neq L_{md}$), the average magnetizing inductance $\bar{L}_m$, the leakage inductance L_{ls}, the moment of inertia J, and the viscous friction coefficient B_m. The circuitry-electromagnetic dynamics is described by the Kirchhoff law

$$\mathbf{u}_{abcs} = \mathbf{r}_s \mathbf{i}_{abcs} + \frac{d\boldsymbol{\psi}_{abcs}}{dt}, \quad \mathbf{u}_{abcs} = \begin{bmatrix} u_{as} \\ u_{bs} \\ u_{cs} \end{bmatrix}, \quad \mathbf{i}_{abcs} = \begin{bmatrix} i_{as} \\ i_{bs} \\ i_{cs} \end{bmatrix}, \quad \boldsymbol{\psi}_{abcs} = \begin{bmatrix} \psi_{as} \\ \psi_{bs} \\ \psi_{cs} \end{bmatrix}, \quad \mathbf{r}_s = \begin{bmatrix} r_s & 0 & 0 \\ 0 & r_s & 0 \\ 0 & 0 & r_s \end{bmatrix}, \tag{6.2}$$

where u_{as}, u_{bs}, and u_{cs} are the phase voltages applied to the *as*, *bs*, and *cs* stator windings; i_{as}, i_{bs}, and i_{cs} are the phase currents; ψ_{as}, ψ_{bs}, and ψ_{cs} are the flux linkages.

The flux linkages are $\boldsymbol{\psi}_{abcs} = \mathbf{L}_s(\theta_r)\mathbf{i}_{abcs}$. Assuming the sinusoidal variation of the magnetizing inductance $L_m(\theta_r)$, with $\bar{L}_m = \frac{1}{3}(L_{mq} + L_{md})$ and $L_{\Delta m} = \frac{1}{3}(L_{md} - L_{mq})$, one finds the inductance mapping as

$$\mathbf{L}_s(\theta_r) = \begin{bmatrix} L_{ls} + \bar{L}_m - L_{\Delta m}\cos 2\theta_r & -\frac{1}{2}\bar{L}_m - L_{\Delta m}\cos 2\left(\theta_r - \frac{1}{3}\pi\right) & -\frac{1}{2}\bar{L}_m - L_{\Delta m}\cos 2\left(\theta_r + \frac{1}{3}\pi\right) \\ -\frac{1}{2}\bar{L}_m - L_{\Delta m}\cos 2\left(\theta_r - \frac{1}{3}\pi\right) & L_{ls} + \bar{L}_m - L_{\Delta m}\cos 2\left(\theta_r - \frac{2}{3}\pi\right) & -\frac{1}{2}\bar{L}_m - L_{\Delta m}\cos 2\left(\theta_r + \pi\right) \\ -\frac{1}{2}\bar{L}_m - L_{\Delta m}\cos 2\left(\theta_r + \frac{1}{3}\pi\right) & -\frac{1}{2}\bar{L}_m - L_{\Delta m}\cos 2\left(\theta_r + \pi\right) & L_{ls} + \bar{L}_m - L_{\Delta m}\cos 2\left(\theta_r + \frac{2}{3}\pi\right) \end{bmatrix}.$$

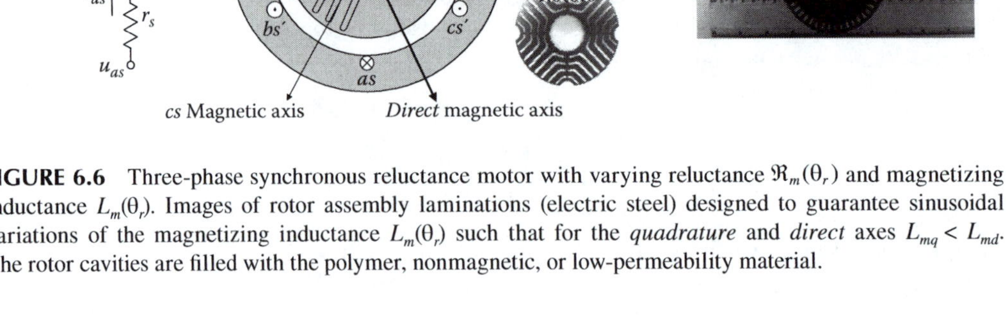

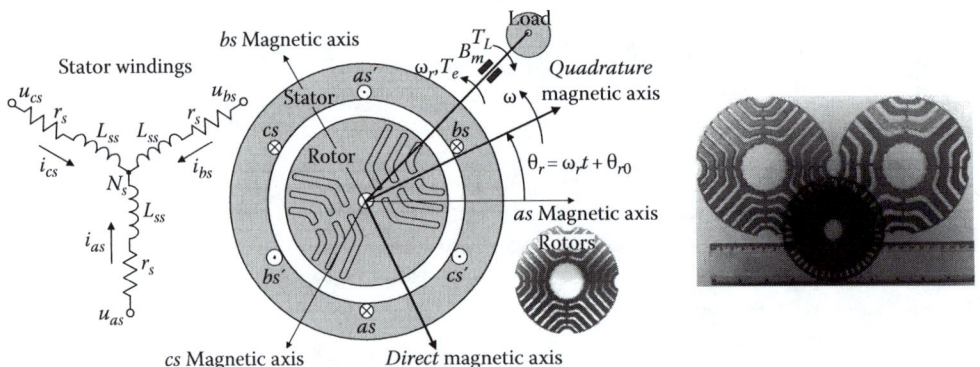

FIGURE 6.6 Three-phase synchronous reluctance motor with varying reluctance $\Re_m(\theta_r)$ and magnetizing inductance $L_m(\theta_r)$. Images of rotor assembly laminations (electric steel) designed to guarantee sinusoidal variations of the magnetizing inductance $L_m(\theta_r)$ such that for the *quadrature* and *direct* axes $L_{mq} < L_{md}$. The rotor cavities are filled with the polymer, nonmagnetic, or low-permeability material.

The evolutions of electrical angular velocity ω_r and displacement θ_r are described by

$$J\frac{2}{P}\frac{d\omega_r}{dt}=T_e-B_m\frac{2}{P}\omega_r-T_L,$$
$$\frac{d\theta_r}{dt}=\omega_r. \tag{6.3}$$

Using the inductance mapping $\mathbf{L}_s(\theta_r)$ and phase currents, the coenergy is

$$W_c=\frac{1}{2}\begin{bmatrix} i_{as} & i_{bs} & i_{cs}\end{bmatrix}\mathbf{L}_s\begin{bmatrix} i_{as}\\ i_{bs}\\ i_{cs}\end{bmatrix}.$$

The electromagnetic torque T_e is

$$T_e=\frac{P}{2}\frac{\partial W_c}{\partial\theta_r}=\frac{P}{2}\frac{1}{2}\begin{bmatrix} i_{as} & i_{bs} & i_{cs}\end{bmatrix}\begin{bmatrix} 2L_{\Delta m}\sin 2\theta_r & 2L_{\Delta m}\sin 2\left(\theta_r-\frac{1}{3}\pi\right) & 2L_{\Delta m}\sin 2\left(\theta_r+\frac{1}{3}\pi\right)\\ 2L_{\Delta m}\sin 2\left(\theta_r-\frac{1}{3}\pi\right) & 2L_{\Delta m}\sin 2\left(\theta_r-\frac{2}{3}\pi\right) & 2L_{\Delta m}\sin 2\theta_r\\ 2L_{\Delta m}\sin 2\left(\theta_r+\frac{1}{3}\pi\right) & 2L_{\Delta m}\sin 2\theta_r & 2L_{\Delta m}\sin 2\left(\theta_r+\frac{2}{3}\pi\right)\end{bmatrix}\begin{bmatrix} i_{as}\\ i_{bs}\\ i_{cs}\end{bmatrix}.$$

One obtains

$$T_e=\frac{P}{2}L_{\Delta m}\left[i_{as}^2\sin 2\theta_r+2i_{as}i_{bs}\sin 2\left(\theta_r-\frac{1}{3}\pi\right)+2i_{as}i_{cs}\sin 2\left(\theta_r+\frac{1}{3}\pi\right)\right.$$
$$\left.+i_{bs}^2\sin 2\left(\theta_r-\frac{2}{3}\pi\right)+2i_{bs}i_{cs}\sin 2\theta_r+i_{cs}^2\sin 2\left(\theta_r+\frac{2}{3}\pi\right)\right]. \tag{6.4}$$

For three-phase synchronous reluctance motors,

$$L_{mq}=\frac{3}{2}(\bar{L}_m-L_{\Delta m})\quad\text{and}\quad L_{md}=\frac{3}{2}(\bar{L}_m+L_{\Delta m}).$$

Therefore,

$$\bar{L}_m=\frac{1}{3}(L_{mq}+L_{md})\quad\text{and}\quad L_{\Delta m}=\frac{1}{3}(L_{md}-L_{mq}).$$

Using the trigonometric identities, we have

$$T_e = \frac{P(L_{md}-L_{mq})}{6}\left[\left(i_{as}^2 - \frac{1}{2}i_{bs}^2 - \frac{1}{2}i_{cs}^2 - i_{as}i_{bs} - i_{as}i_{cs} + 2i_{bs}i_{cs}\right)\sin 2\theta_r + \frac{\sqrt{3}}{2}\left(i_{bs}^2 - i_{cs}^2 - 2i_{as}i_{bs} + 2i_{as}i_{cs}\right)\cos 2\theta_r\right].$$

To maximize the electromagnetic torque, we find the balanced current set

$$i_{as} = \sqrt{2}i_M \sin\left(\theta_r + \frac{1}{3}\varphi_i\pi\right),\quad i_{bs} = \sqrt{2}i_M \sin\left(\theta_r - \frac{1}{3}(2-\varphi_i)\pi\right),\quad i_{cs} = \sqrt{2}i_M \sin\left(\theta_r + \frac{1}{3}(2+\varphi_i)\pi\right),$$

$$\varphi_i = 0.3245$$

which ensures $T_e = \sqrt{2}PL_{\Delta m}i_M^2$. The analytic and numeric results are reported in Example 6.5.

Using PWM power amplifiers, one controls the phase voltages u_{as}, u_{bs}, and u_{cs}. The balanced voltage set is

$$u_{as} = \sqrt{2}u_M \sin\left(\theta_r + \frac{1}{3}\varphi_u\pi\right),\quad u_{bs} = \sqrt{2}u_M \sin\left(\theta_r - \frac{1}{3}(2-\varphi_u)\pi\right),\quad u_{cs} = \sqrt{2}u_M \sin\left(\theta_r + \frac{1}{3}(2+\varphi_u)\pi\right).$$

From $T_e = \sqrt{2}PL_{\Delta m}i_M^2$, torque T_e is controlled by changing the magnitude of the phase currents i_M or voltage u_M. The angular rotor displacement θ_r must be measured or observed. One may ensure a near-optimal operation of synchronous reluctance motors. Analytic results indicate that there is no torque ripple. In practice, there are torque ripple, cogging, eccentricity, and other effects due to electromagnetic field and material nonuniformity, PWM, and other phenomena. Even the high-fidelity analysis using three-dimensional Maxwell's equations, tensor calculus, and heterogeneous simulations do not allow one to ensure an absolute consistency.

Example 6.5

Using (6.4), we calculate and plot the electromagnetic torque for the balanced current set

$$i_{as} = \sqrt{2}i_M \sin\left(\theta_r + \frac{1}{3}\varphi_i\pi\right),\, i_{bs} = \sqrt{2}i_M \sin\left(\theta_r - \frac{1}{3}(2-\varphi_i)\pi\right),\, i_{cs} = \sqrt{2}i_M \sin\left(\theta_r + \frac{1}{3}(2+\varphi_i)\pi\right),$$

$$\varphi_i = 0.3245.$$

Let i_M= 10 A, P = 4 and $L_{\Delta m}$= 0.05 H. For the balanced three-phase current set, the MATLAB m-file is

```
th=0:0.01:4*pi; % angular rotor displacement
IM=10; P=4; LDm=0.05; phi=0.3245;
% Balanced three-phase current set
Ias=sqrt(2)*IM*sin(th+phi*pi/3); Ibs=sqrt(2)*IM*sin(th-(2-phi)*pi/3);
Ics=sqrt(2)*IM*sin(th+(2+phi)*pi/3);
% Calculation of the electromagnetic torque
Te=P*LDm*(Ias.*(sin(2*th).*Ias+2*sin(2*th-2*pi/3).*Ibs+2*sin(2*th+2*pi/3).*Ics)...
+Ibs.*(sin(2*th-4*pi/3).*Ibs+2*sin(2*th).*Ics)+Ics.*sin(2*th+4*pi/3).*Ics)/2;
```

```
% Plot the currents applied to the abc windings
plot(th,Ias,'k-',th,Ibs,'b--',th,Ics,'r:','LineWidth',2.5); axis([0,4*pi,-15,15]);
xlabel('Angular Displacement, \theta_r [rad]','FontSize',18);
title('Phase Currents, {\iti_a_s}, {\iti_b_s} and {\iti_c_s} [A]','FontSize',18); pause;
% Plot of the torque developed versus the angular displacement
plot(th,Te,'k-','LineWidth',2.5); axis([0,4*pi,0,30]);
xlabel('Angular Displacement, \theta_r [rad]','FontSize',18);
title('Electromagnetic Torque, {\itT_e} [N-m]','FontSize',18);
```

The evolutions of the phase currents and T_e are reported in Figures 6.7a and b. The analysis shows that the displacement-independent electromagnetic torque is developed when the balanced current set is used. The numerics correspond to $T_e = \sqrt{2}PL_{\Delta m}i_M^2$. For $i_M = 10$ A, $P = 4$, and $L_{\Delta m} = 0.05$ H, one finds $T_e = \sqrt{2}20$ N-m. ■

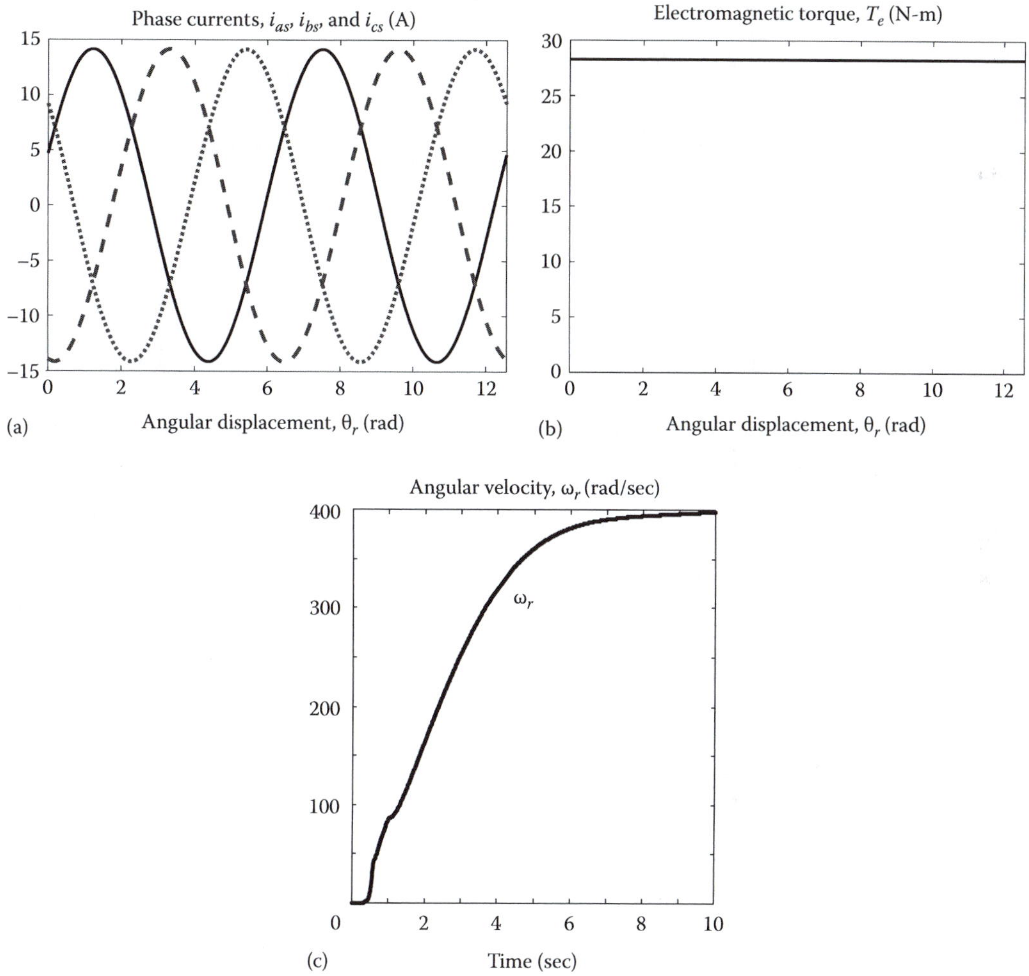

FIGURE 6.7 (a) Phase currents i_{as}, i_{bs}, and i_{cs}; (b) Electromagnetic torque T_e with the balanced current set; (c) Angular velocity dynamics for Example 6.6.

The equations of motion for synchronous reluctance motors are derived in non-Cauchy's and Cauchy's forms. These models in the *abc* variables are found using Kirchhoff's second law (6.2) and Newton's second law of motion (6.3). The electromagnetic torque is expressed by (6.4) using the inductance mapping $\mathbf{L}_s(\theta_r)$. From (6.2) and (6.3), the resulting differential equations in non-Cauchy's form are

$$\frac{di_{as}}{dt}=\frac{1}{L_{ls}+\bar{L}_m-L_{\Delta m}\cos 2\theta_r}\left[-r_s i_{as}+u_{as}+\left(\frac{1}{2}\bar{L}_m+L_{\Delta m}\cos 2\left(\theta_r-\frac{1}{3}\pi\right)\right)\frac{di_{bs}}{dt}\right.$$
$$+\left(\frac{1}{2}\bar{L}_m+L_{\Delta m}\cos 2\left(\theta_r+\frac{1}{3}\pi\right)\right)\frac{di_{cs}}{dt}$$
$$\left.-2L_{\Delta m}\omega_r\left(i_{as}\sin 2\theta_r+i_{bs}\sin 2\left(\theta_r-\frac{1}{3}\pi\right)+i_{cs}\sin 2\left(\theta_r+\frac{1}{3}\pi\right)\right)\right],$$

$$\frac{di_{bs}}{dt}=\frac{1}{L_{ls}+\bar{L}_m-L_{\Delta m}\cos 2\left(\theta_r-\frac{2}{3}\pi\right)}\left[-r_s i_{bs}+u_{bs}+\left(\frac{1}{2}\bar{L}_m+L_{\Delta m}\cos 2\left(\theta_r-\frac{1}{3}\pi\right)\right)\frac{di_{as}}{dt}\right.$$
$$\left.+\left(\frac{1}{2}\bar{L}_m+L_{\Delta m}\cos 2\theta_r\right)\frac{di_{cs}}{dt}-2L_{\Delta m}\omega_r\left(i_{as}\sin 2\left(\theta_r-\frac{1}{3}\pi\right)+i_{bs}\sin 2\left(\theta_r-\frac{2}{3}\pi\right)+i_{cs}\sin 2\theta_r\right)\right],$$

$$\frac{di_{cs}}{dt}=\frac{1}{L_{ls}+\bar{L}_m-L_{\Delta m}\cos 2\left(\theta_r+\frac{2}{3}\pi\right)}\left[-r_s i_{cs}+u_{cs}+\left(\frac{1}{2}\bar{L}_m+L_{\Delta m}\cos 2\left(\theta_r+\frac{1}{3}\pi\right)\right)\frac{di_{as}}{dt}\right.$$
$$\left.+\left(\frac{1}{2}\bar{L}_m+L_{\Delta m}\cos 2\theta_r\right)\frac{di_{bs}}{dt}-2L_{\Delta m}\omega_r\left(i_{as}\sin 2\left(\theta_r+\frac{1}{3}\pi\right)+i_{bs}\sin 2\theta_r+i_{cs}\sin 2\left(\theta_r+\frac{2}{3}\pi\right)\right)\right],$$

$$\frac{d\omega_r}{dt}=\frac{P^2}{4J}L_{\Delta m}\left[i_{as}^2\sin 2\theta_r+2i_{as}i_{bs}\sin 2\left(\theta_r-\frac{1}{3}\pi\right)+2i_{as}i_{cs}\sin 2\left(\theta_r+\frac{1}{3}\pi\right)+i_{bs}^2\sin 2\left(\theta_r-\frac{2}{3}\pi\right)\right.$$
$$\left.+2i_{bs}i_{cs}\sin 2\theta_r+i_{cs}^2\sin 2\left(\theta_r+\frac{2}{3}\pi\right)\right]-\frac{B_m}{J}\omega_r-\frac{P}{2J}T_L,$$

$$\frac{d\theta_r}{dt}=\omega_r. \tag{6.5}$$

Using the Symbolic Toolbox, we find the differential equations in Cauchy's form. The notations are

$$\text{Lbm}=\bar{L}_m,\quad \text{Ldm}=L_{\Delta m},\quad \text{Lls}=L_{ls},\quad \text{rs}=r_s,\quad \text{Bm}=B_m,$$

$$\text{S1}=\sin 2\theta_r,\ \text{S2}=\sin 2\left(\theta_r-\frac{1}{3}\pi\right),\ \text{S3}=\sin 2\left(\theta_r+\frac{1}{3}\pi\right),\ \text{S4}=\sin 2\left(\theta_r-\frac{2}{3}\pi\right),\ \text{S5}=\sin 2(\theta_r+\pi),$$

$$\text{S6}=\sin 2\left(\theta_r+\frac{2}{3}\pi\right),\ \text{and C1}=\cos 2\theta_r,\ \text{C2}=\cos 2\left(\theta_r-\frac{1}{3}\pi\right),\text{C3}=\cos 2\left(\theta_r+\frac{1}{3}\pi\right),\text{C4}=\cos 2\left(\theta_r-\frac{2}{3}\pi\right),$$

$$\text{C5}=\cos 2(\theta_r+\pi),\quad \text{C6}=\cos 2\left(\theta_r+\frac{2}{3}\pi\right).$$

The equations of motion are found by using the following MATLAB file

```
L=sym('[Lls+Lbm-Ldm*C1,-Lbm/2-Ldm*C2,-Lbm/2-Ldm*C3,0;-Lbm/2-Ldm*C2,Lls+Lbm-Ldm*C4,-Lbm/2-Ldm*C5,0;
   -Lbm/2-Ldm*C3,-Lbm/2-Ldm*C5,Lls+Lbm-Ldm*C6,0;0,0,0,2*J/P] ');
R=sym('[-rs,0,0,0;0,-rs,0,0;0,0,-rs,0;0,0,0,-2*Bm/P]');
I=sym('[Ias; Ibs; Ics; Wr] '); V=sym('[vas;vbs;vcs;-TL] ');
K=sym('[Ldm*2*Wr*(S1*Ias+S2*Ibs+S3*Ics); Ldm*2*Wr*(S2*Ias+S4Ibs+S5*Ics);
Ldm*2*Wr*(S3*Ias+S5*Ibs+S6*Ics);Te] '); L1=inv(L); L2=simplify(L1);
FS1=L2*R*I; FS2=simplify(FS1)
FS3=L2*V; FS4=simplify(FS3)
FS5=L2*K; FS6=simplify(FS5)
FS7=FS2+FS4-FS6; FS=simplify(FS7)
```

Applying trigonometric identities, the resulting nonlinear differential equations are

$$
\begin{aligned}
\frac{di_{as}}{dt} = \frac{1}{L_D}\Bigg[& \left(r_s i_{as} - u_{as}\right)\left(4L_{ls}^2 + 3\bar{L}_m^2 - 3L_{\Delta m}^2 + 8\bar{L}_m L_{ls} - 4L_{ls}L_{\Delta m}\cos 2\theta_r\right) \\
& + \left(r_s i_{bs} - u_{bs}\right)\left(3\bar{L}_m^2 - 3L_{\Delta m}^2 + 2\bar{L}_m L_{ls} + 4L_{ls}L_{\Delta m}\cos 2\left(\theta_r - \frac{1}{3}\pi\right)\right) \\
& + \left(r_s i_{cs} - u_{cs}\right)\left(3\bar{L}_m^2 - 3L_{\Delta m}^2 + 2\bar{L}_m L_{ls} + 4L_{ls}L_{\Delta m}\cos 2\left(\theta_r + \frac{1}{3}\pi\right)\right) + 6\sqrt{3}L_{\Delta m}^2 L_{ls}\omega_r\left(i_{cs} - i_{bs}\right) \\
& + \left(8L_{\Delta m}L_{ls}^2\omega_r + 12L_{\Delta m}\bar{L}_m L_{ls}\omega_r\right)\left(\sin 2\theta_r i_{as} + \sin 2\left(\theta_r - \frac{1}{3}\pi\right) i_{bs} + \sin 2\left(\theta_r + \frac{1}{3}\pi\right) i_{cs}\right)\Bigg], \\
\frac{di_{bs}}{dt} = \frac{1}{L_D}\Bigg[& \left(r_s i_{as} - u_{as}\right)\left(3\bar{L}_m^2 - 3L_{\Delta m}^2 + 2\bar{L}_m L_{ls} + 4L_{ls}L_{\Delta m}\cos 2\left(\theta_r - \frac{1}{3}\pi\right)\right) \\
& + \left(r_s i_{bs} - u_{bs}\right)\left(4L_{ls}^2 + 3\bar{L}_m^2 - 3L_{\Delta m}^2 + 8\bar{L}_m L_{ls} - 4L_{ls}L_{\Delta m}\cos 2\left(\theta_r + \frac{1}{3}\pi\right)\right) \\
& + \left(r_s i_{cs} - u_{cs}\right)\left(3\bar{L}_m^2 - 3L_{\Delta m}^2 + 2\bar{L}_m L_{ls} + 4L_{ls}L_{\Delta m}\cos 2\theta_r\right) + 6\sqrt{3}L_{\Delta m}^2 L_{ls}\omega_r\left(i_{as} - i_{cs}\right) \\
& + \left(8L_{\Delta m}L_{ls}^2\omega_r + 12L_{\Delta m}\bar{L}_m L_{ls}\omega_r\right)\left(\sin 2\left(\theta_r - \frac{1}{3}\pi\right) i_{as} + \sin 2\left(\theta_r + \frac{1}{3}\pi\right) i_{bs} + \sin 2\theta_r i_{cs}\right)\Bigg], \\
\frac{di_{cs}}{dt} = \frac{1}{L_D}\Bigg[& \left(r_s i_{as} - u_{as}\right)\left(3\bar{L}_m^2 - 3L_{\Delta m}^2 + 2\bar{L}_m L_{ls} + 4L_{ls}L_{\Delta m}\cos 2\left(\theta_r + \frac{1}{3}\pi\right)\right) \\
& + \left(r_s i_{bs} - u_{bs}\right)\left(3\bar{L}_m^2 - 3L_{\Delta m}^2 + 2\bar{L}_m L_{ls} + 4L_{ls}L_{\Delta m}\cos 2\theta_r\right) \\
& + \left(r_s i_{cs} - u_{cs}\right)\left(4L_{ls}^2 + 3\bar{L}_m^2 - 3L_{\Delta m}^2 + 8\bar{L}_m L_{ls} - 4L_{ls}L_{\Delta m}\cos 2\left(\theta_r - \frac{1}{3}\pi\right)\right) + 6\sqrt{3}L_{\Delta m}^2 L_{ls}\omega_r\left(i_{bs} - i_{as}\right) \\
& + \left(8L_{\Delta m}L_{ls}^2\omega_r + 12L_{\Delta m}\bar{L}_m L_{ls}\omega_r\right)\left(\sin 2\left(\theta_r + \frac{1}{3}\pi\right) i_{as} + \sin 2\theta_r i_{bs} + \sin 2\left(\theta_r - \frac{1}{3}\pi\right) i_{cs}\right)\Bigg], \\
\frac{d\omega_r}{dt} = \frac{P^2}{4J}L_{\Delta m}\Bigg(& i_{as}^2 \sin 2\theta_r + 2i_{as}i_{bs}\sin 2\left(\theta_r - \frac{1}{3}\pi\right) + 2i_{as}i_{cs}\sin 2\left(\theta_r + \frac{1}{3}\pi\right) \\
& + i_{bs}^2 \sin 2\left(\theta_r - \frac{2}{3}\pi\right) + 2i_{bs}i_{cs}\sin 2\theta_r + i_{cs}^2 \sin 2\left(\theta_r + \frac{2}{3}\pi\right)\Bigg) - \frac{B_m}{J}\omega_r - \frac{P}{2J}T_L, \\
\frac{d\theta_r}{dt} = \omega_r, &
\end{aligned}
\tag{6.6}
$$

where $\bar{L}_m = \frac{1}{3}\left(L_{mq} + L_{md}\right)$, $L_{\Delta m} = \frac{1}{3}\left(L_{md} - L_{mq}\right)$, $L_D = L_{ls}\left(9L_{\Delta m}^2 - 4L_{ls}^2 - 12\bar{L}_m L_{ls} - 9\bar{L}_m^2\right)$.

Recall that the mechanical angular velocity and displacement are $\omega_{rm} = 2\omega_r/P$ and $\theta_{rm} = 2\theta_r/P$.

Example 6.6

The nonlinear simulations are performed by using the differential equations of three-phase synchronous reluctance motors in the *machine* variables (6.6). The four-pole, 220 V, 400 rad/sec, 40 kW motor parameters are $r_s = 0.01$ ohm, $L_{md} = 0.0012$ H, $L_{mq} = 0.0002$ H, $J = 0.6$ kg-m^2, and $B_m = 0.003$ N-m-sec/rad.

The three-phase voltage set

$$u_{as} = \sqrt{2}u_M \sin\left(\theta_r + \frac{1}{3}\varphi_u\pi\right),\ u_{bs} = \sqrt{2}u_M \sin\left(\theta_r - \frac{1}{3}(2-\varphi_u)\pi\right),\ u_{cs} = \sqrt{2}u_M \sin\left(\theta_r + \frac{1}{3}(2+\varphi_u)\pi\right)$$

is supplied. Figure 6.7c illustrates the transient dynamics for the angular velocity if the motor accelerates from the stall, and the load torque 10 N-m is applied at 1 sec. The magnitude of the phase voltages is $u_M = 110$ V and $\varphi_u = 0.3882$. The settling time is 10 sec and the steady-state angular velocity is 400 rad/sec. ■

6.2.2.2 Synchronous Reluctance Motors in the Rotor and Synchronous Reference Frames

Nonlinear differential equations (6.5) and (6.6) describe the dynamics of three-phase synchronous reluctance motors in the *machine* variables. Synchronous reluctance machines can be described using the *quadrature*, *direct*, and *zero* quantities. In the rotor and synchronous reference frames, one applies the Park transformations covered in Chapter 5. From Table 5.1, using the *quadrature* (*q*), *direct* (*d*), and *zero* (0) components of voltages (u_{qs}, u_{ds}, u_{0s}), currents (i_{qs}, i_{ds}, i_{0s}), and flux linkages (ψ_{qs}, ψ_{ds}, ψ_{0s}), for $\theta = \theta_r = \theta_e$ we have

1. Rotor reference frame quantities $\mathbf{u}^r_{qd0s} = \mathbf{K}^r_s\mathbf{u}_{abcs}$, $\mathbf{i}^r_{qd0s} = \mathbf{K}^r_s\mathbf{i}_{abcs}$, $\psi^r_{qd0s} = \mathbf{K}^r_s\psi_{abcs}$;
2. Synchronous reference frame quantitates $\mathbf{u}^e_{qd0s} = \mathbf{K}^e_s\mathbf{u}_{abcs}$, $\mathbf{i}^e_{qd0s} = \mathbf{K}^e_s\mathbf{i}_{abcs}$, $\psi^r_{qd0s} = \mathbf{K}^e_s\psi_{abcs}$.

The superscripts *r* and *e* indicate the rotor and synchronous reference frames with $\omega = \omega_r$ and $\omega_r = \omega_e$. The transformations and resulting models are identical because for the normal operation of synchronous machines, $\theta = \theta_r = \theta_e$ and $\omega = \omega_r = \omega_e$. The stator Park transformation matrix

$$\mathbf{K}_s = \frac{2}{3}\begin{bmatrix} \cos\theta & \cos\left(\theta - \frac{2}{3}\pi\right) & \cos\left(\theta + \frac{2}{3}\pi\right) \\ \sin\theta & \sin\left(\theta - \frac{2}{3}\pi\right) & \sin\left(\theta + \frac{2}{3}\pi\right) \\ \frac{1}{2} & \frac{1}{2} & \frac{1}{2} \end{bmatrix}$$

yields

$$\mathbf{K}^r_s = \mathbf{K}^e_s = \frac{2}{3}\begin{bmatrix} \cos\theta_r & \cos\left(\theta_r - \frac{2}{3}\pi\right) & \cos\left(\theta_r + \frac{2}{3}\pi\right) \\ \sin\theta_r & \sin\left(\theta_r - \frac{2}{3}\pi\right) & \sin\left(\theta_r + \frac{2}{3}\pi\right) \\ \frac{1}{2} & \frac{1}{2} & \frac{1}{2} \end{bmatrix} = \frac{2}{3}\begin{bmatrix} \cos\theta_e & \cos\left(\theta_e - \frac{2}{3}\pi\right) & \cos\left(\theta_e + \frac{2}{3}\pi\right) \\ \sin\theta_e & \sin\left(\theta_e - \frac{2}{3}\pi\right) & \sin\left(\theta_e + \frac{2}{3}\pi\right) \\ \frac{1}{2} & \frac{1}{2} & \frac{1}{2} \end{bmatrix}.$$

Using the circuitry-electromagnetic dynamics (6.2), *torsional–mechanical* equations (6.3), and the expression for T_e (6.4), one finds

$$\begin{aligned}
\frac{di_{qs}^r}{dt} &= -\frac{r_s}{L_{ls}+L_{mq}} i_{qs}^r - \frac{L_{ls}+L_{md}}{L_{ls}+L_{mq}} i_{ds}^r \omega_r + \frac{1}{L_{ls}+L_{mq}} u_{qs}^r,\\
\frac{di_{ds}^r}{dt} &= -\frac{r_s}{L_{ls}+L_{md}} i_{ds}^r + \frac{L_{ls}+L_{mq}}{L_{ls}+L_{md}} i_{qs}^r \omega_r + \frac{1}{L_{ls}+L_{md}} u_{ds}^r,\\
\frac{di_{0s}^r}{dt} &= -\frac{r_s}{L_{ls}} i_{0s}^r + \frac{1}{L_{ls}} u_{0s}^r,\\
\frac{d\omega_r}{dt} &= \frac{3P^2}{8J}\left(L_{md}-L_{mq}\right) i_{qs}^r i_{ds}^r - \frac{B_m}{J}\omega_r - \frac{P}{2J} T_L,\\
\frac{d\theta_r}{dt} &= \omega_r.
\end{aligned} \tag{6.7}$$

To attain the balanced operation, the $qd0$ components of currents and voltages are derived by applying the *direct* Park transformations $\mathbf{i}_{qd0s}^r = \mathbf{K}_s^r \mathbf{i}_{abcs}$ and $\mathbf{u}_{qd0s}^r = \mathbf{K}_s^r \mathbf{u}_{abcs}$. By using the three-phase balanced current and voltage sets, we find $i_{qs}^r = \sqrt{2} i_M$, $i_{ds}^r = 0$, $i_{0s}^r = 0$ and $u_{qs}^r = \sqrt{2} u_M$, $u_{ds}^r = 0$, $u_{0s}^r = 0$.

The model in the synchronous reference frame (superscript e) is identical to (6.7) because $\omega = \omega_r = \omega_e$. The AC electromechanical motion devices are controlled by applying the phase voltages u_{as}, u_{bs}, and u_{cs}. Therefore, the analysis in $qd0$ quantities should be applied with this understanding. Control of AC machines using the stationary, rotor, and synchronous reference frames requires computing Park transformations in real time using additional hardware and software routines. There are no practical advantages of the qd control schemes.

6.3 RADIAL TOPOLOGY TWO-PHASE PERMANENT-MAGNET SYNCHRONOUS MACHINES

Consider radial topology permanent-magnet synchronous machines. The studied brushless synchronous motion devices guarantee high efficiency, high power density, high torque density, overloading capabilities, and robustness. The permanent-magnet synchronous AC motors are frequently called the "brushless DC motors". The device physics and the operating principles of permanent-magnet DC and synchronous AC machines are fundamentally distinct.

6.3.1 Two-Phase Permanent-Magnet Synchronous Machines and Stepper Motors

Consider two-phase permanent-magnet synchronous machines. The excitation field is produced by permanent magnets placed on the rotor. Figures 6.8 depict images of synchronous machines used in computer hard-disk drives and servos.

FIGURE 6.8 Images of two- and three-phase permanent-magnet synchronous machines.

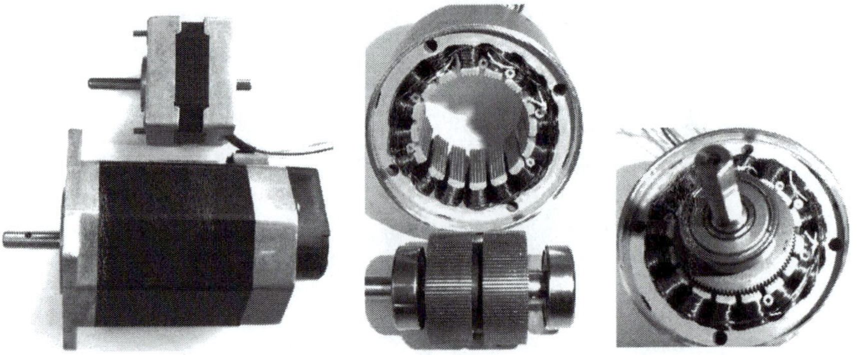

FIGURE 6.9 NEMA 18 and NEMA 23 size stepper motors. There are various stepper motors with different designs, topologies, winding configurations, sizes, etc.

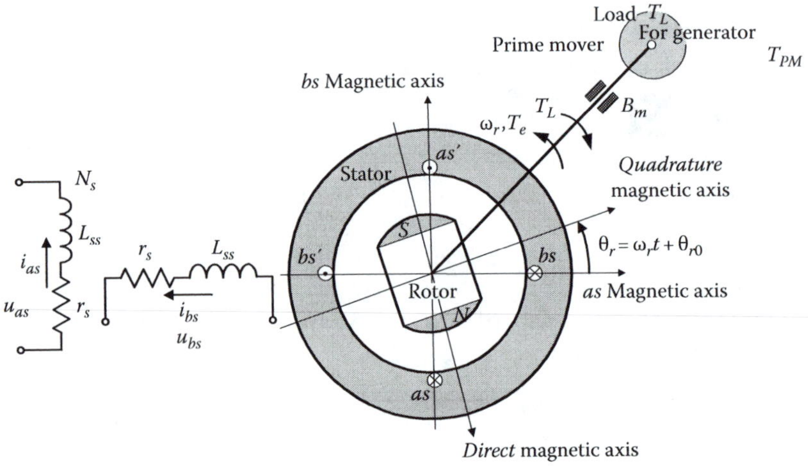

FIGURE 6.10 Two-phase two-pole permanent-magnet synchronous machine.

In servos and electric drives, two- and three-phase permanent-magnet synchronous machines are used. Stepper motors are illustrated in Figure 6.9. Consider the radial topology two-phase permanent-magnet synchronous motors depicted in Figure 6.10.

Using Kirchhoff's voltage law, we have

$$u_{as} = r_s i_{as} + \frac{d\psi_{as}}{dt}, \quad \psi_{as} = L_{asas} i_{as} + L_{asbs} i_{bs} + \psi_{asm},$$
$$u_{bs} = r_s i_{bs} + \frac{d\psi_{bs}}{dt}, \quad \psi_{bs} = L_{bsas} i_{as} + L_{bsbs} i_{bs} + \psi_{bsm}, \tag{6.8}$$

where u_{as} and u_{bs} are the phase voltages applied to the stator windings *as* and *bs*; i_{as} and i_{bs} are the phase currents; ψ_{as} and ψ_{bs} are the stator flux linkages; r_s is the resistances of the stator windings; L_{asas} and L_{bsbs} are the self-inductances; L_{asbs} and L_{bsas} are the mutual inductances.

The flux linkages ψ_{asm} and ψ_{bsm} are periodic functions of the rotor angular displacement θ_r with respect to the stator. Assume an ideal sinusoidal stator–rotor magnetic coupling, linear magnetic system, sinusoidal winding distributions, sinusoidal *mmf* waveforms, etc. Let the rotor with segmented magnets be positioned with respect to the stator such that

$$\psi_{asm} = \psi_m \sin\theta_r \quad \text{and} \quad \psi_{bsm} = -\psi_m \cos\theta_r.$$

The self-inductances of the stator windings are L_{ss}. The stator windings are displaced by 90 electrical degrees. Hence, the mutual inductances between the stator windings are $L_{asbs} = L_{bsas} = 0$. From (6.8), using $\psi_{as} = L_{ss}i_{as} + \psi_m\sin\theta_r$ and $\psi_{bs} = L_{ss}i_{bs} - \psi_m\cos\theta_r$, one finds

$$\begin{aligned} u_{as} &= r_s i_{as} + \frac{d\left(L_{ss}i_{as} + \psi_m \sin\theta_r\right)}{dt} = r_s i_{as} + L_{ss}\frac{di_{as}}{dt} + \psi_m \cos\theta_r\omega_r, \\ u_{bs} &= r_s i_{bs} + \frac{d\left(L_{ss}i_{bs} - \psi_m \cos\theta_r\right)}{dt} = r_s i_{bs} + L_{ss}\frac{di_{bs}}{dt} - \psi_m \sin\theta_r\omega_r. \end{aligned} \tag{6.9}$$

The Newton second law yields

$$\begin{aligned} \frac{d\omega_{rm}}{dt} &= \frac{1}{J}\left(T_e - B_m\omega_{rm} - T_L\right), \\ \frac{d\theta_{rm}}{dt} &= \omega_{rm}. \end{aligned} \tag{6.10}$$

The electromagnetic torque is obtained by using the coenergy

$$W_c = \frac{1}{2}\left(L_{ss}i_{as}^2 + L_{ss}i_{bs}^2\right) + \begin{bmatrix} i_{as} & i_{bs} \end{bmatrix}\begin{bmatrix} \psi_{asm} \\ \psi_{bsm} \end{bmatrix} + W_{PM} = \frac{1}{2}\left(L_{ss}i_{as}^2 + L_{ss}i_{bs}^2\right) + i_{as}\psi_m \sin\theta_r - i_{bs}\psi_m \cos\theta_r + W_{PM}.$$

Here, the inductance L_{ss} and the energy stored in permanent magnets W_{PM} are not functions of θ_r. Hence

$$T_e = \frac{\partial W_c}{\partial\theta_r} = \frac{P\psi_m}{2}\left(i_{as}\cos\theta_r + i_{bs}\sin\theta_r\right). \tag{6.11}$$

Using the circuitry-electromagnetic equations (6.9), *torsional–mechanical* dynamics (6.10), and expression (6.11), we obtain

$$\begin{aligned} \frac{di_{as}}{dt} &= \frac{1}{L_{ss}}\left(-r_s i_{as} - \psi_m \cos\theta_r\omega_r + u_{as}\right), \\ \frac{di_{bs}}{dt} &= \frac{1}{L_{ss}}\left(-r_s i_{bs} + \psi_m \sin\theta_r\omega_r + u_{bs}\right), \\ \frac{d\omega_r}{dt} &= \frac{P^2\psi_m}{4J}\left(i_{as}\cos\theta_r + i_{bs}\sin\theta_r\right) - \frac{B_m}{J}\omega_r - \frac{P}{2J}T_L, \\ \frac{d\theta_r}{dt} &= \omega_r. \end{aligned} \tag{6.12}$$

From (6.11), to guarantee the balanced operation, one finds the balanced current set

$$i_{as} = \sqrt{2}i_M \cos\theta_r, \quad i_{bs} = \sqrt{2}i_M \sin\theta_r.$$

The balanced phase voltages are $u_{as}=\sqrt{2}u_M\cos\theta_r$ and $u_{bs}=\sqrt{2}u_M\sin\theta_r$.
Using the balanced current set, the electromagnetic torque is

$$T_e=\frac{P\psi_m}{2}\sqrt{2}i_M\left(\cos^2\theta_r+\sin^2\theta_r\right)=\frac{P\psi_m}{\sqrt{2}}i_M.$$

The mechanical angular displacement (physical rotor displacement) θ_{rm} is related to the electrical angular displacement, and $\theta_{rm}=2\theta_r/P$. The resulting differential equations (6.12) can be refined using ω_{rm} and θ_{rm}. The mechanical angular displacement θ_{rm} is commonly used in the analysis of stepper motors that usually operate as open-loop devices without measuring θ_r. The stepper motors are rotated step by step by *energizing* the windings by sequentially applying u_{as} and u_{bs}. Permanent-magnet synchronous motors with high P develop high electromagnetic torque, while the mechanical angular velocity is low because $\omega_{rm}=2\omega_r/P$. These motors are effectively used as *direct* drives and servos. The *direct* motor-kinematic connection without gears and couplings ensures a high level of efficiency, reliability, and performance.

6.3.2 Two-Phase Permanent-Magnet Stepper Motors

6.3.2.1 Permanent-Magnet Stepper Motors

For stepper motors, illustrated in Figure 6.9, one *energizes* the stator windings applying an adequate sequence of u_{as} and u_{bs}. The rotor rotates *counterclockwise* or *clockwise* by changing direction of T_e. Applying u_{as} and u_{bs}, one may achieve the incremental rotor displacement equal to a full or half step, or ensure microstepping. The rotor repositioning rate (number of steps per second) is regulated by changing the frequency of u_{as} and u_{bs}. If stepper motors operate in an open-loop mode, the motor can *miss* the step or steps if: (1) Instantaneous torque $T_{e\ instantaneous}$ is not sufficient; (2) $T_{e\ instanteneous}<T_L$; (3) u_{as} and u_{bs} are supplied at high frequency with respect to the motor dynamics that depend on L_{ss} and equivalent moment of inertia J, which may vary. The stepper motor can *pass* the step or steps if: (1) $T_e\gg T_L$; (2) Load T_L is bidirectional; (3) High kinetic energy is stored in the moving rotor with the attached kinematics due to high J. Other factors contribute to *missing* or *passing* steps such as varying J, fast-varying bidirectional T_L, disturbances, parameter variations, etc. Open-loop stepper motors are used if $T_L\approx$ const and $J\approx$ const. The phase voltages switching frequency is found by examining the system dynamics, loads, disturbances, etc. The stepper motors are designed with a high number of rotor teeth RT, which are an analogy to poles P, $RT=P/2$. To achieve a 1.8 degree rotation using the full-step operation, $RT=100$. The full-, half-, and micro-stepping can be achieved by using the electromagnetic torque that comprises two components,

$$T_e=T_{e\ permanent\ magnets}+T_{e\ reluctance}.$$

The reluctance ($T_e=T_{e\ reluctance}$) and *hybrid* ($T_e=T_{e\ permanent\ magnets}+T_{e\ reluctance}$) stepper motors with $L_{md}\neq L_{mq}$, $L_{\Delta m}\neq 0$ are widely used. The synchronous reluctance motors are covered in Section 6.2.

Consider stepper motors, depicted in Figure 6.9, with the number of rotor teeth RT. The electrical angular velocity and displacement are $\omega_r=RT\omega_{rm}$ and $\theta_r=RT\theta_{rm}$. The flux linkages are the functions of the number of RT and displacement. Using (6.8), letting $\psi_{asm}=\psi_m\cos(RT\theta_{rm})$ and $\psi_{bsm}=\psi_m\sin(RT\theta_{rm})$, for a round-rotor with $L_{md}=L_{mq}$ ($L_{\Delta m}=0$), we have

$$u_{as}=r_si_{as}+\frac{d\psi_{as}}{dt},\quad \psi_{as}=L_{asas}i_{as}+L_{asbs}i_{bs}+\psi_{asm},\quad \psi_{asm}=\psi_m\cos\left(RT\theta_{rm}\right),$$

$$u_{bs}=r_si_{bs}+\frac{d\psi_{bs}}{dt},\quad \psi_{bs}=L_{bsas}i_{as}+L_{bsbs}i_{bs}+\psi_{bsm},\quad \psi_{bsm}=\psi_m\sin\left(RT\theta_{rm}\right).$$

One finds the *total emfs*, and

$$u_{as} = r_s i_{as} + \frac{d\left(L_{ss} i_{as} + \psi_m \cos\left(RT\theta_{rm}\right)\right)}{dt} = r_s i_{as} + L_{ss}\frac{di_{as}}{dt} - RT\psi_m \sin\left(RT\theta_{rm}\right)\omega_{rm},$$

$$u_{bs} = r_s i_{bs} + \frac{d\left(L_{ss} i_{bs} + \psi_m \sin\left(RT\theta_{rm}\right)\right)}{dt} = r_s i_{bs} + L_{ss}\frac{di_{bs}}{dt} + RT\psi_m \cos\left(RT\theta_{rm}\right)\omega_{rm}.$$

The governing equations are

$$\frac{di_{as}}{dt} = \frac{1}{L_{ss}}\left[-r_s i_{as} + RT\psi_m \sin\left(RT\theta_{rm}\right)\omega_{rm} + u_{as}\right],$$
$$\frac{di_{bs}}{dt} = \frac{1}{L_{ss}}\left[-r_s i_{bs} - RT\psi_m \cos\left(RT\theta_{rm}\right)\omega_{rm} + u_{bs}\right]. \tag{6.13}$$

Using the coenergy

$$W_c = \frac{1}{2}\left(L_{ss} i_{as}^2 + L_{ss} i_{bs}^2\right) + \psi_m \cos\left(RT\theta_{rm}\right) i_{as} + \psi_m \sin\left(RT\theta_{rm}\right) i_{bs} + W_{PM},$$

one finds

$$T_e = \frac{\partial W_c}{\partial \theta_{rm}} = RT\psi_m\left[-\sin\left(RT\theta_{rm}\right) i_{as} + \cos\left(RT\theta_{rm}\right) i_{bs}\right]. \tag{6.14}$$

From Newton's second law (6.10), (6.13), and (6.14), we have

$$\frac{di_{as}}{dt} = \frac{1}{L_{ss}}\left[-r_s i_{as} + RT\psi_m \sin\left(RT\theta_{rm}\right)\omega_{rm} + u_{as}\right],$$
$$\frac{di_{bs}}{dt} = \frac{1}{L_{ss}}\left[-r_s i_{bs} - RT\psi_m \cos\left(RT\theta_{rm}\right)\omega_{rm} + u_{bs}\right],$$
$$\frac{d\omega_{rm}}{dt} = \frac{1}{J}\left[RT\psi_m\left[-\sin\left(RT\theta_{rm}\right) i_{as} + \cos\left(RT\theta_{rm}\right) i_{bs}\right] - B_m\omega_{rm} - T_L\right], \tag{6.15}$$
$$\frac{d\theta_{rm}}{dt} = \omega_{rm}.$$

Having found T_e (6.14), the stepper motors can be rotated sequentially supplying the phase voltages u_{as} and u_{bs} (energizing windings) without measuring θ_{rm}. The balanced current and voltage sets can be derived if θ_{rm} is measured. The balanced current and voltage sets are $i_{as} = -\sqrt{2} i_M \sin\left(RT\theta_{rm}\right)$, $i_{bs} = \sqrt{2} i_M \cos\left(RT\theta_{rm}\right)$ and $u_{as} = -\sqrt{2} u_M \sin\left(RT\theta_{rm}\right)$, $u_{bs} = \sqrt{2} u_M \cos\left(RT\theta_{rm}\right)$. One finds $T_e = \sqrt{2} RT\psi_m i_M$. To implement the displacement-dependent current or voltage sets, one measures θ_{rm} by using the Hall sensors, or, estimates θ_{rm} by using observers. The observers can be designed using the phase currents.

Stepper motors usually operate in the open-loop configuration without measurements on θ_{rm}. The motors are controlled by power amplifiers by *energizing* the *as* and *bs* windings. The reluctance variations in the *quadrature* and *direct* axes result in the reluctance torque $T_{e\ reluctance}$. The rotor rotates to minimize the reluctance. The *hybrid* stepper motors ensure microstepping, and $T_e = T_{e\ permanent\ magnets} + T_{e\ reluctance}$.

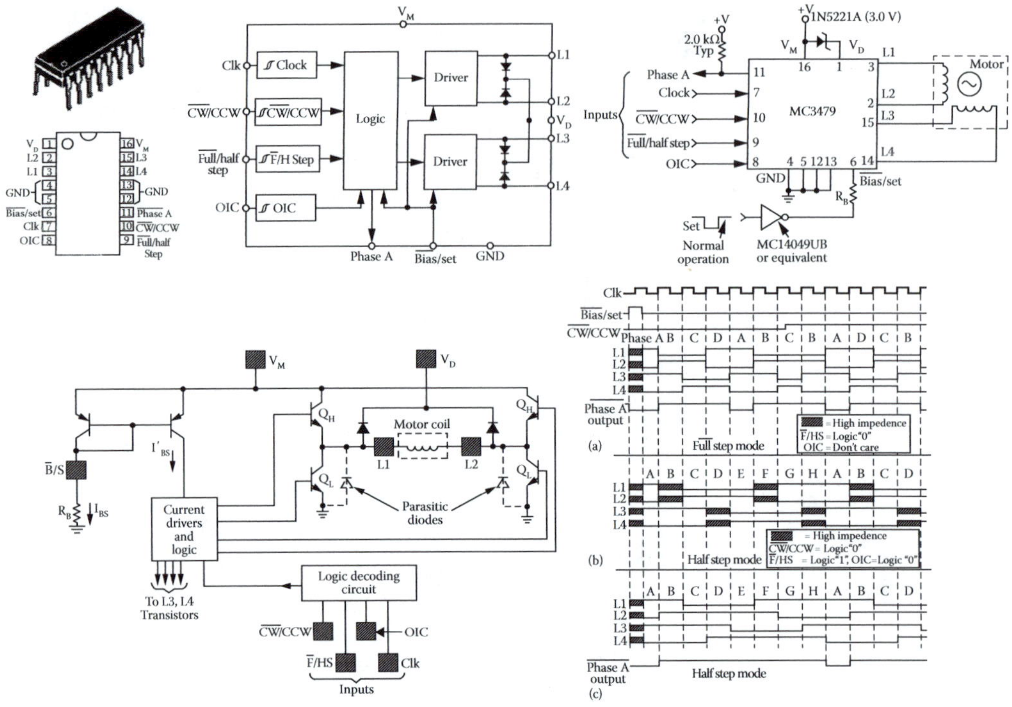

FIGURE 6.11 Motorola MC3479 stepper motor driver. (From Lyshevski, S.E., *Electromechanical Systems, Electric Machines, and Applied Mechatronics*, CRC Press, Boca Raton, FL, 1999; Copyright of Motorola. With permission.)

The Motorola MC3479 stepper motor driver can be used. This driver may drive two-phase stepper motors bidirectionally. The representative block diagrams, circuitry, timing-output diagrams, and connections are shown in Figure 6.11. A high-performance L9942 stepper motor driver ensures the full-, half-, and mini-stepping modes for a well-defined operating envelope if there is no significant bidirectional load.

This MC3479 driver is designed to drive stepper minimotors in various applications such as positioning tables, disk drives, small robots, servos, etc. As illustrated in Figure 6.11, the H-bridge topology power stage supplies the phase voltages u_{as} and u_{bs} to the windings. One winding with terminals L1 and L2 is shown in Figure 6.11. The applied voltage polarity depends on which transistor (Q_H or Q_L) is *on*. These transistors are driven by the signal-level voltages from the logic circuitry. The maximum sink current is a function of the resistor between pin 6 and ground.

When the outputs are in a high impedance state, both transistors (Q_H or Q_L) are *off*. The pin V_D provides a current path for the phase winding ("motor coil") current during transients (switching) to attenuate the *back emf* spikes. Pin V_D is normally connected to V_M (pin 16) through a diode or a resistor, or directly. The instantaneous peak voltage at the outputs must not exceed V_M, which is 6 V. The diodes across Q_L of each output provide a circuit path for the current. When the input is at a Logic "0" (less than 0.8 V), the outputs correspond to a full step operation with each clock cycle. The direction depends on the CW/CCW input. There are four switching phases for each cycle of the sequencing logic. As the phase voltage is applied, current i_{as} or i_{bs} flows in the motor windings. For a Logic "1" (more than 2 V), the outputs change a half step during each clock cycle. Eight switching phases result for each complete cycle of the sequencing logic. The output sequences and timing diagrams are reported in Figure 6.11. A complete description and application notes are reported by Motorola. Stepper motors and PWM controllers/drivers are depicted in Figures 6.12.

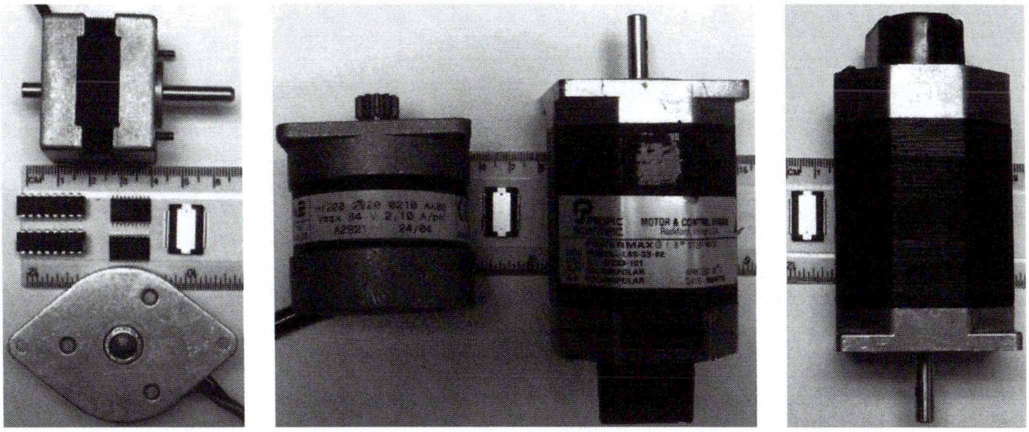

FIGURE 6.12 Permanent-magnet stepper motors and controllers/drivers: Phillips Semiconductor SAA 1027, TI DRV8804DW, and TI DRV8832DKDR, respectively. The Phillips Semiconductor SAA 1027 stepper motor driver may be used for two-stator stepper motors. The SAA 1027 consists of a bidirectional four-state counter and a code converter to vary four output voltages (18 V, 0.5 A per phase). The TI DRV8804DW (5.6 V, 0.8 A per phase) driver is used for unipolar stepper motors. The TI DRV8832DKDR dual full-bridge 100–500 kHz PWM driver may be applied for bipolar four-lead stepper motors. The H-bridge MOSFETs' efficiency is ~90% at 500 kHz, and the peak current per phase is 12 A for 24 V phase voltage.

Example 6.7

Permanent-magnet stepper motors are designed optimizing an electromagnetic system performing three-dimensional electromagnetic, thermal, and mechanical designs. Let the stepper motor be designed such that the flux linkages are $\psi_{asm} = \psi_m \arcsin\left(\sin\left(RT\theta_{rm}\right)\right)$ and $\psi_{bsm} = \psi_m \arcsin\left(\sin\left(RT\theta_{rm} + \pi\right)\right)$. The plots for $\psi_{asm}(\theta_{rm})$ and $\psi_{bsm}(\theta_{rm})$ are depicted in Figure 6.13a if $\psi_m = 1$ and $RT = 100$. The MATLAB statements are

```
psim=1; RT=100; th=0:1e-5:0.1; psiasm=psim*asin(sin(RT*th))/2; psibsm=psim*asin(sin(RT*th+pi))/2;
plot(th, psiasm,'k-', th, psibsm,'b--','LineWidth',2.5);
xlabel('{\it\theta_r_m}','FontSize',18);
title('Flux Linkages {\it\psi_a_s_m}({\it\theta_r_m}) and {\it\psi_b_s_m}({\it\theta_r_m})','FontSize',18);
```

The torque is found using the coenergy

$$W_c = \frac{1}{2}\left(L_{ss}i_{as}^2 + L_{ss}i_{bs}^2\right) + \psi_{asm}i_{as} + \psi_{bsm}i_{bs} + W_{PM}.$$

The derivative of $f(x) = \arcsin(\sin x)$ was found in Example 6.3. For

$$\psi_{asm} = \psi_m \arcsin\left(\sin\left(RT\theta_{rm}\right)\right) \quad \text{and} \quad \psi_{bsm} = \psi_m \arcsin\left(\sin\left(RT\theta_{rm} + \pi\right)\right),$$

we have

$$T_e(\theta_r) = \frac{\partial W_c(\theta_{rm})}{\partial \theta_{rm}} = RT\psi_m \left[\frac{\cos\left(RT\theta_{rm}\right)}{\left|\cos\left(RT\theta_{rm}\right)\right|} i_{as} + \frac{\cos\left(RT\theta_{rm} + \pi\right)}{\left|\cos\left(RT\theta_{rm} + \pi\right)\right|} i_{bs} \right].$$

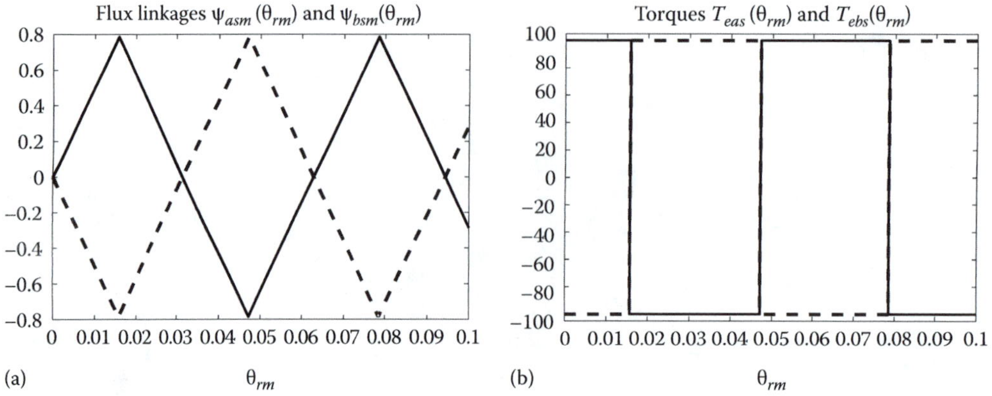

FIGURE 6.13 (a) Flux linkages $\psi_{asm} = \psi_m \arcsin\left(\sin\left(RT\theta_{rm}\right)\right)$ and $\psi_{bsm} = \psi_m \arcsin\left(\sin\left(RT\theta_{rm} + \pi\right)\right)$ depicted using the solid and dotted lines;
(b) Components T_{eas} and T_{ebs} of the electromagnetic torque $T_e(\theta_r) = \underbrace{\frac{1}{2} RT\psi_m \frac{\cos(RT\theta_{rm})}{|\cos(RT\theta_{rm})|} i_{as}}_{T_{eas}} + \underbrace{\frac{1}{2} RT\psi_m \frac{\cos(RT\theta_{rm}+\pi)}{|\cos(RT\theta_{rm}+\pi)|} i_{bs}}_{T_{ebs}}$ if the phase currents are fed $i_{as} = 1$ and $i_{bs} = 1$ A. To ensure rotation, the phase currents i_{as} and i_{bs} must be fed sequentially using the sequential logics. Calculations and plotting are made using

```
psim=1; RT=100; ias=1; ibs=1; th=0:1e-5:0.1;
Teas=ias*psim*RT*cos(RT*th)./abs(cos(RT*th));
Tebs=ibs*psim*RT*cos(RT*th+pi)./abs(cos(RT*th+pi));
plot(th, Teas,'k-', th, Tebs,'b--','LineWidth',2.5); axis([0 0.10 -105 105]); xlabel('{\it\theta_r_m}','FontSize',18);
title('Torques {\itT_e_a_s}({\it\theta_r_m}) and {\itT_e_b_s}({\it\theta_r_m})','FontSize',18);
```

The plots for the electromagnetic torque components $T_{eas}(\theta_{rm})$ and $T_{ebs}(\theta_{rm})$ are shown in Figure 6.13b.

For stepper motor, the phase currents i_{as} and i_{bs} should be fed sequentially with magnitude i_M. We have

$$T_e(\theta_r) = \begin{cases} RT\psi_m \dfrac{\cos\left(RT\theta_{rm}\right)}{\left|\cos\left(RT\theta_{rm}\right)\right|} i_{as}, & i_{as} = \begin{cases} i_M, & \dfrac{d\left[\arcsin\left(\sin\left(RT\theta_{rm}\right)\right)\right]}{d\theta_{rm}} > 0, \text{ with } i_{bs} = 0 \\ 0, & \dfrac{d\left[\arcsin\left(\sin\left(RT\theta_{rm}\right)\right)\right]}{d\theta_{rm}} \le 0, \text{ with } i_{bs} = i_M \end{cases} \\ RT\psi_m \dfrac{\cos\left(RT\theta_{rm}+\pi\right)}{\left|\cos\left(RT\theta_{rm}+\pi\right)\right|} i_{bs}, & i_{bs} = \begin{cases} i_M, & \dfrac{d\left[\arcsin\left(\sin\left(RT\theta_{rm}+\pi\right)\right)\right]}{d\theta_{rm}} > 0, \text{ with } i_{as} = 0 \\ 0, & \dfrac{d\left[\arcsin\left(\sin\left(RT\theta_{rm}+\pi\right)\right)\right]}{d\theta_{rm}} \le 0, \text{ with } i_{as} = i_M. \end{cases} \end{cases}$$

This results in $T_e = RT\psi_m i_M$. The phase voltages u_{as} and u_{bs} must be applied sequentially. The polarity of u_{as} and u_{bs} can be changed. The developed $T_e = RT\psi_m i_M$ and $T_e = -RT\psi_m i_M$ guarantee clockwise or counterclockwise rotation of the rotor without ripple. ■

Note: The triangle wave with range –1 to +1 and period $2T$ can be described as

$$x=\frac{2}{T}\left[t-T\left[\frac{t}{T}+\frac{1}{2}\right]\right](-1)^{\left[\frac{t}{T}+\frac{1}{2}\right]},$$

where the symbol $[n]$ represents the *floor function.*

The saw wave is $x=\left|2\left(\frac{t}{T}-\left[\frac{t}{T}+\frac{1}{2}\right]\right)\right|$.

For a given range $\left[-1 \quad +1\right]$, $x=2\left|2\left(\frac{t}{T}-\left[\frac{t}{T}+\frac{1}{2}\right]\right)\right|-1$.

One recalls that

$$\operatorname{sgn}(x)=\frac{x}{|x|}=\frac{|x|}{x}, \quad x\neq 0.$$

The square periodic function with period T and amplitude A is

$$y=A\operatorname{sgn}\left(\sin\left(\frac{2\pi}{T}x\right)\right), \quad \text{or} \quad y=A\csc\left(\frac{2\pi}{T}x\right)\left|\sin\left(\frac{2\pi}{T}x\right)\right|.$$

One finds $\frac{\partial y}{\partial x}$. In particular,

$$\frac{\partial y}{\partial x}=\frac{\partial}{\partial x}A\operatorname{sgn}\left(\sin\left(\frac{2\pi}{T}x\right)\right)=2A\frac{2\pi}{T}x\cos\left(\frac{2\pi}{T}x\right)\delta\left(\sin\left(\frac{2\pi}{T}x\right)\right).$$

To find the derivative for

$$y=A\csc\left(\frac{2\pi}{T}x\right)\left|\sin\left(\frac{2\pi}{T}x\right)\right|, \quad \csc(x)=\frac{1}{\sin x},$$

one applies the product rule $(fg)'=f'g+fg'$.

Here,

$$\frac{d}{dx}\csc x=\frac{d}{dx}\frac{1}{\sin x}=\sin x\frac{d}{dx}1-1\frac{d}{dx}\frac{\sin x}{\sin^2 x}=-\frac{\cos x}{\sin^2 x}=-\csc x\cdot\cot x.$$

Using

$$|\sin x|=\sqrt{\sin^2 x},$$

one finds

$$\frac{d|\sin x|}{dx}=\frac{\sin x\cos x}{|\sin x|}=\operatorname{sgn}(\sin x)\cos x, \quad \sin x\neq 0.$$ ■

6.3.2.2 Analysis of Permanent-Magnet Stepper Motors Using The Quadrature and Direct Quantities

We examined two-phase permanent-magnet synchronous motors in the *machine* variable. Using the *quadrature* and *direct* components for physical variables, stepper motors can be modeled in the *arbitrary* ω, stationary (ω = 0), rotor ($\omega = \omega_r$), and synchronous ($\omega = \omega_e$) reference frames. Analysis in the rotor and synchronous frames is identical because $\omega = \omega_e = \omega_r$ [2–5]. For (6.13), in the synchronous reference frame, we apply the *direct* Park transformation [2–5]

$$\begin{bmatrix} u_{qs}^e \\ u_{ds}^e \end{bmatrix} = \begin{bmatrix} -\sin(RT\theta_{rm}) & \cos(RT\theta_{rm}) \\ \cos(RT\theta_{rm}) & \sin(RT\theta_{rm}) \end{bmatrix} \begin{bmatrix} u_{as} \\ u_{bs} \end{bmatrix}, \quad \begin{bmatrix} i_{qs}^e \\ i_{ds}^e \end{bmatrix} = \begin{bmatrix} -\sin(RT\theta_{rm}) & \cos(RT\theta_{rm}) \\ \cos(RT\theta_{rm}) & \sin(RT\theta_{rm}) \end{bmatrix} \begin{bmatrix} i_{as} \\ i_{bs} \end{bmatrix}.$$

From (6.13), the following differential equations result

$$u_{qs}^e = r_s i_{qs}^e + L_{ss}\frac{di_{qs}^e}{dt} + RT\psi_m\omega_{rm} + RTL_{ss}i_{ds}^e\omega_{rm}, \quad u_{ds}^e = r_s i_{ds}^e + L_{ss}\frac{di_{ds}^e}{dt} - RTL_{ss}i_{qs}^e\omega_{rm}.$$

Hence,

$$\begin{aligned} \frac{di_{qs}^e}{dt} &= -\frac{r_s}{L_{ss}}i_{qs}^e - \frac{RT\psi_m}{L_{ss}}\omega_{rm} - RTi_{ds}^e\omega_{rm} + \frac{1}{L_{ss}}u_{qs}^e, \\ \frac{di_{ds}^e}{dt} &= -\frac{r_s}{L_{ss}}i_{ds}^e + RTi_{qs}^e\omega_{rm} + \frac{1}{L_{ss}}u_{ds}^e. \end{aligned} \tag{6.16}$$

In (6.16), the phase currents (i_{as}, i_{bs}) are related to (i_{qs}, i_{ds}) as

$$\begin{bmatrix} i_{as} \\ i_{bs} \end{bmatrix} = \begin{bmatrix} -\sin(RT\theta_{rm}) & \cos(RT\theta_{rm}) \\ \cos(RT\theta_{rm}) & \sin(RT\theta_{rm}) \end{bmatrix} \begin{bmatrix} i_{qs}^e \\ i_{ds}^e \end{bmatrix}.$$

From

$$T_e = RT\psi_m\left[-\sin\left(RT\theta_{rm}\right)i_{as} + \cos\left(RT\theta_{rm}\right)i_{bs}\right],$$

we obtain $T_e = RT\psi_m i_{qs}^e$. Using (6.10) and (6.16), one finds

$$\begin{aligned} \frac{di_{qs}^e}{dt} &= -\frac{r_s}{L_{ss}}i_{qs}^e - \frac{RT\psi_m}{L_{ss}}\omega_{rm} - RTi_{ds}^e\omega_{rm} + \frac{1}{L_{ss}}u_{qs}^e, \\ \frac{di_{ds}^e}{dt} &= -\frac{r_s}{L_{ss}}i_{ds}^e + RTi_{qs}^e\omega_{rm} + \frac{1}{L_{ss}}u_{ds}^e, \\ \frac{d\omega_{rm}}{dt} &= \frac{1}{J}\left(RT\psi_m i_{qs}^e - B_m\omega_{rm} - T_L\right), \\ \frac{d\theta_{rm}}{dt} &= \omega_{rm}. \end{aligned} \tag{6.17}$$

Consider the balanced current and voltage sets $i_{as} = -\sqrt{2}i_M \sin\left(RT\theta_{rm}\right)$, $i_{bs} = \sqrt{2}i_M \cos\left(RT\theta_{rm}\right)$, and, $u_{as} = -\sqrt{2}u_M \sin\left(RT\theta_{rm}\right)$, $u_{bs} = \sqrt{2}u_M \cos\left(RT\theta_{rm}\right)$. Applying the *direct* Park transformation, from $i_{qs}^e = -i_{as}\sin(RT\theta_{rm}) + i_{bs}\cos(RT\theta_{rm})$, $i_{ds}^e = i_{as}\cos(RT\theta_{rm}) + i_{bs}\sin(RT\theta_{rm})$, one finds

$$i_{qs}^e = \sqrt{2}i_M \sin^2\left(RT\theta_{rm}\right) + \sqrt{2}i_M \cos^2\left(RT\theta_{rm}\right) = \sqrt{2}i_M$$

and

$$i_{ds}^e = -\sqrt{2}i_M \sin\left(RT\theta_{rm}\right)\cos\left(RT\theta_{rm}\right) + \sqrt{2}i_M \sin\left(RT\theta_{rm}\right)\cos\left(RT\theta_{rm}\right) = 0.$$

Hence, the *quadrature* and *direct* current components are $i_{qs}^e = \sqrt{2}i_M$ and $i_{ds}^e = 0$.

Similarly, the *quadrature* and *direct* voltage components are $u_{qs}^e = \sqrt{2}u_M$ and $u_{ds}^e = 0$. In industrial applications, stepper motors have not been controlled by using the *qd* quantities. In fact, the *as* and *bs* phase voltages u_{as} and u_{bs} must be supplied.

6.3.2.3 Control of Stepper Motors

Experiments are performed for a 8-lead P22NSXA Pacific Scientific permanent-magnet stepper motor with two 100-teeth-each rotor-stack (1.8° full-step), 2.7 V and 4.6 A (unipolar), 1 N-m (rated), $r_s = 0.5$ ohm, $L_{ss} = 7.5 \times 10^{-4}$ H, $\psi_m = 4.9 \times 10^{-3}$ N-m/A, $B_m = 9.2 \times 10^{-4}$ N-m-sec/rad, and $J = 1 \times 10^{-4}$ kg-m^2. A shaft-mount incremental encoder measures θ_{rm}. The angular velocity ω_{rm} can be estimated by counting the encoder pulses. The servo schematics and hardware are reported in Figure 6.14. The clockwise and counterclockwise rotation and precision repositioning should be guaranteed despite bidirectional loading $T_L(t)$. The servo consists of a microcontroller, power electronics, stepper motor, and kinematics. Using the reference r and the mechanical angular displacements θ_{rm}, the controller develops the PWM signals that drive high-frequency MOSFETs. The phase voltages are applied from a dual full-H-bridge topology MOSFET *driver* with up to 500 kHz switching frequency, ~90% efficiency, 50 V, 3 A rated, and 6 A peak. The TI DRV8412 *driver* varies the *average* phase voltages $(u_{as}, u_{bs}, \bar{u}_{as}, \bar{u}_{bs})$ applied to the motor windings. The *driver* is controlled by a TMS320F28035 32-bit fixed-point microcontroller with processing, control, interfacing, peripheral, programming, and other capabilities.

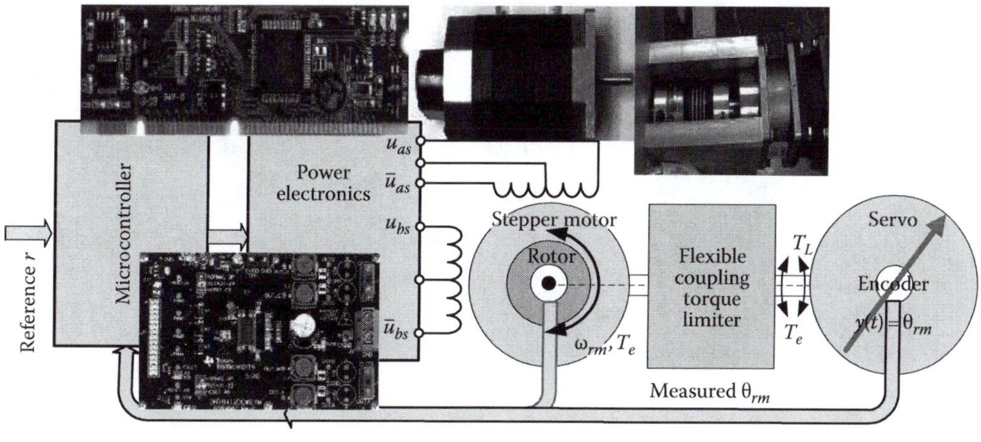

FIGURE 6.14 Schematics and hardware of a bidirectional repositioning closed-loop servo.

The full-, half-, quarter-, and micro-stepping can be ensured. For the bipolar stepper motors with the phase currents (i_{as}, i_{bs}) and voltages (u_{as}, u_{bs}), the full-step displacement is $2\pi/RT$. For the electromagnetic torque, depending on the electromagnetic design and winding configuration, one obtains the electromagnetic toque T_e similar to (6.14) and that reported in Example 6.7. The sequences of the phase voltages for different winding configurations are depicted in Figure 6.11. For a full step repositioning in the open-loop operation without measurements of θ_{rm}, the phase voltages u_{as} and u_{bs} are sequentially applied within the allowed frequency,

$$u_{as} = \begin{cases} u_M, \forall\theta_{rm} \in \left[0 \quad \dfrac{\pi}{RT}\right] \\ 0, \forall\theta_{rm} \in \left[\dfrac{\pi}{RT} \quad \dfrac{2\pi}{RT}\right] \end{cases}, \quad u_{bs} = \begin{cases} u_M, \forall\theta_{rm} \in \left[\dfrac{\pi}{RT} \quad \dfrac{2\pi}{RT}\right] \\ 0, \forall\theta_{rm} \in \left[0 \quad \dfrac{\pi}{RT}\right] \end{cases}.$$

The switching frequency is defined by the motor transients that are affected by L_{ss} and J.

For the unipolar motor with $(i_{as}, i_{bs}, \bar{i}_{as}, \bar{i}_{bs})$ and $(u_{as}, u_{bs}, \bar{u}_{as}, \bar{u}_{bs})$, the expressions for full-, half-, and micro-stepping are found. For a full-step,

$$u_{as} = \begin{cases} u_M, \forall\theta_{rm} \in \left[0 \quad \dfrac{\pi}{2RT}\right] \\ 0, \forall\theta_{rm} \in \left[\dfrac{\pi}{2RT} \quad \dfrac{2\pi}{RT}\right] \end{cases}, u_{bs} = \begin{cases} u_M, \forall\theta_{rm} \in \left[\dfrac{\pi}{2RT} \quad \dfrac{\pi}{RT}\right] \\ 0, \forall\theta_{rm} \in \left[0 \quad \dfrac{\pi}{2RT}\right], \left[\dfrac{\pi}{RT} \quad \dfrac{2\pi}{RT}\right] \end{cases}, \bar{u}_{as} = \begin{cases} u_M, \forall\theta_{rm} \in \left[\dfrac{\pi}{RT} \quad \dfrac{3\pi}{2RT}\right] \\ 0, \forall\theta_{rm} \in \left[0 \quad \dfrac{\pi}{RT}\right], \left[\dfrac{3\pi}{2RT} \quad \dfrac{2\pi}{RT}\right] \end{cases},$$

$$\bar{u}_{bs} = \begin{cases} u_M, \forall\theta_{rm} \in \left[\dfrac{3\pi}{2RT} \quad \dfrac{2\pi}{RT}\right] \\ 0, \forall\theta_{rm} \in \left[0 \quad \dfrac{3\pi}{2RT}\right] \end{cases}.$$

The experimental results, voltages $(u_{as}, u_{bs}, \bar{u}_{as}, \bar{u}_{bs})$ phase currents and displacement are documented in Figure 6.15. The time- and spatially dependent $(u_{as}, u_{bs}, \bar{u}_{as}, \bar{u}_{bs})$ are

$$\begin{cases} u_{as} = u_M \varphi_{as}(\theta_{rm}), \forall\theta_{rm} \in \Theta_{as} \\ u_{bs} = u_M \varphi_{bs}(\theta_{rm}), \forall\theta_{rm} \in \Theta_{bs} \\ \bar{u}_{as} = u_M \bar{\varphi}_{as}(\theta_{rm}), \forall\theta_{rm} \in \bar{\Theta}_{as} \\ \bar{u}_{bs} = u_M \bar{\varphi}_{bs}(\theta_{rm}), \forall\theta_{rm} \in \bar{\Theta}_{bs} \end{cases}.$$

For bidirectional loads, the closed-loop system is designed to prevent repositioning inadequacy and guarantee accuracy. A proportional-integral control law and switching algorithm with switching functions $\varphi_{as}(\cdot)$, $\varphi_{bs}(\cdot)$, $\bar{\varphi}_{as}(\cdot)$, $\bar{\varphi}_{bs}(\cdot)$ are given as

$$u = \mathrm{sat}_0^{u_M}\left(k_p e_\theta + k_i \int e_\theta \, dt\right), \quad e_\theta = r - \theta_{rm}, \quad \begin{cases} u_{as} = u(e_\theta)\varphi_{as}(e_\theta, \theta_{rm}), \forall\theta_{rm} \in \Theta_{as} \\ u_{bs} = u(e_\theta)\varphi_{bs}(e_\theta, \theta_{rm}), \forall\theta_{rm} \in \Theta_{bs} \\ \bar{u}_{as} = u(e_\theta)\bar{\varphi}_{as}(e_\theta, \theta_{rm}), \forall\theta_{rm} \in \bar{\Theta}_{as} \\ \bar{u}_{bs} = u(e_\theta)\bar{\varphi}_{bs}(e_\theta, \theta_{rm}), \forall\theta_{rm} \in \bar{\Theta}_{bs} \end{cases}.$$

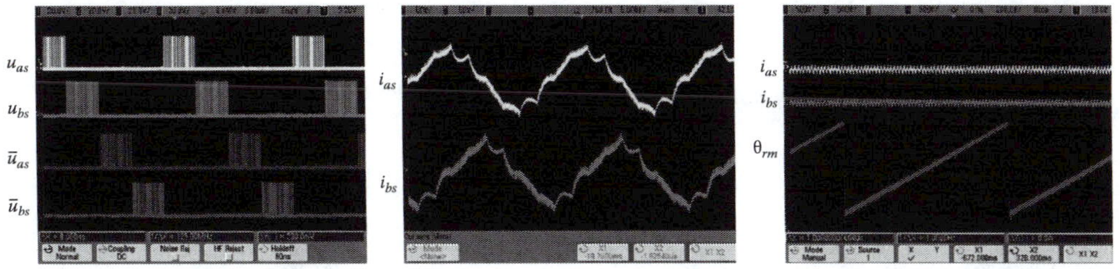

FIGURE 6.15 1.8-degree full-step positioning with 200 steps per a revolution. The PWM phase voltages $(u_{as}, u_{bs}, \bar{u}_{as}, \bar{u}_{bs})$ are applied. The phase currents and measured θ_{rm} are depicted.

For the full-step repositioning, the PWM phase voltages and currents are documented in Figure 6.15. The robust repositioning is ensured in an expanded operating envelope, including bidirectional load T_L. The half-step and 1/128-stepping are studied. The phase voltages $(u_{as}, u_{bs}, \bar{u}_{as}, \bar{u}_{bs})$ for the 1/2- and 1/128 microstepping as well as the phase and currents (i_{as}, i_{bs}) are illustrated in Figures 6.16. The derived switching functions $\varphi_{as}(\cdot)$, $\varphi_{bs}(\cdot)$, $\bar{\varphi}_{as}(\cdot)$, $\bar{\varphi}_{bs}(\cdot)$, which are functions of θ_{rm}, are found from the illustrated spatiotemporal phase voltages $(u_{as}, u_{bs}, \bar{u}_{as}, \bar{u}_{bs})$. The magnitude u_M of $(u_{as}, u_{bs}, \bar{u}_{as}, \bar{u}_{bs})$ is regulated using the PWM concept.

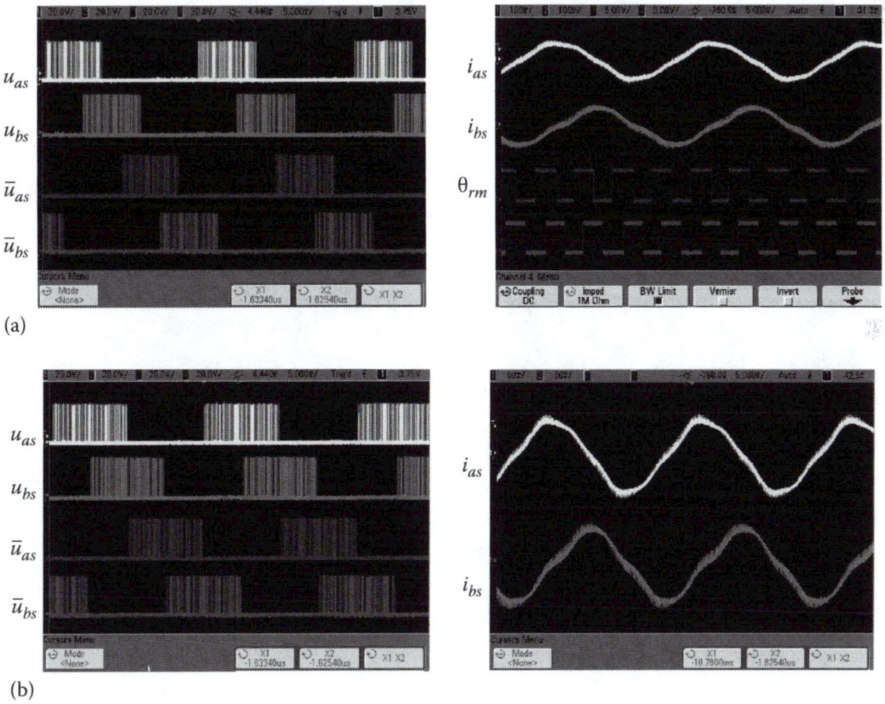

FIGURE 6.16 (a) Half-step motor operation with 400 steps per a revolution. The spatiotemporal phase voltages $(u_{as}, u_{bs}, \bar{u}_{as}, \bar{u}_{bs})$ are applied. The phase currents i_{as} and i_{bs} in the motor windings are measured and illustrated. The measured angular displacement θ_{rm} is reported;

(b) 1/128 stepping: 25,600 steps per a revolution is achieved with the PWM phase voltages $(u_{as}, u_{bs}, \bar{u}_{as}, \bar{u}_{bs})$. The measured phase currents i_{as} and i_{bs} are depicted.

6.4 RADIAL TOPOLOGY THREE-PHASE PERMANENT-MAGNET SYNCHRONOUS MACHINES

A three-phase two-pole permanent-magnet synchronous machine (motor and generator) is depicted in Figure 6.17a. The image of the permanent-magnet synchronous machine is shown in Figure 6.17b. Motors are tested using loads, while generators are rotated by prime movers, as shown in Figures 6.17a and c. In a generator, the motional *emfs* yield the induced phase voltages at the terminals, as shown in Figure 6.17d.

6.4.1 ANALYSIS OF THREE-PHASE PERMANENT-MAGNET SYNCHRONOUS MOTORS

The Kirchhoff second law yields three differential equations for the *as*, *bs*, and *cs* stator phases

$$u_{as} = r_s i_{as} + \frac{d\psi_{as}}{dt}, \quad u_{bs} = r_s i_{bs} + \frac{d\psi_{bs}}{dt}, \quad u_{cs} = r_s i_{cs} + \frac{d\psi_{cs}}{dt}, \tag{6.18}$$

$$\mathbf{u}_{abcs} = \mathbf{r}_s \mathbf{i}_{abcs} + \frac{d\boldsymbol{\psi}_{abcs}}{dt}, \quad \begin{bmatrix} u_{as} \\ u_{bs} \\ u_{cs} \end{bmatrix} = \begin{bmatrix} r_s & 0 & 0 \\ 0 & r_s & 0 \\ 0 & 0 & r_s \end{bmatrix} \begin{bmatrix} i_{as} \\ i_{bs} \\ i_{cs} \end{bmatrix} + \begin{bmatrix} \frac{d\psi_{as}}{dt} \\ \frac{d\psi_{bs}}{dt} \\ \frac{d\psi_{cs}}{dt} \end{bmatrix}.$$

In (6.18), the flux linkages are

$$\psi_{as} = L_{asas} i_{as} + L_{asbs} i_{bs} + L_{ascs} i_{cs} + \psi_{asm},$$

$$\psi_{bs} = L_{bsas} i_{as} + L_{bsbs} i_{bs} + L_{bscs} i_{cs} + \psi_{bsm},$$

$$\psi_{cs} = L_{csas} i_{as} + L_{csbs} i_{bs} + L_{cscs} i_{cs} + \psi_{csm}.$$

The flux linkages ψ_{asm}, ψ_{bsm}, and ψ_{csm} are the periodic functions of θ_r with period 2π. The stator windings are displaced by 120 electrical degrees. Denoting the magnitude of the flux linkages as ψ_m, let

$$\psi_{asm} = \psi_m \sin\theta_r, \quad \psi_{bsm} = \psi_m \sin\left(\theta_r - \frac{2}{3}\pi\right), \quad \text{and} \quad \psi_{csm} = \psi_m \sin\left(\theta_r + \frac{2}{3}\pi\right).$$

The *quadrature* and *direct* axes reluctances are $\Re_{mq} > \Re_{md}$, while for the round-rotor $\Re_{mq} = \Re_{md}$. The *quadrature* and *direct* axes magnetizing inductances are

$$L_{mq} = \frac{N_s^2}{\Re_{mq}} \quad \text{and} \quad L_{md} = \frac{N_s^2}{\Re_{md}}, \quad L_{mq} < L_{md}.$$

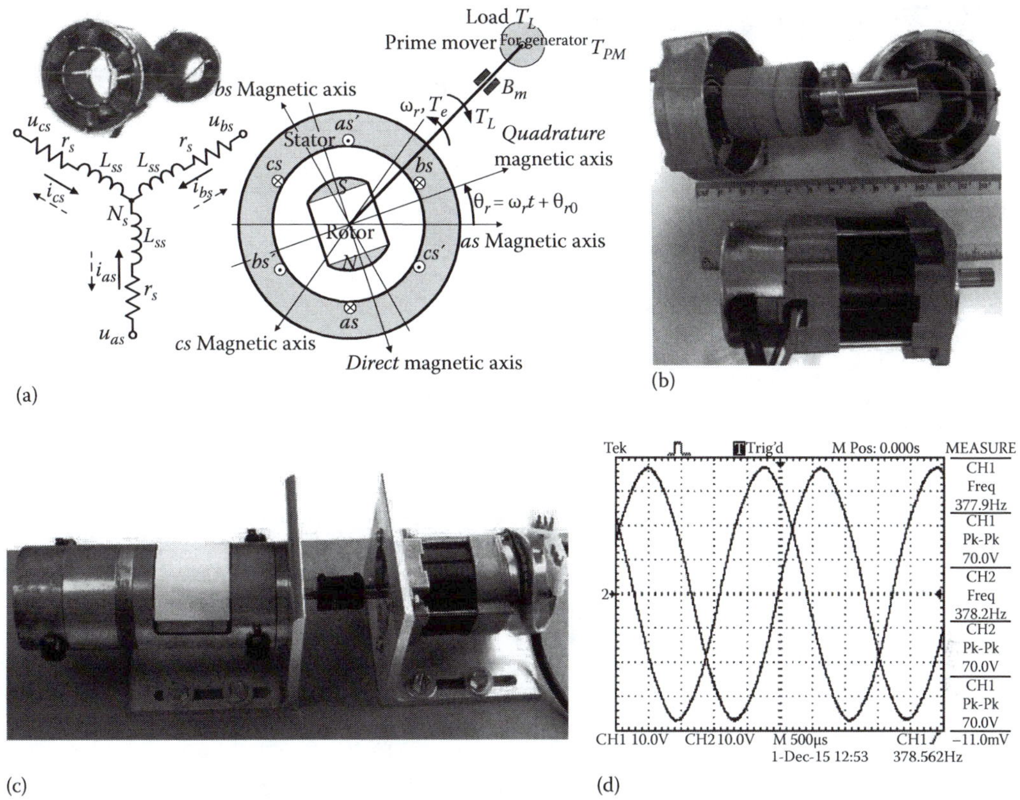

FIGURE 6.17 (a) Two-pole three-phase permanent-magnet synchronous machine. The directions of phase current are shown for motor (solid arrows) and generator (dashed arrows);
(b) Image of the NEMA 23 size permanent-magnet synchronous machine;
(c) Permanent-magnet synchronous machines—motor with a load, or a generator with a prime mover;
(d) Induced phase voltages in a three-phase permanent-magnet synchronous generator.

One has $L_{ls} + L_{mq} \le L_{asas} \le L_{ls} + L_{md}$, and, $L_{asas}(\theta_r)$ is a periodic function of θ_r. Assuming ideal sinusoidal variations, we have

$$L_{asas} = L_{ls} + \bar{L}_m - L_{\Delta m}\cos 2\theta_r,$$

$$\bar{L}_m = \frac{1}{3}\left(L_{mq} + L_{md}\right) = \frac{1}{3}\left(\frac{N_s^2}{\Re_{mq}} + \frac{N_s^2}{\Re_{md}}\right),$$

$$L_{\Delta m} = \frac{1}{3}\left(L_{md} - L_{mq}\right) = \frac{1}{3}\left(\frac{N_s^2}{\Re_{md}} - \frac{N_s^2}{\Re_{mq}}\right),$$

where $\bar{L}_m$ is the average value of the magnetizing inductance; $L_{\Delta m}$ is the half of amplitude of the sinusoidal variation of the magnetizing inductance.

The following equation for the flux linkages results

$$\psi_{abcs} = \mathbf{L}_s \mathbf{i}_{abcs} + \psi_m$$

$$= \begin{bmatrix} L_{ls} + \bar{L}_m - L_{\Delta m}\cos 2\theta_r & -\frac{1}{2}\bar{L}_m - L_{\Delta m}\cos 2\left(\theta_r - \frac{1}{3}\pi\right) & -\frac{1}{2}\bar{L}_m - L_{\Delta m}\cos 2\left(\theta_r + \frac{1}{3}\pi\right) \\ -\frac{1}{2}\bar{L}_m - L_{\Delta m}\cos 2\left(\theta_r - \frac{1}{3}\pi\right) & L_{ls} + \bar{L}_m - L_{\Delta m}\cos 2\left(\theta_r - \frac{2}{3}\pi\right) & -\frac{1}{2}\bar{L}_m - L_{\Delta m}\cos 2\theta_r \\ -\frac{1}{2}\bar{L}_m - L_{\Delta m}\cos 2\left(\theta_r + \frac{1}{3}\pi\right) & -\frac{1}{2}\bar{L}_m - L_{\Delta m}\cos 2\theta_r & L_{ls} + \bar{L}_m - L_{\Delta m}\cos 2\left(\theta_r + \frac{2}{3}\pi\right) \end{bmatrix} \begin{bmatrix} i_{as} \\ i_{bs} \\ i_{cs} \end{bmatrix}$$

$$+ \psi_m \begin{bmatrix} \sin\theta_r \\ \sin\left(\theta_r - \frac{2}{3}\pi\right) \\ \sin\left(\theta_r + \frac{2}{3}\pi\right) \end{bmatrix}.$$

Usually, three-phase permanent-magnet synchronous machines are round-rotor machines. Hence, $\Re_{mq} = \Re_{md}$ and $L_{mq} = L_{md}$. Thus,

$$\bar{L}_m = \frac{2N_s^2}{3\Re_{mq}} = \frac{2N_s^2}{3\Re_{md}} \quad \text{and} \quad L_{\Delta m} = 0.$$

Denoting $L_{ss} = L_{ls} + \bar{L}_m$, the inductance matrix is

$$\mathbf{L}_s = \begin{bmatrix} L_{ls} + \bar{L}_m & -\frac{1}{2}\bar{L}_m & -\frac{1}{2}\bar{L}_m \\ -\frac{1}{2}\bar{L}_m & L_{ls} + \bar{L}_m & -\frac{1}{2}\bar{L}_m \\ -\frac{1}{2}\bar{L}_m & -\frac{1}{2}\bar{L}_m & L_{ls} + \bar{L}_m \end{bmatrix} = \begin{bmatrix} L_{ss} & -\frac{1}{2}\bar{L}_m & -\frac{1}{2}\bar{L}_m \\ -\frac{1}{2}\bar{L}_m & L_{ss} & -\frac{1}{2}\bar{L}_m \\ -\frac{1}{2}\bar{L}_m & -\frac{1}{2}\bar{L}_m & L_{ss} \end{bmatrix}.$$

The expressions for the flux linkages are

$$\psi_{as} = \left(L_{ls} + \bar{L}_m\right) i_{as} - \frac{1}{2}\bar{L}_m i_{bs} - \frac{1}{2}\bar{L}_m i_{cs} + \psi_m \sin\theta_r,$$

$$\psi_{bs} = -\frac{1}{2}\bar{L}_m i_{as} + \left(L_{ls} + \bar{L}_m\right) i_{bs} - \frac{1}{2}\bar{L}_m i_{cs} + \psi_m \sin\left(\theta_r - \frac{2}{3}\pi\right),$$

$$\psi_{cs} = -\frac{1}{2}\bar{L}_m i_{as} - \frac{1}{2}\bar{L}_m i_{bs} + \left(L_{ls} + \bar{L}_m\right) i_{cs} + \psi_m \sin\left(\theta_r + \frac{2}{3}\pi\right), \tag{6.19}$$

$$\psi_{abcs} = \mathbf{L}_s \mathbf{i}_{abcs} + \psi_m = \begin{bmatrix} L_{ls} + \bar{L}_m & -\frac{1}{2}\bar{L}_m & -\frac{1}{2}\bar{L}_m \\ -\frac{1}{2}\bar{L}_m & L_{ls} + \bar{L}_m & -\frac{1}{2}\bar{L}_m \\ -\frac{1}{2}\bar{L}_m & -\frac{1}{2}\bar{L}_m & L_{ls} + \bar{L}_m \end{bmatrix} \begin{bmatrix} i_{as} \\ i_{bs} \\ i_{cs} \end{bmatrix} + \psi_m \begin{bmatrix} \sin\theta_r \\ \sin\left(\theta_r - \frac{2}{3}\pi\right) \\ \sin\left(\theta_r + \frac{2}{3}\pi\right) \end{bmatrix}.$$

Using (6.18) and (6.19), we have

$$\mathbf{u}_{abcs} = \mathbf{r}_s\mathbf{i}_{abcs} + \frac{d\boldsymbol{\psi}_{abcs}}{dt} = \mathbf{r}_s\mathbf{i}_{abcs} + \mathbf{L}_s\frac{d\mathbf{i}_{abcs}}{dt} + \frac{d\boldsymbol{\psi}_m}{dt}, \quad \frac{d\boldsymbol{\psi}_m}{dt} = \psi_m\begin{bmatrix}\cos\theta_r\omega_r \\ \cos\left(\theta_r - \frac{2}{3}\pi\right)\omega_r \\ \cos\left(\theta_r + \frac{2}{3}\pi\right)\omega_r\end{bmatrix}.$$

Cauchy's form of differential equations can be found by using $\mathbf{L}_s^{-1}$. In particular,

$$\frac{d\mathbf{i}_{abcs}}{dt} = -\mathbf{L}_s^{-1}\mathbf{r}_s\mathbf{i}_{abcs} - \mathbf{L}_s^{-1}\frac{d\boldsymbol{\psi}_m}{dt} + \mathbf{L}_s^{-1}\mathbf{u}_{abcs}.$$

The circuitry-electromagnetic dynamics is

$$\begin{aligned}
\frac{di_{as}}{dt} = &-\frac{r_s\left(2L_{ss}-\bar{L}_m\right)}{2L_{ss}^2 - L_{ss}\bar{L}_m - \bar{L}_m^2}i_{as} - \frac{r_s\bar{L}_m}{2L_{ss}^2 - L_{ss}\bar{L}_m - \bar{L}_m^2}i_{bs} - \frac{r_s\bar{L}_m}{2L_{ss}^2 - L_{ss}\bar{L}_m - \bar{L}_m^2}i_{cs} \\
&-\frac{\psi_m\left(2L_{ss}-\bar{L}_m\right)}{2L_{ss}^2 - L_{ss}\bar{L}_m - \bar{L}_m^2}\omega_r\cos\theta_r - \frac{\psi_m\bar{L}_m}{2L_{ss}^2 - L_{ss}\bar{L}_m - \bar{L}_m^2}\omega_r\cos\left(\theta_r - \frac{2}{3}\pi\right) \\
&-\frac{\psi_m\bar{L}_m}{2L_{ss}^2 - L_{ss}\bar{L}_m - \bar{L}_m^2}\omega_r\cos\left(\theta_r + \frac{2}{3}\pi\right) \\
&+\frac{2L_{ss}-\bar{L}_m}{2L_{ss}^2 - L_{ss}\bar{L}_m - \bar{L}_m^2}u_{as} + \frac{\bar{L}_m}{2L_{ss}^2 - L_{ss}\bar{L}_m - \bar{L}_m^2}u_{bs} + \frac{\bar{L}_m}{2L_{ss}^2 - L_{ss}\bar{L}_m - \bar{L}_m^2}u_{cs}, \\
\frac{di_{bs}}{dt} = &-\frac{r_s\bar{L}_m}{2L_{ss}^2 - L_{ss}\bar{L}_m - \bar{L}_m^2}i_{as} - \frac{r_s\left(2L_{ss}-\bar{L}_m\right)}{2L_{ss}^2 - L_{ss}\bar{L}_m - \bar{L}_m^2}i_{bs} - \frac{r_s\bar{L}_m}{2L_{ss}^2 - L_{ss}\bar{L}_m - \bar{L}_m^2}i_{cs} \\
&-\frac{\psi_m\bar{L}_m}{2L_{ss}^2 - L_{ss}\bar{L}_m - \bar{L}_m^2}\omega_r\cos\theta_r - \frac{\psi_m\left(2L_{ss}-\bar{L}_m\right)}{2L_{ss}^2 - L_{ss}\bar{L}_m - \bar{L}_m^2}\omega_r\cos\left(\theta_r - \frac{2}{3}\pi\right) \\
&-\frac{\psi_m\bar{L}_m}{2L_{ss}^2 - L_{ss}\bar{L}_m - \bar{L}_m^2}\omega_r\cos\left(\theta_r + \frac{2}{3}\pi\right) \\
&+\frac{\bar{L}_m}{2L_{ss}^2 - L_{ss}\bar{L}_m - \bar{L}_m^2}u_{as} + \frac{2L_{ss}-\bar{L}_m}{2L_{ss}^2 - L_{ss}\bar{L}_m - \bar{L}_m^2}u_{bs} + \frac{\bar{L}_m}{2L_{ss}^2 - L_{ss}\bar{L}_m - \bar{L}_m^2}u_{cs}, \\
\frac{di_{cs}}{dt} = &-\frac{r_s\bar{L}_m}{2L_{ss}^2 - L_{ss}\bar{L}_m - \bar{L}_m^2}i_{as} - \frac{r_s\bar{L}_m}{2L_{ss}^2 - L_{ss}\bar{L}_m - \bar{L}_m^2}i_{bs} - \frac{r_s\left(2L_{ss}-\bar{L}_m\right)}{2L_{ss}^2 - L_{ss}\bar{L}_m - \bar{L}_m^2}i_{cs} \\
&-\frac{\psi_m\bar{L}_m}{2L_{ss}^2 - L_{ss}\bar{L}_m - \bar{L}_m^2}\omega_r\cos\theta_r - \frac{\psi_m\bar{L}_m}{2L_{ss}^2 - L_{ss}\bar{L}_m - \bar{L}_m^2}\omega_r\cos\left(\theta_r - \frac{2}{3}\pi\right) \\
&-\frac{\psi_m\left(2L_{ss}-\bar{L}_m\right)}{2L_{ss}^2 - L_{ss}\bar{L}_m - \bar{L}_m^2}\omega_r\cos\left(\theta_r + \frac{2}{3}\pi\right) \\
&+\frac{\bar{L}_m}{2L_{ss}^2 - L_{ss}\bar{L}_m - \bar{L}_m^2}u_{as} + \frac{\bar{L}_m}{2L_{ss}^2 - L_{ss}\bar{L}_m - \bar{L}_m^2}u_{bs} + \frac{2L_{ss}-\bar{L}_m}{2L_{ss}^2 - L_{ss}\bar{L}_m - \bar{L}_m^2}u_{cs}.
\end{aligned} \tag{6.20}$$

The electromagnetic torque is found using the coenergy

$$W_c = \frac{1}{2}\begin{bmatrix} i_{as} & i_{bs} & i_{cs} \end{bmatrix} \mathbf{L}_s \begin{bmatrix} i_{as} \\ i_{bs} \\ i_{cs} \end{bmatrix} + \begin{bmatrix} i_{as} & i_{bs} & i_{cs} \end{bmatrix} \begin{bmatrix} \psi_m \sin\theta_r \\ \psi_m \sin\left(\theta_r - \frac{2}{3}\pi\right) \\ \psi_m \sin\left(\theta_r + \frac{2}{3}\pi\right) \end{bmatrix} + W_{PM}, \quad \mathbf{L}_s = \begin{bmatrix} L_{ss} & -\frac{1}{2}\bar{L}_m & -\frac{1}{2}\bar{L}_m \\ -\frac{1}{2}\bar{L}_m & L_{ss} & -\frac{1}{2}\bar{L}_m \\ -\frac{1}{2}\bar{L}_m & -\frac{1}{2}\bar{L}_m & L_{ss} \end{bmatrix},$$

where W_{PM} is the energy stored in permanent magnets.

For round-rotor synchronous machines, $\mathbf{L}_s$ and W_{PM} are not functions of θ_r. One obtains the electromagnetic torque for P-pole three-phase permanent-magnet synchronous motors

$$T_e = \frac{P}{2}\frac{\partial W_c}{\partial \theta_r} = \frac{P\psi_m}{2}\left[i_{as}\cos\theta_r + i_{bs}\cos\left(\theta_r - \frac{2}{3}\pi\right) + i_{cs}\cos\left(\theta_r + \frac{2}{3}\pi\right)\right]. \tag{6.21}$$

With T_e found as (6.21), the Newton second law yields

$$\frac{d\omega_{rm}}{dt} = \frac{P\psi_m}{2J}\left[i_{as}\cos\theta_r + i_{bs}\cos\left(\theta_r - \frac{2}{3}\pi\right) + i_{cs}\cos\left(\theta_r + \frac{2}{3}\pi\right)\right] - \frac{B_m}{J}\omega_{rm} - \frac{1}{J}T_L,$$

$$\frac{d\theta_{rm}}{dt} = \omega_{rm}.$$

The electrical angular velocity ω_r and displacement θ_r are related to the mechanical angular velocity and displacement,

$$\omega_{rm} = \frac{2}{P}\omega_r \quad \text{and} \quad \theta_{rm} = \frac{2}{P}\theta_r.$$

The differential equations of the *torsional–mechanical* dynamics are

$$\frac{d\omega_r}{dt} = \frac{P^2\psi_m}{4J}\left[i_{as}\cos\theta_r + i_{bs}\cos\left(\theta_r - \frac{2}{3}\pi\right) + i_{cs}\cos\left(\theta_r + \frac{2}{3}\pi\right)\right] - \frac{B_m}{J}\omega_r - \frac{P}{2J}T_L,$$

$$\frac{d\theta_r}{dt} = \omega_r. \tag{6.22}$$

A nonlinear mathematical model of permanent-magnet synchronous motors in Cauchy's form is given by a system of five differential equations (6.20) and (6.22).

To control motors, one regulates the *abs* phase currents or voltages. Neglecting the viscous friction coefficient, the analysis of Newton's second law

$$J\frac{d\omega_{rm}}{dt} = T_{em} - T_L, \quad T_{em} = \frac{P}{2}T_e$$

indicates that: (1) The angular velocity ω_{rm} increases (motor accelerates) if $T_{em} > T_L$; (2) The angular velocity ω_{rm} decreases (motor decelerates) if $T_{em} < T_L$; (3) The angular velocity ω_{rm} is constant if $T_e = T_L$. To regulate motion devices, the electromagnetic torque (6.21) must be changed. A balanced three-phase current set is

$$i_{as} = \sqrt{2} i_M \cos\theta_r, \quad i_{bs} = \sqrt{2} i_M \cos\left(\theta_r - \frac{2}{3}\pi\right), \quad i_{cs} = \sqrt{2} i_M \cos\left(\theta_r + \frac{2}{3}\pi\right).$$

From the trigonometric identity

$$\cos^2\theta_r + \cos^2\left(\theta_r - \frac{2}{3}\pi\right) + \cos^2\left(\theta_r + \frac{2}{3}\pi\right) = \frac{3}{2},$$

one yields

$$T_e = \frac{P\psi_m}{2}\sqrt{2} i_M \left(\cos^2\theta_r + \cos^2\left(\theta_r - \frac{2}{3}\pi\right) + \cos^2\left(\theta_r + \frac{2}{3}\pi\right)\right) = \frac{3P\psi_m}{2\sqrt{2}} i_M.$$

The angular displacement θ_r is measured by the Hall-effect sensors. If the PWM amplifiers are used, one changes the magnitude u_M of the phase voltages u_{as}, u_{bs}, and u_{cs}. The balanced voltage set is

$$u_{as} = \sqrt{2} u_M \cos\theta_r, \quad u_{bs} = \sqrt{2} u_M \cos\left(\theta_r - \frac{2}{3}\pi\right), \quad u_{cs} = \sqrt{2} u_M \cos\left(\theta_r + \frac{2}{3}\pi\right).$$

The balanced voltage set can be implemented by the "control logic", which regulates the switching of power transistors. The angular displacement is measured by the Hall-effect sensors. The motor windings *as*, *bs* (for two-phase motors) and *as*, *bs*, *cs* (for three-phase motors) are connected to the power stage outputs. Synchronous electric machines are designed with different rated power, voltage, current, torque, speed, etc. The matching high-performance PWM power amplifiers are available. For ~100 W rated synchronous motors, the schematics of a B15A8 servo amplifier (20–80 V, 7.5 A continuous, 15 A peak, 2.5 kHz bandwidth, 129 × 76 × 25 mm dimensions) is documented in Figure 6.18a. The motor phase windings are connected to P2-1, P2-2, and P2-3. One connects the Hall-effect sensor outputs to P1-12, P1-13, and P1-14. The "control logic" uses the measured rotor angular displacement to generate the PWM phase voltages u_{as}, u_{bs}, and u_{cs} by driving the MOSFETs. The proportional–integral analog controller is used. The reference (command) voltage is supplied to P1-4. The tachometer voltage (proportional to the motor angular velocity) is supplied to P1-6. The reference and measured angular velocities are compared to obtain the tracking error $e(t)$. This $e(t)$ is used by the analog proportional–integral controller to develop the control signals that turn the MOSFETs *on* and *off*. One can change the proportional and integral feedback gains by adjusting the potentiometers. Figure 6.18b documents high-performance machines with the PWM controllers/drivers. The evaluation boards may be effectively used. For example, the EVAL6235 with the L6235 driver is designed for three-phase synchronous motors.

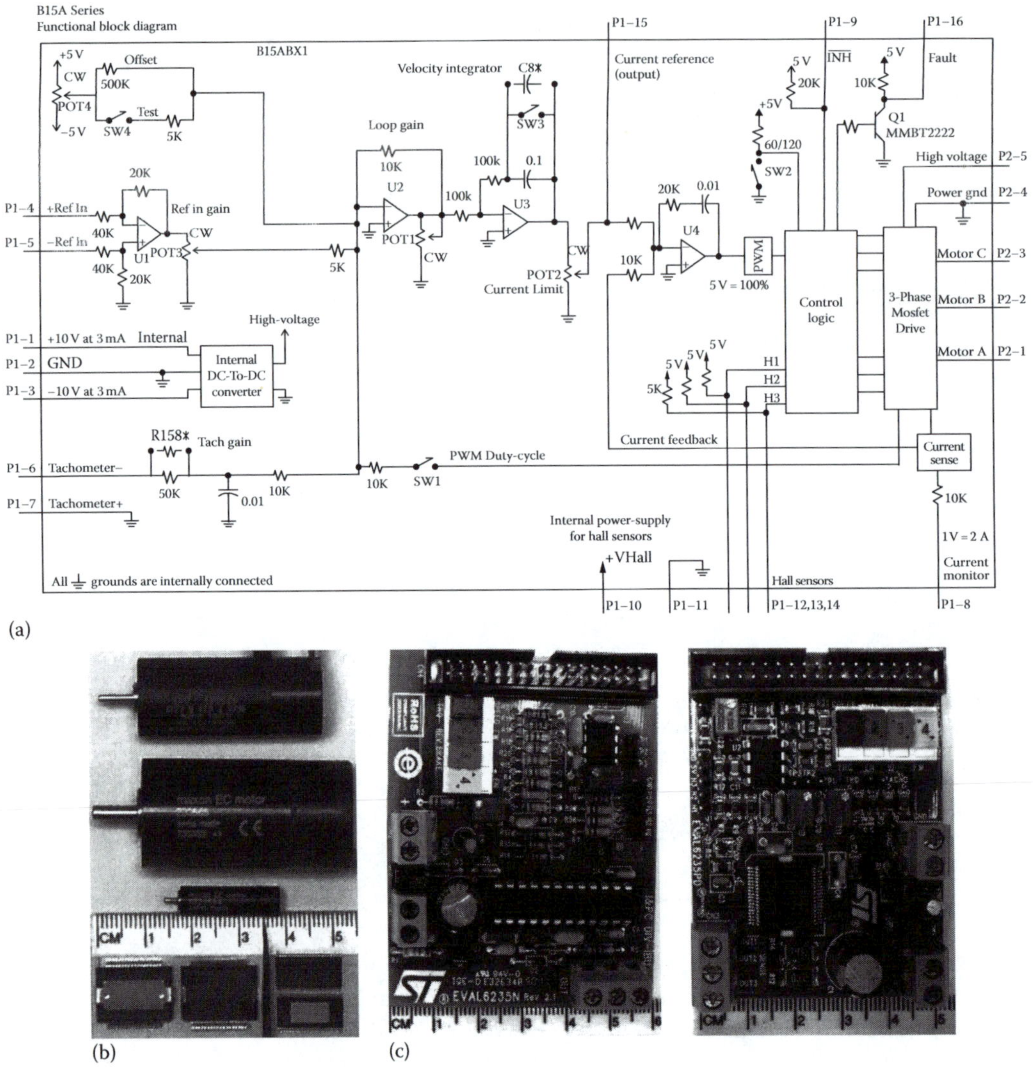

FIGURE 6.18 (a) B15A8 PWM servo amplifier; (From Lyshevski, S.E., *Electromechanical Systems, Electric Machines, and Applied Mechatronics*, CRC Press, Boca Raton, FL, 1999; Courtesy of Advanced Motion Controls, Camarillo, CA, www.a-m-c.com.)
(b) Images of three high-performance radial topology permanent-magnet synchronous motors with Hall-effect sensors—(1) Faulhaber 1628 024B motor with SmCo magnets (Ø15 mm, two-pole, 17 W, 24 V, 0.5 A rated, 1.5 A peak, 3000 rad/sec, 7000 rad/sec maximum, 3.3 mN-m rated, 11 mN-m peak, 15.2 ohm, 0.517 mH, 0.000000054 kg-m^2, up to 70% efficiency. The motor is equipped with the IE2-512 magnetic encoder (512 pulses per revolution); (2) Maxon EC 200685 motor (Ø22 mm, two-pole, 12 V, 2500 rad/sec, 5.75 A and 21 mN-m rated, ~85% efficiency); (3) Maxon EC motor (Ø6 mm, two-pole, 12 V, 4500 rad/sec, 0.27 A and 0.41 mN-m rated, ~65% efficiency). Images of DRV8312DDWR (52.5 V and 3.5 A rated) and DRV8332DKDR (52.5 V and 8 A rated) PWM controllers/drivers;
(c) EVAL6235N and EVAL6235PD evaluation board with the L6235 DMOS driver for three-phase synchronous motors, 52 V, 2.8 A (rated), 5.6 A (peak), up to 100 kHz. The double-diffused metal-oxide-semiconductor (DMOS) power FETs are used. The bipolar-CMOS-DMOS STMicroelectronics technology combines isolated DMOS power transistors with CMOS and bipolar circuits on the same chip. The L6235 has a three-phase DMOS bridge. The current and voltage controllers use the current sensors as well as Hall-effect sensors displacement measurements. The closed-loop ensures control capabilities.

Small permanent-magnet synchronous motors (from mW to ~10 W) are used in rotating and positioning stages, hard drives, robots, appliances, etc. Permanent-magnet ~10 W rated synchronous motors can be controlled by the 30 V, 1 A MC33035 PWM driver [4], the L6235 STMicroelectronics driver, etc. The phase voltages u_{as}, u_{bs}, and u_{cs} are obtained using the rotor angular displacement measured by the Hall-effect sensors, as reported in Figures 6.19a and b. The MC33035 can be used to drive power MOSFETs, see Figure 6.19b. The representative block diagrams provide the functional schematics. The three-phase, six-step full-wave converter topology are implemented. The closed-loop configuration is realized using proportional or proportional-integral controllers. The "error amplifier" yields the error. The reader is referred to the Motorola application notes for detailed information. The images of the PWM Motorola, STMicroelectronics, and Texas Instrument drivers and evaluation boards are documented in Figures 6.2 and 6.18.

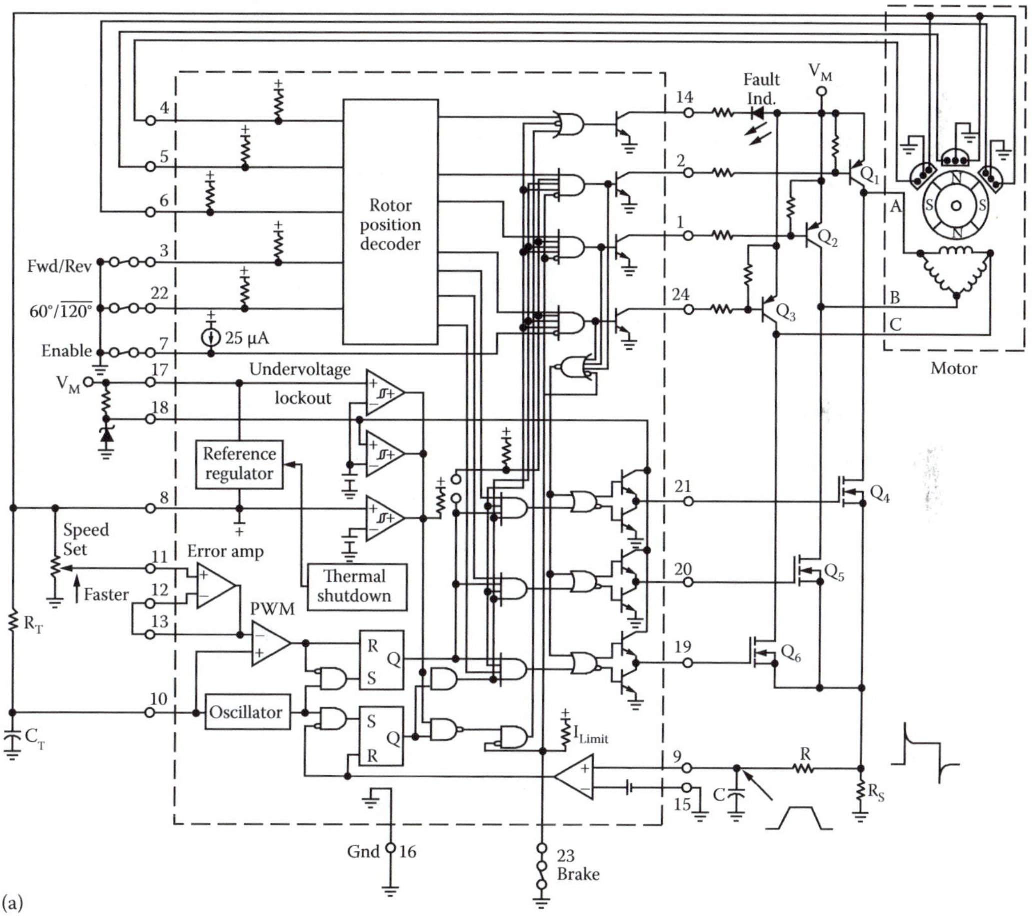

(a)

FIGURE 6.19 Schematics of (a) the MC33035 *Brushless DC Motor Controller* to drive and control synchronous minimotors; *(Continued)*

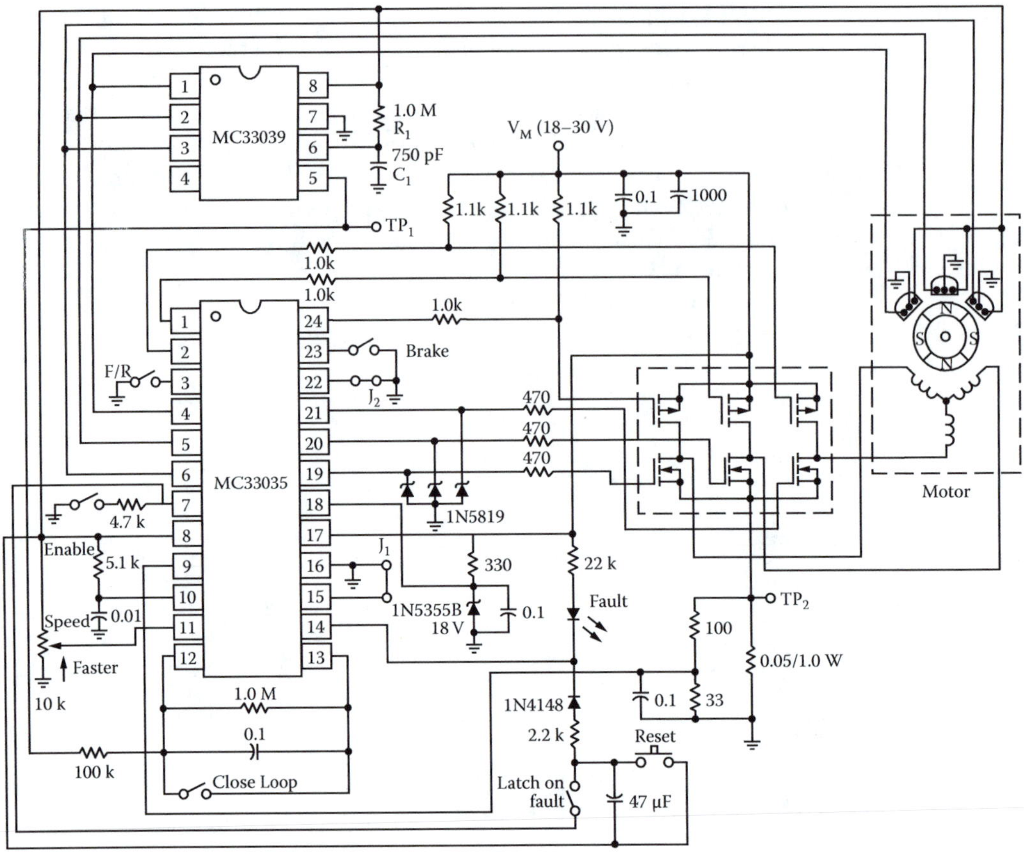

(b)

FIGURE 6.19 (*Continued*) (b) The MC33039 and MC33035 controller/driver to drive the power MOSFETs controlling synchronous motors. (From Lyshevski, S.E., *Electromechanical Systems, Electric Machines, and Applied Mechatronics*, CRC Press, Boca Raton, FL, 1999; Copyright of Motorola. With permission.)

6.4.2 Lagrange Equations of Motion and the Dynamics of Permanent-Magnet Synchronous Motors

We derived mathematical models for synchronous motors using Kirchhoff's voltage law, Faraday's law of electromagnetic induction, Newtonian mechanics, Lorenz force, and the coenergy concept. The coenergy was used to find the electromagnetic torque.

We apply the Lagrange equations

$$\frac{d}{dt}\left(\frac{\partial \Gamma}{\partial \dot{q}_i}\right) - \frac{\partial \Gamma}{\partial q_i} + \frac{\partial D}{\partial \dot{q}_i} + \frac{\partial \Pi}{\partial q_i} = Q_i,$$

where the kinetic Γ, potential Π, and dissipation D energies are found by using the *generalized* coordinates q_i and *generalized* forces Q_i.

The *generalized* coordinates are the electric charges in the *abc* stator windings

$$q_1 = \int i_{as}\,dt,\ \dot{q}_1 = i_{as},\quad q_2 = \int i_{bs}\,dt,\ \dot{q}_2 = i_{bs},\quad q_3 = \int i_{cs}\,dt,\ \dot{q}_3 = i_{cs},$$

and the angular displacement $q_4 = \theta_r$, $\dot{q}_4 = \omega_r$.

The *generalized* forces are the applied voltages to the *abc* windings $Q_1 = u_{as}$, $Q_2 = u_{bs}$, $Q_3 = u_{cs}$ and the load torque $Q_4 = -T_L$.

One yields four Lagrange equations

$$\frac{d}{dt}\left(\frac{\partial \Gamma}{\partial \dot{q}_1}\right) - \frac{\partial \Gamma}{\partial q_1} + \frac{\partial D}{\partial \dot{q}_1} + \frac{\partial \Pi}{\partial q_1} = Q_1, \quad \frac{d}{dt}\left(\frac{\partial \Gamma}{\partial \dot{q}_2}\right) - \frac{\partial \Gamma}{\partial q_2} + \frac{\partial D}{\partial \dot{q}_2} + \frac{\partial \Pi}{\partial q_2} = Q_2,$$

$$\frac{d}{dt}\left(\frac{\partial \Gamma}{\partial \dot{q}_3}\right) - \frac{\partial \Gamma}{\partial q_3} + \frac{\partial D}{\partial \dot{q}_3} + \frac{\partial \Pi}{\partial q_3} = Q_3, \quad \frac{d}{dt}\left(\frac{\partial \Gamma}{\partial \dot{q}_4}\right) - \frac{\partial \Gamma}{\partial q_4} + \frac{\partial D}{\partial \dot{q}_4} + \frac{\partial \Pi}{\partial q_4} = Q_4.$$

The total kinetic energy includes kinetic energies of electrical and mechanical systems

$$\Gamma = \Gamma_E + \Gamma_M = \frac{1}{2}L_{asas}\dot{q}_1^2 + \frac{1}{2}\left(L_{asbs} + L_{bsas}\right)\dot{q}_1\dot{q}_2 + \frac{1}{2}\left(L_{ascs} + L_{csas}\right)\dot{q}_1\dot{q}_3 + \frac{1}{2}L_{bsbs}\dot{q}_2^2$$

$$+\frac{1}{2}\left(L_{bscs} + L_{csbs}\right)\dot{q}_2\dot{q}_3 + \frac{1}{2}L_{cscs}\dot{q}_3^2 + \psi_m\dot{q}_1 \sin q_4 + \psi_m\dot{q}_2 \sin\left(q_4 - \frac{2}{3}\pi\right) + \psi_m\dot{q}_3 \sin\left(q_4 + \frac{2}{3}\pi\right) + \frac{1}{2}J\dot{q}_4^2.$$

Therefore,

$$\frac{\partial \Gamma}{\partial q_1} = 0, \quad \frac{\partial \Gamma}{\partial \dot{q}_1} = L_{asas}\dot{q}_1 + \frac{1}{2}\left(L_{asbs} + L_{bsas}\right)\dot{q}_2 + \frac{1}{2}\left(L_{ascs} + L_{csas}\right)\dot{q}_3 + \psi_m \sin q_4,$$

$$\frac{\partial \Gamma}{\partial q_2} = 0, \quad \frac{\partial \Gamma}{\partial \dot{q}_2} = \frac{1}{2}\left(L_{asbs} + L_{bsas}\right)\dot{q}_1 + L_{bsbs}\dot{q}_2 + \frac{1}{2}\left(L_{bscs} + L_{csbs}\right)\dot{q}_3 + \psi_m \sin\left(q_4 - \frac{2}{3}\pi\right),$$

$$\frac{\partial \Gamma}{\partial q_3} = 0, \quad \frac{\partial \Gamma}{\partial \dot{q}_3} = \frac{1}{2}\left(L_{ascs} + L_{csas}\right)\dot{q}_1 + \frac{1}{2}\left(L_{bscs} + L_{csbs}\right)\dot{q}_2 + L_{cscs}\dot{q}_3 + \psi_m \sin\left(q_4 + \frac{2}{3}\pi\right),$$

$$\frac{\partial \Gamma}{\partial q_4} = \psi_m\dot{q}_1 \cos q_4 + \psi_m\dot{q}_2 \cos\left(q_4 - \frac{2}{3}\pi\right) + \psi_m\dot{q}_3 \cos\left(q_4 + \frac{2}{3}\pi\right), \quad \frac{\partial \Gamma}{\partial \dot{q}_4} = J\dot{q}_4.$$

The total potential energy is $\Pi = 0$.

The total dissipated energy is a sum of the heat energy dissipated by the electrical system and the heat energy dissipated by the mechanical system. That is,

$$D = D_E + D_M = \frac{1}{2}\left(r_s\dot{q}_1^2 + r_s\dot{q}_2^2 + r_s\dot{q}_3^2 + B_m\dot{q}_4^2\right).$$

The differentiation of D with respect to the *generalized* coordinates yields

$$\frac{\partial D}{\partial \dot{q}_1} = r_s\dot{q}_1, \quad \frac{\partial D}{\partial \dot{q}_2} = r_s\dot{q}_2, \quad \frac{\partial D}{\partial \dot{q}_3} = r_s\dot{q}_3, \quad \text{and} \quad \frac{\partial D}{\partial \dot{q}_4} = B_m\dot{q}_4.$$

The Lagrange equations result in the following differential equations

$$L_{asas}\frac{di_{as}}{dt}+\frac{1}{2}\left(L_{asbs}+L_{bsas}\right)\frac{di_{bs}}{dt}+\frac{1}{2}\left(L_{ascs}+L_{csas}\right)\frac{di_{cs}}{dt}+\psi_m\omega_r\cos\theta_r+r_si_{as}=u_{as},$$

$$\frac{1}{2}\left(L_{asbs}+L_{bsas}\right)\frac{di_{as}}{dt}+L_{bsbs}\frac{di_{bs}}{dt}+\frac{1}{2}\left(L_{bscs}+L_{csbs}\right)\frac{di_{cs}}{dt}+\psi_m\omega_r\cos\left(\theta_r-\frac{2}{3}\pi\right)+r_si_{bs}=u_{bs},$$

$$\frac{1}{2}\left(L_{ascs}+L_{csas}\right)\frac{di_{as}}{dt}+\frac{1}{2}\left(L_{bscs}+L_{csbs}\right)\frac{di_{bs}}{dt}+L_{cscs}\frac{di_{cs}}{dt}+\psi_m\omega_r\cos\left(\theta_r+\frac{2}{3}\pi\right)+r_si_{cs}=u_{cs},$$

$$J\frac{d^2\theta_r}{dt^2}-\psi_mi_{as}\cos\theta_r-\psi_mi_{bs}\cos\left(\theta_r-\frac{2}{3}\pi\right)-\psi_mi_{cs}\cos\left(\theta_r+\frac{2}{3}\pi\right)+B_m\frac{d\theta_r}{dt}=-T_L.$$

For round-rotor permanent-magnet synchronous motors, $\Re_{mq}=\Re_{md}$ and $L_{mq}=L_{md}$. Thus,

$$\bar{L}_m=\frac{2N_s^2}{3\Re_{mq}}=\frac{2N_s^2}{3\Re_{md}}\quad\text{and}\quad L_{\Delta m}=0.$$

We obtain

$$\left(L_{ls}+\bar{L}_m\right)\frac{di_{as}}{dt}-\frac{1}{2}\bar{L}_m\frac{di_{bs}}{dt}-\frac{1}{2}\bar{L}_m\frac{di_{cs}}{dt}+\psi_m\omega_r\cos\theta_r+r_si_{as}=u_{as},$$

$$-\frac{1}{2}\bar{L}_m\frac{di_{as}}{dt}+\left(L_{ls}+\bar{L}_m\right)\frac{di_{bs}}{dt}-\frac{1}{2}\bar{L}_m\frac{di_{cs}}{dt}+\psi_m\omega_r\cos\left(\theta_r-\frac{2}{3}\pi\right)+r_si_{bs}=u_{bs},$$

$$-\frac{1}{2}\bar{L}_m\frac{di_{as}}{dt}-\frac{1}{2}\bar{L}_m\frac{di_{bs}}{dt}+\left(L_{ls}+\bar{L}_m\right)\frac{di_{cs}}{dt}+\psi_m\omega_r\cos\left(\theta_r+\frac{2}{3}\pi\right)+r_si_{cs}=u_{cs},$$

$$J\frac{d\omega_r}{dt}+B_m\omega_r-\psi_m\left[i_{as}\cos\theta_r+i_{bs}\cos\left(\theta_r-\frac{2}{3}\pi\right)+i_{cs}\cos\left(\theta_r+\frac{2}{3}\pi\right)\right]=-T_L,$$

$$\frac{d\theta_r}{dt}=\omega_r,$$

where $L_{ss}=L_{ls}+\bar{L}_m$.

In equation

$$\frac{d}{dt}\left(\frac{\partial\Gamma}{\partial\dot{q}_4}\right)-\frac{\partial\Gamma}{\partial q_4}+\frac{\partial D}{\partial\dot{q}_4}+\frac{\partial\Pi}{\partial q_4}=Q_4$$

we found

$$\frac{\partial\Gamma}{\partial q_4}=\psi_m\dot{q}_1\cos q_4+\psi_m\dot{q}_2\cos\left(q_4-\frac{2}{3}\pi\right)+\psi_m\dot{q}_3\cos\left(q_4+\frac{2}{3}\pi\right).$$

Hence, the electromagnetic torque is

$$T_e = \frac{\partial W_c}{\partial \theta_r} = \frac{\partial \Gamma}{\partial q_4} = \psi_m \left[i_{as} \cos\theta_r + i_{bs} \cos\left(\theta_r - \frac{2}{3}\pi\right) + i_{cs} \cos\left(\theta_r + \frac{2}{3}\pi\right)\right].$$

For P-pole permanent-magnet synchronous motors, differential equations in Cauchy's form, as given by (6.20) and (6.22), result.

6.4.3 Three-Phase Permanent-Magnet Synchronous Generators

For synchronous generators, shown in Figure 6.17a, the mathematical model can be found. The image of a round-rotor permanent-magnet synchronous generator is depicted in Figure 6.17b. If rotated by a prime mover, as shown in Figures 6.17a and c, the phase voltages are the induced *emfs* in the *as*, *bs*, and *cs* phases. Using the direction of currents depicted in Figure 6.17a, we have

$$\mathbf{u}_{abcs} = -\mathbf{r}_s \mathbf{i}_{abcs} + \frac{d\psi_{abcs}}{dt}, \quad \begin{bmatrix} u_{as} \\ u_{bs} \\ u_{cs} \end{bmatrix} = -\begin{bmatrix} r_s & 0 & 0 \\ 0 & r_s & 0 \\ 0 & 0 & r_s \end{bmatrix}\begin{bmatrix} i_{as} \\ i_{bs} \\ i_{cs} \end{bmatrix} + \begin{bmatrix} \frac{d\psi_{as}}{dt} \\ \frac{d\psi_{bs}}{dt} \\ \frac{d\psi_{cs}}{dt} \end{bmatrix}, \tag{6.23}$$

$$\psi_{abcs} = -\mathbf{L}_s \mathbf{i}_{abcs} + \psi_m = -\begin{bmatrix} L_{ls} + \bar{L}_{mr} & -\frac{1}{2}\bar{L}_m & -\frac{1}{2}\bar{L}_m \\ -\frac{1}{2}\bar{L}_m & L_{ls} + \bar{L}_m & -\frac{1}{2}\bar{L}_m \\ -\frac{1}{2}\bar{L}_m & -\frac{1}{2}\bar{L}_m & L_{ls} + \bar{L}_m \end{bmatrix}\begin{bmatrix} i_{as} \\ i_{bs} \\ i_{cs} \end{bmatrix} + \psi_m \begin{bmatrix} \sin\theta_r \\ \sin\left(\theta_r - \frac{2}{3}\pi\right) \\ \sin\left(\theta_r + \frac{2}{3}\pi\right) \end{bmatrix}$$

$$= -\begin{bmatrix} L_{ss} & -\frac{1}{2}\bar{L}_m & -\frac{1}{2}\bar{L}_m \\ -\frac{1}{2}\bar{L}_m & L_{ss} & -\frac{1}{2}\bar{L}_m \\ -\frac{1}{2}\bar{L}_m & -\frac{1}{2}\bar{L}_m & L_{ss} \end{bmatrix}\begin{bmatrix} i_{as} \\ i_{bs} \\ i_{cs} \end{bmatrix} + \psi_m \begin{bmatrix} \sin\theta_r \\ \sin\left(\theta_r - \frac{2}{3}\pi\right) \\ \sin\left(\theta_r + \frac{2}{3}\pi\right) \end{bmatrix}.$$

Using the Newton second law of motion

$$J\frac{d^2\theta_{rm}}{dt^2} = -T_e - B_m\omega_{rm} + T_{PM},$$

one obtains

$$\frac{d\omega_{rm}}{dt} = \frac{1}{J}\left(-T_e - B_m\omega_{rm} + T_{PM}\right), \quad \frac{d\theta_{rm}}{dt} = \omega_{rm},$$

where the loading generator electromagnetic torque T_e is given by (6.21). Using the results of Section 6.4.1, we derive differential equations

$$\frac{di_{as}}{dt}=-\frac{r_s\left(2L_{ss}-\bar{L}_m\right)}{2L_{ss}^2-L_{ss}\bar{L}_m-\bar{L}_m^2}i_{as}-\frac{r_s\bar{L}_m}{2L_{ss}^2-L_{ss}\bar{L}_m-\bar{L}_m^2}i_{bs}-\frac{r_s\bar{L}_m}{2L_{ss}^2-L_{ss}\bar{L}_m-\bar{L}_m^2}i_{cs}$$

$$+\frac{\psi_m\left(2L_{ss}-\bar{L}_m\right)}{2L_{ss}^2-L_{ss}\bar{L}_m-\bar{L}_m^2}\omega_r\cos\theta_r+\frac{\psi_m\bar{L}_m}{2L_{ss}^2-L_{ss}\bar{L}_m-\bar{L}_m^2}\omega_r\cos\left(\theta_r-\frac{2}{3}\pi\right)+\frac{\psi_m\bar{L}_m}{2L_{ss}^2-L_{ss}\bar{L}_m-\bar{L}_m^2}\omega_r\cos\left(\theta_r+\frac{2}{3}\pi\right)$$

$$-\frac{2L_{ss}-\bar{L}_m}{2L_{ss}^2-L_{ss}\bar{L}_m-\bar{L}_m^2}u_{as}-\frac{\bar{L}_m}{2L_{ss}^2-L_{ss}\bar{L}_m-\bar{L}_m^2}u_{bs}-\frac{\bar{L}_m}{2L_{ss}^2-L_{ss}\bar{L}_m-\bar{L}_m^2}u_{cs},$$

$$\frac{di_{bs}}{dt}=-\frac{r_s\bar{L}_m}{2L_{ss}^2-L_{ss}\bar{L}_m-\bar{L}_m^2}i_{as}-\frac{r_s\left(2L_{ss}-\bar{L}_m\right)}{2L_{ss}^2-L_{ss}\bar{L}_m-\bar{L}_m^2}i_{bs}-\frac{r_s\bar{L}_m}{2L_{ss}^2-L_{ss}\bar{L}_m-\bar{L}_m^2}i_{cs}$$

$$+\frac{\psi_m\bar{L}_m}{2L_{ss}^2-L_{ss}\bar{L}_m-\bar{L}_m^2}\omega_r\cos\theta_r+\frac{\psi_m\left(2L_{ss}-\bar{L}_m\right)}{2L_{ss}^2-L_{ss}\bar{L}_m-\bar{L}_m^2}\omega_r\cos\left(\theta_r-\frac{2}{3}\pi\right)+\frac{\psi_m\bar{L}_m}{2L_{ss}^2-L_{ss}\bar{L}_m-\bar{L}_m^2}\omega_r\cos\left(\theta_r+\frac{2}{3}\pi\right)$$

$$-\frac{\bar{L}_m}{2L_{ss}^2-L_{ss}\bar{L}_m-\bar{L}_m^2}u_{as}-\frac{2L_{ss}-\bar{L}_m}{2L_{ss}^2-L_{ss}\bar{L}_m-\bar{L}_m^2}u_{bs}-\frac{\bar{L}_m}{2L_{ss}^2-L_{ss}\bar{L}_m-\bar{L}_m^2}u_{cs},$$

$$\frac{di_{cs}}{dt}=-\frac{r_s\bar{L}_m}{2L_{ss}^2-L_{ss}\bar{L}_m-\bar{L}_m^2}i_{as}-\frac{r_s\bar{L}_m}{2L_{ss}^2-L_{ss}\bar{L}_m-\bar{L}_m^2}i_{bs}-\frac{r_s\left(2L_{ss}-\bar{L}_m\right)}{2L_{ss}^2-L_{ss}\bar{L}_m-\bar{L}_m^2}i_{cs}$$

$$+\frac{\psi_m\bar{L}_m}{2L_{ss}^2-L_{ss}\bar{L}_m-\bar{L}_m^2}\omega_r\cos\theta_r+\frac{\psi_m\bar{L}_m}{2L_{ss}^2-L_{ss}\bar{L}_m-\bar{L}_m^2}\omega_r\cos\left(\theta_r-\frac{2}{3}\pi\right)+\frac{\psi_m\left(2L_{ss}-\bar{L}_m\right)}{2L_{ss}^2-L_{ss}\bar{L}_m-\bar{L}_m^2}\omega_r\cos\left(\theta_r+\frac{2}{3}\pi\right)$$

$$-\frac{\bar{L}_m}{2L_{ss}^2-L_{ss}\bar{L}_m-\bar{L}_m^2}u_{as}-\frac{\bar{L}_m}{2L_{ss}^2-L_{ss}\bar{L}_m-\bar{L}_m^2}u_{bs}-\frac{2L_{ss}-\bar{L}_m}{2L_{ss}^2-L_{ss}\bar{L}_m-\bar{L}_m^2}u_{cs},$$

$$\frac{d\omega_r}{dt}=-\frac{P^2\psi_m}{4J}\left(i_{as}\cos\theta_r+i_{bs}\cos\left(\theta_r-\frac{2}{3}\pi\right)+i_{cs}\cos\left(\theta_r+\frac{2}{3}\pi\right)\right)-\frac{B_m}{J}\omega_r+\frac{P}{2J}T_{PM},$$

$$\frac{d\theta_r}{dt}=\omega_r. \tag{6.24}$$

For a balanced Y-Y source-load system with the three-phase load impedances $\mathbf{Z}_{Las}=\mathbf{Z}_{Lbs}=\mathbf{Z}_{Lcs}=R_L$, using (6.24), the state-space mathematical model of a synchronous generator with a balanced Y-load is

$$\begin{bmatrix} \frac{di_{as}}{dt} \\ \frac{di_{bs}}{dt} \\ \frac{di_{cs}}{dt} \\ \frac{d\omega_r}{dt} \\ \frac{d\theta_r}{dt} \end{bmatrix} = \begin{bmatrix} -\frac{(r_s+R_L)(2L_{ss}-\bar{L}_m)}{2L_{ss}^2-L_{ss}\bar{L}_m-\bar{L}_m^2} & -\frac{(r_s+R_L)\bar{L}_m}{2L_{ss}^2-L_{ss}\bar{L}_m-\bar{L}_m^2} & -\frac{(r_s+R_L)\bar{L}_m}{2L_{ss}^2-L_{ss}\bar{L}_m-\bar{L}_m^2} & 0 & 0 \\ -\frac{(r_s+R_L)\bar{L}_m}{2L_{ss}^2-L_{ss}\bar{L}_m-\bar{L}_m^2} & -\frac{(r_s+R_L)(2L_{ss}-\bar{L}_m)}{2L_{ss}^2-L_{ss}\bar{L}_m-\bar{L}_m^2} & -\frac{(r_s+R_L)\bar{L}_m}{2L_{ss}^2-L_{ss}\bar{L}_m-\bar{L}_m^2} & 0 & 0 \\ -\frac{(r_s+R_L)\bar{L}_m}{2L_{ss}^2-L_{ss}\bar{L}_m-\bar{L}_m^2} & -\frac{(r_s+R_L)\bar{L}_m}{2L_{ss}^2-L_{ss}\bar{L}_m-\bar{L}_m^2} & -\frac{(r_s+R_L)(2L_{ss}-\bar{L}_m)}{2L_{ss}^2-L_{ss}\bar{L}_m-\bar{L}_m^2} & 0 & 0 \\ 0 & 0 & 0 & -\frac{B_m}{J} & 0 \\ 0 & 0 & 0 & 1 & 0 \end{bmatrix} \begin{bmatrix} i_{as} \\ i_{bs} \\ i_{cs} \\ \omega_r \\ \theta_r \end{bmatrix}$$

$$+ \begin{bmatrix} \frac{\psi_m(2L_{ss}-\bar{L}_m)}{2L_{ss}^2-L_{ss}\bar{L}_m-\bar{L}_m^2}\omega_r & \frac{\psi_m\bar{L}_m}{2L_{ss}^2-L_{ss}\bar{L}_m-\bar{L}_m^2}\omega_r & \frac{\psi_m\bar{L}_m}{2L_{ss}^2-L_{ss}\bar{L}_m-\bar{L}_m^2}\omega_r \\ \frac{\psi_m\bar{L}_m}{2L_{ss}^2-L_{ss}\bar{L}_m-\bar{L}_m^2}\omega_r & \frac{\psi_m(2L_{ss}-\bar{L}_m)}{2L_{ss}^2-L_{ss}\bar{L}_m-\bar{L}_m^2}\omega_r & \frac{\psi_m\bar{L}_m}{2L_{ss}^2-L_{ss}\bar{L}_m-\bar{L}_m^2}\omega_r \\ \frac{\psi_m\bar{L}_m}{2L_{ss}^2-L_{ss}\bar{L}_m-\bar{L}_m^2}\omega_r & \frac{\psi_m\bar{L}_m}{2L_{ss}^2-L_{ss}\bar{L}_m-\bar{L}_m^2}\omega_r & \frac{\psi_m(2L_{ss}-\bar{L}_m)}{2L_{ss}^2-L_{ss}\bar{L}_m-\bar{L}_m^2}\omega_r \\ -\frac{P^2\psi_m}{4J}i_{as} & -\frac{P^2\psi_m}{4J}i_{bs} & -\frac{P^2\psi_m}{4J}i_{cs} \\ 0 & 0 & 0 \end{bmatrix} \begin{bmatrix} \cos\theta_r \\ \cos\left(\theta_r-\frac{2}{3}\pi\right) \\ \cos\left(\theta_r+\frac{2}{3}\pi\right) \end{bmatrix} + \begin{bmatrix} 0 \\ 0 \\ 0 \\ \frac{P}{2J} \\ 0 \end{bmatrix} T_{PM}.$$

Example 6.8: Experimental Studies of a Permanent-Magnet Synchronous Generator

Synchronous machines can be examined, tested, and analyzed. Consider a synchronous generator rotated by a prime mover. The induced phase voltages (*motional emfs*) in the *as*, *bs*, and *cs* phases constitute a three-phase Y- or Δ-connected voltage source. There are balanced and unbalanced Y- or Δ-connected *RLC* loads. The balanced and unbalanced Y- or Δ-source to a Y- or Δ-load can be examined. The three-phase system analysis can be performed in time and frequency domains. In the steady state, for an unbalanced Y-connected load with the phase load impedances

$$\mathbf{Z}_{Las} \neq \mathbf{Z}_{Lbs} \neq \mathbf{Z}_{Lcs},$$

we have

$$\mathbf{I}_{bs} = \frac{\mathbf{U}_{bsN}}{\mathbf{Z}_{Lbs}}, \quad \mathbf{I}_{as} = \frac{\mathbf{U}_{asN}}{\mathbf{Z}_{Las}}, \quad \text{and} \quad \mathbf{I}_{cs} = \frac{\mathbf{U}_{csN}}{\mathbf{Z}_{Lcs}}.$$

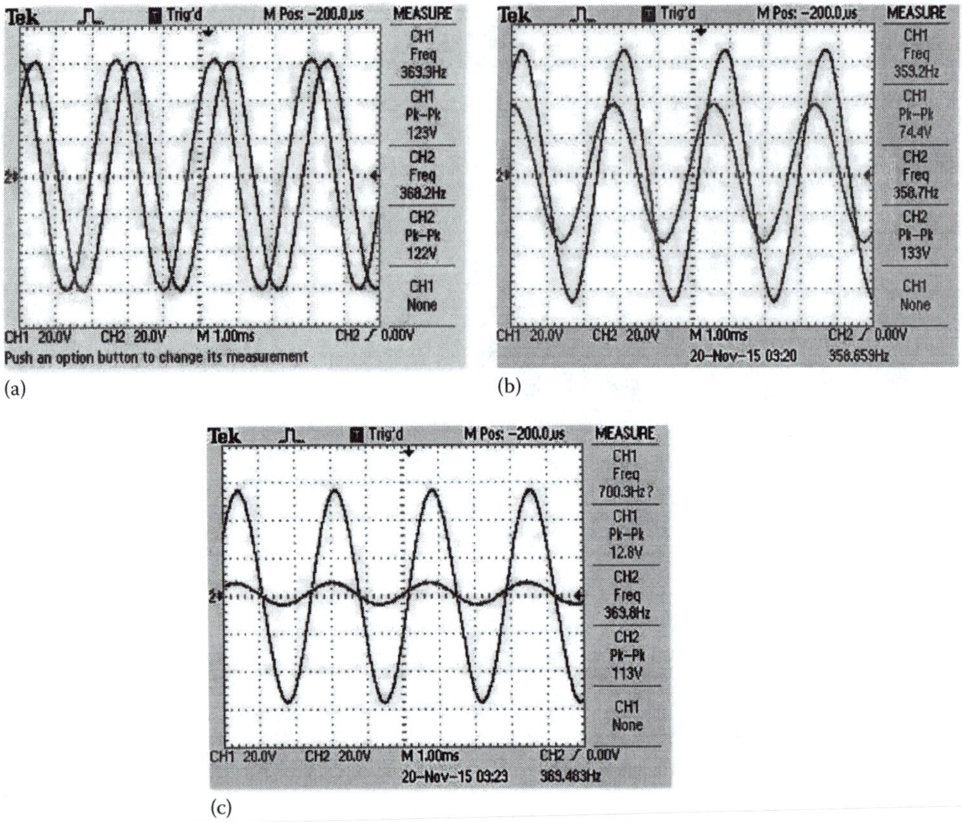

FIGURE 6.20 Induced phase voltages in a three-phase permanent-magnet synchronous generator when
(a) $R_{Las} = R_{Lbs} = R_{Lcs} = 100$ ohm;
(b) $R_{Las} = R_{Lbs} = 100$ ohm and $R_{Lcs} = 10$ ohm;
(c) $R_{Las} = R_{Lbs} = 100$ ohm and $R_{Lcs} = 1$ ohm.

The complex power per phase is $\mathbf{S}_i = P_i + jQ_i = \mathbf{U}_i\mathbf{I}_i^*$. For the balanced system with $\mathbf{Z}_{Las} = \mathbf{Z}_{Lbs} = \mathbf{Z}_{Lcs}$, using the rms values of phase currents I_p and voltages V_p, the total complex power is

$$\mathbf{S} = 3\mathbf{S}_p = 3\mathbf{U}_p\mathbf{I}_p^* = 3I_p^2\mathbf{Z}_p = 3I_p^2 Z_p \angle\phi = 3\frac{V_p^2}{\mathbf{Z}_p^*} = 3\frac{V_p^2}{Z_p\angle\phi}.$$

One performs an analysis for the Y-Y, Y-Δ, Δ-Y, or Δ-Δ systems using the load impedances $\mathbf{Z}_{Las}$, $\mathbf{Z}_{Lbs}$, and $\mathbf{Z}_{Lcs}$.

Generators may operate in balanced and unbalanced operating envelopes. For a generator, when $R_{Las} = R_{Lbs} = R_{Lcs}$ and $R_{Las} \neq R_{Lbs} \neq R_{Lcs}$, the experimental results are reported in Figures 6.20. ■

Example 6.9: Experimental Studies and Analysis of a Permanent-Magnet Synchronous Machine

The radial topology permanent-magnet synchronous motors and generators are described by five differential equations (6.20)–(6.22), and, (6.23)–(6.24). For different steady-state ω_r and loads, the induced *emfs* are illustrated in Figure 6.21a.

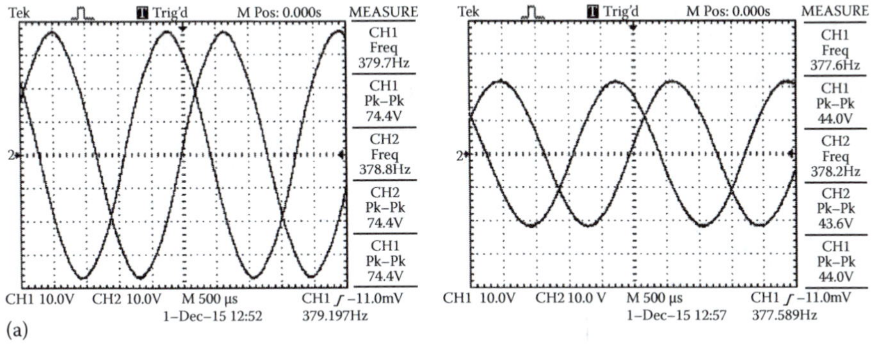

(a)

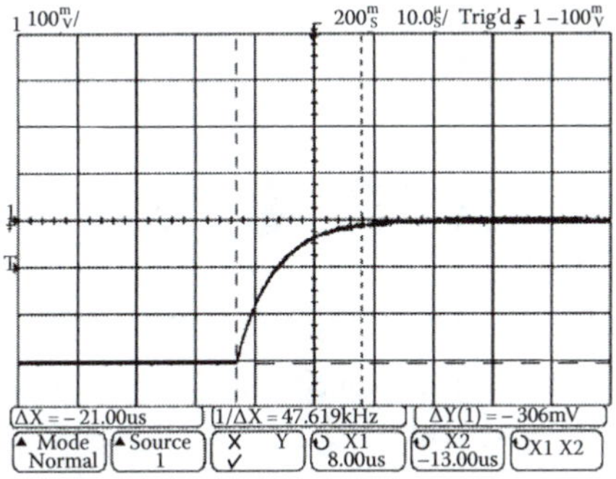

(b)

FIGURE 6.21 (a) Unloaded and loaded generator: Induced $emf_{as\omega} = \psi_m \cos\theta_r\omega_r$ if $\omega_r = \text{const}$; (b) Current in the phase winding for the step voltage applied (motor at stall).

In steady state, $emf_{as\omega} = \psi_m \cos\theta_r\omega_r$. The measured magnitude $\psi_m\omega_r$ yields the unknown ψ_m. This ψ_m varies in an operating envelope due to nonlinear electromagnetic system, saturation, etc. One obtains the self-inductance. At the stall motor, the phase windings are the *RL* Y- or Δ-circuits. The current *i*(*t*) dynamics in the *RL* circuit depends on the values of *R* and *L*. By measuring the current as the voltage across the resistor r_R for not rotating motor, one obtains *i*(*t*) applying the voltage. For the step voltage,

$$i(t) = \frac{u}{R}\left(1 - e^{-\frac{R}{L}t}\right).$$

The time constant is $\tau = L/R$, $L = 2L_{ss}$ and $R = 2r_s + r_R$ because we test two Y-connected phase windings in series. For the first-order *RL* circuit, the time delay corresponds to $0.632i_{steady\text{-}state}$. Hence, $L_{ss} = \tau(2r_s + r_R)/2$. The oscilloscope data for the voltage across the resistor r_R is shown in Figure 6.21b. For $r_R = 1.97 \times 10^3$ ohm, one obtains $\tau = 5.2 \times 10^{-6}$ sec. Hence, $L_{ss} = 0.0051$ H.

The electric machine parameters can be directly measured and derived by performing experiments. One can measure the stator resistance r_s. The constant ψ_m is found using the induced *emfs,* e.g. examining the induced phase voltage when machine used as a generator at the steady-state ω_r. The viscous friction coefficient B_m is found by measuring the magnitude of the phase currents at no load. For motors, $T_e = T_{friction} = B_m\omega_{rm}$ at $T_L = 0$, $B_m = T_{em}/\omega_{rm}$. The moment of inertia *J* can be estimated as $\frac{1}{2}m_{rotor}R_{rotor}^2$, or derived by using the deceleration or acceleration rate. During deceleration, if $u_M = 0$ and $T_L = 0$, we have

$$\frac{d\omega_{rm}}{dt} = -\frac{1}{J}B_m\omega_{rm}.$$

For a four-pole permanent-magnet synchronous machine, the parameters are experimentally found to be as follows: $r_s = 1$ ohm, $L_{ls} = 0.0005$ H, $L_{ss} = 0.005$ H, $\psi_m = 0.15$ V-sec/rad (N-m/A), $B_m = 0.0005$ N-m-sec/rad, and $J = 0.00025$ kg-m^2.

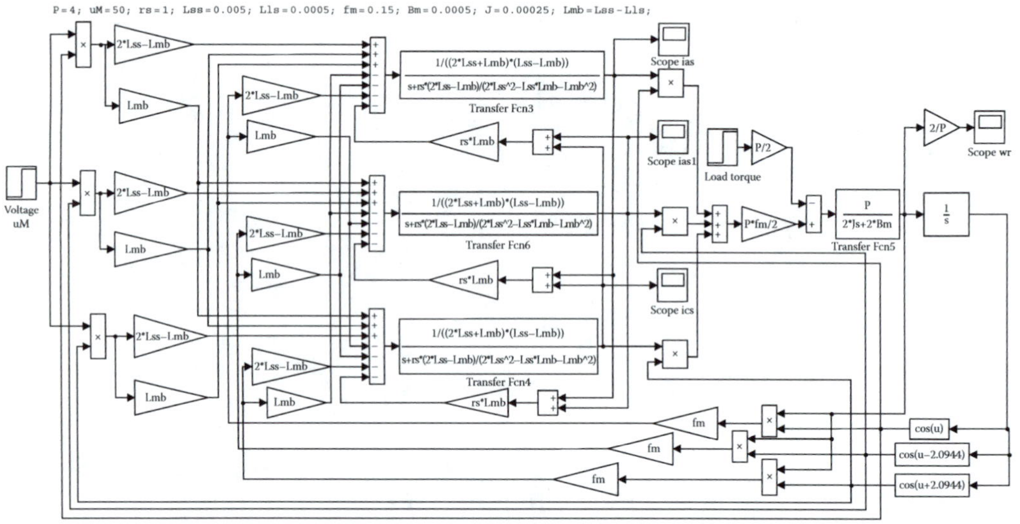

FIGURE 6.22 Simulink® diagram to simulate permanent-magnet synchronous motors.

The Simulink diagram to simulate permanent-magnet synchronous motors is documented in Figure 6.22. The balanced voltage set is

$$u_{as} = \sqrt{2}u_M \cos\theta_r, \quad u_{bs} = \sqrt{2}u_M \cos\left(\theta_r - \frac{2}{3}\pi\right), \quad u_{cs} = \sqrt{2}u_M \cos\left(\theta_r + \frac{2}{3}\pi\right).$$

The motor dynamics is studied if motor accelerates with the rated voltage applied, $u_M = 50$ V. The motor parameters are uploaded as

```
P=4; uM=50; rs=1; Lss=0.005; Lls=0.0005; fm=0.15; Bm=0.0005; J=0.00025; Lmb=Lss-Lls;
```

The motor accelerates from stall at load $T_{L0|t=0} = 0.25$ N-m, $t \in [0 \quad 0.125)$ sec. The load $T_L = 0.5$ N-m is applied at $t = 0.125$ sec. Figures 6.23 illustrate the evolution of the phase currents and mechanical angular velocity. The motor reaches the steady-state mechanical angular velocity with the load $T_{L0|t=0} = 0.25$ N-m. The angular velocity reduces and the phase currents magnitude increases as motor is loaded. One examines motor dynamics, acceleration, starting capabilities, etc. ■

6.4.4 Mathematical Models of Permanent-Magnet Synchronous Machines in the Arbitrary, Rotor, and Synchronous Reference Frames

6.4.4.1 Arbitrary Reference Frame

Our goal is to derive the governing equations using the *quadrature*, *direct*, and *zero* (*qd*0) components of stator currents, voltages, and flux linkages. We find a mathematical model in the *arbitrary* reference frame when the frame angular velocity ω is not specified. Using the *direct* Park transformation, we have

$$\mathbf{u}_{qd0s} = \mathbf{K}_s \mathbf{u}_{abcs}, \quad \mathbf{i}_{qd0s} = \mathbf{K}_s \mathbf{i}_{abcs}, \quad \psi_{qd0s} = \mathbf{K}_s \psi_{abcs}, \quad \mathbf{K}_s = \frac{2}{3}\begin{bmatrix} \cos\theta & \cos\left(\theta - \frac{2}{3}\pi\right) & \cos\left(\theta + \frac{2}{3}\pi\right) \\ \sin\theta & \sin\left(\theta - \frac{2}{3}\pi\right) & \sin\left(\theta + \frac{2}{3}\pi\right) \\ \frac{1}{2} & \frac{1}{2} & \frac{1}{2} \end{bmatrix}. \tag{6.25}$$

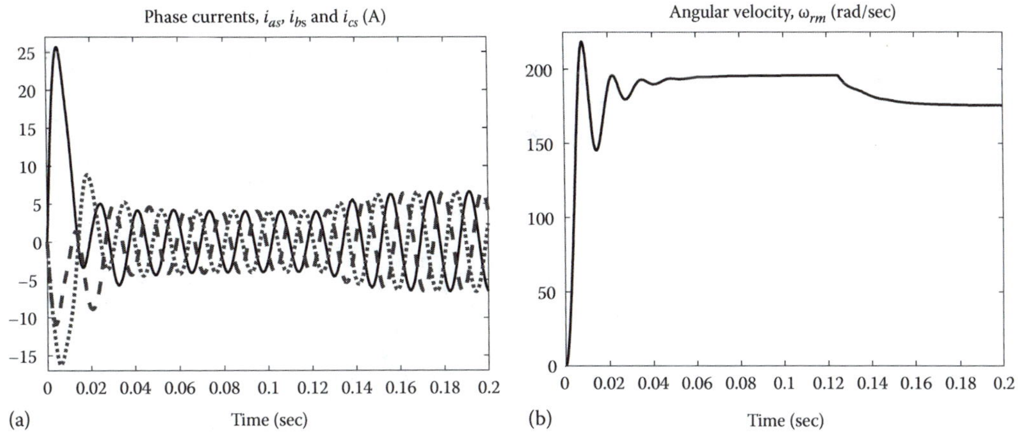

FIGURE 6.23 (a) Transient dynamics of the phase currents $i_{as}(t)$, $i_{bs}(t)$, and $i_{cs}(t)$ (solid, dashed, and dotted lines);

(b) Dynamics of the angular velocity $\Omega_r(t)$, $T_L = \begin{cases} 0.25, 0 \le t \le 0.125 \text{ s} \\ 0.5, 0.125 < t \le 0.2 \text{ s} \end{cases}$.

The $qd0$ components of stator phase voltages $\mathbf{u}_{qd0s}$, currents $\mathbf{i}_{qd0s}$, and flux linkages $\boldsymbol{\psi}_{qd0s}$ are found using the *machine* variables and $\mathbf{K}_s$. From (6.18) $\mathbf{u}_{abcs} = \mathbf{r}_s \mathbf{i}_{abcs} + \dfrac{d\boldsymbol{\psi}_{abcs}}{dt}$ with (6.25), one has

$$\mathbf{K}_s^{-1}\mathbf{u}_{qd0s} = \mathbf{r}_s\mathbf{K}_s^{-1}\mathbf{i}_{qd0s} + \frac{d\left(\mathbf{K}_s^{-1}\boldsymbol{\psi}_{qd0s}\right)}{dt}, \quad \mathbf{K}_s^{-1} = \begin{bmatrix} \cos\theta & \sin\theta & 1 \\ \cos\left(\theta - \frac{2}{3}\pi\right) & \sin\left(\theta - \frac{2}{3}\pi\right) & 1 \\ \cos\left(\theta + \frac{2}{3}\pi\right) & \sin\left(\theta + \frac{2}{3}\pi\right) & 1 \end{bmatrix}. \tag{6.26}$$

Multiplication of the left and right sides in (6.26) by $\mathbf{K}_s$ yields

$$\mathbf{K}_s\mathbf{K}_s^{-1}\mathbf{u}_{qdos} = \mathbf{K}_s\mathbf{r}_s\mathbf{K}_s^{-1}\mathbf{i}_{qd0s} + \mathbf{K}_s\frac{d\mathbf{K}_s^{-1}}{dt}\boldsymbol{\psi}_{qd0s} + \mathbf{K}_s\mathbf{K}_s^{-1}\frac{d\boldsymbol{\psi}_{qd0s}}{dt}. \tag{6.27}$$

The matrix $\mathbf{r}_s$ is diagonal, and $\mathbf{K}_s\mathbf{r}_s\mathbf{K}_s^{-1} = \mathbf{r}_s$.
From

$$\frac{d\mathbf{K}_s^{-1}}{dt} = \omega\begin{bmatrix} -\sin\theta & \cos\theta & 0 \\ -\sin\left(\theta - \frac{2}{3}\pi\right) & \cos\left(\theta - \frac{2}{3}\pi\right) & 0 \\ -\sin\left(\theta + \frac{2}{3}\pi\right) & \cos\left(\theta + \frac{2}{3}\pi\right) & 0 \end{bmatrix},$$

we have

$$\mathbf{K}_s \frac{d\mathbf{K}_s^{-1}}{dt} = \omega \begin{bmatrix} 0 & 1 & 0 \\ -1 & 0 & 0 \\ 0 & 0 & 0 \end{bmatrix}.$$

Hence, the vector-matrix equation (6.27) gives

$$\mathbf{u}_{qd0s} = \mathbf{r}_s \mathbf{i}_{qd0s} + \omega \begin{bmatrix} \psi_{ds} \\ -\psi_{qs} \\ 0 \end{bmatrix} + \frac{d\psi_{qd0s}}{dt}, \tag{6.28}$$

where

$$\psi_{qd0s} = \mathbf{K}_s \psi_{abcs}, \quad \psi_{abcs} = \mathbf{L}_s \mathbf{i}_{abcs} + \psi_m = \begin{bmatrix} L_{ls} + \bar{L}_m & -\frac{1}{2}\bar{L}_m & -\frac{1}{2}\bar{L}_m \\ -\frac{1}{2}\bar{L}_m & L_{ls} + \bar{L}_m & -\frac{1}{2}\bar{L}_m \\ -\frac{1}{2}\bar{L}_m & -\frac{1}{2}\bar{L}_m & L_{ls} + \bar{L}_m \end{bmatrix} \begin{bmatrix} i_{as} \\ i_{bs} \\ i_{cs} \end{bmatrix} + \psi_m \begin{bmatrix} \sin\theta_r \\ \sin\left(\theta_r - \frac{2}{3}\pi\right) \\ \sin\left(\theta_r + \frac{2}{3}\pi\right) \end{bmatrix}.$$

Hence,

$$\psi_{qd0s} = \mathbf{K}_s \mathbf{L}_s \mathbf{K}_s^{-1} \mathbf{i}_{qd0s} + \mathbf{K}_s \psi_m = \begin{bmatrix} L_{ls} + \frac{3}{2}\bar{L}_m & 0 & 0 \\ 0 & L_{ls} + \frac{3}{2}\bar{L}_m & 0 \\ 0 & 0 & L_{ls} \end{bmatrix} \mathbf{i}_{qd0s} + \psi_m \begin{bmatrix} -\sin(\theta - \theta_r) \\ \cos(\theta - \theta_r) \\ 0 \end{bmatrix},$$

because

$$\mathbf{K}_s \mathbf{L}_s \mathbf{K}_s^{-1} = \begin{bmatrix} L_{ls} + \frac{3}{2}\bar{L}_m & 0 & 0 \\ 0 & L_{ls} + \frac{3}{2}\bar{L}_m & 0 \\ 0 & 0 & L_{ls} \end{bmatrix} \text{ and}$$

$$\mathbf{K}_s \psi_m = \frac{2}{3} \begin{bmatrix} \cos\theta & \cos\left(\theta - \frac{2}{3}\pi\right) & \cos\left(\theta + \frac{2}{3}\pi\right) \\ \sin\theta & \sin\left(\theta - \frac{2}{3}\pi\right) & \sin\left(\theta + \frac{2}{3}\pi\right) \\ \frac{1}{2} & \frac{1}{2} & \frac{1}{2} \end{bmatrix} \psi_m \begin{bmatrix} \sin\theta_r \\ \sin\left(\theta_r - \frac{2}{3}\pi\right) \\ \sin\left(\theta_r + \frac{2}{3}\pi\right) \end{bmatrix} = \psi_m \begin{bmatrix} -\sin(\theta - \theta_r) \\ \cos(\theta - \theta_r) \\ 0 \end{bmatrix}.$$

Having derived $\boldsymbol{\psi}_{qd0}$ one finds $d\boldsymbol{\psi}_{qd0}/dt$ in (6.28). Using (6.28), a model for the permanent-magnet synchronous motors circuitry in the *arbitrary* reference frame is

$$\mathbf{u}_{qd0s} = \mathbf{r}_s\mathbf{i}_{qd0s} + \omega\begin{bmatrix} \psi_{ds} \\ -\psi_{qs} \\ 0 \end{bmatrix} + \begin{bmatrix} L_{ls}+\frac{3}{2}\bar{L}_m & 0 & 0 \\ 0 & L_{ls}+\frac{3}{2}\bar{L}_m & 0 \\ 0 & 0 & L_{ls} \end{bmatrix}\frac{d\mathbf{i}_{qd0s}}{dt} + \psi_m \frac{d}{dt}\begin{bmatrix} -\sin(\theta-\theta_r) \\ \cos(\theta-\theta_r) \\ 0 \end{bmatrix}. \quad (6.29)$$

In model (6.29), one specifies the frame angular velocity ω and θ. The stationary ω = 0, rotor $\omega = \omega_r$, and synchronous $\omega = \omega_e$ reference frames are commonly used.

6.4.4.2 Synchronous Motors in the Rotor and Synchronous Reference Frames

The electrical angular velocity is equal to the synchronous angular velocity. The angular velocity of the reference frame is specified as $\omega = \omega_e = \omega_r$. From $\theta = \theta_e = \theta_r$, the Park transformations matrix for stator variables is

$$\mathbf{K}_s^r = \mathbf{K}_s^e = \frac{2}{3}\begin{bmatrix} \cos\theta_r & \cos\left(\theta_r - \frac{2}{3}\pi\right) & \cos\left(\theta_r + \frac{2}{3}\pi\right) \\ \sin\theta_r & \sin\left(\theta_r - \frac{2}{3}\pi\right) & \sin\left(\theta_r + \frac{2}{3}\pi\right) \\ \frac{1}{2} & \frac{1}{2} & \frac{1}{2} \end{bmatrix}.$$

The last term in (6.29) is defined. In the expanded form, for $\omega = \omega_r$ and $\theta = \theta_r$, differential equations (6.29) yield

$$\frac{di_{qs}^r}{dt} = -\frac{r_s}{L_{ls}+\frac{3}{2}\bar{L}_m} i_{qs}^r - \frac{\psi_m}{L_{ls}+\frac{3}{2}\bar{L}_m}\omega_r - i_{ds}^r\omega_r + \frac{1}{L_{ls}+\frac{3}{2}\bar{L}_m}u_{qs}^r,$$

$$\frac{di_{ds}^r}{dt} = -\frac{r_s}{L_{ls}+\frac{3}{2}\bar{L}_m} i_{ds}^r + i_{qs}^r\omega_r + \frac{1}{L_{ls}+\frac{3}{2}\bar{L}_m}u_{ds}^r, \quad (6.30)$$

$$\frac{di_{0s}^r}{dt} = -\frac{r_s}{L_{ls}} i_{0s}^r + \frac{1}{L_{ls}}u_{0s}^r.$$

We apply the Park transformation $\mathbf{i}_{abcs} = \mathbf{K}_s^{r^{-1}}\mathbf{i}_{qd0s}^r$ to the phase current in the electromagnetic torque equation (6.21). Using

$$\begin{bmatrix} i_{as} \\ i_{bs} \\ i_{cs} \end{bmatrix} = \begin{bmatrix} \cos\theta_r & \sin\theta_r & 1 \\ \cos\left(\theta_r - \frac{2}{3}\pi\right) & \sin\left(\theta_r - \frac{2}{3}\pi\right) & 1 \\ \cos\left(\theta_r + \frac{2}{3}\pi\right) & \sin\left(\theta_r + \frac{2}{3}\pi\right) & 1 \end{bmatrix}\begin{bmatrix} i_{qs}^r \\ i_{ds}^r \\ i_{0s}^r \end{bmatrix},$$

we substitute the resulting

$$i_{as} = \cos\theta_r i_{qs}^r + \sin\theta_r i_{ds}^r + i_{0s}^r,\quad i_{bs} = \cos\left(\theta_r - \frac{2}{3}\pi\right)i_{qs}^r + \sin\left(\theta_r - \frac{2}{3}\pi\right)i_{ds}^r + i_{0s}^r, \quad \text{and}$$

$$i_{cs} = \cos\left(\theta_r + \frac{2}{3}\pi\right)i_{qs}^r + \sin\left(\theta_r + \frac{2}{3}\pi\right)i_{ds}^r + i_{0s}^r$$

$$\text{in } T_e = \frac{P\psi_m}{2}\left[i_{as}\cos\theta_r + i_{bs}\cos\left(\theta_r - \frac{2}{3}\pi\right) + i_{cs}\cos\left(\theta_r + \frac{2}{3}\pi\right)\right].$$

One obtains

$$T_e = \frac{3P\psi_m}{4} i_{qs}^r. \tag{6.31}$$

Using (6.30), the Newtonian *torsional–mechanical* dynamics and (6.31), we have

$$\begin{aligned}
\frac{di_{qs}^r}{dt} &= -\frac{r_s}{L_{ls} + \frac{3}{2}\bar{L}_m} i_{qs}^r - \frac{\psi_m}{L_{ls} + \frac{3}{2}\bar{L}_m}\omega_r - i_{ds}^r\omega_r + \frac{1}{L_{ls} + \frac{3}{2}\bar{L}_m} u_{qs}^r,\\
\frac{di_{ds}^r}{dt} &= -\frac{r_s}{L_{ls} + \frac{3}{2}\bar{L}_m} i_{ds}^r + i_{qs}^r\omega_r + \frac{1}{L_{ls} + \frac{3}{2}\bar{L}_m} u_{ds}^r,\\
\frac{di_{0s}^r}{dt} &= -\frac{r_s}{L_{ls}} i_{0s}^r + \frac{1}{L_{ls}} u_{0s}^r,\\
\frac{d\omega_r}{dt} &= \frac{3P^2\psi_m}{8J} i_{qs}^r - \frac{B_m}{J}\omega_r - \frac{P}{2J}T_L,\\
\frac{d\theta_r}{dt} &= \omega_r.
\end{aligned} \tag{6.32}$$

A balanced three-phase current set is

$$i_{as}(t) = \sqrt{2}i_M\cos\theta_r,\quad i_{bs}(t) = \sqrt{2}i_M\cos\left(\theta_r - \frac{2}{3}\pi\right),\quad i_{cs}(t) = \sqrt{2}i_M\cos\left(\theta_r + \frac{2}{3}\pi\right).$$

Using the *direct* Park transformation

$$\begin{bmatrix} i_{qs}^r \\ i_{ds}^r \\ i_{0s}^r \end{bmatrix} = \frac{2}{3}\begin{bmatrix} \cos\theta_r & \cos\left(\theta_r - \frac{2}{3}\pi\right) & \cos\left(\theta_r + \frac{2}{3}\pi\right) \\ \sin\theta_r & \sin\left(\theta_r - \frac{2}{3}\pi\right) & \sin\left(\theta_r + \frac{2}{3}\pi\right) \\ \frac{1}{2} & \frac{1}{2} & \frac{1}{2} \end{bmatrix}\begin{bmatrix} i_{as} \\ i_{bs} \\ i_{cs} \end{bmatrix},$$

one obtains the $qd0$-current components

$$\begin{bmatrix} i_{qs}^r \\ i_{ds}^r \\ i_{0s}^r \end{bmatrix} = \frac{2}{3}\begin{bmatrix} \cos\theta_r & \cos\left(\theta_r - \frac{2}{3}\pi\right) & \cos\left(\theta_r + \frac{2}{3}\pi\right) \\ \sin\theta_r & \sin\left(\theta_r - \frac{2}{3}\pi\right) & \sin\left(\theta_r + \frac{2}{3}\pi\right) \\ \frac{1}{2} & \frac{1}{2} & \frac{1}{2} \end{bmatrix}\begin{bmatrix} \sqrt{2}i_M\cos\theta_r \\ \sqrt{2}i_M\cos\left(\theta_r - \frac{2}{3}\pi\right) \\ \sqrt{2}i_M\cos\left(\theta_r + \frac{2}{3}\pi\right) \end{bmatrix} = \begin{bmatrix} \sqrt{2}i_M \\ 0 \\ 0 \end{bmatrix}.$$

Hence, $i_{qs}^r(t) = \sqrt{2}i_M,\ i_{ds}^r(t) = 0,\ i_{0s}^r(t) = 0$.

Due to the self-inductances, the *abc* voltages may be supplied with advanced phase shifting

$$u_{as}(t) = \sqrt{2}u_M\cos(\theta_r + \varphi_u),\quad u_{bs}(t) = \sqrt{2}u_M\cos\left(\theta_r - \frac{2}{3}\pi + \varphi_u\right),\quad u_{cs}(t) = \sqrt{2}u_M\cos\left(\theta_r + \frac{2}{3}\pi + \varphi_u\right).$$

From

$$\begin{bmatrix} u_{qs}^r \\ u_{ds}^r \\ u_{0s}^r \end{bmatrix} = \frac{2}{3}\begin{bmatrix} \cos\theta_r & \cos\left(\theta_r - \frac{2}{3}\pi\right) & \cos\left(\theta_r + \frac{2}{3}\pi\right) \\ \sin\theta_r & \sin\left(\theta_r - \frac{2}{3}\pi\right) & \sin\left(\theta_r + \frac{2}{3}\pi\right) \\ \frac{1}{2} & \frac{1}{2} & \frac{1}{2} \end{bmatrix}\begin{bmatrix} u_{as} \\ u_{bs} \\ u_{cs} \end{bmatrix},$$

one finds

$$\begin{bmatrix} u_{qs}^r \\ u_{ds}^r \\ u_{0s}^r \end{bmatrix} = \frac{2}{3}\begin{bmatrix} \cos\theta_r & \cos\left(\theta_r - \frac{2}{3}\pi\right) & \cos\left(\theta_r + \frac{2}{3}\pi\right) \\ \sin\theta_r & \sin\left(\theta_r - \frac{2}{3}\pi\right) & \sin\left(\theta_r + \frac{2}{3}\pi\right) \\ \frac{1}{2} & \frac{1}{2} & \frac{1}{2} \end{bmatrix}\begin{bmatrix} \sqrt{2}u_M\cos(\theta_r + \varphi_u) \\ \sqrt{2}u_M\cos\left(\theta_r - \frac{2}{3}\pi + \varphi_u\right) \\ \sqrt{2}u_M\cos\left(\theta_r + \frac{2}{3}\pi + \varphi_u\right) \end{bmatrix} = \begin{bmatrix} \sqrt{2}u_M\cos\varphi_u \\ -\sqrt{2}u_M\sin\varphi_u \\ 0 \end{bmatrix}.$$

Using the trigonometric identities, we obtain

$$u_{qs}^r(t) = \sqrt{2}u_M\cos\varphi_u, \quad u_{ds}^r(t) = -\sqrt{2}u_M\sin\varphi_u, \quad u_{0s}^r(t) = 0.$$

If the self-inductance is small,

$$u_{as}(t) = \sqrt{2}u_M\cos\theta_r, \quad u_{bs}(t) = \sqrt{2}u_M\cos\left(\theta_r - \frac{2}{3}\pi\right), \quad u_{cs}(t) = \sqrt{2}u_M\cos\left(\theta_r + \frac{2}{3}\pi\right).$$

For $\varphi_u = 0$, we have $u_{qs}^r(t) = \sqrt{2}u_M$, $u_{ds}^r(t) = 0$, $u_{0s}^r(t) = 0$.

In the synchronous reference frame, one finds the identical model because $\mathbf{u}_{qd0s}^e = \mathbf{u}_{qd0s}^r$, $\mathbf{i}_{qd0s}^e = \mathbf{i}_{qd0s}^r$, and $\psi_{qd0s}^e = \psi_{qd0s}^r$. From $\omega = \omega_e = \omega_r$ and $\theta = \theta_e = \theta_r$, differential equations (6.32) result in the state-space model in the synchronous reference frame

$$\begin{bmatrix} \frac{di_{qs}^e}{dt} \\ \frac{di_{ds}^e}{dt} \\ \frac{di_{0s}^e}{dt} \\ \frac{d\omega_r}{dt} \\ \frac{d\theta_r}{dt} \end{bmatrix} = \begin{bmatrix} -\frac{r_s}{L_{ls} + \frac{3}{2}\bar{L}_m} & 0 & 0 & -\frac{\psi_m}{L_{ls} + \frac{3}{2}\bar{L}_m} & 0 \\ 0 & -\frac{r_s}{L_{ls} + \frac{3}{2}\bar{L}_m} & 0 & 0 & 0 \\ 0 & 0 & -\frac{r_s}{L_{ls}} & 0 & 0 \\ \frac{3P^2\psi_m}{8J} & 0 & 0 & -\frac{B_m}{J} & 0 \\ 0 & 0 & 0 & 1 & 0 \end{bmatrix}\begin{bmatrix} i_{qs}^e \\ i_{ds}^e \\ i_{0s}^e \\ \omega_r \\ \theta_r \end{bmatrix} + \begin{bmatrix} -i_{ds}^e\omega_r \\ i_{qs}^e\omega_r \\ 0 \\ 0 \\ 0 \end{bmatrix}$$

$$+ \begin{bmatrix} \frac{1}{L_{ls} + \frac{3}{2}\bar{L}_m} & 0 & 0 \\ 0 & \frac{1}{L_{ls} + \frac{3}{2}\bar{L}_m} & 0 \\ 0 & 0 & \frac{1}{L_{ls}} \\ 0 & 0 & 0 \\ 0 & 0 & 0 \end{bmatrix}\begin{bmatrix} u_{qs}^e \\ u_{ds}^e \\ u_{0s}^e \end{bmatrix} - \begin{bmatrix} 0 \\ 0 \\ 0 \\ \frac{P}{2J} \\ 0 \end{bmatrix} T_L.$$

The balanced current set is $i_{qs}^e(t) = \sqrt{2}i_M$, $i_{ds}^e(t) = 0$, $i_{0s}^e(t) = 0$.

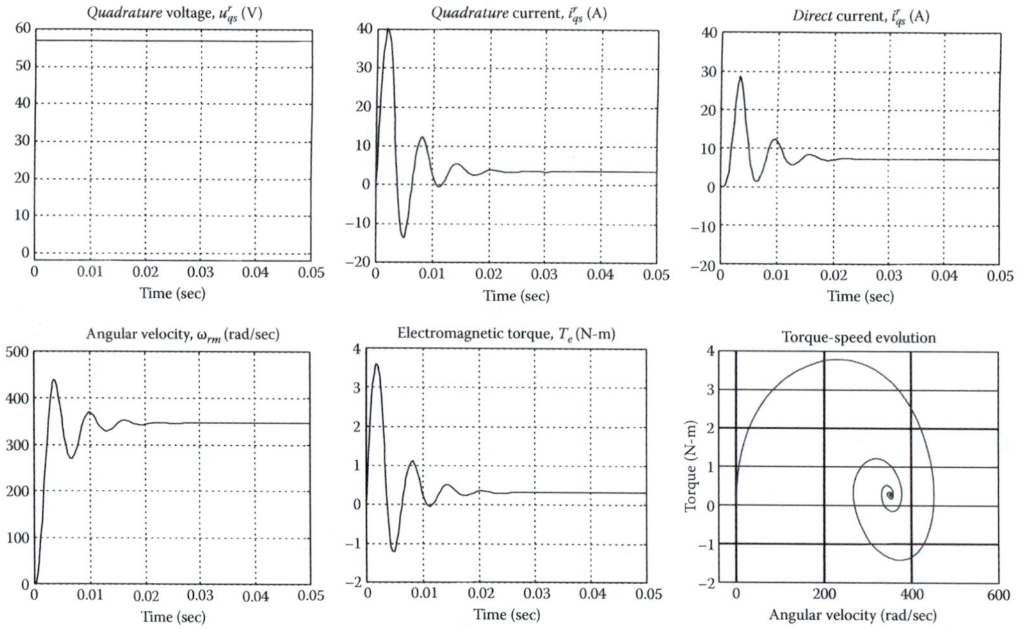

FIGURE 6.24 Acceleration of a three-phase permanent-magnet synchronous motor from stall if $u_M = 50$ V and $T_L = 0.3$ N-m.

Example 6.10

Using differential equations (6.32), we model the permanent-magnet synchronous motor. The parameters are given in Example 6.9. The *qd*0 voltage components that guarantee the balanced operation are $u_{qs}^r(t) = \sqrt{2}u_M$, $u_{ds}^r(t) = 0$, $u_{0s}^r(t) = 0$. Let $u_M = 40$ V. The dynamics are illustrated in Figure 6.24 if $T_L = 0.3$ N-m. The motor starts at stall and reaches the angular velocity 350 rad/sec. Two-dimensional torque-speed evolution is depicted in the last plot. ■

6.4.4.3 Synchronous Generators in the Rotor and Synchronous Reference Frames

We use the results reported in Section 6.4.3. Equations (6.23) $\mathbf{u}_{abcs} = -(\mathbf{r}_s + \mathbf{R}_L)\mathbf{i}_{abcs} + \dfrac{d\boldsymbol{\psi}_{abcs}}{dt}$, $\boldsymbol{\psi}_{abcs} = -\mathbf{L}_s\mathbf{i}_{abcs} + \boldsymbol{\psi}_m$ and (6.24) are used. We apply the Park transformation to find

$$\mathbf{u}_{qd0s}^r = \mathbf{K}_s^r\mathbf{u}_{abcs},\ \mathbf{i}_{qd0s}^r = \mathbf{K}_s^r\mathbf{i}_{abcs}, \text{ and } \boldsymbol{\psi}_{qd0s}^r = \mathbf{K}_s^r\boldsymbol{\psi}_{abcs},\ \mathbf{K}_s^r = \frac{2}{3}\begin{bmatrix} \cos\theta_r & \cos\left(\theta_r - \frac{2}{3}\pi\right) & \cos\left(\theta_r + \frac{2}{3}\pi\right) \\ \sin\theta_r & \sin\left(\theta_r - \frac{2}{3}\pi\right) & \sin\left(\theta_r + \frac{2}{3}\pi\right) \\ \frac{1}{2} & \frac{1}{2} & \frac{1}{2} \end{bmatrix}.$$

As reported in Sections 6.4.4.1 and 6.4.4.2, one finds the resulting models. For the balanced resistive load $R_L = R_{Las} = R_{Lbs} = R_{Lcs}$, we have

$$\frac{di_{qs}^r}{dt} = -\frac{r_s + R_L}{L_{ls} + \frac{3}{2}\bar{L}_m} i_{qs}^r + \frac{\psi_m}{L_{ls} + \frac{3}{2}\bar{L}_m}\omega_r - i_{ds}^r\omega_r,$$

$$\frac{di_{ds}^r}{dt} = -\frac{r_s + R_L}{L_{ls} + \frac{3}{2}\bar{L}_m} i_{ds}^r + i_{qs}^r\omega_r,$$

$$\frac{di_{0s}^r}{dt} = -\frac{r_s + R_L}{L_{ls}} i_{0s}^r.$$

The generator is rotated by the prime mover with the mechanical angular velocity $\omega_{rm} = \frac{2}{P}\omega_r$. The circuitry-electromagnetic and *torsional–mechanical* dynamics is

$$\frac{di_{qs}^r}{dt} = -\frac{r_s + R_L}{L_{ls} + \frac{3}{2}\bar{L}_m} i_{qs}^r + \frac{\psi_m}{L_{ls} + \frac{3}{2}\bar{L}_m}\frac{P}{2}\omega_{rm} - \frac{P}{2} i_{ds}^r\omega_{rm},$$

$$\frac{di_{ds}^r}{dt} = -\frac{r_s + R_L}{L_{ls} + \frac{3}{2}\bar{L}_m} i_{ds}^r + \frac{P}{2} i_{qs}^r\omega_{rm},$$

$$\frac{d\omega_{rm}}{dt} = -\frac{3P\psi_m}{4J} i_{qs}^r - \frac{B_m}{J}\omega_{rm} + \frac{1}{J}T_{PM}.$$

The induced voltage is proportional to ω_{rm}. The generator electromagnetic torque $T_e = \frac{3P\psi_m}{4} i_{qs}^r$ acts against T_{PM}. One may perform analysis in the $qd0$ quantities, and then obtain the abc voltages and currents using the *inverse* Park transformation

$$\mathbf{u}_{abcs} = \mathbf{K}_s^{r^{-1}}\mathbf{u}_{qd0s}^r, \quad \mathbf{i}_{abcs} = \mathbf{K}_s^{r^{-1}}\mathbf{i}_{qd0s}^r, \quad \mathbf{K}_s^{r^{-1}} = \begin{bmatrix} \cos\theta_r & \sin\theta_r & 1 \\ \cos\left(\theta_r - \frac{2}{3}\pi\right) & \sin\left(\theta_r - \frac{2}{3}\pi\right) & 1 \\ \cos\left(\theta_r + \frac{2}{3}\pi\right) & \sin\left(\theta_r + \frac{2}{3}\pi\right) & 1 \end{bmatrix}.$$

The model in the synchronous reference frame is identical.

6.5 ADVANCED TOPICS IN THE ANALYSIS OF PERMANENT-MAGNET SYNCHRONOUS MACHINES

We documented analysis, modeling, and simulation of permanent-magnet synchronous machines. The designer can achieve near-optimal design and performance in the specified operating envelope. The nonlinear magnetic system, magnetic field nonuniformity, nonlinear B–H characteristic, saturation, spacing between magnets, eccentricity, and other effects significantly affect device capabilities. The circuitry-electromagnetic dynamics, energy conversion, and torque production were studied assuming a linear magnetic system, ideal sinusoidal winding–magnets coupling, etc. Under these assumptions, the magnetizing inductance and flux linkages can be described by the odd sine and even cosine functions

$$\boldsymbol{\psi}_{abcm} \equiv \psi_m \sin(\theta_r + \phi) \quad \text{or} \quad \boldsymbol{\psi}_{abcm} \equiv \psi_m \cos(\theta_r + \phi).$$

In (6.19), we assumed that

$$\psi_m = \begin{bmatrix} \psi_{asm} \\ \psi_{bsm} \\ \psi_{csm} \end{bmatrix} = \psi_m \begin{bmatrix} \sin\theta_r \\ \sin\left(\theta_r - \frac{2}{3}\pi\right) \\ \sin\left(\theta_r + \frac{2}{3}\pi\right) \end{bmatrix}.$$

The function is even if $f(x) = f(-x)$, and $f(x)$ is odd if $-f(x) = f(-x)$. In general,

$$\boldsymbol{\psi}_{abcm} \neq \psi_m \sin(\theta_r + \phi) \quad \text{and} \quad \boldsymbol{\psi}_{abcm} \neq \psi_m \cos(\theta_r + \phi).$$

One experimentally finds flux linkages $\boldsymbol{\psi}_{abcm}$ in the full operating envelope. The flux linkages ψ_{asm}, ψ_{bsm}, and ψ_{csm} in the specified operating envelope (loads, angular velocity, etc.) can be obtained experimentally by examining the induced *emfs* by operating synchronous machines as generators. For example, one may find

$$\psi_m = \begin{bmatrix} \psi_{asm} \\ \psi_{bsm} \\ \psi_{csm} \end{bmatrix},$$

$$\psi_{asm} = \psi_{m1}\sum_{n=1}^{\infty}\left(a_{asn}\sin^{2n-1}\theta_r + b_{asn}\cos^{2n-1}\theta_r\right)$$
$$+\psi_{m2}\sum_{k,l=1}^{\infty}\left(a_{asl,k}\,\mathrm{sgn}\left(\sin\theta_r\right)\left|\sin^{\frac{2l-1}{2k-1}}\theta_r\right| + b_{asl,k}\,\mathrm{sgn}\left(\cos\theta_r\right)\left|\cos^{\frac{2l-1}{2k-1}}\theta_r\right|\right), \tag{6.33}$$

$$\psi_{bsm} = \psi_{m1}\sum_{n=1}^{\infty}\left(a_{bsn}\sin^{2n-1}\left(\theta_r - \frac{2}{3}\pi\right) + b_{bsn}\cos^{2n-1}\left(\theta_r - \frac{2}{3}\pi\right)\right)$$
$$+\psi_{m2}\sum_{k,l=1}^{\infty}\left(a_{bsl,k}\,\mathrm{sgn}\left(\sin\left(\theta_r - \frac{2}{3}\pi\right)\right)\left|\sin^{\frac{2l-1}{2k-1}}\left(\theta_r - \frac{2}{3}\pi\right)\right| + b_{bsl,k}\,\mathrm{sgn}\left(\cos\left(\theta_r - \frac{2}{3}\pi\right)\right)\left|\cos^{\frac{2l-1}{2k-1}}\left(\theta_r - \frac{2}{3}\pi\right)\right|\right),$$

$$\psi_{csm} = \psi_{m1}\sum_{n=1}^{\infty}\left(a_{csn}\sin^{2n-1}\left(\theta_r + \frac{2}{3}\pi\right) + b_{csn}\cos^{2n-1}\left(\theta_r + \frac{2}{3}\pi\right)\right)$$
$$+\psi_{m2}\sum_{k,l=1}^{\infty}\left(a_{csl,k}\,\mathrm{sgn}\left(\sin\left(\theta_r + \frac{2}{3}\pi\right)\right)\left|\sin^{\frac{2l-1}{2k-1}}\left(\theta_r + \frac{2}{3}\pi\right)\right| + b_{csl,k}\,\mathrm{sgn}\left(\cos\left(\theta_r + \frac{2}{3}\pi\right)\right)\left|\cos^{\frac{2l-1}{2k-1}}\left(\theta_r + \frac{2}{3}\pi\right)\right|\right).$$

Here, a_n, $a_{l,k}$, and b_n, $b_{l,k}$ are the coefficients or functions that depend on the electromagnetic system, operating envelope, machine design, materials, fabrication technology, sizing, and other factors, e.g.,

$$a_i(\mathbf{E}, \mathbf{D}, \mathbf{B}, \mathbf{H}, \mathbf{i}_{abcs}, \boldsymbol{\omega}_r, \mathbf{T}_L, \boldsymbol{\varepsilon}, \boldsymbol{\mu}) \quad \text{and} \quad b_i(\mathbf{E}, \mathbf{D}, \mathbf{B}, \mathbf{H}, \mathbf{i}_{abcs}, \boldsymbol{\omega}_r, \mathbf{T}_L, \boldsymbol{\varepsilon}, \boldsymbol{\mu}).$$

Other expressions experimentally found are

$$\psi_m = \begin{bmatrix} \psi_{asm} \\ \psi_{bsm} \\ \psi_{csm} \end{bmatrix}, \quad \psi_{asm} = d_{as0} + \psi_{m3}e^{\sum_{p=1}^{\infty}\left(a_{asp}\sin^p\frac{1}{c_p}\theta_r + b_{asp}\cos^p\frac{1}{c_p}\theta_r\right)}, \tag{6.34}$$

$$\psi_{bsm} = d_{bs0} + \psi_{m3}e^{\sum_{p=1}^{\infty}\left(a_{bsp}\sin^p\left(\frac{1}{c_p}\theta_r - \frac{2}{3}\pi\right) + b_{bsp}\cos^p\left(\frac{1}{c_p}\theta_r - \frac{2}{3}\pi\right)\right)},$$

$$\psi_{csm} = d_{cs0} + \psi_{m3}e^{\sum_{p=1}^{\infty}\left(a_{csp}\sin^p\left(\frac{1}{c_p}\theta_r + \frac{2}{3}\pi\right) + b_{csp}\cos^p\left(\frac{1}{c_p}\theta_r + \frac{2}{3}\pi\right)\right)}.$$

One combines (6.33) and (6.34) to obtain consistent expressions for the flux linkages ψ_{asm}, ψ_{bsm}, and ψ_{csm} in the full operating envelope. For the symmetric machines $a_{asn} = a_{bsn} = a_{csn}$ and $b_{asn} = b_{bsn} = b_{csn}$, $d_{as0} = d_{bs0} = d_{cs0}$, etc. Using the element-by-element product, we have

$$\psi_m = \begin{bmatrix} \psi_{asm} \\ \psi_{bsm} \\ \psi_{csm} \end{bmatrix},$$

$$\psi_{asm} = \left[\psi_{m0} + \psi_{m1}\sum_{n=1}^{\infty} a_n \sin^{2n-1}\theta_r + \psi_{m2}\sum_{k,l=1}^{\infty} a_{l,k}\,\mathrm{sgn}(\sin\theta_r)\left|\sin^{\frac{2l-1}{2k-1}}\theta_r\right|\right]$$

$$\circ\left[d_0 + \psi_{m3} e^{\sum_{p=1}^{\infty}\left(a_p \sin^p \frac{1}{c_p}\theta_r + b_p \cos^p \frac{1}{c_p}\theta_r\right)}\right], \quad (6.35)$$

$$\psi_{bsm} = \left[\psi_{m0} + \psi_{m1}\sum_{n=1}^{\infty} a_n \sin^{2n-1}\left(\theta_r - \frac{2}{3}\pi\right) + \psi_{m2}\sum_{k,l=1}^{\infty} a_{l,k}\,\mathrm{sgn}\left(\sin\left(\theta_r - \frac{2}{3}\pi\right)\right)\left|\sin^{\frac{2l-1}{2k-1}}\left(\theta_r - \frac{2}{3}\pi\right)\right|\right]$$

$$\circ\left[d_0 + \psi_{m3} e^{\sum_{p=1}^{\infty}\left(a_p \sin^p\left(\frac{1}{c_p}\theta_r - \frac{2}{3}\pi\right) + b_p \cos^p\left(\frac{1}{c_p}\theta_r - \frac{2}{3}\pi\right)\right)}\right],$$

$$\psi_{csm} = \left[\psi_{m0} + \psi_{m1}\sum_{n=1}^{\infty} a_n \sin^{2n-1}\left(\theta_r + \frac{2}{3}\pi\right) + \psi_{m2}\sum_{k,l=1}^{\infty} a_{l,k}\,\mathrm{sgn}\left(\sin\left(\theta_r + \frac{2}{3}\pi\right)\right)\left|\sin^{\frac{2l-1}{2k-1}}\left(\theta_r + \frac{2}{3}\pi\right)\right|\right]$$

$$\circ\left[d_0 + \psi_{m3} e^{\sum_{p=1}^{\infty}\left(a_p \sin^p\left(\frac{1}{c_p}\theta_r + \frac{2}{3}\pi\right) + b_p \cos^p\left(\frac{1}{c_p}\theta_r + \frac{2}{3}\pi\right)\right)}\right].$$

Example 6.11

Using (6.33)–(6.35), with the corresponding coefficients ψ_{mi}, a_i, b_i, c_i, and d_0, consider:

1. $\psi_{asm} = \sin\theta_r + \sin^5\theta_r$, $\psi_{asm} = \sin^3\theta_r + \sin^5\theta_r$, $\psi_{asm} = \frac{1}{2}\sin\theta_r + \mathrm{sgn}(\sin\theta_r)\left|\sin^{1/3}\theta_r\right|$, and $\psi_{asm} = \frac{1}{2}\sin\theta_r + \mathrm{sgn}(\sin\theta_r)\left|\sin^{1/9}\theta_r\right|$;
2. $\psi_{asm} = e^{\left(-\sin\frac{1}{2}\theta_r + \cos^{12}\frac{1}{2}\theta_r\right)}$, $\psi_{asm} = e^{\left(\sin^4\frac{1}{4}\theta_r + \cos^4\frac{1}{4}\theta_r\right)}$, and $\psi_{asm} = \sin\theta_r e^{-\cos\theta_r}$;
3. $\psi_{asm} = \sin\theta_r e^{-a\cos^2\theta_r}$, $\psi_{bsm} = \cos\theta_r e^{-a\sin^2\theta_r}$, and $\psi_{asm} = \sin^3\theta_r e^{-a\cos^4\theta_r}$, $\psi_{bsm} = \cos^3\theta_r e^{-a\sin^4\theta_r}$.

Figures 6.25 document the resulting plots. ■

Consistent analysis can be performed. For adequately designed permanent-magnet synchronous machines in the rated operating envelope, let (6.33) be simplified to

$$\psi_{asm} = \psi_m\sum_{n=1}^{\infty} a_n \sin^{2n-1}\theta_r,\quad \psi_{bsm} = \psi_m\sum_{n=1}^{\infty} a_n \sin^{2n-1}\left(\theta_r - \frac{2}{3}\pi\right),\ \text{and}\ \psi_{csm} = \psi_m\sum_{n=1}^{\infty} a_n \sin^{2n-1}\left(\theta_r + \frac{2}{3}\pi\right).$$

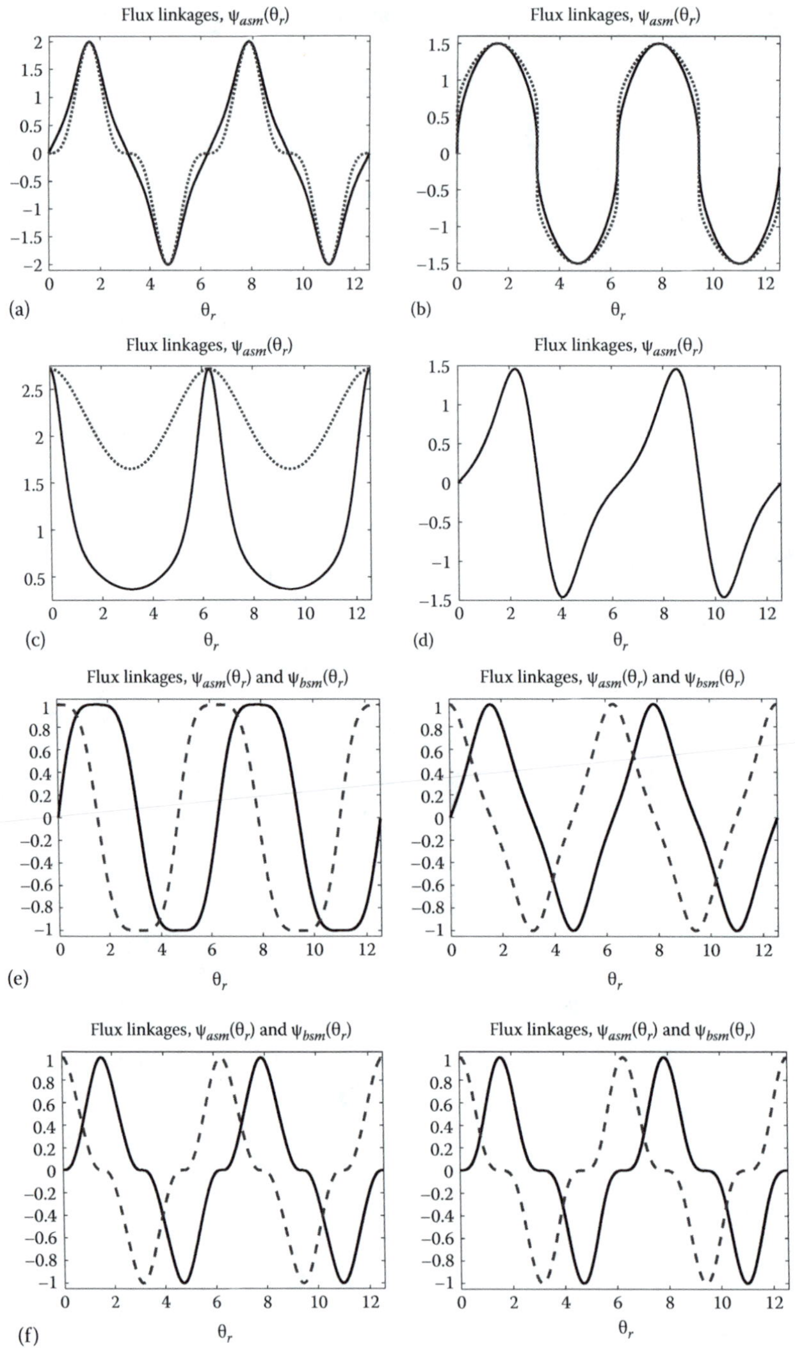

FIGURE 6.25 (a) Plots of $\psi_{asm} = \sin\theta_r + \sin^5\theta_r$ and $\psi_{asm} = \sin^3\theta_r + \sin^5\theta_r$ (solid and dotted lines);
(b) $\psi_{asm} = \frac{1}{2}\sin\theta_r + \mathrm{sgn}(\sin\theta_r)\left|\sin^{1/3}\theta_r\right|$ and $\psi_{asm} = \frac{1}{2}\sin\theta_r + \mathrm{sgn}(\sin\theta_r)\left|\sin^{1/9}\theta_r\right|$;
(c) $\psi_{asm} = e^{\left(-\sin\frac{1}{2}\theta_r + \cos^{12}\frac{1}{2}\theta_r\right)}$ and $\psi_{asm} = e^{\left(\sin^4\frac{1}{4}\theta_r + \cos^4\frac{1}{4}\theta_r\right)}$;
(d) $\psi_{asm} = \sin\theta_r e^{-\cos\theta_r}$;
(e) $\psi_{asm} = \sin\theta_r e^{-a\cos^2\theta_r}$, $\psi_{bsm} = \cos\theta_r e^{-a\sin^2\theta_r}$ for $a = 0.5$ and $a = -0.5$, respectively;
(f) Plots of $\psi_{asm} = \sin^3\theta_r e^{-a\cos^4\theta_r}$, $\psi_{bsm} = \cos^3\theta_r e^{-a\sin^4\theta_r}$ for $a = 0.5$ and $a = -0.5$, respectively.

From (6.18), one yields

$$\mathbf{u}_{abcs} = \mathbf{r}_s \mathbf{i}_{abcs} + \frac{d\boldsymbol{\psi}_{abcs}}{dt}, \boldsymbol{\psi}_{abcs} = \begin{bmatrix} \psi_{as} \\ \psi_{bs} \\ \psi_{cs} \end{bmatrix} = \begin{bmatrix} L_{ls} + \bar{L}_m & -\frac{1}{2}\bar{L}_m & -\frac{1}{2}\bar{L}_m \\ -\frac{1}{2}\bar{L}_m & L_{ls} + \bar{L}_m & -\frac{1}{2}\bar{L}_m \\ -\frac{1}{2}\bar{L}_m & -\frac{1}{2}\bar{L}_m & L_{ls} + \bar{L}_m \end{bmatrix} \begin{bmatrix} i_{as} \\ i_{bs} \\ i_{cs} \end{bmatrix} + \psi_m \begin{bmatrix} \sum_{n=1}^{\infty} a_n \sin^{2n-1} \theta_r \\ \sum_{n=1}^{\infty} a_n \sin^{2n-1}\left(\theta_r - \frac{2}{3}\pi\right) \\ \sum_{n=1}^{\infty} a_n \sin^{2n-1}\left(\theta_r + \frac{2}{3}\pi\right) \end{bmatrix}, \tag{6.36}$$

$$u_{as} = r_s i_{as} + \frac{d\psi_{as}}{dt}, \quad u_{bs} = r_s i_{bs} + \frac{d\psi_{bs}}{dt}, \quad u_{cs} = r_s i_{cs} + \frac{d\psi_{cs}}{dt}.$$

In (6.36), the total derivatives $d\psi_{as}/dt$, $d\psi_{bs}/dt$, and $d\psi_{cs}/dt$ can be derived yielding the circuitry-electromagnetic equations.

The electromagnetic torque is

$$T_e = \frac{\partial W_c}{\partial \theta_r} = \frac{P\psi_m}{2}\left[i_{as}\sum_{n=1}^{\infty}(2n-1)a_n \cos\theta_r \sin^{2n-2}\theta_r + i_{bs}\sum_{n=1}^{\infty}(2n-1)a_n \cos\left(\theta_r - \frac{2}{3}\pi\right)\sin^{2n-2}\left(\theta_r - \frac{2}{3}\pi\right) \right.$$
$$\left. + i_{cs}\sum_{n=1}^{\infty}(2n-1)a_n \cos\left(\theta_r + \frac{2}{3}\pi\right)\sin^{2n-2}\left(\theta_r + \frac{2}{3}\pi\right)\right]. \tag{6.37}$$

The Newton second law (6.22) yields

$$\frac{d\omega_r}{dt} = \frac{P^2\psi_m}{4J}\left[i_{as}\sum_{n=1}^{\infty}(2n-1)a_n \cos\theta_r \sin^{2n-2}\theta_r + i_{bs}\sum_{n=1}^{\infty}(2n-1)a_n \cos\left(\theta_r - \frac{2}{3}\pi\right)\sin^{2n-2}\left(\theta_r - \frac{2}{3}\pi\right) \right.$$
$$\left. + i_{cs}\sum_{n=1}^{\infty}(2n-1)a_n \cos\left(\theta_r + \frac{2}{3}\pi\right)\sin^{2n-2}\left(\theta_r + \frac{2}{3}\pi\right)\right] - \frac{B_m}{J}\omega_r - \frac{P}{2J}T_L,$$

$$\frac{d\theta_r}{dt} = \omega_r. \tag{6.38}$$

The balanced current and voltage sets are derived using the expression for T_e (6.37). We have

$$i_{as} = \sqrt{2}i_M \cos\theta_r \left(\sum_{n=1}^{\infty}(2n-1)a_n \sin^{2n-2}\theta_r\right)^{-1},$$

$$i_{bs} = \sqrt{2}i_M \cos\left(\theta_r - \frac{2}{3}\pi\right)\left(\sum_{n=1}^{\infty}(2n-1)a_n \sin^{2n-2}\left(\theta_r - \frac{2}{3}\pi\right)\right)^{-1}, \tag{6.39}$$

$$i_{cs} = \sqrt{2}i_M \cos\left(\theta_r + \frac{2}{3}\pi\right)\left(\sum_{n=1}^{\infty}(2n-1)a_n \sin^{2n-2}\left(\theta_r + \frac{2}{3}\pi\right)\right)^{-1}, \quad |i_{as}, i_{bs}, i_{cs}| \le i_{\max},$$

and

$$u_{as}=\sqrt{2}u_M\cos\theta_r\left(\sum_{n=1}^{\infty}(2n-1)a_n\sin^{2n-2}\theta_r\right)^{-1},$$

$$u_{bs}=\sqrt{2}u_M\cos\left(\theta_r-\frac{2}{3}\pi\right)\left(\sum_{n=1}^{\infty}(2n-1)a_n\sin^{2n-2}\left(\theta_r-\frac{2}{3}\pi\right)\right)^{-1}, \tag{6.40}$$

$$u_{cs}=\sqrt{2}u_M\cos\left(\theta_r+\frac{2}{3}\pi\right)\left(\sum_{n=1}^{\infty}(2n-1)a_n\sin^{2n-2}\left(\theta_r+\frac{2}{3}\pi\right)\right)^{-1}, \quad |u_{as},u_{bs},u_{cs}|\le u_{\max}.$$

The phase currents and voltages are constrained. The current and voltage limits should be examined. The singularity problem can be resolved. The balanced current and voltage sets (6.39) and (6.40) should be implemented. Using power MOSFET output stages (usually six- or twelve-step) and converter topologies (hard-switching or *passive* and *active* soft-switching), one strives to ensure efficiency, adequateness, and consistency. The hardware solutions largely define the voltage waveforms. Advanced DSPs are used to identify unknown a_n and b_n applying the estimation algorithms. Filtering, processing, nonlinear analysis, conditional logics, look-up tables, and other routines are implemented. Within the existing converter topologies, it is not always possible to ensure ideal sinusoidal voltage waveforms. Furthermore, the PWM concept implies the voltage *averaging*. The rated solid-state device voltage, current, switching frequency, and other characteristics affect current and voltage waveforms. The hardware-dependent phase voltages must be used in the performance analysis. We enabled the analysis tasks providing justifications and foundations for advanced studies. One may apply the derived basic electromagnetics, control concepts and hardware solutions ensuring near-optimal performance.

Example 6.12

For two-phase permanent-magnet synchronous machines, it is desired to ensure the design which leads to $\psi_{asm}=\psi_m\sin\theta_r$ and $\psi_{bsm}=\psi_m\cos\theta_r$. For the electromagnetic torque $T_e=\frac{P\psi_m}{2}\left(\cos\theta_r i_{as}-\sin\theta_r i_{bs}\right)$, one finds the balanced current set $i_{as}=i_M\cos\theta_r$ and $i_{bs}=-i_M\sin\theta_r$.

Let the flux linkages, established by the permanent magnets as viewed from the windings, be

$$\psi_{asm}=\psi_m\sum_{n=1}^{\infty}a_n\sin^{2n-1}\theta_r \quad \text{and} \quad \psi_{bsm}=\psi_m\sum_{n=1}^{\infty}a_n\cos^{2n-1}\theta_r.$$

The electromagnetic torque is

$$T_e=\frac{P\psi_m}{2}\left[i_{as}\sum_{n=1}^{\infty}(2n-1)a_n\cos\theta_r\sin^{2n-2}\theta_r-i_{bs}\sum_{n=1}^{\infty}(2n-1)a_n\sin\theta_r\cos^{2n-2}\theta_r\right].$$

For $a_1\neq 1$, $a_2\neq 0$ and $\forall a_n=0$, $n>2$,

$$T_e=\frac{P\psi_m}{2}\left[i_{as}\cos\theta_r\left(a_1+3a_2\sin^2\theta_r\right)-i_{bs}\sin\theta_r\left(a_1+3a_2\cos^2\theta_r\right)\right].$$

The phase voltages u_{as} and u_{bs}, which ensure the near-balanced operating conditions are

$$u_{as}=u_M\frac{\cos\theta_r}{a_1+3a_2\sin^2\theta_r},\ |u_{as}|\le u_{\max},\quad u_{bs}=-u_M\frac{\sin\theta_r}{a_1+3a_2\cos^2\theta_r},\ |u_{bs}|\le u_{\max}.$$

If $a_1\gg a_2$, one may use $u_{as}=u_M\cos\theta_r$ and $u_{bs}=-u_M\sin\theta_r$. ■

Example 6.13

We study a radial topology three-phase synchronous motor with $P = 4$, $u_M = 50$, $r_s = 1$ ohm, $L_{ls} = 0.0002$ H, $\bar{L}_m = 0.0018$ H, $L_{ss} = 0.002$ H, $\psi_m = 0.1$ V-sec/rad (N-m/A), $B_m = 0.00008$ N-m-sec/rad, and $J = 0.00004$ kg-m^2. For no load and light load conditions, the constants in (6.33) may be $a_1 \neq 0$ and $\forall a_n = 0$, $n > 1$. For the loaded motor, $a_1 = 1$, $a_2 = 0.05$, $a_3 = 0.02$, and $\forall a_n = 0$, $n > 3$. The permanent-magnet synchronous motor is described by five nonlinear differential equations (6.36) and (6.38). For $a_1 \neq 0$, $a_2 \neq 0$, $a_3 \neq 0$, and $\forall a_n = 0$, $n > 3$, the flux linkages are

$$\psi_{as} = L_{ss}i_{as} - \frac{1}{2}\bar{L}_m i_{bs} - \frac{1}{2}\bar{L}_m i_{cs} + \psi_m\left(a_1 \sin\theta_r + a_2 \sin^3\theta_r + a_3 \sin^5\theta_r\right),$$

$$\psi_{bs} = -\frac{1}{2}\bar{L}_m i_{as} + L_{ss}i_{bs} - \frac{1}{2}\bar{L}_m i_{cs} + \psi_m\left(a_1 \sin\left(\theta_r + \frac{2}{3}\pi\right) + a_2 \sin^3\left(\theta_r + \frac{2}{3}\pi\right) + a_3 \sin^5\left(\theta_r + \frac{2}{3}\pi\right)\right),$$

$$\psi_{cs} = -\frac{1}{2}\bar{L}_m i_{as} - \frac{1}{2}\bar{L}_m i_{bs} + L_{ss}i_{cs} + \psi_m\left(a_1 \sin\left(\theta_r - \frac{2}{3}\pi\right) + a_2 \sin^3\left(\theta_r - \frac{2}{3}\pi\right) + a_3 \sin^5\left(\theta_r - \frac{2}{3}\pi\right)\right).$$

The Kirchhoff second law equations (6.36) yield

$$u_{as} = r_s i_{as} + L_{ss}\frac{di_{as}}{dt} - \frac{1}{2}\bar{L}_m\frac{di_{bs}}{dt} - \frac{1}{2}\bar{L}_m\frac{di_{cs}}{dt} + \psi_m \cos\theta_r\left(a_1 + 3a_2\sin^2\theta_r + 5a_3\sin^4\theta_r\right)\omega_r,$$

$$u_{bs} = r_s i_{bs} - \frac{1}{2}\bar{L}_m\frac{di_{as}}{dt} + L_{ss}\frac{di_{bs}}{dt} - \frac{1}{2}\bar{L}_m\frac{di_{cs}}{dt} + \psi_m\cos\left(\theta_r + \frac{2}{3}\pi\right)\left(a_1 + 3a_2\sin^2\left(\theta_r + \frac{2}{3}\pi\right) + 5a_3\sin^4\left(\theta_r + \frac{2}{3}\pi\right)\right)\omega_r,$$

$$u_{cs} = r_s i_{cs} - \frac{1}{2}\bar{L}_m\frac{di_{as}}{dt} - \frac{1}{2}\bar{L}_m\frac{di_{bs}}{dt} + L_{ss}\frac{di_{cs}}{dt} + \psi_m\cos\left(\theta_r - \frac{2}{3}\pi\right)\left(a_1 + 3a_2\sin^2\left(\theta_r - \frac{2}{3}\pi\right) + 5a_3\sin^4\left(\theta_r - \frac{2}{3}\pi\right)\right)\omega_r.$$

The circuitry-electromagnetic dynamics is

$$\frac{di_{as}}{dt} = \frac{1}{L_{ss}}\left[-r_s i_{as} + \frac{1}{2}\bar{L}_m\frac{di_{bs}}{dt} + \frac{1}{2}\bar{L}_m\frac{di_{cs}}{dt} - \psi_m\cos\theta_r\left(a_1 + 3a_2\sin^2\theta_r + 5a_3\sin^4\theta_r\right)\omega_r + u_{as}\right], \tag{6.41}$$

$$\frac{di_{bs}}{dt} = \frac{1}{L_{ss}}\left[-r_s i_{bs} + \frac{1}{2}\bar{L}_m\frac{di_{as}}{dt} + \frac{1}{2}\bar{L}_m\frac{di_{cs}}{dt} - \psi_m\cos\left(\theta_r + \frac{2}{3}\pi\right)\left(a_1 + 3a_2\sin^2\left(\theta_r + \frac{2}{3}\pi\right) + 5a_3\sin^4\left(\theta_r + \frac{2}{3}\pi\right)\right)\omega_r + u_{bs}\right],$$

$$\frac{di_{cs}}{dt} = \frac{1}{L_{ss}}\left[-r_s i_{cs} + \frac{1}{2}\bar{L}_m\frac{di_{as}}{dt} + \frac{1}{2}\bar{L}_m\frac{di_{bs}}{dt} - \psi_m\cos\left(\theta_r - \frac{2}{3}\pi\right)\left(a_1 + 3a_2\sin^2\left(\theta_r - \frac{2}{3}\pi\right) + 5a_3\sin^4\left(\theta_r - \frac{2}{3}\pi\right)\right)\omega_r + u_{cs}\right].$$

From (6.37), the expression for the electromagnetic torque is

$$T_e = \frac{P\psi_m}{2}\left[i_{as}\cos\theta_r\left(a_1 + 3a_2\sin^2\theta_r + 5a_3\sin^4\theta_r\right)\right.$$

$$+ i_{bs}\cos\left(\theta_r + \frac{2}{3}\pi\right)\left(a_1 + 3a_2\sin^2\left(\theta_r + \frac{2}{3}\pi\right) + 5a_3\sin^4\left(\theta_r + \frac{2}{3}\pi\right)\right)$$

$$\left. + i_{cs}\cos\left(\theta_r - \frac{2}{3}\pi\right)\left(a_1 + 3a_2\sin^2\left(\theta_r - \frac{2}{3}\pi\right) + 5a_3\sin^4\left(\theta_r - \frac{2}{3}\pi\right)\right)\right].$$

The *torsional–mechanical* equations of motion (6.38) are

$$\frac{d\omega_r}{dt} = \frac{P^2\psi_m}{4J}\left[i_{as}\cos\theta_r\left(a_1 + 3a_2\sin^2\theta_r + 5a_3\sin^4\theta_r\right)\right.$$

$$+ i_{bs}\cos\left(\theta_r + \frac{2}{3}\pi\right)\left(a_1 + 3a_2\sin^2\left(\theta_r + \frac{2}{3}\pi\right) + 5a_3\sin^4\left(\theta_r + \frac{2}{3}\pi\right)\right)$$

$$\left. + i_{cs}\cos\left(\theta_r - \frac{2}{3}\pi\right)\left(a_1 + 3a_2\sin^2\left(\theta_r - \frac{2}{3}\pi\right) + 5a_3\sin^4\left(\theta_r - \frac{2}{3}\pi\right)\right)\right] - \frac{B_m}{J}\omega_r - \frac{P}{2J}T_L,$$

$$\frac{d\theta_r}{dt} = \omega_r. \tag{6.42}$$

Using (6.41) and (6.42), the Simulink diagram to simulate permanent-magnet synchronous motors ($a_1 \neq 0$, $a_2 \neq 0$, $a_3 \neq 0$ and $\forall a_n = 0$, $n > 3$) is developed as reported in Figure 6.26. The following phase voltages are supplied

$$u_{as} = \sqrt{2}u_M\cos\theta_r,\quad u_{bs} = \sqrt{2}u_M\cos\left(\theta_r + \frac{2}{3}\pi\right),\quad \text{and}\quad u_{cs} = \sqrt{2}u_M\cos\left(\theta_r - \frac{2}{3}\pi\right).$$

This voltage set can be used because $a_1 \gg (2n-1)a_n$, $\forall n > 1$.

The motor dynamics is studied as the motor accelerates from stall with the rated voltage applied $u_M = 50$ V at no load. The load torque $T_L = 0.1$ N-m is applied at $t = 0.025$ sec. Figures 6.27 illustrate the evolution of the phase currents and electrical angular velocity. The motor reaches the steady state $\omega_r = 500$ rad/sec with no load and 490 rad/sec if $T_L = 0.1$ N-m. The angular velocity reduces and phase currents magnitude increases as T_L is applied.

For the loaded motor, a_n coefficients are $a_1 = 1$, $a_2 = 0.05$, $a_3 = 0.02$, and $\forall a_n = 0$, $n > 3$. The motor accelerates with the rated voltage applied, for example, $u_M = 50$ V. At $t = 0$ sec, the load torque is $T_{L0} = 0.2$ N-m. At $t = 0.025$ sec, the load increases to $T_L = 0.5$ N-m. The evolutions of the phase currents and ω_r are documented in Figures 6.28. The motor reaches the steady-state operation within 0.02 sec. One observes the phase current chattering and the electromagnetic torque ripple. This leads to the reduction of efficiency, losses, vibration, noise, heating, etc. There are many other secondary effects that degrade the motor performance. For example, in overheated motors, ψ_m reduces leading to the reduction of T_e and torque density. Using the conditional statements and look-up table, the near-balanced voltage sets can be implemented in a full operating envelope. The closed-loop electromechanical systems must be designed to ensure optimal performance, guarantying *achievable* capabilities. ■

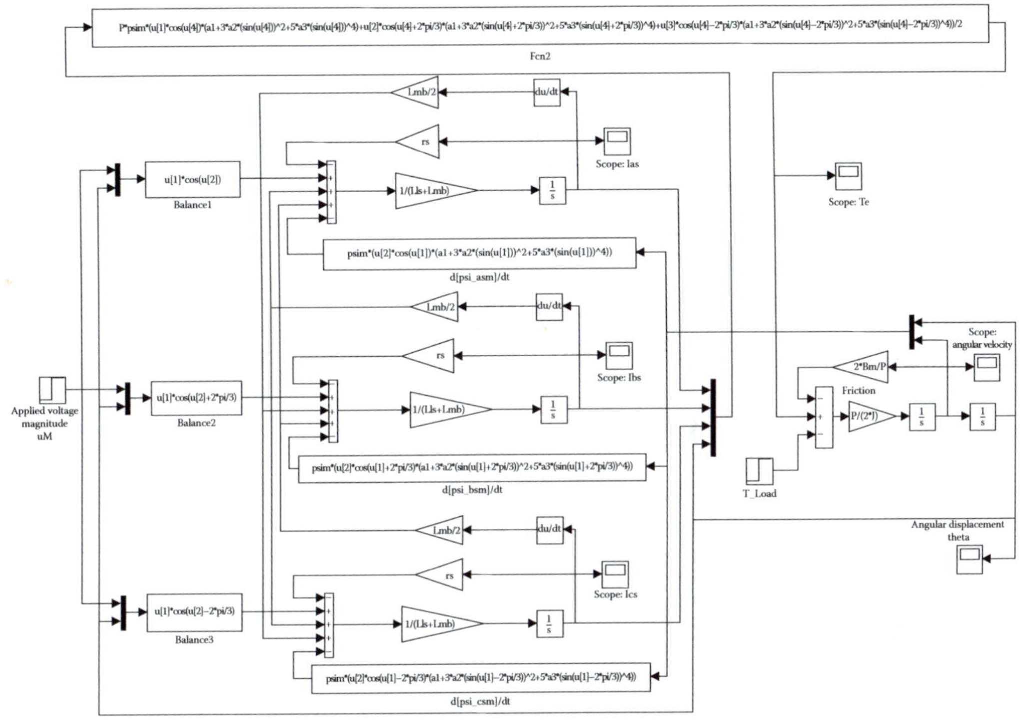

FIGURE 6.26 Simulink® diagram to simulate permanent-magnet synchronous motors when $a_1 \neq 0$, $a_2 \neq 0$, $a_3 \neq 0$ and $\forall a_n = 0$, $n > 3$.

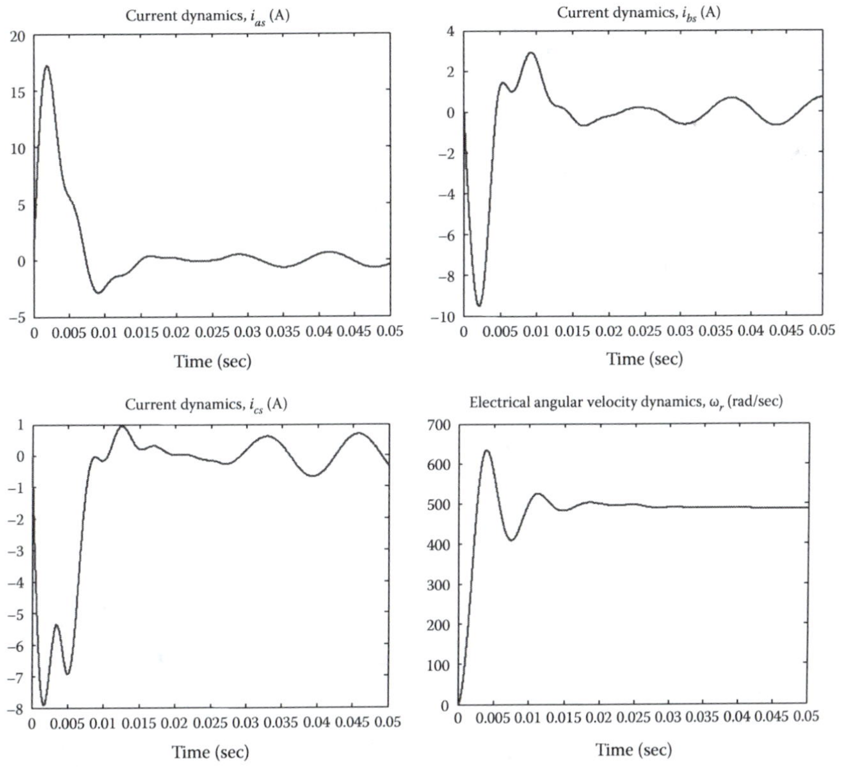

FIGURE 6.27 Transient dynamics for light loads ($T_L = 0.1$ N-m at $t = 0.025$ sec).

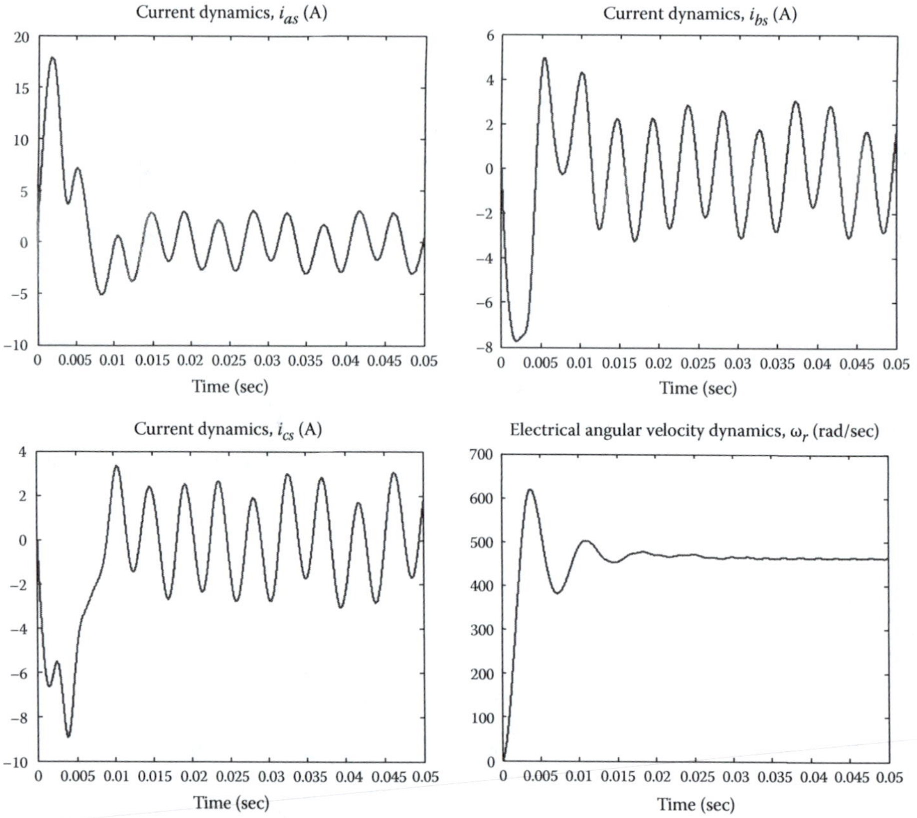

FIGURE 6.28 Transient dynamics for the rated load (T_{L0} = 0.2 N-m and T_L = 0.5 N-m at t = 0.025 sec).

Example 6.14: Modeling, Analysis, and Simulation of a Permanent-Magnet Synchronous Generator

The permanent-magnet synchronous machine considered in Example 6.13 is used as a generator. In an operating envelope $a_1 \neq 0, a_2 \neq 0, a_3 \neq 0$, and $\forall a_n = 0, n > 3$. With a Y-connected balanced resistive load with R_L, the equations of motion for a permanent-magnet synchronous generator is developed using (6.36)

$$\frac{di_{as}}{dt} = \frac{1}{L_{ss}}\left[-(r_s + R_L)i_{as} + \frac{1}{2}\bar{L}_m\frac{di_{bs}}{dt} + \frac{1}{2}\bar{L}_m\frac{di_{cs}}{dt} - \psi_m\cos\theta_r\left(a_1 + 3a_2\sin^2\theta_r + 5a_3\sin^4\theta_r\right)\omega_r\right],$$

$$\frac{di_{bs}}{dt} = \frac{1}{L_{ss}}\left[-(r_s + R_L)i_{bs} + \frac{1}{2}\bar{L}_m\frac{di_{as}}{dt} + \frac{1}{2}\bar{L}_m\frac{di_{cs}}{dt}\right.$$
$$\left.-\psi_m\cos\left(\theta_r + \frac{2}{3}\pi\right)\left(a_1 + 3a_2\sin^2\left(\theta_r + \frac{2}{3}\pi\right) + 5a_3\sin^4\left(\theta_r + \frac{2}{3}\pi\right)\right)\omega_r\right],$$

$$\frac{di_{cs}}{dt} = \frac{1}{L_{ss}}\left[-(r_s + R_L)i_{cs} + \frac{1}{2}\bar{L}_m\frac{di_{as}}{dt} + \frac{1}{2}\bar{L}_m\frac{di_{bs}}{dt}\right.$$
$$\left.-\psi_m\cos\left(\theta_r - \frac{2}{3}\pi\right)\left(a_1 + 3a_2\sin^2\left(\theta_r - \frac{2}{3}\pi\right) + 5a_3\sin^4\left(\theta_r - \frac{2}{3}\pi\right)\right)\omega_r\right].$$

The prime mover rotates the generator. The *torsional–mechanical* dynamics is

$$\frac{d\omega_r}{dt} = -\frac{P^2\psi_m}{4J}\left[i_{as}\cos\theta_r\left(a_1 + 3a_2\sin^2\theta_r + 5a_3\sin^4\theta_r\right)\right.$$

$$+ i_{bs}\cos\left(\theta_r + \frac{2}{3}\pi\right)\left(a_1 + 3a_2\sin^2\left(\theta_r + \frac{2}{3}\pi\right) + 5a_3\sin^4\left(\theta_r + \frac{2}{3}\pi\right)\right)$$

$$\left. + i_{cs}\cos\left(\theta_r - \frac{2}{3}\pi\right)\left(a_1 + 3a_2\sin^2\left(\theta_r - \frac{2}{3}\pi\right) + 5a_3\sin^4\left(\theta_r - \frac{2}{3}\pi\right)\right)\right] - \frac{B_m}{J}\omega_r + \frac{P}{2J}T_{PM},$$

$$\frac{d\theta_r}{dt} = \omega_r.$$

For no load and light loads when $R_L \in [100 \quad \infty]$ ohm, one has $a_1 = 1$ and $\forall a_n = 0, n > 1$. For the heavily loaded generator with $R_L \in [25 \quad 75]$ ohm, $a_1 = 1$, $a_2 = 0.05$, $a_3 = 0.02$, and $\forall a_n = 0$, $n > 3$. The Simulink model to simulate permanent-magnet synchronous generators is illustrated in Figure 6.29.

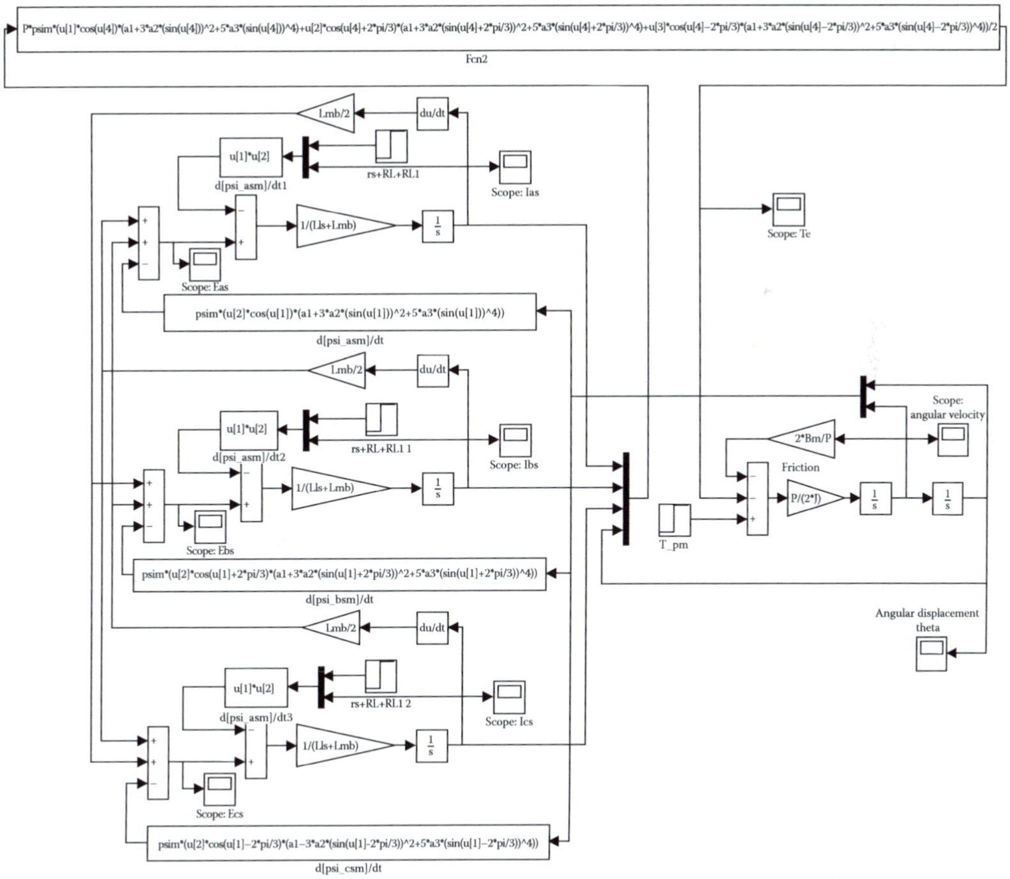

FIGURE 6.29 Simulink® diagram for permanent-magnet synchronous generators if $a_1 \neq 0$, $a_2 \neq 0$, $a_3 \neq 0$ and $\forall a_n = 0$, $n > 3$.

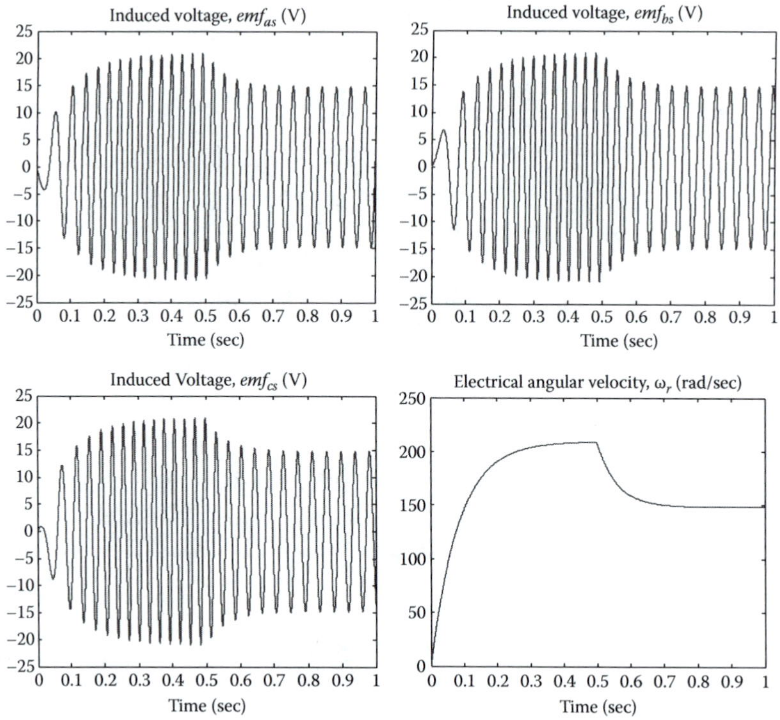

FIGURE 6.30 Generator dynamics: $T_{PM} = 0.05$ N-m, $R_L = 150$ ohm at $t = 0$ sec, and $R_L = 100$ ohm at $t = 0.5$ sec.

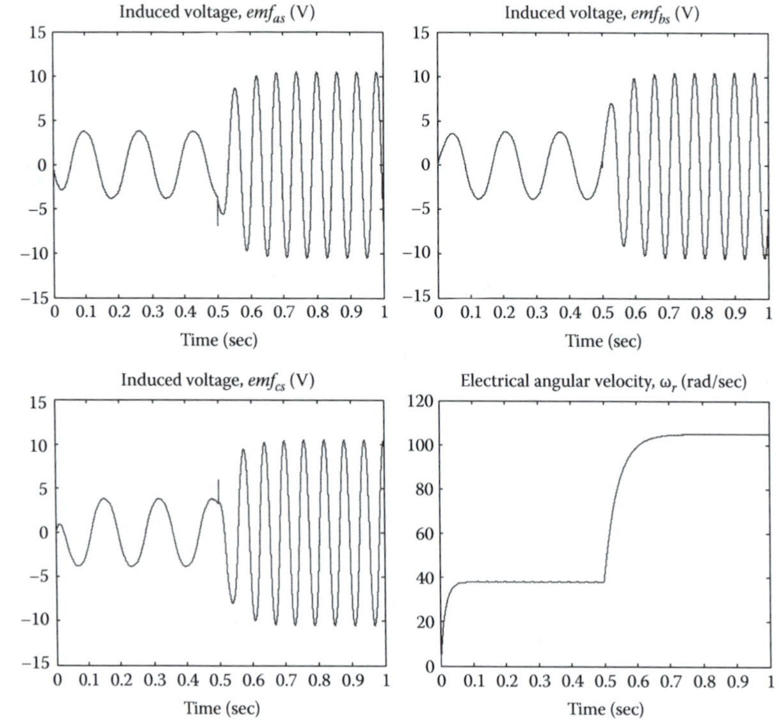

FIGURE 6.31 Generator dynamics: $T_{PM} = 0.05$ N-m, $R_L = 25$ ohm at $t = 0$ sec, and $R_L = 75$ ohm at $t = 0.5$ sec.

The generator dynamics and voltage generation are studied as the generator accelerates from stall with $T_{PM} = 0.05$ N-m. The resistors $R_L = 150$ ohm and $R_L = 100$ ohm are inserted in the balanced Y-load at $t = 0$ sec and $t = 0.5$ sec, respectively. Figures 6.30 illustrate the evolution of the induced *motional emfs*. The electrical angular velocity is depicted. The generator reaches the steady-state ω_r when $T_{PM} = T_e + T_{friction}$. The generator is loaded by the balanced Y-load with R_L. As the load increases (R_L reduces), the angular velocity and induced *emfs* decrease due to prime mover torque constraints, $T_{PM} = 0.05$ N-m. We assess the lightly loaded generator performance when $a_1 = 1$ and $\forall a_n = 0$, $n > 1$.

For the loaded generator, when $R_L \in \begin{bmatrix} 25 & 75 \end{bmatrix}$ ohm (peak and rated loads, respectively), we have $a_1 = 1$, $a_2 = 0.05$, $a_3 = 0.02$, and $\forall a_n = 0$, $n > 3$. The generator performance is studied if $R_L = 25$ ohm at $t = 0$ sec and $R_L = 75$ ohm at $t = 0.5$ sec. The evolution of the induced *emfs* and electrical angular velocity are documented in Figures 6.31. As R_L increases at $t = 0.5$ sec, the load reduces, and the angular velocity increases. The terminal phase voltage increases due to increase of ω_r. Permanent-magnet synchronous machines are designed with the attempt to guarantee near-optimal design with $a_1 \gg a_n$, $\forall n > 1$ within the rated operating envelope. This can be achieved by applying a consistent electromagnetic design, advanced technologies, enhanced processes, adequate magnets, appropriate materials, etc. ■

6.6 AXIAL TOPOLOGY PERMANENT-MAGNET SYNCHRONOUS MACHINES

In automotive, aerospace, biotechnology, consumer electronics, energy systems, marine, medical, robotics, and other applications, axial topology permanent-magnet synchronous machines could be a preferable solution. The axial topology permanent-magnet DC electromechanical motion devices were examined in Chapter 4. In synchronous machines, the stationary magnetic field is established by magnets placed on the rotor, and the AC phase voltages are applied to the stator windings as functions of θ_r. The images of axial topology permanent-magnet synchronous machines are shown in Figures 6.32.

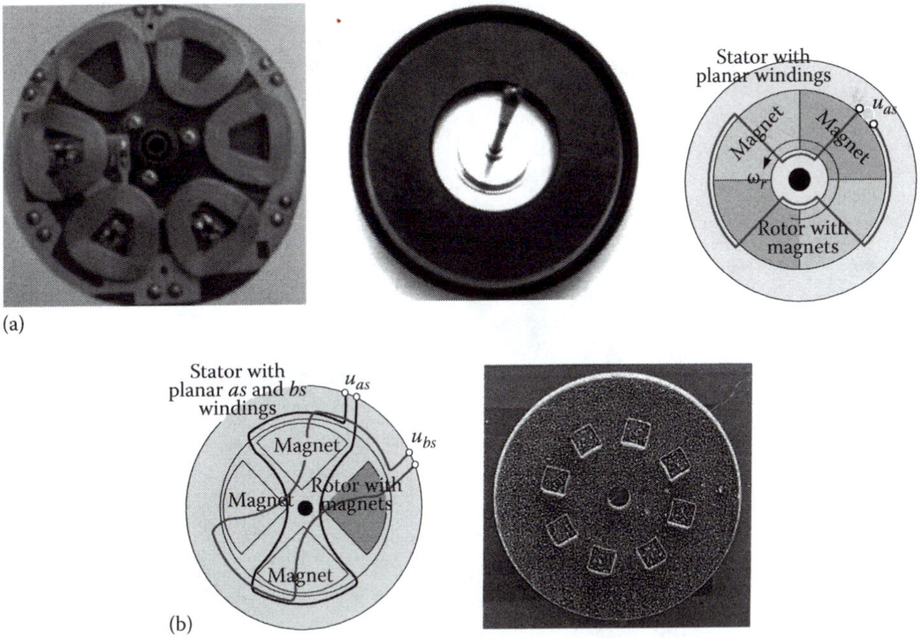

FIGURE 6.32 (a) Three-phase axial topology permanent-magnet synchronous machine—stator with planar windings and rotor with segmented magnets;
(b) Single- and two-phase axial topology synchronous machine and micromachined rotor with magnets.

Axial topology synchronous machines are fabricated in different sizes. Conventional and micromachining technologies are applied as illustrated in Figures 6.32. The advantages of axial topology electric machines are affordability, high-yield fabrication, robustness, assembly, and packaging simplicity. These advantages are mainly due to: (1) Segmented arrays of three-dimensionally optimized planar magnets, which can be pre- and post-assembly magnetized; (2) Precise three-dimensional winding and magnet fabrication and assembly ensuring optimal electromagnetics; (3) Rotor back ferromagnetic material is not required; (4) Simplicity of placing planar windings on the stator which is made from polymers or ceramics; (5) Optimal mechanical and thermal designs, ensuring robustness, heat exchange, etc.

The torque on a planar current loop of any size and shape in the uniform magnetic field is $\vec{T} = i\vec{s} \times \vec{B} = \vec{m} \times \vec{B}$. In permanent-magnet DC motion devices, the voltage is supplied to the armature windings on rotor using brushes and commutator. In permanent-magnet synchronous machines, the phase windings are on the stator. Hence, brushes and commutator are not needed. To develop the electromagnetic torque, the phase voltages u_{as}, u_{bs}, and u_{cs} are supplied as functions of the angular displacement θ_r.

The *effective as*, *bs*, and *cs* phase flux densities vary as a function of θ_r due to the angular displacement of the rotor with magnets relative to the stator windings. Depending on the topology, magnet magnetization, geometry, and shape, one finds distinct expressions for the *effective* $B_{as}(\theta_r)$, $B_{bs}(\theta_r)$, and $B_{cs}(\theta_r)$, which are the periodic functions of θ_r, $\theta_{rm} = 2\theta_r/P$. The number of poles P depends on the number of magnets N_m. If an optimal electromagnetic design is accomplished, planar windings are ideally placed, and magnets are ideally magnetized, one may ensure

$$B_{as}(\theta_r) = B_{\max}\sin(\theta_r),\quad B_{bs}(\theta_r) = B_{\max}\sin\left(\theta_r - \frac{2}{3}\pi\right),\quad \text{and}\quad B_{cs}(\theta_r) = B_{\max}\sin\left(\theta_r + \frac{2}{3}\pi\right),$$

where $B_{\max}$ is the *effective* flux density produced by the magnets as viewed from the winding ($B_{\max}$ depends on the magnets used, magnet-winding separation, number of coil layers, temperature, etc.).

For the segmented magnet topologies and configurations, the design-dependent *effective* phase flux densities $B_{as}(\theta_r)$, $B_{bs}(\theta_r)$, and $B_{cs}(\theta_r)$ are the periodic functions. For example, $B_{as}(\theta_r) = B_{\max}\,\mathrm{sgn}(\sin)$, $B_{bs}(\theta_r) = B_{\max}\,\mathrm{sgn}\left(\sin\left(\theta_r - \frac{2}{3}\pi\right)\right)$, $B_{cs}(\theta_r) = B_{\max}\,\mathrm{sgn}\left(\sin\left(\theta_r + \frac{2}{3}\pi\right)\right)$.

Using (6.33), let

$$B_{as} = B_{\max}\sum_{n=1}^{\infty} a_{Bn}\sin^{2n-1}\theta_r,\quad B_{bs} = B_{\max}\sum_{n=1}^{\infty} a_{Bn}\sin^{2n-1}\left(\theta_r - \frac{2}{3}\pi\right),\quad B_{cs} = B_{\max}\sum_{n=1}^{\infty} a_{Bn}\sin^{2n-1}\left(\theta_r + \frac{2}{3}\pi\right). \tag{6.43}$$

The stator is made from nonferromagnetic materials. The stator and rotor reluctances are zero, and the mutual inductances between the planar windings are negligibly small. Using (6.43), applying the number of turns and using the flux linkages $\psi = N\int_s \vec{B}\cdot d\vec{s}$, one obtains

$$\begin{bmatrix}\psi_{as}\\ \psi_{bs}\\ \psi_{cs}\end{bmatrix} = \begin{bmatrix}L_{ss} & 0 & 0\\ 0 & L_{ss} & 0\\ 0 & 0 & L_{ss}\end{bmatrix}\begin{bmatrix}i_{as}\\ i_{bs}\\ i_{cs}\end{bmatrix} + \psi_m\begin{bmatrix}\sum_{n=1}^{\infty} a_n\sin^{2n-1}\theta_r\\ \sum_{n=1}^{\infty} a_n\sin^{2n-1}\left(\theta_r - \frac{2}{3}\pi\right)\\ \sum_{n=1}^{\infty} a_n\sin^{2n-1}\left(\theta_r + \frac{2}{3}\pi\right)\end{bmatrix}. \tag{6.44}$$

Using the Kirchhoff second law, the differential equations are

$$u_{as} = r_s i_{as} + \frac{d\psi_{as}}{dt}, \quad u_{bs} = r_s i_{bs} + \frac{d\psi_{bs}}{dt}, \quad u_{cs} = r_s i_{cs} + \frac{d\psi_{cs}}{dt}, \tag{6.45}$$

$$\mathbf{u}_{abcs} = \mathbf{r}_s \mathbf{i}_{abcs} + \frac{d\psi_{abcs}}{dt}, \quad \begin{bmatrix} u_{as} \\ u_{bs} \\ u_{cs} \end{bmatrix} = \begin{bmatrix} r_s & 0 & 0 \\ 0 & r_s & 0 \\ 0 & 0 & r_s \end{bmatrix} \begin{bmatrix} i_{as} \\ i_{bs} \\ i_{cs} \end{bmatrix} + \begin{bmatrix} \dfrac{d\psi_{as}}{dt} \\ \dfrac{d\psi_{bs}}{dt} \\ \dfrac{d\psi_{cs}}{dt} \end{bmatrix}.$$

The electromagnetic torque is

$$T_e = \frac{\partial W_c}{\partial \theta_r} = \frac{P\psi_m}{2} \left[i_{as} \sum_{n=1}^{\infty} (2n-1) a_n \cos\theta_r \sin^{2n-2}\theta_r + i_{bs} \sum_{n=1}^{\infty} (2n-1) a_n \cos\left(\theta_r - \frac{2}{3}\pi\right) \sin^{2n-2}\left(\theta_r - \frac{2}{3}\pi\right) \right.$$
$$\left. + i_{cs} \sum_{n=1}^{\infty} (2n-1) a_n \cos\left(\theta_r + \frac{2}{3}\pi\right) \sin^{2n-2}\left(\theta_r + \frac{2}{3}\pi\right) \right]. \tag{6.46}$$

The *torsional–mechanical* dynamics is given by

$$\frac{d\omega_r}{dt} = \frac{P^2\psi_m}{4J} \left[i_{as} \sum_{n=1}^{\infty} (2n-1) a_n \cos\theta_r \sin^{2n-2}\theta_r + i_{bs} \sum_{n=1}^{\infty} (2n-1) a_n \cos\left(\theta_r - \frac{2}{3}\pi\right) \sin^{2n-2}\left(\theta_r - \frac{2}{3}\pi\right) \right.$$
$$\left. + i_{cs} \sum_{n=1}^{\infty} (2n-1) a_n \cos\left(\theta_r + \frac{2}{3}\pi\right) \sin^{2n-2}\left(\theta_r + \frac{2}{3}\pi\right) \right] - \frac{B_m}{J}\omega_r - \frac{P}{2J} T_L,$$

$$\frac{d\theta_r}{dt} = \omega_r. \tag{6.47}$$

The differential equations are found using (6.45) and (6.47). The near-balanced current and voltage sets can be derived. Examples 6.15 through 6.17 illustrate the applications of our findings.

Example 6.15: Single-Phase Axial Topology Permanent-Magnet Synchronous Machine

Consider a single-phase permanent-magnet synchronous motor, depicted in Figure 6.32b. Let

$$B_{as}(\theta_r) = B_{\max}\sin\theta_r.$$

Using the coenergy

$$W_c\left(i_{as}, \theta_r\right) = \frac{1}{2}\psi_{as}(\theta_r) i_{as},$$

we have

$$T_e = \frac{\partial W_c\left(i_{as}, \theta_r\right)}{\partial \theta_r} = \frac{P\psi_m}{2}\cos\theta_r i_{as}.$$

If i_{as} is the DC current, the electromagnetic torque is zero on average.

For the phase current $i_{as} = i_M \cos \theta_r$, one has

$$T_e = \frac{1}{2} P \psi_m i_M \cos^2 \theta_r, \quad T_{e\,average} \neq 0.$$

However, there is a torque ripple. The equations of motion are found using the Kirchhoff law $u_{as} = r_s i_{as} + \frac{d\psi_{as}}{dt}$, $\psi_{as} = L_{ss} i_{as} + \psi_m \sin \theta_r$ and Newtonian mechanics. We have

$$\frac{di_{as}}{dt} = \frac{1}{L_{ss}} \left(-r_s i_{as} - \psi_m \cos \theta_r \omega_r + u_{as} \right),$$

$$\frac{d\omega_r}{dt} = \frac{P^2 \psi_m}{4J} \cos \theta_r i_{as} - \frac{B_m}{J} \omega_r - \frac{P}{2J} T_L,$$

$$\frac{d\theta_r}{dt} = \omega_r.$$

■

Example 6.16: Two-Phase Axial Topology Permanent-Magnet Synchronous Motors

Consider two-phase permanent-magnet synchronous machines, as depicted in Figure 6.32b. Let

$$\psi_{asm} = \psi_m \sin \theta_r \quad \text{and} \quad \psi_{bsm} = \psi_m \cos \theta_r.$$

The electromagnetic torque is

$$T_e = \frac{1}{2} P \psi_m \left(i_{as} \cos \theta_r - i_{bs} \sin \theta_r \right).$$

The balanced current set is $i_{as} = i_M \cos \theta_r$, $i_{bs} = -i_M \sin \theta_r$. The T_e is maximized, and $T_e = \frac{1}{2} P \psi_m i_M$.

Consider a case when $\psi_{asm} = \psi_m \sin^5 \theta_r$ and $\psi_{bsm} = \psi_m \cos^5 \theta_r$, $a_3 = 1$ and all others $\forall a_n = 0$. The electromagnetic torque is

$$T_e = \frac{\partial W_c}{\partial \theta_r} = \frac{5}{2} P \psi_m \left(i_{as} \cos \theta_r \sin^4 \theta_r - i_{bs} \sin \theta_r \cos^4 \theta_r \right).$$

Let $N = 20$, $A_{eq} = 0.001$, $B_{\max} = 1$ and $P = 8$. Assume that the phase currents are $i_{as} = i_M \cos \theta_r$ and $i_{bs} = -i_M \sin \theta_r$, $i_M = 2$ A. Differentiation, calculations, and plotting are performed using the Symbolic Toolbox. The following MATLAB file with comments is used

```
x=sym('x'); % To use a symbolic variable, create an object of type SYM
N=20; Aeq=0.001; Bmax=1; psim=N*Aeq*Bmax; P=8; iM=2;
psias=psim*sin(x)^5; psibs=psim*cos(x)^5;
% Differentiate y1 and y2 using the DIFF command
dpsias=diff(psias); dpsibs=diff(psibs);
ias=iM*cos(x); ibs=-iM*sin(x); % Phase currents
% Derive and plot the electromagnetic torque
Te=P*(dpsias*ias+dpsibs*ibs)/2, Te=simplify(Te), ezplot(Te)
```

The results of the calculations are reported in the Command Window as

```
Te = (4*cos(x)^2*sin(x)^4)/5 + (4*cos(x)^4*sin(x)^2)/5
Te = 1/10 - cos(4*x)/10
```

The electromagnetic torque is

$$T_e = \frac{1}{10}\left(1 - \cos 4\theta_r\right) \text{ N-m.}$$

The plot for T_e is illustrated in Figure 6.33a. The electromagnetic torque varies. The torque ripple is an undesirable phenomenon due to losses, noise, vibration, etc. One may derive the balance current and voltage sets. From

$$T_e = \frac{5}{2} P\psi_m \left(i_{as}\cos\theta_r \sin^4\theta_r - i_{bs}\sin\theta_r \cos^4\theta_r\right),$$

the balanced current set is

$$i_{as} = i_M \frac{\cos\theta_r}{\sin^4\theta_r}, \quad i_{bs} = -i_M \frac{\sin\theta_r}{\cos^4\theta_r}\sin^4\theta_r, \quad \left|i_{as}\right| \le i_{\max}, \left|i_{bs}\right| \le i_{\max}.$$

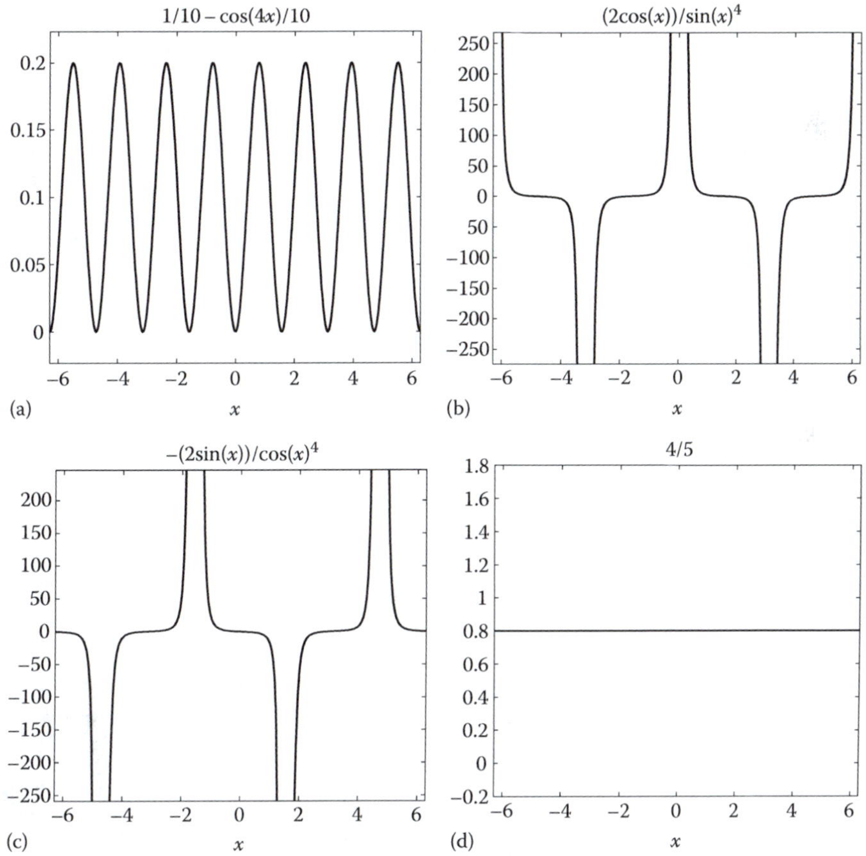

FIGURE 6.33 (a) Electromagnetic torque $T_e = \frac{4}{5}\left(1 - \cos^2\theta_r\right)\cos^2\theta_r$ N-m;

(b) Phase current $i_{as} = i_M \frac{\cos\theta_r}{\sin^4\theta_r}$;

(c) Phase current $i_{bs} = -i_M \frac{\sin\theta_r}{\cos^4\theta_r}\sin^4\theta_r$;

(d) Plot for $T_e = 0.8$ N-m.

If the phase current limits are not reached, one has

```
ias=iM*cos(x)/sin(x)^4; ibs=-iM*sin(x)/cos(x)^4; % Phase currents
ezplot(ias); pause; ezplot(ibs); pause;
% Derive and plot the electromagnetic torque
Te=P*(dpsias*ias+dpsibs*ibs)/2, Te=simplify(Te), ezplot(Te)
```

We obtain

```
Te = (4*cos(x)^2)/5 + (4*sin(x)^2)/5
Te = 4/5
```

That is, $T_e = 0.8$ N-m. The phase currents i_{as} and i_{bs} are depicted in Figures 6.33b and c. The T_e is plotted in Figure 6.33d. Due to singularity and saturation, it is not always possible to guarantee the balanced current and voltage sets. ■

Example 6.17: Three-Phase Axial Topology Permanent-Magnet Synchronous Machines

For a three-phase axial topology machine, depicted in Figure 6.32a, one may find the unknown parameters. The parameters r_a, L_{ss}, and J are directly measured. The ψ_m and a_n are found experimentally measuring the *motional emfs* by rotating the synchronous generator at different loads. The experimental results are illustrated in Figure 6.34. At no load, for the steady-state angular velocities $\omega_r = 956$ rad/sec and $\omega_r = 1382$ rad/sec, the output voltages are 29.9 and 39.7 V. One finds ideal sinusoidal *motional emfs* (induced phase voltages) $emf_{as\omega} = \psi_m \cos\theta_r\omega_r$. We have $\psi_m = 0.0288$ and $\psi_m = 0.0313$ V-sec/rad, $a_1 = 1$ ($a_1 \neq 0$), $\forall a_n = 0$, $n > 1$. There are variations $\psi_m \in \left[\psi_{m\min} \quad \psi_{m\max}\right]$ due to nonlinearities, temperature variations, etc.

For the loaded generator, using (6.44) for the *as* phase,

$$\psi_{asm} = \psi_m \sum_{n=1}^{\infty} a_n \sin^{2n-1}\theta_r.$$

The *motional emf* is

$$emf_{as\omega} = \psi_m \left(\sum_{n=1}^{\infty} (2n-1) a_n \cos\theta_r \sin^{2n-2}\theta_r \right) \omega_r.$$

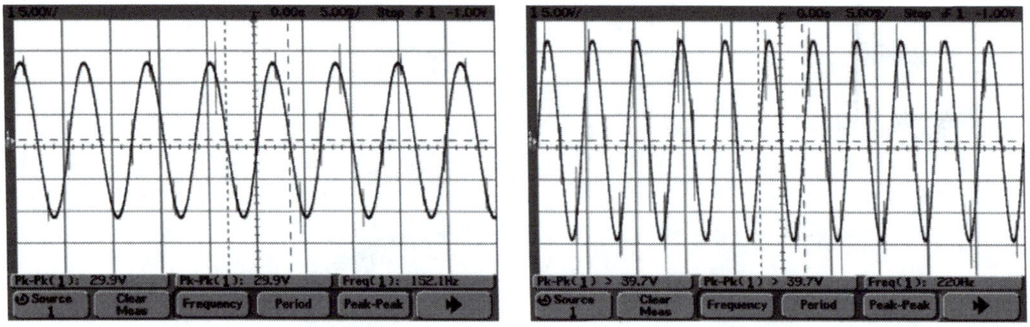

FIGURE 6.34 Induced $emf_{as\omega} = \psi_m \cos\theta_r\omega_r$ if the generator is rotated at 956 and 1382 rad/sec.

The experimental results and numeric analysis yield

$$emf_{as\omega} = \psi_m \cos\theta_r (a_1 + 3a_2 \sin^2\theta_r + 5a_3 \sin^4\theta_r)\omega_r,$$

e.g., $a_1 \neq 0$, $a_2 \neq 0$, $a_3 \neq 0$ while $\forall a_n = 0$, $n > 3$.

For the rated load, we found $\psi_m = 0.03$ V-sec/rad, $a_1 = 0.85$, $a_2 = 0.06$, $a_3 = 0.04$, and $\forall a_n = 0$, $n > 3$.

Using (6.44), the experimental results yield

$$\psi_{as} = L_{ss} i_{as} + \psi_m (a_1 \sin\theta_r + a_2 \sin^3\theta_r + a_3 \sin^5\theta_r),$$

$$\psi_{bs} = L_{ss} i_{bs} + \psi_m \left(a_1 \sin\left(\theta_r + \frac{2}{3}\pi\right) + a_2 \sin^3\left(\theta_r + \frac{2}{3}\pi\right) + a_3 \sin^5\left(\theta_r + \frac{2}{3}\pi\right)\right)$$

and

$$\psi_{cs} = L_{ss} i_{cs} + \psi_m \left(a_1 \sin\left(\theta_r - \frac{2}{3}\pi\right) + a_2 \sin^3\left(\theta_r - \frac{2}{3}\pi\right) + a_3 \sin^5\left(\theta_r - \frac{2}{3}\pi\right)\right).$$

For the motor, applying (6.45), we have

$$\frac{di_{as}}{dt} = \frac{1}{L_{ss}} \left[-r_s i_{as} - \psi_m \cos\theta_r \left(a_1 + 3a_2 \sin^2\theta_r + 5a_3 \sin^4\theta_r \right)\omega_r + u_{as} \right],$$

$$\frac{di_{bs}}{dt} = \frac{1}{L_{ss}} \left[-r_s i_{bs} - \psi_m \cos\left(\theta_r + \frac{2}{3}\pi\right)\left(a_1 + 3a_2 \sin^2\left(\theta_r + \frac{2}{3}\pi\right) + 5a_3 \sin^4\left(\theta_r + \frac{2}{3}\pi\right)\right)\omega_r + u_{bs} \right],$$

$$\frac{di_{cs}}{dt} = \frac{1}{L_{ss}} \left[-r_s i_{cs} - \psi_m \cos\left(\theta_r - \frac{2}{3}\pi\right)\left(a_1 + 3a_2 \sin^2\left(\theta_r - \frac{2}{3}\pi\right) + 5a_3 \sin^4\left(\theta_r - \frac{2}{3}\pi\right)\right)\omega_r + u_{cs} \right]. \tag{6.48}$$

The expression for the electromagnetic torque (6.46) is

$$\begin{aligned} T_e = \frac{1}{2} P \psi_m \Bigg[& \cos\theta_r \left(a_1 + 3a_2 \sin^2\theta_r + 5a_3 \sin^4\theta_r \right) i_{as} \\ & + \cos\left(\theta_r + \frac{2}{3}\pi\right)\left(a_1 + 3a_2 \sin^2\left(\theta_r + \frac{2}{3}\pi\right) + 5a_3 \sin^4\left(\theta_r + \frac{2}{3}\pi\right)\right) i_{bs} \\ & + \cos\left(\theta_r - \frac{2}{3}\pi\right)\left(a_1 + 3a_2 \sin^2\left(\theta_r - \frac{2}{3}\pi\right) + 5a_3 \sin^4\left(\theta_r - \frac{2}{3}\pi\right)\right) i_{cs} \Bigg]. \end{aligned}$$

The *torsional–mechanical* equations (6.47) are

$$\begin{aligned} \frac{d\omega_r}{dt} = \frac{P^2 \psi_m}{4J} \Bigg[& \cos\theta_r \left(a_1 + 3a_2 \sin^2\theta_r + 5a_3 \sin^4\theta_r \right) i_{as} \\ & + \cos\left(\theta_r + \frac{2}{3}\pi\right)\left(a_1 + 3a_2 \sin^2\left(\theta_r + \frac{2}{3}\pi\right) + 5a_3 \sin^4\left(\theta_r + \frac{2}{3}\pi\right)\right) i_{bs} \\ & + \cos\left(\theta_r - \frac{2}{3}\pi\right)\left(a_1 + 3a_2 \sin^2\left(\theta_r - \frac{2}{3}\pi\right) + 5a_3 \sin^4\left(\theta_r - \frac{2}{3}\pi\right)\right) i_{cs} \Bigg] - \frac{B_m}{J}\omega_r - \frac{P}{2J} T_L, \end{aligned}$$

$$\frac{d\theta_r}{dt} = \omega_r. \tag{6.49}$$

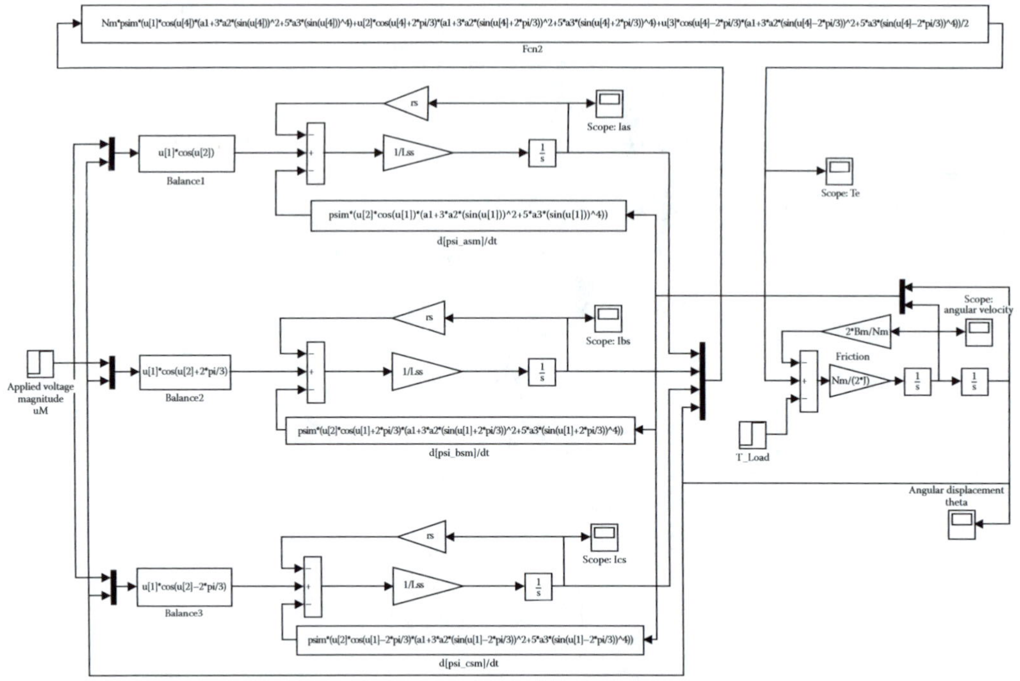

FIGURE 6.35 Simulink® diagram for axial topology permanent-magnet synchronous motors ($a_1 \neq 0$, $a_2 \neq 0$, $a_3 \neq 0$ while $\forall a_n = 0$, $n > 3$).

Using (6.48) and (6.49), the Simulink model to simulate axial topology permanent-magnet synchronous motors ($a_1 \neq 0, a_2 \neq 0, a_3 \neq 0, \forall a_n = 0, n > 3$) is depicted in Figure 6.35. The phase voltages are

$$u_{as} = \sqrt{2}u_M \cos\theta_r, \quad u_{bs} = \sqrt{2}u_M \cos\left(\theta_r + \frac{2}{3}\pi\right), \quad \text{and} \quad u_{cs} = \sqrt{2}u_M \cos\left(\theta_r - \frac{2}{3}\pi\right).$$

The motor parameters are $P = 8$, $r_s = 13.5$ ohm, $L_{ss} = 0.035$ H, $B_m = 0.0000005$ N-m-sec/rad, and $J = 0.00001$ kg-m^2. We found ψ_m=0.03 V-sec/rad with the following a_i: (1) $a_1 = 1$ and $\forall a_n = 0$, $n > 1$ (no load); (2) $a_1 = 0.85$, $a_2 = 0.06$, $a_3 = 0.04$, and $\forall a_n = 0$, $n > 3$ (rated load). The motor parameters are uploaded as

```
P=8; uM=50; rs=13.5; Lss=0.035; Bm=0.0000005; J=0.00001;
psim=0.03; a1=1; a2=0; a3=0;               % No load
psim=0.03; a1=0.85; a2=0.06; a3=0.04; % Rated load
```

The motor dynamics is studied as the motor accelerates from stall with the rated voltage applied ($u_M = 50$ V) if $T_{L0} = 0.005$ N-m. The load torque $T_L = 0.01$ N-m is applied at $t = 0.5$ sec. Figure 6.36a illustrates the dynamics of $i_{as}(t)$, $i_{bs}(t)$, $i_{cs}(t)$, and $\omega_r(t)$. At $T_L = 0.005$ N-m, the steady-state ω_r is 1470 rad/sec, and the angular velocity decreases as the load applied at $t = 0.5$ sec.

For the loaded motor, $a_1 = 0.85$, $a_2 = 0.06$, $a_3 = 0.04$, and $\forall a_n = 0$, $n > 3$. The motor accelerates with the rated voltage applied. At $t = 0$ sec, the load torque is $T_{L0} = 0.015$ N-m. Load $T_L = 0.03$ N-m is applied at $t = 0.5$ sec. The evolutions of the motor variables are reported in Figure 6.36b. One observes the torque ripple. These undesirable phenomena can be minimized applying the balanced voltages set.

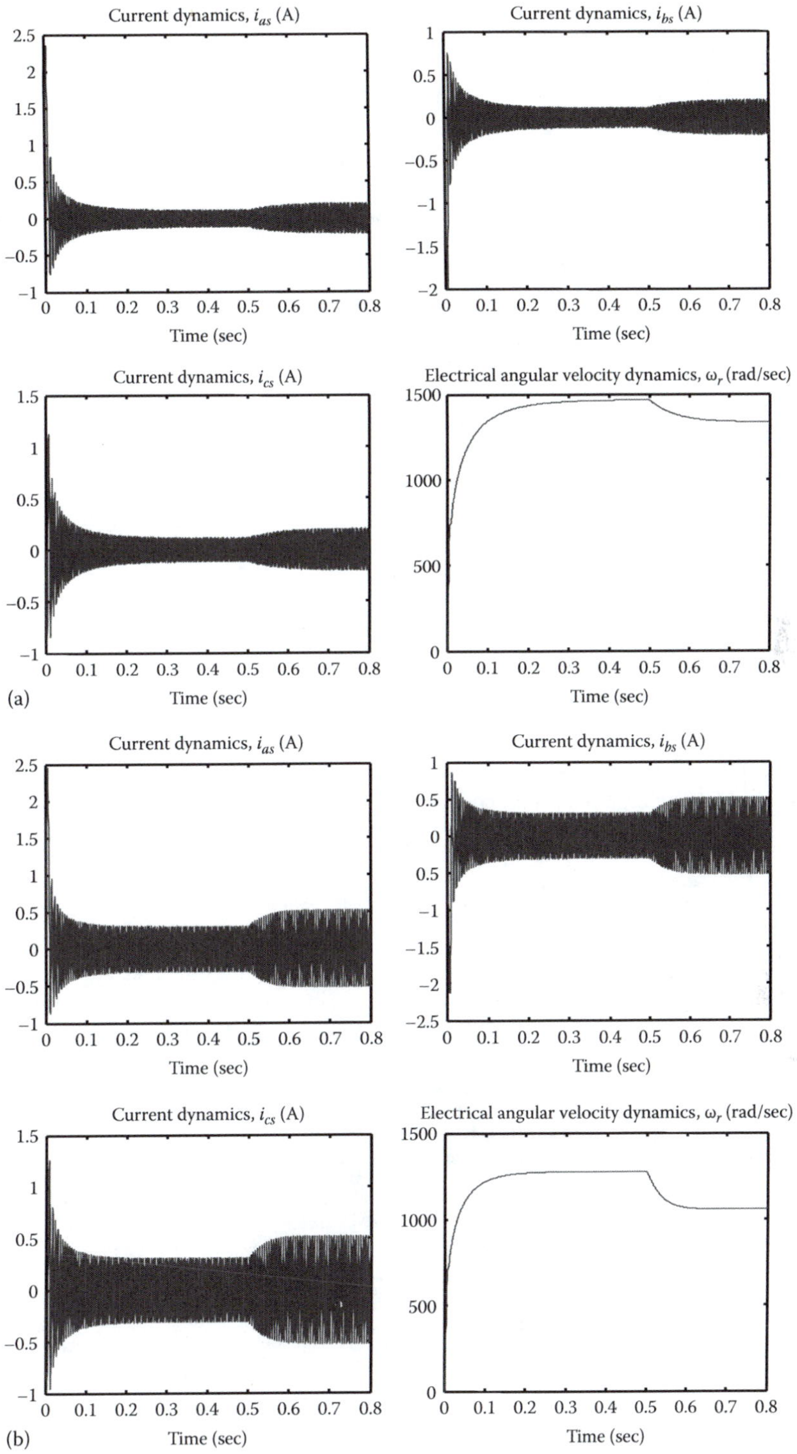

FIGURE 6.36 (a) Dynamics of an axial topology permanent-magnet synchronous motor, $a_1 = 1$, $\forall a_n = 0$, $n > 1$;
(b) Dynamics of an axial topology permanent-magnet synchronous motor ($a_1 = 0.85$, $a_2 = 0.06$, $a_3 = 0.04$, and $\forall a_n = 0$, $n > 3$), $T_{L0} = 0.015$ N-m and $T_L = 0.03$ N-m at $t = 0.5$ sec.

Using (6.48) and (6.49), for the generator, one finds

$$\frac{di_{as}}{dt}=\frac{1}{L_{ss}}\left[-r_s i_{as}-\psi_m\cos\theta_r\left(a_1+3a_2\sin^2\theta_r+5a_3\sin^4\theta_r\right)\omega_r\right],$$

$$\frac{di_{bs}}{dt}=\frac{1}{L_{ss}}\left[-r_s i_{bs}-\psi_m\cos\left(\theta_r+\frac{2}{3}\pi\right)\left(a_1+3a_2\sin^2\left(\theta_r+\frac{2}{3}\pi\right)+5a_3\sin^4\left(\theta_r+\frac{2}{3}\pi\right)\right)\omega_r\right],$$

$$\frac{di_{cs}}{dt}=\frac{1}{L_{ss}}\left[-r_s i_{cs}-\psi_m\cos\left(\theta_r-\frac{2}{3}\pi\right)\left(a_1+3a_2\sin^2\left(\theta_r-\frac{2}{3}\pi\right)+5a_3\sin^4\left(\theta_r-\frac{2}{3}\pi\right)\right)\omega_r\right],$$

$$\frac{d\omega_r}{dt}=-\frac{P^2\psi_m}{4J}\left[\cos\theta_r\left(a_1+3a_2\sin^2\theta_r+5a_3\sin^4\theta_r\right)i_{as}\right.$$
$$+\cos\left(\theta_r+\frac{2}{3}\pi\right)\left(a_1+3a_2\sin^2\left(\theta_r+\frac{2}{3}\pi\right)+5a_3\sin^4\left(\theta_r+\frac{2}{3}\pi\right)\right)i_{bs}$$
$$\left.+\cos\left(\theta_r-\frac{2}{3}\pi\right)\left(a_1+3a_2\sin^2\left(\theta_r-\frac{2}{3}\pi\right)+5a_3\sin^4\left(\theta_r-\frac{2}{3}\pi\right)\right)i_{cs}\right]-\frac{B_m}{J}\omega_r+\frac{P}{2J}T_{PM},$$

$$\frac{d\theta_r}{dt}=\omega_r.$$

■

6.7 CONVENTIONAL DC-CURRENT EXITED THREE-PHASE SYNCHRONOUS MACHINES

Consider three-phase synchronous machines with the stator windings and the rotor field *fr* winding. Applying the DC field voltage u_{fr}, one forms the electromagnet with the north and south poles. We analyze synchronous machines as depicted in Figure 6.37.

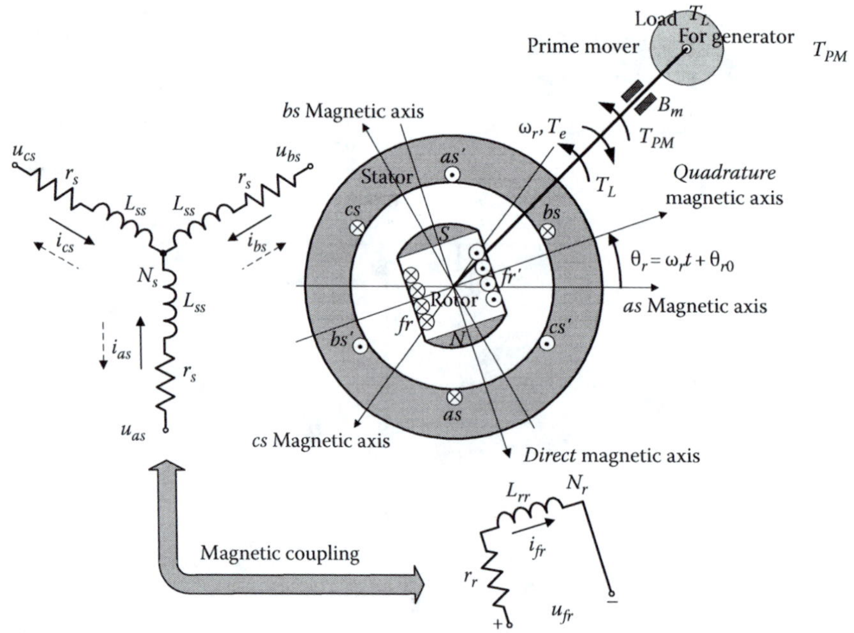

FIGURE 6.37 Three-phase Y-connected DC-current exited synchronous machine (directions of the stator currents i_{as}, i_{bs}, and i_{cs} are shown by the solid arrows for motor operation and dashed arrows for the generator operation).

6.7.1 Dynamics of Synchronous Motors in the Machine Variables

The application of Kirchhoff's voltage law yields four differential equations for the stator (*as*, *bs*, and *cs*) and rotor *fr* windings. For motor, we have

$$u_{as} = r_s i_{as} + \frac{d\psi_{as}}{dt}, \quad u_{bs} = r_s i_{bs} + \frac{d\psi_{bs}}{dt}, \quad u_{cs} = r_s i_{cs} + \frac{d\psi_{cs}}{dt}, \quad u_{fr} = r_r i_{fr} + \frac{d\psi_{fr}}{dt}, \tag{6.50}$$

$$\mathbf{u}_{abcs} = \mathbf{r}_s \mathbf{i}_{abcs} + \frac{d\psi_{abcs}}{dt}, \quad u_{fr} = r_r i_{fr} + \frac{d\psi_{fr}}{dt}, \quad \text{or} \quad \begin{bmatrix} u_{as} \\ u_{bs} \\ u_{cs} \\ u_{fr} \end{bmatrix} = \begin{bmatrix} r_s & 0 & 0 & 0 \\ 0 & r_s & 0 & 0 \\ 0 & 0 & r_s & 0 \\ 0 & 0 & 0 & r_r \end{bmatrix} \begin{bmatrix} i_{as} \\ i_{bs} \\ i_{cs} \\ i_{fr} \end{bmatrix} + \begin{bmatrix} \frac{d\psi_{as}}{dt} \\ \frac{d\psi_{bs}}{dt} \\ \frac{d\psi_{cs}}{dt} \\ \frac{d\psi_{fr}}{dt} \end{bmatrix},$$

where the flux linkages are

$$\psi_{as} = L_{asas} i_{as} + L_{asbs} i_{bs} + L_{ascs} i_{cs} + L_{asfr} i_{fr}, \quad \psi_{bs} = L_{bsas} i_{as} + L_{bsbs} i_{bs} + L_{bscs} i_{cs} + L_{bsfr} i_{fr},$$

$$\psi_{cs} = L_{csas} i_{as} + L_{csbs} i_{bs} + L_{cscs} i_{cs} + L_{csfr} i_{fr}, \quad \psi_{fr} = L_{fras} i_{as} + L_{frbs} i_{bs} + L_{frcs} i_{cs} + L_{frfr} i_{fr}.$$

For synchronous reluctance machines, the self- and mutual inductances were found in Section 6.2.2. We use the following expressions for the stator and rotor flux linkages

$$\begin{aligned}
\psi_{as} &= \left(L_{ls} + \bar{L}_m - L_{\Delta m}\cos 2\theta_r\right) i_{as} + \left(-\frac{1}{2}\bar{L}_m - L_{\Delta m}\cos 2\left(\theta_r - \frac{1}{3}\pi\right)\right) i_{bs} \\
&\quad + \left(-\frac{1}{2}\bar{L}_m - L_{\Delta m}\cos 2\left(\theta_r + \frac{1}{3}\pi\right)\right) i_{cs} + L_{md}\sin\theta_r i_{fr}, \\
\psi_{bs} &= \left(-\frac{1}{2}\bar{L}_m - L_{\Delta m}\cos 2\left(\theta_r - \frac{1}{3}\pi\right)\right) i_{as} + \left(L_{ls} + \bar{L}_m - L_{\Delta m}\cos 2\left(\theta_r - \frac{2}{3}\pi\right)\right) i_{bs} \\
&\quad + \left(-\frac{1}{2}\bar{L}_m - L_{\Delta m}\cos 2\theta_r\right) i_{cs} + L_{md}\sin\left(\theta_r - \frac{2}{3}\pi\right) i_{fr}, \\
\psi_{cs} &= \left(-\frac{1}{2}\bar{L}_m - L_{\Delta m}\cos 2\left(\theta_r + \frac{1}{3}\pi\right)\right) i_{as} + \left(-\frac{1}{2}\bar{L}_m - L_{\Delta m}\cos 2\theta_r\right) i_{bs} \\
&\quad + \left(L_{ls} + \bar{L}_m - L_{\Delta m}\cos 2\left(\theta_r + \frac{2}{3}\pi\right)\right) i_{cs} + L_{md}\sin\left(\theta_r + \frac{2}{3}\pi\right) i_{fr}, \\
\psi_{fr} &= L_{md}\sin\theta_r i_{as} + L_{md}\sin\left(\theta_r - \frac{2}{3}\pi\right) i_{bs} + L_{md}\sin\left(\theta_r + \frac{2}{3}\pi\right) i_{cs} + \left(L_{lf} + L_{mf}\right) i_{fr}.
\end{aligned} \tag{6.51}$$

where $L_{md} = \dfrac{N_r N_s}{\Re_{md}}$ and $L_{mf} = \dfrac{N_r^2}{\Re_{md}}$.

The self- and mutual inductances mapping $\mathbf{L}_{abcs/fr}$ is

$$\mathbf{L}_{abcs/fr} = \begin{bmatrix} L_{ls} + \bar{L}_m - L_{\Delta m}\cos 2\theta_r & -\frac{1}{2}\bar{L}_m - L_{\Delta m}\cos 2\left(\theta_r - \frac{1}{3}\pi\right) & -\frac{1}{2}\bar{L}_m - L_{\Delta m}\cos 2\left(\theta_r + \frac{1}{3}\pi\right) & L_{md}\sin\theta_r \\ -\frac{1}{2}\bar{L}_m - L_{\Delta m}\cos 2\left(\theta_r - \frac{1}{3}\pi\right) & L_{ls} + \bar{L}_m - L_{\Delta m}\cos 2\left(\theta_r - \frac{2}{3}\pi\right) & -\frac{1}{2}\bar{L}_m - L_{\Delta m}\cos 2\theta_r & L_{md}\sin\left(\theta_r - \frac{2}{3}\pi\right) \\ -\frac{1}{2}\bar{L}_m - L_{\Delta m}\cos 2\left(\theta_r + \frac{1}{3}\pi\right) & -\frac{1}{2}\bar{L}_m - L_{\Delta m}\cos 2\theta_r & L_{ls} + \bar{L}_m - L_{\Delta m}\cos 2\left(\theta_r + \frac{2}{3}\pi\right) & L_{md}\sin\left(\theta_r + \frac{2}{3}\pi\right) \\ L_{md}\sin\theta_r & L_{md}\sin\left(\theta_r - \frac{2}{3}\pi\right) & L_{md}\sin\left(\theta_r + \frac{2}{3}\pi\right) & L_{lf} + L_{mf} \end{bmatrix}.$$

One obtains

$$\boldsymbol{\psi}_{abcsfr} = \mathbf{L}_{abcs/fr}\mathbf{i}_{abcsfr}$$

$$= \begin{bmatrix} L_{ls} + \bar{L}_m - L_{\Delta m}\cos 2\theta_r & -\frac{1}{2}\bar{L}_m - L_{\Delta m}\cos 2\left(\theta_r - \frac{1}{3}\pi\right) & -\frac{1}{2}\bar{L}_m - L_{\Delta m}\cos 2\left(\theta_r + \frac{1}{3}\pi\right) & L_{md}\sin\theta_r \\ -\frac{1}{2}\bar{L}_m - L_{\Delta m}\cos 2\left(\theta_r - \frac{1}{3}\pi\right) & L_{ls} + \bar{L}_m - L_{\Delta m}\cos 2\left(\theta_r - \frac{1}{3}\pi\right) & -\frac{1}{2}\bar{L}_m - L_{\Delta m}\cos 2\theta_r & L_{md}\sin\left(\theta_r - \frac{2}{3}\pi\right) \\ -\frac{1}{2}\bar{L}_m - L_{\Delta m}\cos 2\left(\theta_r + \frac{1}{3}\pi\right) & -\frac{1}{2}\bar{L}_m - L_{\Delta m}\cos 2\theta_r & L_{ls} + \bar{L}_m - L_{\Delta m}\cos 2\left(\theta_r + \frac{2}{3}\pi\right) & L_{md}\sin\left(\theta_r + \frac{2}{3}\pi\right) \\ L_{md}\sin\theta_r & L_{md}\sin\left(\theta_r - \frac{2}{3}\pi\right) & L_{md}\sin\left(\theta_r + \frac{2}{3}\pi\right) & L_{lf} + L_{mf} \end{bmatrix}\begin{bmatrix} i_{as} \\ i_{bs} \\ i_{cs} \\ i_{fr} \end{bmatrix}.$$

From (6.50) and (6.51), we have the following set of four differential equations

$$\frac{d\left[\left(L_{ls} + \bar{L}_m - L_{\Delta m}\cos 2\theta_r\right)i_{as} + \left(-\frac{1}{2}\bar{L}_m - L_{\Delta m}\cos 2\left(\theta_r - \frac{1}{3}\pi\right)\right)i_{bs} + \left(-\frac{1}{2}\bar{L}_m - L_{\Delta m}\cos 2\left(\theta_r + \frac{1}{3}\pi\right)\right)i_{cs} + L_{md}\sin\theta_r i_{fr}\right]}{dt}$$
$$= -r_s i_{as} + u_{as},$$

$$\frac{d\left[\left(-\frac{1}{2}\bar{L}_m - L_{\Delta m}\cos 2\left(\theta_r - \frac{1}{3}\pi\right)\right)i_{as} + \left(L_{ls} + \bar{L}_m - L_{\Delta m}\cos 2\left(\theta_r - \frac{2}{3}\pi\right)\right)i_{bs} + \left(-\frac{1}{2}\bar{L}_m - L_{\Delta m}\cos 2\theta_r\right)i_{cs} + L_{md}\sin\left(\theta_r - \frac{2}{3}\pi\right)i_{fr}\right]}{dt}$$
$$= -r_s i_{bs} + u_{bs},$$

$$\frac{d\left[\left(-\frac{1}{2}\bar{L}_m - L_{\Delta m}\cos 2\left(\theta_r + \frac{1}{3}\pi\right)\right)i_{as} + \left(-\frac{1}{2}\bar{L}_m - L_{\Delta m}\cos 2\theta_r\right)i_{bs} + \left(L_{ls} + \bar{L}_m - L_{\Delta m}\cos 2\left(\theta_r + \frac{2}{3}\pi\right)\right)i_{cs} + L_{md}\sin\left(\theta_r + \frac{2}{3}\pi\right)i_{fr}\right]}{dt}$$
$$= -r_s i_{cs} + u_{cs},$$

$$\frac{d\left[L_{md}\sin\theta_r i_{as} + L_{md}\sin\left(\theta_r - \frac{2}{3}\pi\right)i_{bs} + L_{md}\sin\left(\theta_r + \frac{2}{3}\pi\right)i_{cs} + \left(L_{lf} + L_{mf}\right)i_{fr}\right]}{dt} = -r_r i_{fr} + u_{fr}.$$

(6.52)

Using (6.52), Cauchy's form of differential equations for the circuitry-electromagnetic dynamics of conventional three-phase synchronous motors is found using the Symbolic Toolbox. The notations are

$$\text{S1} = \sin\theta_r,\ \text{S2} = \sin\left(\theta_r - \frac{2}{3}\pi\right),\ \text{S3} = \sin\left(\theta_r + \frac{2}{3}\pi\right),\ \text{C1} = \cos\theta_r,\ \text{C2} = \cos\left(\theta_r - \frac{2}{3}\pi\right),$$

$$\text{C3} = \cos\left(\theta_r + \frac{2}{3}\pi\right),\ \text{Lls} = L_{ls},\ \text{Lmb} = \bar{L}_m,\ \text{Lmd} = L_{md},\ \text{Lmf} = L_{mf},\ \text{and Llf} = L_{lf}.$$

We have

```
L=sym('[Lls+Lmb,-Lmb/2,-Lmb/2,Lmd*S1;-Lmb/2,Lls+Lmb,-Lmb/2,Lmd*S2;-Lmb/2,-Lmb/2,Lls+Lmb,Lmd*S3;
   Lmd*S1,Lmd*S2,Lmd*S3,Llf+Lmf]');
R=sym('[-rs 0  0   0; 0 -rs 0   0; 0  0 -rs 0; 0  0  0  -rr]');
I=sym('[ias; ibs; ics; ifr]');
V=sym('[uas;ubs;ucs;ufr]');
K=sym('[Lmd*C1*wr*ifr; Lmd*C2*wr*ifr; Lmd*C3*wr*ifr; Lmd*wr*(ias*C1+ibs*C2+ics*C3)]');
L1=inv(L);  L2=simplify(L1)
FS1=L2*R*I; FS2=simplify(FS1)
FS3=L2*V;   FS4=simplify(FS3)
FS5=L2*K;   FS6=simplify(FS5)
FS7=FS2+FS4-FS6; FS=simplify(FS7)
```

Using the derived results, and applying the trigonometric identities, the following nonlinear differential equations are obtained

$$
\begin{aligned}
\frac{di_{as}}{dt}=&\frac{1}{\left(2L_{ls}+3\bar{L}_m\right)\left(3L_{md}^2L_{ls}-\left(2L_{ls}^2+3L_{ls}\bar{L}_m\right)L_{ff}\right)}\Bigg[\left(3L_{md}^2\bar{L}_m+4L_{ls}L_{md}^2-\left(3\bar{L}_m^2+4L_{ls}^2+8L_{ls}\bar{L}_m\right)L_{ff}+2L_{md}^2L_{ls}\cos 2\theta_r\right)\left(-r_si_{as}+u_{as}\right)\\
&+\left(3L_{md}^2\bar{L}_m+L_{ls}L_{md}^2-\left(3\bar{L}_m^2+2L_{ls}\bar{L}_m\right)L_{ff}+2L_{md}^2L_{ls}\cos 2\left(\theta_r-\frac{1}{3}\pi\right)\right)\left(-r_si_{bs}+u_{bs}\right)\\
&+\left(3L_{md}^2\bar{L}_m+L_{ls}L_{md}^2-\left(3\bar{L}_m^2+2L_{ls}\bar{L}_m\right)L_{ff}+2L_{md}^2L_{ls}\cos 2\left(\theta_r+\frac{1}{3}\pi\right)\right)\left(-r_si_{cs}+u_{cs}\right)+\left(6L_{ls}L_{md}\bar{L}_m+4L_{ls}^2L_{md}\right)\sin\theta_r\left(-r_ri_{fr}+u_{fr}\right)\\
&-\left(6L_{ls}L_{md}^2\bar{L}_m+4L_{ls}^2L_{md}^2\right)\left(i_{as}\cos\theta_r+i_{bs}\cos\left(\theta_r-\frac{2}{3}\pi\right)+i_{cs}\cos\left(\theta_r+\frac{2}{3}\pi\right)\right)\omega_r\sin\theta_r\\
&+\left(\left(6L_{md}L_{ls}\bar{L}_m+4L_{md}L_{ls}^2\right)L_{ff}-6L_{md}^3L_{ls}\right)i_{fr}\omega_r\cos\theta_r\Bigg],\\
\frac{di_{bs}}{dt}=&\frac{1}{\left(2L_{ls}+3\bar{L}_m\right)\left(3L_{md}^2L_{ls}-\left(2L_{ls}^2+3L_{ls}\bar{L}_m\right)L_{ff}\right)}\Bigg[\left(3L_{md}^2\bar{L}_m+L_{ls}L_{md}^2-\left(3\bar{L}_m^2+2L_{ls}\bar{L}_m\right)L_{ff}+2L_{md}^2L_{ls}\cos 2\left(\theta_r-\frac{1}{3}\pi\right)\right)\left(-r_si_{as}+u_{as}\right)\\
&+\left(3L_{md}^2\bar{L}_m+4L_{ls}L_{md}^2-\left(3\bar{L}_m^2+4L_{ls}^2+8L_{ls}\bar{L}_m\right)L_{ff}+2L_{md}^2L_{ls}\cos 2\left(\theta_r-\frac{2}{3}\pi\right)\right)\left(-r_si_{bs}+u_{bs}\right)\\
&+\left(3L_{md}^2\bar{L}_m+L_{ls}L_{md}^2-\left(3\bar{L}_m^2+2L_{ls}\bar{L}_m\right)L_{ff}+2L_{md}^2L_{ls}\cos 2\theta_r\right)\left(-r_si_{cs}+u_{cs}\right)+\left(6L_{ls}L_{md}\bar{L}_m+4L_{ls}^2L_{md}\right)\sin\left(\theta_r-\frac{2}{3}\pi\right)\left(-r_ri_{fr}+u_{fr}\right)\\
&-\left(6L_{ls}L_{md}^2\bar{L}_m+4L_{ls}^2L_{md}^2\right)\left(i_{as}\cos\theta_r+i_{bs}\cos\left(\theta_r-\frac{2}{3}\pi\right)+i_{cs}\cos\left(\theta_r+\frac{2}{3}\pi\right)\right)\omega_r\sin\left(\theta_r-\frac{2}{3}\pi\right)\\
&+\left(\left(6L_{md}L_{ls}\bar{L}_m+4L_{md}L_{ls}^2\right)L_{ff}-6L_{md}^3L_{ls}\right)i_{fr}\omega_r\cos\left(\theta_r-\frac{2}{3}\pi\right)\Bigg],\\
\frac{di_{cs}}{dt}=&\frac{1}{\left(2L_{ls}+3\bar{L}_m\right)\left(3L_{md}^2L_{ls}-\left(2L_{ls}^2+3L_{ls}\bar{L}_m\right)L_{ff}\right)}\Bigg[\left(3L_{md}^2\bar{L}_m+L_{ls}L_{md}^2-\left(3\bar{L}_m^2+2L_{ls}\bar{L}_m\right)L_{ff}+2L_{md}^2L_{ls}\cos 2\left(\theta_r+\frac{1}{3}\pi\right)\right)\left(-r_si_{as}+u_{as}\right)\\
&+\left(3L_{md}^2\bar{L}_m+L_{ls}L_{md}^2-\left(3\bar{L}_m^2+2L_{ls}\bar{L}_m\right)L_{ff}+2L_{md}^2L_{ls}\cos 2\theta_r\right)\left(-r_si_{bs}+u_{bs}\right)\\
&+\left(3L_{md}^2\bar{L}_m+4L_{ls}L_{md}^2-\left(3\bar{L}_m^2+4L_{ls}^2+8L_{ls}\bar{L}_m\right)L_{ff}+2L_{md}^2L_{ls}\cos 2\left(\theta_r+\frac{2}{3}\pi\right)\right)\left(-r_si_{cs}+u_{cs}\right)\\
&+\left(6L_{ls}L_{md}\bar{L}_m+4L_{ls}^2L_{md}\right)\sin\left(\theta_r+\frac{2}{3}\pi\right)\left(-r_ri_{fr}+u_{fr}\right)\\
&-\left(6L_{ls}L_{md}^2\bar{L}_m+4L_{ls}^2L_{md}^2\right)\left(i_{as}\cos\theta_r+i_{bs}\cos\left(\theta_r-\frac{2}{3}\pi\right)+i_{cs}\cos\left(\theta_r+\frac{2}{3}\pi\right)\right)\omega_r\sin\left(\theta_r+\frac{2}{3}\pi\right)\\
&+\left(\left(6L_{md}L_{ls}\bar{L}_m+4L_{md}L_{ls}^2\right)L_{ff}-6L_{md}^3L_{ls}\right)i_{fr}\omega_r\cos\left(\theta_r+\frac{2}{3}\pi\right)\Bigg],\\
\frac{di_{fr}}{dt}=&\frac{1}{3L_{md}^2L_{ls}-\left(2L_{ls}^2+3L_{ls}\bar{L}_m\right)L_{ff}}\Bigg[2L_{md}L_{ls}\left(\sin\theta_r\left(-r_si_{as}+u_{as}\right)+\sin\left(\theta_r-\frac{2}{3}\pi\right)\left(-r_si_{bs}+u_{bs}\right)+\sin\left(\theta_r+\frac{2}{3}\pi\right)\left(-r_si_{cs}+u_{cs}\right)\right)\\
&-\left(2L_{ls}^2+3\bar{L}_mL_{ls}\right)\left(-r_ri_{fr}+u_{fr}\right)+\left(3L_{ls}L_{md}\bar{L}_m+2L_{ls}^2L_{md}\right)\left(i_{as}\omega_r\cos\theta_r+i_{bs}\omega_r\cos\left(\theta_r-\frac{2}{3}\pi\right)+i_{cs}\omega_r\cos\left(\theta_r+\frac{2}{3}\pi\right)\right)\Bigg].
\end{aligned}
\tag{6.53}
$$

Here, $L_{ff}=L_{lf}+L_{mf}$.

In matrix form we have

$$
\begin{bmatrix} \frac{di_{as}}{dt} \\ \frac{di_{bs}}{dt} \\ \frac{di_{cs}}{dt} \\ \frac{di_{fr}}{dt} \end{bmatrix}
=
\begin{bmatrix}
-\frac{r_s L_{Ds}}{L_{\Sigma s}} & -\frac{r_s L_{Ms}}{L_{\Sigma s}} & -\frac{r_s L_{Ms}}{L_{\Sigma s}} & 0 \\
-\frac{r_s L_{Ms}}{L_{\Sigma s}} & -\frac{r_s L_{Ds}}{L_{\Sigma s}} & -\frac{r_s L_{Ms}}{L_{\Sigma s}} & 0 \\
-\frac{r_s L_{Ms}}{L_{\Sigma s}} & -\frac{r_s L_{Ms}}{L_{\Sigma s}} & -\frac{r_s L_{Ds}}{L_{\Sigma s}} & 0 \\
0 & 0 & 0 & \frac{r_r\left(2L_{ls}^2 + 3\bar{L}_m L_{ls}\right)}{L_{\Sigma f}}
\end{bmatrix}
\begin{bmatrix} i_{as} \\ i_{bs} \\ i_{cs} \\ i_{fr} \end{bmatrix}
$$

$$
+\begin{bmatrix}
-\frac{2r_s L_{md}^2 L_{ls}}{L_{\Sigma s}}\cos 2\theta_r & -\frac{2r_s L_{md}^2 L_{ls}}{L_{\Sigma s}}\cos 2\left(\theta_r - \frac{1}{3}\pi\right) & -\frac{2r_s L_{md}^2 L_{ls}}{L_{\Sigma s}}\cos 2\left(\theta_r + \frac{1}{3}\pi\right) & -\frac{r_r L_{Mf}}{L_{\Sigma s}}\sin\theta_r \\
-\frac{2r_s L_{md}^2 L_{ls}}{L_{\Sigma s}}\cos 2\left(\theta_r - \frac{1}{3}\pi\right) & -\frac{2r_s L_{md}^2 L_{ls}}{L_{\Sigma s}}\cos 2\left(\theta_r - \frac{2}{3}\pi\right) & -\frac{2r_s L_{md}^2 L_{ls}}{L_{\Sigma s}}\cos 2\theta_r & -\frac{r_r L_{Mf}}{L_{\Sigma s}}\sin\left(\theta_r - \frac{2}{3}\pi\right) \\
-\frac{2r_s L_{md}^2 L_{ls}}{L_{\Sigma s}}\cos 2\left(\theta_r + \frac{1}{3}\pi\right) & -\frac{2r_s L_{md}^2 L_{ls}}{L_{\Sigma s}}\cos 2\theta_r & -\frac{2r_s L_{md}^2 L_{ls}}{L_{\Sigma s}}\cos 2\left(\theta_r + \frac{2}{3}\pi\right) & -\frac{r_r L_{Mf}}{L_{\Sigma s}}\sin\left(\theta_r + \frac{2}{3}\pi\right) \\
-\frac{2r_r L_{md} L_{ls}}{L_{\Sigma f}}\sin\theta_r & -\frac{2r_r L_{md} L_{ls}}{L_{\Sigma f}}\sin\left(\theta_r - \frac{2}{3}\pi\right) & -\frac{2r_r L_{md} L_{ls}}{L_{\Sigma f}}\sin\left(\theta_r + \frac{2}{3}\pi\right) & 0
\end{bmatrix}
\begin{bmatrix} i_{as} \\ i_{bs} \\ i_{cs} \\ i_{fr} \end{bmatrix}
$$

$$
+\begin{bmatrix}
-\frac{6L_{ls}L_{md}^2\bar{L}_m + 4L_{ls}^2 L_{md}^2}{L_{\Sigma s}}\sin\theta_r & -\frac{6L_{ls}L_{md}^2\bar{L}_m + 4L_{ls}^2 L_{md}^2}{L_{\Sigma s}}\sin\theta_r & -\frac{6L_{ls}L_{md}^2\bar{L}_m + 4L_{ls}^2 L_{md}^2}{L_{\Sigma s}}\sin\theta_r \\
-\frac{6L_{ls}L_{md}^2\bar{L}_m + 4L_{ls}^2 L_{md}^2}{L_{\Sigma s}}\sin\left(\theta_r - \frac{2}{3}\pi\right) & -\frac{6L_{ls}L_{md}^2\bar{L}_m + 4L_{ls}^2 L_{md}^2}{L_{\Sigma s}}\sin\left(\theta_r - \frac{2}{3}\pi\right) & -\frac{6L_{ls}L_{md}^2\bar{L}_m + 4L_{ls}^2 L_{md}^2}{L_{\Sigma s}}\sin\left(\theta_r - \frac{2}{3}\pi\right) \\
-\frac{6L_{ls}L_{md}^2\bar{L}_m + 4L_{ls}^2 L_{md}^2}{L_{\Sigma s}}\sin\left(\theta_r + \frac{2}{3}\pi\right) & -\frac{6L_{ls}L_{md}^2\bar{L}_m + 4L_{ls}^2 L_{md}^2}{L_{\Sigma s}}\sin\left(\theta_r + \frac{2}{3}\pi\right) & -\frac{6L_{ls}L_{md}^2\bar{L}_m + 4L_{ls}^2 L_{md}^2}{L_{\Sigma s}}\sin\left(\theta_r + \frac{2}{3}\pi\right) \\
\frac{3L_{ls}L_{md}\bar{L}_m + 2L_{ls}^2 L_{md}}{L_{\Sigma f}} & \frac{3L_{ls}L_{md}\bar{L}_m + 2L_{ls}^2 L_{md}}{L_{\Sigma f}} & \frac{3L_{ls}L_{md}\bar{L}_m + 2L_{ls}^2 L_{md}}{L_{\Sigma f}}
\end{bmatrix}
$$

$$
\times\begin{bmatrix}
i_{as}\omega_r\cos\theta_r \\
i_{bs}\omega_r\cos\left(\theta_r - \frac{2}{3}\pi\right) \\
i_{cs}\omega_r\cos\left(\theta_r + \frac{2}{3}\pi\right)
\end{bmatrix}
+\begin{bmatrix}
\frac{\left(6L_{md}L_{ls}\bar{L}_m + 4L_{md}L_{ls}^2\right)L_{ff} - 6L_{md}^3 L_{ls}}{L_{\Sigma s}} i_{fr}\omega_r\cos\theta_r \\
\frac{\left(6L_{md}L_{ls}\bar{L}_m + 4L_{md}L_{ls}^2\right)L_{ff} - 6L_{md}^3 L_{ls}}{L_{\Sigma s}} i_{fr}\omega_r\cos\left(\theta_r - \frac{2}{3}\pi\right) \\
\frac{\left(6L_{md}L_{ls}\bar{L}_m + 4L_{md}L_{ls}^2\right)L_{ff} - 6L_{md}^3 L_{ls}}{L_{\Sigma s}} i_{fr}\omega_r\cos\left(\theta_r + \frac{2}{3}\pi\right) \\
0
\end{bmatrix}
$$

$$
+\begin{bmatrix}
\frac{L_{Ds} + 2L_{md}^2 L_{ls}\cos 2\theta_r}{L_{\Sigma s}} & \frac{L_{Ms} + 2L_{md}^2 L_{ls}\cos 2\left(\theta_r - \frac{1}{3}\pi\right)}{L_{\Sigma s}} & \frac{L_{Ms} + 2L_{md}^2 L_{ls}\cos 2\left(\theta_r + \frac{1}{3}\pi\right)}{L_{\Sigma s}} & \frac{L_{Mf}}{L_{\Sigma s}}\sin\theta_r \\
\frac{L_{Ms} + 2L_{md}^2 L_{ls}\cos 2\left(\theta_r - \frac{1}{3}\pi\right)}{L_{\Sigma s}} & \frac{L_{Ds} + 2L_{md}^2 L_{ls}\cos 2\left(\theta_r - \frac{2}{3}\pi\right)}{L_{\Sigma s}} & \frac{L_{Ms} + 2L_{md}^2 L_{ls}\cos 2\theta_r}{L_{\Sigma s}} & \frac{L_{Mf}}{L_{\Sigma s}}\sin\left(\theta_r - \frac{2}{3}\pi\right) \\
\frac{L_{Ms} + 2L_{md}^2 L_{ls}\cos 2\left(\theta_r + \frac{1}{3}\pi\right)}{L_{\Sigma s}} & \frac{L_{Ms} + 2L_{md}^2 L_{ls}\cos 2\theta_r}{L_{\Sigma s}} & \frac{L_{Ds} + 2L_{md}^2 L_{ls}\cos 2\left(\theta_r + \frac{2}{3}\pi\right)}{L_{\Sigma s}} & \frac{L_{Mf}}{L_{\Sigma s}}\sin\left(\theta_r + \frac{2}{3}\pi\right) \\
\frac{2L_{md}L_{ls}\sin\theta_r}{L_{\Sigma f}} & \frac{2L_{md}L_{ls}\sin\left(\theta_r - \frac{2}{3}\pi\right)}{L_{\Sigma f}} & \frac{2L_{md}L_{ls}\sin\left(\theta_r + \frac{2}{3}\pi\right)}{L_{\Sigma f}} & -\frac{2L_{ls}^2 + 3\bar{L}_m L_{ls}}{L_{\Sigma f}}
\end{bmatrix}
\begin{bmatrix} u_{as} \\ u_{bs} \\ u_{cs} \\ u_{fr} \end{bmatrix}
$$

where

$$L_{Ds} = 3L_{md}^2\bar{L}_m + 4L_{ls}L_{md}^2 - \left(3\bar{L}_m^2 + 4L_{ls}^2 + 8L_{ls}\bar{L}_m\right)\left(L_{lf} + L_{mf}\right),$$
$$L_{Ms} = 3L_{md}^2\bar{L}_m + L_{ls}L_{md}^2 - \left(3\bar{L}_m^2 + 2L_{ls}\bar{L}_m\right)\left(L_{lf} + L_{mf}\right),$$
$$L_{\Sigma s} = \left(2L_{ls} + 3\bar{L}_m\right)\left[3L_{md}^2L_{ls} - \left(2L_{ls}^2 + 3L_{ls}\bar{L}_m\right)\left(L_{lf} + L_{mf}\right)\right],$$
$$L_{\Sigma f} = 3L_{md}^2L_{ls} - \left(2L_{ls}^2 + 3L_{ls}\bar{L}_m\right)\left(L_{lf} + L_{mf}\right),$$
$$L_{Mf} = 6L_{ls}L_{md}\bar{L}_m + 4L_{ls}^2L_{md}.$$

The *torsional–mechanical* equations are

$$\frac{d\omega_{rm}}{dt} = \frac{1}{J}\left(T_e - B_m\omega_{rm} - T_L\right), \quad \frac{d\theta_{rm}}{dt} = \omega_{rm},$$

where the mechanical angular velocity ω_{rm} and displacement θ_{rm} are

$$\omega_{rm} = \frac{2}{P}\omega_r, \quad \theta_{rm} = \frac{2}{P}\theta_r.$$

Using the electrical angular velocity and displacement, one has

$$\begin{aligned}\frac{d\omega_r}{dt} &= \frac{P}{2J}T_e - \frac{B_m}{J}\omega_r - \frac{P}{2J}T_L,\\ \frac{d\theta_r}{dt} &= \omega_r.\end{aligned} \tag{6.54}$$

For a P-pole synchronous motor, the electromagnetic torque is

$$T_e = \frac{P}{2}\frac{\partial W_c\left(i_{as}, i_{bs}, i_{cs}, i_{fr}, \theta_r\right)}{\partial\theta_r} = \frac{P}{2}\frac{\partial}{\partial\theta_r}\frac{1}{2}\begin{bmatrix} i_{as} & i_{bs} & i_{cs} & i_{fr}\end{bmatrix}\mathbf{L}_{abcs/fr}\begin{bmatrix} i_{as} \\ i_{bs} \\ i_{cs} \\ i_{fr}\end{bmatrix}.$$

We have

$$\begin{aligned}T_e = \frac{P}{2}\Bigg[& L_{\Delta m}\left(i_{as}^2 \sin 2\theta_r + 2i_{as}i_{bs}\sin 2\left(\theta_r - \frac{1}{3}\pi\right) + 2i_{as}i_{cs}\sin 2\left(\theta_r + \frac{1}{3}\pi\right) + i_{bs}^2\sin 2\left(\theta_r - \frac{2}{3}\pi\right)\right.\\ & \left. + 2i_{bs}i_{cs}\sin 2\theta_r + i_{cs}^2\sin 2\left(\theta_r + \frac{2}{3}\pi\right)\right) + L_{md}i_{fr}\left(i_{as}\cos\theta_r + i_{bs}\cos\left(\theta_r - \frac{2}{3}\pi\right) + i_{cs}\cos\left(\theta_r + \frac{2}{3}\pi\right)\right)\Bigg].\end{aligned} \tag{6.55}$$

Using trigonometric identities, from (6.55), one also finds

$$\begin{aligned}T_e = \frac{P}{2}\Bigg[& \frac{L_{md} - L_{mq}}{3}\left(\left(i_{as}^2 - \frac{1}{2}i_{bs}^2 - \frac{1}{2}i_{cs}^2 - i_{as}i_{bs} - i_{as}i_{cs} + 2i_{bs}i_{cs}\right)\sin 2\theta_r + \frac{\sqrt{3}}{2}\left(i_{bs}^2 - i_{cs}^2 - 2i_{as}i_{bs} + 2i_{as}i_{cs}\right)\cos 2\theta_r\right)\\ & + L_{md}i_{fr}\left(\left(i_{as} - \frac{1}{2}i_{bs} - \frac{1}{2}i_{cs}\right)\cos\theta_r + \frac{\sqrt{3}}{2}\left(i_{bs} - i_{cs}\right)\sin\theta_r\right)\Bigg].\end{aligned}$$

The *torsional–mechanical* differential equations of motion (6.54) with (6.55) become

$$\frac{d\omega_r}{dt}=\frac{P^2}{4J}\left[L_{\Delta m}\left(i_{as}^2\sin 2\theta_r+2i_{as}i_{bs}\sin 2\left(\theta_r-\frac{1}{3}\pi\right)+2i_{as}i_{cs}\sin 2\left(\theta_r+\frac{1}{3}\pi\right)+i_{bs}^2\sin 2\left(\theta_r-\frac{2}{3}\pi\right)\right.\right.$$
$$\left.\left.+2i_{bs}i_{cs}\sin 2\theta_r+i_{cs}^2\sin 2\left(\theta_r+\frac{2}{3}\pi\right)\right)+L_{md}i_{fr}\left(i_{as}\cos\theta_r+i_{bs}\cos\left(\theta_r-\frac{2}{3}\pi\right)+i_{cs}\cos\left(\theta_r+\frac{2}{3}\pi\right)\right)\right]$$
$$-\frac{B_m}{J}\omega_r-\frac{P}{2J}T_L,$$
$$\frac{d\theta_r}{dt}=\omega_r. \qquad (6.56)$$

The resulting equations (6.53) and (6.56) give a set of six nonlinear differential equations. In the operating envelope, the electrical angular velocity is equal to the synchronous angular velocity, and

$$\omega_r=\omega_e=\frac{4\pi f}{P}.$$

The angular velocity is regulated by changing the frequency of the phase voltages applied. For round-rotor synchronous machines $L_{\Delta m}=0$. Hence equation (6.55) becomes

$$T_e=\frac{PL_{md}}{2}i_{fr}\left[i_{as}\cos\theta_r+i_{bs}\cos\left(\theta_r-\frac{2}{3}\pi\right)+i_{cs}\cos\left(\theta_r+\frac{2}{3}\pi\right)\right].$$

The DC voltage u_{fr} is supplied. This u_{fr} can be regulated. The balanced current and voltage sets are

$$i_{as}=\sqrt{2}i_M\cos\theta_r,\quad i_{bs}=\sqrt{2}i_M\cos\left(\theta_r-\frac{2}{3}\pi\right),\quad i_{cs}=\sqrt{2}i_M\cos\left(\theta_r+\frac{2}{3}\pi\right),$$

$$u_{as}=\sqrt{2}u_M\cos\theta_r,\quad u_{bs}=\sqrt{2}u_M\cos\left(\theta_r-\frac{2}{3}\pi\right),\quad u_{cs}=\sqrt{2}u_M\cos\left(\theta_r+\frac{2}{3}\pi\right).$$

We have

$$T_e=\frac{PL_{md}}{2}i_{fr}\sqrt{2}i_M\left(\cos^2\theta_r+\cos^2\left(\theta_r-\frac{2}{3}\pi\right)+\cos^2\left(\theta_r+\frac{2}{3}\pi\right)\right)=\frac{3PL_{md}}{2\sqrt{2}}i_{fr}i_M.$$

Example 6.18

Consider a three-phase two-pole synchronous motor with r_s = 0.25 ohm, r_r = 0.47 ohm, L_{ls} = 0.0001 H, L_{mq} = 0.00095 H, L_{md} = 0.00095 H, L_{lf} = 0.00022 H, L_{mf} = 0.0001 H, L_{md} = 0.00035 H, J = 0.003 kg-m², and B_m = 0.00072 N-m-sec/rad. The phase voltages are

$$u_{as}(t)=\sqrt{2}150\cos(377t),\quad u_{bs}(t)=\sqrt{2}150\cos\left(377t-\frac{2}{3}\pi\right),\quad u_{cs}(t)=\sqrt{2}150\cos\left(377t+\frac{2}{3}\pi\right).$$

The field voltage is $u_{fr} = 5$ V.
Let

$$T_L = \begin{cases} 0 \text{ N-m}, \forall t \in [0 \quad 0.2) \text{ s} \\ 1 \text{ N-m}, \forall t \in [0.2 \quad 0.5] \text{ s} \end{cases}.$$

The differential equations in Cauchy's form (6.53) and (6.56) are used to model the conventional synchronous motors. Two MATLAB files are developed.

MATLAB script 1

```
tspan = [0 0.5]; y0=[0 0 0 0 0 0]';
options = odeset('RelTol',5e-3,'AbsTol',[1e-4 1e-4 1e-4 1e-4 1e-4 1e-4]);
[t,y] = ODE45('crm_02',tspan,y0,options);
subplot(3,2,1); plot(t,y(:,1)); ylabel('ias'); axis([0,0.5,-500,500]); grid;
subplot(3,2,3); plot(t,y(:,2)); ylabel('ibs'); axis([0,0.5,-500,500]); grid;
subplot(3,2,5); plot(t,y(:,3)); ylabel('ics'); axis([0,0.5,-500,500]); grid;
subplot(3,2,2); plot(t,y(:,4)); ylabel('ifr'); axis([0,0.5,-200,200]); grid;
subplot(3,2,4); plot(t,y(:,5)); ylabel('wr'); axis([0,0.5,0,400]); grid;
subplot(3,2,6); plot(t,y(:,6)); ylabel('Q'); axis([0,0.5,0,200]); grid;
```

MATLAB script 2

```
function yprime=difer(t,y);
Lls=0.0001; Lmqs=0.00095; Lmds=0.00095; % Stator leakage and magnetizing inductances
Llf=0.00022; Lmf=0.0001; % Rotor leakage and magnetizing inductances
Lmd=0.00035; % Mutual inductance
Lmb=(Lmqs+Lmds)/3; % Average value of the magnetizing inductance
rs=0.25; rr=0.47; % Resistances of the stator and rotor windings
J=0.003; Bm=0.00072; % Equivalent moment of inertia and viscous friction coefficient
P=2; % Number of poles
um=sqrt(2)*150; w=377; % Magnitude of the applied phase voltages and angular frequency
ufr=5; % DC voltage applied to the rotor field winding
% Load torque, applied at time tTl sec
if t<=0.2
   Tl=0;
else
   Tl=1;
end
% Applied phase voltages to the abc windings
uas=um*cos(w*t); ubs=um*cos(w*t-2*pi/3); ucs=um*cos(w*t+2*pi/3);
% Expressions used
Ld=2*Lls*Lmd^2; Ldr=2*Lls*Lmd;
Lss=(-3*Lmb^2-4*Lls^2-8*Lls*Lmb)*(Lmf+Llf)+3*Lmd^2*Lmb+4*Lls*Lmd^2;
Ldens=(2*Lls+3*Lmb)*((-2*Lls^2-3*Lmb*Lls)*(Llf+Lmf)+3*Lmd^2*Lls);
Ldenr=(-2*Lls^2-3*Lmb*Lls)*(Llf+Lmf)+3*Lmd^2*Lls;
Lms=(-3*Lmb^2-2*Lls*Lmb)*(Lmf+Llf)+3*Lmd^2*Lmb+Lls*Lmd^2;
Lrr=6*Lls*Lmd*Lmb+4*Lls^2*Lmd; Llr=-(2*Lls^2+3*Lmb*Lls);
% Variables used
S1=sin(y(6,:)); S2=sin(y(6,:)-2*pi/3); S3=sin(y(6,:)+2*pi/3);
IUas=-rs*y(1,:)+uas; IUbs=-rs*y(2,:)+ubs; IUcs=-rs*y(3,:)+ucs; IUfr=-rr*y(4,:)+ufr;
C1=cos(y(6,:)); C2=cos(y(6,:)-2*pi/3); C3=cos(y(6,:)+2*pi/3);
C12=cos(2*y(6,:)); C22=cos(2*y(6,:)-2*pi/3); C32=cos(2*y(6,:)+2*pi/3);
Nsts=(-6*Lmd^2*Lls*Lmb-4*Lmd^2*Lls^2)*y(5,:).*(C1.*y(1,:)+C2.*y(2,:)+C3.*y(3,:));
Nstr=(3*Lmd*Lmb*Lls+2*Lmd*Lls^2)*y(5,:).*(C1.*y(1,:)+C2.*y(2,:)+C3.*y(3,:));
Nct=((6*Lmd*Lls*Lmb+4*Lmd*Lls^2)*(Llf+Lmf)-6*Lmd^3*Lls)*y(5,:).*y(4,:);
Te=P*(Lmd*y(4,:).*(y(1,:).*C1+y(2,:).*C2+y(3,:).*C3))/2;
% Differential equations
yprime=[((Lss+Ld*C12).*IUas+(Lms+Ld*C22).*IUbs+(Lms+Ld*C32).*IUcs+Lrr*S1.*IUfr+Nsts.*S1+Nct.*C1)/Ldens;...
((Lms+Ld*C22).*IUas+(Lss+Ld*C32).*IUbs+(Lms+Ld*C12).*IUcs+Lrr*S2.*IUfr+Nsts.*S2+Nct.*C2)/Ldens;...
((Lms+Ld*C32).*IUas+(Lms+Ld*C12).*IUbs+(Lss+Ld*C22).*IUcs+Lrr*S3.*IUfr+Nsts.*S3+Nct.*C3)/Ldens;...
(Ldr*S1.*IUas+Ldr*S2.*IUbs+Ldr*S3.*IUcs+Llr*IUfr+Nstr)/Ldenr;...
(P*Te-2*Bm*y(5,:)-P*Tl)/(2*J);...
y(5,:)];
```

The transient dynamics are illustrated in Figures 6.38. A two-pole motor starts from stall and reaches the synchronous angular velocity 377 rad/sec. The angular velocity of the motor is locked with the synchronous velocity. ■

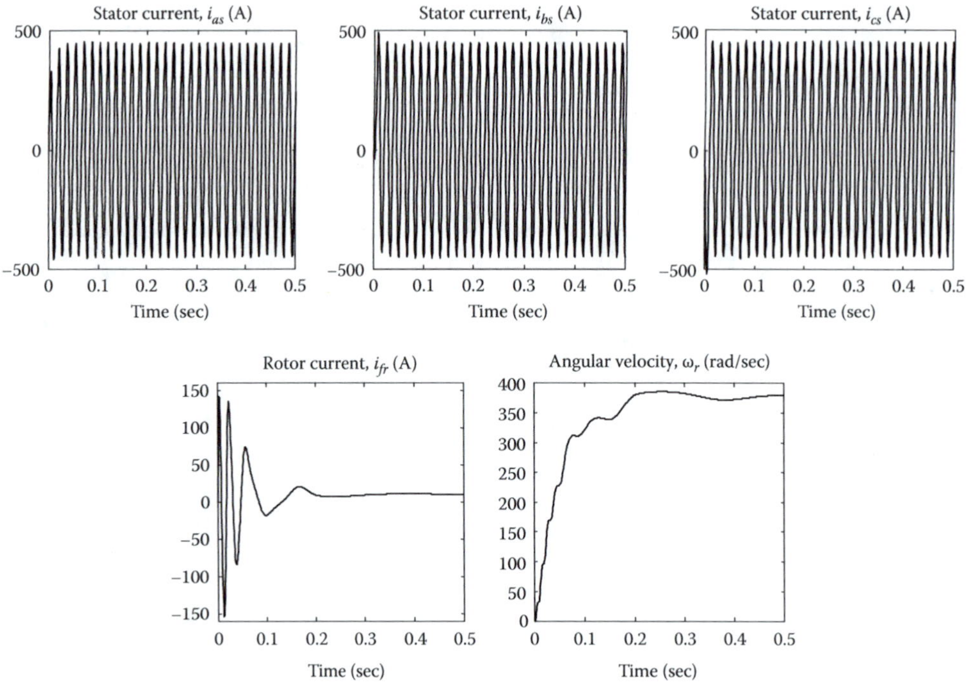

FIGURE 6.38 Dynamics of a conventional three-phase synchronous motor, $T_L = \begin{cases} 0 \text{ N-m}, \forall t \in [0 \quad 2)\text{ s} \\ 1 \text{ N-m}, \forall t \in [0.2 \quad 0.5]\text{ s} \end{cases}$.

6.7.2 Three-Phase DC-Current Exited Synchronous Generators

Three-phase synchronous generators are driven by the prime mover, see Figure 6.37. Using Kirchhoff's second law, we have four equations

$$u_{as} = -r_s i_{as} + \frac{d\psi_{as}}{dt}, \quad u_{bs} = -r_s i_{bs} + \frac{d\psi_{bs}}{dt}, \quad u_{cs} = -r_s i_{cs} + \frac{d\psi_{cs}}{dt}, \quad u_{fr} = r_r i_{fr} + \frac{d\psi_{fr}}{dt}, \tag{6.57}$$

where the flux linkages are

$$\begin{aligned}
\psi_{as} &= -\left(L_{ls} + \bar{L}_m - L_{\Delta m}\cos 2\theta_r\right)i_{as} - \left(-\frac{1}{2}\bar{L}_m - L_{\Delta m}\cos 2\left(\theta_r - \frac{1}{3}\pi\right)\right)i_{bs} \\
&\quad - \left(-\frac{1}{2}\bar{L}_m - L_{\Delta m}\cos 2\left(\theta_r + \frac{1}{3}\pi\right)\right)i_{cs} + L_{md}\sin\theta_r i_{fr}, \\
\psi_{bs} &= -\left(-\frac{1}{2}\bar{L}_m - L_{\Delta m}\cos 2\left(\theta_r - \frac{1}{3}\pi\right)\right)i_{as} - \left(L_{ls} + \bar{L}_m - L_{\Delta m}\cos 2\left(\theta_r - \frac{2}{3}\pi\right)\right)i_{bs} \\
&\quad - \left(-\frac{1}{2}\bar{L}_m - L_{\Delta m}\cos 2\theta_r\right)i_{cs} + L_{md}\sin\left(\theta_r - \frac{2}{3}\pi\right)i_{fr}, \\
\psi_{cs} &= \left(-\frac{1}{2}\bar{L}_m - L_{\Delta m}\cos 2\left(\theta_r + \frac{1}{3}\pi\right)\right)i_{as} + \left(-\tfrac{1}{2}\bar{L}_m - L_{\Delta m}\cos 2\theta_r\right)i_{bs} \\
&\quad + \left(L_{ls} + \bar{L}_m - L_{\Delta m}\cos 2\left(\theta_r + \frac{2}{3}\pi\right)\right)i_{cs} + L_{md}\sin\left(\theta_r + \frac{2}{3}\pi\right)i_{fr}, \\
\psi_{fr} &= -L_{md}\sin\theta_r i_{as} - L_{md}\sin\left(\theta_r - \frac{2}{3}\pi\right)i_{bs} - L_{md}\sin\left(\theta_r + \frac{2}{3}\pi\right)i_{cs} + \left(L_{lf} + L_{mf}\right)i_{fr}.
\end{aligned} \tag{6.58}$$

Hence,

$$\psi_{abcs/fr} = \mathbf{L}_{abcs/fr}\mathbf{i}_{abcs/fr}$$

$$= \begin{bmatrix} L_{ls}+\bar{L}_m - L_{\Delta m}\cos 2\theta_r & -\tfrac{1}{2}\bar{L}_m - L_{\Delta m}\cos 2\left(\theta_r - \frac{1}{3}\pi\right) & -\frac{1}{2}\bar{L}_m - L_{\Delta m}\cos 2\left(\theta_r + \frac{1}{3}\pi\right) & L_{md}\sin\theta_r \\ -\frac{1}{2}\bar{L}_m - L_{\Delta m}\cos 2\left(\theta_r - \frac{1}{3}\pi\right) & L_{ls}+\bar{L}_m - L_{\Delta m}\cos 2\left(\theta_r - \frac{2}{3}\pi\right) & -\frac{1}{2}\bar{L}_m - L_{\Delta m}\cos 2\theta_r & L_{md}\sin\left(\theta_r - \frac{2}{3}\pi\right) \\ -\frac{1}{2}\bar{L}_m - L_{\Delta m}\cos 2\left(\theta_r + \frac{1}{3}\pi\right) & -\frac{1}{2}\bar{L}_m - L_{\Delta m}\cos 2\theta_r & L_{ls}+\bar{L}_m - L_{\Delta m}\cos 2\left(\theta_r + \frac{2}{3}\pi\right) & L_{md}\sin\left(\theta_r + \frac{2}{3}\pi\right) \\ L_{md}\sin\theta_r & L_{md}\sin\left(\theta_r - \frac{2}{3}\pi\right) & L_{md}\sin\left(\theta_r + \frac{2}{3}\pi\right) & L_{lf}+L_{mf} \end{bmatrix}$$

$$\times \begin{bmatrix} -i_{as} \\ -i_{bs} \\ -i_{cs} \\ i_{fr} \end{bmatrix}$$

From (6.57) and (6.58), for round-rotor synchronous machines with $L_{\Delta m} = 0$, we have

$$\frac{di_{as}}{dt} = \frac{1}{L_{ls}+\bar{L}_m}\left(-r_s i_{as} + \frac{d\left(\frac{1}{2}\bar{L}_m i_{bs} + \frac{1}{2}\bar{L}_m i_{cs} + L_{md}\sin\left(\frac{P}{2}\theta_{rm}\right)i_{fr}\right)}{dt}\right),$$

$$\frac{di_{bs}}{dt} = \frac{1}{L_{ls}+\bar{L}_m}\left(-r_s i_{bs} + \frac{d\left(\frac{1}{2}\bar{L}_m i_{as} + \frac{1}{2}\bar{L}_m i_{bs} + L_{md}\sin\left(\frac{P}{2}\theta_{rm} - \frac{2}{3}\pi\right)i_{fr}\right)}{dt}\right),$$

$$\frac{di_{cs}}{dt} = \frac{1}{L_{ls}+\bar{L}_m}\left(-r_s i_{cs} + \frac{d\left(\frac{1}{2}\bar{L}_m i_{as} + \frac{1}{2}\bar{L}_m i_{bs} + L_{md}\sin\left(\frac{P}{2}\theta_{rm} + \frac{2}{3}\pi\right)i_{fr}\right)}{dt}\right),$$

$$\frac{di_{fr}}{dt} = \frac{1}{L_{lf}+L_{mf}}\left(-r_s i_{fr} + \frac{d\left(L_{md}\sin\left(\frac{P}{2}\theta_{rm}\right)i_{as} + L_{md}\sin\left(\frac{P}{2}\theta_{rm} - \frac{2}{3}\pi\right)i_{bs} + L_{md}\sin\left(\frac{P}{2}\theta_{rm} + \frac{2}{3}\pi\right)i_{cs}\right)}{dt} + u_{fr}\right).$$

Cauchy's form of the circuitry dynamics for conventional synchronous generator is found as given by the first four equations in (6.59). The electromagnetic torque of the generator is found to be as given by (6.55). With the *torsional–mechanical* dynamics, the resulting mathematical model for three-phase synchronous generators is

$$
\begin{aligned}
\frac{di_{as}}{dt} = {} & \frac{1}{\left(2L_{ls}+3\bar{L}_m\right)\left(3L_{md}^2L_{ls}-\left(2L_{ls}^2+3L_{ls}\bar{L}_m\right)L_{ff}\right)}\Bigg[-\left(3L_{md}^2\bar{L}_m+4L_{ls}L_{md}^2-\left(3\bar{L}_m^2+4L_{ls}^2+8L_{ls}\bar{L}_m\right)L_{ff}+2L_{md}^2L_{ls}\cos 2\theta_r\right)r_s i_{as} \\
& -\left(3L_{md}^2\bar{L}_m+L_{ls}L_{md}^2-\left(3\bar{L}_m^2+2L_{ls}\bar{L}_m\right)L_{ff}+2L_{md}^2L_{ls}\cos 2\left(\theta_r-\frac{1}{3}\pi\right)\right)r_s i_{bs} \\
& -\left(3L_{md}^2\bar{L}_m+L_{ls}L_{md}^2-\left(3\bar{L}_m^2+2L_{ls}\bar{L}_m\right)L_{ff}+2L_{md}^2L_{ls}\cos 2\left(\theta_r+\frac{1}{3}\pi\right)\right)r_s i_{cs}-\left(6L_{ls}L_{md}\bar{L}_m+4L_{ls}^2L_{md}\right)\sin\theta_r\left(-r_r i_{fr}+u_{fr}\right) \\
& -\left(6L_{ls}L_{md}^2\bar{L}_m+4L_{ls}^2L_{md}^2\right)\left(i_{as}\cos\theta_r+i_{bs}\cos\left(\theta_r-\frac{2}{3}\pi\right)+i_{cs}\cos\left(\theta_r+\frac{2}{3}\pi\right)\right)\omega_r\sin\theta_r \\
& -\left(\left(6L_{md}L_{ls}\bar{L}_m+4L_{md}L_{ls}^2\right)L_{ff}-6L_{md}^3L_{ls}\right)i_{fr}\omega_r\cos\theta_r\Bigg],
\end{aligned}
$$

$$
\begin{aligned}
\frac{di_{bs}}{dt} = {} & \frac{1}{\left(2L_{ls}+3\bar{L}_m\right)\left(3L_{md}^2L_{ls}-\left(2L_{ls}^2+3L_{ls}\bar{L}_m\right)L_{ff}\right)}\Bigg[-\left(3L_{md}^2\bar{L}_m+L_{ls}L_{md}^2-\left(3\bar{L}_m^2+2L_{ls}\bar{L}_m\right)L_{ff}+2L_{md}^2L_{ls}\cos 2\left(\theta_r-\frac{1}{3}\pi\right)\right)r_s i_{as} \\
& -\left(3L_{md}^2\bar{L}_m+4L_{ls}L_{md}^2-\left(3\bar{L}_m^2+4L_{ls}^2+8L_{ls}\bar{L}_m\right)L_{ff}+2L_{md}^2L_{ls}\cos 2\left(\theta_r-\frac{2}{3}\pi\right)\right)r_s i_{bs} \\
& -\left(3L_{md}^2\bar{L}_m+L_{ls}L_{md}^2-\left(3\bar{L}_m^2+2L_{ls}\bar{L}_m\right)L_{ff}+2L_{md}^2L_{ls}\cos 2\theta_r\right)r_s i_{cs}-\left(6L_{ls}L_{md}\bar{L}_m+4L_{ls}^2L_{md}\right)\sin\left(\theta_r-\frac{2}{3}\pi\right)\left(-r_r i_{fr}+u_{fr}\right) \\
& -\left(6L_{ls}L_{md}^2\bar{L}_m+4L_{ls}^2L_{md}^2\right)\left(i_{as}\cos\theta_r+i_{bs}\cos\left(\theta_r-\frac{2}{3}\pi\right)+i_{cs}\cos\left(\theta_r+\frac{2}{3}\pi\right)\right)\omega_r\sin\left(\theta_r-\frac{2}{3}\pi\right) \\
& -\left(\left(6L_{md}L_{ls}\bar{L}_m+4L_{md}L_{ls}^2\right)L_{ff}-6L_{md}^3L_{ls}\right)i_{fr}\omega_r\cos\left(\theta_r-\frac{2}{3}\pi\right)\Bigg],
\end{aligned}
$$

$$
\begin{aligned}
\frac{di_{cs}}{dt} = {} & \frac{1}{\left(2L_{ls}+3\bar{L}_m\right)\left(3L_{md}^2L_{ls}-\left(2L_{ls}^2+3L_{ls}\bar{L}_m\right)L_{ff}\right)}\Bigg[-\left(3L_{md}^2\bar{L}_m+L_{ls}L_{md}^2-\left(3\bar{L}_m^2+2L_{ls}\bar{L}_m\right)L_{ff}+2L_{md}^2L_{ls}\cos 2\left(\theta_r+\frac{1}{3}\pi\right)\right)r_s i_{as} \\
& -\left(3L_{md}^2\bar{L}_m+L_{ls}L_{md}^2-\left(3\bar{L}_m^2+2L_{ls}\bar{L}_m\right)L_{ff}+2L_{md}^2L_{ls}\cos 2\theta_r\right)r_s i_{bs} \\
& -\left(3L_{md}^2\bar{L}_m+4L_{ls}L_{md}^2-\left(3\bar{L}_m^2+4L_{ls}^2+8L_{ls}\bar{L}_m\right)L_{ff}+2L_{md}^2L_{ls}\cos 2\left(\theta_r+\frac{2}{3}\pi\right)\right)r_s i_{cs} \\
& -\left(6L_{ls}L_{md}\bar{L}_m+4L_{ls}^2L_{md}\right)\sin\left(\theta_r+\frac{2}{3}\pi\right)\left(-r_r i_{fr}+u_{fr}\right) \\
& -\left(6L_{ls}L_{md}^2\bar{L}_m+4L_{ls}^2L_{md}^2\right)\left(i_{as}\cos\theta_r+i_{bs}\cos\left(\theta_r-\frac{2}{3}\pi\right)+i_{cs}\cos\left(\theta_r+\frac{2}{3}\pi\right)\right)\omega_r\sin\left(\theta_r+\frac{2}{3}\pi\right) \\
& -\left(\left(6L_{md}L_{ls}\bar{L}_m+4L_{md}L_{ls}^2\right)L_{ff}-6L_{md}^3L_{ls}\right)i_{fr}\omega_r\cos\left(\theta_r+\frac{2}{3}\pi\right)\Bigg],
\end{aligned}
$$

$$
\begin{aligned}
\frac{di_{fr}}{dt} = {} & \frac{1}{3L_{md}^2L_{ls}-\left(2L_{ls}^2+3L_{ls}\bar{L}_m\right)L_{ff}}\Bigg[2L_{md}L_{ls}\left(\sin\theta_r r_s i_{as}+\sin\left(\theta_r-\frac{2}{3}\pi\right)r_s i_{bs}+\sin\left(\theta_r+\frac{2}{3}\pi\right)r_s i_{cs}\right) \\
& -\left(2L_{ls}^2+3\bar{L}_mL_{ls}\right)\left(-r_r i_{fr}+u_{fr}\right)-\left(3L_{ls}L_{md}\bar{L}_m+2L_{ls}^2L_{md}\right)\left(i_{as}\omega_r\cos\theta_r+i_{bs}\omega_r\cos\left(\theta_r-\frac{2}{3}\pi\right)+i_{cs}\omega_r\cos\left(\theta_r+\frac{2}{3}\pi\right)\right)\Bigg].
\end{aligned}
$$

$$
\frac{d\omega_r}{dt} = -\frac{P^2L_{md}}{4J}i_{fr}\left[i_{as}\cos\theta_r+i_{bs}\cos\left(\theta_r-\frac{2}{3}\pi\right)+i_{cs}\cos\left(\theta_r+\frac{2}{3}\pi\right)\right]-\frac{B_m}{J}\omega_r+\frac{P}{2J}T_{PM},
$$

$$
\frac{d\theta_r}{dt} = \omega_r. \tag{6.59}
$$

Example 6.19

We model a synchronous generator, driven by a finite-power and finite-torque prime mover. The generator parameters are documented in Example 6.18. One may develop MATLAB files, similar to those in Example 6.18. Figures 6.39 depict the induced voltages u_{as}, u_{bs} and u_{cs} (*motional emfs*), and phase currents for the varying balanced Y-load with R_L. The frequency and magnitude of u_{as}, u_{bs}, and u_{cs} depend on the angular velocity ω_{rm} ($\omega_{rm} = \omega_{PM}$) and load. The induced voltages decrease at low angular velocity of the prime mover. ■

6.7.3 Mathematical Models of Synchronous Machines in the Rotor and Synchronous Reference Frames

To analyze machines in the *qd*0 quantities, one specifies the angular velocity of the reference frame. In the *arbitrary* reference frame, ω is not specified. If $\omega = \omega_r$ or $\omega = \omega_e$, one studies the rotor and synchronous frames. We recall that $\omega_r = \omega_e$, and the resulting models are identical. The *direct* and *inverse* Park transformations are applied.

One finds the *qd*0 components of voltages, currents, and flux linkages using the *machine* variables. In the rotor reference frame

$$\mathbf{u}_{qd0s}^r = \mathbf{K}_s^r \mathbf{u}_{abcs},\quad \mathbf{i}_{qd0s}^r = \mathbf{K}_s^r \mathbf{i}_{abcs},\quad \boldsymbol{\psi}_{qd0s}^r = \mathbf{K}_s^r \boldsymbol{\psi}_{abcs},\quad \mathbf{K}_s^r = \frac{2}{3}\begin{bmatrix} \cos\theta_r & \cos\left(\theta_r - \frac{2}{3}\pi\right) & \cos\left(\theta_r + \frac{2}{3}\pi\right) \\ \sin\theta_r & \sin\left(\theta_r - \frac{2}{3}\pi\right) & \sin\left(\theta_r + \frac{2}{3}\pi\right) \\ \frac{1}{2} & \frac{1}{2} & \frac{1}{2} \end{bmatrix}.$$

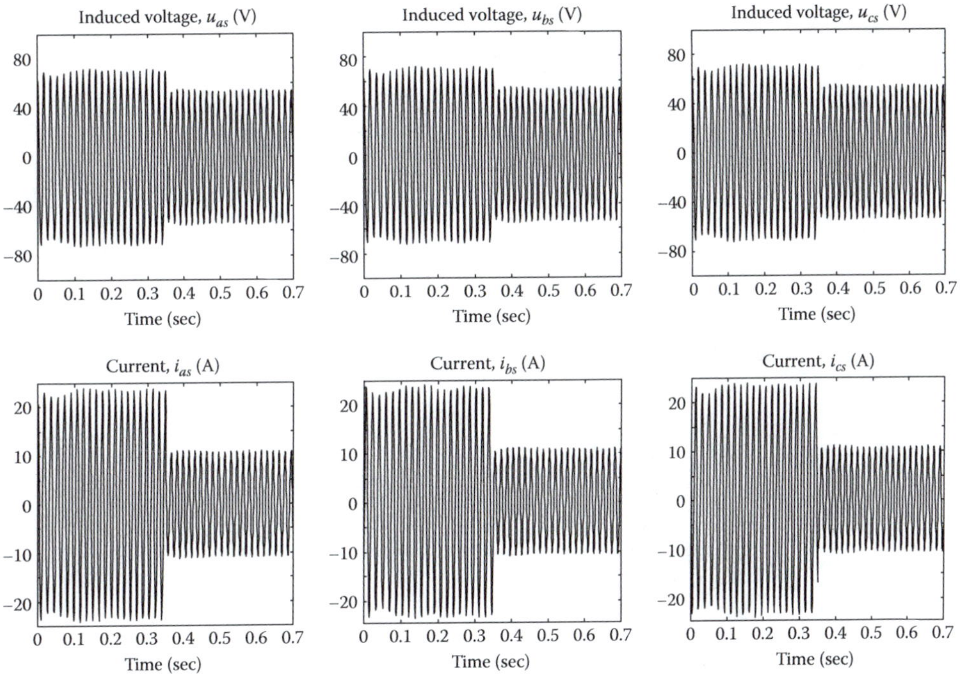

FIGURE 6.39 Induced phase voltages and phase currents if $R_L = \begin{cases} 3 \text{ ohm}, t \in [0 \quad 0.35)\text{ s} \\ 5 \text{ ohm}, t \in [0.35 \quad 0.7]\text{ s} \end{cases}$.

For synchronous motors and generators, we found

$$\mathbf{u}_{abcs} = \mathbf{r}_s \mathbf{i}_{abcs} + \frac{d\boldsymbol{\psi}_{abcs}}{dt}, \quad \text{and,} \quad \mathbf{u}_{abcs} = -\mathbf{r}_s \mathbf{i}_{abcs} + \frac{d\boldsymbol{\psi}_{abcs}}{dt}. \tag{6.60}$$

For motors, the application of the Park transformations to (6.60) results in

$$u_{qs}^r = r_s i_{qs}^r + \omega_r \psi_{ds}^r + \frac{d\psi_{qs}^r}{dt}, \quad u_{qs}^r = r_s i_{ds}^r - \omega_r \psi_{qs}^r + \frac{d\psi_{ds}^r}{dt}, \quad u_{0s}^r = r_s i_{0s}^r + \frac{d\psi_{0s}^r}{dt}, \quad u_{fr} = r_r i_{fr} + \frac{d\psi_{fr}}{dt}. \tag{6.61}$$

where the equations for the flux linkages are

$$\psi_{qs}^r = \left(L_{ls} + L_{mq}\right) i_{qs}^r, \quad \psi_{ds}^r = \left(L_{ls} + L_{md}\right) i_{ds}^r + L_{md} i_{fr}, \quad \psi_{0s}^r = L_{ls} i_{0s}^r, \quad \text{and} \quad \psi_{fr} = L_{md} i_{ds}^r + \left(L_{lf} + L_{md}\right) i_{fr}.$$

From (6.60) and (6.61), one finds

$$\begin{aligned}
u_{qs}^r &= r_s i_{qs}^r + \left(L_{ls} + L_{mq}\right)\frac{di_{qs}^r}{dt} + \left(L_{ls} + L_{md}\right) i_{ds}^r \omega_r + L_{md} i_{fr} \omega_r, \\
u_{ds}^r &= r_s i_{ds}^r + \left(L_{ls} + L_{md}\right)\frac{di_{ds}^r}{dt} - \left(L_{ls} + L_{mq}\right) i_{qs}^r \omega_r + L_{md}\frac{di_{fr}}{dt}, \\
u_{0s}^r &= r_s i_{0s}^r + L_{ls}\frac{di_{0s}^r}{dt}, \\
u_{fr} &= r_r i_{fr} + L_{md}\frac{di_{ds}^r}{dt} + \left(L_{lf} + L_{md}\right)\frac{di_{fr}}{dt}.
\end{aligned} \tag{6.62}$$

For the round-rotor synchronous machines $L_{\Delta m} = 0$. From (6.55), the electromagnetic torque is

$$T_e = \frac{PL_{md}}{2} i_{fr}\left[i_{as}\cos\theta_r + i_{bs}\cos\left(\theta_r - \frac{2}{3}\pi\right) + i_{cs}\cos\left(\theta_r + \frac{2}{3}\pi\right)\right].$$

Applying the Park transformation, one obtains

$$T_e = \frac{3P}{4}\left[L_{mq} i_{qs}^r i_{ds}^r - L_{md} i_{qs}^r \left(i_{ds}^r - i_{fr}\right)\right].$$

The *torsional–mechanical* differential equations are

$$\begin{aligned}
\frac{d\omega_r}{dt} &= \frac{3P^2}{8J}\left[L_{mq} i_{qs}^r i_{ds}^r - L_{md} i_{qs}^r \left(i_{ds}^r - i_{fr}\right)\right] - \frac{B_m}{J}\omega_r - \frac{P}{2J}T_L, \\
\frac{d\theta_r}{dt} &= \omega_r.
\end{aligned} \tag{6.63}$$

We derived the mathematical model for conventional synchronous motors in the rotor reference frame as given by (6.62) and (6.63). The models in the rotor and synchronous reference frames are identical.

For synchronous generators, one obtains

$$0=-r_s i_{qs}^r+\omega_r\psi_{ds}^r+\frac{d\psi_{qs}^r}{dt},\quad 0=-r_s i_{ds}^r-\omega_r\psi_{qs}^r+\frac{d\psi_{ds}^r}{dt},\quad 0=-r_s i_{0s}^r+\frac{d\psi_{0s}^r}{dt},\quad u_{fr}=r_r i_{fr}+\frac{d\psi_{fr}}{dt}.$$

where the flux linkages are

$$\psi_{qs}^r=-\left(L_{ls}+L_{mq}\right)i_{qs}^r,\ \psi_{ds}^r=-\left(L_{ls}+L_{md}\right)i_{ds}^r+L_{md}i_{fr},\ \psi_{0s}^r=-L_{ls}i_{0s}^r,\text{ and }\psi_{fr}=-L_{md}i_{ds}^r+\left(L_{lf}+L_{md}\right)i_{fr}.$$

Hence, the circuitry-electromagnetic and *torsional–mechanical* dynamics is

$$\frac{di_{qs}^r}{dt}=-\frac{r_s}{L_{ls}+L_{mq}}i_{qs}^r-\frac{L_{ls}+L_{md}}{L_{ls}+L_{mq}}i_{ds}^r\omega_r+\frac{L_{md}}{L_{ls}+L_{mq}}i_{fr}\omega_r,$$

$$\frac{di_{ds}^r}{dt}=-\frac{r_s}{L_{ls}+L_{md}}i_{ds}^r+\frac{L_{ls}+L_{mq}}{L_{ls}+L_{md}}i_{qs}^r\omega_r+\frac{L_{md}}{L_{ls}+L_{md}}\frac{di_{fr}}{dt},$$

$$\frac{di_{0s}^r}{dt}=-\frac{r_s}{L_{ls}}i_{0s}^r,$$

$$\frac{di_{fr}}{dt}=-\frac{r_r}{L_{lf}+L_{md}}i_{fr}+\frac{L_{md}}{L_{lf}+L_{md}}\frac{di_{ds}^r}{dt}+u_{fr},$$

$$\frac{d\omega_r}{dt}=-\frac{3P^2}{8J}\left[L_{mq}i_{qs}^r i_{ds}^r-L_{md}i_{qs}^r\left(i_{ds}^r-i_{fr}\right)\right]-\frac{B_m}{J}\omega_r+\frac{P}{2J}T_{PM},$$

$$\frac{d\theta_r}{dt}=\omega_r,$$

where $L_{mq}=L_{md}$.

In the synchronous reference frame, synchronous machines are analyzed using the same equations. The superscript e is used for the *quadrature*, *direct*, and *zero* voltages and currents components.

PRACTICE PROBLEMS

6.1 *Synchronous Reluctance Motors*

Consider single-phase synchronous reluctance actuators (limited-angle rotational relay and motor) as illustrated in Figures 6.40.

For a limited-angle rotational relay, consider the magnetizing inductance

$$L_m(\theta_r)=\bar{L}_m-L_{\Delta m}\cos^3\left(4\theta_r-\frac{\pi}{16}\right).$$

For motor, let

$$L_m(\theta_r)=\bar{L}_m-L_{\Delta m}\cos^3 4\theta_r.$$

For a limited-angle rotational relay, using the coenergy, one finds

$$T_e(\theta_r)=\frac{\partial W_c}{\partial\theta_r}=\frac{\partial W_c}{\partial\theta_r}\frac{1}{2}L_m(\theta_r)i_{as}^2=6L_{\Delta m}\sin\left(4\theta_r-\frac{\pi}{16}\right)\cos^2\left(4\theta_r-\frac{\pi}{16}\right)i_{as}^2.$$

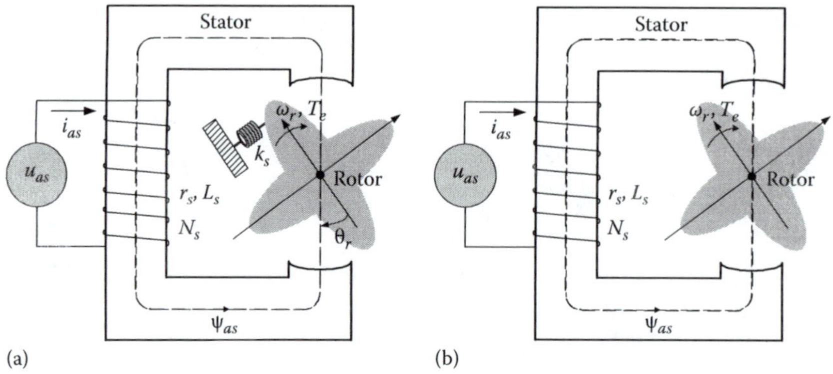

FIGURE 6.40 (a) Limited-angle rotational relay; (b) Single-phase synchronous reluctance motor.

For a limited displacement angle θ_r within ±0.1 rad, T_e is developed if a DC voltage u_{as} is applied. The rotor rotates to minimize the reluctance, which is minimum if $\theta_r = 0$. The relay remains in the equilibrium angular position if the DC voltage is supplied when $T_e = T_{spring}$ or $T_e > T_{spring}$. The governing differential equations are

$$\frac{di_{as}}{dt} = \frac{1}{L_{ls} + \bar{L}_m - L_{\Delta m}\cos^3\left(4\theta_r - \frac{\pi}{16}\right)}\left[-r_s i_{as} - 12L_{\Delta m}\sin\left(4\theta_r - \frac{\pi}{16}\right)\cos^2\left(4\theta_r - \frac{\pi}{16}\right)i_{as}\omega_r + u_{as}\right],$$

$$\frac{d\omega_r}{dt} = \frac{1}{J}\left(6L_{\Delta m}\sin\left(4\theta_r - \frac{\pi}{16}\right)\cos^2\left(4\theta_r - \frac{\pi}{16}\right)i_{as}^2 - B_m\omega_r - k_s\theta_r - T_L\right),$$

$$\frac{d\theta_r}{dt} = \omega_r.$$

For motor,

$$L_m(\theta_r) = \bar{L}_m - L_{\Delta m}\cos^3 4\theta_r$$

and

$$W_c(i_{as}, \theta_r) = \frac{1}{2}\left(L_{ls} + \bar{L}_m - L_{\Delta m}\cos^3 4\theta_r\right)i_{as}^2.$$

One finds the electromagnetic torque

$$T_e = \frac{\partial W_c(i_{as}, \theta_r)}{\partial\theta_r} = \frac{\partial}{\partial\theta_r}\frac{1}{2}i_{as}^2\left(L_{ls} + \bar{L}_m - L_{\Delta m}\cos^3 4\theta_r\right) = 6L_{\Delta m}i_{as}^2\sin 4\theta_r\cos^2 4\theta_r.$$

The total *emf* $d\psi_{as}/dt$ is found by using the expression $\psi_{as} = \left(L_{ls} + \bar{L}_m - L_{\Delta m}\cos^3 4\theta_r\right)i_{as}$. We obtain a set of three first-order nonlinear differential equations

$$\frac{di_{as}}{dt} = \frac{1}{L_{ls} + \bar{L}_m - L_{\Delta m}\cos^3 4\theta_r}\left[-r_s i_{as} - 12L_{\Delta m}\sin 4\theta_r\cos^2 4\theta_r i_{as}\omega_r + u_{as}\right],$$

$$\frac{d\omega_r}{dt} = \frac{1}{J}\left(6L_{\Delta m}\sin 4\theta_r\cos^2 4\theta_r i_{as}^2 - B_m\omega_r - T_L\right),$$

$$\frac{d\theta_r}{dt} = \omega_r.$$

To rotate the motor, the phase current i_{as} should be fed as a function of the angular displacement. For example,

$$i_{as} = \begin{cases} i_M \dfrac{1}{\cos 4\theta_r \sqrt{\sin 4\theta_r}} & \text{if } \cos 4\theta_r \sqrt{\sin 4\theta_r} > 0 \\ 0 & \text{if } \cos 4\theta_r \sqrt{\sin 4\theta_r} \le 0 \end{cases}, \quad \|i_{as}\| \le i_{\max}.$$

This results in T_e with $T_{e\,average} \neq 0$.

6.2 Consider a two-phase permanent-magnet synchronous motor.
Let

$$\psi_{asm} = \psi_m \cos^5\theta_r \quad \text{and} \quad \psi_{bsm} = \psi_m \sin^5\theta_r.$$

Using the flux linkages

$$\psi_{as} = L_{ss}i_{as} + \psi_{asm} = L_{ss}i_{as} + \psi_m \cos^5\theta_r \quad \text{and} \quad \psi_{bs} = L_{ss}i_{bs} + \psi_{bsm} = L_{ss}i_{bs} + \psi_m \sin^5\theta_r,$$

the Kirchhoff voltage law yields

$$\frac{di_{as}}{dt} = \frac{1}{L_{ss}}\left(-r_s i_{as} + 5\psi_m \sin\theta_r \cos^4\theta_r\omega_r + u_{as}\right),$$

$$\frac{di_{bs}}{dt} = \frac{1}{L_{ss}}\left(-r_s i_{bs} - 5\psi_m \cos\theta_r \sin^4\theta_r\omega_r + u_{bs}\right).$$

In the *as* and *bs* phases, the *motional emfs* are $5\psi_m \sin\theta_r \cos^4\theta_r\omega_r$ and $5\psi_m \cos\theta_r \sin^4\theta_r\omega_r$. The electromagnetic torque is

$$T_e = \frac{P}{2}\frac{\partial W_c}{\partial\theta_r} = \frac{P}{2}\frac{\partial}{\partial\theta_r}\begin{bmatrix}\psi_{asm} & \psi_{bsm}\end{bmatrix}\begin{bmatrix} i_{as} \\ i_{bs}\end{bmatrix} = \frac{P}{2}\frac{\partial}{\partial\theta_r}\psi_m\left(\cos^5\theta_r i_{as} + \sin^5\theta_r i_{bs}\right)$$

$$= \frac{5P}{2}\psi_m\left(-\sin\theta_r \cos^4\theta_r i_{as} + \cos\theta_r \sin^4\theta_r i_{bs}\right).$$

The balanced current and voltage sets are

$$i_{as} = -i_M \frac{\sin\theta_r}{\cos^4\theta_r}, \quad i_{bs} = i_M \frac{\cos\theta_r}{\sin^4\theta_r}, \quad |i_{as}| \le i_{\max}, |i_{abs}| \le i_{\max},$$

$$u_{as} = -u_M \frac{\sin\theta_r}{\cos^4\theta_r}, \quad u_{bs} = u_M \frac{\cos\theta_r}{\sin^4\theta_r}, \quad |u_{as}| \le u_{\max}, |u_{abs}| \le u_{\max}.$$

6.3 Consider a two-phase permanent-magnet machine with the flux linkages

$$\psi_{as} = L_{ss}i_{as} + \psi_{asm} = L_{ss}i_{as} + \psi_{m1}\cos\theta_r + \psi_{m4}\cos^7\theta_r \quad \text{and}$$
$$\psi_{bs} = L_{ss}i_{bs} + \psi_{bsm} = L_{ss}i_{bs} + \psi_{m1}\sin\theta_r + \psi_{m4}\sin^7\theta_r.$$

The electromagnetic torque is

$$T_e = \frac{P}{2}\frac{\partial W_c}{\partial \theta_r} = \frac{P}{2}\frac{\partial}{\partial \theta_r}\begin{bmatrix}\psi_{asm} & \psi_{bsm}\end{bmatrix}\begin{bmatrix} i_{as} \\ i_{bs}\end{bmatrix}$$

$$= \frac{P}{2}\frac{\partial}{\partial \theta_r}\left(\psi_{m1}\cos\theta_r i_{as} + \psi_{m4}\cos^7\theta_r i_{as} + \psi_{m1}\sin\theta_r i_{bs} + \psi_{m4}\sin^7\theta_r i_{bs}\right)$$

$$= \frac{P}{2}\left(-\psi_{m1}\sin\theta_r i_{as} - 7\psi_{m4}\sin\theta_r\cos^6\theta_r i_{as} + \psi_{m1}\cos\theta_r i_{bs} + 7\psi_{m4}\cos\theta_r\sin^6\theta_r i_{bs}\right).$$

The governing equations are

$$\frac{di_{as}}{dt} = \frac{1}{L_{ss}}\left[-r_s i_{as} + \psi_{m1}\sin\theta_r\omega_r + 7\psi_{m4}\sin\theta_r\cos^6\theta_r\omega_r + u_{as}\right],$$

$$\frac{di_{bs}}{dt} = \frac{1}{L_{ss}}\left[-r_s i_{bs} - \psi_{m1}\cos\theta_r\omega_r - 7\psi_{m4}\cos\theta_r\sin^6\theta_r\omega_r + u_{bs}\right],$$

$$\frac{d\omega_r}{dt} = \frac{1}{J}\left[\frac{P^2}{4}\left(-\psi_{m1}\sin\theta_r i_{as} - 7\psi_{m4}\sin\theta_r\cos^6\theta_r i_{as} + \psi_{m1}\cos\theta_r i_{bs} + 7\psi_{m4}\cos\theta_r\sin^6\theta_r i_{bsL}\right) - B_m\omega_r - \frac{P}{2}T_L\right],$$

$$\frac{d\theta_r}{dt} = \omega_r.$$

The balanced current and voltage sets are

$$i_{as} = -i_M\frac{\sin\theta_r}{1+\dfrac{7\psi_{m4}}{\psi_{m1}}\cos^6\theta_r},\quad i_{bs} = i_M\frac{\cos\theta_r}{1+\dfrac{7\psi_{m4}}{\psi_{m1}}\sin^6\theta_r},\quad |i_{as}|\le i_{\max},\ |i_{bs}|\le i_{\max},$$

$$u_{as} = -u_M\frac{\sin\theta_r}{1+\dfrac{7\psi_{m4}}{\psi_{m1}}\cos^6\theta_r},\quad u_{bs} = u_M\frac{\cos\theta_r}{1+\dfrac{7\psi_{m4}}{\psi_{m1}}\sin^6\theta_r},\quad |u_{as}|\le u_{\max},\ |u_{bs}|\le u_{\max}.$$

Assuming there are not bounds on the phase current, the electromagnetic torque is $T_e = \frac{1}{2}P\psi_{m1}i_M$.

6.4 Consider a two-phase permanent-magnet synchronous motor. The experimentally measured flux linkages, which are nonlinear functions of θ_r, may be parametrized using (6.35). The mixed exponential-trigonometric nonlinear interpolation of the measured *motional emfs* results in

$$\psi_{asm} = \psi_m\sin\theta_r e^{-\cos\theta_r} \quad\text{and}\quad \psi_{bsm} = \psi_m\cos\theta_r e^{\sin\theta_r}.$$

Figure 6.25d documents the plot for ψ_{asm}. The electromagnetic torque is

$$T_e = \frac{P}{2}\frac{\partial W_c}{\partial \theta_r} = \frac{P}{2}\frac{\partial}{\partial \theta_r}\begin{bmatrix}\psi_{asm} & \psi_{bsm}\end{bmatrix}\begin{bmatrix} i_{as} \\ i_{bs}\end{bmatrix} = \frac{P}{2}\frac{\partial}{\partial \theta_r}\left(\psi_m\sin\theta_r e^{-\cos\theta_r} i_{as} + \psi_m\cos\theta_r e^{\sin\theta_r} i_{bs}\right)$$

$$= \frac{P}{2}\psi_m\left[\left(\sin^2\theta_r + \cos\theta_r\right)e^{-\cos\theta_r}i_{as} + \left(\cos^2\theta_r - \sin\theta_r\right)e^{\sin\theta_r}i_{bs}\right].$$

In the Kirchhoff voltage law, the *total emfs* $d\psi_{as}/dt$ and $d\psi_{bs}/dt$ are found using $\psi_{as} = L_{ss}i_{as} + \psi_m \sin\theta_r e^{-\cos\theta_r}$ and $\psi_{bs} = L_{ss}i_{bs} + \psi_m \cos\theta_r e^{\sin\theta_r}$.

Applying the Kirchhoff law and Newton's second law, we find

$$\frac{di_{as}}{dt} = \frac{1}{L_{ss}}\left[-r_s i_{as} - \psi_m\left(\sin^2\theta_r + \cos\theta_r\right)e^{-\cos\theta_r}\omega_r + u_{as}\right],$$

$$\frac{di_{bs}}{dt} = \frac{1}{L_{ss}}\left[-r_s i_{as} - \psi_m\left(\cos^2\theta_r - \sin\theta_r\right)e^{\sin\theta_r}\omega_r + u_{bs}\right],$$

$$\frac{d\omega_r}{dt} = \frac{1}{J}\left[\frac{P^2}{4}\psi_m\left[\left(\sin^2\theta_r + \cos\theta_r\right)e^{-\cos\theta_r}i_{as} + \left(\cos^2\theta_r - \sin\theta_r\right)e^{\sin\theta_r}i_{bs}\right] - B_m\omega_r - \frac{P}{2}T_L\right]$$

$$\frac{d\theta_r}{dt} = \omega_r$$

From

$$T_e = \frac{P}{2}\psi_m\left[\left(\sin^2\theta_r + \cos\theta_r\right)e^{-\cos\theta_r}i_{as} + \left(\cos^2\theta_r - \sin\theta_r\right)e^{\sin\theta_r}i_{bs}\right],$$

the balanced current sets are

$$\begin{cases} i_{as} = i_M \dfrac{\sin^2\theta_r}{\left(\sin^2\theta_r + \cos\theta_r\right)e^{-\cos\theta_r}}, & |i_{as}| \le i_{max} \\ i_{bs} = i_M \dfrac{\cos^2\theta_r}{\left(\cos^2\theta_r - \sin\theta_r\right)e^{\sin\theta_r}}, & |i_{bs}| \le i_{max} \end{cases} \quad \text{or} \quad \begin{cases} i_{as} = i_M \dfrac{e^{\cos\theta_r}}{\sin^2\theta_r + \cos\theta_r}, & |i_{as}| \le i_{max} \\ i_{bs} = i_M \dfrac{e^{-\sin\theta_r}}{\cos^2\theta_r - \sin\theta_r}, & |i_{bs}| \le i_{max}. \end{cases}$$

For the aforementioned current sets, if the phase currents are not bounded,

$$T_e = \frac{P}{2}\psi_m i_M.$$

However, in the full operating envelope, it is impossible to guarantee the balanced operation due to the current and voltage limits.

HOMEWORK PROBLEMS

6.1 Consider a two-phase permanent-magnet synchronous motor. Let $\psi_{asm} = \psi_m \cos^7\theta_r$ and $\psi_{bsm} = \psi_m \sin^7\theta_r$. Solve the following problems:

a. Using Kirchhoff's voltage law, derive the differential equation for the phase current i_{as} and i_{bs}, that is, obtain and report the circuitry-electromagnetic differential equations.

b. Report the *emf*$_{as}$ induced, as found in (a). Plot the derived *emf*$_{as}$ as a function of θ_r at the steady-state operation if $\omega_r = 100$ rad/sec and $\psi_m = 0.1$. Report the MATLAB statement to calculate and plot *emf*$_{as}$.

c. Derive an explicit expression for the electromagnetic torque T_e.

d. For T_e found, derive the balanced voltage and current sets with the goal to eliminate the torque ripple and current chattering. Report the problems one might face.

6.2 Consider a two-phase axial topology permanent-magnet synchronous motor. The *effective* flux with respect to the *as* and *bs* windings are $B_{as} = B_{M\,\max} \sin^7 \theta_r$ and $B_{bs} = B_{\max} \cos^7 \theta_r$. The motor parameters are $N = 100$, $A_{ag} = 0.0001$, $B_{\max} = 0.75$ and $N_m = 10$. Let the phase currents be $i_{as} = i_M \cos \theta_r$ and $i_{bs} = -i_M \sin \theta_r$. Solve the following problems:

a. Drive the expression for the electromagnetic torque.
b. Examine and document how the electromagnetic toque varies as a function of the rotor displacement. Plot a torque versus displacement curve.
c. Make the conclusions on how to improve the motor's performance and enhance its capabilities. For example, how to maximize the torque and ensure that T_e does not have the ripple.
d. Document how to use MATLAB to solve Problems (a) through (c).

6.3 Simulate and analyze an electromechanical system actuated by a NEMA 23 size, two-phase 1.8° full-step, 5.4 V (*rms*), 1.4 N-m permanent-magnet stepper motor. The parameters are as follows: $RT = 50$, $r_s = 1.68$ ohm, $L_{ss} = 0.0057$ H, $\psi_m = 0.0064$ V-sec/rad (N-m/A), $B_m = 0.000074$ N-m-sec/rad, and $J = 0.000024$ kg-m^2. Study the motor performance:

(1) $u_{as} = -u_M \operatorname{sgn}\left(\sin\left(RT\theta_{rm}\right)\right)$ and $u_{bs} = u_M \operatorname{sgn}\left(\cos\left(RT\theta_{rm}\right)\right)$;

(2) $u_{as} = -\sqrt{2}u_M \sin\left(RT\theta_{rm}\right)$ and $u_{bs} = \sqrt{2}u_M \cos\left(RT\theta_{rm}\right)$;

(3) Phase voltages u_{as} and u_{bs} with the magnitude u_M are applied as sequences of pulses with different frequency to ensure a step-by-step operation.

Explain why stepper motors can be a favorable solution in some *direct* drives and *servo* applications. Discuss the possibility to use stepper motors in the open-loop systems without Hall-effect sensors to measure θ_r. Report the challenges one faces using the stepper motors in an open-loop configuration and in not using the rotor displacement sensor.

REFERENCES

1. S. J. Chapman, *Electric Machinery Fundamentals*, McGraw-Hill, New York, 2011.
2. P. C. Krause and O. Wasynczuk, *Electromechanical Motion Devices*, McGraw-Hill, New York, 1989.
3. P. C. Krause, O. Wasynczuk, S. D. Sudhoff, and S. Pekarek, *Analysis of Electric Machinery*, Wiley-IEEE Press, New York, 2013.
4. S. E. Lyshevski, *Electromechanical Systems, Electric Machines, and Applied Mechatronics*, CRC Press, Boca Raton, FL, 1999.
5. S. E. Lyshevski, *Electromechanical Systems and Devices*, CRC Press, Boca Raton, FL, 2008.
6. G. R. Slemon, *Electric Machines and Drives*, Addison-Wesley Publishing Company, Reading, MA, 1992.

7 Electronics and Power Electronics

Signal Processing, Filtering, Data Analysis, and Data Analytics

7.1 MICROELECTRONICS, OPERATIONAL AMPLIFIERS, AND INTEGRATED CIRCUITS

Signal processing, signal conditioning, and other tasks are accomplished by integrated circuits (ICs). The analog and digital controllers and filters are implemented using analog, *hybrid*, and digital controllers. Electromechanical systems are predominantly continuous. Analog controllers and filters can be implemented using operational amplifiers and specialized ICs. These controllers and filters are integrated with high-switching-frequency power amplifiers. The solid-state semiconductor devices (diodes, transistors, thyristors, and others) are continuous. The use of analog and digital controllers, filters and sensors results in *hybrid* closed-loop electromechanical systems. Low-power analog electronics ensures compliance, effectiveness, and simplicity.

To ensure data acquisition, one performs sensing, data processing, and data analysis to ensure control, decision-making, hierarchical management, etc. Data processing must be performed to enable descriptive, predictive, and prescriptive data analytics. Predictive data analytics is needed to ensure data reduction using statistical models, data mining, predictive analysis, etc. The prescriptive deterministic analysis may not be sufficient. The descriptive data analytics by means of consistent data analysis is implemented using microelectronics, and sensing hardware is studied in this chapter.

We examine the use of operational amplifiers to implement analog controller and filters. Various physical quantities (displacement, velocity, acceleration, force, torque, pressure, temperature, and others) can be directly measured. The sensor's output voltage or current is a function of the measured physical quantity. The signal-level sensor outputs, used to implement controllers, must be filtered. A single operational amplifier has *noninverting* and *inverting* inputs (pins 3 and 2) as well as an output terminal (pin 6), see Figures 7.1. The DC voltage is supplied. The terminal 7 is connected to a positive voltage u_+, while a negative voltage (or ground) u_- is supplied to the terminal 4. The pin connections of the single, dual, and quad low-power operational amplifiers MC33171, MC33172, and MC33174 are reported in Figure 7.1a. There are various packages, including surface mount. Operational amplifiers, which consist of dozens of field-effect transistors, are fabricated using the CMOS or biCMOS technology [1]. Figures 7.1a depict the representative schematics. There are general-purpose, instrumental, precision, high-speed, differential, power, and other operational amplifiers shown in Figure 7.1b. Using analog microelectronics, one may perform summation, subtraction, multiplication, and division of input signals. Images of the AD534JDZ and AD734 four-quadrant multipliers-dividers are documented in Figure 7.1c.

The operational amplifier output is the difference between two input voltages $[u_1(t) - u_2(t)]$ applied to the *inverting* input terminal and *noninverting* input terminal, multiplied by the differential open-loop coefficient k_0. The resulting output voltage is $u_0(t) = k_0[u_2(t) - u_1(t)]$. The differential open-loop coefficient is positive. The gain k_0 is very large, $k_0 \in \left[1\times10^5 \;\; 1\times10^7\right]$. The general-purpose operational amplifiers have input and output resistances $\sim\left[1\times10^5 \;\; 1\times10^{12}\right]$ and $\sim\left[10 \;\; 1000\right]$ ohm, respectively. The *inverting* and *noninverting* input terminals are distinguished by the "–"

and "+" signs. Supplying the signal-level input voltage $u_1(t)$ to the *inverting* input terminal using external resistor R_1, and grounding the *noninverting* input terminal, one can find the differential closed-loop coefficient k_0 if a negative feedback with R_2 is used. The output terminal is connected to the *inverting* input terminal, and the resistor R_2 is inserted as depicted in Figure 7.2a. The inverting summing amplifier (weighted summer) is shown in Figure 7.2b. The instrumentational amplifiers are a preferable choice. The images of AD524ADZ-ND are documented in Figure 7.2c.

As illustrated in Figure 7.3a, we use the input impedance $Z_1(s)$ and the feedback path impedance $Z_2(s)$. The impedance is the ratio of the phasor voltage to the phasor current. The impedances of the resistor, capacitor, and inductor are

$$Z_R(s) = R, Z_R(j\omega) = R, \quad Z_C(s) = \frac{1}{sC}, \quad Z_C(j\omega) = \frac{1}{j\omega C} = -\frac{j}{\omega C}, \text{ and } Z_L(s) = sL, Z_L(j\omega) = j\omega L.$$

The transfer function of the closed-loop amplifier configuration is $G(s) = \frac{U_0(s)}{U_1(s)} = -\frac{Z_2(s)}{Z_1(s)}$.

For the operational amplifier, shown in Figure 7.3b, the transfer function is found using

$$Z_1(s) = R_1 \quad \text{and} \quad Z_2(s) = \frac{R_2}{R_2C_2s+1}.$$

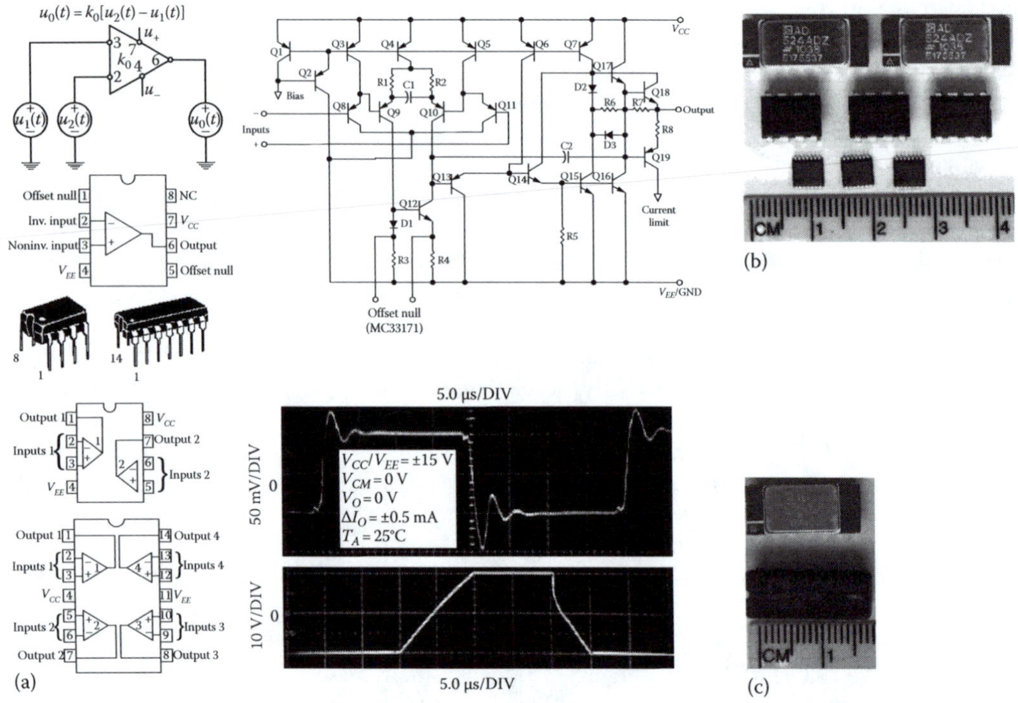

FIGURE 7.1 (a) General-purpose operational amplifiers, pin connections, packages, schematics, and transient responses; (From Lyshevski, S.E., *Electromechanical Systems, Electric Machines, and Applied Mechatronics*, CRC Press, Boca Raton, FL, 1999; Copyright of Motorola. With permission.)
(b) Images of Analog Devices AD524ADZ-ND 16-CDIP package instrumentation amplifiers, AD620ANZ low-power instrumentation amplifiers, and AD8612ARUZ 14-lead ultrafast 4 nsec single supply comparators, respectively;
(c) Images of AD534JDZ and AD734 four-quadrant multipliers-dividers with a multiplication error ~ ±0.25%. Summation, subtraction, division, and multiplication ($XY + Z$) operations can be implemented. The AD734AQ is a high-speed (10 MHz bandwidth, 200 nsec settling time) four-quadrant analog multiplier that computes XY/Z with −80 dB distortion. The low capacitance X, Y, and Z inputs are differential.

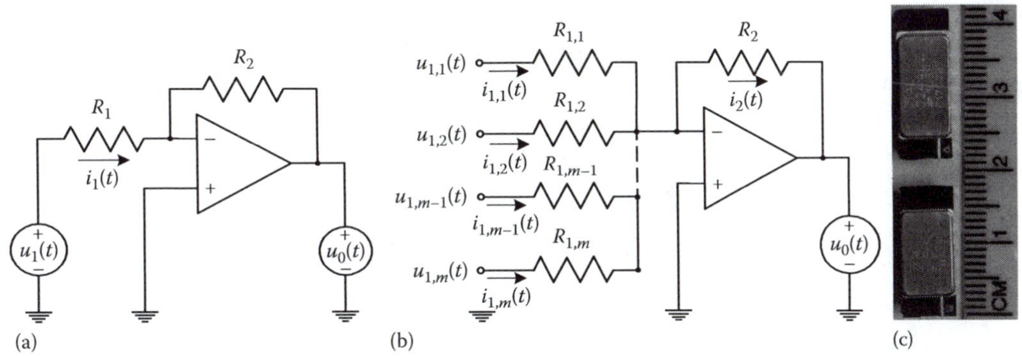

FIGURE 7.2 (a) Inverting configuration of the operational amplifier;

(b) A summing amplifier with m inputs. The current in the feedback path is $i_2(t) = i_{1,1}(t) + \cdots + i_{1,m}(t)$, $i_{1,1}(t) = u_{1,1}(t)/R_{1,1},\ldots, i_{1,m}(t) = u_{1,m}(t)/R_{1,m}$. The amplifier output is $u_0(t) = -\left(\frac{R_2}{R_{1,1}}u_{1,1}(t) + \cdots + \frac{R_2}{R_{1,m}}u_{1,m}(t)\right)$;

(c) Image of the Analog Devices AD524ADZ-ND 16-CDIP package instrumentation amplifiers.

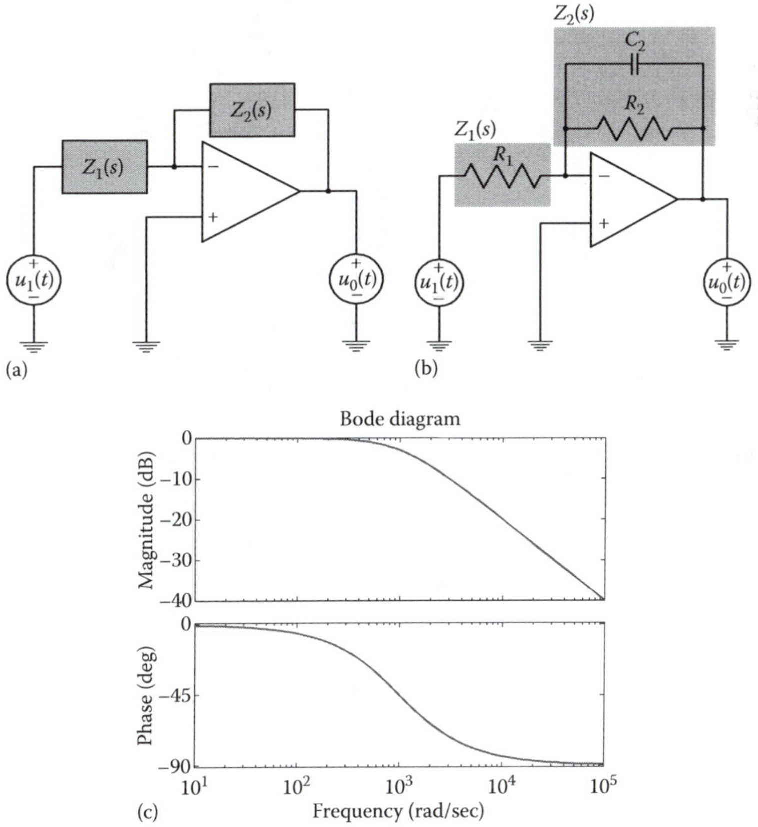

FIGURE 7.3 (a) Inverting configuration of the operational amplifier with $Z_1(s)$ and $Z_2(s)$, $G(s) = -\frac{Z_2(s)}{Z_1(s)}$;

(b) Inverting operational amplifier with $Z_1(s) = R_1$ and $Z_2(s) = \frac{R_2}{R_2C_2s+1}$, $G(s) = -\frac{Z_2(s)}{Z_1(s)} = -\frac{R_2/R_1}{R_2C_2s+1}$;

(c) Bode plots for a low-pass filter $G(j\omega) = \frac{R_2}{R_1}\frac{1}{R_2C_2j\omega+1}$. The MATLAB® statement is

```
R1=1e5;R2=1e5;C2=1e-8;num=[R2/R1];den=[R2*C2 1];bode(num,den)
```

We have

$$G(s)=\frac{U_0(s)}{U_1(s)}=-\frac{Z_2(s)}{Z_1(s)}=-\frac{R_2/R_1}{R_2C_2s+1}.$$

The closed-loop gain coefficient for the inverting operational amplifier is $-R_2/R_1$, while the time constant is R_2C_2. In the frequency domain, substituting $s=j\omega$, we have

$$G(j\omega)=\frac{U_0(j\omega)}{U_1(j\omega)}=-\frac{Z_2(j\omega)}{Z_1(j\omega)}=-\frac{R_2}{R_1}\frac{1}{R_2C_2j\omega+1}.$$

Consider a low-pass filter

$$G(j\omega)=\frac{R_2}{R_1}\frac{1}{R_2C_2j\omega+1}$$

with the gain $k=1$ and the cutoff frequency $\omega=1/R_2C_2=1000$ rad/sec. One finds $R_1=R_2=1\times10^5$ ohm and $C_2=1\times10^{-8}$ C. The Bode plots are calculated and plotted in Figure 7.3c.

For input and feedback impedances $Z_1(s)$ and $Z_2(s)$, Table 7.1 reports the transfer functions of the inverting operational amplifier configurations.

7.2 ANALOG FILTERS

Operational amplifiers perform the arithmetic operations (addition, subtraction, and multiplication), signal conditioning, and processing to implement filters, control algorithms, etc. Sensor signals contain noise which have different origins. The low-, medium-, and high-frequency noise can be attenuated by filters that must be consistently designed and implemented.

The system bandwidth, noise characteristics, robustness, immunity, effectiveness, simplicity, compliance, and other quantities are considered. Using the system bandwidth, noise spectra, and control algorithms, one finds the frequencies to be preserved (system bandwidth) and attenuated (noise frequency). Using system requirements, one finds the cutoff frequencies and attenuation. The elliptical, Butterworth, Chebyshev, Bessel, Cauer, notch, and other filters can be used. We focus on the low-pass Butterworth and notch filters, which guarantee no passband and stopband ripples. These filters ensure the preferable overdamped transient response. In the Butterworth and notch filters, the magnitude $|G(\omega)|$ is a constant, monotonically decreasing, or monolithically decreasing function of frequency at all frequencies.

The filters are designed by specifying the gain. Using the system bandwidth (frequencies to be preserved) and noise or disturbance frequencies to attenuate, one finds the cutoff frequencies, $\omega_{ci}=1/R_iC_i$. Other specifications imposed are the phase lag, filter compliance, complexity, and sensitivity.

Example 7.1: First-Order Notch Filters

For the filter depicted in Figure 7.4a, the impedances are

$$Z_1(s)=\frac{R_1}{R_1C_1s+1}\quad\text{and}\quad Z_2(s)=\frac{R_2}{R_2C_2s+1}.$$

One finds

$$G(s)=\frac{U_0(s)}{U_1(s)}=-\frac{Z_2(s)}{Z_1(s)}=\frac{N(s)}{D(s)}=-\frac{R_2}{R_1}\frac{\left(R_1C_1s+1\right)}{\left(R_2C_2s+1\right)}.$$

The order of the numerator $N(s)$ is the same as the order of the denominator $D(s)$. One has the notch filter with a band-stop phase $\phi\to0$ as $\omega\to\infty$.

TABLE 7.1
Transfer Functions of the Inverting Amplifier Configurations

Input Circuit with Impedance $Z_1(s)$	Feedback Circuit with Impedance $Z_2(s)$	Transfer Function
$Z_1(s)$ 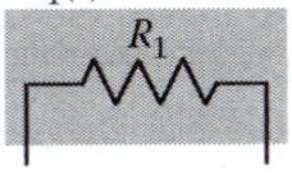 $Z_1(s)=R_1$	$Z_2(s)$ 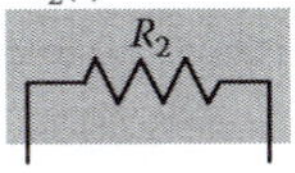 $Z_2(s)=R_2$	$G(s)=\frac{U_0(s)}{U_1(s)}=-\frac{R_2}{R_1}$
$Z_1(s)$ 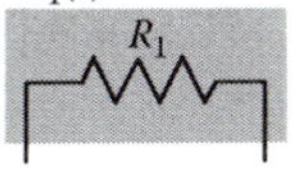 $Z_1(s)=R_1$	$Z_2(s)$ 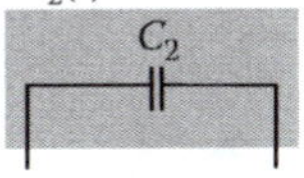 $Z_2(s)=\frac{1}{C_2s}$	$G(s)=\frac{U_0(s)}{U_1(s)}=-\frac{1}{R_1C_2s}$
$Z_1(s)$ 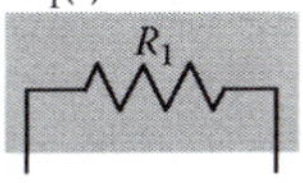 $Z_1(s)=R_1$	$Z_2(s)$ 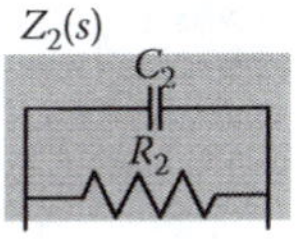 $Z_2(s)=\frac{R_2}{R_2C_2s+1}$	$G(s)=\frac{U_0(s)}{U_1(s)}=-\frac{\frac{R_2}{R_1}}{R_2C_2s+1}$
$Z_1(s)$ 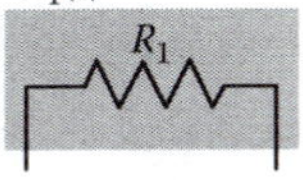 $Z_1(s)=R_1$	$Z_2(s)$ 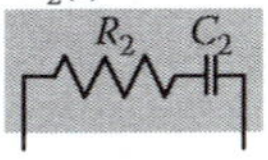 $Z_2(s)=\frac{R_2C_2s+1}{C_2s}$	$G(s)=\frac{U_0(s)}{U_1(s)}=-\frac{R_2C_2s+1}{R_1C_2s}$
$Z_1(s)$ 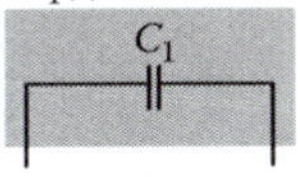 $Z_1(s)=\frac{1}{C_1s}$	$Z_2(s)$ 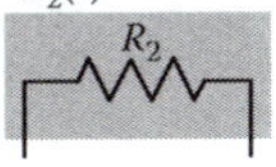 $Z_2(s)=R_2$	$G(s)=\frac{U_0(s)}{U_1(s)}=-R_1C_2s$
$Z_1(s)$ 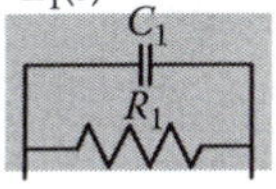 $Z_1(s)=\frac{R_1}{R_1C_1s+1}$	$Z_2(s)$ 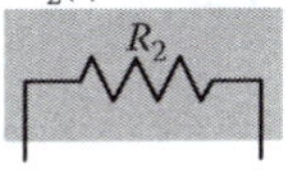 $Z_2(s)=R_2$	$G(s)=\frac{U_0(s)}{U_1(s)}=-\frac{R_1R_2C_1s+R_2}{R_1}$
$Z_1(s)$ 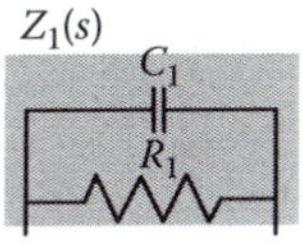 $Z_1(s)=\frac{R_1}{R_1C_1s+1}$	$Z_2(s)$ 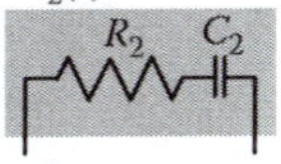 $Z_2(s)=\frac{R_2C_2s+1}{C_2s}$	$G(s)=\frac{U_0(s)}{U_1(s)}=-\frac{(R_1C_1s+1)(R_2C_2s+1)}{R_1C_2s}$
$Z_1(s)$ 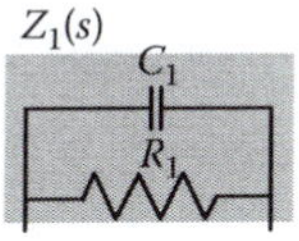 $Z_1(s)=\frac{R_1}{R_1C_1s+1}$	$Z_2(s)$ 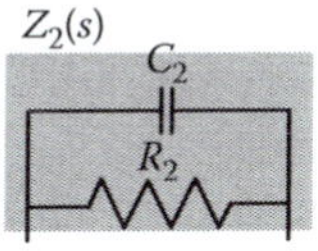 $Z_2(s)=\frac{R_2}{R_2C_2s+1}$	$G(s)=\frac{U_0(s)}{U_1(s)}=-\frac{R_2}{R_1}\frac{(R_1C_1s+1)}{(R_2C_2s+1)}$
$Z_1(s)$ 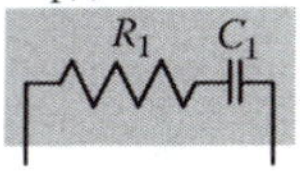 $Z_1(s)=\frac{R_1C_1s+1}{C_1s}$	$Z_2(s)$ 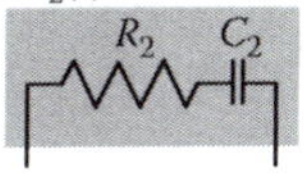 $Z_2(s)=\frac{R_2C_2s+1}{C_2s}$	$G(s)=\frac{U_0(s)}{U_1(s)}=-\frac{C_1}{C_2}\frac{(R_2C_2s+1)}{(R_1C_1s+1)}$

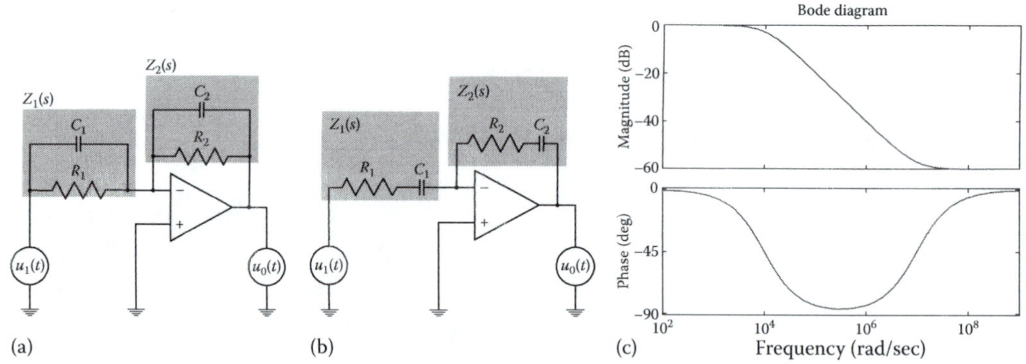

FIGURE 7.4 Analog first-order notch filters implemented using an inverting operational amplifier: (a) $G(s) = -\dfrac{R_2}{R_1}\dfrac{(R_1C_1s+1)}{(R_2C_2s+1)}$ $Z_1(s) = \dfrac{R_1}{R_1C_1s+1}$ and $Z_2(s) = \dfrac{R_2}{R_2C_2s+1}$; (b) $G(s) = -\dfrac{C_1}{C_2}\dfrac{(R_2C_2s+1)}{(R_1C_1s+1)}$ $Z_1(s) = \dfrac{R_1C_1s+1}{C_1s}$ and $Z_2(s) = \dfrac{R_2C_2s+1}{C_2s}$; (c) Bode plots for a notch filter $G(s) = \dfrac{R_2}{R_1}\dfrac{(R_1C_1s+1)}{(R_2C_2s+1)}$ ($R_1 = R_2 = 1000$ ohm, $C_1 = 1 \times 10^{-10}$ F, $C_2 = 1 \times 10^{-7}$ F). The MATLAB® statement is

```
R1=1e3; R2=1e3; C1=1e-10; C2=1e-7; num=[R2*R1*C1 R2];den=[R1*R2*C2 R1]; bode(num,den)
```

The filter has a zero $(R_1C_1s + 1)$ and a pole $(R_2C_2s + 1)$. Let the system bandwidth be 1000 Hz. The ~100000 Hz noise should be attenuated by ~100 times, while at frequencies within the system bandwidth, a unit gain $k = 1$ should be ensured. Hence, $k = R_2/R_1 = 1$. From $\omega = 2\pi f$, one finds the cutoff frequencies $\omega_{ci} = 1/R_iC_i$. For the numerator and denominator, $\omega_{cN} = \dfrac{1}{R_1C_1}$ and $\omega_{cD} = \dfrac{1}{R_2C_2}$. At the corner angular frequencies ω_{ci}, the $\Delta\,|G|_{\text{dB}} = -3$ dB of the nominal passband value. To ensure $k = 1$ up to $f = 1000$ Hz, we let $\omega_{cD} = 10000$ rad/sec. One has $R_1 = R_2 = 1000$ ohm, $C_1 = 1 \times 10^{-10}$ F and $C_2 = 1 \times 10^{-7}$ F. The Bode plot is illustrated in Figure 7.4c. The design specifications are met.

For the notch filter shown in Figure 7.4b

$$Z_1(s) = \frac{R_1C_1s+1}{C_1s} \quad \text{and} \quad Z_2(s) = \frac{R_2C_2s+1}{C_2s}.$$

The transfer function is

$$G(s) = \frac{U_0(s)}{U_1(s)} = -\frac{C_1}{C_2}\frac{(R_2C_2s+1)}{(R_1C_1s+1)}.$$

The filter design is similar as reported. ■

Example 7.2: Second-Order Notch Filter

Consider the notch filter represented in Figure 7.5a, which composes a series (cascade) configuration of two inverting operational amplifiers with input and feedback impedances.

The input and feedback impedances are

$$Z_{11}(s) = \frac{R_{11}}{R_{11}C_{11}s+1},\ Z_{12}(s) = \frac{R_{12}}{R_{12}C_{12}s+1},\ Z_{21}(s) = \frac{R_{21}}{R_{21}C_{21}s+1} \quad \text{and} \quad Z_{22}(s) = \frac{R_{22}}{R_{22}C_{22}s+1}.$$

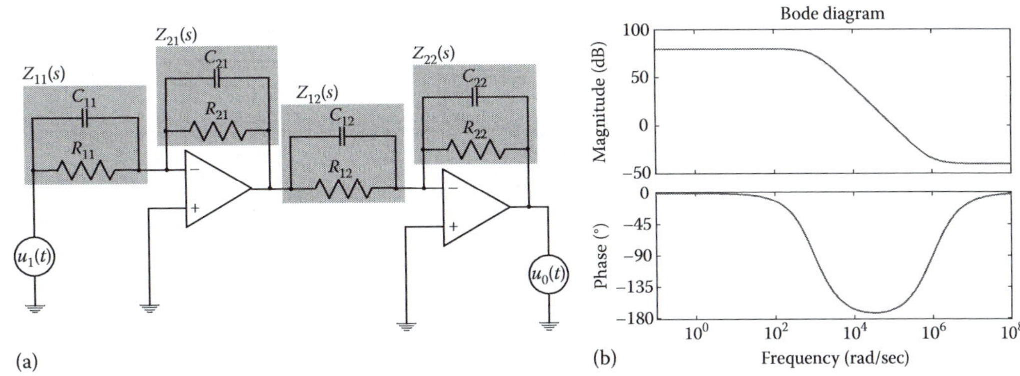

FIGURE 7.5 (a) Notch filter schematics with $G(s) = \dfrac{Z_{21}(s)}{Z_{11}(s)}\dfrac{Z_{22}(s)}{Z_{12}(s)} = \dfrac{R_{21}R_{22}}{R_{11}R_{12}}\dfrac{(R_{11}C_{11}s+1)(R_{12}C_{12}s+1)}{(R_{21}C_{21}s+1)(R_{22}C_{22}s+1)}$; (b) Bode plots. The MATLAB® statements are

```
R11=100; R12=100; R21=10000; R22=10000; C11=10e-9; C12=10e-9; C21=100e-9; C22=100e-9;
num1=[R11*C11 1]; num2=[R12*C12 1]; num=(R21*R22)/(R11*R12)*conv(num1,num2);
den1=[R21*C21 1]; den2=[R22*C22 1]; den=conv(den1, den2); bode(num,den,{0.1,1e8})
```

The transfer function of the notch filter is

$$G(s) = \frac{Z_{21}(s)}{Z_{11}(s)}\frac{Z_{22}(s)}{Z_{12}(s)} = \frac{\frac{R_{21}}{R_{11}}(R_{11}C_{11}s+1)}{R_{21}C_{21}s+1}\frac{\frac{R_{22}}{R_{12}}(R_{12}C_{12}s+1)}{R_{22}C_{22}s+1} = \frac{\frac{R_{21}R_{22}}{R_{11}R_{12}}(R_{11}C_{11}s+1)(R_{12}C_{12}s+1)}{(R_{21}C_{21}s+1)(R_{22}C_{22}s+1)}.$$

The values of resistors and capacitors are found for the specified cutoff frequencies, gain, and attenuation at the specified frequencies. Let the system bandwidth be ~10 Hz, which should be preserved. The noise frequency is ~100000 Hz. The noise should be attenuated at least 1000 times. The filter gain at low frequency should be 10000 or 80 dB. The corner angular frequencies are $1/(R_{ij}C_{ij})$. We have two poles with $\omega_{cD1} = \dfrac{1}{R_{21}C_{21}}$ and $\omega_{cD2} = \dfrac{1}{R_{22}C_{22}}$, and, two zeros with $\omega_{cN1} = \dfrac{1}{R_{11}C_{11}}$ and $\omega_{cN2} = \dfrac{1}{R_{12}C_{12}}$. Let $\omega_{cD1} = \omega_{cD2} = 1000$ rad/sec and $\omega_{cN1} = \omega_{cN2} = 1 \times 10^6$ rad/sec. Recall that $\omega = 2\pi f$. The system bandwidth $f = 10$ Hz gives $2\pi f = 62.8$ rad/sec. However, at the corner angular frequencies $1/(R_{ij}C_{ij})$, one has $\Delta|G|_{dB} = -3$ dB at the nominal passband value.

Using the specified low-frequency gain 80 dB, one has $(R_{21}R_{22})/(R_{11}R_{12}) = 10000$. Using the cutoff frequencies $1/(R_{ij}C_{ij})$, we have $R_{11} = R_{12} = 100$ ohm, $R_{21} = R_{22} = 10000$ ohm, $C_{11} = C_{12} = 10$ nF and, $C_{21} = C_{22} = 100$ nF. The Bode plots are documented in Figure 7.5b. ■

Example 7.3: Second- and Third Order Filters

Consider the filters reported in Figures 7.6. For the schematic in Figure 7.6a, one has

$$Z_1(s) = \frac{\left(R_{11}+\frac{1}{C_{11}s}\right)\left(R_{12}+\frac{1}{C_{12}s}\right)}{R_{11}+\frac{1}{C_{11}s}+R_{12}+\frac{1}{C_{12}s}} = \frac{(R_{11}C_{11}s+1)(R_{12}C_{12}s+1)}{s(R_{11}C_{11}C_{12}s+R_{12}C_{11}C_{12}s+C_{11}+C_{12})}$$

and

$$Z_2(s) = \frac{(R_{21}C_{21}s+1)(R_{22}C_{22}s+1)}{s(R_{21}C_{21}C_{22}s+R_{22}C_{21}C_{22}s+C_{21}+C_{22})}.$$

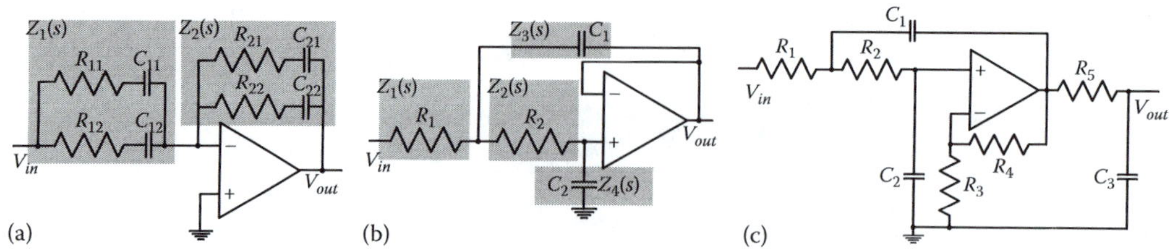

FIGURE 7.6 (a) Notch filter with $G(s)=\dfrac{(R_{21}C_{21}s+1)(R_{22}C_{22}s+1)(R_{11}C_{11}C_{12}s+R_{12}C_{11}C_{12}s+C_{11}+C_{12})}{(R_{11}C_{11}s+1)(R_{12}C_{12}s+1)(R_{21}C_{21}C_{22}s+R_{22}C_{21}C_{22}s+C_{21}+C_{22})}$; (b) Analog Sallen–Key low-pass filter with $G(s)=\dfrac{V_{out}(s)}{V_{in}(s)}=\dfrac{Z_3(s)Z_4(s)}{Z_1(s)Z_2(s)+[Z_1(s)+Z_2(s)]Z_3(s)+Z_3(s)Z_4(s)}$ and $G(s)=\dfrac{V_{out}(s)}{V_{in}(s)}=\dfrac{1}{R_1R_2C_1C_2s^2+(R_1+R_2)C_2s+1}$; (c) Low-pass third-order Butterworth filter.

The resulting transfer function is

$$G(s)=-\frac{Z_2(s)}{Z_1(s)}=\frac{(R_{21}C_{21}s+1)(R_{22}C_{22}s+1)(R_{11}C_{11}C_{12}s+R_{12}C_{11}C_{12}s+C_{11}+C_{12})}{(R_{11}C_{11}s+1)(R_{12}C_{12}s+1)(R_{21}C_{21}C_{22}s+R_{22}C_{21}C_{22}s+C_{21}+C_{22})}.$$

This transfer function $G(s)$ corresponds to the third-order notch filter.

The Sallen–Key low-pass filter is reported in Figure 7.6b. Using the node analysis reported in Example 7.4, one finds the resulting transfer function

$$G(s)=\frac{V_{out}(s)}{V_{in}(s)}=\frac{1}{R_1R_2C_1C_2s^2+(R_1+R_2)C_2s+1}.$$

The third-order Butterworth filter is depicted in Figure 7.6c. ■

Example 7.4: Single Operational Amplifier Filters

There are many single-amplifier active filter schemes that implement the second- and third-order filter's transfer functions. As documented, the gain and cutoff frequencies are found using the system bandwidth (reciprocal of settling time), noise frequency, attenuation requirements, complexity, control laws used, and other specifications. A multiple feedback scheme, which implements low-pass, high-pass, and band-pass second-order filters, is illustrated in Figure 7.7a. The passive elements (resistors and capacitors) are used. While the impedances $Z(s) = Z(j\omega)$ are commonly used, the admittances can also be applied. The admittance $Y(s) = Y(j\omega)$ is the reciprocal of impedance $Z(s)$. For a capacitor $Y_C(s) = sC$, and for a resistor $Y_R(s) = 1/R$.

One finds the node equations (Kirchhoff's current law) at node 1 and at the summing node 2. We have $(Y_1 + Y_2 + Y_3 + Y_4)V_1 - Y_1V_{in} - Y_4V_{out} = 0$ and $-Y_3V_1 - Y_5V_{out} = 0$. Eliminating V_1 and grouping the terms yield the transfer function

$$G(s)=\frac{V_{out}(s)}{V_{in}(s)}=\frac{-Y_1Y_3}{(Y_1+Y_2+Y_3+Y_4)Y_5+Y_3Y_4}.$$

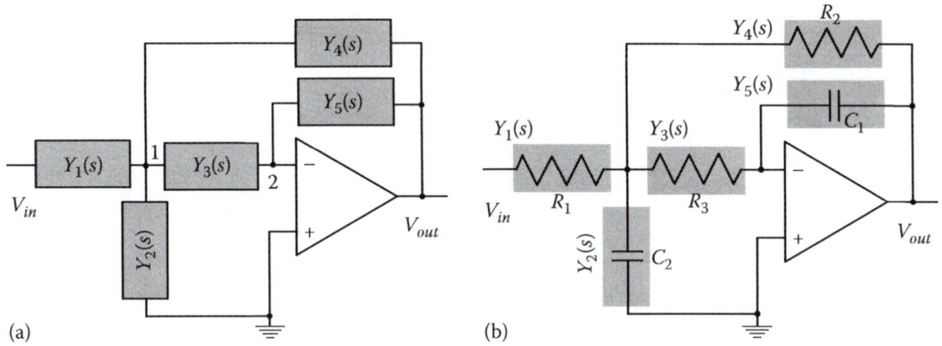

FIGURE 7.7 (a) A multiple-feedback analog filter schematics with $G(s)=\frac{V_{out}(s)}{V_{in}(s)}=\frac{-Y_1Y_3}{(Y_1+Y_2+Y_3+Y_4)Y_5+Y_3Y_4}$; (b) The second-order low pass filter with transfer function $G(s)=\frac{V_{out}(s)}{V_{in}(s)}=-\frac{\frac{R_2}{R_1}\frac{1}{R_2R_3C_1C_2}}{s^2+\frac{1}{C_2}\left(\frac{1}{R_1}+\frac{1}{R_2}+\frac{1}{R_3}\right)s+\frac{1}{R_2R_3C_1C_2}}$.

Various filters may be implemented by using five admittances $Y_i(s)$. In Figure 7.7b, $Y_1=1/R_1$, $Y_2=sC_2$, $Y_3=1/R_3$, $Y_4=1/R_2$, and $Y_5=sC_1$. The resulting transfer function is

$$G(s)=\frac{V_{out}(s)}{V_{in}(s)}=\frac{-\frac{R_2}{R_1}\frac{1}{R_2R_3C_1C_2}}{s^2+\frac{1}{C_2}\left(\frac{1}{R_1}+\frac{1}{R_2}+\frac{1}{R_3}\right)s+\frac{1}{R_2R_3C_1C_2}}=\frac{-ka_0}{s^2+a_1s+a_0}.$$ ■

Example 7.5: Butterworth Filters

The elliptical, Chebyshev, Bessel, Cauer, and others filters can be used. In the notch and Butterworth filters, the magnitude $|G(\omega)|$ is constant, monotonically decreasing or monolithically decreasing function at all frequencies. One specifies the pass band gain $|G|_{\max\,\mathrm{dB}}$ at the pass band frequency ω_p, and the minimum stop band gain $|G|_{\min\,\mathrm{dB}}$ at the stop band frequency ω_s. The gain, specified attenuation, the pass band frequency, and stop band frequency define the filter order. Using the cutoff frequency ω_c, one expresses the filter transfer function as $G(s)=\frac{k_0}{B_n\left(\frac{1}{\omega_c}s\right)}$. The n-degree Butterworth polynomials, normalized for $\omega_c=1$, are

$$B_n(s)=\begin{cases}\prod_{k=1}^{\frac{n}{2}}\left[s^2-2s\cos\left(\frac{2k+n-1}{2n}\pi\right)+1\right], & n \text{ even}\\ (s+1)\prod_{k=1}^{\frac{n-1}{2}}\left[s^2-2s\cos\left(\frac{2k+n-1}{2n}\pi\right)+1\right], & n \text{ odd}\end{cases}.$$

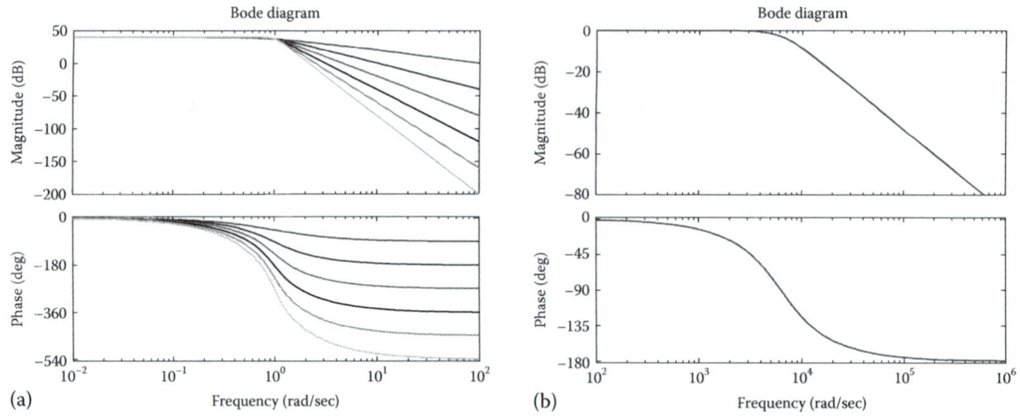

FIGURE 7.8 (a) Bode plots for the low-pass Butterworth filters $G(s)=\dfrac{k_0}{B_n\left(\frac{1}{\omega_c}s\right)}$, $k_0=100$, $\omega_c=1$, $n=1$ to $n=6$; (b) Bode plots for the second-order Butterworth filter $G(s)=\dfrac{3.95\times10^7}{s^2+8886s+3.95\times10^7}$.

The high-order filters are formed by cascading in series the first- and second-order stages, which implement

$$G(s)=\frac{k_0}{B_n\left(\frac{1}{\omega_c}s\right)}.$$

For example, three second-order low pass filters can be cascaded to implement a sixth-order filter with the specified gain and cutoff frequencies. The normalized n-degree Butterworth polynomials are $B_1(s)=(s+1)$, $B_2(s)=(s^2+1.414s+1)$, $B_3(s)=(s+1)(s^2+s+1)$,

$B_4(s)=(s^2+0.765s+1)(s^2+1.848s+1)$, $B_5(s)=(s+1)(s^2+0.618s+1)(s^2+1.618s+1)$,

$B_6(s)=(s^2+0.518s+1)(s^2+1.414s+1)(s^2+1.932s+1)$,

$B_7(s)=(s+1)(s^2+0.445s+1)(s^2+1.247s+1)(s^2+1.802s+1)$,

$B_8(s)=(s^2+0.39s+1)(s^2+1.111s+1)(s^2+1.663s+1)(s^2+1.962s+1)$, etc.

The Bode plots for $G(s)$ with $B_n(s)$, $n=1,\ldots,6$ of the normalized Butterworth filters with $k_0=100$ are documented in Figure 7.8a.

The filters can be designed in MATLAB®. Using the butter command, specifying the filter order $n=2$ and cutoff frequency to be 1000 Hz, one has

```
n=2; f=1000; [num,den]=butter(n,2*pi*f,'low','s'); filter=tf(num,den), bode(num,den)
```

The resulting transfer function is $G(s)=\dfrac{3.95\times10^7}{s^2+8886s+3.95\times10^7}$. The Bode plots are depicted in Figure 7.8b. ■

7.3 DESCRIPTIVE ANALYSIS, DATA ANALYTICS, AND STATISTICAL MODELS

To ensure data acquisition, control, and decision making, sensing, data processing and data analysis are performed by sensing microelectronic hardware and software. Predictive and prescriptive data analytics is needed to ensure data reduction using statistical models, data mining, predictive analysis, etc. The descriptive data analytics can be ensured by consistent statistical models. This section focuses on solution of the aforementioned problems.

Probability theory and statistical analysis are applied to examine noise to design filters, closed-loop systems, decision-and-control systems, management systems, etc. To ensure a cognizant overall system design, the probability theory is used in the yield, failure, redundancy, effectiveness, and other analyses. The sample space is finite, and one finds the corresponding probability model. A statistical model is given as a pair $(\mathcal{S}, \mathcal{P})$, where $\mathcal{S}$ is the set of possible observations (sample space), and, $\mathcal{P}$ is a set of probability distributions on $\mathcal{S}$. The set $\mathcal{P}$ is parameterized as $\mathcal{P} = \{f_\phi: \phi \in \Phi\}$, where the set $\Phi \in \mathbb{R}$ defines the model parameters. Parameterization is *identifiable* if $f_{\phi 1} = f_{\phi 2} | \phi_1 = \phi_2$.

Example 7.6

Consider a statistical model $(\mathcal{S}, \mathcal{P})$ with $\mathcal{P} = \{f_\phi: \phi \in \Phi\}$. The model is parametric if Φ has a finite dimension, $\Phi \subseteq \mathbb{R}^d$ where d is a positive integer, and, d is a model dimension. Assuming that data evolves from a single-variable Gaussian distribution, $\mathcal{P} = \left\{ f_{\mu,\sigma}(x) = \frac{1}{\sigma\sqrt{2\pi}} e^{-\frac{(x-\mu)^2}{2\sigma^2}} : \mu \in \mathbb{R}, \sigma > 0 \right\}$. The model dimension d is 2. ■

The cumulative distribution function (cdf) $F_X(x)$ of a random variable X is a real-valued continuous function $F_X(\cdot): \mathbb{R} \to \mathbb{R}$, defined as $F_X(x) = \Pr\{\omega \in \Omega: X(\omega) \le x\}$. Omitting the argument ω, $F_X(x) = \Pr\{X \le x\}$. If there exists a function f_X, such that for all $x \in \mathbb{R}$, the cdf satisfies $F_X(x) = \int_{-\infty}^{x} f_X(y)dy$, then, the random variable is said to be continuous. Here, $f_X(x)$ is the probability density function (pdf). The cdf of a continuous random variable X is an integral of f_X, and

$$F_X(x) = \int_{-\infty}^{x} f_X(y)dy, \quad f_X(x) = \frac{d}{dx} F_X(x) \quad \text{with} \quad \Pr\left[a \le X \le b\right] = \int_{a}^{b} f_X(x)dx,$$

$$\int_{-\infty}^{\infty} f_X(x)dx = 1, \quad f_X(x) \ge 0, \forall x.$$

As documented in the Illustrative Example 7.2, the cdf $F_X(x)$ defines the probability for a real-valued variable X, $F_X(x) = \Pr(X \le x)$, $\Pr(a < X \le b) = F_X(b) - F_X(a)$. The cdf of a continuous X is a nondecreasing, right-continuous function such that $\lim_{x \to -\infty} F_X(x) \to 0$, $\lim_{x \to \infty} F_X(x) \to 1$. The $F_X(x)$ and $f_X(x)$ are obtained from the measured data that can be mapped by histograms. The histogram represents the probability distribution by using the tabulated frequencies within the equally spaced discrete intervals (*bins*) within the data range.

To find a statistical model $(\mathcal{S}, \mathcal{P})$, the cdf and pdf should be parametrized. For the normal (Gaussian) distribution, the pdf is

$$f_X(x) = \frac{1}{\sigma\sqrt{2\pi}} e^{-\frac{(x-\mu)^2}{2\sigma^2}},$$

while the cdf is

$$F_X(x) = \frac{1}{2}\left[1 + \operatorname{erf}\left(\frac{x-\mu}{\sigma\sqrt{2}}\right)\right].$$

These $F_X(x)$ and $f_X(x)$ are parametrized by finding the mean μ and variance σ^2, yielding $f_X(x; \mu, \sigma^2)$, $\mu \in \mathbb{R}$, $\sigma > 0$.

Sensors, data quality, and data integrity: The physical quantities are measured by sensors. The data quality and data integrity are affected by the sensing physics, sensors used, measurement technology, precision, nonlinearities, errors, sensitivity, noise, etc. In aerospace, automotive, electromechanical, electronic, mechanical, robotic, and other systems one must ensure data quality and data integrity. The data conformity, data consistency, data completeness, and data validity must be guaranteed to ensure data acquisition, adequate control, effectiveness analysis, etc.

Example 7.7: MEMS Accelerometers, Gyroscopes, and Internal Measurement Unit

The MEMS accelerometers, gyroscopes, and internal measurement units (IMU) are widely used. The Analog Devices iMEMS® accelerometer and gyroscope and the InvenSense MPU-6500 and MPU-9250 multi-axis accelerometers and gyroscopes are documented in Figures 7.11a and b. In aerospace, automotive, electronic, manufacturing, medical, naval, and robotic systems, the linear (a_x, a_y, a_z) and angular $(\alpha_\theta, \alpha_\phi, \alpha_\psi)$ accelerations are measured by IMUs. The measured output tuples $(\hat{a}_x, \hat{a}_y, \hat{a}_z)$ and $(\hat{\alpha}_\theta, \hat{\alpha}_\phi, \hat{\alpha}_\psi)$ depend on acting physical accelerations (a_x, a_y, a_z) and $(\alpha_\theta, \alpha_\phi, \alpha_\psi)$, alignment A_i, ICs processing error e_i, nonlinearity N_i, bias B_{i0}, noise n_i, etc. For the measured linear and angular accelerations, we have

$$\hat{a}_i = f(a_x, a_y, a_z, A_i, e_i, N_i, B_{i0}, n_i)_{i=x,y,z} \quad \text{and} \quad \hat{\alpha}_j = f(\alpha_\theta, \alpha_\phi, \alpha_\psi, A_j, e_j, N_j, B_{j0}, n_j)_{j=\theta,\phi,\psi}.$$

For example, the measured linear and angular accelerations are

$$\hat{a}_x = P_x(a_x) + P_y(a_y) + P_z(a_z) + P_{xyz}(a_x, a_y, a_z) + B_{x0} + n_x \text{ and}$$
$$\hat{\alpha}_\theta = P_\theta(\alpha_\theta) + P_\phi(\alpha_\phi) + P_\psi(\alpha_\psi) + P_{\theta\phi\psi}(\alpha_\theta, \alpha_\phi, \alpha_\psi) + B_{\theta 0} + n_\theta.$$

Here, $P_x(a_x)$ is the polynomial that maps nonlinearities and errors; $P_y(a_y)$ and $P_z(a_z)$ are the cross-coupling polynomials; $P_{xyz}(a_x, a_y, a_z)$ is the cross-coupling nonhomogeneous polynomial; B_{x0} is the measurement zero-offset bias; n_x is the noise. For a white noise $n_x = \xi_x$, and, real-valued n_x is characterized by the finite variance σ_x^2, covariance $E[n_x(t_1) \quad n_x(t_2)] = 0$, $\forall t_1 \neq t_2$, etc.

Within the sensing axis, the measured accelerations $\hat{a}_i$ and $\hat{\alpha}_j$ depend on the accelerations acting along other axes. Ideally, for accelerometers and gyroscopes

$$\begin{bmatrix} \hat{a}_x \\ \hat{a}_y \\ \hat{a}_z \end{bmatrix} = \begin{bmatrix} a_x & 0 & 0 \\ 0 & a_y & 0 \\ 0 & 0 & a_z \end{bmatrix} + \begin{bmatrix} B_{0x} \\ B_{0y} \\ B_{0z} \end{bmatrix} + \begin{bmatrix} n_x \\ n_y \\ n_z \end{bmatrix}, \quad \begin{bmatrix} \hat{\alpha}_\theta \\ \hat{\alpha}_\phi \\ \hat{\alpha}_\psi \end{bmatrix} = \begin{bmatrix} \alpha_\theta & 0 & 0 \\ 0 & \alpha_\phi & 0 \\ 0 & 0 & \alpha_\psi \end{bmatrix} + \begin{bmatrix} B_{0\theta} \\ B_{0\phi} \\ B_{0\psi} \end{bmatrix} + \begin{bmatrix} n_\theta \\ n_\phi \\ n_\psi \end{bmatrix}.$$

A concurrent statistical analysis of noise is needed. The adequate pdfs must be found, and the parameters must be estimated. The experimental results are reported in Example 7.10. ■

The statistical characteristics of noise must be found using the experimental data. The probabilistic analysis results in descriptive quantitative models. The statistical models $(\mathcal{S}, \mathcal{P})$ can be found for noise, perturbations, errors, failures, etc. The normal (Gaussian), extreme value, and generalized extreme value distributions are commonly used. For the normal distribution $\mathcal{N}(\mu, \sigma^2)$, the single-variable pdf and cdf are

$$f_X(x; \mu, \sigma^2) = \frac{1}{\sigma\sqrt{2\pi}} e^{-\frac{(x-\mu)^2}{2\sigma^2}}, \quad F_X(x) = \frac{1}{2}\left[1 + \text{erf}\left(\frac{x-\mu}{\sigma\sqrt{2}}\right)\right], \quad \mu \in \mathbb{R}, \quad \sigma > 0,$$

where μ is the means; σ is the standard deviation.

The variance σ^2 is

$$\mathrm{var}(X) = \sigma^2 = \int_{-\infty}^{\infty} \frac{(x-\mu)^2}{\sigma\sqrt{2\pi}} e^{-\frac{(x-\mu)^2}{2\sigma^2}} dx.$$

The pdfs for the extreme value $\mathcal{EV}(\mu, \sigma)$ and generalized extreme value $\mathcal{GEV}(\mu, \sigma, k)$ distributions are

$$f_X(x;\mu,\sigma) = \frac{1}{\sigma} e^{-\frac{x-\mu}{\sigma}} e^{-e^{-\frac{x-\mu}{\sigma}}}, \quad f_X(x;\mu,\sigma,k) = \frac{1}{\sigma} e^{-\left(1+k\frac{x-\mu}{\sigma}\right)^{-\frac{1}{k}}} \left(1+k\frac{x-\mu}{\sigma}\right)^{-1-\frac{1}{k}}, \quad \mu \in \mathbb{R},\ \sigma > 0,\ k \in \mathbb{R}$$

where μ, σ, and k are the location, scale, and shape parameters, $k \neq 0$.

The continuous two- and three-parameter Weibull distributions $W(\cdot)$ are mapped by

$$f_X(x;a,b) = \begin{cases} \dfrac{b}{a}\left(\dfrac{x}{a}\right)^{b-1} e^{-\left(\frac{x}{a}\right)^b}, & x \geq 0 \\ 0, & x < 0 \end{cases}$$

and

$$f_X(x;a,b,c) = \begin{cases} \dfrac{b}{a}\left(\dfrac{x-c}{a}\right)^{b-1} e^{-\left(\frac{x-c}{a}\right)^b}, & x \geq c \\ 0, & x < c \end{cases},$$

where a, b, and c are the scale, shape, and location parameters, $a > 0$, $b > 0$.

For a two-parameter distribution $\mathcal{W}(a, b)$,

$$F_X(x) = \begin{cases} 1 - e^{-\left(\frac{x}{a}\right)^b}, & x \geq 0. \\ 0, & x < 0 \end{cases}$$

Example 7.8: Failure Analysis

Failure analysis is very important in electronic, electromechanical, energy, mechanical, power, and other systems. For example, the maintenance and reliability analyses can be performed using consistent failure models. These analyses determine the maintenance schedule, reduce maintenance cost, and improve safety.

The time between failures of devices, components, modules, and systems can be measured. Let us measure the operating time to failure for identical motion devices. The probabilistic analysis is performed and a failure model is found. We use the time-to-failure data set as reported in the two first lines of the MATLAB file reported below. Examining the histogram reported in Figure 7.9a and the corresponding cdf, one concludes that the Weibull distribution can be used. We apply a function

$$f_X(x;a,b,d) = d\left(\frac{x}{a}\right)^{b-1} e^{-\left(\frac{x}{a}\right)^b}, \quad x \geq 0, \quad \int_{-\infty}^{\infty} f_X(x)dx = 1, \quad a > 0, \quad b > 0, \quad d > 0.$$

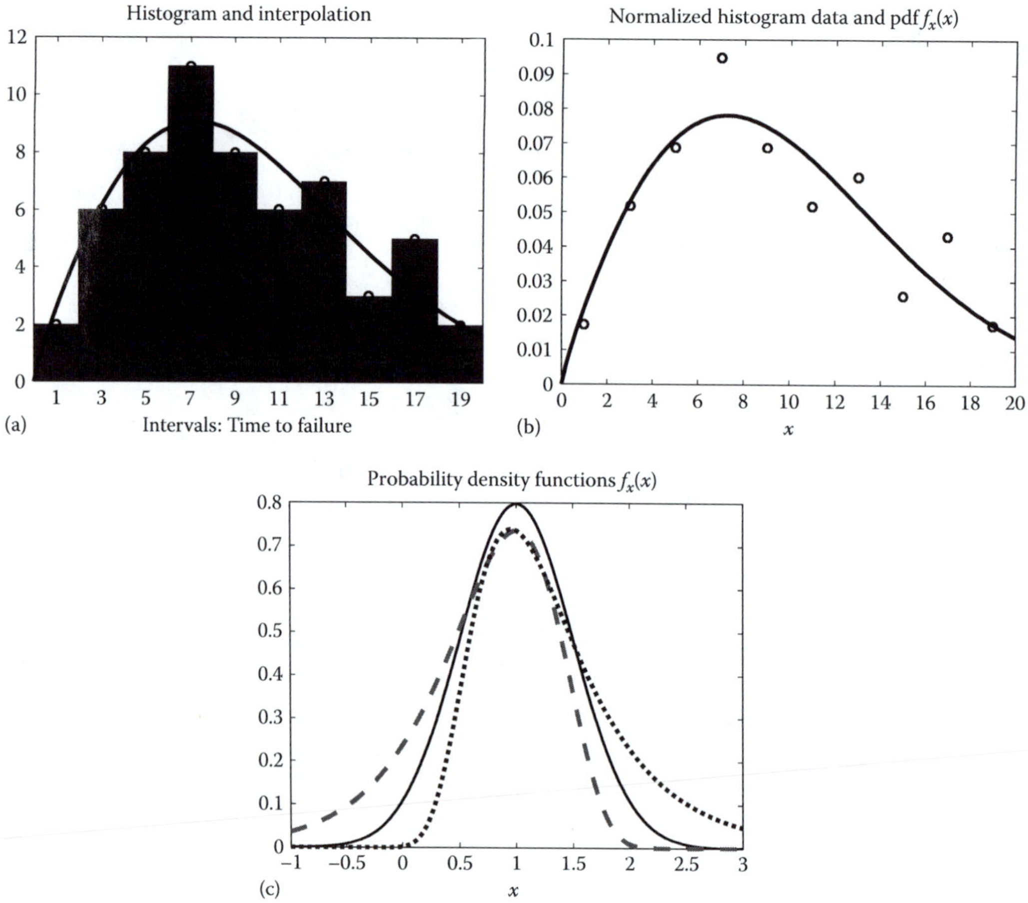

FIGURE 7.9 (a) Failure histogram and interpolation;

(b) Weibull pdf $f_X(x;a,b,d) = d\left(\frac{x}{a}\right)^{b-1} e^{-\left(\frac{x}{a}\right)^b}$, $x \geq 0$, $a = 10.97$, $b = 1.864$ and $d = 0.178$;

(c) Normal $\mathcal{N}(\mu, \sigma^2)$, extreme value $\mathcal{EV}(\mu, \sigma)$ and generalized extreme value $\mathcal{GEV}(\mu, \sigma, k)$ distributions: Probability density functions $f_X(x)$ for $\mathcal{N}(\mu, \sigma^2)$, $\mathcal{EV}(\mu, \sigma)$ and $\mathcal{GEV}(\mu, \sigma, k)$, $\mu = 1$, $\sigma = 0.5$ and $k = 0.1$ (solid, dashed, and dotted lines) for Example 7.9.

The MATLAB is used to plot the histogram and perform parametrization. For a histogram, we have $a = 10.97$ and, $b = 1.864$. Figure 7.9a documents the histogram and the corresponding interpolation.

The Kolmogorov axiom states that

$$\int_{-\infty}^{\infty} f_X(x)dx = 1, \quad \text{or} \quad \sum_x f_X(x) = 1.$$

One ensures

$$\int_{-\infty}^{\infty} f_X(x)dx = 1$$

by normalizing the histogram, cdf, or pdf within the data range. The cdf and pdf can be parametrized. To find the unknown parameters, the interpolation on $f_X(x)$ must converge. We examine the Weibull pdfs

$$f_X(x;a,b)=\frac{b}{a}\left(\frac{x}{a}\right)^{b-1}e^{-\left(\frac{x}{a}\right)^b},$$

and

$$f_X(x;a,b,d)=d\left(\frac{x}{a}\right)^{b-1}e^{-\left(\frac{x}{a}\right)^b},\quad x\ge 0.$$

The unknown parameters are $a = 10.97$, $b = 1.864$, and $d = 0.178$. The plot for the resulting $f_X(x)$ is depicted in Figure 7.9b. The MATLAB file is

```
OperatingTime=[0.9 1.2 2.3 2.5 2.9 3 3.4 3.9 4.4 4.7 5.1 5.2 5.4 5.7 5.7 5.9 6.2 6.4 6.5 6.5 6.7 7.1
   7.3 7.4 7.9 7.9 8 8.2 8.3 8.7 8.8  8.8 8.9 9.3 9.8 10.4 11 11.1 11.4 11.5 11.9 12.1 12.2 12.4 12.7
   13 13.2 13.5 14.1 14.8 15.9 16.1 16.2 16.7 16.9 17.2 18.1 18.7];
BinWidth =2; bin=1:BinWidth:19; hist(OperatingTime,bin);
xlabel('Intervals: Time to Failure','FontSize',18); title('Histogram','FontSize',18); hold on;
counts=hist(OperatingTime,bin); plot(bin,counts,'ko','linewidth',3); hold on
Interpolation=@(p,x) p(3).* (x ./ p(1)).^(p(2)-1) .* exp(-(x ./ p(1)).^p(2)); StartingValues=[5 2 10];
Coefficients=nlinfit(bin,counts, Interpolation,StartingValues)
xgrid=linspace(0,20,100);line(xgrid, Interpolation(Coefficients,xgrid),'Color','k','linewidth',4);
xlabel('Intervals: Time to Failure ','FontSize',18); title('Histogram and Interpolation','FontSize',18);
   pause; hold off;
S=sum(BinWidth*counts); plot(bin,counts./S,'ko','linewidth',3); hold on
InterpolationfX= @(p,x) p(3).* (x ./ p(1)).^(p(2)-1) .* exp(-(x ./ p(1)).^p(2)); StartingValuesfX=[5 2 10];
CoefficientsfX=nlinfit(bin,counts./S, InterpolationfX,StartingValuesfX)
xgrid=linspace(0,20,100); line(xgrid, InterpolationfX(CoefficientsfX,xgrid),'Color','k','linewidth',4);
xlabel('{\itx}','FontSize',18); title('Normalized Histogram Data and pdf {\itf_X}({\itx})','FontSize',18);
```

■

Example 7.9

The pdfs for normal $\mathcal{N}(\mu, \sigma^2)$, extreme value $\mathcal{EV}(\mu, \sigma)$ and generalized extreme value $\mathcal{GEV}(\mu, \sigma, k)$ distributions are $f_X(x;\mu,\sigma^2)=\frac{1}{\sigma\sqrt{2\pi}}e^{-\frac{(x-\mu)^2}{2\sigma^2}}$, $f_X(x;\mu,\sigma)=\frac{1}{\sigma}e^{\frac{x-\mu}{\sigma}}e^{-e^{\frac{x-\mu}{\sigma}}}$, and $f_X(x;\mu,\sigma,k)=\frac{1}{\sigma}e^{-\left(1+k\frac{x-\mu}{\sigma}\right)^{-\frac{1}{k}}}\left(1+k\frac{1}{\sigma}(x-\mu)\right)^{-1-\frac{1}{k}}$. Let $\mu = 1$, $\sigma = 0.5$ and $k = 0.1$. The plots for $f_X(x)$ are reported in Figure 7.9c. The MATLAB statements to calculate and plot $f_X(x)$ are

```
x=-1:1e-4:3; mu=1; sigma=0.5; k=0.1;
fXN=(1/(sigma*sqrt(2*pi)))*exp(-((x-mu).^2)/(2*sigma^2)); plot(x,fXN,'k','linewidth',3); hold on;
fXev=(1/sigma)*exp((x-mu)./sigma).*exp(-exp((x-mu)./sigma)); plot(x,fXev,'b--','linewidth',3); hold on;
fXgev=(1/sigma)*exp(-(1+k*(x-mu)./sigma).^(-1/k)).*(1+k*(x-mu)./sigma).^(-1-1/k);
plot(x,fXgev,'r:','linewidth',3);
xlabel('{\itx}','FontSize',18); title('Probability Density Functions {\itf_X}({\itx})','FontSize',18);
```

■

The normal $\mathcal{N}(\mu, \sigma^2)$, extreme value $\mathcal{EV}(\mu, \sigma)$, and Weibull $\mathcal{W}(a, b, c)$ distributions may be physics-consistent. One also applies the lognormal distrubution ln $\mathcal{N}(\mu, \sigma^2)$ with

$$f_X(x;\mu,\sigma^2)=\frac{1}{x\sigma\sqrt{2\pi}}e^{-\frac{(\ln x-\mu)^2}{2\sigma^2}},\quad F_X(x)=\frac{1}{2}+\frac{1}{2}\text{erf}\left(\frac{\ln x-\mu}{\sqrt{2}\sigma}\right),\quad x\ge 0,\quad \mu\in\mathbb{R},\quad \sigma>0.$$

TABLE 7.2
Conventional Distributions and Generalized Distributions

Distributions	Conventional pdfs	Generalized pdfs $f_X(x)$, f_X: $\mathbb{R} \to \mathbb{R}$, $f_X(x) \geq 0$, $\int_{-\infty}^{\infty} f_X(x)dx = 1$
Normal $\mathcal{N}(\cdot)$ and multimodal normal $\mathcal{GN}(\cdot)$	$f_X(x;\mu,\sigma^2)=\frac{1}{\sigma\sqrt{2\pi}}e^{-\frac{(x-\mu)^2}{2\sigma^2}}$	$f_X(x;\mu,\sigma^2,b,a_n)=\frac{1}{\sigma b\sqrt{2\pi}}e^{-\frac{1}{2\sigma^2}\sum_{n=0}^{\infty}a_n(x-\mu)^n}$, $\mu\in\mathbb{R}, \sigma>0, b>0, a_n\in\mathbb{R}, \forall n$
Extreme value $\mathcal{EV}(\cdot)$ and multimodal extreme value $\mathcal{GEV}(\cdot)$	$f_X(x;\mu,\sigma)=\frac{1}{\sigma}e^{-\frac{x-\mu}{\sigma}}e^{-e^{-\frac{x-\mu}{\sigma}}}$	$f_X(x;\mu,\sigma,b,a_n)=\frac{1}{\sigma b}e^{-\frac{1}{\sigma}\sum_{n=0}^{\infty}a_n(x-\mu)^n}e^{-e^{-\frac{1}{\sigma}\sum_{n=0}^{\infty}a_n(x-\mu)^n}}$, $\mu\in\mathbb{R}, \sigma>0, b>0, a_n\in\mathbb{R}, \forall n$
Lognormal ln $\mathcal{N}(\cdot)$ and generalized lognormal ln $\mathcal{GN}(\cdot)$	$f_X(x;\mu,\sigma^2)=\frac{1}{x\sigma\sqrt{2\pi}}e^{-\frac{(\ln x-\mu)^2}{2\sigma^2}}$	$f_X(x;\mu,\sigma^2,b,a_n)=\frac{1}{x\sigma b\sqrt{2\pi}}e^{-\frac{1}{2\sigma^2}\sum_{n=0}^{\infty}a_n(\ln x-\mu)^n}$, $\mu\in\mathbb{R}, \sigma>0, b>0, a_n\in\mathbb{R}, \forall n$
Rayleigh $\mathcal{R}(\cdot)$ and generalized Rayleigh $\mathcal{GR}(\cdot)$	$f_X(x;\sigma)=\frac{x}{\sigma^2}e^{-\frac{x^2}{2\sigma^2}}$	$f_X(x;\sigma,b,a_n)=\frac{x}{\sigma^2 b}e^{-\frac{1}{2\sigma^2}\sum_{n=0}^{\infty}a_n x^n}$, $\sigma>0, b>0, a_n\in\mathbb{R}, \forall n$
Inverse Gaussian $\mathcal{IN}(\cdot)$ and generalized inverse Gaussian $\mathcal{GIN}(\cdot)$	$f_X(x;\mu,\lambda)=\sqrt{\frac{\lambda}{2\pi x^3}}e^{-\frac{\lambda(x-\mu)^2}{2\mu^2 x}}$	$f_X(x;\mu,\lambda,b,a_n)=\sqrt{\frac{\lambda}{2b\pi x^3}}e^{-\frac{\lambda}{2\mu^2 x}\sum_{n=0}^{\infty}a_n(x-\mu)^n}$, $\mu>0, \lambda>0, b>0, a_n\in\mathbb{R}, \forall n$
Other generalized continuous and piecewise continuous distributions	See *Illustrative Example* 7.1.	The homogeneous or nonhomogeneous polynomials are used in $f_X(x)$.

Rayleigh distribution $\mathcal{R}(\sigma)$ with

$$f_X(x;\sigma)=\frac{x}{\sigma^2}e^{-\frac{x^2}{2\sigma^2}},\quad F_X(x)=1-e^{-\frac{x^2}{2\sigma^2}},\quad x\geq 0,\quad \sigma>0,$$

inverse Gaussian distribution $\mathcal{IN}(\mu, \lambda)$ with

$$f_X(x;\mu,\lambda)=\sqrt{\frac{\lambda}{2\pi x^3}}e^{-\frac{\lambda(x-\mu)^2}{2\mu^2 x}},\quad x>0,\quad \mu>0,\quad \lambda>0,$$

and other distributions are examined. The conventional distributions not always guarantee consistency, conformity, and adequateness. Generalized multimodal distributions are introduced, as given in Table 7.2.

Illustrative Example 7.1

There are a great number of distributions, including multivariate, matrix-valued, etc. These and other continuous and piecewise continuous distributions, such as the Maxwell–Boltzmann, Cauchy, and others, can be refined by applying the proposed concept. Homogeneous or nonhomogeneous polynomials are used. For example, consider the Maxwell–Boltzmann pdf

$$f_X(x;a)=\sqrt{\frac{2}{\pi}}\frac{1}{a^3}x^2 e^{-\frac{1}{2a^2}x^2},\quad x\in(0,\infty),\quad a>0.$$

Using the single-variable polynomials, we define the generalized Maxwell–Boltzmann distribution $\mathcal{GM}(\cdot)$ with

$$f_X(x;a,b_m,a_n) = \sqrt{\frac{2}{\pi}}\frac{1}{a^3}\left(\sum_{m=1}^{\infty} b_m x^m\right) \circ e^{-\frac{1}{2a^2}\sum_{n=0}^{\infty} a_n x^n},$$

$$\int_{-\infty}^{\infty} f_X(x)dx = 1,\ x \in (0,\infty), \quad a > 0, \quad b_m \in \mathbb{R}, \quad a_n \in \mathbb{R}, \quad \forall (m,n).$$ ■

Illustrative Example 7.2

The nondecreasing, right-continuous function $F_X(x)$ is an antiderivative of a pdf $f_X(x)$, and the derivative of $F_X(x)$ is $f_X(x)$. The pdf $f_X(x)$ and cdf $F_X(x)$ are related as

$$F_X(x) = \int_{-\infty}^{x} f_X(y)dy \quad \text{and} \quad f_X(x) = \frac{d}{dx}F_X(x)$$

with $\lim_{x\to-\infty} F_X(x) \to 0$ and $\lim_{x\to\infty} F_X(x) \to 1$. Furthermore, $f_X(x) \geq 0$, $\forall x$, $f_X(x)$ has an absolute maximum at x_0 such that $f_X(x_0) \geq f_X(x)$, $\forall x$, and $\int_{-\infty}^{\infty} f_X(x)dx = 1$.

For the generalized multimodal distributions proposed, the conditions on the pdfs $f_X(x)$ and cdfs $F_X(x)$ are guaranteed. That is, real-valued continuous and monolithic cdfs and pdfs exist, such that

1. F_X: $\mathbb{R} \to \mathbb{R}$, $F_X(x) = \int_{-\infty}^{x} f_X(y)dy$, $\lim_{x\to-\infty} F_X(x) \to 0$, $\lim_{x\to\infty} F_X(x) \to 1$
2. f_X: $\mathbb{R} \to \mathbb{R}$, $f_X(x) = \frac{d}{dx}F_X(x)$, $f_X(x) \geq 0$, $\exists f_X(x_0)_{\max} \geq f_X(x)$, $\forall x$, $\int_{-\infty}^{\infty} f_X(x)dx = 1.$

For the normal distribution $\mathcal{N}(\mu, \sigma^2)$ with domain $-\infty \leq x \leq \infty$, the single variable pdf and cdf are

$$f_X(x;\mu,\sigma^2) = \frac{1}{\sigma\sqrt{2\pi}}e^{-\frac{(x-\mu)^2}{2\sigma^2}}, \quad F_X(x) = \frac{1}{2}\left[1+\operatorname{erf}\left(\frac{x-\mu}{\sigma\sqrt{2}}\right)\right], \quad \mu \in \mathbb{R}, \quad \sigma > 0.$$

Considering $\phi = e^{-x^2}$, $\frac{\partial}{\partial x}e^{-x^2} = -2xe^{-x^2}$, $\int e^{-x^2}dx = \frac{1}{2}\sqrt{\pi}\operatorname{erf}(x)$, while for $\phi = e^{-x-x^2}$, $\frac{\partial}{\partial x}e^{-x-x^2} = -(1+2x)e^{-x-x^2}$, $\int e^{-x-x^2}dx = \frac{1}{2}\sqrt[4]{e}\sqrt{\pi}\operatorname{erf}\left(\frac{1}{2}+x\right)$. For $\phi = e^{-x-x^2-x^3-x^4-\cdots}$, the global maximum, derivatives, and indefinite and definite integrals exist. ■

Example 7.10

Consider the normal $\mathcal{N}(\mu, \sigma^2)$ and multimodal normal $\mathcal{GN}(\mu, \sigma^2, b, a_n)$ distributions with the corresponding pdfs

$$f_X(x;\mu,\sigma^2)=\frac{1}{\sigma\sqrt{2\pi}}e^{-\frac{(x-\mu)^2}{2\sigma^2}}$$

and

$$f_X(x;\mu,\sigma^2,b,a_n)=\frac{1}{\sigma b\sqrt{2\pi}}e^{-\frac{1}{2\sigma^2}\sum_{n=0}^{\infty}a_n(x-\mu)^n},\quad \int_{-\infty}^{\infty}f_X(x)dx=1.$$

For $\mu = 0$, $\sigma = 1$, $b > 0$ and $a_{n=0,1,2,3,4} \neq 0$ ($\forall a_n = 0$ for $a_n \geq 5$), the calculated pdfs are reported in Figure 7.10a.

The extreme value distribution $\mathcal{EV}(\mu, \sigma)$, $f_X(x;\mu,\sigma)=\frac{1}{\sigma}e^{-\frac{x-\mu}{\sigma}}e^{-e^{-\frac{x-\mu}{\sigma}}}$ and the multimodal extreme value distribution

$$\mathcal{GEV}(\mu, \sigma, b, a_n),\ f_X(x;\mu,\sigma,b,a_n)=\frac{1}{\sigma b}e^{-\frac{1}{\sigma}\sum_{n=0}^{\infty}a_n(x-\mu)^n}e^{-e^{-\frac{1}{\sigma}\sum_{n=0}^{\infty}a_n(x-\mu)^n}},\quad \int_{-\infty}^{\infty}f_X(x)dx=1$$

are studied. Figure 7.10b reports the resulting pdfs for $\mu = 0$, $\sigma = 1$, $b > 0$ and $a_{n=0,1,2,3,4} \neq 0$ ($\forall a_n = 0$ for $a_n \geq 5$). ■

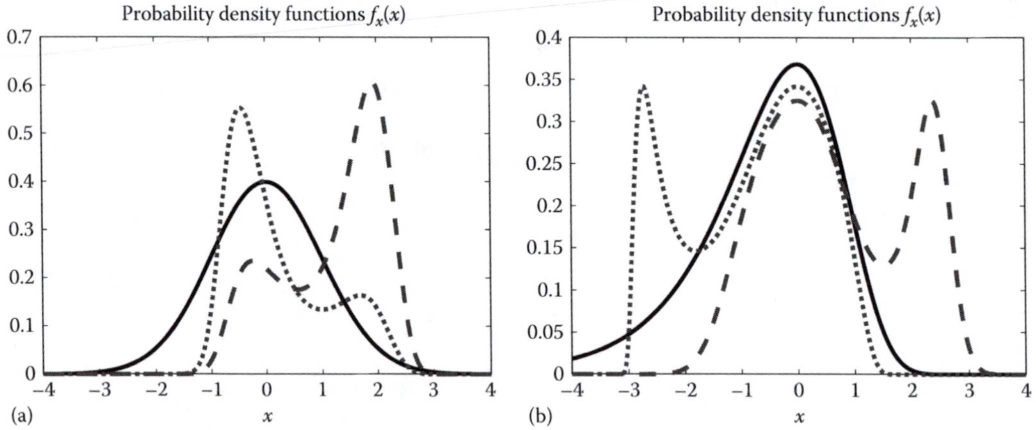

FIGURE 7.10 (a) Normal distribution $\mathcal{N}(\mu, \sigma^2)$, $f_X(x;\mu,\sigma^2)=\frac{1}{\sigma\sqrt{2\pi}}e^{-\frac{(x-\mu)^2}{2\sigma^2}}$, $\mu = 0$, $\sigma = 1$ (solid line), and multimodal normal distribution $\mathcal{GN}(\mu,\sigma^2,b,a_n)$, $f_X(x;\mu,\sigma^2,b,a_n)=\frac{1}{\sigma b\sqrt{2\pi}}e^{-\frac{a_0+a_1x+a_2x^2+a_3x^3+a_4x^4}{2\sigma^2}}$, $\int_{-\infty}^{\infty}f_X(x)dx=1$, $\sigma = 1$, $b = 1.094$, $a_0 = 0$, $a_1 = 3$, $a_2 = 1$, $a_3 = -3$, $a_4 = 1$ (dotted line) and $b = 1.82$, $a_0 = 0$, $a_1 = 1$, $a_2 = 1$, $a_3 = -3$, $a_4 = 1$ (dashed line); (b) Extreme value distribution $\mathcal{EV}(\mu, \sigma)$, $f_X(x;\mu,\sigma)=\frac{1}{\sigma}e^{-\frac{x-\mu}{\sigma}}e^{-e^{-\frac{x-\mu}{\sigma}}}$, $\mu = 0$, $\sigma = 1$ (solid line), and multimodal extreme value distribution $\mathcal{GEV}(\mu, \sigma, b, a_n)$, $f_X(x;\mu,\sigma,b,a_n)=\frac{1}{\sigma b}e^{-\frac{a_0+a_1x+a_2x^2+a_3x^3+a_4x^4}{\sigma}}e^{-e^{\frac{a_0+a_1x+a_2x^2+a_3x^3+a_4x^4}{\sigma}}}$, $\sigma = 1$, $b = 0.883$, $a_0 = 0$, $a_1 = 1$, $a_2 = -0.1$, $a_3 = 0.1$, $a_4 = -0.1$ (dashed line), and, $b = 0.929$, $a_0 = 0$, $a_1 = 1$, $a_2 = -0.1$, $a_3 = 0.1$, $a_4 = 0.1$ (dotted line).

Example 7.11

The internal measurement unit (IMU) was discussed in Example 7.7. The Analog Devices iMEMS accelerometer and gyroscope on evaluation boards are documented in Figure 7.11a. We examine the InvenSense MPU-6500 and MPU-9250 multiaxis accelerometers and gyroscopes shown in Figure 7.11b. The experiments are performed. At equilibrium (rest or steady motion), the measured linear ($\hat{a}_x$, $\hat{a}_y$, $\hat{a}_z$) and angular ($\hat{\alpha}_\theta$, $\hat{\alpha}_\phi$, $\hat{\alpha}_\psi$) accelerations are depicted in Figures 7.11c and d. We examine the noise tuples (n_{ax}, n_{ay}, n_{az}) and ($n_{\alpha\theta}$, $n_{\alpha\phi}$, $n_{\alpha\psi}$) in the channels (a_x, a_y, a_z) and (α_θ, α_ϕ, α_ψ).

The linear accelerations, measured by the (a_x, a_y, a_z) channel accelerometers are ($\hat{a}_x$, $\hat{a}_y$, $\hat{a}_z$). The angular accelerations are measured by the (α_θ, α_ϕ, α_ψ) channel gyroscopes with the outputs ($\hat{\alpha}_\theta$, $\hat{\alpha}_\phi$, $\hat{\alpha}_\psi$). One has the measured tuples ($\hat{a}_x$, $\hat{a}_y$, $\hat{a}_z$) and ($\hat{\alpha}_\theta$, $\hat{\alpha}_\phi$, $\hat{\alpha}_\psi$). Measurements at rest yield the noise tuples (n_{ax}, n_{ay}, n_{az}) and ($n_{\alpha\theta}$, $n_{\alpha\phi}$, $n_{\alpha\psi}$). The histograms to characterize noise in the (α_θ, α_ϕ, α_ψ) channels are documented in Figure 7.12. The statistical models ($\mathcal{S}$, $\mathcal{P}$), $\mathcal{P} = \{f_\phi: \phi \in \Phi\}$ are procured using the normal $\mathcal{N}(\mu, \sigma^2)$ and multimodal-normal $\mathcal{GN}(\mu, \sigma^2, b, a_n)$ distributions with the corresponding cdfs and pdfs. One finds the corresponding models with

$$\mathcal{P} = \left\{ f_{\mu,\sigma}(x) = \frac{1}{\sigma\sqrt{2\pi}} e^{-\frac{(x-\mu)^2}{2\sigma^2}} : \mu \in \mathbb{R}, \sigma > 0 \right\}$$

and

$$\mathcal{P} = \left\{ f_{\mu,\sigma,b,a_n}(x) = \frac{1}{\sigma b\sqrt{2\pi}} e^{-\frac{1}{2\sigma^2}\sum_{n=0}^{\infty} a_n (x-\mu)^2} : \mu \in \mathbb{R}, \sigma > 0, b > 0, a_n \in \mathbb{R}, \forall n \right\}.$$

The pdfs are parametrized. For $\mathcal{N}_{\alpha i}(\cdot)$ and $\mathcal{GN}_{\alpha i}(\cdot)$, the resulting pdfs are reported in Figure 7.12 by dashed and solid lines. The multimodal normal distribution

$$\mathcal{GN}(\mu, \sigma^2, b, a_n) \text{ with } f_X(x; \mu, \sigma^2, b, a_n) = \frac{1}{\sigma b\sqrt{2\pi}} e^{-\frac{1}{2\sigma^2}\sum_{n=0}^{\infty} a_n (x-\mu)^2}, \quad \int_{-\infty}^{\infty} f_X(x)dx = 1$$

guarantee accuracy and consistency. Usually, the normal distribution model is assumed. It is frequently supposed that the channels are symmetric. We found that the Gaussian, extreme value, and other commonly used distributions do not guarantee model consistency and accuracy. The multimodal normal distribution ensures consistency and data conformity. One may find the statistical model ($\mathcal{S}$, $\mathcal{P}$), $\mathcal{P} = \{f_\phi: \phi \in \Phi\}$ using the multimodal extreme value $\mathcal{GEV}(\cdot)$ distribution with

$$f_X(x; \mu, \sigma, b, a_n) = \frac{1}{\sigma b} e^{-\frac{1}{\sigma}\sum_{n=0}^{\infty}(a_n x - \mu)^n} e^{-e^{-\frac{1}{\sigma}\sum_{n=0}^{\infty} a_n (x-\mu)^n}}, \quad \mu \in \mathbb{R}, \quad \sigma > 0, \quad b > 0, \quad a_n \in \mathbb{R}, \quad \forall n$$

For $n_{\alpha\theta}$, we have $\mathcal{N}(\mu = 0.008, \sigma = 0.167)$ and $\mathcal{GN}(\mu = 0.228, \sigma = 0.0878, a_1 = 0.228, a_2 = 2.23, a_3 = 7.44, a_4 = 8, b = 2.56)$.

For $n_{\alpha\phi}$, one obtains $\mathcal{N}(\mu = 0.0192, \sigma = 0.447)$ and $\mathcal{GN}(\mu = 0.351, \sigma = 0.29, a_1 = -0.158, a_2 = 0.924, a_3 = 3.17, a_4 = 2.22, b = 1.24)$.

For $n_{\alpha\psi}$, we find $\mathcal{N}(\mu = 0.0076, \sigma = 0.184)$ and $\mathcal{GN}(\mu = 0.341, \sigma = 0.182, a_1 = 3.82, a_2 = 20.3, a_3 = 43, a_4 = 31.3, b = 32.4)$. ■

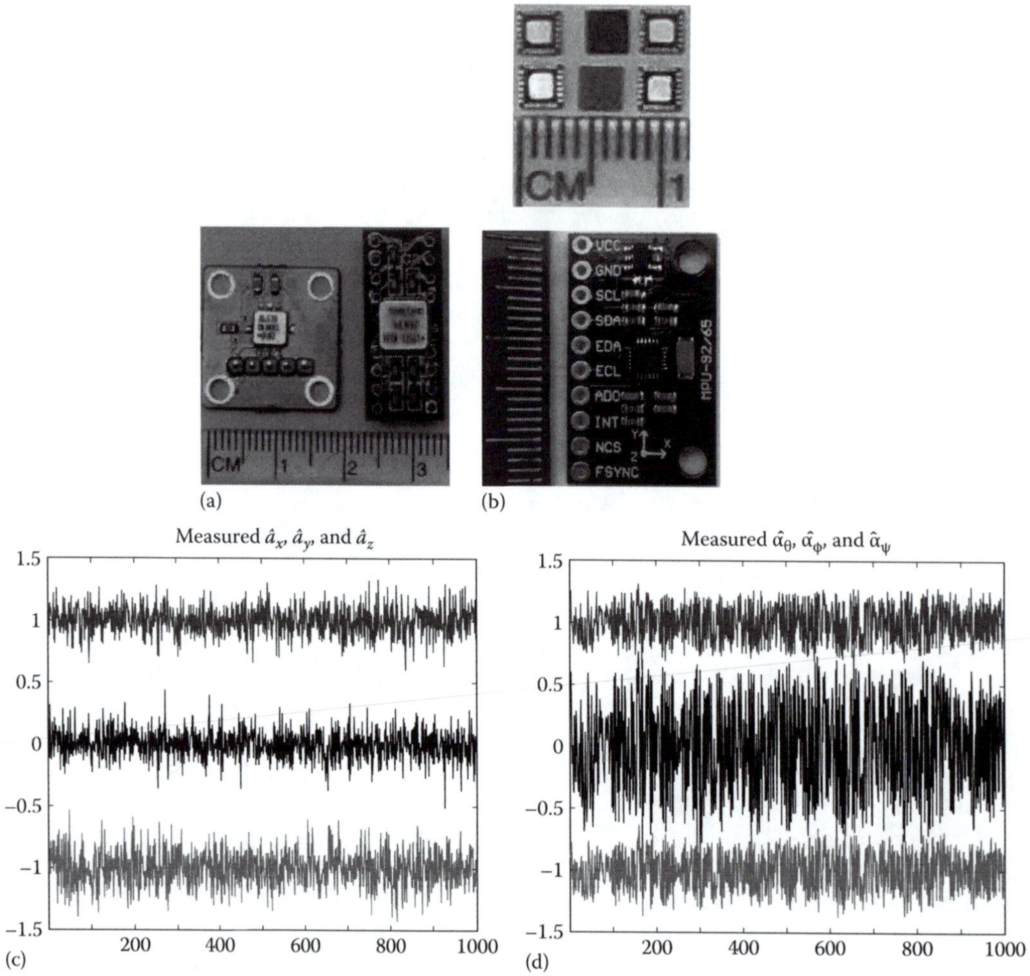

FIGURE 7.11 (a) ADXL203 and ADXRS300 high-precision iMEMS® accelerometer and gyroscope on evaluation boards. The ADXL203 dual-axis high-precision iMEMS accelerometer (5 × 5 × 2 mm LCC package) measures ±2*g* accelerations with low error and noise. The analog signal conditioned voltage is linearly proportional to the acceleration. The iMEMS surface micromachining technology ADXRS300 gyroscope measures ±300°/s angular acceleration. The signal conditioning is ensured by on-chip ICs;

(b) Images of MEMS MPU-6500 and MPU-9250 multiaxis accelerometers and gyroscopes evaluation board, and 3 × 3 × 0.9 mm QFN packages. The MPU-6500 and MPU-9250 three-axis accelerometer and three-axis gyroscope are used in various applications. This IMU includes an on-chip 16-bit processor, operates at 1.8 V, and consumes ~6 mW. Three-axis gyroscope, accelerometer, and compass are in MPU-9250. The on-chip 16-bit digital motion processor performs signal processing to reduce the noise and ensure accuracy. Images of the IMU evaluation boards with MPU-6550 and MPU-9250 are also shown;

(c) Measured linear accelerations in the *x*, *y*, and *z* axes $\hat{a}_x$, $\hat{a}_y$, and $\hat{a}_z$ (top, center, and bottom plots) at equilibrium. For plotting, the off-sets for $\hat{a}_x$, $\hat{a}_y$, and $\hat{a}_z$ are 1, 0, and −1;

(d) Measured angular accelerations in the pitch, roll, and yaw axes ($\hat{\alpha}_\theta$, $\hat{\alpha}_\phi$, $\hat{\alpha}_\psi$) at equilibrium.

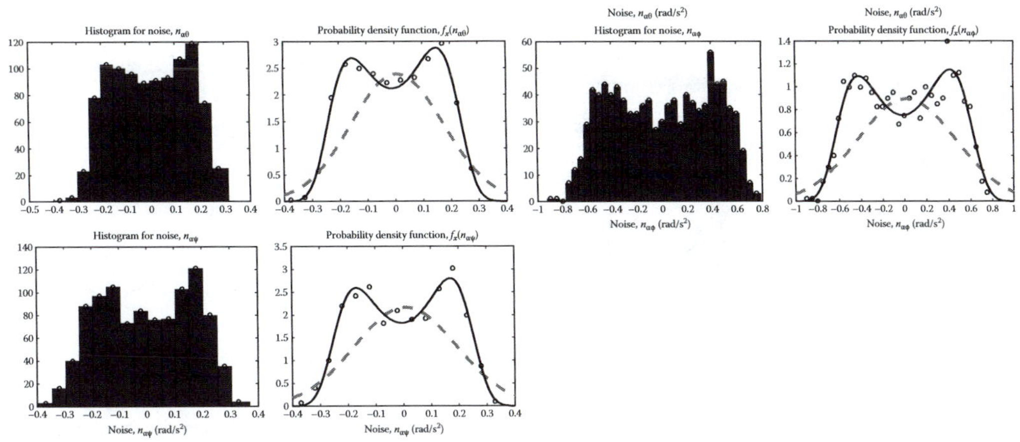

FIGURE 7.12 Histograms and pdfs of noise in the (α_θ, α_ϕ, α_ψ) gyroscope channels. The pdfs for the $\mathcal{N}_i(\mu_i, \sigma_i)$ and $\mathcal{GN}_i(\mu_i, \sigma_i^2, b_i, a_{ni})$ distributions are depicted by the dashed and solid lines respectively.

7.4 POWER AMPLIFIERS AND PWM CONVERTERS

7.4.1 Analog Controllers and PWM Amplifiers

Different transfer functions are implemented by operational amplifiers using passive elements. Operational amplifiers are widely used to implement analog control laws. An inverting integrator is obtained by placing a capacitor C_2 in the feedback path with $Z_2(s) = 1/C_2 s$, see Figure 7.13a. The resulting transfer function is

$$G(s) = -\frac{Z_2(s)}{Z_1(s)} = -\frac{1}{R_1 C_2 s}.$$

Denoting the initial value of the capacitor voltage as $u_C(t_0)$, the amplifier output voltage is

$$u_0(t) = -u_C(t_0) - \frac{1}{R_1 C_2}\int_{t_0}^{t_f} u_1(\tau)d\tau.$$

The operational differentiator performs the differentiation of the input signal. The current through the input capacitor is $C_1 \frac{du_1(t)}{dt}$, see Figure 7.13b. The output voltage is proportional to the derivative of the input voltage. Hence,

$$u_0(t) = -R_2 C_1 \frac{du_1(t)}{dt}.$$

The transfer function is

$$G(s) = -\frac{Z_2(s)}{Z_1(s)} = -R_2 C_1 s.$$

The transfer function of the proportional–integral–derivative (PID) control law is

$$G_{PID}(s) = k_p + \frac{k_i}{s} + k_d s = \frac{k_d s^2 + k_p s + k_i}{s}, \quad k_p > 0, \quad k_i > 0 \quad \text{and} \quad k_d > 0.$$

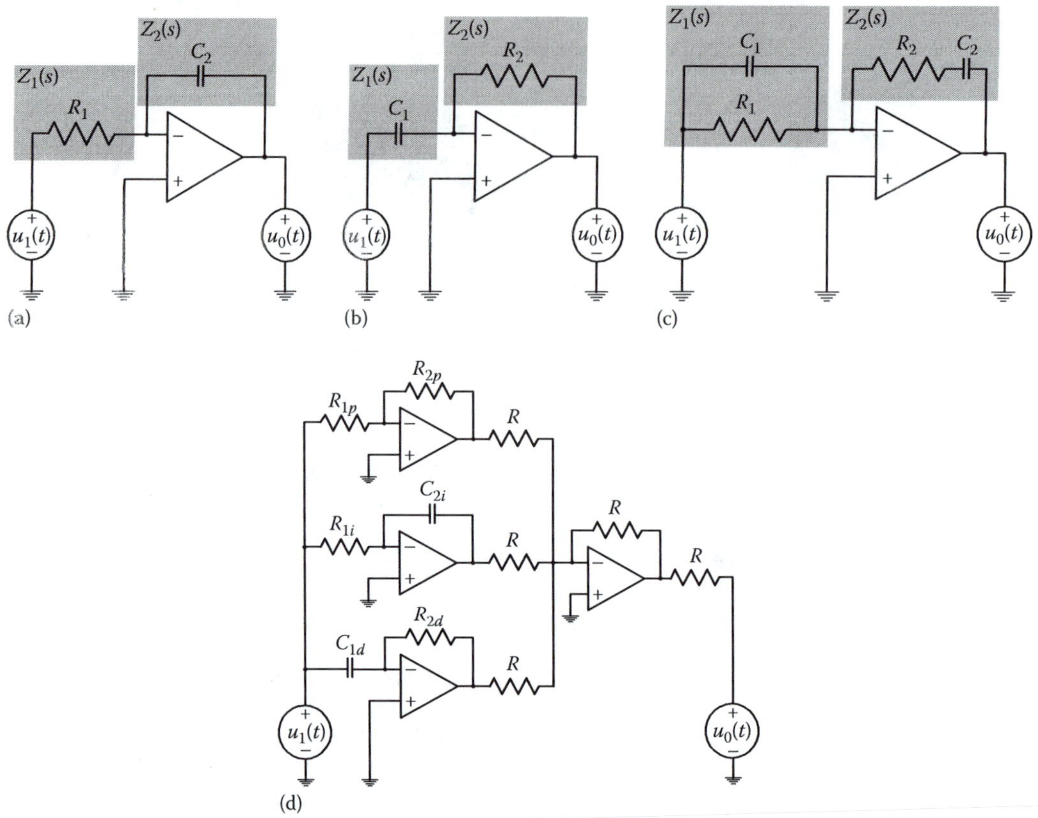

FIGURE 7.13 (a) Inverting integrator, $G(s) = -\dfrac{1}{R_1C_2s}$;
(b) Inverting differentiator, $G(s) = -R_2C_1s$;
(c) An inverting operational amplifier implements an analog PID controller with a transfer function
$G(s) = \dfrac{U_0(s)}{U_1(s)} = -\dfrac{(R_1C_1s+1)(R_2C_2s+1)}{R_1C_2s} = -\dfrac{R_2C_1s^2 + \dfrac{R_1C_1+R_2C_2}{R_1C_2}s + \dfrac{1}{R_1C_2}}{s}$;
(d) Analog PID controller $G(s) = \dfrac{U_0(s)}{U_1(s)} = \dfrac{R_{2p}}{R_{1p}} + \dfrac{1}{R_{1i}C_{2i}s} + R_{2d}C_{1d}s$ implementation.

The PID controller can be implemented using the configuration depicted in Figure 7.13c. The transfer function of an inverting operational amplifier is

$$G(s) = \frac{U_0(s)}{U_1(s)} = -\frac{(R_1C_1s+1)(R_2C_2s+1)}{R_1C_2s} = -\frac{R_2C_1s^2 + \dfrac{R_1C_1+R_2C_2}{R_1C_2}s + \dfrac{1}{R_1C_2}}{s}.$$

The feedback gains k_p, k_i, and k_d are $k_p = -\dfrac{R_1C_1+R_2C_2}{R_1C_2}$, $k_i = -\dfrac{1}{R_1C_2}$, and $k_d = -R_2C_1$. The positive definiteness of k_p, k_i, and k_d are ensured by the use of another inverting amplifier.

The designer needs to vary feedback gains. The configuration illustrated in Figure 7.13d is commonly used to implement the PID controller. One has

$$G(s) = \frac{U_0(s)}{U_1(s)} = \frac{R_{2p}}{R_{1p}} + \frac{1}{R_{1i}C_{2i}s} + R_{2d}C_{1d}s, \quad k_p = \frac{R_{2p}}{R_{1p}}, \quad k_i = \frac{1}{R_{1i}C_{2i}}, \quad \text{and} \quad k_d = R_{2d}C_{1d}.$$

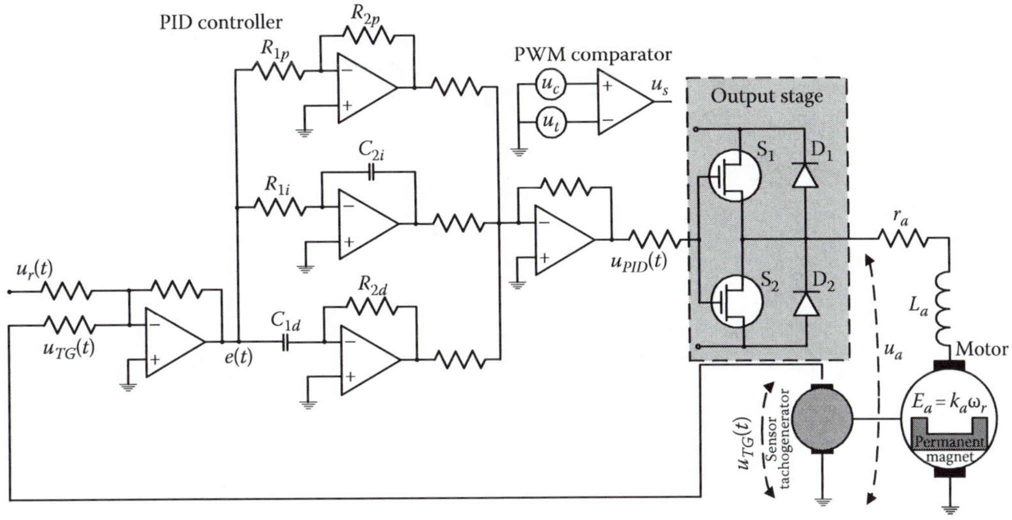

FIGURE 7.14 Application of the D-class power amplifier to control an electric drive with a permanent-magnet DC motor: Closed-loop configuration with a PID controller and sensor (tachogenerator).

Power electronic solutions are application specific. For example, for 100 W (rated), 50 V permanent-magnet motors, the rated current is ~2 A, while the peak current may reach ~20 A. Power amplifiers are used to ensure the needed voltage and current [2–5]. High-frequency pulse-width-modulation (PWM) amplifiers with one-, two-, or four-quadrant output stages are used. The power dissipated in the output stage power transistors should be minimized. The output stages are classified as A, B, AB, C, and D classes. In electromechanical systems, D-class PWM switching amplifiers are commonly used due to high efficiency, simplicity, reliability, low harmonic distortion, etc. A simplified typifying configuration with an output stage to control the permanent-magnet DC motor is illustrated in Figure 7.14. Using the PID controller, the MOSFETs switching is controlled. As will be reported, the comparators are used to drive the MOSFETs. The output voltage u_a, applied to the motor winding, is regulated. The angular velocity ω_r is measured by a tachogenerator. The measured angular velocity is compared with the desired velocity. The error $e(t) = u_r(t) - u_{TG}(t)$ is used as an input to the PID controller. The error amplifier compares the desired angular velocity $u_r(t)$ with the measured angular velocity $u_{TG}(t)$. If the signal-level voltage u_{PID} of the output inverting operational amplifier of the PID controller is positive, transistor S_1 is *off*. Transistor S_2 is *on*, and the negative voltage u_a is applied to the motor. If the voltage u_{PID} is negative, the S_1 is *on*, and S_2 is *off*. The applied voltage u_a is positive. The diodes D_1 and D_2 prevent damage of MOSFETs by the *back emf*. This simple schematics can be modified to ensure the PWM concept to vary the average value of u_a. Usually, PI controllers are used due to the sensitivity of the derivative feedback to the noise that may not be sufficiently attenuated by filters. The output stage topology and circuitry are much more complex, see Figures 7.15 and 7.16. The back *emf*, current ripple, PWM switching frequency, inductance matching, filtering, and efficiency are considered.

The high-switching frequency PWM DC–DC converters are used in electromechanical systems. As documented in Figure 7.15, the MC33030 DC servo-motor controller/driver integrates on-chip operational amplifier and comparator, driving logics, PWM four-quadrant converter, etc. The rated (peak) output voltage and current are 36 V and 1 A. Hence, one can use MC33030 for small ~10 W DC motors and actuators. Electric machines can operate at high voltage and current for short periods. For motors, $T_{e\,peak}/T_{e\,rated}$, i_{peak}/i_{rated}, and P_{peak}/P_{rated} could be ~10. The power electronics, with $i_{peak}/i_{rated} \sim 2$, should accommodate the peak motor current within the specific operating envelope. The MC33030 servo-motor driver contains 119 active transistors. The difference between the reference and actual angular velocity or displacement is compared by the error amplifier. Two comparators are used as shown in Figure 7.15.

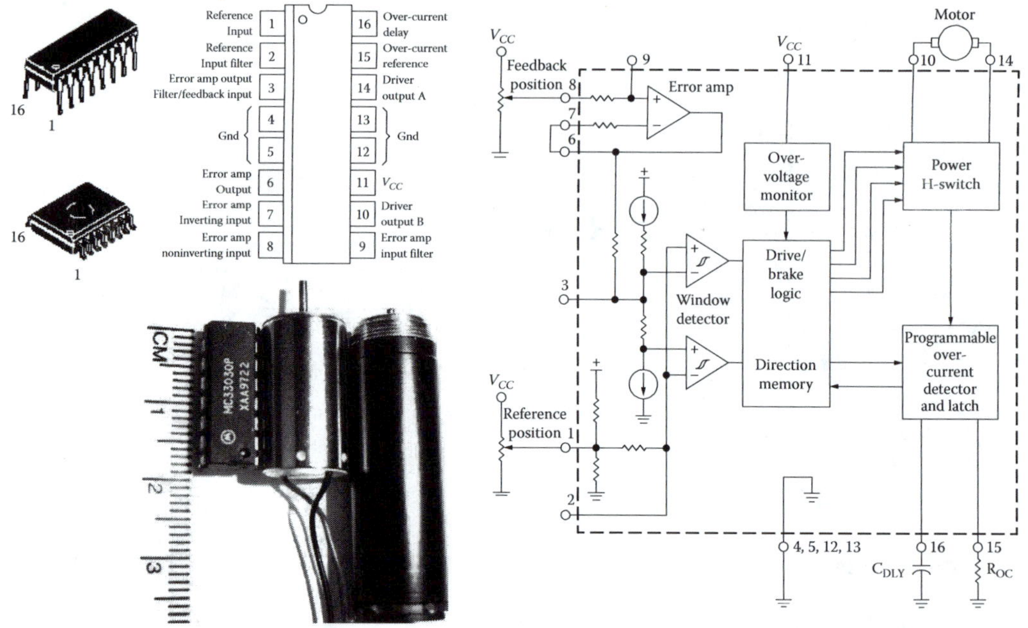

FIGURE 7.15 Pin connection and block diagram of the MC33030 DC servo-motor controller/driver for permanent-magnet DC motors and actuators. (From Lyshevski, S.E., *Electromechanical Systems, Electric Machines, and Applied Mechatronics*, CRC Press, Boca Raton, FL, 1999; Copyright of Motorola. With permission.)

A *pnp* differential output power stage ensures driving and braking capabilities. The four-quadrant H-configured power stage guarantees high performance and efficiency. A schematics of a servo system with MC33030 is illustrated in Figure 7.16.

One specifies a voltage on the reference input (pin 1). The velocity or displacement sensor measures the output velocity or displacement. The sensor's output voltage is supplied to pin 3. The reference voltage $r(t)$ is compared with the measured output $y(t)$. The tracking error is $e(t) = r(t) - y(t)$. The "window detector" is implemented by two comparators with hysteresis. The proportional controller controls comparators that drive the MOSFET drivers. The four-quadrant power stage outputs the PWM voltage. The permanent-magnet DC motor is connected to pins 10 and 14. The current limit is set on pin 15. The voltage protection is ensured by the "overvoltage monitor", which is important due to the *back emf*. This schematics can be modified by adding additional filters, control, and data acquisitions circuitry.

The dual-power operational amplifiers can be used. Figure 7.17a documents the image of a 7-pin, 40 V, 1.5 A, heatsink-mount dual-power operational amplifier that can be used in half- and full-bridge motor drivers. The application-specific electronics are used. Advanced-technology miniscale electric machines with enabling CMOS microelectronics are widely deployed. The size of minimachines can be less than electronics. Images of 2 and 4 mm diameter electric machines and power electronic board are illustrated in Figure 7.17b.

7.4.2 Switching Converter: Buck Converter

Using a pulse-width-modulation (PWM) switching concept, the voltage at the load terminal can be effectively regulated. The schematics for a high switching frequency DC–DC *buck* (*step-down*) converter is shown in Figure 7.18a. The converter components are MOSFET *S*, diode *D* and *LC* filter. The *RL* load with r_a and L_a is illustrated. The contact resistances, parasitic resistances, ohmic losses, and the inherent switch and inductor resistances r_s, r_L, and r_c are considered. The images of the Texas Instruments TPS544C20RVFT and TPS5410D regulators are illustrated in Figure 7.18b.

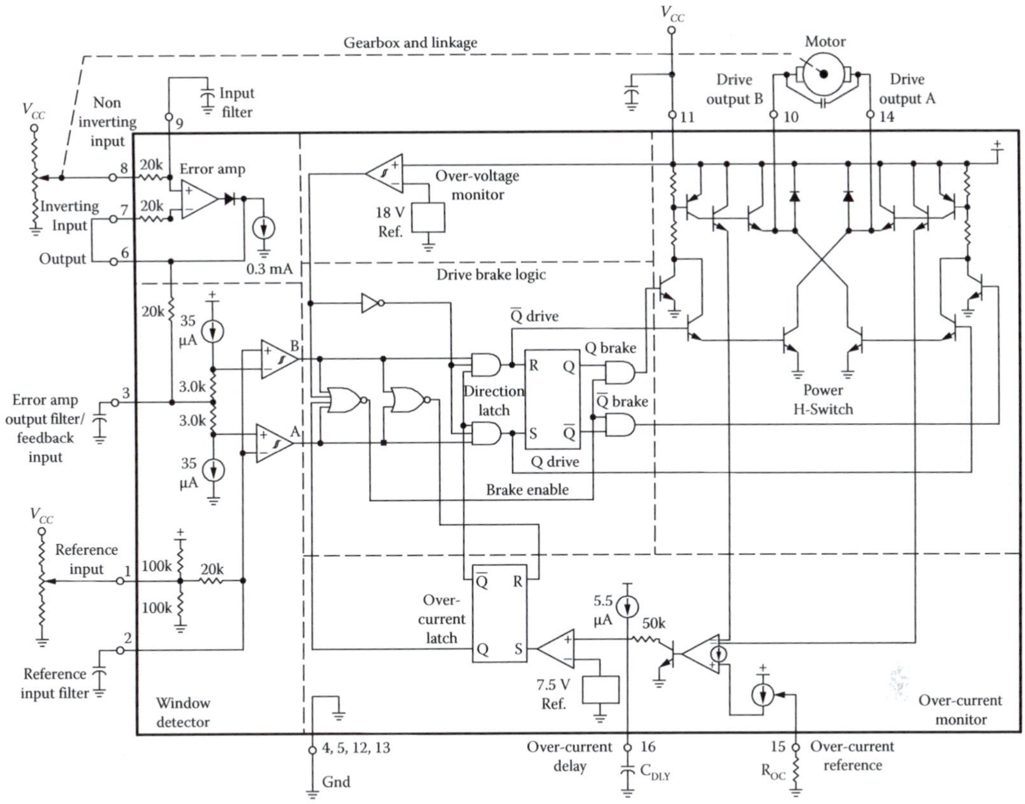

FIGURE 7.16 Schematics of a servosystem with MC33030 DC servo-motor controller/driver. (From Lyshevski, S.E., *Electromechanical Systems, Electric Machines, and Applied Mechatronics*, CRC Press, Boca Raton, FL, 1999; Copyright of Motorola. With permission.)

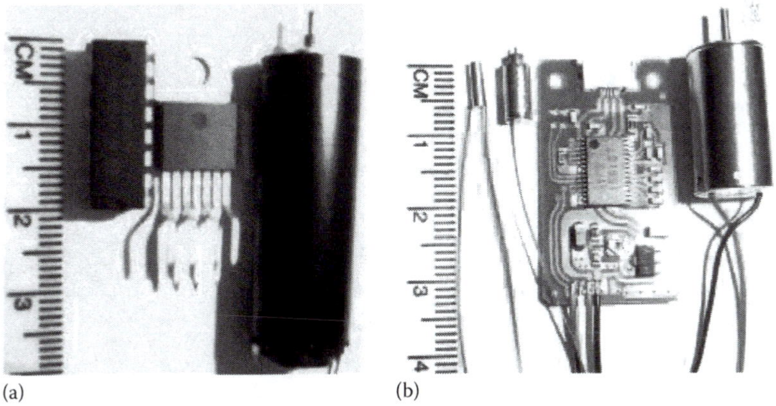

FIGURE 7.17 Power electronics hardware and permanent-magnet electric machines:
(a) MC30330 servo-motor controller/driver and dual power amplifier (~30 W peak) to control permanent-magnet DC motors and actuators (~3 W rated and ~30 W peak);
(b) Application-specific closed-loop PWM amplifier, 2 and 4 mm diameter permanent-magnet synchronous motors, and a 10 mm diameter permanent-magnet DC motor.

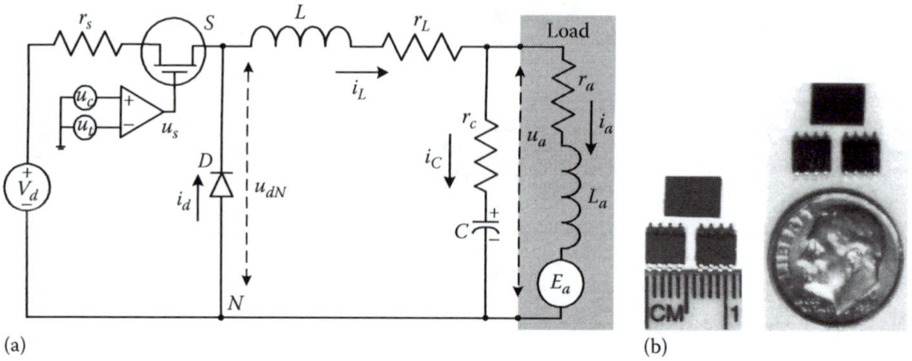

FIGURE 7.18 (a) *Step-down* switching converter;
(b) Texas Instruments TPS544C20RVFT 40LQFN-CLIP package (4.5–18 V input voltage, 0.6–5.5 V output voltage, 30 A output current, 1 MHz) and TPS5410D 8SOIC package (5.5–36 V input voltage, 1.23–31 V output voltage, 1 A output current, 500 kHz) *buck* regulators.

In the *step-down* converter, the switch S is open and closed. The switching frequency is

$$f = \frac{1}{t_{on} + t_{off}},$$

where t_{on} and t_{off} are the switching *on* and *off* durations. Assuming that the switch is lossless, the voltage u_{dN} is equal to the supplied voltage V_d when the switch is closed. The output voltage is zero if the switch is open, see Figure 7.19a.

The voltage u_{dN} and the voltage applied to the load u_a are regulated by controlling the switching *on* and *off* durations, denoted as t_{on} and t_{off}. The average voltage, applied to the load, depends on t_{on} and t_{off}. In steady state

$$u_{dN\,av} = \frac{t_{on}}{t_{on} + t_{off}} V_d = d_D V_d, \quad d_D = \frac{t_{on}}{t_{on} + t_{off}}, \quad d_D \in \begin{bmatrix} 0 & 1 \end{bmatrix},$$

where d_D is the duty ratio (duty cycle).

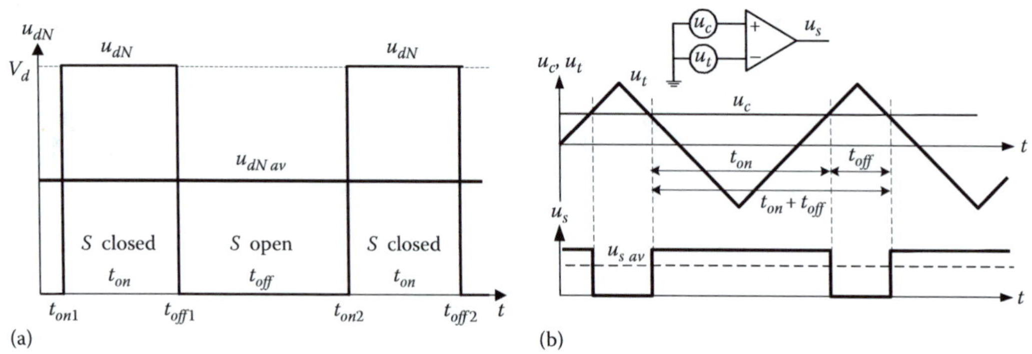

FIGURE 7.19 (a) Voltage waveforms;
(b) A comparator ensures the PWM transistor switching: The signals u_c, u_t, and u_s are shown.

The duty ratio d_D is a function of the switching frequency and the time during which the switch is *on*. One has $d_D \in \begin{bmatrix} 0 & 1 \end{bmatrix}$, and $d_D = 0$ if $t_{on} = 0$, while $d_D = 1$ if $t_{off} = 0$. By changing d_D one controls the transistor switching activity. The average voltage, supplied to the load u_a, is regulated. To establish PWM switching, a control-triangle concept is used. The switching signal u_s, which drives the switch, is generated by comparing a signal-level control voltage u_c with a repetitive triangular u_t. The comparators are shown in Figures 7.1 and 7.20. The duration of the output pulses u_s represents the *weighted* value between the triangular voltage u_t with the assigned switching frequency and control signal u_c. The output voltage of the comparator u_s drives the switch S. The *on* and *off* switching is accomplished by comparing u_c and u_t. The Motorola dual operational amplifier and dual comparator MC3405 are reported in Figure 7.20. Figures 7.19b and 7.20 illustrate the voltage waveforms.

For the *step-down* converter, illustrated in Figures 7.18, a low-pass first-order LC filter with inductance L and capacitance C ensures the specified voltage ripple. Neglecting the small resistances r_s, r_L, and r_c, we have the expressions for voltage ripple $\dfrac{\Delta u_a}{u_a} = \dfrac{1-d_D}{8LCf^2}$.

The minimum value for the LC filter inductor is $L_{\min} = \dfrac{(1-d_D)r_a}{2f}$.

Two circuits when the switch is closed and open are illustrated in Figures 7.21.

Using Kirchhoff's laws, one finds the differential equations to describe the converter dynamics. If the switch is closed, the diode D is reverse biased. For the circuit, shown in Figure 7.21a, we have

$$\frac{du_C}{dt} = \frac{1}{C}(i_L - i_a), \quad \frac{di_L}{dt} = \frac{1}{L}\left(-u_C - (r_L + r_c)i_L + r_c i_a - r_s i_L + V_d\right), \quad \frac{di_a}{dt} = \frac{1}{L_a}\left(u_C + r_c i_L - (r_a + r_c)i_a - E_a\right).$$

If the switch is open, the diode D is forward biased, and $i_d = i_L$, see Figure 7.21b. One obtains

$$\frac{du_C}{dt} = \frac{1}{C}(i_L - i_a), \quad \frac{di_L}{dt} = \frac{1}{L}\left(-u_C - (r_L + r_c)i_L + r_c i_a\right), \quad \frac{di_a}{dt} = \frac{1}{L_a}\left(u_C + r_c i_L - (r_a + r_c)i_a - E_a\right).$$

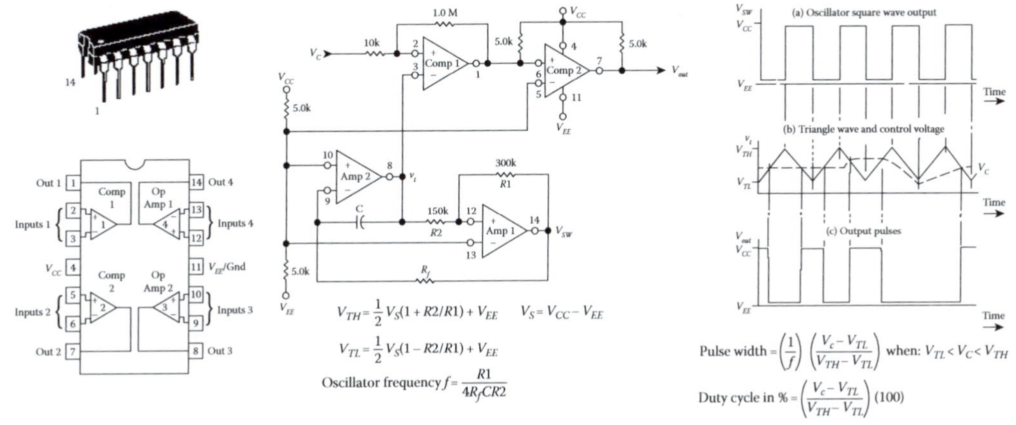

FIGURE 7.20 MC3405 comparator pin connections, schematics, and waveforms. (From Lyshevski, S.E., *Electromechanical Systems, Electric Machines, and Applied Mechatronics*, CRC Press, Boca Raton, FL, 1999; Copyright of Motorola. With permission.)

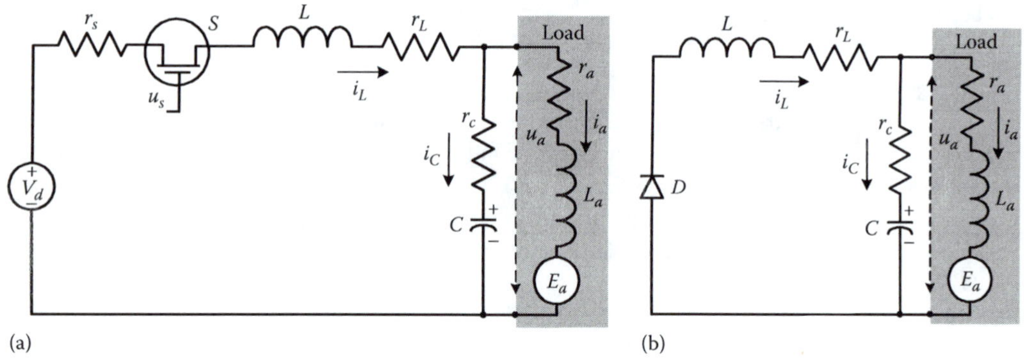

FIGURE 7.21 Circuits of the *buck* DC–DC converter:
(a) When the switch is closed;
(b) When the switch is open.

When the switch is closed, the duty ratio is $d_D = 1$. If the switch is open, the duty ratio is zero, $d_D = 0$. By using the *averaging* concept, from two sets of differential equations derived, one obtains the resulting nonlinear differential equations for the *buck* switching converter

$$\frac{du_C}{dt} = \frac{1}{C}\left(i_L - i_a\right),$$

$$\frac{di_L}{dt} = \frac{1}{L}\left(-u_C - \left(r_L + r_c\right)i_L + r_c i_a - r_s i_L d_D + V_d d_D\right),$$

$$\frac{di_a}{dt} = \frac{1}{L_a}\left(u_C + r_c i_L - \left(r_a + r_c\right)i_a - E_a\right).$$

The duty ratio is regulated by the signal-level control voltage u_c. We have

$$d_D = \frac{u_c}{u_{t\max}} \in \left[0 \quad 1\right], \quad u_c \in \left[0 \quad u_{c\max}\right], \quad u_{c\max} = u_{t\max}.$$

Neglecting small r_s, r_L, and r_c, the analysis of the steady-state performance yields $\frac{u_{a\ average}}{V_d} = d_D$. The converter output is the voltage applied to the load u_a. We have $u_a = u_C + r_c i_L - r_c i_a$. From $d_D = \frac{u_c}{u_{t\max}}$, a nonlinear term $\frac{r_s}{L} i_L d_D = \frac{r_s}{L u_{t\max}} i_L u_c$ is the multiplication of the state variable i_L and control u_c. The control limit is $0 \le u_c \le u_{c\max}$, $u_c \in \left[0 \quad u_{c\max}\right]$.

Example 7.12: Simulation and Experimental Studies of a Closed-Loop System

We simulate and examine the *step-down* converter. The converter parameters are $r_s = 0.025$ ohm, $r_L = 0.02$ ohm, $r_c = 0.15$ ohm, $r_a = 3$ ohm, $C = 0.003$ F, $L = 0.0007$ H, and $L_a = 0.005$ H. In simulations, let $d_D = 0.5$. The supplied DC voltage is $V_d = 50$ V and $E_a = 10$ V. Using the differential equations derived, the following m-files are developed to perform the simulations using the ode45 differential equation solver.

MATLAB file (ch7 _ 01.m)

```
t0=0; tfinal=0.03; tspan=[t0 tfinal]; y0=[0 0 0]';
[t,y]=ode45('ch7_02',tspan,y0);
subplot(2,2,1); plot(t,y); xlabel('Time (seconds)','FontSize',10);
title('Transient Dynamics of State Variables','FontSize',10);
subplot(2,2,2); plot(t,y(:,1),'-'); xlabel('Time (seconds)','FontSize',10);
title('Voltage u_C, [V]','FontSize',10);
subplot(2,2,3); plot(t,y(:,2),'-'); xlabel('Time (seconds)','FontSize',10);
title('Current i_L, [A]','FontSize',10);
subplot(2,2,4); plot(t,y(:,3),'-'); xlabel('Time (seconds)','FontSize',10);
title('Current i_a, [A]','FontSize',10);
```

MATLAB file (ch7 _ 02.m)

```
% Dynamics of the buck converter
function yprime=difer(t,y);
% parameters
Vd=50; Ea=10; rs=0.025; rl=0.02; rc=0.15; ra=3; C=0.003; L=0.0007; La=0.005; D=0.5;
% Differential equations for a buck converter
yprime=[(y(2,:)-y(3,:))/C;...
(-y(1,:)-(rl+rc)*y(2,:)+rc*y(3,:)-rs*y(2,:)*D+Vd*D)/L;...
(y(1,:)+rc*y(2,:)-(rc+ra)*y(3,:)-Ea)/La];
```

The transient dynamics for the state variables $u_C(t)$, $i_L(t)$, and $i_a(t)$ are illustrated in Figure 7.22a. The settling time is 0.025 sec. The steady-state value of the output voltage is 25 V because the applied voltage is 50 V and $d_D = 0.5$. Using $u_a = u_C + r_c i_L - r_a i_a$, the voltage at the load terminal u_a is calculated and plotted. The plot for $u_a(t)$ is illustrated in Figure 7.22b. One types in the Command Window

```
plot(t,y(:,1)+rc*y(:,2)-rc*y(:,3),'-');
xlabel('Time (seconds)','FontSize',14); title('Voltage u_a, [V]','FontSize',14);
```

The experimental studies are performed. The output voltage is stabilized for the resistive load. In particular, we set the output voltage to be 3.95 V. Using the tracking error $e = (r - u_a)$, the proportional-integral control law is $u = k_p e + k_i \int e\, dt$. The PI control law is implemented using a single operational amplifier. The feedback gains are $k_p = 0.52$ and $k_i = 0.11$. The experimental results are reported in Figure 7.22c. For different resistances $r_a = 14$ ohm and $r_a = 9.5$ ohm, the terminal voltage u_a is stabilized. The reference voltage is 3.95 V. ■

7.4.3 Boost Converter

A typical configuration of a one-quadrant *boost* (*step-up*) DC–DC switching converter is documented in Figure 7.23a. When the switch S is closed, the diode D is reverse biased. One finds

$$\frac{du_C}{dt} = -\frac{1}{C} i_a, \quad \frac{di_L}{dt} = \frac{1}{L}\left(-(r_L + r_s) i_L + V_d\right), \quad \frac{di_a}{dt} = \frac{1}{L_a}\left(u_C - (r_a + r_c) i_a - E_a\right).$$

If the switch is open, the diode is forward biased because the direction of the current in the inductor i_L does not change instantly. Hence,

$$\frac{du_C}{dt} = \frac{1}{C}(i_L - i_a), \quad \frac{di_L}{dt} = \frac{1}{L}\left(-u_C - (r_L + r_c) i_L + r_c i_a + V_d\right), \quad \frac{di_a}{dt} = \frac{1}{L_a}\left(u_C + r_c i_L - (r_a + r_c) i_a - E_a\right).$$

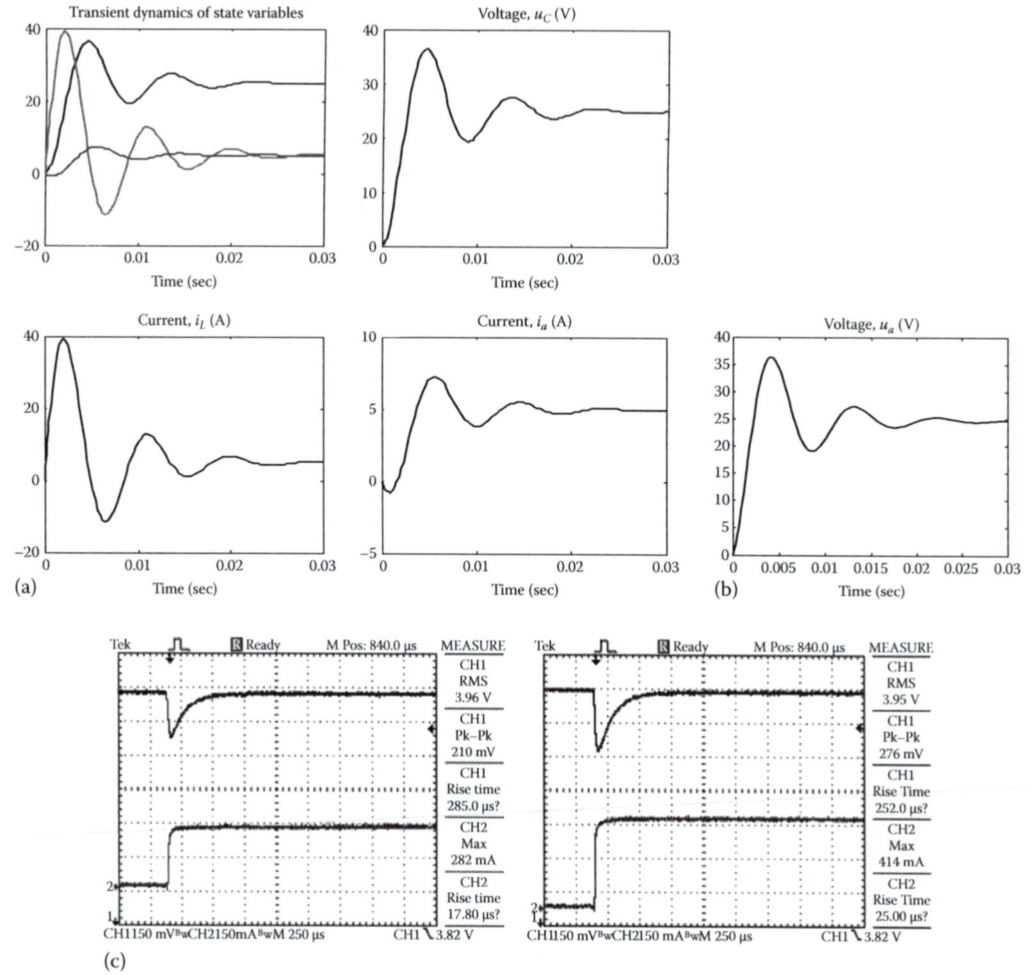

FIGURE 7.22 (a) Transient dynamics of the *buck* converter, $V_d = 50$ V and $d_D = 0.5$;
(b) Voltage u_a on the load terminal;
(c) Closed-loop *buck* regulator: Stabilization of the terminal voltage u_a if $r = 3.95$ V by using an analog proportional-integral control law. The experimental results are reported for $r_a = 14$ ohm and $r_a = 9.5$ ohm at which $i_{a\,steady\text{-}state} = 0.282$ A and $i_{a\,steady\text{-}state} = 0.414$ A.

Applying the *averaging* concept, using d_D, one finds

$$\frac{du_C}{dt} = \frac{1}{C}\left(i_L - i_a - i_L d_D\right),$$

$$\frac{di_L}{dt} = \frac{1}{L}\left(-u_C - \left(r_L + r_c\right)i_L + r_c i_a + u_C d_D + \left(r_c - r_s\right)i_L d_D - r_c i_a d_D + V_d\right),$$

$$\frac{di_a}{dt} = \frac{1}{L_a}\left(u_C + r_c i_L - \left(r_a + r_c\right)i_a - r_c i_L d_D - E_a\right).$$

The steady-state analysis results in $\dfrac{u_{a\,average}}{V_d} = \dfrac{1}{1 - d_D}$.

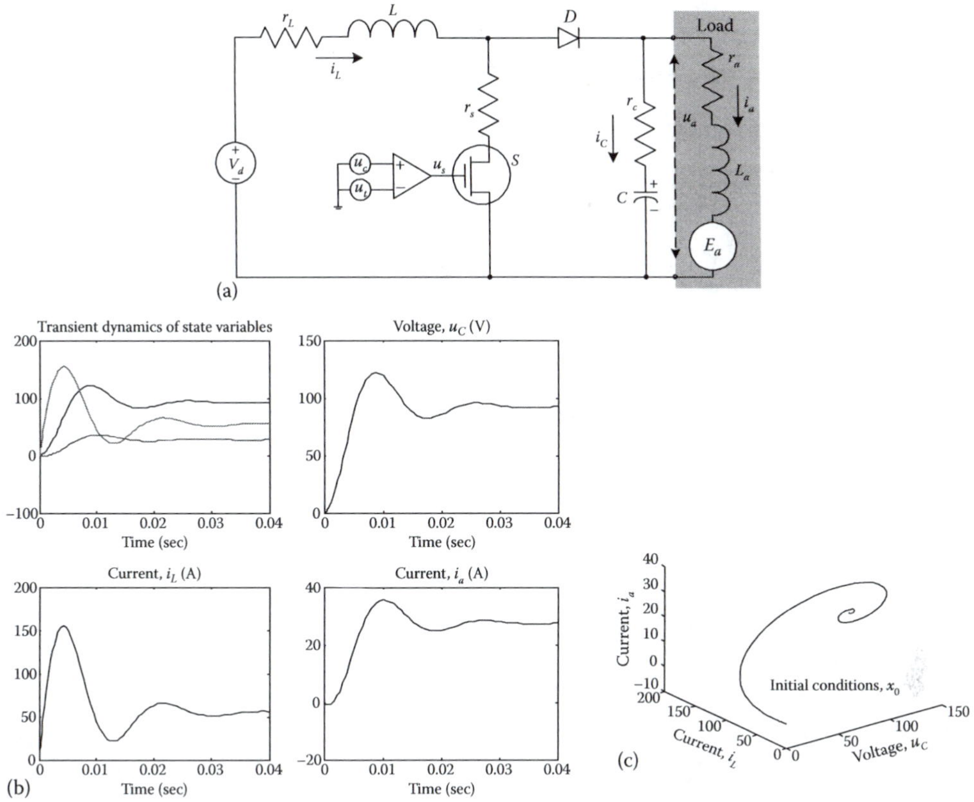

FIGURE 7.23 (a) High-frequency *boost* converter;
(b) Transient dynamics of the *boost* converter;
(c) Evolution of the state variables.

The voltage ripple is

$$\frac{\Delta u_a}{u_a} = \frac{d_D}{r_a C f^2}.$$

The minimum value of the inductance depends on the switching frequency and load resistance, and

$$L_{\min} = \frac{d_D(1-d_D)^2 r_a}{2f}.$$

The compliance, matching, and compatibility or power electronics and electromechanical motion devices must be ensured.

Example 7.13

We perform simulations of the *boost* converter if $r_s = 0.025$ ohm, $r_L = 0.02$ ohm, $r_c = 0.15$ ohm, $r_a = 3$ ohm, $C = 0.003$ F, $L = 0.0007$ H, and $L_a = 0.005$ H. Assume $d_D = 0.5$, $V_d = 50$ V, and $E_a = 10$ V. Using the differential equations found, simulations are conducted. The transients for $u_C(t)$, $i_L(t)$, and $i_a(t)$ are plotted in Figure 7.23b. The settling time is 0.038 sec. The transient dynamics of the *boost* and other DC–DC switching converters is fast. Therefore, the converter equations of motion are not usually used to analyze electromechanical system dynamics where mechanical dynamics dominate. Power electronic solutions affect the overall system performance, including efficiency, loading capabilities, stability, etc. The applied voltage to the motor or actuator windings u_a may not be considered as the control input. The three-dimensional plot, depicted in Figure 7.23c, illustrates the evolution of states $u_C(t)$, $i_L(t)$, and $i_a(t)$. ■

We applied the Kirchhoff's laws to derive the mathematical model in the form of nonlinear differential equations for a one-quadrant *boost* DC–DC converter illustrated in Figure 7.23a. The Lagrange concept can be applied. The Lagrange equations of motion are

$$\frac{d}{dt}\left(\frac{\partial \Gamma}{\partial \dot{q}_1}\right)-\frac{\partial \Gamma}{\partial q_1}+\frac{\partial D}{\partial \dot{q}_1}+\frac{\partial \Pi}{\partial q_1}=Q_1 \quad \text{and} \quad \frac{d}{dt}\left(\frac{\partial \Gamma}{\partial \dot{q}_2}\right)-\frac{\partial \Gamma}{\partial q_2}+\frac{\partial D}{\partial \dot{q}_2}+\frac{\partial \Pi}{\partial q_2}=Q_2.$$

The electric charges in the first and the second loops are q_1 and q_2. That is, $i_L=\dot{q}_1$ and $i_a=\dot{q}_2$. The *generalized* forces are $Q_1 = V_d$ and $Q_2 = -E_a$.

When the switch is closed, the total kinetic Γ, potential Π, and dissipation D energies are

$$\Gamma=\frac{1}{2}\left(L\dot{q}_1^2+L_a\dot{q}_2^2\right),\quad \Pi=\frac{1}{2}\frac{q_2^2}{C},\quad \text{and} \quad D=\frac{1}{2}\left((r_L+r_s)\dot{q}_1^2+(r_c+r_a)\dot{q}_2^2\right).$$

Assume that the resistances, inductances, and capacitance are constant. We have

$$\frac{\partial \Gamma}{\partial q_1}=0,\quad \frac{\partial \Gamma}{\partial q_2}=0,\quad \frac{\partial \Gamma}{\partial \dot{q}_1}=L\dot{q}_1,\quad \frac{\partial \Gamma}{\partial \dot{q}_2}=L_a\dot{q}_2,\quad \frac{d}{dt}\left(\frac{\partial \Gamma}{\partial \dot{q}_1}\right)=L\ddot{q}_1,\quad \frac{d}{dt}\left(\frac{\partial \Gamma}{\partial \dot{q}_2}\right)=L_a\ddot{q}_2,$$

$$\frac{\partial \Pi}{\partial q_1}=0,\quad \frac{\partial \Pi}{\partial q_2}=\frac{q_2}{C},\quad \text{and,}\quad \frac{\partial D}{\partial \dot{q}_1}=(r_L+r_s)\dot{q}_1,\quad \frac{\partial D}{\partial \dot{q}_2}=(r_c+r_a)\dot{q}_2.$$

The Lagrange equations of motion yield

$$L\ddot{q}_1+(r_L+r_s)\dot{q}_1=Q_1,\quad L_a\ddot{q}_2+(r_c+r_a)\dot{q}_2+\frac{1}{C}q_2=Q_2.$$

One obtains

$$\ddot{q}_1=\frac{1}{L}\left(-(r_L+r_s)\dot{q}_1+Q_1\right),\quad \ddot{q}_2=\frac{1}{L_a}\left(-(r_c+r_a)\dot{q}_2-\frac{1}{C}q_2+Q_2\right),$$

If the switch is open

$$\Gamma=\frac{1}{2}\left(L\dot{q}_1^2+L_a\dot{q}_2^2\right),\quad \Pi=\frac{1}{2}\frac{(q_1-q_2)^2}{C},\quad \text{and} \quad D=\frac{1}{2}\left(r_L\dot{q}_1^2+r_c(\dot{q}_1-\dot{q}_2)^2+r_a\dot{q}_2^2\right).$$

Hence, $\frac{\partial \Gamma}{\partial q_1}=0, \frac{\partial \Gamma}{\partial q_2}=0, \frac{\partial \Gamma}{\partial \dot{q}_1}=L\dot{q}_1, \frac{\partial \Gamma}{\partial \dot{q}_2}=L_a\dot{q}_2, \frac{d}{dt}\left(\frac{\partial \Gamma}{\partial \dot{q}_1}\right)=L\ddot{q}_1, \frac{d}{dt}\left(\frac{\partial \Gamma}{\partial \dot{q}_2}\right)=L_a\ddot{q}_2,$

$$\frac{\partial \Pi}{\partial q_1}=\frac{q_1-q_2}{C}, \quad \frac{\partial \Pi}{\partial q_2}=-\frac{q_1-q_2}{C}, \quad \text{and,} \quad \frac{\partial D}{\partial \dot{q}_1}=(r_L+r_c)\dot{q}_1-r_c\dot{q}_2, \quad \frac{\partial D}{\partial \dot{q}_2}=-r_c\dot{q}_1+(r_c+r_a)\dot{q}_2.$$

The resulting equations are

$$L\ddot{q}_1+(r_L+r_c)\dot{q}_1-r_c\dot{q}_2+\frac{q_1-q_2}{C}=Q_1, \quad L_a\ddot{q}_2-r_c\dot{q}_1+(r_c+r_a)\dot{q}_2-\frac{q_1-q_2}{C}=Q_2.$$

Therefore,

$$\ddot{q}_1=\frac{1}{L}\left(-(r_L+r_c)\dot{q}_1+r_c\dot{q}_2-\frac{q_1-q_2}{C}+Q_1\right), \quad \ddot{q}_2=\frac{1}{L_a}\left(r_c\dot{q}_1-(r_c+r_a)\dot{q}_2+\frac{q_1-q_2}{C}+Q_2\right).$$

From the differential equations derived when the switch is closed and open, Cauchy's form of differential equations are found by using $i_L=\dot{q}_1$ and $i_a=\dot{q}_2$. That is, $\frac{dq_1}{dt}=i_L$ and $\frac{dq_2}{dt}=i_a$. The voltage across the capacitor u_C can be expressed using the charges. When the switch is closed $u_C=-\frac{q_2}{C}$, while if the switch is open $u_C=\frac{q_1-q_2}{C}$. The differential equations, found using Kirchhoff's voltage law and the Lagrange equations of motion, result in the identical models where the physical variables u_C, i_L, i_a and q_1, i_L, q_2, i_a are used.

7.4.4 Buck-Boost Converters

The *buck-boost* switching converter is illustrated in Figure 7.24a. The images of the TPS63060DSCR *buck-boost* regulator are shown in Figure 7.24b.

If the switch is closed, the diode is reverse biased. When the switch is open, the diode is forward biased. One derives a set of differential equations using Kirchhoff's law or Lagrange equations. The steady-state relationship between the supplied and terminal voltage is

$$\frac{u_{a\ average}}{V_d}=\frac{-d_D}{1-d_D}.$$

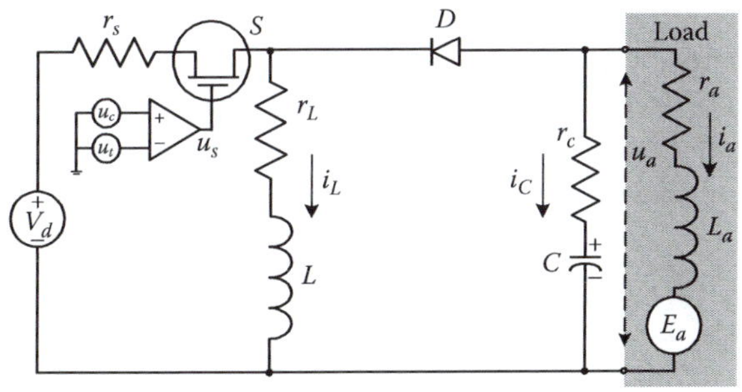

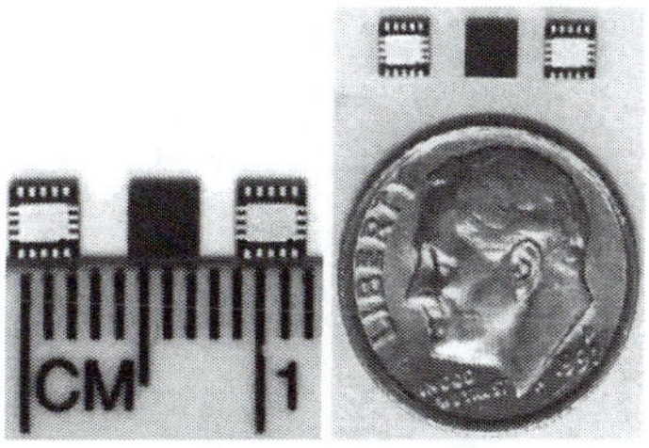

FIGURE 7.24 (a) High-frequency *buck-boost* switching converter;
(b) Texas Instruments TPS63060DSCR *buck-boost* regulator (2.5–12 V input voltage, 2.5–8 V output voltage, output current 2 A in the *buck* mode and 1.3 A in the *boost* mode, up to 93 % efficiency).

The expressions for the voltage ripple and minimum inductance are

$$\frac{\Delta u_a}{u_a}=\frac{d_D}{r_a Cf} \quad \text{and} \quad L_{\min}=\frac{(1-d_D)^2 r_a}{2f}.$$

When the switch S is closed and open, one finds the following differential equations

$$\frac{du_C}{dt}=-\frac{1}{C}i_a,\quad \frac{di_L}{dt}=\frac{1}{L}\left(-\left(r_L+r_s\right)i_L+V_d\right),\quad \frac{di_a}{dt}=\frac{1}{L_a}\left(u_C-\left(r_a+r_c\right)i_a-E_a\right),$$

and

$$\frac{du_C}{dt}=-\frac{1}{C}\left(i_L+i_a\right),\quad \frac{di_L}{dt}=\frac{1}{L}\left(u_C-\left(r_L+r_c\right)i_L-r_c i_a\right),\quad \frac{di_a}{dt}=\frac{1}{L_a}\left(u_C-r_c i_L-\left(r_a+r_c\right)i_a-E_a\right).$$

Applying the *averaging* concept, using d_D, one finds

$$\frac{du_C}{dt}=-\frac{1}{C}\left(i_L+i_a-i_L d_D\right),$$

$$\frac{di_L}{dt}=\frac{1}{L}\left(u_C-(r_L+r_c)i_L-r_c i_a-u_C d_D-r_s i_L d_D+r_c i_L d_D+r_c i_a d_D+V_d d_D\right),$$

$$\frac{di_a}{dt}=\frac{1}{L_a}\left(u_C-r_c i_L-(r_a+r_c)i_a+r_c i_L d_D-E_a\right).$$

Example 7.14

We perform simulations of the *buck-boost* converter. The parameters are r_s = 0.025 ohm, r_L = 0.02 ohm, r_c = 0.15 ohm, r_a = 3 ohm, C = 0.003 F, L = 0.0007 H, and L_a = 0.005 H. Assume that V_d = 50 V and E_a = 10 V.

Solving the differential equations, the transient dynamics for $u_C(t)$, $i_L(t)$, and $i_a(t)$ are plotted in Figure 7.25 for different values of d_D. The results derived correspond to the expected values for the steady-state solution. ■

Example 7.15: Modeling, Simulation, and Control of Open- and Closed-Loop Buck-Boost Converter

Consider the *buck-boost* converter shown in Figure 7.26. When the MOSFET is closed, using the currents i_L and i_{RL} and voltage u_C, one has

$$\frac{du_C}{dt}=-\frac{1}{C}i_{RL},\quad \frac{di_L}{dt}=\frac{1}{L}\left[-\left(r_L+r_s\right)i_L+V_d\right],\quad \frac{di_{RL}}{dt}=\frac{1}{L_L}\left[u_C-\left(R_L+r_c\right)i_{RL}\right].$$

When the MOSFET is open, the capacitor C is charged by the voltage source. The differential equations are

$$\frac{du_C}{dt}=\frac{1}{C}\left(-i_L-i_{RL}\right),\ \frac{di_L}{dt}=\frac{1}{L}\left[u_C-\left(r_L+r_c\right)i_L-r_c i_{RL}\right],\ \frac{di_{RL}}{dt}=\frac{1}{L_L}\left[u_C-r_c i_L-\left(R_L+r_c\right)i_{RL}\right].$$

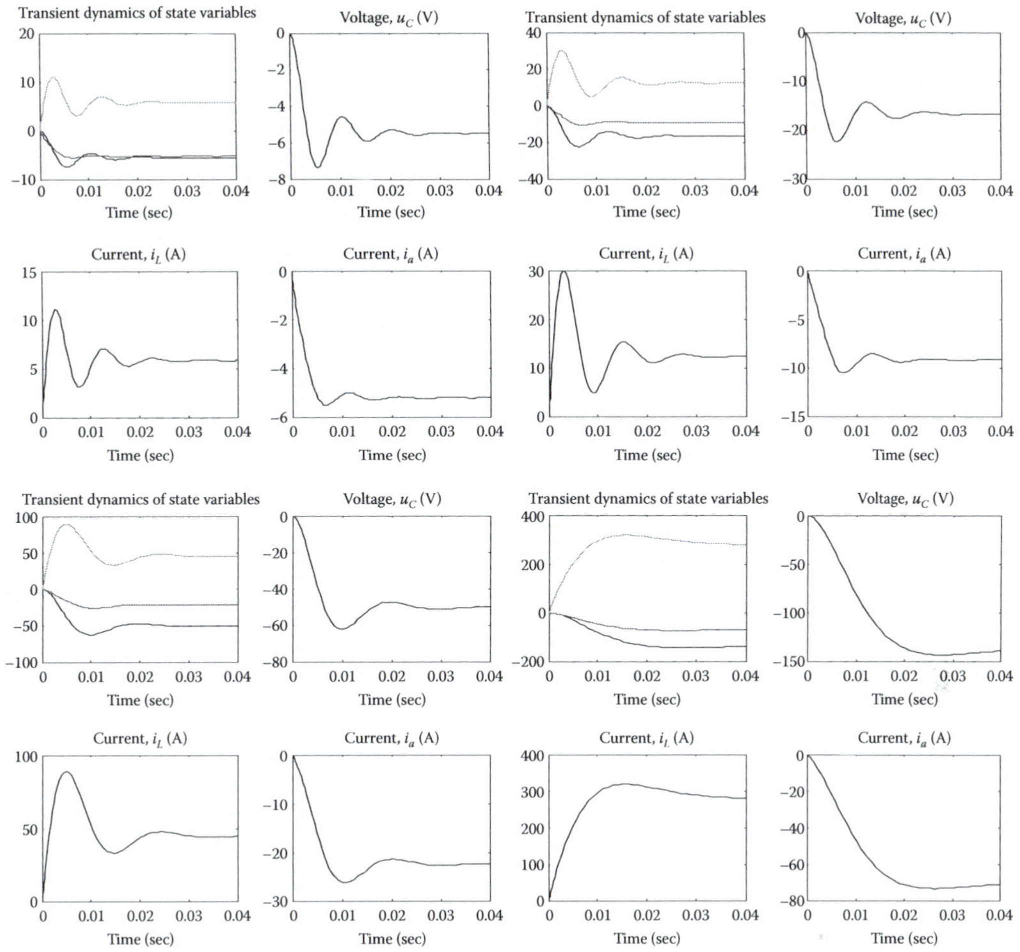

FIGURE 7.25 Transient dynamics of the *buck-boost* converter when $d_D = 0.1$, $d_D = 0.25$, $d_D = 0.5$, $d_D = 0.75$.

The voltage, applied to the load u_{RL}, is regulated by controlling the switching *on* and *off* durations t_{on} and t_{off}. The duty cycle $d_D = \dfrac{t_{on}}{t_{on}+t_{off}} \in [0 \quad 1]$ varies, and $0 \le d_D \le 1$.

Using the *averaging* concept, we have

$$\frac{du_C}{dt} = \frac{1}{C}\left(-i_L - i_{RL} + i_L d_D\right),$$

$$\frac{di_L}{dt} = \frac{1}{L}\left[u_C - \left(r_L + r_c\right)i_L - r_c i_{RL} - u_C d_D - \left(r_s - r_c\right)i_L d_D + r_c i_{RL} d_D + V_d d_D\right],$$

$$\frac{di_{RL}}{dt} = \frac{1}{L_L}\left[u_C - r_c i_L - \left(R_L + r_c\right)i_{RL} + r_c i_L d_D\right].$$

We simulate the open-loop *buck-boost* converter dynamics. One has $r_s = 0.025$ ohm, $r_L = 0.02$ ohm, $r_c = 0.15$ ohm, $R_L = 5$ ohm, $C = 0.000022$ F, $L = 0.001$ H, and $L_L = 0.005$ H. Figure 7.27 reports the Simulink® model to simulate the open- and closed-loop system with PID control law.

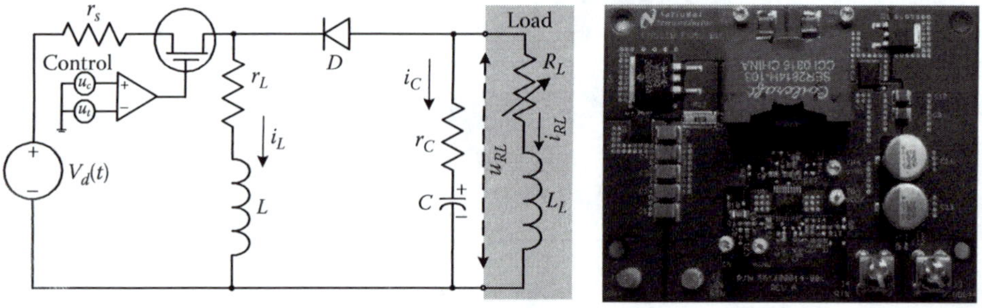

FIGURE 7.26 Controlled *buck-boost* converter with the varying *RL* load.

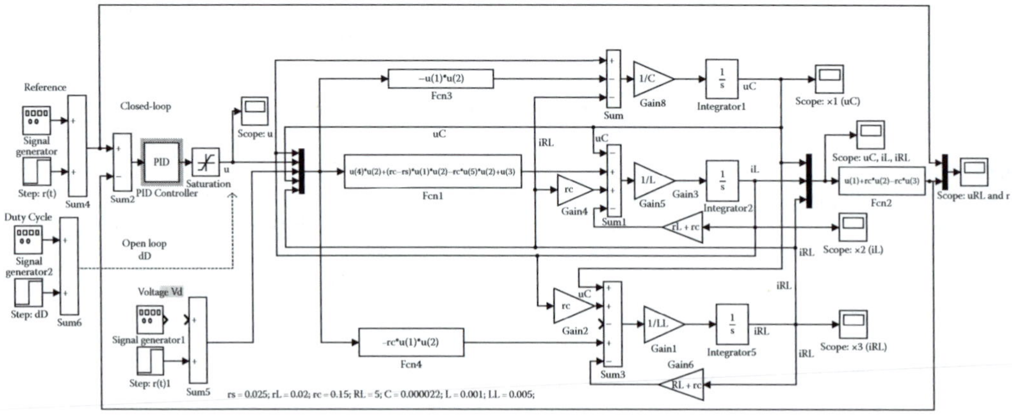

FIGURE 7.27 Simulink® model for open- and closed-loop *buck-boost* converters.

In an open-loop system, let the duty ratio changes as $d_D = \begin{cases} 0.5 \\ 0.7 \end{cases}$ with frequency 100 rad/sec. The plots for the u_C, i_L, and i_{RL} are documented in Figure 7.28a. The open-loop evolution of u_{RL} is shown in Figure 7.28a.

The closed-loop *buck-boost* converter is studied with the PID control law. We apply the proportional control law $u = k_p e$, $u \in [0 \quad 1]$, $e = (r - u_{RL})$, $k_p = 10$. The reference input (command) $r = \begin{cases} 20\text{ V} \\ 40\text{ V} \end{cases}$ varies with frequency 1000 rad/sec. The dynamics of u_C, i_L and i_{RL}, as well as evolution of u_{RL}, are reported in Figure 7.28b. The output voltage u_{RL} precisely follows the step references $r = \begin{cases} 20\text{ V} \\ 40\text{ V} \end{cases}$. ■

7.4.5 Cuk Converter

The Cuk converter is based on a capacitive-inductive energy transfer, while in the *buck*, many *buck-boost*, *boost*, and *flyback* converter topologies are based on the inductive energy transfer. The DC–DC PWM Cuk regulator is a *buck-boost* high-power-factor converter. If the switch is *on* or *off*, the currents in the input and output inductors L_1 and L are continuous. The output voltage, applied to the load can be either smaller or greater than the supplied voltage V_d. When the switch is turned *off*, the diode is forward biased. The voltage V_d is supplied, and the capacitor C_1 is charged through the inductor L_1, see Figure 7.29a. To study how the converter operates, assume that the switch is turned *on*. The current through the inductor L_1 rises, the voltage of capacitor C_1 reverse-biases diode D and turns it *off*. The capacitor C_1 discharges the stored energy through the circuit formed by capacitors C_1, C, the load $r_a - L_a$, and the inductor L. If the switch is turned *off*, the voltage V_d is applied. The capacitor C_1 charges. The energy, stored in the inductor L, transfers to the load. The diode and switch provide a synchronous switching, and the capacitor C_1 is a key element for transferring energy from the energy source to the load. Figure 7.29b reports the images of the Texas Instruments LM2611AMF/NOPB Cuk regulators.

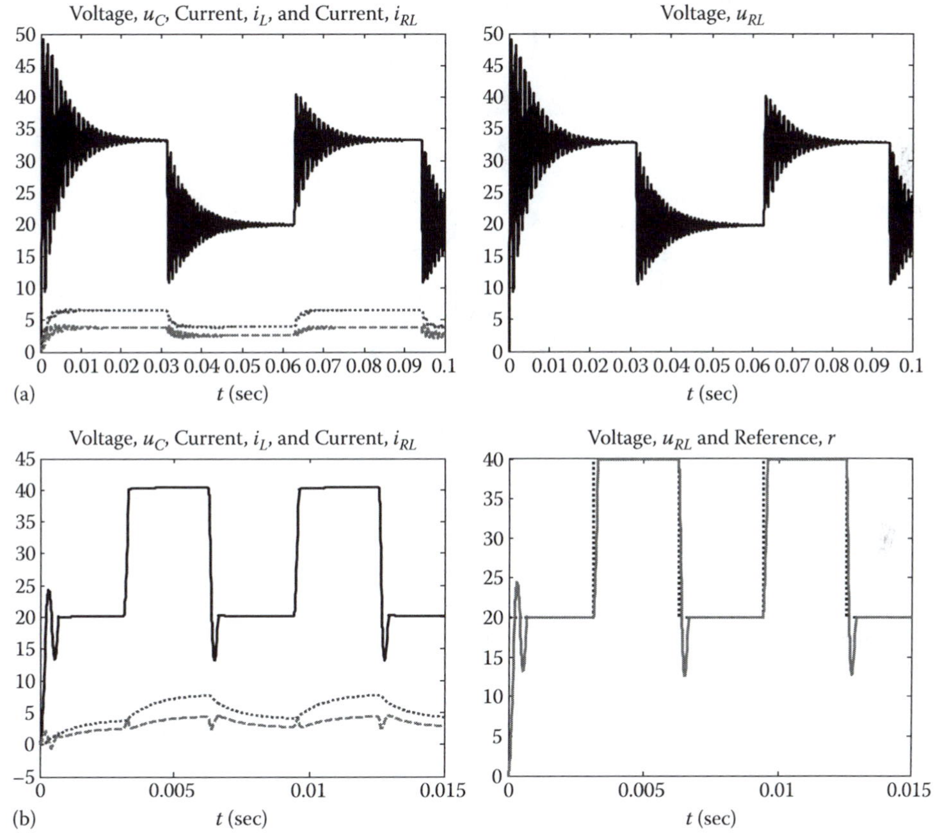

FIGURE 7.28 (a) Open-loop dynamics for u_C (solid line), i_L (dashed line), and i_{RL} (dotted line) if $d_D = \begin{cases} 0.5 \\ 0.7 \end{cases}$ changes at frequency 100 rad/sec. Evolution of the terminal voltage u_{RL}; (b) Closed-loop dynamics for u_C (solid line), i_L (dashed line), and i_{RL} (dotted line) for the step reference $r = \begin{cases} 20\text{ V} \\ 40\text{ V} \end{cases}$, which changes at frequency 1000 rad/sec. Closed-loop converter evolution of the terminal voltage u_{RL} (solid line) for varying r (dashed line).

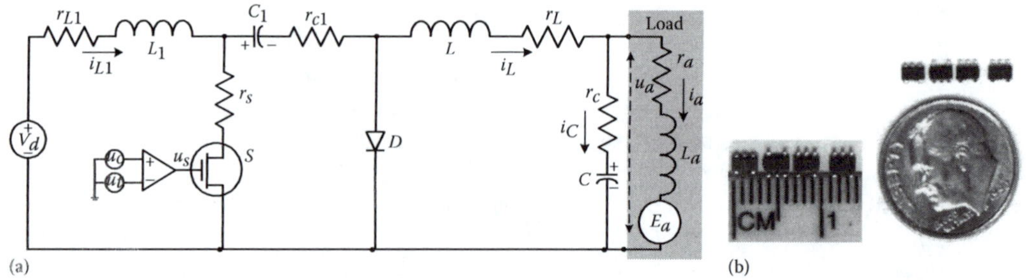

FIGURE 7.29 (a) High-frequency switching Cuk converter;
(b) Image of a Texas Instruments LM2611AMF/NOPB SOT-23-5 package Cuk regulator (2.7–14 V input voltage, −1.2 to −27 V output voltage, 900 mA, 1.4 MHz).

From Kirchhoff's laws, examining, when the switch is open and closed, one finds the following set of differential equations

$$\frac{du_{C1}}{dt} = \frac{1}{C_1}\left(i_{L1} - i_{L1}d_D + i_L d_D\right),$$

$$\frac{du_C}{dt} = \frac{1}{C}\left(i_L - i_a\right),$$

$$\frac{di_{L1}}{dt} = \frac{1}{L_1}\left(-u_{C1} - \left(r_{L1} + r_{c1}\right)i_{L1} + u_{C1}d_D + \left(r_{c1} - r_s\right)i_{L1}d_D + r_s i_L d_D + V_d\right),$$

$$\frac{di_L}{dt} = \frac{1}{L}\left(-u_C - \left(r_L + r_c\right)i_L + r_c i_a - u_{C1}d_D + r_s i_{L1}d_D - \left(r_{c1} + r_s\right)i_L d_D\right),$$

$$\frac{di_a}{dt} = \frac{1}{L_a}\left(u_C + r_c i_L - \left(r_c + r_a\right)i_a - E_a\right).$$

The steady-state equations, which are important in the converter design and converter-load matching, are

$$\frac{u_{a\,average}}{V_d} = \frac{-d_D}{1-d_D}, \quad \frac{\Delta u_a}{u_a} = \frac{1-d_D}{8LCf^2}, \quad L_{1\min} = \frac{(1-d_D)^2 r_a}{2d_D f}, \quad \text{and} \quad L_{\min} = \frac{(1-d_D)r_a}{2f}.$$

Example 7.16

Simulations are performed for the Cuk converter if $V_d = 50$ V, $E_a = 10$ V, $r_{L1} = 0.035$ ohm, $r_L = 0.02$ ohm, $r_c = 0.15$ ohm, $r_s = 0.03$ ohm, $r_{c1} = 0.018$ ohm, $r_a = 3$ ohm, $C_1 = 2 \times 10^{-5}$ F, $C = 3.5 \times 10^{-6}$ F, $L_1 = 5 \times 10^{-6}$ H, $L = 7 \times 10^{-6}$ H, and $L_a = 0.005$ H. Using the differential equations found, the simulation is performed letting $d_D = 0.5$. The transient dynamics for voltages and currents $u_{C1}(t)$, $u_C(t)$, $i_{L1}(t)$, $i_L(t)$, and $i_a(t)$ are illustrated in Figure 7.30. The settling time is 0.0005 sec. ■

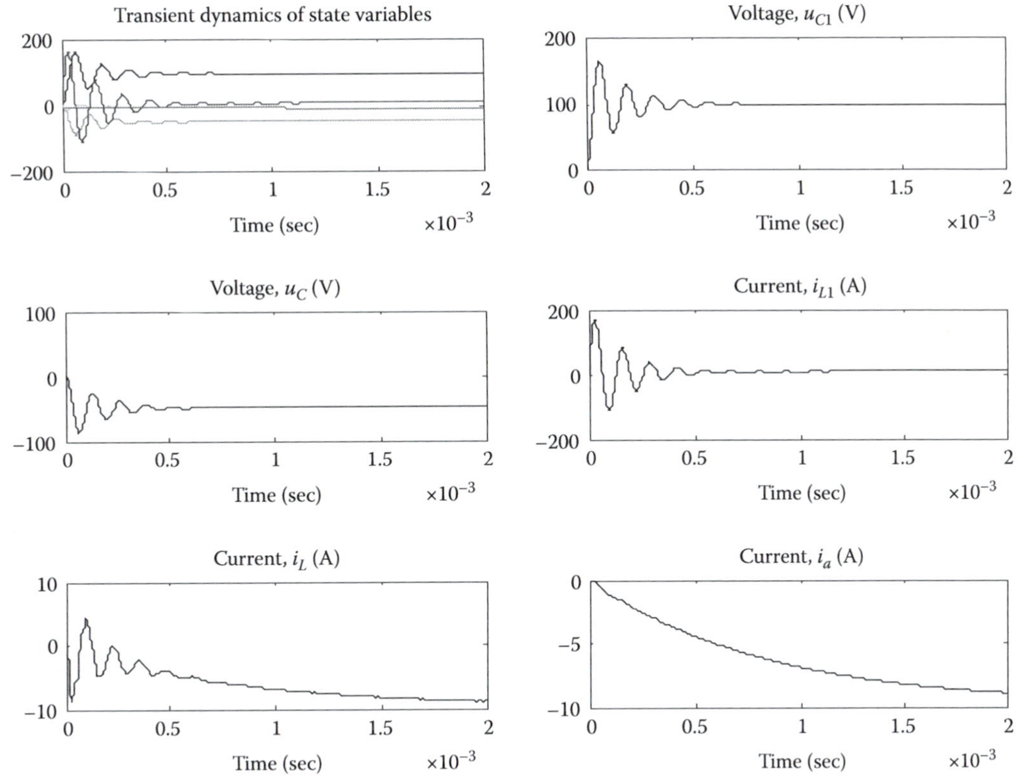

FIGURE 7.30 Transient dynamics of the Cuk converter.

Example 7.17: Analysis, Modeling, and Simulation of a Cuk Converter Using Simulink

We examine a high-performance PWM DC–DC inverting Cuk converter as illustrated in Figure 7.31. The output voltage u_{RL} should be regulated. In many applications, u_{RL} should be stabilized at the specified value despite the RL load variations. Depending on d_D, the output voltage u_{RL} can be smaller or greater than V_d.

Using Kirchhoff's laws and the *averaging* concept, one finds the following set of differential equations

$$\frac{du_{C1}}{dt} = \frac{1}{C_1}\left(i_{L1} - i_{L1}d_D + i_L d_D\right),$$

$$\frac{du_C}{dt} = \frac{1}{C}\left(i_L - i_{RL}\right),$$

$$\frac{di_{L1}}{dt} = \frac{1}{L_1}\left[-u_{C1} - \left(r_{L1} + r_{C1}\right)i_{L1} + u_{C1}d_D + \left(r_{C1} - r_s\right)i_{L1}d_D + r_s i_L d_D + V_d\right],$$

$$\frac{di_L}{dt} = \frac{1}{L}\left[-u_C - \left(r_L + r_C\right)i_L + r_C i_{RL} - u_{C1}d_D + r_s i_{L1}d_D - \left(r_{C1} + r_s\right)i_L d_D\right],$$

$$\frac{di_{RL}}{dt} = \frac{1}{L_L}\left[u_C + r_C i_L - \left(r_C + R_L\right)i_{RL}\right].$$

The bounded duty cycle is

$$d_D = \frac{t_{on}}{t_{on} + t_{off}}, \quad d_D \in \begin{bmatrix} 0 & 1 \end{bmatrix}.$$

Let $V_d = 10$ V, $r_{L1} = 0.035$ ohm, $r_L = 0.03$ ohm, $r_C = 0.01$ ohm, $r_s = 0.03$ ohm, $r_{c1} = 0.01$ ohm, $C_1 = 1 \times 10^{-6}$ F, $C = 22 \times 10^{-6}$ F, $L_1 = 15 \times 10^{-6}$ H, $L = 47 \times 10^{-6}$ H, and $L_L = 0.001$ H. The load varies as $R_L = 30$ and $R_L = 10$ ohm, $R_L = \begin{cases} 10 \\ 30 \end{cases}$ ohm with $f = 50$ Hz. The Simulink model is reported in Figure 7.32.

Let the duty ratio $d_D = \begin{cases} 0.75 \\ 0.25 \end{cases}$ varies with $f = 50$ Hz. The plots for u_{C1}, u_C, i_{L1}, i_L, and i_{RL} are documented in Figures 7.33, which also depicts the evolution of u_{RL}. ■

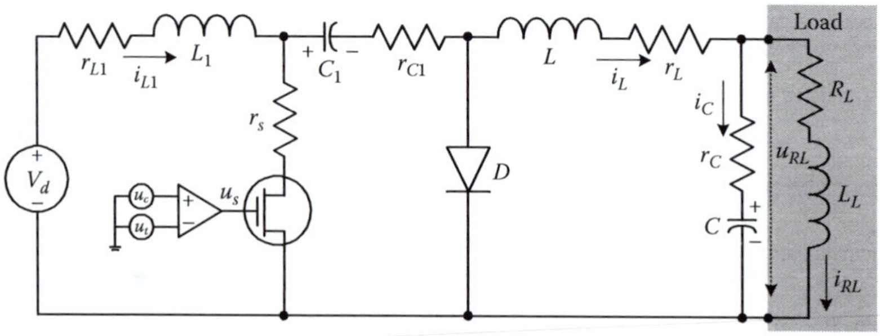

FIGURE 7.31 Cuk high-frequency DC–DC switching converter.

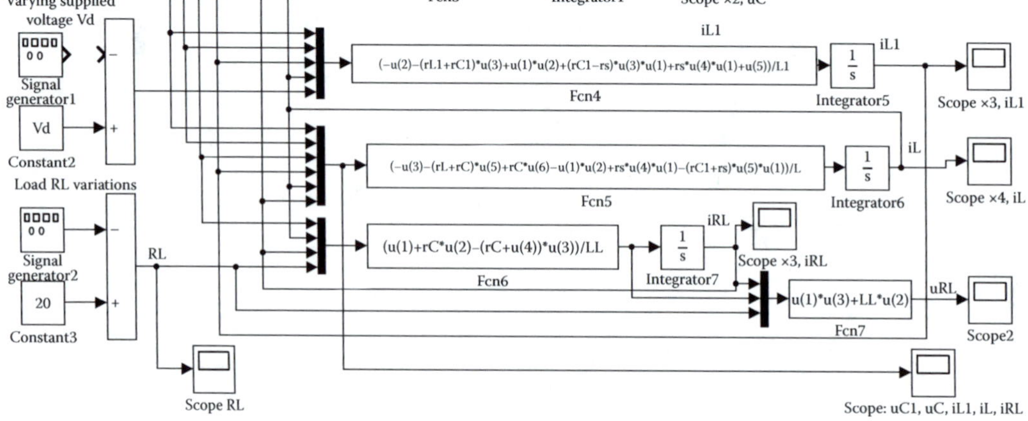

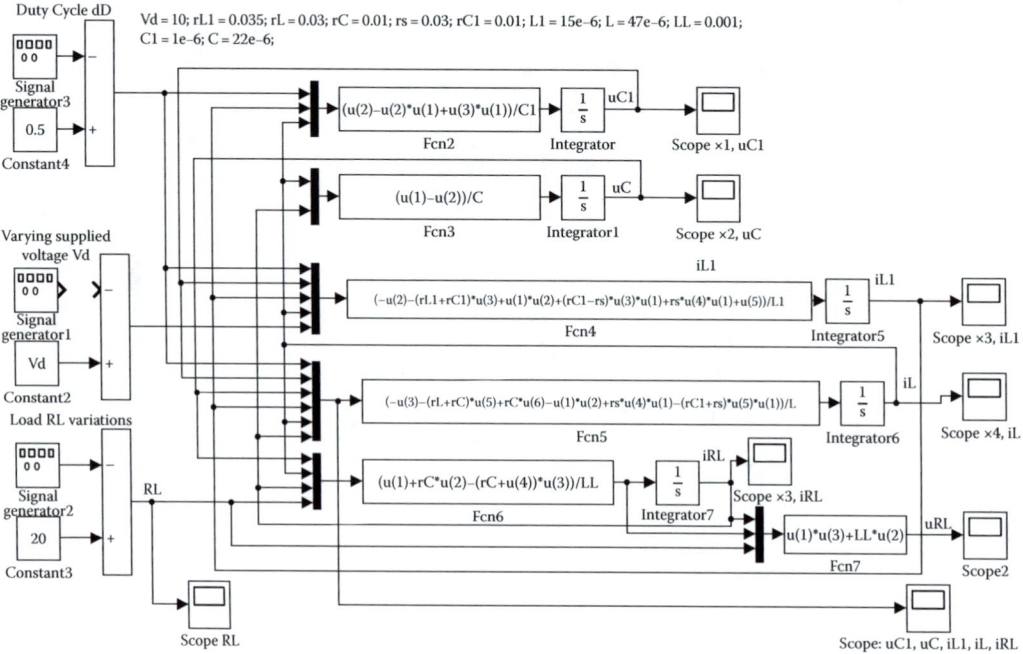

FIGURE 7.32 Simulink® model for open-loop Cuk converter.

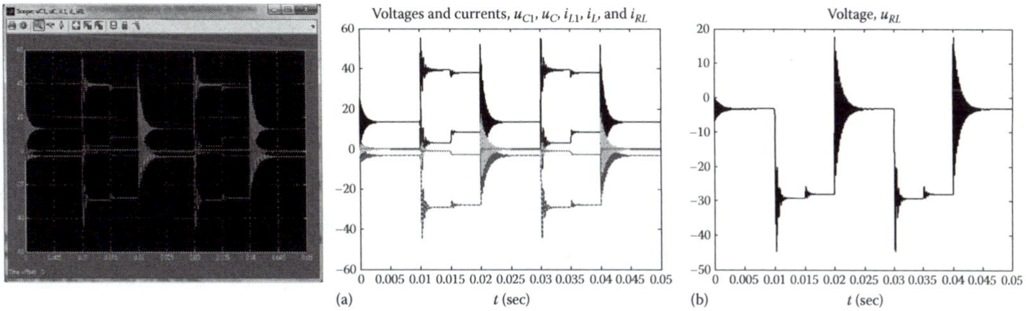

FIGURE 7.33 Dynamics of u_{C1} (solid line), u_C (dashed line), i_{L1} (solid line), i_L (dashed line), and i_{RL} (solid line) if $d_D = \begin{cases} 0.75 \\ 0.25 \end{cases}$ varies with the frequency 50 Hz. The load is $R_L = \begin{cases} 10 \\ 30 \end{cases}$ ohm with $f = 100$ Hz.

7.4.6 Flyback and Forward Converters

To decouple the input and output stages, *flyback* and *forward* converters are used. These DC–DC regulators magnetically isolate the input and output stages using transformers that support the switching schemes. The application of transformers increases the size and cost. The energy is stored in the inductor when the switch is closed. The stored energy is transformed to the load when the switch is open. The *flyback* and *forward* magnetically coupled DC–DC converters are illustrated in Figures 7.34.

In the *flyback* DC–DC converter, when the switch is closed, the diode is reverse biased. If the switch is open, the diode is forward biased. The switch is closed for time $\frac{d_D}{f}$ and open for $\frac{1-d_D}{f}$. For the open switch $i_d = \frac{N_1}{N_2} i_L$. Using the duty ratio d_D, the differential equations can be found.

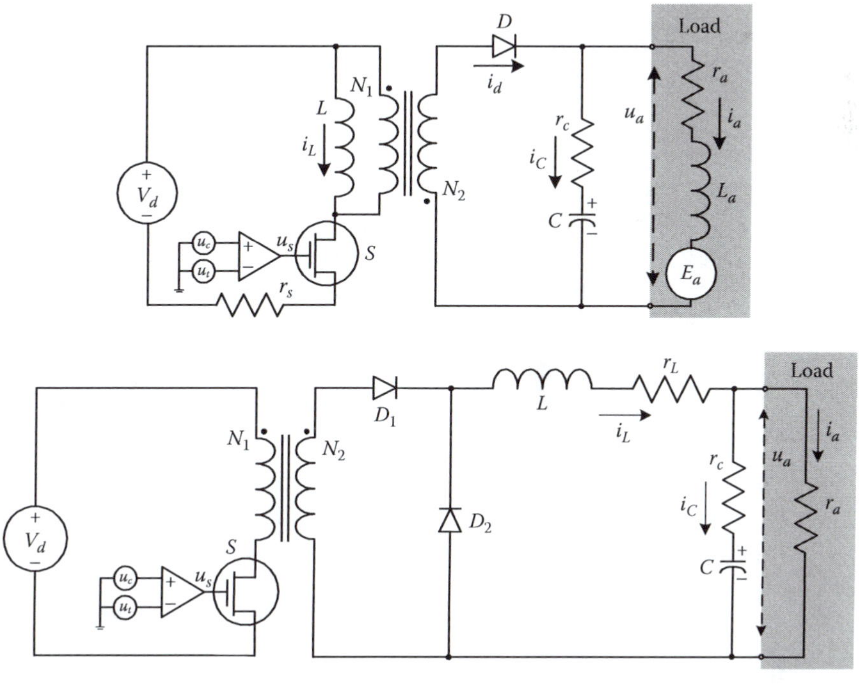

FIGURE 7.34 *Flyback* DC–DC converter and the *forward* DC–DC regulator.

Consider a *forward* DC–DC regulator. The differential equations when the switch is closed are

$$\frac{du_C}{dt}=\frac{1}{C(r_c+r_a)}(-u_C+r_ai_L),\quad \frac{di_L}{dt}=\frac{1}{L}\left(-\frac{r_a}{r_c+r_a}u_C+\left(\frac{r_a^2}{r_c+r_a}-r_L-r_a\right)i_L+\frac{N_2}{N_1}V_d\right).$$

If the switch is open,

$$\frac{du_C}{dt}=\frac{1}{C(r_c+r_a)}(-u_C+r_ai_L),\quad \frac{di_L}{dt}=\frac{1}{L}\left(-\frac{r_a}{r_c+r_a}u_C+\left(\frac{r_a^2}{r_c+r_a}-r_L-r_a\right)i_L\right).$$

The resulting differential equations are

$$\frac{du_C}{dt}=\frac{1}{C(r_c+r_a)}(-u_C+r_ai_L),$$

$$\frac{di_L}{dt}=\frac{1}{L}\left(-\frac{r_a}{r_c+r_a}u_C+\left(\frac{r_a^2}{r_c+r_a}-r_L-r_a\right)i_L+\frac{N_2}{N_1}V_dd_D\right).$$

Example 7.18

Using a set of differential equations derived, we simulate the *forward* converter if $V_d = 50$ V and $d_D = 0.5$. The parameters are $r_L = 0.02$ ohm, $r_C = 0.01$ ohm, $r_a = 3$ ohm, $L = 0.000005$ H, $C = 0.003$ F, and $N_2/N_1 = 1$.

The transient dynamics for the state variables $u_C(t)$ and $i_L(t)$ are documented in Figures 7.35. The settling time is 0.0015 sec. The steady-state voltage, applied to the load terminal is 25 V. The three-dimensional plot for $u_C(t)$, $i_L(t)$, and time t is depicted in the last plot in Figures 7.35. ■

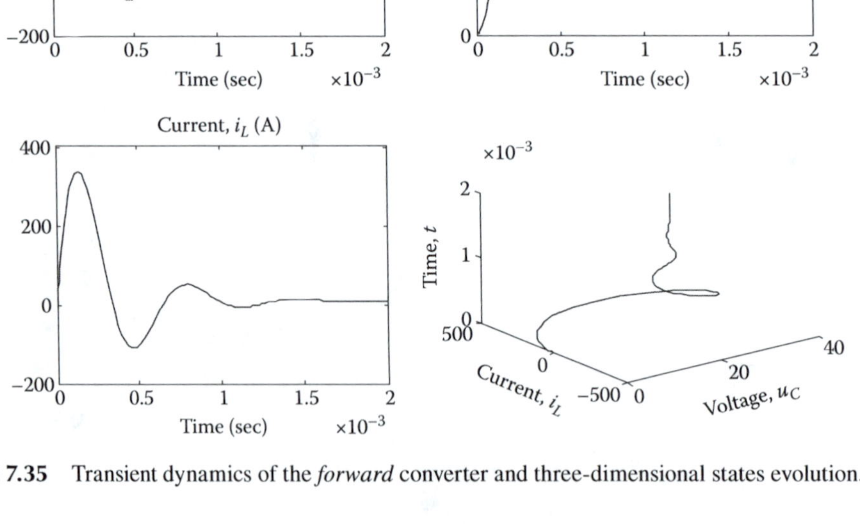

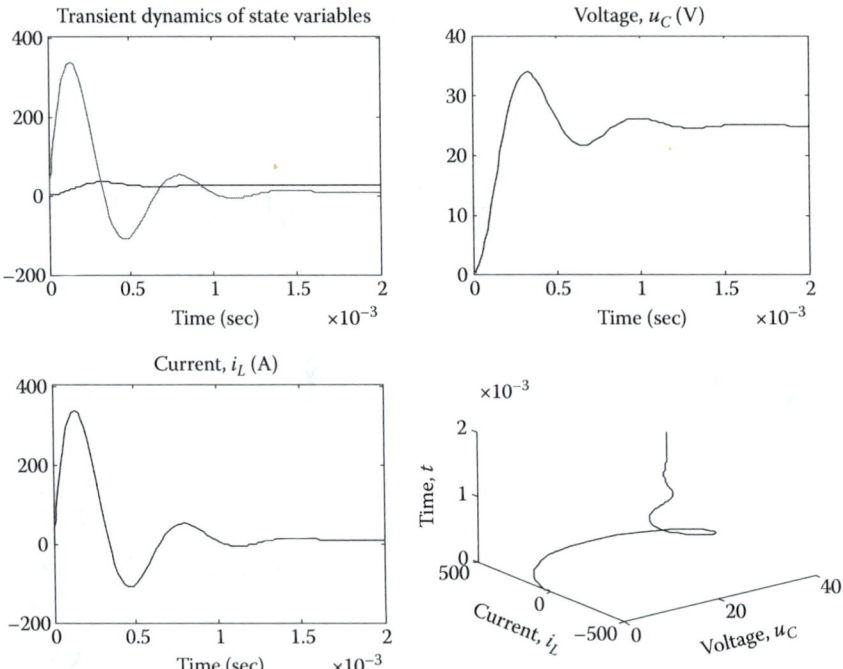

FIGURE 7.35 Transient dynamics of the *forward* converter and three-dimensional states evolution.

7.4.7 Resonant and Switching Converters

The developments of advanced switching converters focus on topology design, nonlinear analysis, optimization, and control. To attain high efficiency and power density, new topologies were developed. Nonlinear analysis and design must be performed to guarantee specifications and requirements imposed. To enable converter performance (minimize settling time, ensure stability and robustness, minimize losses, etc.), resonant converters are studied. The zero-voltage and zero-current switching are enabling concepts that can improve power density, efficiency, reliability, switching frequency, and other performance characteristics. Advanced converter topologies and control algorithms are proposed. Nonlinear analysis, optimization, and control problems should be solved to improve the steady-state and dynamic characteristics. A variety of resonant converter topologies and filters are used. A resonant converter is shown in Figure 7.36a.

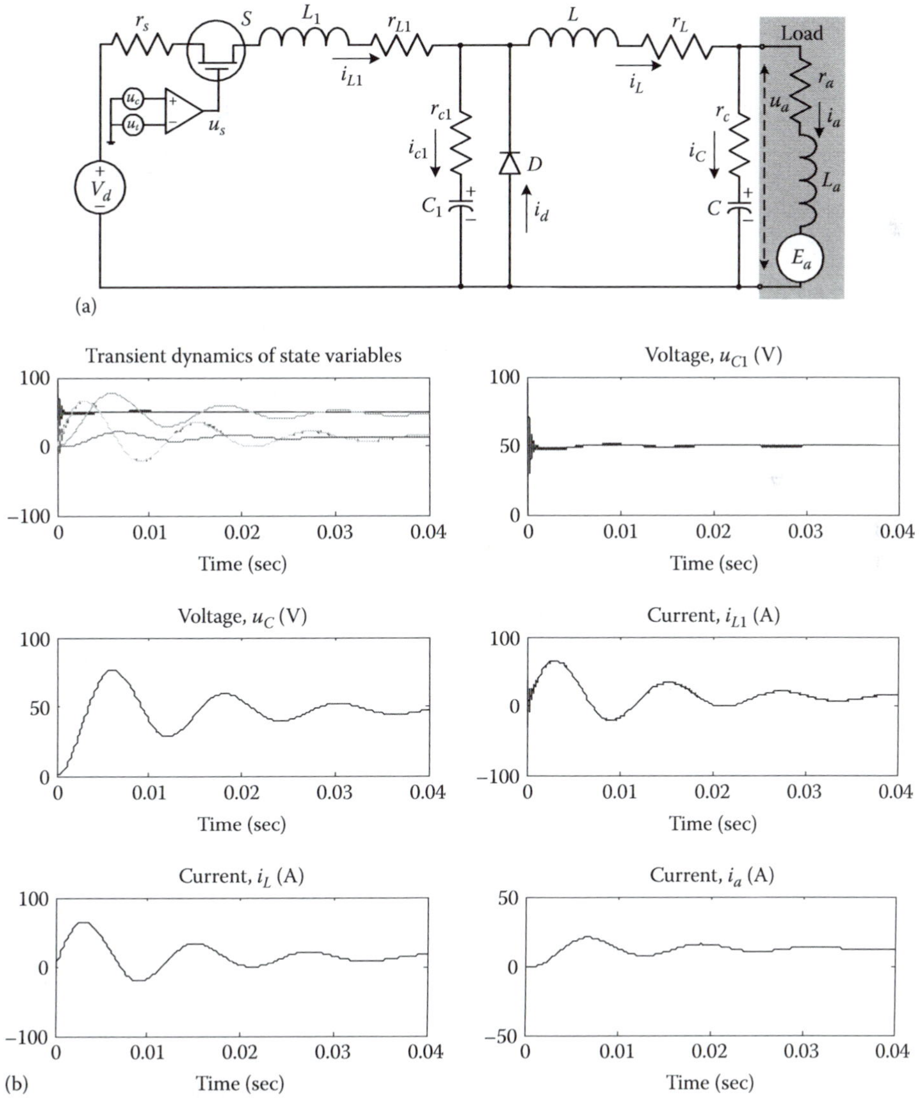

FIGURE 7.36 (a) Resonant converter with zero-current switching; (b) Transient dynamics of the resonant converter.

The output voltage at the load terminal (formed by a resistor r_a and inductor L_a) is regulated by controlling the switching *on* and *off* durations. When the switch is open, the diode D is forward biased to carry the output inductor current i_L. The voltage across the capacitor C_1 is zero. When the switch is closed, the diode remains forward biased while $i_{L1} < i_L$. As i_{L1} reaches i_L, the diode turns *off*. Hence, the switch turns *off* and *on* at zero-current, and i_{L1} is zero. A comparator implements the PWM switching. The switching signal u_s, which drives the switch, is generated by comparing a signal-level control voltage u_c with a repetitive triangular signal u_t. In resonant converters, the frequency may be controlled to regulate the output voltage. A set of differential equations to describe the resonant converter dynamics is

$$\frac{du_{C1}}{dt} = \frac{1}{C_1}\left(i_{L1} - i_L\right)d_D,$$

$$\frac{du_C}{dt} = \frac{1}{C}\left(i_L - i_a\right),$$

$$\frac{di_{L1}}{dt} = \frac{1}{L_1}\left[-u_{C1} - \left(r_s + r_{L1} + r_{c1}\right)i_{L1} + r_{c1}i_L + V_d\right]d_D,$$

$$\frac{di_L}{dt} = \frac{1}{L}\left[u_{C1} - u_C + r_{c1}i_{L1} - \left(r_{c1} + r_L + r_c\right)i_L + r_c i_a\right]d_D,$$

$$\frac{di_a}{dt} = \frac{1}{L_a}\left(u_C + r_c i_L - \left(r_c + r_a\right)i_a - E_a\right).$$

A nonlinear mathematical model results due to the multiplication of state variables $u_{C1}(t)$, $u_C(t)$, $i_{L1}(t)$, $i_L(t)$, $i_a(t)$ and the control variable d_D. We have the following nonlinear state-space model

$$\begin{bmatrix} \frac{du_{C1}}{dt} \\ \frac{du_C}{dt} \\ \frac{di_{L1}}{dt} \\ \frac{di_L}{dt} \\ \frac{di_a}{dt} \end{bmatrix} = \begin{bmatrix} 0 & 0 & 0 & 0 & 0 \\ 0 & 0 & 0 & \frac{1}{C} & -\frac{1}{C} \\ 0 & 0 & 0 & 0 & 0 \\ 0 & 0 & 0 & 0 & 0 \\ 0 & \frac{1}{L_a} & 0 & \frac{r_c}{L_a} & -\frac{r_c + r_a}{L_a} \end{bmatrix} \begin{bmatrix} u_{C1} \\ u_C \\ i_{L1} \\ i_L \\ i_a \end{bmatrix} + \begin{bmatrix} \frac{1}{C_1}\left(i_{L1} - i_L\right) \\ 0 \\ \frac{1}{L_1}\left(-u_{C1} - \left(r_s + r_{L1} + r_{c1}\right)i_{L1} + r_{c1}i_L + V_d\right) \\ \frac{1}{L}\left(u_{C1} - u_C + r_{c1}i_{L1} - \left(r_{c1} + r_L + r_c\right)i_L + r_c i_a\right) \\ 0 \end{bmatrix} d_D - \begin{bmatrix} 0 \\ 0 \\ 0 \\ 0 \\ \frac{1}{L_a}E_a \end{bmatrix}.$$

Example 7.19

We perform simulations and examine the resonant converter illustrated in Figure 7.36a. Let $V_d = 50$ V, $E_a = 10$ V, and $d_D = 0.5$. The parameters are $r_s = 0.025$ ohm, $r_{L1} = 0.01$ ohm, $r_{c1} = 0.04$ ohm, $r_L = 0.02$ ohm, $r_c = 0.02$ ohm, $L_1 = 0.000005$ H, $L = 0.0007$ H, $C_1 = 0.000003$ F, and $C = 0.003$ F. The *RL* load is formed by $r_a = 3$ ohm and $L_a = 0.005$ H. Using the derived set of differential equations, the simulations are performed. The transient responses for the state variables $u_{C1}(t)$, $u_C(t)$, $i_{L1}(t)$, $i_L(t)$, and $i_a(t)$ are plotted in Figure 7.36b. ■

There are a great variety of high-performance PWM switching and resonant converters. We examined one-quadrant converters, and reported two- and four-quadrant regulators. In electromechanical systems, high-performance PWM regulators are used. As the converter topology, filters, and controllers are found, the corresponding data-intensive analysis is performed by addressing, solving, and substantiating findings in analysis and design. For various converters, nonlinear analyses and design tasks were performed.

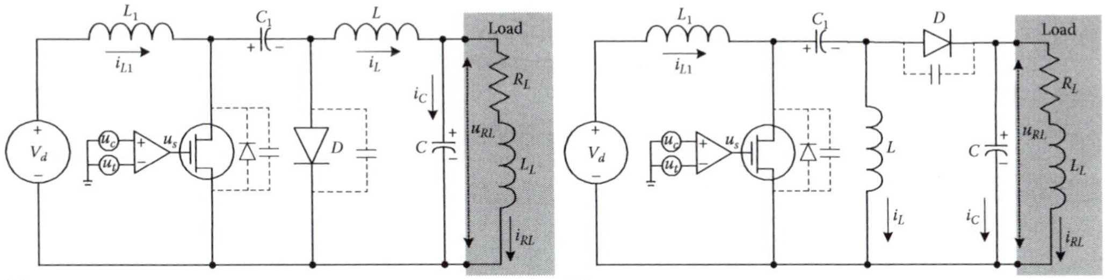

FIGURE 7.37 (a) PWM *soft*-switching Cuk converter; (b) PWM *soft*-switching SEPIC converter.

Consider a high-performance DC–DC inverting Cuk and single-ended primary-inductor converter (SEPIC) converters with output filters, see Figures 7.37. These PWM regulators accomplish inductive-capacitive energy conversion and transfer. The inverted *buck-boost* regulator topologies ensure high efficiency, minimal losses, and overloading capabilities due to the use of high-performance MOSFETs, high-Q ceramic capacitors, and low-loss inductors. Depending on applications and operating envelopes, the input voltage $V_d(t)$ can be time-invariant or time-varying. The output voltage at the load terminal u_{RL} should be stabilized at the specified value. Alternatively, the voltage u_{RL} may be regulated in the operating envelope defined by the reference command $r(t)$, load, etc. Stability, robustness, time-optimal dynamics, minimal dynamic error, and zero steady-state error must be guaranteed. With the specified ripple, the output voltage u_{RL} is applied to time-varying *RLC* or *RL* loads. If the MOSFET is *on* or *off*, it is desirable that the currents in the input and output inductors L_1 and L be smooth. Depending on the application, specified voltage, and current, the MOSFET switching frequency $f = 1/(t_{on} + t_{off})$ may vary from ~100 kHz to 10 GHz. In Cuk and SEPIC regulators, illustrated in Figures 7.37, the output voltage u_{RL} is varied by changing the comparator control voltage u_c.

The switching signal u_s drives the MOSFET driver or MOSFET. The passive *soft*-switching schemes may be used. *Soft*-switching converters use: (1) Zero-current inductors L_{ZC}, which enable zero-current turn-on of active switches *S*; (2) Zero-voltage capacitors C_{ZV}, which ensure zero-voltage turn-off of switches; (3) Snubber inductors; (4) Voltage-storage capacitors; (5) Diodes. In *soft*-switching converters, clamp circuit ensures MOSFET operation at near-zero-voltage switching turn-on. This enables operation at a high frequency, thereby reducing the size of the components. The clamp capacitor–diode circuit is connected in parallel with the switch. The MOSFET turns on at the near-zero voltage.

HOMEWORK PROBLEMS

7.1 Using operational amplifiers, develop the schematics to implement the analog proportional, PI, and PID controllers. Report at least two possible schematics for PI and PID controllers. Derive the transfer functions $G(s)$ and express the feedback gains k_p and k_i using the circuitry parameters (resistors and capacitors).

7.2 Explain why converters and power amplifiers should be used in electromechanical systems.

7.3 Study the PWM switching concept applying the sinusoidal-like (not triangular u_t) signal to the comparator. The switching signal u_s, which drives the switch *S*, is generated by comparing a signal-level control voltage u_c with a repetitive sinusoidal signal u_{sin}. Propose the schematics to generate u_{sin} (various oscillators schemes and circuitries can be used). Report how to define and vary (if needed) the frequency of u_{sin}. Report the voltage waveforms.

7.4 Report the four-quadrant H-configured power stage schematics. Explain how it operates.

REFERENCES

1. A. S. Sedra and K. C. Smith, *Microelectronic Circuits*, Oxford University Press, New York, 2014.
2. S. E. Lyshevski, *Electromechanical Systems, Electric Machines, and Applied Mechatronics*, CRC Press, Boca Raton, FL, 1999.
3. D. W. Hart, *Power Electronics*, McGraw-Hill, New York, 2010.
4. J. G. Kassakian, M. F. Schlecht, and G. C. Verghese, *Principles of Power Electronics*, Addison-Wesley Publishing Company, Reading, MA, 1991.
5. N. Mohan, T. M. Undeland, and W. P. Robbins, *Power Electronics: Converters, Applications, and Design*, John Wiley & Sons, New York, 2002.

8 Control of Electromechanical Systems

8.1 INTRODUCTION TO CONTROL AND OPTIMIZATION

Optimized and controlled electromechanical systems should ensure best *achievable* performance and capabilities, such as functionality, stability, robustness, and disturbance rejection. Consistent control laws should be designed and implemented as analog or digital controllers. One designs closed-loop systems to optimize dynamics and guarantee stability. Optimal control laws should be consistent with device physics and hardware. One applies stability criteria, assigns performance goals, and solves the optimization problems by minimizing performance functionals. We document various design methods. Some methods result in control laws that assume that all state variables are measurable or observable. Many variables cannot be measured. The *minimal-complexity* control laws are synthesized to: (1) Guarantee near-optimal performance; (2) Minimize system complexity; (3) Ensure hardware consistency and simplicity.

There are a large variety of subsystems, components, and devices in electromechanical systems. Due to the fast dynamics of microelectronic components and sensors, the overall behavior of electromechanical systems is predefined by the dynamics of electromechanical motion devices with the attached kinematics. In Chapters 2 through 7, we found consistent mathematical models in the form of nonlinear differential equations. These models will be used to solve control problems.

Closed-loop systems are designed to ensure the best performance as measured against a spectrum of specifications and requirements. The hardware solutions predefine system performance and capabilities, while control laws affect the system performance and capabilities. For example, for a high-performance ~30 kW electric drive in automotive application, a permanent-magnet synchronous motor with a *soft-switching* PWM amplifier ensures the best performance. This electric drive is open-loop stable. The tracking control is designed to guarantee cruise and tracking control. Inadequate designs may lead to control laws that may destabilize the stable system. Maximum efficiency, stability, robustness, accuracy, and disturbance attenuation are obvious criteria among other requirements. Abstract problem formulation and minimization of settling time may result in high feedback gains or discontinuous relay-type control. The chattering phenomena, oscillatory dynamics, losses, low efficiency, and other unacceptable phenomena result. Hence, high-gain, discontinuous, and other theoretical control laws may be inadequate. Practical control methods result in *minimal-complexity* hardware and near-optimal overall performance. The specifications imposed on closed-loop systems are examined. The criteria under the consideration can be:

- Microelectronics, power electronics, and electromagnetics control consistency and efficiency. Control laws should be designed using device physics examining energy conversion, torque production, etc.;
- Stability with the desired stability margins in the full operating envelope;
- Robustness to parameter variations, structural, and kinematic changes;
- Tracking accuracy, minimal time dynamic, and acceptable steady-state errors;
- Disturbance, perturbations, and load and noise attenuation;
- Transient response specifications, such as settling times, overshoot, etc.

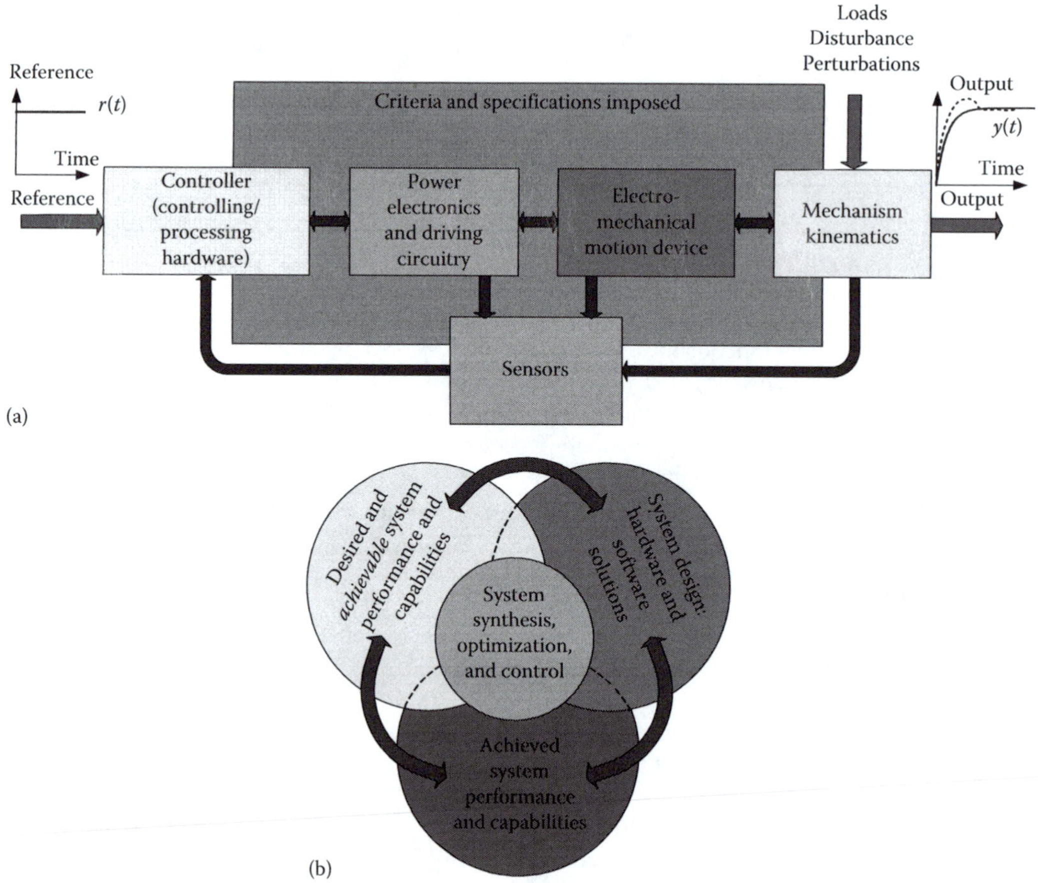

FIGURE 8.1 (a) Electromechanical system with reference (command) input $r(t)$ and output $y(t)$; (b) Synthesis taxonomy to ensure optimal performance.

The system performance is measured and assessed against multiple criteria, such as stability, robustness, disturbance rejection, and steady-state error. The requirements and specifications are defined in the full operating envelope. The performance characteristics can be assigned and assessed using criteria and metrics given by the performance functionals, performance indexes, equalities, inequalities, etc. The system can be controlled by using different control laws $u(t)$, which are found by applying performance and stability criteria. The electromechanical system with the reference $r(t)$ and output $y(t)$ is illustrated in Figure 8.1a. We control and optimize dynamics ensuring adequate steady-state characteristics. The structural design is performed by finding system organization, using advanced hardware, such as motion devices, sensors, power electronics, digital signal processors (DSPs), integrated circuits (ICs), etc. By using the synthesis taxonomy reported in Figure 8.1b, the optimal and near-optimal capabilities should be assured.

The high-level functional diagram and a possible closed-loop system organization are reported in Figure 8.2. Control laws must be designed, examined, substantiated, and implemented. To implement analog and digital control laws, analog microelectronics, microcontrollers, and DSPs are used.

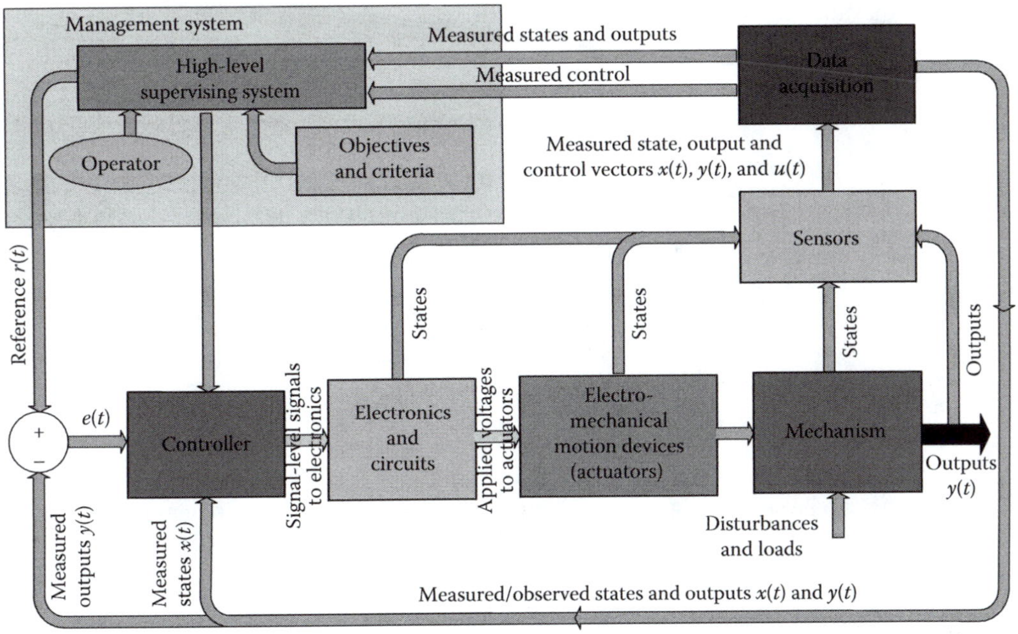

FIGURE 8.2 High-level functional diagram of closed-loop electromechanical systems.

Steady-state and dynamic behavior can be optimized using different methods [1–5]. The frequency- and s-domain methods (Laplace and Fourier transforms, transfer function, etc.) are applicable to linear systems. The majority of electromechanical systems are nonlinear. Using the laws of physics, nonlinear differential equations are derived using state, output, control, reference, disturbance, and other variables. Linear, nonlinear, and bounded proportional-integral-derivative (PID) control laws will be designed and examined. To implement PID control laws, the directly measured tracking error $e(t)$ is used, $e(t) = r(t) - y(t)$. System performance and capabilities, which are largely defined by hardware, can be improved by control laws. The departure from PID control laws may result in additional hardware components, such as sensors, microelectronics, etc. If the state variables $x(t)$ are used to implement controllers, these $x(t)$ must be measured or observed. For some systems, one may achieve a substantial performance improvement using advanced control laws. For other systems, PID control laws may guarantee *near-optimal* performance ensuring hardware and software simplicity. For many open-loop stable and unstable systems, PID controllers have been successfully used for centuries. Due to inherent physical limits and system complexity, complex control laws may ensure a quite moderate improvement. The advanced methods in design of control laws are applied if strengthened specifications are imposed and if these specifications are supported by hardware. For multi-objective problems (high accuracy, efficiency maximization, disturbance attenuation, vibration, and noise minimization, etc.), advanced concepts are applied. Advanced control laws are used in high-accuracy pointing systems, multi-degree-of-freedom robots (manipulators), high-precision positioning systems, sound and audio systems, high-performance drives, etc.

Example 8.1: System Performance and Evaluation Using Performance Functionals

Control laws can be designed by evaluating system performance using performance functionals and indexes, which represent and assess specifications and requirements. These performance functionals, which must be physics-consistent, predefine control laws u, which must be implemented by controllers.

By using the tracking error $e(t)=r(t)-y(t)$ and settling time, the performance functional can be expressed as $J=\min_{e}\int_0^{\infty}|e|\,dt$, $J=\min_{e}\int_0^{\infty}e^2\,dt$, $J=\min_{e}\int_0^{\infty}\left(|e|+e^2+e^6\right)dt$, $J=\min_{t,e}\int_0^{\infty}t|e|\,dt$, etc. If other performance requirements are imposed, such as electric losses P_{loss} and control energy E_u, recalling that $P_{loss}\equiv i^2$ and $E_u\equiv u^2$, one may use $J=\min_{t,e,i,u}\int_0^{\infty}(t|e|+e^2+i^2+u^2)dt$. We use various performance quantities, such as control efforts, state transient evolutions, etc. The control efforts can be assessed using the positive-definite integrands u^{2n} (n = 1, 2, 3,…) or $|u|$. The control rate is evaluated as $(du/dt)^{2n}$ or $|du/dt|$. One uses the physics-consistent and device-specific specifications. For example, the torque ripple can be assessed by ΔT_e^2 or $|\Delta T_e|$, $\Delta T_e=(T_e-T_{e\,average})$. The analysis of the torque ripple leads to quantitative assessment of efficiency, heating, vibration, noise, wearing, etc. One may use $J=\min_{e,\Delta T_e,u}\int_0^{\infty}(e^2+\Delta T_e^2+u^2)dt$.

The *first principles* of electromechanical devices and control methods must be consistently applied. The variables x, e, and u are used in the performance functionals to mathematically define and assess the dissipated energy, torque ripple, and other criteria. ■

A great number of specifications are imposed. Stability, efficiency, robustness, accuracy, and settling time are usually prioritized because these measures affect safety, effectiveness, value, commercialization, etc. Consider the output transient dynamics and two evolution envelopes. The output response is illustrated in Figure 8.3 for the step reference $r(t)$ = const. The system dynamics is stable because output is bounded and converges to the steady-state value $y_{steady\text{-}state}$. That is, $\lim_{t\to\infty}y(t)=y_{steady\text{-}state}$ and $\lim_{t\to\infty}e(t)\to 0$. The output dynamics $y(t)$ is studied within the evolution envelopes. Two evolution envelopes (I and II) can be assigned specifying the desired accuracy, settling time, and overshoot as shown in Figure 8.3. The systems dynamics is within the *achievable* evolution envelope. The *desired* evolution envelope I may or may be achieved and the specifications imposed may not be met. Due to the limits (maximum torque, force, power, voltage, current, acceleration, etc.), the designer may not be able to achieve the *desired* performance while guaranteeing the best *achievable* performance. The system hardware redesign may be needed if necessary, and, the solutions may or may not exist. The hardware physical limits may not be surpassed and overcome by software or control solutions. The designer must be aware on hardware physical limits, device specificity, technological constraints, affordability, and other factors.

The settling time is the time needed for the system output $y(t)$ to reach and stay within the steady-state value $y_{steady\text{-}state}$. The allowable difference between $y(t)$ and $y_{steady\text{-}state}$ is used to specify the settling time. This difference Δy may vary from ~5% to 0.1% or less. For example, in high-accuracy pointing systems, the required repositioning accuracy can be milliradians for the radian-range repositioning. The settling time is the minimum time after which the system response remains within the desired accuracy, taking into account the steady-state value $y_{steady\text{-}state}$ and command $r(t)$. The maximum overshoot is the difference between the maximum peak value of the systems output $y(t)$ and the steady-state $y_{steady\text{-}state}$ divided by $y_{steady\text{-}state}$, e.g.,

$$\text{Overshoot}=\frac{y_{\max}-y_{steady\text{-}state}}{y_{steady\text{-}state}}\times 100\%.$$

As depicted in Figure 8.3, the peak time is the time required for the system output $y(t)$ to reach the first peak of the overshoot.

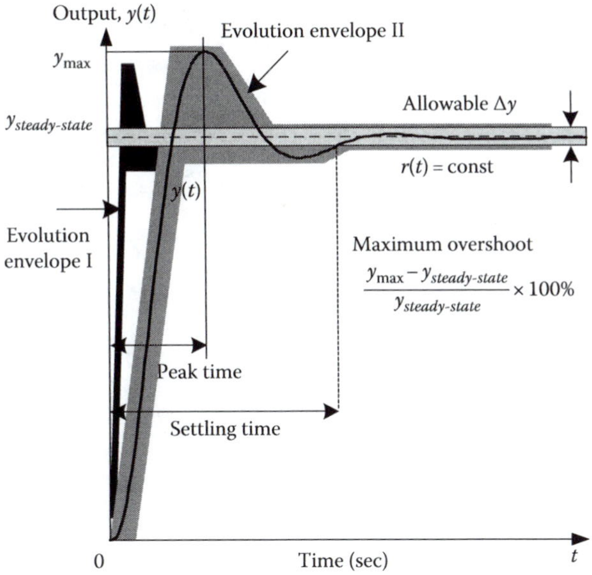

FIGURE 8.3 Output transient response and evolution envelopes.

8.2 STATE-SPACE EQUATIONS OF MOTION AND TRANSFER FUNCTIONS

Device physics and dynamics of electromechanical motion devices, amplifiers, converters, and other components are modeled by nonlinear differential equations. For a few electromechanical devices, under assumptions and simplifications, linear differential equations were obtained from which transfer functions can be found. Linear systems can be described in the s and z domains using transfer functions $G_{sys}(s)$ and $G_{sys}(z)$. The Laplace operator $s = d/dt$ and Laplace transforms are used for continuous-time systems. The transfer functions can be used to design control laws. For nonlinear systems, one cannot effectively apply linear control theory. However, many nonlinear electromechanical systems can be controlled using PID control laws that ensure adequate near-optimal performance.

For linear systems in the state-space form, we use n states $x \in \mathbb{R}^n$ and m controls $u \in \mathbb{R}^m$. The transient dynamics of linear systems is described by a set of n linear first-order differential equations

$$\frac{dx_1}{dt} = a_{11}x_1 + a_{12}x_2 + \cdots + a_{1n-1}x_{n-1} + a_{1n}x_n + b_{11}u_1 + b_{12}u_2 + \cdots + b_{1m-1}u_{m-1} + b_{1m}u_m,\ x_1(t_0) = x_{10},$$

$$\vdots$$

$$\frac{dx_n}{dt} = a_{n1}x_1 + a_{n2}x_2 + \cdots + a_{nn-1}x_{n-1} + a_{nn}x_n + b_{n1}u_1 + b_{n2}u_2 + \cdots + b_{nm-1}u_{m-1} + b_{nm}u_m,\ x_n(t_0) = x_{n0}.$$

In the matrix form, we have

$$\frac{dx}{dt}=\begin{bmatrix}\frac{dx_1}{dt}\\ \frac{dx_2}{dt}\\ \vdots\\ \frac{dx_{n-1}}{dt}\\ \frac{dx_n}{dt}\end{bmatrix}=\begin{bmatrix}a_{11} & a_{12} & \cdots & a_{1n-1} & a_{1n}\\ a_{21} & a_{22} & \cdots & a_{2n-1} & a_{2n}\\ \vdots & \vdots & \ddots & \vdots & \vdots\\ a_{n-1\,1} & a_{n-1\,2} & \cdots & a_{n-1\,n-1} & a_{n-1\,n}\\ a_{n\,1} & a_{n\,2} & \cdots & a_{nn-1} & a_{nn}\end{bmatrix}\begin{bmatrix}x_1\\ x_2\\ \vdots\\ x_{n-1}\\ x_n\end{bmatrix}$$

$$+\begin{bmatrix}b_{11} & b_{12} & \cdots & b_{1m-1} & b_{1m}\\ b_{21} & b_{22} & \cdots & b_{2m-1} & b_{2m}\\ \vdots & \vdots & \ddots & \vdots & \vdots\\ b_{n-1\,1} & b_{n-1\,2} & \cdots & b_{n-1\,m-1} & b_{n-1\,m}\\ b_{n1} & b_{n2} & \cdots & b_{nm-1} & b_{nm}\end{bmatrix}\begin{bmatrix}u_1\\ u_2\\ \vdots\\ u_{m-1}\\ u_m\end{bmatrix}$$

$$= Ax+Bu,\quad x(t_0)=x_0.$$

If system parameters are constant, matrices $A \in \mathbb{R}^{n\times n}$ and $B \in \mathbb{R}^{n\times m}$ are constant-coefficients. One finds the characteristic equation using the determinant

$$\det\left(sI-A\right)=0 \quad \text{or} \quad \left|sI-A\right|=0,$$

where $I \in \mathbb{R}^{n\times n}$ is the identity matrix.

For an $n \times n$ matrix A, the characteristic polynomial of A, denoted by $p_A(s)$, is found using the determinant, $p_A(s) = \det(sI - A)$. From the characteristic equation $\det(sI - A) = \left|sI - A\right| = 0$ or $a_n s^n + a_{n-1}s^{n-1} + \cdots + a_1 s + a_0 = 0$, we obtain the eigenvalues, which are also called the characteristic roots or characteristic poles. The system is stable if the real parts of all eigenvalues are negative. The stability analysis using the eigenvalues is valid only for linear systems.

The transfer function can be found using the state-space equations. Consider the linear time-invariant systems

$$\frac{dx}{dt}=Ax+Bu,\quad y=Hx.$$

The output vector $y \in Y \subset \mathbb{R}^b$ is expressed using the output equation $y = Hx$, where $H \in \mathbb{R}^{b\times n}$ is the matrix of the constant coefficients. The Laplace transform for the state-space model $\dot{x} = Ax + Bu$, $y = Hx$, yields

$$sX(s)-x(t_0)=AX(s)+BU(s),\quad Y(s)=HX(s).$$

Assuming that the initial conditions are zero, we have $X(s) = (sI–A)^{-1}BU(s)$.

Matrix $A \in \mathbb{R}^{n\times n}$ is invertible if 0 is not an eigenvalue of A, and, $\det A \neq 0$.

Using the system output $y(t)$, $Y(s) = HX(s) = H(sI - A)^{-1}BU(s)$.

The transfer function is

$$G(s)=\frac{Y(s)}{U(s)}=H\left(sI-A\right)^{-1}B=H\Phi(s)B,\quad \Phi\left(s\right)=\left(sI-A\right)^{-1},$$

where $\Phi(s)$ is the state transition matrix, $\Phi(s) = (sI - A)^{-1}$ which yields

$$\Phi(t)=L^{-1}[(sI-A)^{-1}],\ \Phi(t)=e^{At},\ \Phi(t)=e^{At}=I+At+\frac{1}{2!}A^2t^2+\frac{1}{3!}A^3t^3+\cdots.$$

Assuming that the initial conditions are zero, one may apply the Laplace transform to both sides of the n-order differential equation

$$\sum_{i=0}^{n} a_i \frac{d^i y(t)}{dt^i} = \sum_{i=0}^{m} b_i \frac{d^i u(t)}{dt^i}.$$

From

$$\left(\sum_{i=0}^{n} a_i s^i\right) Y(s) = \left(\sum_{i=0}^{m} b_i s^i\right) U(s),$$

We have

$$G(s) = \frac{Y(s)}{U(s)} = \frac{b_m s^m + b_{m-1} s^{m-1} + \cdots + b_1 s + b_0}{a_n s^n + a_{n-1} s^{n-1} + \cdots + a_1 s + a_0}.$$

The stability of linear time-invariant systems is guaranteed if all eigenvalues, obtained by solving the characteristic equation $a_n s^n + a_{n-1} s^{n-1} + \cdots + a_1 s + a_0 = 0$ have negative real parts.

Using the state transition matrix $\Phi(t) = e^{At}$, the solution due to the initial conditions x_0 and input u are

$$x(t) = e^{A(t-t_0)} x_0 = \Phi(t - t_0) x_0$$

and $$x(t) = e^{A(t-t_0)} x_0 + \int_{t_0}^{t} e^{A(t-\tau)} Bu(\tau) d\tau, \quad t > t_0.$$

The output dynamics is $y(t) = He^{A(t-t_0)} x_0 + H \int_{t_0}^{t} e^{A(t-\tau)} Bu(\tau) d\tau.$

Example 8.2

Consider an nth-order linear system described by

$$\frac{d^n y}{dt^n} + a_{n-1} \frac{d^{n-1} y}{dt^{n-1}} + \cdots + a_1 \frac{dy}{dt} + a_0 y = u.$$

The state-space model is found defining the state variables as

$$x_1 = y,\ x_2 = dy/dt, \ldots,\ x_n = d^{n-1}y/dt^{n-1}.$$

Hence, $\frac{dx_1}{dt} = x_2, \frac{dx_2}{dt} = x_3, \cdots, \frac{dx_n}{dt} = -a_0 x_1 - a_1 x_2 - \cdots - a_{n-1} x_n + u.$

The state-space equation is

$$\frac{dx}{dt} = \begin{bmatrix} \frac{dx_1}{dt} \\ \frac{dx_2}{dt} \\ \vdots \\ \frac{dx_{n-1}}{dt} \\ \frac{dx_n}{dt} \end{bmatrix} = \begin{bmatrix} 0 & 1 & 0 & \cdots & 0 & 0 \\ 0 & 0 & 1 & \cdots & 0 & 0 \\ \vdots & \vdots & \vdots & \ddots & \vdots & \vdots \\ 0 & 0 & 0 & \cdots & 0 & 1 \\ -a_0 & -a_1 & -a_2 & \cdots & -a_{n-2} & -a_{n-1} \end{bmatrix} \begin{bmatrix} x_1 \\ x_2 \\ \vdots \\ x_{n-1} \\ x_n \end{bmatrix} + \begin{bmatrix} 0 \\ 0 \\ 0 \\ 0 \\ 1 \end{bmatrix} u = Ax + Bu.$$

From $x_1 = y$, the output equation is $y = \begin{bmatrix} 1 & 0 & \cdots & 0 & 0 \end{bmatrix} x = Hx$, $H = \begin{bmatrix} 1 & 0 & \cdots 0 & 0 \end{bmatrix}$. ■

Example 8.3

Consider a system

$$\frac{dx}{dt}=\begin{bmatrix}\frac{dx_1}{dt}\\ \frac{dx_2}{dt}\end{bmatrix}=\begin{bmatrix}0 & 1\\ -2 & -3\end{bmatrix}\begin{bmatrix}x_1\\ x_2\end{bmatrix}+\begin{bmatrix}0\\ 1\end{bmatrix}u=Ax+Bu.$$

The state transition matrix is

$$\Phi(s)=\left(sI-A\right)^{-1}=\begin{bmatrix}s & -1\\ 2 & s+3\end{bmatrix}^{-1}=\frac{1}{s^2+3s+2}\begin{bmatrix}s+3 & 1\\ -2 & s\end{bmatrix}.$$

Here, $A^{-1}=\begin{bmatrix}a_{11} & a_{12}\\ a_{21} & a_{22}\end{bmatrix}^{-1}=\frac{1}{a_{11}a_{22}-a_{12}a_{21}}\begin{bmatrix}a_{22} & -a_{12}\\ -a_{21} & a_{11}\end{bmatrix}.$

The characteristic equation

$$|sI-A|=0,\ s^2+3s+2=0$$

yields two eigenvalues $s_1=-1$ and $s_2=-2$.

Using the partial fractioning and Laplace transforms, from

$$\Phi(s)=\left(sI-A\right)^{-1}=\begin{bmatrix}\frac{s+3}{s^2+3s+2} & \frac{1}{s^2+3s+2}\\ -\frac{2}{s^2+3s+2} & \frac{s}{s^2+3s+2}\end{bmatrix},$$

one finds $\Phi(t)=e^{At}=\begin{bmatrix}2e^{-t}-e^{-2t} & e^{-t}-e^{-2t}\\ -2e^{-t}+2e^{-2t} & -e^{-t}+2e^{-2t}\end{bmatrix}.$

Using the Symbolic Toolbox, one finds A^{-1} and $\Phi(t)$ as

```
>> A=[0 1; -2 -3]; t=sym('t'); eAt=expm(A*t)
eAt =
[  2*exp(-t) - exp(-2*t),        exp(-t) - exp(-2*t)]
[ 2*exp(-2*t) - 2*exp(-t), 2*exp(-2*t) - exp(-t)]
or
>> A=[0 1; -2 -3]; s=sym('s'); [n,n]=size(A); Ainv=inv(s*eye(n)-A), eAt=ilaplace(Ainv)
Ainv =
[ (s + 3)/(s^2 + 3*s + 2), 1/(s^2 + 3*s + 2)]
[       -2/(s^2 + 3*s + 2), s/(s^2 + 3*s + 2)]
eAt =
[   2*exp(-t) - exp(-2*t),         exp(-t) - exp(-2*t)]
[ 2*exp(-2*t) - 2*exp(-t), 2*exp(-2*t) - exp(-t)]
```

With $y=x_1$, the output equation is $y=Hx+Du=\begin{bmatrix}1 & 0\end{bmatrix}x+\begin{bmatrix}0\end{bmatrix}u$. The transfer function is

$$G(s)=\frac{Y(s)}{U(s)}=H\left(sI-A\right)^{-1}B=H\Phi(s)B=\begin{bmatrix}1 & 0\end{bmatrix}\begin{bmatrix}\frac{s+3}{s^2+3s+2} & \frac{1}{s^2+3s+2}\\ -\frac{2}{s^2+3s+2} & \frac{s}{s^2+3s+2}\end{bmatrix}\begin{bmatrix}0\\ 1\end{bmatrix}=\frac{1}{s^2+3s+2}.$$

The MATLAB® solution is

```
>> A=[0 1; -2 -3]; B=[0; 1]; H=[1 0]; D=0; [num, den]=ss2tf(A,B,H,D)
num =
 0   0   1
den =
 1   3   2
```

■

Electromechanical systems are described by nonlinear differential equations with bounds imposed on state and control variables. In the state-space form, we have

$$\dot{x}(t) = F(x,r,d) + B(x)u, \quad y = H\left(x\right), \quad u_{\min} \le u \le u_{\max}, \quad x\left(t_0\right) = x_0,$$

where $x \in X \subset \mathbb{R}^n$ is the state vector (displacement, position, velocity, current, voltage, etc.) which evolves in X; $u \in U \subset \mathbb{R}^m$ is the bounded control vector (voltage, duty cycle, signal-level voltage to the comparator, etc.) which evolves in U; $r \in R \subset \mathbb{R}^b$ and $y \in Y \subset \mathbb{R}^b$ are the reference and output vectors; $d \in D \subset \mathbb{R}^v$ is the disturbance vector (load, noise and perturbations); $F(\cdot)$: $\mathbb{R}^n \times \mathbb{R}^b \times \mathbb{R}^v \to \mathbb{R}^n$ and $B(\cdot)$: $\mathbb{R}^n \to \mathbb{R}^{n\times m}$ are the nonlinear maps; $H(\cdot)$: $\mathbb{R}^n \to \mathbb{R}^b$ is the nonlinear map defined in the neighborhood of the origin, $H(0) = 0$.

The system output $y(t)$ is a nonlinear function of the state variables $x(t)$ and $y = H(x)$.

The majority of electromechanical motion devices are continuous, and, are described by differential equations. For discrete systems, or, if digital control laws should be designed, one studies discrete systems. The differential equations can be discretized. For n-dimensional state, m-dimensional control, and b-dimensional output vectors, the state-space difference equation is

$$x_{k+1} = \begin{bmatrix} x_{k+1,1} \\ x_{k+1,2} \\ \vdots \\ x_{k+1,n-1} \\ x_{k+1,n} \end{bmatrix} = \begin{bmatrix} a_{k11} & a_{k12} & \cdots & a_{k1n-1} & a_{k1n} \\ a_{k21} & a_{k22} & \cdots & a_{k2n-1} & a_{k2n} \\ \vdots & \vdots & \ddots & \vdots & \vdots \\ a_{kn-11} & a_{kn-12} & \cdots & a_{kn-1\,n-1} & a_{kn-1n} \\ a_{kn1} & a_{kn2} & \cdots & a_{knn-1} & a_{knn} \end{bmatrix} \begin{bmatrix} x_{k1} \\ x_{k2} \\ \vdots \\ x_{kn-1} \\ x_{kn} \end{bmatrix}$$

$$+ \begin{bmatrix} b_{k11} & b_{k12} & \cdots & b_{k1m-1} & b_{k1\,m} \\ b_{k21} & b_{k22} & \cdots & b_{k2m-1} & b_{k2\,m} \\ \vdots & \vdots & \ddots & \vdots & \vdots \\ b_{kn-11} & b_{kn-12} & \cdots & b_{kn-1\,m-1} & b_{kn-1\,m} \\ b_{kn1} & b_{kn2} & \cdots & b_{knm-1} & b_{knm} \end{bmatrix} \begin{bmatrix} u_{k1} \\ u_{k2} \\ \vdots \\ u_{km-1} \\ u_{km} \end{bmatrix} = A_k x_k + B_k u_k, \quad x_{k=k_0} = x_{k0},$$

where $A_k \in \mathbb{R}^{n\times n}$ and $B_k \in \mathbb{R}^{n\times m}$ are the matrices of coefficients.

The output equation is $y_k = H_k x_k$, where $H_k \in \mathbb{R}^{b\times n}$ is the matrix of the constant coefficients.

Consider an n-order linear difference equation

$$\sum_{i=0}^{n} a_i y_{n-i} = \sum_{i=0}^{m} b_i u_{n-i}, \quad n \ge m.$$

Assuming that the coefficients are time-invariant, using the z-transform and letting the initial conditions be zero, we have

$$\left(\sum_{i=0}^{n} a_i z^i \right) Y(z) = \left(\sum_{i=0}^{m} b_i z^i \right) U(z).$$

The transfer function is

$$G(z) = \frac{Y(z)}{U(z)} = \frac{b_m z^m + b_{m-1} z^{m-1} + \cdots + b_1 z + b_0}{a_n z^n + a_{n-1} z^{n-1} + \cdots + a_1 z + a_0}.$$

Nonlinear discrete electromechanical systems are described using nonlinear difference equations

$$x_{k+1} = F(x_k, r_k, d_k) + B(x_k)u_k, \quad y_k = H(x_k), \quad u_{k\min} \le u \le u_{k\max}.$$

8.3 ANALOG AND DIGITAL PROPORTIONAL–INTEGRAL–DERIVATIVE CONTROL

8.3.1 Analog Proportional–Integral–Derivative Control Laws

The majority of electromechanical motion devices, solid-state devices, and power electronics are continuous. Simple and effective proportional–integral–derivative (PID) control laws have been used for centuries. The classical analog PID control law is

$$u(t) = \underbrace{k_p e(t)}_{\text{Proportional Feedback}} + \underbrace{k_i \int e(t)dt}_{\text{Integral Feedback}} + \underbrace{k_d \frac{de(t)}{dt}}_{\text{Derivative Feedback}}, \quad e(t) = r(t) - y(t), \tag{8.1}$$

where $e(t)$ is the error between the reference and the system output, $e(t) = r(t) - y(t)$; k_p, k_i, and k_d are the proportional, integral, and derivative feedback gains, $k_p > 0$, $k_i > 0$, and $k_d > 0$.

The diagram of the analog PID control law (8.1) is shown in Figure 8.4. The Laplace operator $s = d/dt$ is used, yielding

$$U(s) = \left(k_p + \frac{k_i}{s} + k_d s\right)E(s).$$

The transfer function is

$$G_{PID}(s) = \frac{U(s)}{E(s)} = \frac{k_d s^2 + k_p s + k_i}{s}.$$

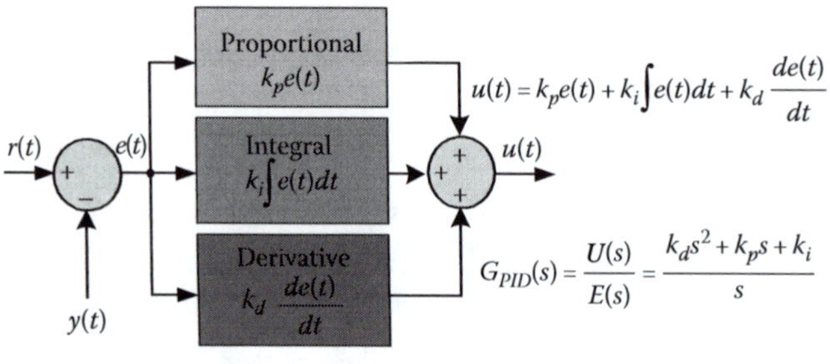

FIGURE 8.4 Linear analog PID control law with $G_{PID}(s) = \frac{U(s)}{E(s)} = \frac{k_d s^2 + k_p s + k_i}{s}$.

The proportional, proportional–integral (PI), and proportional–derivative (PD) control laws are

$$u(t) = k_p e(t), \quad u(t) = k_p e(t) + k_i \int e(t)dt, \quad \text{and} \quad u(t) = k_p e(t) + k_d \frac{de(t)}{dt}.$$

Different linear and nonlinear analog PID control laws can be designed and implemented. The closed-loop electromechanical system with PID control law in the time- and s-domains is shown in Figures 8.5a and b. If the system is linear or can be linearized, the transfer function $G_{sys}(s)$ can be used. The closed-loop electromechanical systems, described by $G_{sys}(s)$ or nonlinear differential equations, are depicted in Figures 8.5a and b. The control bound is shown. For linear systems with no bounds, shown in Figure 8.5b, the transfer function algebra, Laplace transforms, frequency-domain analysis, final-value theorem, and other concepts of linear control theory can be used.

If the system output $y(t)$ is bounded and $y(t)$ converges to the bounded $r(t)$ as time approaches infinity, the tracking of the reference input is accomplished. Tracking is achieved if the tracking error $e(t) = [r(t) - y(t)]$ is bounded. Ideally, $\lim_{t\to\infty} e(t) = 0$. For physical systems, one specifies $\lim_{t\to\infty} |e(t)| \le \delta$, $\delta > 0$. The physical systems always exhibit the tracking error due to the sensor errors, digital-to-analog and analog-to-digital conversion errors, processing errors, etc. An absolute accuracy is assumed in theoretical analysis assuming ideal measurements of $y(t)$, $r(t)$, and $x(t)$.

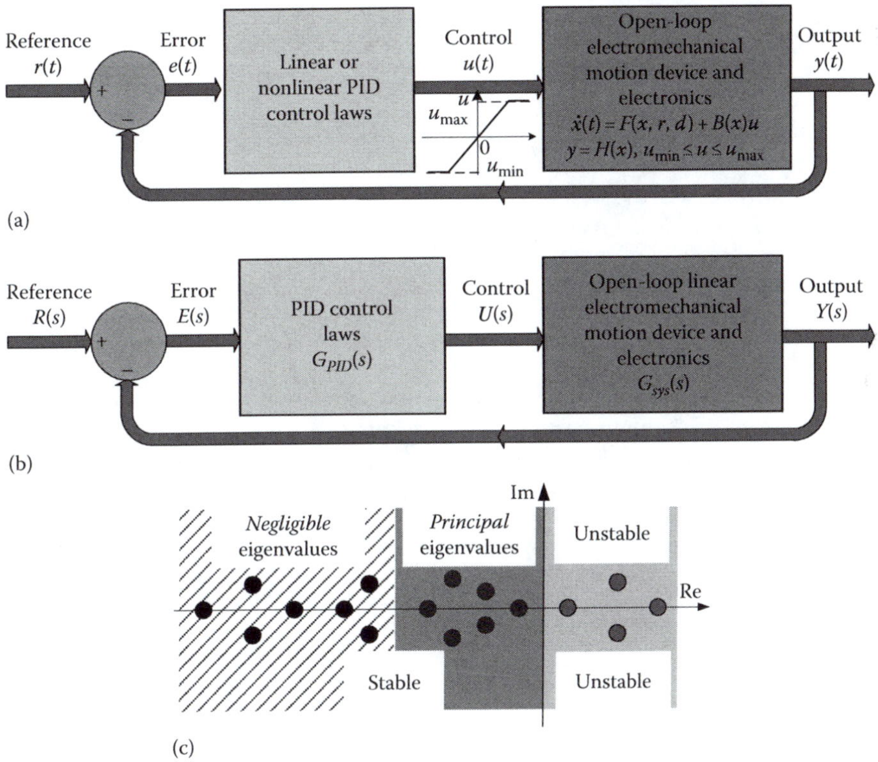

FIGURE 8.5 (a) Time-domain diagram of a nonlinear closed-loop system with a PID control law; (b) s-Domain diagrams of the closed-loop system, $G(s) = \frac{Y(s)}{R(s)} = \frac{G_{PID}(s)G_{sys}(s)}{1 + G_{PID}(s)G_{sys}(s)}$, $\lim_{t\to\infty} e(t) = \lim_{s\to 0} \frac{sR(s)}{1 + G_{PID}G_{sys}}$;
(c) Linear systems analysis: Eigenvalues in the complex plane.

In the time domain, the tracking error is $e(t) = r(t) - y(t)$. The Laplace transform is $E(s) = R(s) - Y(s)$. For a linear closed-loop system, as given in Figure 8.5b, the Laplace transform of the output $y(t)$ is $Y(s) = G_{sys}(s)U(s) = G_{sys}(s)G_{PID}(s)E(s) = G_{sys}(s)G_{PID}(s)[R(s)-Y(s)]$. The transfer function of the closed-loop linear electromechanical systems with a linear PID control law is

$$G(s)=\frac{Y(s)}{R(s)}=\frac{G_{PID}(s)G_{sys}(s)}{1+G_{PID}(s)G_{sys}(s)},\quad G(j\omega)=\frac{Y(j\omega)}{R(j\omega)}=\frac{G_{PID}(j\omega)G_{sys}(j\omega)}{1+G_{PID}(j\omega)G_{sys}(j\omega)},$$

and,

$$E(s)=\frac{R(s)}{1+G_{PID}(s)G_{sys}(s)}.$$

The final value theorem yields the steady-state error

$$\lim_{t\to\infty} e(t)=\lim_{s\to 0}\frac{sR(s)}{1+G_{PID}G_{sys}}.$$

The characteristic equation can be found. Stability and performance depend on control laws with $G_{PID}(s)$. One may refine the proportional, integral, and derivative feedback gains that affect the eigenvalues and closed-loop system dynamics. The current, angular velocity, and other feedback can be implemented. Using the constant factor k, eigenvalues at the origin, as well as real and complex-conjugate eigenvalues and zeros, one has

$$G(s)=\frac{k\left(T_{n1}s+1\right)\left(T_{n2}s+1\right)\cdots\left(T_{n,l-1}^2s^2+2\xi_{n,l-1}T_{n,l-1}s+1\right)\left(T_{n,l}^2s^2+2\xi_{n,l}T_{n,l}s+1\right)}{s^M\left(T_{d1}s+1\right)\left(T_{d2}s+1\right)\cdots\left(T_{d,p-1}^2s^2+2\xi_{d,p-1}T_{d,p-1}s+1\right)\left(T_{d,p}^2s^2+2\xi_{d,p}T_{d,p}s+1\right)},$$

where T_i and ξ_i are the time constants and damping coefficients; M is the order of the eigenvalues at the origin.

The control law transfer function $G_{PID}(s)$ can be found specifying $G(s)$, assigning zeros and eigenvalues that affect the settling time, stability margins, tracking error, overshoot, etc. If the PID control law (8.1) is used, the feedback gains k_p, k_i, and k_d can be derived to attain the specific *principal* characteristic eigenvalues because other poles can be located far left in the complex plane, see Figure 8.5c. Many textbooks [1–5] cover *modal* control assigning the eigenvalues and tracking error. These methods are applicable only to linear systems with no constraints.

Example 8.4

Study the so-called *force* or *torque control* for a one-dimensional mechanical motion of frictionless mass when the disturbances and electromagnetic dynamics are neglected. Only translational or rotational equations of motion due to the Newton second law are considered. One-dimensional motion of a frictionless not-perturbed rigid-body mechanical system is described by

$$\frac{dx_1}{dt}=x_2,\quad \frac{dx_2}{dt}=u.$$

Here, the state variables $x_1(t)$ and $x_2(t)$ denote the displacement and velocity, while u is the force or torque to be applied to control the system motion.

The transfer function of the open-loop system is

$$G_{sys}(s) = 1/s^2.$$

Consider the PD control law

$$u(t) = k_p e(t) + k_d(de(t)/dt) \text{ with } G_{PD}(s) = U(s)/E(s) = k_p + k_d s.$$

The transfer function of the closed-loop system is

$$G(s)=\frac{Y(s)}{R(s)}=\frac{G_{PID}(s)G_{sys}(s)}{1+G_{PID}(s)G_{sys}(s)}=\frac{(k_p+k_ds)\frac{1}{s^2}}{1+(k_p+k_ds)\frac{1}{s^2}}=\frac{k_ds+k_p}{s^2+k_ds+k_p}.$$

The characteristic equation is

$$s^2 + k_d s + k_p = 0.$$

One can specify the settling time that results in the desired characteristic eigenvalues. Let the desired eigenvalues be −1 and −1.

The specified characteristic equation is

$$(s + 1)(s + 1) = s^2 + 2s + 1 = 0.$$

From the obtained characteristic equations $s^2 + k_d s + k_p = 0$ and $s^2 + 2s + 1 = 0$, one finds the feedback coefficients $k_p = 1$ and $k_d = 2$.

The steady-state error for the unit step reference $r(t) = 1(t)$ with $R(s) = 1/s$ is

$$\lim_{t\to\infty} e(t) = \lim_{s\to 0}\frac{sR(s)}{1+G_{PID}G_{sys}} = \lim_{s\to 0}\frac{s\dfrac{1}{s}}{1+\left(k_p+k_d s\right)\dfrac{1}{s^2}} = \lim_{s\to 0}\frac{s^2}{s^2+k_d s+k_p} = 0.$$

The steady-state error for the unit ramp function $r(t) = t$ with $R(s) = 1/s^2$ is

$$\lim_{t\to\infty} e(t) = \lim_{s\to 0}\frac{sR(s)}{1+G_{PID}G_{sys}} = \lim_{s\to 0}\frac{s\dfrac{1}{s^2}}{1+\left(k_p+k_d s\right)\dfrac{1}{s^2}} = \lim_{s\to 0}\frac{s}{s^2+k_d s+k_p} = 0.$$

The system is stable. The analytic expressions for evolutions of $x_1(t)$ and $x_2(t)$ can be found by using the Laplace transform for given $r(t)$ and initial conditions. The use of derivative feedback results in the system being sensitive to the noise and dependent on $r(t)$. ■

Example 8.5

Consider an electric drive with a permanent-magnet DC motor. The output is the angular velocity ω_r. The motor is open-loop stable, and the fundamental and experimental findings were reported in Chapter 4. The torque-speed characteristics are $\omega_r = (u_a - r_a i_a)/k_a = (u_a/k_a) - (r_a/k_a^2)T$. As the armature voltage u_a is applied, the motor rotates at the particular angular velocity ω_r defined by u_a, k_a, r_a, B_m, and T_L. We study the dynamics and stability using the linear control system theory. Using Kirchhoff's voltage law and Newton's second law of motion, we found

$$\frac{di_a}{dt} = -\frac{r_a}{L_a}i_a - \frac{k_a}{L_a}\omega_r + \frac{1}{L_a}u_a,$$

$$\frac{d\omega_r}{dt} = \frac{k_a}{J}i_a - \frac{B_m}{J}\omega_r - \frac{1}{J}T_L.$$

Denoting $x_1 = i_a$, $x_2 = \omega_r$, and $u = u_a$, we have

$$\frac{dx}{dt} = \begin{bmatrix}\dfrac{dx_1}{dt}\\ \dfrac{dx_2}{dt}\end{bmatrix} = \begin{bmatrix}-\dfrac{r_a}{L_a} & -\dfrac{k_a}{L_a}\\ \dfrac{k_a}{J} & -\dfrac{B_m}{J}\end{bmatrix}\begin{bmatrix}x_1\\ x_2\end{bmatrix} + \begin{bmatrix}\dfrac{1}{L_a}\\ 0\end{bmatrix}u = Ax + Bu.$$

The state transition matrix is $\Phi(s) = (sI-A)^{-1}$. The characteristic equation is $|sI-A| = 0$. The transfer function is found using the results reported in Examples 8.2 and 8.3. The transfer function algebra also can be applied. For an open-loop electric drive (the output is ω_r and $T_L = 0$), we have

$$G_{sys}(s)=\frac{Y(s)}{U(s)}=\frac{\Omega_r(s)}{U(s)}=\frac{k_a}{L_aJs^2+\left(r_aJ+L_aB_m\right)s+r_aB_m+k_a^2}.$$

The characteristic equation is

$$L_aJs^2+\left(r_aJ+L_aB_m\right)s+r_aB_m+k_a^2=0.$$

Consider the second-order quadratic equation

$$as^2 + bs + c = 0,$$

where $a = L_aJ$, $b = (r_aJ + L_aB_m)$, and $c=\left(r_aB_m+k_a^2\right)$.

The solution of the quadratic equation is $s_{1,2}=\left(-b\pm\sqrt{b^2-4ac}\right)/2a$.

All motor parameters are positive. Hence, $a > 0$, $b > 0$, and $c > 0$. For any possible a, b, and c, the real parts of eigenvalues are negative. Hence, the open-loop system (electric drive) is stable.

Any PID control laws will guarantee stability of the closed-loop system.

Consider the proportional tracking control law

$$u = k_pe,\ k_p > 0 \text{ with } G_p(s) = k_p.$$

With $y = \omega_r$, the tracking error is $e = r - y = r - \omega_r$.

The transfer function of the closed-loop system is

$$G(s)=\frac{Y(s)}{R(s)}=\frac{\Omega_r(s)}{R(s)}=\frac{G_{PID}(s)G_{sys}(s)}{1+G_{PID}(s)G_{sys}(s)}=\frac{k_pk_a}{L_aJs^2+\left(r_aJ+L_aB_m\right)s+r_aB_m+k_a^2+k_pk_a}.$$

The characteristic equation is

$$L_aJs^2+\left(r_aJ+L_aB_m\right)s+r_aB_m+k_a^2+k_pk_a=0.$$

For the closed-loop system $c = (r_aB_m + k_a^2 + k_pk_a)$, $c > 0$. The stability is guaranteed because the real parts of all characteristic eigenvalues $s_{1,2}=\left(-b\pm\sqrt{b^2-4ac}\right)/2a$ are negative. The proportional control law changes the eigenvalues. One may find the proportional feedback gain k_p of the control law $u = k_pe$ to ensure the specified tracking error. The steady-state error is $\lim_{t\to\infty}e(t)=\lim_{s\to 0}\frac{sR(s)}{1+G_{PID}G_{sys}}$.

For the unit step reference $r(t) = 1$, and $R(s) = 1/s$. Hence

$$\lim_{t\to\infty}e(t)=\lim_{s\to 0}\frac{sR(s)}{1+k_pG_{sys}}=\lim_{s\to 0}\frac{1}{1+k_p\dfrac{k_a}{L_aJs^2+\left(r_aJ+L_aB_m\right)s+r_aB_m+k_a^2}}$$

$$=\lim_{s\to 0}\frac{L_aJs^2+\left(r_aJ+L_aB_m\right)s+r_aB_m+k_a^2}{L_aJs^2+\left(r_aJ+L_aB_m\right)s+r_aB_m+k_a^2+k_pk_a}.$$

Let the sensor is measuring of ω_r with accuracy 0.1%. Correspondingly, one may specify the tracking error to be 0.1%. From $\lim_{t\to\infty} e(t) = e(\infty) = 0.001$, we have

$$\lim_{t\to\infty}e(t)=\lim_{s\to 0}\frac{L_aJs^2+\left(r_aJ+L_aB_m\right)s+r_aB_m+k_a^2}{L_aJs^2+\left(r_aJ+L_aB_m\right)s+r_aB_m+k_a^2+k_pk_a}=0.001.$$

One finds k_p. As an illustrative example, letting all the motor parameters be equal to 1 ($r_a = 1$ ohm, $L_a = 1$ H, $k_a = 1$ V-sec/rad, $B_m = 1$ N-m-sec/rad and $J = 1$ kg-m^2), one has $k_p = 1998$.

The characteristic equation $L_aJs^2+(r_aJ+L_aB_m)s+r_aB_m+k_a^2+k_pk_a=s^2+2s+2000=0$ yields the characteristic eigenvalues. Solving $s_{1,2}=\left(-b\pm\sqrt{b^2-4ac}\right)/2a$, one finds the eigenvalues $s_{1,2}=-1\pm 44.7i$. ■

Example 8.6

Consider the electric drive with a permanent-magnet DC motor in Example 8.5. The output is the angular velocity ω_r. That is, $y=\omega_r$. The tracking error is $e=r-y=r-\omega_r$.

Our goal is to find the feedback gains of a proportional-integral control with the current feedback

$$u(t)=-k_Ii_a+k_pe+k_i\int e\,dt$$

to guarantee the desired eigenvalues $s_1=s_2=s_3=-1$. Let all motor parameters be equal to 1.

The transfer function for an open-loop system is

$$G_{sys}(s)=\frac{\Omega(s)}{U(s)}=\frac{k_a}{L_aJs^2+(r_aJ+L_aB_m)s+r_aB_m+k_a^2}.$$

We find the transfer function of the closed-loop system. Assuming that there is no saturation, using $G_{sys}(s)$ with the current feedback $-k_Ii_a$, one finds

$$G_{sys}^*(s)=\frac{k_a}{L_aJs^2+(r_aJ+k_IJ+L_aB_m)s+(r_a+k_I)B_m+k_a^2}.$$

For a closed-loop system, we have

$$G(s)=\frac{\Omega(s)}{R(s)}=\frac{G_{PID}(s)G_{sys}^*(s)}{1+G_{PID}(s)G_{sys}^*(s)}=\frac{\dfrac{k_ps+k_i}{s}G_{sys}^*(s)}{1+\dfrac{k_ps+k_i}{s}G_{sys}^*(s)}$$

$$=\frac{(k_ps+k_i)k_a}{L_aJs^3+(r_aJ+k_IJ+L_aB_m)s^2+(r_aB_m+k_IB_m+k_a^2+k_pk_a)s+k_ik_a}$$

The characteristic equation is

$$L_aJs^3+(r_aJ+k_IJ+L_aB_m)s^2+(r_aB_m+k_IB_m+k_a^2+k_pk_a)s+k_ik_a=0.$$

The characteristic equation with the specified eigenvalues is

$$(s+1)(s+1)(s+1)=s^3+3s^2+3s+1=0.$$

The comparison of these characteristic equations yields

$$\frac{r_aJ+k_IJ+L_aB_m}{L_aJ}=3,\quad \frac{r_aB_m+k_IB_m+k_a^2+k_pk_a}{L_aJ}=3,\quad\text{and}\quad \frac{k_ik_a}{L_aJ}=1.$$

Hence, the feedback gains are $k_I=1$, $k_p=0$, and $k_i=1$.

The steady-state error is found by using the final value theorem. For $r(t) = 1$, one has $R(s) = 1/s$. Hence

$$\lim_{t\to\infty} e(t) = \lim_{s\to 0}\frac{sR(s)}{1+G_{PID}G^*_{sys}} = \lim_{s\to 0}\frac{sR(s)}{1+\dfrac{k_p s+k_i}{s}G^*_{sys}}$$

$$= \lim_{s\to 0}\frac{1}{1+\dfrac{k_p s+k_i}{s}\dfrac{k_a}{L_aJs^2+(r_aJ+k_IJ+L_aB_m)s+r_aB_m+k_IB_m+k_a^2}}$$

$$= \lim_{s\to 0}\frac{s\left[L_aJs^2+(r_aJ+k_IJ+L_aB_m)s+r_aB_m+k_IB_m+k_a^2\right]}{L_aJs^3+(r_aJ+k_IJ+L_aB_m)s^2+(r_aB_m+k_IB_m+k_a^2+k_pk_a)s+k_ik_a} = 0.$$

Therefore, $\lim_{t\to\infty} e(t) = e(\infty) = 0$, and the tracking error is zero. ■

The linear PID control laws with N_i integrals and N_d derivative terms are

$$u(t) = \underbrace{k_pe(t)}_{\text{Proportional}} + \underbrace{\sum_{j=1}^{N_i}\int\cdots\int k_{ij}e\,dt}_{\text{Integral}} + \underbrace{\sum_{j=1}^{N_d}k_{dj}\frac{d^je(t)}{dt^j}}_{\text{Derivative}},$$

$$U(s) = \left(k_P + \sum_{j=1}^{N_i}k_{ij}\frac{1}{s^j} + \sum_{j=1}^{N_d}k_{dj}s^j\right)E(s), \tag{8.2}$$

where N_i and N_d are the positive integers; k_{ij} and k_{dj} are the integral and derivative feedback coefficients.

From (8.2), we obtain the transfer function

$$G_{PID}(s) = U(s)/E(s).$$

For example, for

$$u(t) = k_pe(t) + k_{i1}\int e(t)dt + k_{i2}\iint e(t)dt + k_{i3}\iiint e(t)dt + k_d\frac{de(t)}{dt},$$

one yields

$$G_{PID}(s) = \frac{U(s)}{E(s)} = \frac{k_ds^4+k_ps^3+k_{i1}s^2+k_{i2}s+k_{i3}}{s^3}.$$

Nonlinear PID control laws can be designed and implemented. With the nonlinear tracking feedback mappings, one has

$$u(t) = \underbrace{\sum_{m=1}^{M_p}k_{p(2m-1)}e^{2m-1}(t)}_{\text{Proportional}} + \underbrace{\sum_{j=1}^{N_i}\sum_{m=1}^{M_i}\int\cdots\int k_{ij(2m-1)}e^{2m-1}\,dt}_{\text{Integral}} + \underbrace{\sum_{j=1}^{N_d}\sum_{m=1}^{M_d}k_{dj(2m-1)}\frac{d^j}{dt^j}e^{2m-1}(t)}_{\text{Derivative}}, \tag{8.3}$$

where M_p, M_i, and M_d are the positive integers; $k_{p(2m-1)}$, $k_{ij(2m-1)}$, and $k_{dj(2m-1)}$ are the proportional, integral, and derivative feedback coefficients.

Example 8.7

In (8.3), integers M_p, M_i, and M_d are assigned by the designer defining the power for the tracking error. Setting $N_i = 1$, $N_d = 1$, $M_p = 1$, $M_i = 1$, and $M_d = 1$, we have the PID control law (8.1). Letting $N_i = 2$, $N_d = 1$, $M_p = 3$, $M_i = 2$, and $M_d = 1$, from (8.3), one obtains the nonlinear PID control law

$$u(t) = k_{p1}e(t) + k_{p3}e^3(t) + k_{p5}e^5(t) + k_{i1,1}\int e\,dt + k_{i2,1}\iint e\,dt + k_{i1,3}\int e^3\,dt + k_{i2,3}\iint e^3\,dt + k_{d1,1}\frac{de(t)}{dt}.$$

■

Nonlinear control laws improve system dynamics, enhance stability, ensure robustness, guarantee disturbance attenuation, etc. Using the multi-index notations, the power-series nonlinear PID-type control law is

$$u = \underbrace{\sum_{m=1}^{M_p} k_{p(2m-1)}\,\mathrm{sgn}(e)|e|^{\sum_{p=0}^{L_p}\frac{2m-1}{2p+1}}}_{\text{Proportional}} + \underbrace{\sum_{j=1}^{N_i}\sum_{m=1}^{M_i}\int\cdots\int k_{ij(2m-1)}\,\mathrm{sgn}(e)|e|^{\sum_{p=0}^{L_i}\frac{2m-1}{2p+1}}\,dt}_{\text{Integral}}$$

$$+ \underbrace{\sum_{j=1}^{N_d}\sum_{m=1}^{M_d} k_{dj(2m-1)}\,\mathrm{sgn}(e)\frac{d^j}{dt^j}|e|^{\sum_{p=0}^{L_d}\frac{2m-1}{2p+1}}}_{\text{Derivative}}, \tag{8.4}$$

where L_p, L_i, and L_d are the nonnegative integers.

For example, nonlinear feedback $\mathrm{sgn}(e)|e(t)|^{1/3}$, $\mathrm{sgn}(e)|e(t)|^{1/7}$, $\mathrm{sgn}(e)|e(t)|^{3/7}$, and others ensure a large control signal $u(t)$ for a small tracking error. Nonlinear PID control laws (8.4) ensure stability, high precision, and high accuracy despite control constraints.

The inherent hardware control bounds $u_{\min} \le u \le u_{\max}$ arise because $u(t)$, given by (8.1) through (8.4), exceed the hardware limits. For example, the duty ratio in PWM amplifiers or the applied voltage are constrained as $d_{D\min} \le d_D \le d_{D\max}$, $d_D \in [0\ 1]$ or $d_D \in [-1\ 1]$. For motors, $u_{\min} \le u \le u_{\max}$. If hardware limits are reached, or the nonlinear control laws (8.3) or (8.4) are used, linear methods (transfer function, eigenvalues, pole-placement, etc.) cannot be applied because the closed-loop system is nonlinear. All control laws usually lead to saturation as $u_{\min} \le u \le u_{\max}$. Linear analysis may be inadequate for many problems. For physical systems, the limits imposed on the voltage, current, charge, force, torque, power, and other physical quantities are considered. There are the mechanical limits on the maximum angular and linear displacement, velocities, accelerations, etc. The rated, peak, and *admissible* (maximum allowed) voltages, currents, velocities, and displacements are specified. The closed-loop electromechanical system with a saturated control is shown in Figure 8.6a. The constrained PID-type control laws (8.2), (8.3), and (8.4) with linear and nonlinear feedback mappings are

$$u = \mathrm{sat}_{u_{\min}}^{u_{\max}}\left(k_p e + \sum_{j=1}^{N_i}\int\cdots\int k_{ij}e\,dt + \sum_{j=1}^{N_d}k_{dj}\frac{d^j e}{dt^j}\right), \quad u_{\min} \le u \le u_{\max}, \tag{8.5}$$

$$u = \mathrm{sat}_{u_{\min}}^{u_{\max}}\left(\sum_{m=1}^{M_p}k_{p(2m-1)}e^{2m-1} + \sum_{j=1}^{N_i}\sum_{m=1}^{M_i}\int\cdots\int k_{ij(2m-1)}e^{2m-1}dt + \sum_{j=1}^{N_d}\sum_{m=1}^{M_d}k_{dj(2m-1)}\frac{d^j e^{2m-1}}{dt^j}\right),$$

$$u = \mathrm{sat}_{u_{\min}}^{u_{\max}}\left(\sum_{m=1}^{M_p}k_{p(2m-1)}\,\mathrm{sgn}(e)|e|^{\sum_{p=0}^{L_p}\frac{2m-1}{2p+1}} + \sum_{j=1}^{N_i}\sum_{m=1}^{M_i}\int\cdots\int k_{ij(2m-1)}\,\mathrm{sgn}(e)|e|^{\sum_{p=0}^{L_i}\frac{2m-1}{2p+1}}\,dt\right.$$

$$\left.+\sum_{j=1}^{N_d}\sum_{m=1}^{M_d}k_{dj(2m-1)}\,\mathrm{sgn}(e)\frac{d^j}{dt^j}|e|^{\sum_{p=0}^{L_d}\frac{2m-1}{2p+1}}\right).$$

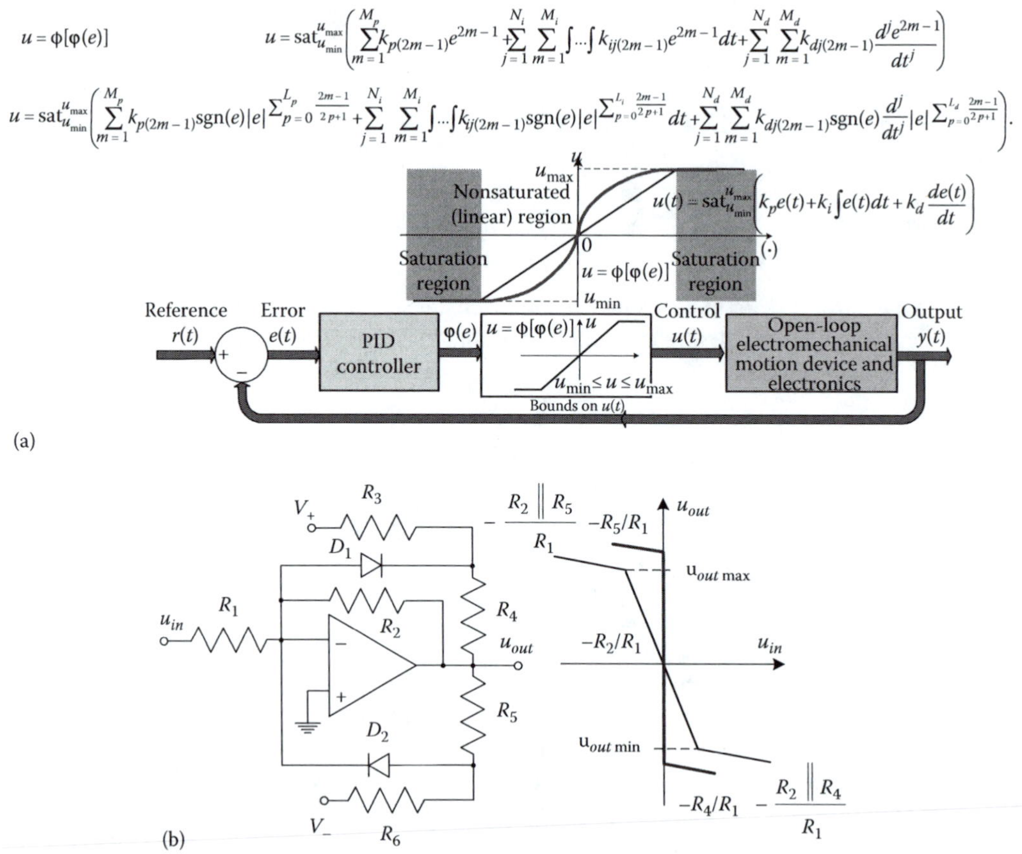

FIGURE 8.6 (a) Electromechanical system with a bounded PID control law $u = \phi[\varphi(e)]$, $u_{\min} \le u \le u_{\max}$; (b) Implementation of the variable-gain feedback using the limiter operational amplifier circuit with diodes.

In general, the continuous, piecewise continuous, or discontinuous bounded function $\phi(\cdot)$ represents the inherent limits in physical systems $u_{\min} \le u \le u_{\max}$. Using $e(t)$, we have

$$u = \phi[\varphi(e)],$$

$$u_{\min} \le u \le u_{\max}.$$

Smooth continuous bounds ϕ are exhibited by electronic, microelectronic, and electromagnetic systems. The hard limits, such saturation, can be exhibited by mechanical systems. With inherent symmetric or asymmetric bounds $\phi(\cdot)$ and nonlinear feedback map $\varphi(e)$, one explicitly defines $u = \phi[\varphi(e)]$ as

$$u = \phi\left(k_p e + \sum_{j=1}^{N_i} \int \cdots \int k_{ij} e \, dt + \sum_{j=1}^{N_d} k_{dj} \frac{d^j e}{dt^j}\right), \quad u_{\min} \le u \le u_{\max},$$

$$u=\phi\left(\sum_{m=1}^{M_p}k_{p(2m-1)}e^{2m-1}+\sum_{j=1}^{N_i}\sum_{m=1}^{M_i}\int\cdots\int k_{ij(2m-1)}e^{2m-1}dt+\sum_{j=1}^{N_d}\sum_{m=1}^{M_d}k_{dj(2m-1)}\frac{d^j e^{2m-1}}{dt^j}\right),\quad u_{\min}\le u\le u_{\max},$$

$$u=\phi\left(\sum_{m=1}^{M_p}k_{p(2m-1)}\,\mathrm{sgn}(e)\left|e\right|^{\sum_{p=0}^{L_p}\frac{2m-1}{2p+1}}+\sum_{j=1}^{N_i}\sum_{m=1}^{M_i}\int\cdots\int k_{ij(2m-1)}\,\mathrm{sgn}(e)\left|e\right|^{\sum_{p=0}^{L_i}\frac{2m-1}{2p+1}}dt\right.$$

$$\left.+\sum_{j=1}^{N_d}\sum_{m=1}^{M_d}k_{dj(2m-1)}\,\mathrm{sgn}(e)\frac{d^j}{dt^j}\left|e\right|^{\sum_{p=0}^{L_d}\frac{2m-1}{2p+1}}\right).$$

As documented in Figure 8.6a, control $u(t)$ varies between the minimum and maximum values, $u_{\min}\le u\le u_{\max}$, $u_{\min}\le 0$ and $u_{\max}>0$. For

$$u(t)=\mathrm{sat}_{u_{\min}}^{u_{\max}}\left(k_pe(t)+k_i\int e(t)dt+k_d\frac{de(t)}{dt}\right),$$

if $k_pe(t)+k_i\int e(t)dt+k_d\frac{de(t)}{dt}>u_{\max}$ and $k_pe(t)+k_i\int e(t)dt+k_d\frac{de(t)}{dt}<u_{\min}$, the control is bounded as $u=u_{\max}$ and $u=u_{\min}$.

Due to control bounds, one must apply nonlinear control theory even if a system is described by linear differential equations. The reported PID control laws are implemented by analog and digital controllers. The reference, output, and error $e(t)$ are directly measured. Various hardware solutions exist to implement control laws, as covered in Chapter 7. The limiter inverted operational amplifier circuit with diodes implements a nonlinear feedback, as illustrated in Figure 8.6b. The slopes of the piecewise–continuous u_{out}–u_{in} transfer characteristic are defined by the input and feedback resistors R_1 and R_2, as well as the resistors connected to $\pm V$. We have

$$u_{out\max}=V\frac{R_5}{R_6}+V_D\left(1+\frac{R_5}{R_6}\right)\quad\text{and}\quad u_{out\min}=-V\frac{R_4}{R_3}-V_D\left(1+\frac{R_4}{R_3}\right),$$

where V_D is the voltage drop at the diodes. If the feedback resistor R_2 is removed, one realizes the comparator with the upper and lower slopes of the u_{out}–u_{in} characteristic $-R_5/R_1$ and $-R_4/R_1$, respectively.

Example 8.8: Control of a Servo with a Permanent-Magnet DC Motor

A permanent-magnet DC machine under many assumptions can be described by linear differential equations. Consider a servosystem with a permanent-magnet DC motor that actuates a rotating stage, as documented in Figure 8.7a. A geared motor with a planetary gearhead is attached to the rotating stage. Our goal is to design and examine control laws. The platform angular displacement is a function of the rotor displacement. Using the gear ratio k_{gear}, the output equation is $y=Hx$, $y(t)=k_{gear}\theta_r(t)$. To change the angular velocity and displacement, one regulates the voltage applied to the armature winding u_a. The rated armature voltage for the motor is $\pm u_{\max}$. The rated current is $i_{a\max}$, and the maximum angular velocity is $\omega_{r\max}$. For a DC motor, $u_{\max}=30$ V ($-30\le u_a\le 30$ V), $i_{a\max}=0.15$ A, $\omega_{r\max}=150$ rad/sec, $r_a=200$ ohm, $L_a=0.002$ H, $k_a=0.2$ V-sec/rad, (N-m/A), $J=0.00000002$ kg-m^2, and $B_m=0.00000005$ N-m-sec/rad. The reduction gear ratio is 100:1.

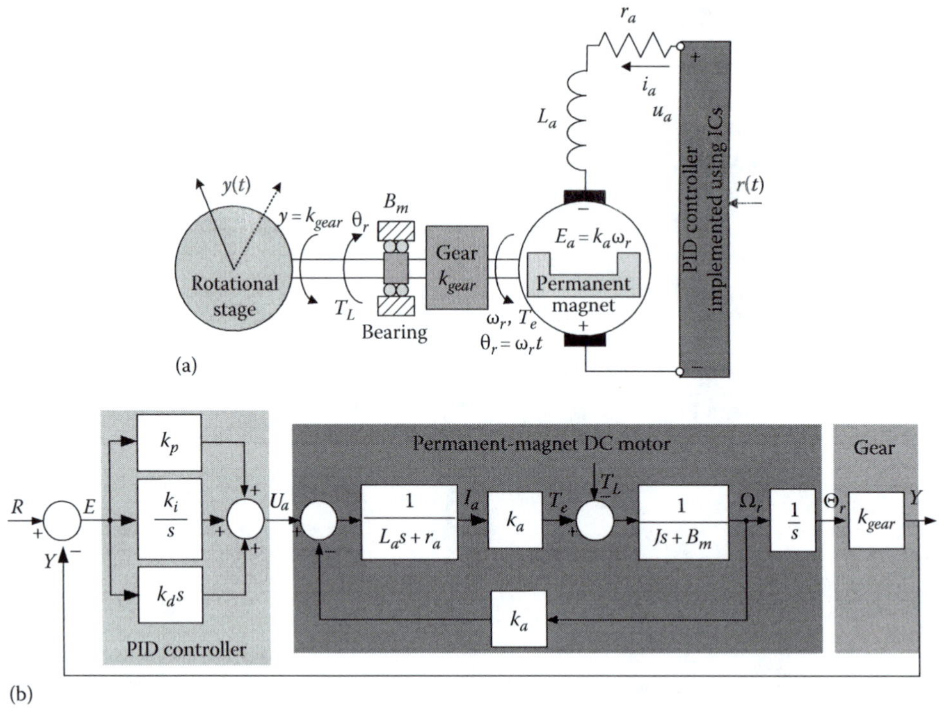

FIGURE 8.7 (a) Schematic diagram of a servosystem with a permanent-magnet DC motor; (b) s-domain diagram of the closed-loop system with an analog PID control law.

The differential equations are

$$\frac{di_a}{dt} = \frac{1}{L_a}(-r_a i_a - k_a\omega_r + u_a), \quad \frac{d\omega_r}{dt} = \frac{1}{J}\left(T_e - T_{viscous} - T_L\right) = \frac{1}{J}\left(k_a i_a - B_m\omega_r - T_L\right), \quad \frac{d\theta_r}{dt} = \omega_r.$$

The output equation is $y = k_{gear}\,\theta_r(t)$, $Y(s) = k_{gear}\,\Theta_r(s)$. One obtains the s-domain diagram of the open- and closed-loop systems, as documented in Figure 8.7b. The transfer function of the open-loop system is

$$G_{sys}(s) = \frac{Y(s)}{U_a(s)} = \frac{k_{gear}k_a}{s\left(L_a J s^2 + \left(r_a J + L_a B_m\right)s + r_a B_m + k_a^2\right)}.$$

For the PID control law

$$u_a(t) = k_p e(t) + k_i\int e(t)dt + k_d\frac{de(t)}{dt},$$

we have $G_{PID}(s) = \dfrac{U_a(s)}{E(s)} = \dfrac{k_d s^2 + k_p s + k_i}{s}$.

The closed-loop transfer function is

$$G(s)=\frac{Y(s)}{R(s)}=\frac{G_{PID}(s)G_{sys}(s)}{1+G_{PID}(s)G_{sys}(s)}=\frac{\left(k_d s^2+k_p s+k_i\right)k_{gear}k_a}{s^2\left(L_aJs^2+\left(r_aJ+L_aB_m\right)s+r_aB_m+k_a^2\right)+\left(k_d s^2+k_p s+k_i\right)k_{gear}k_a}$$

$$=\frac{\frac{k_d}{k_i}s^2+\frac{k_p}{k_i}s+1}{\frac{L_aJ}{k_{gear}k_ak_i}s^4+\frac{\left(r_aJ+L_aB_m\right)}{k_{gear}k_ak_i}s^3+\frac{\left(r_aB_m+k_a^2+k_{gear}k_ak_d\right)}{k_{gear}k_ak_i}s^2+\frac{k_p}{k_i}s+1}.$$

The numerical values of the numerator and denominator coefficients in the transfer function $G_{sys}(s)$ may be found using the following MATLAB statements:

```
format short e; ra=200; La=0.002; ka=0.2; J=0.00000002; Bm=0.00000005; kgear=0.01;
num_s=[ka*kgear]; den_s=[La*J ra*J+La*Bm ra*Bm+ka^2 0]; num_s, den_s
```

For the open-loop system, $G_{sys}(s)=\frac{Y(s)}{U(s)}=\frac{2\times10^{-3}}{s\left(4\times10^{-11}s^2+4\times10^{-6}s+4\times10^{-2}\right)}$.

The open-loop system is unstable because one of the eigenvalues is at origin. Using the `roots` command, we find the eigenvalues:

```
>> Eigenvalues=roots(den_s)
Eigenvalues =
             0
 -8.8729e+004
 -1.1273e+004
```

The characteristic equation of the closed-loop transfer function $G(s)$ is

$$\frac{L_aJ}{k_{gear}k_ak_i}s^4+\frac{\left(r_aJ+L_aB_m\right)}{k_{gear}k_ak_i}s^3+\frac{\left(r_aB_m+k_a^2+k_{gear}k_ak_d\right)}{k_{gear}k_ak_i}s^2+\frac{k_p}{k_i}s+1=0.$$

The proportional k_p, integral k_i, and derivative k_d feedback coefficients affect the location of the eigenvalues and alter the tracking error. Let k_p = 25000, k_i = 250, and k_d = 25. The MATLAB statements are

```
kp=25000; ki=250; kd=25;
% Denominator of the closed-loop transfer function
den_c=[(La*J)/(kgear*ka*ki) (ra*J+La*Bm)/(kgear*ka*ki) ...
(ra*Bm+ka^2+kgear*ka*kd)/(kgear*ka*ki) kp/ki 1];
Eigenvalues_Closed_Loop=roots(den_c) %Eigenvalues of the closed-loop system
```

The eigenvalues of the closed-loop system are found as

```
Eigenvalues_Closed_Loop =
 -6.6393e+004
 -3.3039e+004
 -5.6983e+002
 -1.0000e-002
```

The closed-loop system is stable because the real parts of eigenvalues are negative. The real eigenvalues are preferred because they usually result in minimum or no overshoot. We study three cases: (a) $k_p = 25000$, $k_i = 250$, and $k_d = 25$; (b) $k_p = 2500$, $k_i = 250000$, and $k_d = 25$; (c) $k_p = 25000$, $k_i = 250$, and $k_d = 0$. The following MATLAB file allows one to simulate the closed-loop electromechanical system.

```
ra=200; La=0.002; ka=0.2; J=0.00000002; Bm=0.00000005; kgear=0.01;
kp=25000; ki=250; kd=25; ref=1; % reference (command) displacement is 1 rad
num_c=[kd/ki kp/ki 1]; % Numerator and denominator of the closed-loop system
den_c=[(La*J)/(kgear*ka*ki) (ra*J+La*Bm)/(kgear*ka*ki) ...
(ra*Bm+ka^2+kgear*ka*kd)/(kgear*ka*ki) kp/ki 1];
t=0:0.0001:0.02; u=ref*ones(size(t)); y=lsim(num_c,den_c,u,t);
plot(t,y,'-',y,u,':','LineWidth',2);
title('Angular Displacement, y(t)=0.01\theta_r, r(t)=1 [rad]','FontSize',14);
xlabel('Time (seconds)','FontSize',14);
ylabel('Output y(t) and Reference r(t)','FontSize',14); axis([0 0.02,0 1.2])
```

The closed-loop servo output (angular displacement) and reference $r(t)$ for $r(t) = 1$ rad are illustrated in Figures 8.8.

The Simulink® model to perform simulations is documented in Figure 8.9a. We use the PID controller block. The applied voltage is constrained. Let

$$u_a(t) = \text{sat}_{-30}^{+30}\left(25000e(t) + 250\int e(t)dt\right), \quad -30 \le u_a \le 30 \text{ V}.$$

The Saturation block is used. The transient dynamics of the system variables, as well as the output and voltage evolutions, are documented in Figure 8.9b. The physical limits and constraints significantly increase the settling time. One observes the effect of the load T_L. The reference significantly affects the settling time and system behavior. The control bounds are examined. Nonlinear friction, backlash, dead zone, and other nonlinear phenomena can be studied. The quadratic performance functional

$$J = \int_0^\infty \left(qe^2 + gu_a^2\right)dt, \quad q \ge 0, \quad g \ge 0$$

is used to assess the system dynamics. The evolution of $J(t)$ is found. Simulink solves differential equations, allows interactive analysis, ensures visualization, etc. The motor parameters and feedback gain coefficients are uploaded in the Command Window as

```
ra=200; La=0.002; ka=0.2; J=0.00000002; Bm=0.00000005; kgear=0.01;
kp=25000; ki=250; kd=0; g=1; q=1;
```

■

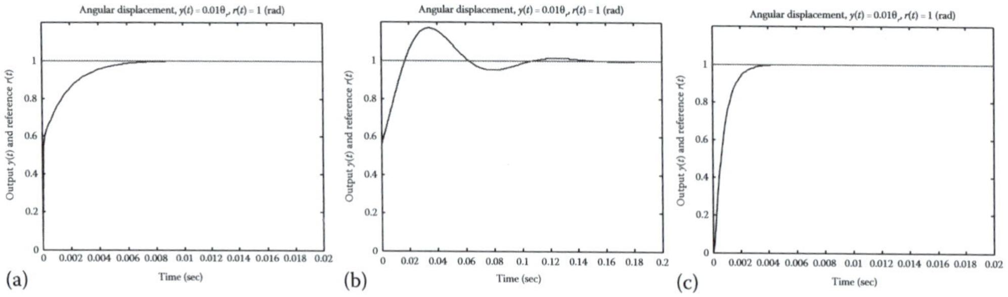

FIGURE 8.8 Dynamics of the closed-loop system with an analog PID control law:
(a) $k_p = 25000$, $k_i = 250$, and $k_d = 25$;
(b) $k_p = 2500$, $k_i = 250000$, and $k_d = 25$;
(c) $k_p = 25000$, $k_i = 250$, and $k_d = 0$.

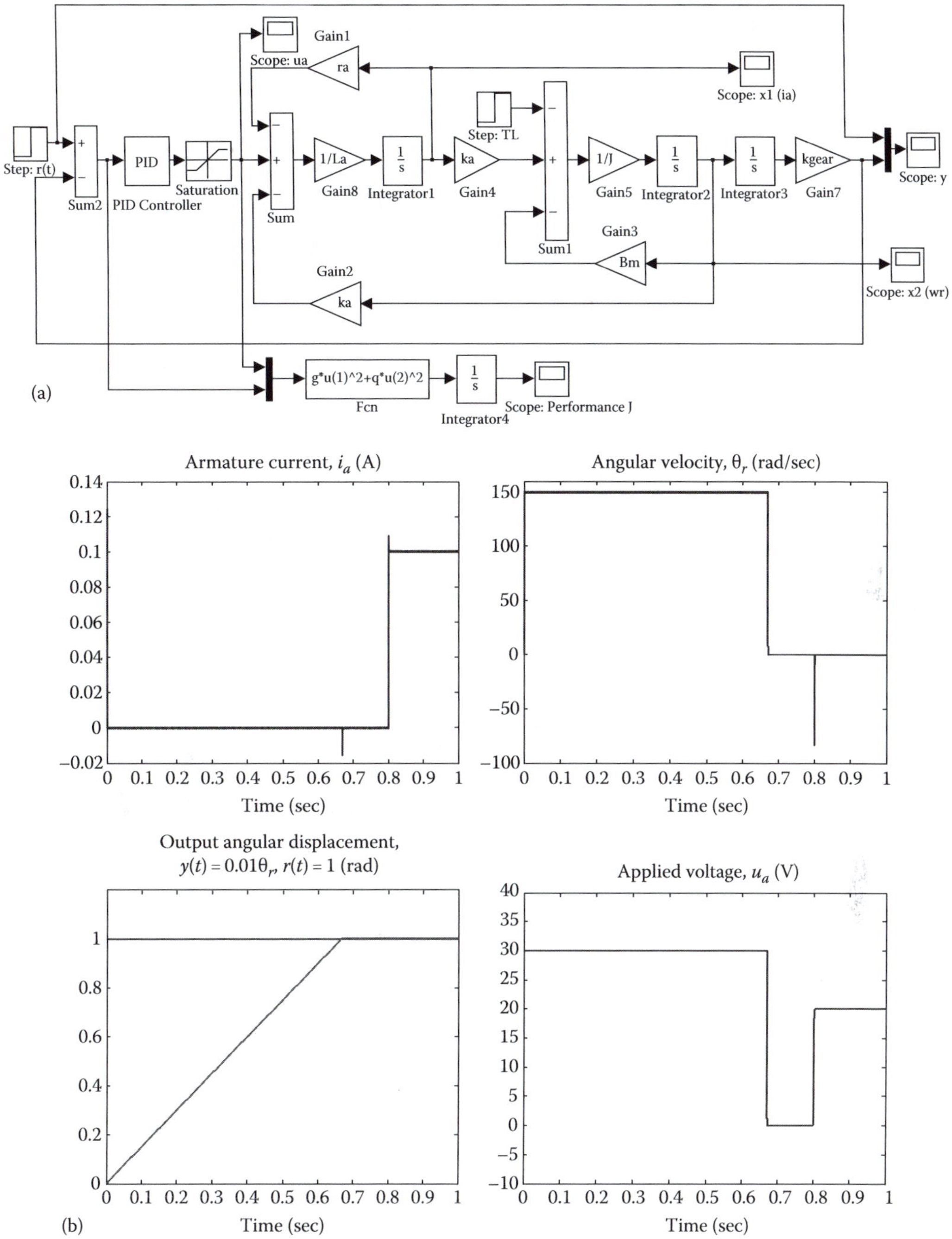

FIGURE 8.9 (a) Simulink® model of the closed-loop system with saturation; (b) Closed-loop system dynamics with the bounded control law $u_a(t) = \mathrm{sat}_{u_{\min}}^{u_{\max}}\left(k_p e(t) + k_i \int e(t)dt + k_d \frac{de(t)}{dt}\right)$, $k_p = 25000$, $k_i = 250$, $k_d = 0$ for $r(t) = 1$ rad and $T_L = \begin{cases} 0, & t \in [0\ \ 0.8)\ \mathrm{s} \\ 0.02\ \mathrm{N\text{-}m}, & t \in [0.8\ \ 1)\ \mathrm{s} \end{cases}$.

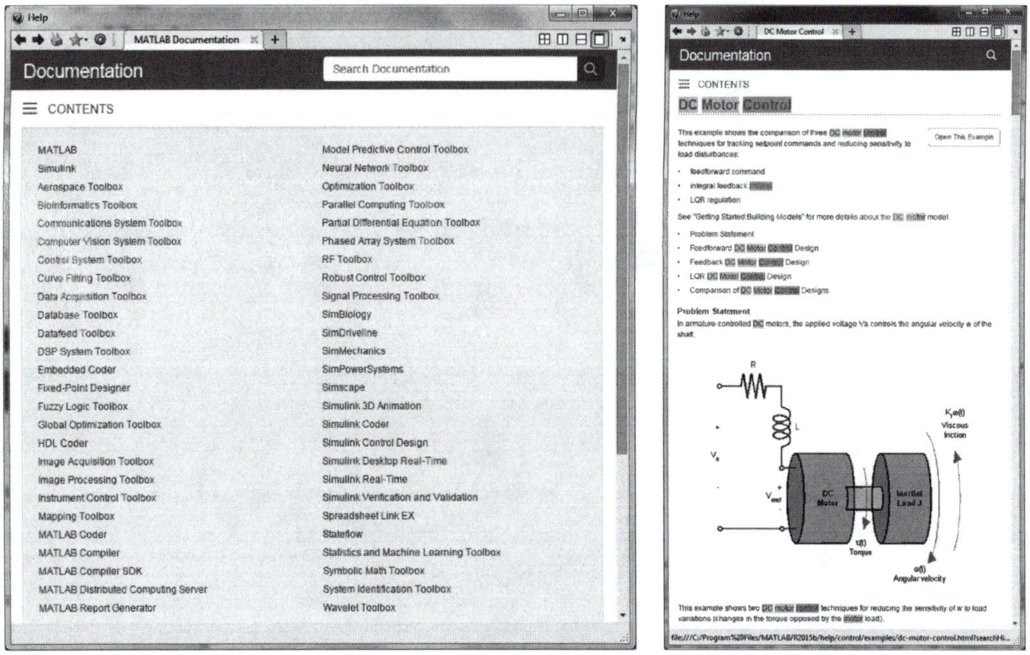

(a) (b)

FIGURE 8.10 (a) MATLAB® toolboxes; (b) Demonstrative DC motor control example.

MATLAB supports basic control law designs, enables simulation, and provides educational examples. The user refines existing or develops new solutions for engineering problems. The MATLAB windows with a list of toolboxes and DC motors control example are documented in Figures 8.10a and b. The device-physics-centric engineering designs are different compared to educational exercises. One may not rely on simplifications, postulates, and assumptions in engineering practice because practitioners deal with electromechanical systems hardware.

Example 8.9: Control of a Magnetic Levitation System

We studied variable-reluctance electromechanical motion devices, such as solenoids, relays, magnetic levitation systems, and others in Chapter 3. Nonlinear differential equations were derived and used to perform design and analysis. Many magnetic levitation systems use the reluctance electromagnetics. To design optimal systems, various control and optimization methods can be used. In addition to Hamilton–Jocobi, Lyapunov, and other methods covered in this chapter, as illustrative educational examples, neural network and fuzzy logic concepts are used. Control methods, control laws, and controllers must be applied and substantiated for physical systems. The designed controllers must be compliant, practical, and *implementable*.

MATLAB offers various educational examples. Consider the so-called NARMA-L2 control problem for a magnetic levitation system from the Neural Network Toolbox. One must refine a "plant" model, derive and use adequate equations of motion for practical levitation systems, consider the voltage applied as a control variable, etc. We use the provided "plant" as a ready-to-use example. No changes were made to the original "plant" model and simulation settings. Our goal is to compare a neural network control law, which is extremely difficult or impossible to implement, with a simplest proportional control. The proportional controller can be implemented by using a single operational amplifier. The Simulink model is reported in Figure 8.11a. The simulation results are documented for a trained "neural controller" and a proportional control law $u = k_p e$, $e = (r - y)$ with $k_p = 1000$. The simulated dynamics is illustrated in Figures 8.11b and c, respectively. A simplest proportional control law $u = k_p e$ guarantees an adequate performance and overperforms a hypothetical neural network solution. In general, PID control laws guarantee compliance, *implementability*, and practicality.

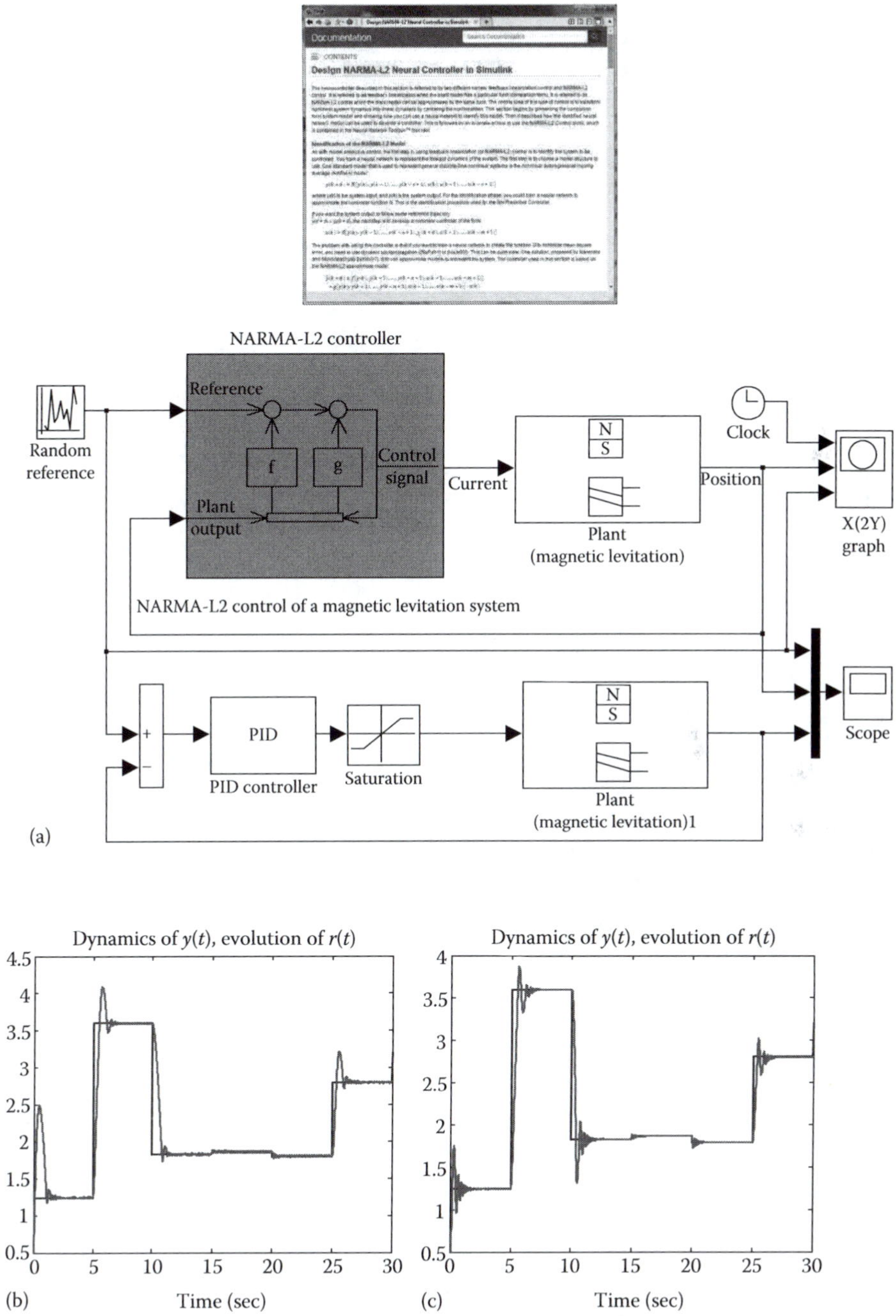

FIGURE 8.11 (a) NARMA-L2 "neural controller" and Simulink® model; (b) "Trained neural controller": Dynamics of displacement $y(t)$ and evolution of reference $r(t)$; (c) Proportional control law $u = k_p e$, $e = (r - y)$, $k_p = 1000$: Dynamics of $y(t)$ and evolution of reference $r(t)$.

The magnetic levitation system was studied in Chapter 3. A consistent mathematical model is

$$\frac{di}{dt}=\frac{1}{L(x)+L_l}\left[-ri-iv\frac{\partial L(x)}{\partial x}+u\right],$$

$$\frac{dv}{dt}=\frac{1}{m}\left[\frac{1}{2}i^2\frac{\partial L(x)}{\partial x}-mg-F_\xi\right],$$

$$\frac{dx}{dt}=v,\quad x_{\min}\le x\le x_{\max}.$$

where $L(x)$ is the magnetizing inductance that was measured and approximated, such as $L(x)=ae^{-bx}$ or $L(x)=\dfrac{a}{b+cx}$; F_ξ is the disturbance and perturbation force.

For a magnetic levitation system, documented in Figure 8.12a, the measured parameters are $r_a = 3.4$ ohm, $L_l = 0.001$ H, $m = 0.0054$ kg, $L(x) = ae^{-bx}$, $a = 1.9$, and $b = 381$. The closed-loop system is designed with the proportional-integral control law as

$$u(t)=\text{sat}_{u_{\min}}^{u_{\max}}\left(k_pe(t)+k_i\int e(t)dt\right),\quad 0\le u\le 20\text{ V}.$$

As documented in Chapter 3, the one-directional electromagnetic force F_e is developed. The stability of the closed-loop system can be proven using the Lyapunov stability theory. The feedback gains are $k_p = 3.1 \times 10^5$ and $k_i = 8.6 \times 10^3$. The image of a closed-loop magnetic levitation system, experimental results on stabilization of a suspended mass, and moving mass steering with $r = 2$ mm and $r = 2.5$ mm are reported in Figures 8.12. ■

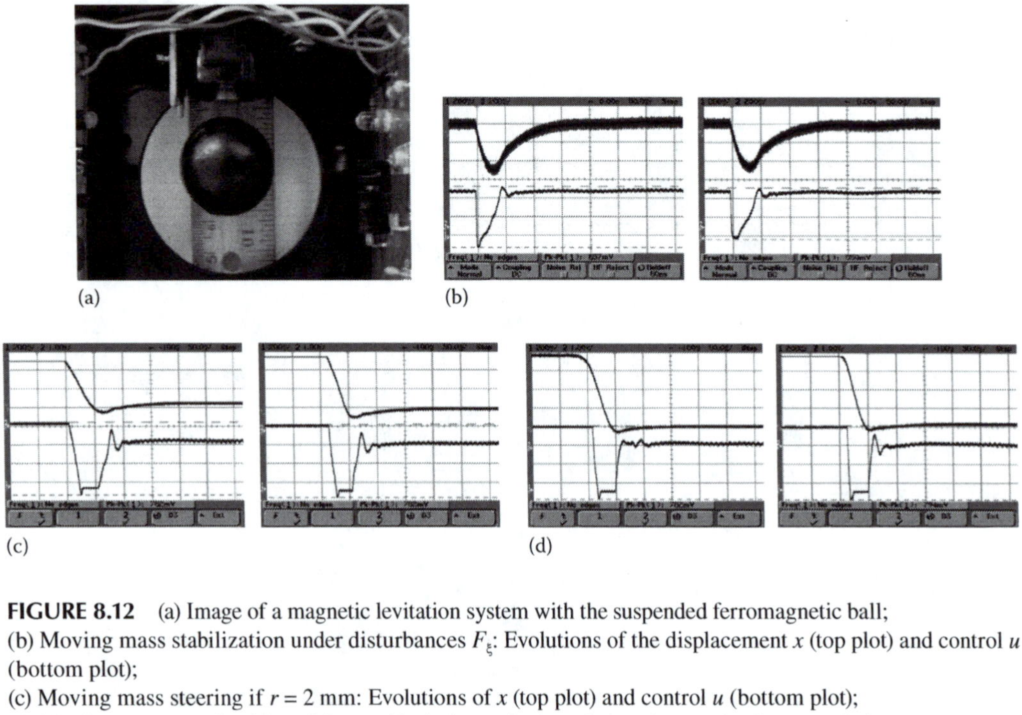

(a) (b) (c) (d)

FIGURE 8.12 (a) Image of a magnetic levitation system with the suspended ferromagnetic ball;
(b) Moving mass stabilization under disturbances F_ξ: Evolutions of the displacement x (top plot) and control u (bottom plot);
(c) Moving mass steering if $r = 2$ mm: Evolutions of x (top plot) and control u (bottom plot);
(d) Moving mass steering if $r = 2.5$ mm: Evolutions of x (top plot) and control u (bottom plot).

8.3.2 Digital Control Laws and Transfer Functions

Microcontrollers and DSPs are used to implement control algorithms. Diagnostics, filtering, data acquisition, and other tasks can be performed using discrete mathematics, processing calculus, and digital processing capabilities of microcontrollers. Digital control algorithms can be designed and discrete-time systems studied. The measured continuous-time $x(t)$, $y(t)$, $e(t)$, $r(t)$, and other quantities can be sampled with the sampling period T_s. The continuous- and discrete-time domains are related. We have $t = kT_s$, where k is the integer. Discrete-time systems are studied using difference equations. In *hybrid* electromechanical systems, continuous-time physical devices and components are controlled using digital controllers. To design digital control laws, the differential equations can be discretized.

Example 8.10

For the first-order linear constant-coefficient differential equation

$$\frac{dx}{dt} = -ax(t) + bu(t),$$

one may find the discrete-time model in the form of a difference equation and a transfer function.

A differential equation $\frac{dx}{dt} = -ax(t) + bu(t)$ is discretized by using $t = kT_s$, yielding

$$\left.\frac{dx}{dt}\right|_{t=kT_s} = -ax(kT_s) + bu(kT_s).$$

With an adequate sampling period T_s, the forward rectangular rule (Euler approximation) gives

$$\frac{dx}{dt} \approx \frac{x(t+T_s) - x(t)}{T_s}.$$

Thus, $\left.\frac{dx}{dt}\right|_{t=kT_s} = \frac{x(kT_s + T_s) - x(kT_s)}{T_s}$.

Using the forward difference, one obtains

$$\frac{x(kT_s + T_s) - x(kT_s)}{T_s} = -ax(kT_s) + bu(kT_s).$$

We denote $x(t)$ and $u(t)$ at discrete instances t_k and t_{k+1} as $x_k = x(t)\big|_{t=kT_s}$, $x_{k+1} = x(t)\big|_{t=(k+1)T_s}$, and $u_k = u(t)\big|_{t=kT_s}$. Hence,

$$\frac{x_{k+1} - x_k}{T_s} = -ax_k + bu_k,$$

where $x_{k+1} = x[(k+1)T_s]$, $x_k = x(kT_s)$, and $u_k = u(kT_s)$.

The resulting difference equation is

$$x_{k+1} = (1 - aT_s)x_k + bT_s u_k,$$

$$\text{or} \quad x_{k+1} = a_k x_k + b_k u_k, \quad a_k = (1 - aT_s), \quad b_k = bT_s.$$

This difference equation can be written as

$$x_k = (1 - aT_s)x_{k-1} + bT_s u_{k-1}.$$

The transfer function is

$$G(z) = \frac{X(z)}{U(z)} = \frac{bT_s z^{-1}}{1 - (1 - aT_s)z^{-1}} = \frac{bT_s}{z - (1 - aT_s)}.$$

■

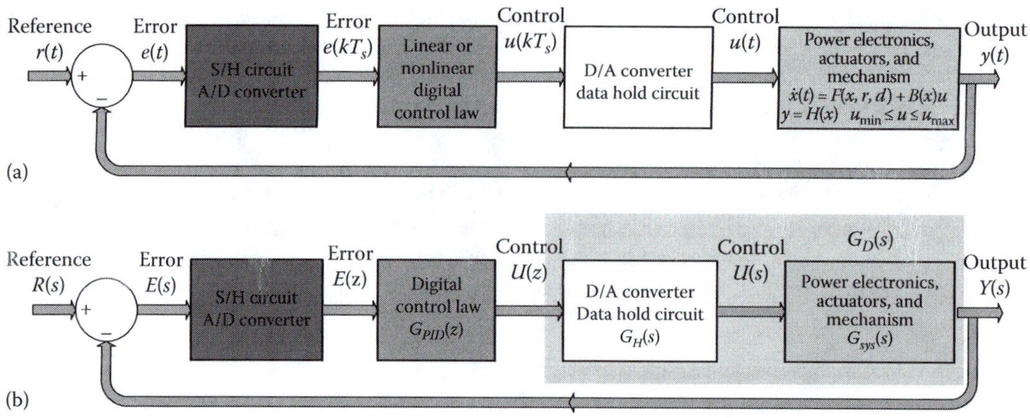

FIGURE 8.13 (a) Block diagrams of a nonlinear *hybrid* system with a digital control law; (b) A linear *hybrid* systems with a digital PID control law.

There are analog and digital components in *hybrid* systems as shown in Figures 8.13. Nonlinear and linear continuous-time systems with digital controllers, *hybrid* circuits (A/D and D/A converters, data hold circuits, etc.), power electronics and analog electromechanical motion devices are represented.

Assume that the electromechanical motion device is described by linear constant-coefficient (time-invariant) differential equations. The closed-loop system is documented in Figure 8.13b using the transfer function for the open-loop system $G_{sys}(s)$, data hold circuit $G_H(s)$, and digital control law $G_{PID}(z)$. To convert the discrete-time signals from microcontrollers to signals which drive transistors in PWM amplifiers, distinct data hold circuits are used. Zero- and first-order data hold circuits are usually implemented to avoid complexity and time delay associated with higher-order data hold circuits. The N-order data hold circuit with the zero-order data hold is documented in Figure 8.14.

For the zero-order data hold circuit, the piecewise continuous data hold output is

$$h(t) = \sum_{k=0}^{\infty} e(kT_s)\left[1\left(t - kT_s\right) - 1\left(t - (k+1)T_s\right)\right].$$

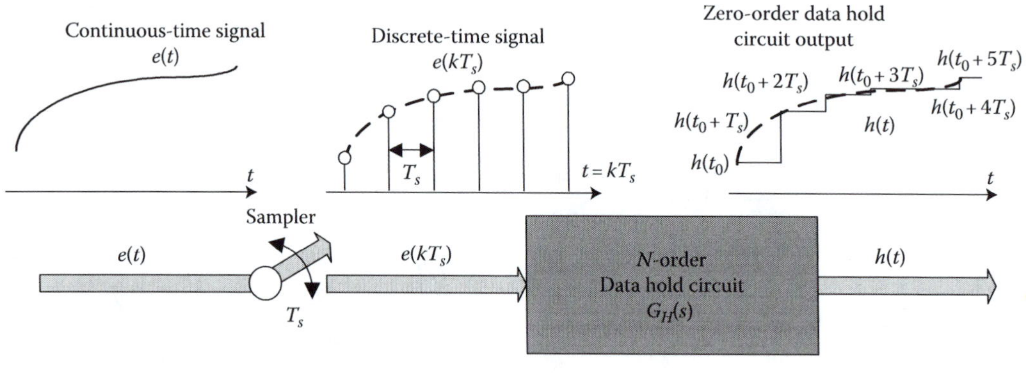

FIGURE 8.14 Sampler and N-order data hold circuit with zero-order data hold.

The transfer function of the zero-order data hold is

$$G_H(s) = (1 - e^{-T_s s})/s.$$

The first-order data hold, which can be used to perform the direct linear extrapolation, is expressed in the time domain as

$$h(t) = 1(t) + \frac{t}{T_s}1(t) - \frac{t - T_s}{T_s}1(t - T_s) - 1(t - T_s).$$

Hence, the transfer function is

$$G_H(s) = \frac{1}{s} + \frac{1}{T_s s^2} - \frac{1}{T_s s^2}e^{-T_s s} - \frac{1}{s}e^{-T_s s} = \left(1 - e^{-T_s s}\right)\frac{T_s s + 1}{T_s s^2}.$$

The system $G_{sys}(s)$ with the data hold circuit $G_H(s)$ is given by the transfer function

$$G_D(s) = G_H(s)G_{sys}(s).$$

One finds $G_D(z)$ for a given T_s.

Example 8.11

We derive the z-domain representations for digital proportional, integral, and derivative terms of the PID control law

$$u(t) = k_p e(t) + k_i \int e(t)dt + k_d \frac{de(t)}{dt}$$

$$\text{with } G_{PID}(s) = \frac{U(s)}{E(s)} = \frac{k_d s^2 + k_p s + k_i}{s}.$$

For the analog proportional control law, one has

$$u_p(t) = k_p e(t), \quad G_P(s) = \frac{U_p(s)}{E(s)} = k_p.$$

The proportional digital control law is $u_p(kT_s) = k_p e(kT_s)$ and

$$G_P(z) = \frac{U_p(z)}{E(z)} = k_p.$$

The integral and derivative terms

$$u_i(t) = k_i \int e(t)dt, G_I(s) = \frac{U_i(s)}{E(s)} = \frac{k_i}{s}$$

and $u_d(t) = k_d \dfrac{de(t)}{dt}$, $G_D(s) = \dfrac{U_d(s)}{E(s)} = k_d s$
can be discretized.
Using the z-transform, for the integral part, applying the Euler approximation, the transfer function is

$$G_I(z) = \frac{U_i(z)}{E(z)} = \frac{T_s}{1 - z^{-1}} = \frac{T_s z}{z - 1}.$$

To find the derivative term, using the trapezoidal approximation the first difference results in

$$G_D(z) = \frac{U_d(z)}{E(z)} = \frac{1-z^{-1}}{T_s} = \frac{z-1}{T_s z}.$$

Performing the summation of the derived terms, the PI, PD, or PID control laws are found. ■

For

$$u(t) = k_p e(t) + k_i \int e(t)dt + k_d \frac{de(t)}{dt} \quad \text{and} \quad U(s) = \frac{k_d s^2 + k_p s + k_i}{s} E(s),$$

one has

$$U(z) = \left(k_{dp} + \frac{k_{di}}{1-z^{-1}} + k_{dd}\left(1-z^{-1}\right) \right) E(z).$$

From $G_{PID}(z) = \frac{U(z)}{E(z)} = k_{dp} + \frac{k_{di}}{1-z^{-1}} + k_{dd}\left(1-z^{-1}\right)$,

we have

$$G_{PID}(z) = \frac{\left(k_{dp} + k_{di} + k_{dd}\right)z^2 - \left(k_{dp} + 2k_{dd}\right)z + k_{dd}}{z^2 - z}.$$

The *reference-output* form of the digital PID control law is

$$U(z) = -k_{dp}Y(z) - k_{di}\frac{Y(z) - R(z)}{1-z^{-1}} - k_{dd}\left(1-z^{-1}\right)Y(z).$$

The z-domain diagrams of the digital PID control laws for the *error* and *reference-output* forms are illustrated in Figure 8.15.

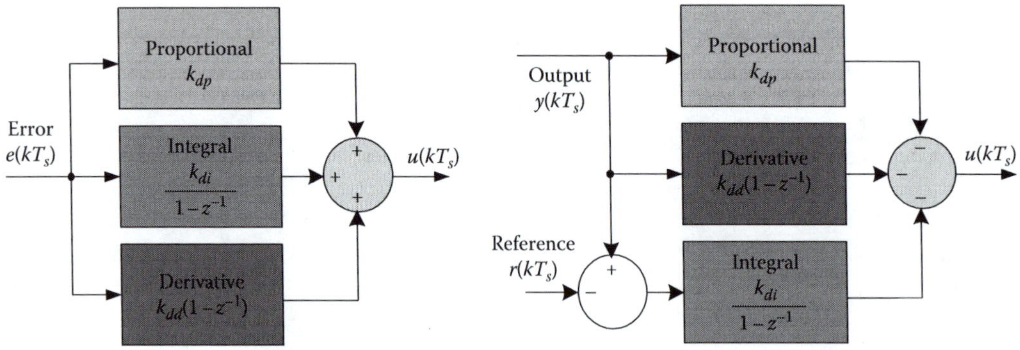

FIGURE 8.15 *Error* and *reference-output* forms of the digital PID control law.

The feedback gains k_{dp}, k_{di}, and k_{dd} of the digital control laws are related to the proportional, integral, and derivative coefficients of the analog PID control law (k_p, k_i, and k_d) as well as the sampling period T_s. The relationships between k_{dp}, k_{di}, k_{dd} and k_p, k_i, k_d can be obtained using different analytical and numerical approaches. There are equations that relate z and s. Approximation of the integral term by the trapezoidal summation and derivative term by a two-point difference yields

$$k_{dp} = k_p - \frac{1}{2}k_{di},\ k_{di} = k_i T_s, \text{ and } k_{dd} = k_d / T_s.$$

Approximating the integration (rectangular, trapezoidal Tustin, bilinear, etc.) and differentiation (Euler, Taylor, backward difference, etc.), one may obtain other $G_{PID}(z)$ and expressions for feedback gains. Using microcontrollers, the PID control law can be implemented as

$$u(kT_s) = \underbrace{k_{dp}e(kT_s)}_{\text{Proportional}} + \underbrace{\frac{1}{2}k_i T_s \sum_{j=1}^{k}\left[e((j-1)T_s) + e(iT_s)\right]}_{\text{Integral}} + \underbrace{\frac{k_d}{T_s}\left[e(kT_s) - e((k-1)T_s)\right]}_{\text{Derivative}}.$$

To find the transfer function for systems and controllers in the z-domain, the Tustin approximation is commonly applied using transfer functions in the s-domain. From

$$z = e^{sT_s},$$

we have

$$s = \frac{1}{T_s}\ln(z).$$

The series expansion of $\ln(z)$ is

$$\ln(z) = 2\left[\frac{z-1}{z+1} + \frac{1}{3}\left(\frac{z-1}{z+1}\right)^3 + \frac{1}{5}\left(\frac{z-1}{z+1}\right)^5 + \cdots\right], \quad z > 0.$$

By truncating this series expansion for $\ln(z)$, one obtains the Tustin approximation

$$\ln(z) \approx 2\frac{z-1}{z+1} = 2\frac{1-z^{-1}}{1+z^{-1}}.$$

This yields

$$s = \frac{1}{T_s}\ln(z) \approx \frac{2}{T_s}\frac{z-1}{z+1} = \frac{2}{T_s}\frac{1-z^{-1}}{1+z^{-1}}.$$

Example 8.12

For a linear PID control law with

$$G_{PID}(s) = \frac{U(s)}{E(s)} = \frac{k_d s^2 + k_p s + k_i}{s},$$

we derive the expression for $G_{PID}(z)$ applying the Tustin approximation.

From $s \approx \frac{2}{T_s}\frac{1-z^{-1}}{1+z^{-1}}$, we have

$$G_{PID}(z) = \frac{U(z)}{E(z)} = \frac{k_d\left(\frac{2}{T_s}\frac{1-z^{-1}}{1+z^{-1}}\right)^2 + k_p\frac{2}{T_s}\frac{1-z^{-1}}{1+z^{-1}} + k_i}{\frac{2}{T_s}\frac{1-z^{-1}}{1+z^{-1}}}$$

$$= \frac{\left(2k_pT_s + k_iT_s^2 + 4k_d\right) + \left(2k_iT_s^2 - 8k_d\right)z^{-1} + \left(-2k_pT_s + k_iT_s^2 + 4k_d\right)z^{-2}}{2T_s\left(1-z^{-2}\right)}.$$

Thus,

$$U(z) - U(z)z^{-2} = k_{e0}E(z) + k_{e1}E(z)z^{-1} + k_{e2}E(z)z^{-2},$$

where $k_{e0} = k_p + \frac{1}{2}k_iT_s + 2\frac{k_d}{T_s}$, $k_{e1} = k_iT_s - 4\frac{k_d}{T_s}$, and $k_{e2} = -k_p + \frac{1}{2}k_iT_s + 2\frac{k_d}{T_s}$.

The expression to implement the digital control law is

$$u(k) = u(k-2) + k_{e0}e(k) + k_{e1}e(k-1) + k_{e2}e(k-2).$$

To implement a digital control law, $e(k)$, $e(k-1)$, $e(k-2)$, and $u(k-2)$ are used. ■

The closed-loop system with a digital control law $G_{PID}(z)$ is illustrated in Figure 8.13b. The transfer function of the closed-loop systems is

$$G(z) = \frac{Y(z)}{R(z)} = \frac{G_{PID}(z)G_D(z)}{1+G_{PID}(z)G_D(z)}.$$

The analysis of linear discrete-time systems is straightforward by applying the methods of linear control theory.

Example 8.13: Digital Electromechanical Servosystem with a Permanent-Magnet DC Motor

Consider a pointing system actuated by a permanent-magnet DC motor. For this system, analog control was examined in Example 8.8. Our goal is to study the digital PID control laws. The objectives are to guarantee stability, attain the fast repositioning, and minimize tracking error. The block diagram of the closed-loop system with a digital PID controller, A/D and D/A converters, and data hold circuit is documented in Figure 8.16.

The transfer function of the open-loop system (permanent-magnet DC motor with a gearhead) is

$$G_{sys}(s) = \frac{Y(s)}{U(s)} = \frac{k_{gear}k_a}{s\left(L_aJs^2 + \left(r_aJ + L_aB_m\right)s + r_aB_m + k_a^2\right)}.$$

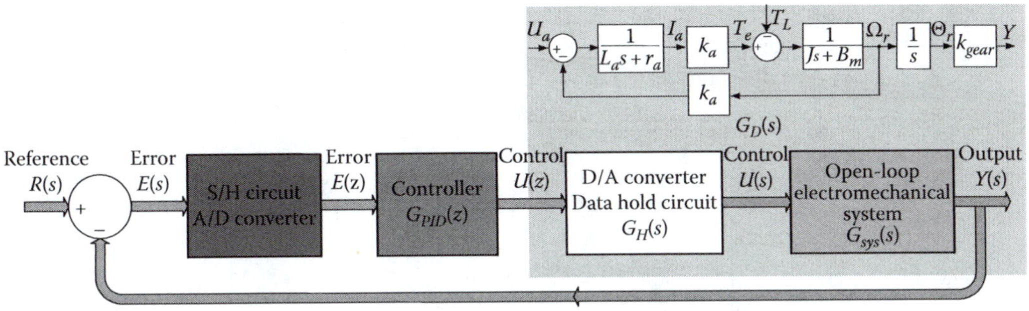

FIGURE 8.16 Block diagrams of the closed-loop system with the digital PID controller.

The transfer function of the zero-order data hold is $G_H(s) = \dfrac{1-e^{-T_s s}}{s}$. Hence,

$$G_D(s) = G_H(s)G_{sys}(s) = \frac{1-e^{-T_s s}}{s}\frac{k_{gear}k_a}{s\left(L_aJs^2 + \left(r_aJ + L_aB_m\right)s + r_aB_m + k_a^2\right)}.$$

The DC motor parameters are r_a = 200 ohm, L_a = 0.002 H, k_a = 0.2 V-sec/rad (N-m/A), J = 0.00000002 kg-m², and B_m = 0.00000005 N-m-sec/rad. The transfer function in the z-domain $G_D(z)$ is found from $G_D(s)$ by using the `c2dm` command. The `filter` command is used to simulate the dynamics. The following MATLAB file discretizes the system, performs simulations, and plots the system evolution:

```
ra=200; La=0.002; ka=0.2; J=0.00000002; Bm=0.00000005; kgear=0.01; % Motor parameters
% Numerator and denominator of the open-loop transfer function
format short e; num_s=[ka*kgear]; den_s=[La*J ra*J+La*Bm ra*Bm+ka^2 0]; num_s, den_s, pause;
% Numerator and denominator of GD(z) with zero-order data hold
Ts=0.0002;      % Sampling time (sampling period) Ts
[num_dz,den_dz]=c2dm(num_s,den_s,Ts,'zoh'); num_dz, den_dz, pause;
kp=25000; ki=250; kd=0.25;  % Feedback coefficient gains of the analog PID control law
kdi=ki*Ts; kdp=kp-kdi/2; kdd=kd/Ts;  % Feedback coefficient gains of the digital PID control law
% Numerator and denominator of the transfer function of the PID controller
num_pidz=[(kdp+kdi+kdd) -(kdp+2*kdd) kdd]; den_pidz=[1 -1 0]; num_pidz, den_pidz, pause;
% Numerator and denominator of the closed-loop transfer function G(z)
num_z=conv(num_pidz,num_dz); den_z=conv(den_pidz,den_dz)+conv(num_pidz,num_dz); num_z, den_z, pause;
k_final=20; k=0:1:k_final; % Samples, t=k*Ts
ref=1; r=ref*ones(1,k_final+1);  % Reference (command) input is r=1 rad
% Modeling of the servo-system output y(k)
y=filter(num_z,den_z,r);
plot(k,y,'o',k,y,'k-',k,r,'b:','LineWidth',3);
title('Angular Displacement, y(t)=0.01\theta_r, r(t)=1 [rad]','FontSize',16);
xlabel('Discrete Time k, Continuous Time t=kT_s [sec]','FontSize',16);
ylabel('Output y(k) and Reference r(k)','FontSize',16); axis([0 20,0 1.2])
```

For the open-loop system $G_{sys}(s) = \dfrac{Y(s)}{U(s)} = \dfrac{2\times10^{-3}}{s\left(4\times10^{-11}s^2 + 4\times10^{-6}s + 4\times10^{-2}\right)}$.

The sampling time is T_s = 0.0002 sec. The transfer function $G_D(z)$ is

$$G_D(z) = \frac{5.53\times10^{-6}z^2 + 3.41\times10^{-6}z + 8.6\times10^{-9}}{z^3 - 1.1z^2 + 0.105z - 2.06\times10^{-9}}.$$

The transfer function of the digital PID control law is

$$G_{PID}(z) = \frac{\left(k_{dp} + k_{di} + k_{dd}\right)z^2 - \left(k_{dp} + 2k_{dd}\right)z + k_{dp}}{z^2 - z},$$

$$k_{dp} = k_p - \frac{1}{2}k_{di}, \quad k_{di} = k_iT_s, \quad k_{dd} = \frac{k_d}{T_s}.$$

Let the feedback gains of the analog PID control law are $k_p = 25000$, $k_i = 250$, and $k_d = 0.25$. The feedback coefficients of the digital control law are found using

$$k_{dp} = k_p - \frac{1}{2}k_{di}, \quad k_{di} = k_i T_s, \quad \text{and} \quad k_{dd} = k_d / T_s.$$

Hence,

$$G_{PID}(z) = \frac{2.63 \times 10^4 z^2 - 2.75 \times 10^4 z + 1.25 \times 10^3}{z^2 - z}.$$

The transfer function of the closed-loop system is

$$G(z) = \frac{Y(z)}{R(z)} = \frac{G_{PID}(z)G_D(z)}{1 + G_{PID}(z)G_D(z)}.$$

We have $G(z) = \dfrac{0.145z^4 - 0.063z^3 - 0.087z^2 + 0.004z + 1.07 \times 10^{-5}}{z^5 - 1.96z^4 + 1.15z^3 - 0.19z^2 + 0.004z + 1.07 \times 10^{-5}}$.

The output dynamics for the reference input $r(kT_s) = 1$ rad, $k \geq 0$ is shown in Figure 8.17a. The settling time is $t_{settling} = 15 \times 0.0002 = 0.003$ sec, and there is no overshoot.

The sampling time, defined by the microcontroller capabilities, significantly affects the system dynamics. Using the MATLAB file reported, for the sampling time $T_s = 0.001$ sec, the closed-loop transfer function is

$$G(z) = \frac{1.14z^4 - 1.02z^3 - 0.12z^2 + 0.0012z + 2.61 \times 10^{-10}}{z^5 - 0.86z^4 - 0.021z^3 - 0.12z^2 + 0.0012z + 2.61 \times 10^{-10}}.$$

The output of the servosystem $y(kT_s)$ is plotted in Figures 8.17b and c for $T_s = 0.001$ sec and $T_s = 0.0015$ sec. If $T_s = 0.001$ sec, the settling time is $t_{settling} = 5 \times 0.001 = 0.005$ sec, and the overshoot is ~14%. For $T_s = 0.0015$ sec, as documented in Figure 8.17c, the overshoot is 77% and the settling time $t_{settling} = 10 \times 0.0015 = 0.015$ sec. If $T_s = 0.0018$ sec, the closed-loop system becomes unstable. Thus, the sampling time significantly affects the closed-loop system's performance and stability. For large T_s, one must refine the feedback coefficients k_{dp}, k_{di}, and k_{dd} to ensure the stability, adequate dynamics, overshoot, etc. However, the linear analysis has a limited practicality. The control bounds $u_{min} \leq u \leq u_{max}$ must be examined. Neglecting nonlinearities, presumably stable systems may become unstable because nonlinearities significantly affect stability. Unstable systems can be stable with inherent system nonlinearities and constraints. Using the `filter` command, simulations were performed assuming that the system was linear, and no constraints were imposed. One must carry out the nonlinear simulations. Various MATLAB commands and Simulink blocks can be used. ■

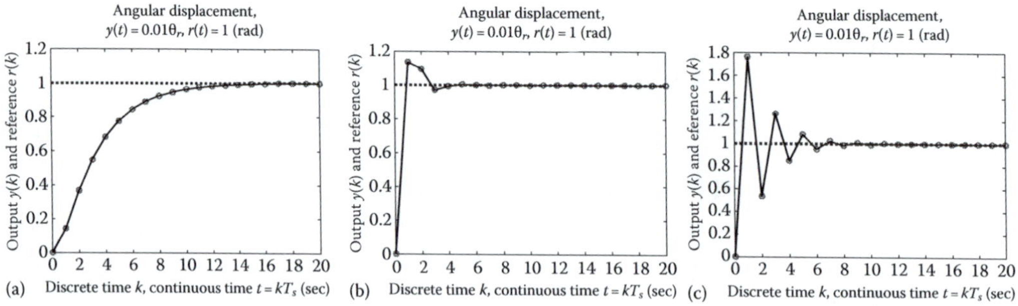

FIGURE 8.17 Output dynamics of the system with a digital PID control law: (a) $T_s = 0.0002$ s; (b) $T_s = 0.001$ s; (c) $T_s = 0.0015$ s.

8.4 CONTROLLABILITY, OBSERVABILITY, OBSERVER DESIGN, AND MODAL CONTROL

Physical electromechanical, electronic, and mechanical systems are controllable. These systems are described by nonlinear differential and partial differential equations. One may examine controllability and observability. For physical electromechanical systems, which are controllable, the analysis of controllability is mainly mathematical exercize. The observability analysis and observer design are studied solving the pole-placement *modal* control problem.

A multi-input/multi-output linear time-invariant system

$$\frac{dx}{dt} = Ax + Bu, \quad u = -K_F x, \quad y = Hx + Du, \quad C = \begin{bmatrix} B & AB & \cdots & A^{n-1}B \end{bmatrix},$$

is controllable if the rank(C) = n. Here, $C \in \mathbb{R}^{n\times n}$ is the controllability matrix $C = \begin{bmatrix} B & AB & \cdots & A^{n-1}B \end{bmatrix}$ and $K_F \in \mathbb{R}^{m\times n}$ is the feedback gain matrix.

If rank(C) = $r < n$, then only r eigenvalues of $(A-BK_F) \in \mathbb{R}^{n\times n}$ can be assigned.

Using control laws $u = -K_F x$, one can alter the eigenvalues of closed-loop systems. The eigenvalues can be precisely placed if a constant gain matrix K_F exists such that the feedback control law

$$u = -K_F x, \quad K_F \in \mathbb{R}^{m\times n}$$

will ensure the specified characteristic eigenvalues.

We examine the eigenvalues of the closed-loop system

$$\frac{dx}{dt} = Ax + Bu = Ax - BK_F x = \left(A - BK_F\right)x$$

using the characteristic equation $\left|sI - \left(A - BK_F\right)\right| = 0$.

The system is observable if a constant estimator gain matrix K_E exists such that the eigenvalues of $(A-K_E H) \in \mathbb{R}^{n\times n}$ can be assigned. A condition for observability is rank(O) = n, where $O \in \mathbb{R}^{n\times n}$ is the observability matrix, $O = \begin{bmatrix} H & HA & \cdots & HA^{n-1} \end{bmatrix}^T$. If rank($O$) = $r < n$, only r eigenvalues of $(A - K_E H)$ can be assigned.

Example 8.14

For a servosystem with a permanent-magnet DC motor, consider a stabilizing control law $u = -K_F x$. Our goal is to find the feedback matrix K_F for a linear servosystem, studied in Example 8.8. To illustrate the concept, let all motor parameters be equal to 1. Hence,

$$\frac{di_a}{dt} = \frac{1}{L_a}(-r_a i_a - k_a \omega_r + u_a) = -i_a - \omega_r + u_a,$$

$$\frac{d\omega_r}{dt} = \frac{1}{J}\left(k_a i_a - B_m \omega_r - T_L\right) = i_a - \omega_r,$$

$$\frac{d\theta_r}{dt} = \omega_r.$$

One has

$$\frac{dx}{dt} = Ax + Bu = \begin{bmatrix} -1 & -1 & 0 \\ 1 & -1 & 0 \\ 0 & 1 & 0 \end{bmatrix} x + \begin{bmatrix} 1 \\ 0 \\ 0 \end{bmatrix} u, \quad y = Hx, \quad H = \begin{bmatrix} 0 & 0 & 1 \end{bmatrix},$$

$$x_1 = i_a, \quad x_2 = \omega_r, \quad x_3 = \theta_r, \quad y = x_3 = \theta_r \quad \text{and} \quad u = u_a.$$

Let the control law $u = -K_F x$ must guarantee the specified eigenvalues $s_1 = -1$, $s_2 = -2$, and $s_3 = -3$.

We find and compare two characteristic equations

$$\left|sI-\left(A-BK_F\right)\right|=0, \text{ and, } (s+1)(s+2)(s+3)=0.$$

One finds

$$\left|sI-\left(A-BK_F\right)\right|=\left|\begin{bmatrix} s & 0 & 0 \\ 0 & s & 0 \\ 0 & 0 & s \end{bmatrix}-\begin{bmatrix} -1 & -1 & 0 \\ 1 & -1 & 0 \\ 0 & 1 & 0 \end{bmatrix}+\begin{bmatrix} 1 \\ 0 \\ 0 \end{bmatrix}\begin{bmatrix} k_{F1} & k_{F2} & k_{F3} \end{bmatrix}\right|=\left|\begin{bmatrix} s+1+k_{F1} & 1+k_{F2} & k_{F3} \\ -1 & s+1 & 0 \\ 0 & -1 & s \end{bmatrix}\right|$$

$$=s^3+\left(2+k_{F1}\right)s^2+\left(2+k_{F1}+k_{F2}\right)s+k_{F3}=0$$

and $(s+1)(s+2)(s+3)=s^3+6s^2+11s+6=0$.

Hence, $k_{F1}=4$, $k_{F2}=5$ and $k_{F3}=6$.

Thus $K_F=\begin{bmatrix} 4 & 5 & 6 \end{bmatrix}$. The stabilizing control law that guarantees $s_1=-1$, $s_2=-2$, $s_3=-3$ is given as

$$u=-K_Fx=-\begin{bmatrix} 4 & 5 & 6 \end{bmatrix}x=-4x_1-5x_2-6x_3.$$

The following MATLAB script uses the place command to solve this problem

```
A=[-1 -1 0; 1 -1 0; 0 1 0]; B=[1; 0; 0]; H=[0 0 1]; % Matrices A, B and H
p=[-1 -2 -3];                  % Desired eigenvalues locations
KF=place(A,B,p); KF            % Computing control gain matrix KF
Aclosed=A-B*KF; Aclosed % Finding closed-loop system matrix
E=eig(Aclosed); E              % Computing closed-loop eigenvalues
```

The controllability and observability matrices, and, their ranks, are found using

```
C=ctrb(A,B); C, Crank=rank(C)  % Controllability matrix C, rank of C
O=obsv(A,H); O, Orank =rank(O) % Observability matrix O, rank of O
```

We have

$$C=\begin{bmatrix} 1 & -1 & 0 \\ 0 & 1 & -2 \\ 0 & 0 & 1 \end{bmatrix},\quad \text{rank}(C)=3,\quad \text{and}\quad O=\begin{bmatrix} 0 & 0 & 1 \\ 0 & 1 & 0 \\ 1 & -1 & 0 \end{bmatrix},\quad \text{rank}(O)=3.$$ ■

Example 8.15

For a servosystem with a permanent-magnet DC motor, we design a stabilizing control law $u=-K_Fx$. The feedback matrix K_F should be found to guarantee $s_1=-100$, $s_2=-500$, and $s_3=-1000$. Adequate eigenvalues must be assigned. Inadequate eigenvalues may result in unstable dynamics, sensitivity, etc. The system coefficients are $r_a=3.15$ ohm, $L_a=0.0066$ H, $k_a=0.16$ V-sec/rad, $J=0.0001$ kg-m², $B_m=0.0001$ N-m-sec/rad, and $k_{gear}=0.1$. The control is bounded as $-30\le u_a\le 30$ V. Hence,

$$\frac{dx}{dt}=Ax+Bu=\begin{bmatrix} -\frac{r_a}{L_a} & -\frac{k_a}{L_a} & 0 \\ \frac{k_a}{J} & -\frac{B_m}{J} & 0 \\ 0 & 1 & 0 \end{bmatrix}x+\begin{bmatrix} \frac{1}{L_a} \\ 0 \\ 0 \end{bmatrix}u,\quad y=Hx,\quad H=\begin{bmatrix} 0 & 0 & k_{gear} \end{bmatrix},$$

$$x_1=i_a,\quad x_2=\omega_r,\quad x_3=\theta_r,\quad y=k_{gear}x_3=k_{gear}\theta_r,\quad \text{and}\quad u=u_a.$$

The control law $u = -K_F x$ should guarantee $s_1 = -100$, $s_2 = -500$ and $s_3 = -1000$.

The MATLAB file to find the controllability, observability, and feedback matrices is

```
ra=3.15; La=0.0066; ka=0.16; J=0.0001; Bm=0.0001; kgear=0.1; % System parameters
% Matrices A, B and H
A=[-ra/La -ka/La 0; ka/J -Bm/J 0; 0 1 0]; B=[1/La; 0; 0]; H=[0 0 kgear];
C=ctrb(A,B); Crank=rank(C), C, Crank          % Controllability matrix
O=obsv(A,H); Orank=rank(O), O, Orank          % Observability matrix
p=[-100 -500 -1000];                          % Desired eigenvalues
KF=place(A,B,p); KF                           % Control gain matrix KF
Aclosed=A-B*KF; Aclosed                       % Closed-loop system matrix
E=eig(Aclosed); E                             % Closed-loop eigenvalues
```

We have rank(C) = 3, rank(O) = 3, and $K_F = \begin{bmatrix} 7.4 & 2.51 & 206.3 \end{bmatrix}$.

The stabilizing control law is

$$u = -K_F x = -\begin{bmatrix} 7.4 & 2.51 & 206.3 \end{bmatrix} x$$

$$= -7.4x_1 - 2.51x_2 - 206.3x_3 = -7.4i_a - 2.51\omega_r - 206.3\theta_r.$$

The Simulink model is reported in Figure 8.18a. With no constraints, the transient dynamics is documented in Figure 8.18b if $\begin{bmatrix} x_{10} \\ x_{20} \\ x_{30} \end{bmatrix} = \begin{bmatrix} 10 \\ 25 \\ 1 \end{bmatrix}$. The stabilization problem is solved, and the eigenvalues are placed at $s_1 = -100$, $s_2 = -500$, and $s_3 = -1000$. Even with an adequate choice of the eigenvalues, control u bounds $-30 \le u \le 30$ are reached even for small initial conditions. The constraints significantly affect the system dynamics, as shown in Figure 8.18c. ■

If a system

$$\frac{dx}{dt} = Ax + Bu, \quad y = Hx + Du, \quad D = 0$$

is completely observable, a state observer can be designed using $u(t)$ and output $y(t)$ to estimate the state vector $x(t)$. The equation of the state observer is

$$\frac{d\hat{x}}{dt} = A\hat{x} + Bu + K_E\left(y - \hat{y}\right) = \left(A - K_E H\right)\hat{x} + Bu + K_E y,$$

where $\hat{x}(t)$ is the observer state, which provides an estimate of $x(t)$; $K_E \in \mathbb{R}^{n \times b}$ is the estimator matrix.

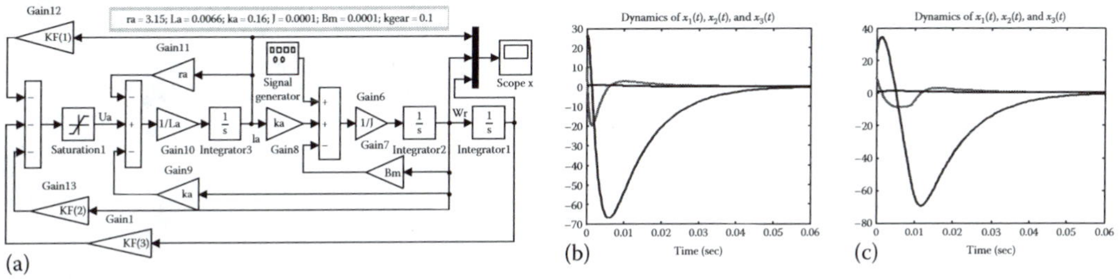

FIGURE 8.18 (a) Simulink® model of the closed-loop system with the state feedback;
(b) The dynamics of the closed-loop system, if u is not constrained. The eigenvalues $s_1 = -100$, $s_2 = -500$, and $s_3 = -1000$ are guaranteed;
(c) The dynamics of the closed-loop system if $-30 \le u \le 30$.

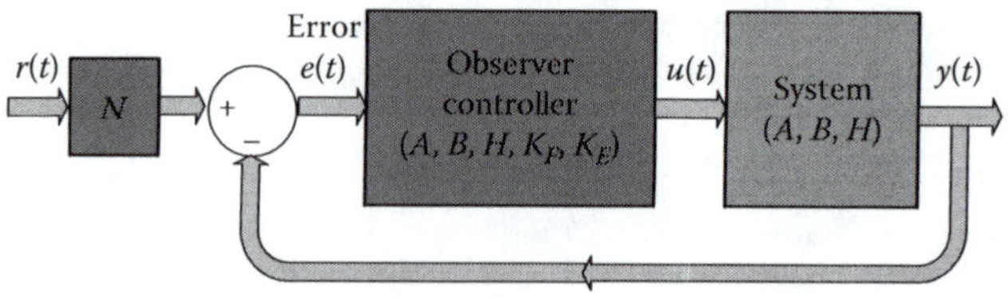

FIGURE 8.19 Observer and controller functional diagram.

The observer error is

$$\bar{e} = x - \hat{x}, \text{ and } \frac{d\bar{e}}{dt} = \left(A - K_E H\right)\bar{e},$$

where $\bar{e} \in \mathbb{R}^n$ is the observer error vector.

The convergence is guaranteed if $\lim_{t\to\infty} \bar{e}(t) = 0$. The eigenvalues of $(A - K_E H)$ must have the negative real parts. If $\lim_{t\to\infty} \bar{e}(t) = 0$, $\hat{x}(t)$ follows $x(t)$ ensuring the asymptotic convergence. The observer gain matrix K_E is found by specifying the eigenvalues of $(A - K_E H)$. The state estimation error $\left[x(t) - \hat{x}(t)\right]$ should converge faster than the system transients. Therefore, the observer eigenvalues must be greater than the closed-loop system's eigenvalues.

For the closed-loop system

$$\frac{dx}{dt} = Ax + Bu = Ax - BK_F x = \left(A - BK_F\right)x, \quad y = Hx,$$

the state-space equation with the observer becomes

$$\frac{d\hat{x}}{dt} = \left(A - BK_F - K_E H\right)\hat{x} - K_E\left(Nr - y\right),$$

where $[Nr(t) - y(t)]$ is the observer and controller input.

The eigenvalues of the closed-loop system are the union of the eigenvalues of $(A - BK_F)$ and $(A - K_E H)$ because

$$A_\Sigma = \begin{bmatrix} A & -BK_F \\ K_E H & A - BK_F - K_E H \end{bmatrix}.$$

The representative functional diagram is reported in Figure 8.19.

Example 8.16

We design the observer for the problem considered in Example 8.14. For a system

$$\frac{dx}{dt} = Ax + Bu = \begin{bmatrix} -1 & -1 & 0 \\ 1 & -1 & 0 \\ 0 & 1 & 0 \end{bmatrix} x + \begin{bmatrix} 1 \\ 0 \\ 0 \end{bmatrix} u, \quad y = Hx, \quad H = \begin{bmatrix} 0 & 0 & 1 \end{bmatrix},$$

the observer should be designed to guarantee the state vector $x(t)$ estimations using the system output y. The observer gain matrix K_E is found specifying the eigenvalues $s_1 = -5$, $s_2 = -10$, and $s_3 = -15$.

We have $K_E = \begin{bmatrix} 477 \\ 217 \\ 28 \end{bmatrix}$.

The following MATLAB script uses the `place` command to find K_E for the simulation. The evolution of $x(t)$ and $\hat{x}(t)$, as well as $y(t)$ and $\hat{y}(t)$, are reported in Figure 8.20a.

```
A=[-1 -1 0; 1 -1 0; 0 1 0]; B=[1; 0; 0]; % State matrices A and B
H=[0  0  1]; D=0;    % Output matrices C and D
p=[-5 -10 -15];      % Desired observer eigenvalues locations
KE=place (A',H',p)';           % Observer gain matrix KE
Aob=A-KE*H; eig(Aob);  % Observer system matrix and eigenvalues
x0=[-1; 1; 0]; t=[0:0.01:5]'; % Initial conditions and time
u=0*t;        % No input
G=ss(A,B,H,D);         % Develop system as a linear time-invariant model
[y,t,x]=lsim(G,u,t,x0);      % Simulate system to find x(t) and y(t)
G_ob=ss(Aob,KE,H,D); % Observer design
[yhat,t,xhat]=lsim(G_ob,y,t); % Simulate observer with zero initial conditions
plot(t,x,'r',t,xhat,'k--','LineWidth',2.5);  % Plot for the states x and xhat
title('Dynamics of x(t) and $\hat{x}$(t)','Interpreter','latex','FontSize',20)
xlabel('Time [seconds]', 'Interpreter','latex','FontSize',20); pause
plot(t,y,'r',t,yhat,'k--','LineWidth',2.5);  % plot for the output y and yhat
title('Dynamics of y(t) and $\hat{y}$(t)','Interpreter','latex','FontSize',20)
xlabel('Time [seconds]', 'Interpreter','latex','FontSize',20);
```

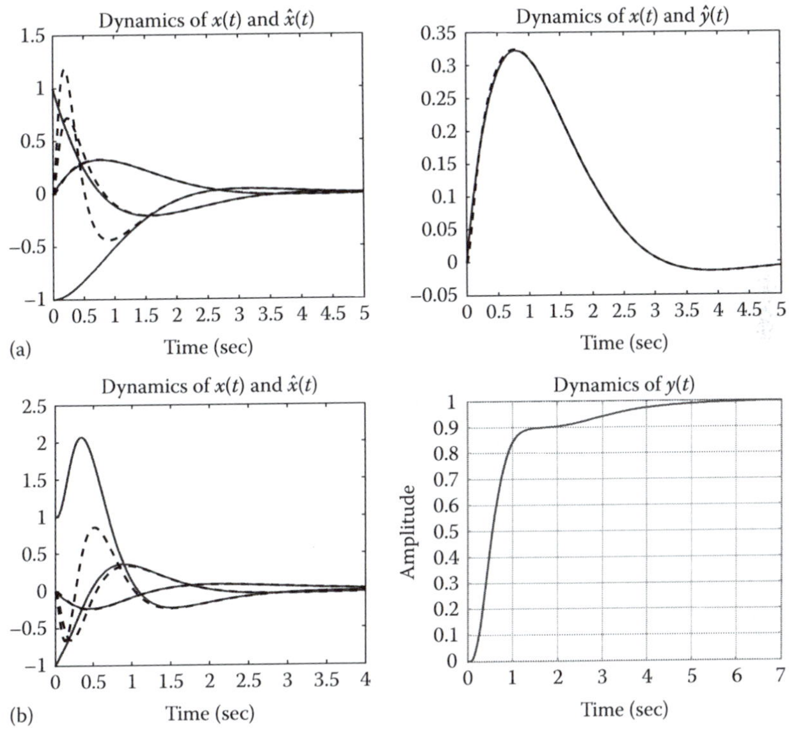

FIGURE 8.20 Dynamics of $x(t)$, $\hat{x}(t)$, and $y(t)$:
(a) Observer design: Dynamics of $x(t)$ (solid lines), $\hat{x}(t)$ (dashed lines), $y(t)$ (solid lines) and $\hat{y}(t)$ (dashed lines);
(b) Control law and observer design: Transient dynamics of $x(t)$ (solid lines) and $\hat{x}(t)$ (dashed lines) due to the initial conditions, and the evolution of the output $y(t)$ for the step reference input.

One designs the controller and observer. As in Example 8.14, we find the stabilizing control law $u = -K_F x$ to guarantee $s_1 = -1$, $s_2 = -2$, and $s_3 = -3$. The simulations are performed using the `lsim` and `step` commands. Figure 8.20b documents the dynamics of $x(t)$, $\hat{x}(t)$, and $y(t)$.

```
A=[-1 -1 0; 1 -1 0; 0 1 0]; B=[1; 0; 0]; H=[0  0  1]; D=0;
% State and output matrices
Gp=ss(A,B,H,D);                    % System as a linear time-invariant model
ps=[-1 -2 -3];               % Desired system eigenvalues
KF=place(A,B,ps);            % Feedback gain matrix KF: Modal control design
po = [-5 -10 -15];           % Desired eigenvalues
KE=place (A',H',po)';        % Observer gain matrix KE
% Design of an observer and control law
As=A-B*KF-KE*H;                    % Observer-controller matrix
Gc=ss(As,-KE,-KF,0);         % Closed-loop system with controller and observer
f_poles=pole(Gc);            % Controller and observer characteristic eignevalues
f_zeros=tzero(Gc);           % Controller and observer zeros
bode(Gc); pause;             % Bode plot
GpGc=Gp*Gc;                  % System with controller and observer
Gcl=feedback(GpGc,1,-1); cl_loop_poles=pole(Gcl);
lfg=dcgain(Gcl); N=1/lfg;          % DC gain & Normalization
Ref=N*Gcl; t=[0:0.01:4]'; r=0*t;          % Reference input
z0=[1 -1 0 0 0 0]'; [y,t,z]=lsim(Ref,r,t,z0); % Closed-loop system simulation
plot (t,z(:,1:3),'r-',t,z(:,4:6),'k--','LineWidth',2.5);
title('Dynamics of x(t) and $\hat{x}$(t)','Interpreter','latex','FontSize',20)
xlabel('Time [seconds]', 'Interpreter','latex','FontSize',20); pause
step(Ref), grid; [ys,t,z]=step(Ref);
title('Dynamics of y(t)','Interpreter','latex','FontSize',20)
xlabel('Time','Interpreter','latex','FontSize',20);
```

■

8.5 OPTIMAL CONTROL

We solve optimization problems and design control laws by minimizing performance functionals that quantitatively describe system performance and capabilities. The state $x \in \mathbb{R}^n$, output $y \in \mathbb{R}^b$, tracking error $e \in \mathbb{R}^b$, and control $u \in \mathbb{R}^m$ are used. The variables tuple (x, e, u) or quadruple (x, y, e, u) quantify and specify the system performance. We need to analytically derive the *admissible* (bounded or unbounded) time-invariant or time-varying control law as a nonlinear function of error $e(t)$ and state $x(t)$ vectors

$$u = \phi(t, x) \quad \text{or} \quad u = \phi(t, e, x), \tag{8.6}$$

by minimizing the performance functional

$$J\big(x(\cdot),u(\cdot)\big) = \int_{t_0}^{t_f} W_{xu}(x,u)dt \quad \text{or} \quad J\big(x(\cdot),e(\cdot),u(\cdot)\big) = \int_{t_0}^{t_f} W_{xeu}(x,e,u)dt, \tag{8.7}$$

subject to the system dynamics and constraints imposed.

In (8.6) and (8.7), $\phi(\cdot)$ is a continuous, piecewise continuous, or discontinuous bounded function that represents the inherent limits in physical systems; $W_{xeu}(\cdot)$: $\mathbb{R}^n \times \mathbb{R}^b \times \mathbb{R}^m \to \mathbb{R}_{\geq 0}$ is the positive-definite, continuous, and differentiable integrand function synthesized by the designer; t_0 and t_f are the initial and final time that define the time horizon.

One may use the quadratic functional

$$J = \int_{t_0}^{t_f} \left(e^2 + u^2\right) dt,$$

where $W_{eu}(e, u) = e^2 + u^2$. This performance functional is positive-definite, and the performance integrands are differentiable ensuring analytic solutions. A positive-definite $W_{eu}(e, u) = |e| + |u|$

may cause complexity. The discontinuous integrands in $W_{xeu}(\cdot)$ can be approximated by continuous differentiable functions, or search, parametric, and numeric optimization can be performed.

One may design *admissible* control law $u = \phi(e, x)$ or *minimal-complexity* control $u = \phi(e, x_m)$ by minimizing the performance functional $J\left(x(\cdot), e(\cdot), u(\cdot)\right) = \int_{t_0}^{t_f} W_{xeu}(x, e, u)\,dt$ subject to the system dynamics and constraints. Here, e and x_m are the directly measured tracking error and states.

Feedback coefficients of linear and nonlinear control laws can be derived.

One can design a great variety of functionals by specifying optimal performance. For example, nonquadratic functionals

$$J = \int_{t_0}^{t_f}\left(e^4 + u^6\right)dt \quad \text{or} \quad J = \int_{t_0}^{t_f}\left(|e|e^2 + e^8 + |u|u^{10}\right)dt$$

use differentiable $W_{eu}(\cdot)$.

The quadratic functionals, such as

$$J = \int_{t_0}^{t_f}\left(x^2 + e^2 + u^2\right)dt,$$

ease optimization, ensuring that the problem is analytically solvable. The application of nonquadratic integrands may result in mathematical complexity. However, the system performance may be improved.

As a performance functional (8.7) is chosen by synthesizing the integrand functions, the design implies the minimization or maximization problem using the Hamilton–Jacobi, dynamic programming, maximum principle, calculus of variations, nonlinear programming, or other concepts.

Consider the electromechanical system dynamics described by nonlinear differential equations

$$\dot{x}(t) = F(x) + B(x)u, \quad x(t_0) = x_0, \tag{8.8}$$

where $F(\cdot)\colon \mathbb{R}^n \to \mathbb{R}^n$ and $B(\cdot)\colon \mathbb{R}^n \to \mathbb{R}^{n\times m}$ are the continuous Lipschitz nonlinear maps, $F(0) = 0$ and $B(0) = 0$.

To find an optimal control, the necessary conditions for optimality are applied. For $J = \int_{t_0}^{t_f} W_{xu}(x, u)\,dt$, the Hamiltonian function is

$$H\left(x, u, \frac{\partial V}{\partial x}\right) = W_{xu}(x, u) + \left(\frac{\partial V}{\partial x}\right)^T \left[F(x) + B(x)u\right], \tag{8.9}$$

where $V(\cdot)\colon \mathbb{R}^n \to \mathbb{R}_{\geq 0}$ is the continuous and differentiable return function, $V(0) = 0$.

Control law (8.6) can be found using the first-order necessary condition for optimality

$$\frac{\partial H\left(x, u, \dfrac{\partial V}{\partial x}\right)}{\partial u} = 0. \tag{8.10}$$

Hamiltonian and Lagrangian: To maximize or minimize a function $W(x, u)$, one uses the Hamiltonian H or Lagrangian Λ. For linear and nonlinear dynamic systems (8.8), one has

$$\Lambda = W(x,u) + \lambda^T\left[Ax + Bu\right], \quad \Lambda = W(x,u) + \lambda^T\left[F(x) + B(x)u\right],$$

where λ is the Lagrange multiplier, $\lambda \geq 0$.

Using the *maximum principle*, the resulting equations are

$$\dot{x}(t) = \frac{\partial H}{\partial \lambda}^T, -\dot{\lambda} = \frac{\partial H}{\partial x}^T, \frac{\partial H}{\partial u} = 0,$$

$$\text{and} \quad \dot{x}(t) = \frac{\partial \Lambda}{\partial \lambda}^T, -\dot{\lambda} = \frac{\partial \Lambda}{\partial x}^T, \frac{\partial \Lambda}{\partial u} = 0.$$

The control function $u(\cdot)$: $[t_0, t_f) \to \mathbb{R}^m$ is obtained from (8.10) applying the performance functional (8.7). Using the quadratic performance functional

$$J = \frac{1}{2}\int_{t_0}^{t_f}\left(x^TQx + u^TGu\right)dt, \quad Q \in \mathbb{R}^{n\times n}, \quad Q \geq 0, \quad G \in \mathbb{R}^{m\times m}, \quad G > 0, \tag{8.11}$$

from (8.9) and (8.10), one finds

$$H = \frac{1}{2}\left(x^TQx + u^TGu\right) + \left(\frac{\partial V}{\partial x}\right)^T\left[F(x) + B(x)u\right]$$

and

$$\frac{\partial H}{\partial u} = u^TG + \left(\frac{\partial V}{\partial x}\right)^T B(x).$$

Hence, the control law is

$$u = -G^{-1}B^T(x)\frac{\partial V}{\partial x}. \tag{8.12}$$

The second-order necessary conditions for optimality

$$\frac{\partial^2 H\left(x, u, \dfrac{\partial V}{\partial x}\right)}{\partial u \times \partial u^T} > 0 \tag{8.13}$$

is satisfied because

$$\frac{\partial^2 H}{\partial u \times \partial u^T} = G > 0.$$

Example 8.17

Consider the *force control* problem for a moving mass assuming that the applied force F_a is a control variable u, and the velocity v is the state variable x, $x \equiv v$. From

$$\dot{x}(t) = \frac{dv}{dt} = \frac{1}{m}\sum F,$$

with the viscous friction, the dynamics is

$$\dot{x}(t) = \frac{1}{m}\sum F = \frac{1}{m}(F_a - B_m x).$$

The first-order differential equation that describes the input–output dynamics is

$$\dot{x}(t) = ax + bu,$$

where $a = -B_m/m$ and $b = 1/m$.

We minimize the quadratic functional (8.11) with the weighting coefficients q and g. The functional

$$J = \frac{1}{2}\int_{t_0}^{t_f}\left(qx^2 + gu^2\right)dt, \quad q \geq 0, \quad g > 0,$$

yields the Hamiltonian function (8.9) as

$$H\left(x, u, \frac{\partial V}{\partial x}\right) = \underbrace{\frac{1}{2}\left(qx^2 + gu^2\right)}_{\substack{\text{Performance Integrand} \\ J(x,u)=\frac{1}{2}\int_{t_0}^{t_f}(qx^2+gu^2)dt}} + \frac{\partial V}{\partial x}\underbrace{\frac{dx}{dt}}_{\substack{\text{System Dynamics} \\ \dot{x}(t)=ax+bu}} = \frac{1}{2}\left(qx^2 + gu^2\right) + \frac{\partial V}{\partial x}\left(ax + bu\right).$$

The Hamiltonian $H(\cdot)$ is minimized using the first-order necessary condition for optimality. From (8.10), one obtains

$$gu + \frac{\partial V}{\partial x}b = 0.$$

Hence,

$$u = -\frac{b}{g}\frac{\partial V}{\partial x} = -g^{-1}b\frac{\partial V}{\partial x}.$$

The continuous and differentiable return function $V(x)$ is given in the quadratic form

$$V(x) = \frac{1}{2}kx^2,$$

where k is the positive-definite unknown coefficient.

From

$$u = -g^{-1}b\frac{\partial V}{\partial x},$$

the control law is

$$u = -g^{-1}bkx, \quad k > 0.$$

The unknown k is found by solving the Riccati differential equation

$$-dk/dt = q + 2ak - g^{-1}b^2k^2.$$

If $t_f = \infty$, one solves the quadratic algebraic equation

$$-q - 2ak + g^{-1}b^2k^2 = 0, \; k > 0.$$

For $\dot{x}(t) = ax + bu$ with $u = -g^{-1}bkx$, the closed-loop system $\dot{x}(t) = \left(a - g^{-1}b^2k\right)x$ is stable if $(a - g^{-1}b^2k) < 0$. This condition for stability is guaranteed because $a < 0$, $g > 0$, and $k > 0$.

The second-order necessary conditions for optimality (8.13) is guaranteed because $\frac{\partial^2 H}{\partial u^2} = g > 0$.

The electromechanical systems are nonlinear. For example, consider nonlinear viscous friction,

$$\dot{x}(t) = \frac{1}{m}\sum F = \frac{1}{m}(F_a - B_m x - B_{m1}\,\mathrm{sgn}(x)x^{1/3}).$$

With $u = -g^{-1}bkx$, $k > 0$, the closed-loop system dynamics is

$$\dot{x}(t) = -\frac{B_m}{m}x - \frac{B_{m1}}{m}\mathrm{sgn}(x)x^{1/3} - g^{-1}b^2kx.$$

This system is stable, which can be proven by applying the Lyapunov stability theory covered in Section 8.14. ■

8.6 OPTIMIZATION OF LINEAR SYSTEMS

Consider a linear time-invariant system described by linear differential equations

$$\dot{x}(t) = Ax + Bu, \quad x(t_0) = x_0, \tag{8.14}$$

where $A \in \mathbb{R}^{n\times n}$ and $B \in \mathbb{R}^{n\times m}$ are the constant-coefficient matrices.

We solve the linear quadratic regulator (LQR) problem. Using the quadratic integrands, the quadratic performance functional is

$$J = \frac{1}{2}\int_{t_0}^{t_f}\left(x^TQx + u^TGu\right)dt, \quad Q \ge 0, \quad G > 0, \tag{8.15}$$

where $Q \in \mathbb{R}^{n\times n}$ is the positive-semidefinite constant-coefficient weighting matrix; $G \in \mathbb{R}^{m\times m}$ is the positive-definite constant-coefficient weighting matrix.

Using (8.14) and (8.15), the Hamiltonian function is

$$H\left(x, u, \frac{\partial V}{\partial x}\right) = \frac{1}{2}\left(x^TQx + u^TGu\right) + \left(\frac{\partial V}{\partial x}\right)^T\left(Ax + Bu\right). \tag{8.16}$$

The Hamilton–Jacobi functional equation is

$$-\frac{\partial V}{\partial t} = \min_u\left[\frac{1}{2}\left(x^TQx + u^TGu\right) + \left(\frac{\partial V}{\partial x}\right)^T\left(Ax + Bu\right)\right]. \tag{8.17}$$

The derivative of the Hamiltonian H exists, and the control function $u(\cdot)$: $[t_0, t_f) \to \mathbb{R}^m$ is found by using the first-order necessary condition for optimality (8.10). From

$$\frac{\partial H}{\partial u} = u^T G + \left(\frac{\partial V}{\partial x}\right)^T B,$$

one finds

$$u = -G^{-1}B^T \frac{\partial V}{\partial x}. \tag{8.18}$$

The second-order necessary condition for optimality (8.13) is guaranteed because for the positive-definite weighting matrix $G > 0$, one yields $\dfrac{\partial^2 H}{\partial u \times \partial u^T} = G > 0$.

Substituting (8.18) in (8.17), the following partial differential equation results

$$-\frac{\partial V}{\partial t} = \frac{1}{2}x^T Q x + \frac{1}{2}\left(\frac{\partial V}{\partial x}\right)^T BG^{-1}B^T \frac{\partial V}{\partial x} + \left(\frac{\partial V}{\partial x}\right)^T Ax - \left(\frac{\partial V}{\partial x}\right)^T BG^{-1}B^T \frac{\partial V}{\partial x}$$

$$= \frac{1}{2}x^T Q x + \left(\frac{\partial V}{\partial x}\right)^T Ax - \frac{1}{2}\left(\frac{\partial V}{\partial x}\right)^T BG^{-1}B^T \frac{\partial V}{\partial x}. \tag{8.19}$$

The solution of (8.19) is satisfied by the quadratic return function

$$V(x) = \frac{1}{2}x^T K x, \tag{8.20}$$

where $K \in \mathbb{R}^{n\times n}$ is the symmetric matrix, $K = K^T$.

The matrix

$$K = \begin{bmatrix} k_{11} & k_{12} & \cdots & k_{1n-1} & k_{1n} \\ k_{21} & k_{22} & \cdots & k_{2n-1} & k_{2n} \\ \vdots & \vdots & \ddots & \vdots & \vdots \\ k_{(n-1)1} & k_{(n-1)2} & \cdots & k_{(n-1)\,n-1} & k_{(n-1)\,n} \\ k_{n1} & k_{n2} & \cdots & k_{nn-1} & k_{nn} \end{bmatrix}, \quad k_{ij} = k_{ji}$$

must be positive-definite because positive-semidefinite and positive-definite constant-coefficient weighting matrices Q and G are used in the quadratic performance functional (8.15) yielding $J > 0$. The positive definiteness of the quadratic return function $V(x)$ can be verified using the Sylvester criterion.

From (8.20), applying the matrix identity $x^T KAx = \frac{1}{2}x^T\left(A^T K + KA\right)x$ in (8.19), one obtains

$$-\frac{\partial\left(\frac{1}{2}x^T K x\right)}{\partial t} = \frac{1}{2}x^T Q x + \frac{1}{2}x^T A^T K x + \frac{1}{2}x^T KAx - \frac{1}{2}x^T KBG^{-1}B^T K x. \tag{8.21}$$

The boundary condition is

$$V(t_f, x) = \frac{1}{2}x^T K(t_f) x = \frac{1}{2}x^T K_f x. \tag{8.22}$$

The following nonlinear differential equation, called the Riccati equation, must be solved to find the unknown symmetric matrix K

$$-\dot{K} = Q + A^T K + K^T A - K^T BG^{-1}B^T K, \quad K(t_f) = K_f. \tag{8.23}$$

From (8.18) and (8.20), the control law is

$$u = -G^{-1}B^T Kx. \tag{8.24}$$

The feedback gain matrix is $K_F = G^{-1}B^TK$. Using (8.14) and (8.24), the closed-loop system is

$$\dot{x}(t) = Ax + Bu = Ax - BG^{-1}B^T Kx = \left(A - BG^{-1}B^T K\right)x = \left(A - BK_F\right)x. \tag{8.25}$$

The eigenvalues of the matrix $\left(A - BG^{-1}B^T K\right) = \left(A - BK_F\right) \in \mathbb{R}^{n \times n}$ have negative real parts ensuring stability.

If in functional (8.15) $t_f = \infty$, the matrix K can be found by solving the nonlinear equation

$$0 = -Q - A^T K - K^T A + K^T BG^{-1}B^T K, \quad K > 0. \tag{8.26}$$

Example 8.18

Consider the system studied in Example 8.17 assuming $m = 1$ and neglecting viscous friction, $B_m = 0$. The differential equation is

$$\frac{dx}{dt} = u.$$

In the state-space model

$$\dot{x}(t) = Ax + Bu, A = [0] \text{ and } B = [1].$$

In the functional $J = \frac{1}{2}\int_{t_0}^{t_f} (qx^2 + gu^2)dt$, let the weighting coefficients be $q = 1$ and $g = 1$. Hence, $Q = [1]$ and $G = [1]$.

The unknown k of the quadratic return function (8.20) $V(x) = ½kx^2$ is found by solving (8.23), which is

$$-\dot{k}(t) = 1 - k^2(t), \quad k(t_f) = 0.$$

The solution of this differential equation is

$$k(t) = \frac{1 - e^{-2(t_f - t)}}{1 + e^{-2(t_f - t)}}.$$

A control law, which guarantees the minimum of the quadratic functional $J = \frac{1}{2}\int_{t_0}^{t_f} (x^2 + u^2)dt$ subject to the system dynamics $\frac{dx}{dt} = u$, is obtained using (8.24). We have

$$u = -k(t)x = -\frac{1 - e^{-2(t_f - t)}}{1 + e^{-2(t_f - t)}}x.$$

For $t_f = \infty$ and $J = \frac{1}{2}\int_{0}^{\infty} (x^2 + u^2)dt$, solving $1 - k^2 = 0$, we have $k = 1$.

Hence $u = -x$.

By applying the lqr MATLAB command, for $t_f = \infty$, one finds the feedback gain, return function coefficient k, and the eigenvalue as

```
[K_feedback,K,Eigenvalues] = lqr(0,1,1,1,0)
```

with the resulting

```
K_feedback =  1
K =           1
Eigenvalues =       -1
```

We conclude that the control law is $u = -x$, and the characteristic eigenvalue is −1. The closed-loop system is stable. The numerical results correspond to the analytic solution found. ■

Example 8.19

Consider a one-dimensional motion of a rigid-body mechanical system described by a set of two first-order differential equations

$$\frac{dx_1}{dt} = x_2, \quad \frac{dx_2}{dt} = u.$$

As considered in Example 8.4, the state variables are the displacement $x_1(t)$ and velocity $x_2(t)$, while u is the force or torque. Using the state-space notations (8.14), we have

$$\dot{x}(t) = Ax + Bu, \quad \begin{bmatrix} \dot{x}_1 \\ \dot{x}_2 \end{bmatrix} = \begin{bmatrix} 0 & 1 \\ 0 & 0 \end{bmatrix} \begin{bmatrix} x_1 \\ x_2 \end{bmatrix} + \begin{bmatrix} 0 \\ 1 \end{bmatrix} u, \quad A = \begin{bmatrix} 0 & 1 \\ 0 & 0 \end{bmatrix}, \quad B = \begin{bmatrix} 0 \\ 1 \end{bmatrix}$$

The quadratic functional (8.15) is

$$J = \frac{1}{2}\int_0^{\infty} \left(q_{11}x_1^2 + q_{22}x_2^2 + gu^2\right) dt, \quad q_{11} \geq 0, \quad q_{22} \geq 0, g > 0,$$

or $$J = \frac{1}{2}\int_0^{\infty} \left(\begin{bmatrix} x_1 & x_2 \end{bmatrix} \begin{bmatrix} q_{11} & 0 \\ 0 & q_{22} \end{bmatrix} \begin{bmatrix} x_1 \\ x_2 \end{bmatrix} + uGu \right) dt, \quad Q = \begin{bmatrix} q_{11} & 0 \\ 0 & q_{22} \end{bmatrix}, \quad G = g.$$

Using the quadratic return function (8.20)

$$V(x) = V(x_1, x_2) = \frac{1}{2}k_{11}x_1^2 + k_{12}x_1x_2 + \frac{1}{2}k_{22}x_2^2 = \frac{1}{2}\begin{bmatrix} x_1 & x_2 \end{bmatrix} \begin{bmatrix} k_{11} & k_{12} \\ k_{21} & k_{22} \end{bmatrix} \begin{bmatrix} x_1 \\ x_2 \end{bmatrix}, \quad k_{12} = k_{21},$$

the control law (8.24) is

$$u = -G^{-1}B^T Kx = -g^{-1}\begin{bmatrix} 0 & 1 \end{bmatrix} \begin{bmatrix} k_{11} & k_{12} \\ k_{21} & k_{22} \end{bmatrix} \begin{bmatrix} x_1 \\ x_2 \end{bmatrix} = -\frac{1}{g}\left(k_{21}x_1 + k_{22}x_2\right).$$

The unknown matrix $K = \begin{bmatrix} k_{11} & k_{12} \\ k_{21} & k_{22} \end{bmatrix}$, $k_{12} = k_{21}$, is found by solving the matrix Riccati equation (8.26). We have

$-Q - A^T K - K^T A + K^T B G^{-1} B^T K = 0,$

and,

$$-\begin{bmatrix} q_{11} & 0 \\ 0 & q_{22} \end{bmatrix} - \begin{bmatrix} 0 & 0 \\ 1 & 0 \end{bmatrix}\begin{bmatrix} k_{11} & k_{12} \\ k_{21} & k_{22} \end{bmatrix} - \begin{bmatrix} k_{11} & k_{21} \\ k_{12} & k_{22} \end{bmatrix}\begin{bmatrix} 0 & 1 \\ 0 & 0 \end{bmatrix} + \begin{bmatrix} k_{11} & k_{21} \\ k_{12} & k_{22} \end{bmatrix}\begin{bmatrix} 0 \\ 1 \end{bmatrix} g^{-1} \begin{bmatrix} 0 & 1 \end{bmatrix}\begin{bmatrix} k_{11} & k_{12} \\ k_{21} & k_{22} \end{bmatrix}$$
$$= \begin{bmatrix} 0 & 0 \\ 0 & 0 \end{bmatrix}.$$

Three algebraic equations to be solved are

$$\frac{k_{12}^2}{g} - q_{11} = 0, \quad -k_{11} + \frac{k_{12}k_{22}}{g} = 0, \quad \text{and} \quad -2k_{12} + \frac{k_{22}^2}{g} - q_{22} = 0.$$

The solution yields $k_{12} = k_{21} = \pm\sqrt{q_{11}g}$, $k_{22} = \pm\sqrt{g\left(q_{22} + 2k_{12}\right)}$, and $k_{11} = \dfrac{k_{12}k_{22}}{g}$.

The performance functional

$$J = \frac{1}{2}\int_0^\infty \left(q_{11}x_1^2 + q_{22}x_2^2 + gu^2\right)dt$$

is positive-definite because the quadratic terms are used, and $q_{11} \geq 0$, $q_{22} \geq 0$, $g > 0$.

Hence, $k_{11} = \sqrt{q_{11}\left(q_{22} + 2\sqrt{q_{11}g}\right)}$, $k_{12} = k_{21} = \sqrt{q_{11}g}$, and $k_{22} = \sqrt{g\left(q_{22} + 2\sqrt{q_{11}g}\right)}$.

The control law is

$$u = -\frac{1}{g}\left(\sqrt{q_{11}g}\,x_1 + \sqrt{g\left(q_{22} + 2\sqrt{q_{11}g}\right)}\,x_2\right) = -\sqrt{\frac{q_{11}}{g}}\,x_1 - \sqrt{\frac{q_{22} + 2\sqrt{q_{11}g}}{g}}\,x_2.$$

We found an analytic solution in the symbolic form. One may find the feedback gains and eigenvalues applying the `lqr` command. Let $q_{11} = 100$, $q_{22} = 10$, and $g = 1$. Using

```
A=[0 1;0 0]; B=[0;1]; Q=[100 0;0 10]; G=[1];
[K_feedback,K,Eigenvalues]= lqr(A,B,Q,G)
```

one finds

```
K_feedback =
   10.0000    5.4772
K =
   54.7723   10.0000
   10.0000    5.4772
Eigenvalues =
  -2.7386 + 1.5811i
  -2.7386 - 1.5811i
```

Hence,

$$K = \begin{bmatrix} k_{11} & k_{12} \\ k_{21} & k_{22} \end{bmatrix} = \begin{bmatrix} 54.77 & 10 \\ 10 & 5.48 \end{bmatrix}, \quad k_{11} = 54.77, \quad k_{12} = k_{21} = 10, \quad k_{22} = 5.48.$$

The control law is $u = -10x_1 - 5.48x_2$.
The stability of the closed-loop system

$$\frac{dx_1}{dt} = x_2,$$

$$\frac{dx_2}{dt} = -10x_1 - 5.48x_2,$$

is guaranteed. The eigenvalues have negative real parts. The complex eigenvalues are $-2.74 \pm 1.58i$.

■

Example 8.20

Consider a system described by the following state-space equation

$$\dot{x} = Ax + Bu = \begin{bmatrix} -10 & 0 & -20 & 0 \\ 0 & -10 & -10 & 0 \\ 10 & 5 & -1 & 0 \\ 0 & 0 & 1 & 0 \end{bmatrix} \begin{bmatrix} x_1 \\ x_2 \\ x_3 \\ x_4 \end{bmatrix} + \begin{bmatrix} 10 & 0 \\ 0 & 10 \\ 0 & 0 \\ 0 & 0 \end{bmatrix} \begin{bmatrix} u_1 \\ u_2 \end{bmatrix}.$$

The output is x_4, e.g., $y = x_4$. Hence, the output equation is

$$y = \begin{bmatrix} 0 & 0 & 0 & 1 \end{bmatrix} \begin{bmatrix} x_1 \\ x_2 \\ x_3 \\ x_4 \end{bmatrix} + \begin{bmatrix} 0 & 0 \end{bmatrix} \begin{bmatrix} u_1 \\ u_2 \end{bmatrix}.$$

The quadratic performance functional (8.15) is

$$J = \frac{1}{2}\int_0^\infty \left(x^T Q x + u^T G u\right) dt$$

$$= \frac{1}{2}\int_0^\infty \left(\begin{bmatrix} x_1 & x_2 & x_3 & x_4 \end{bmatrix} \begin{bmatrix} 0.05 & 0 & 0 & 0 \\ 0 & 0.1 & 0 & 0 \\ 0 & 0 & 0.01 & 0 \\ 0 & 0 & 0 & 1 \end{bmatrix} \begin{bmatrix} x_1 \\ x_2 \\ x_3 \\ x_4 \end{bmatrix} + \begin{bmatrix} u_1 & u_2 \end{bmatrix} \begin{bmatrix} 0.001 & 0 \\ 0 & 0.001 \end{bmatrix} \begin{bmatrix} u_1 \\ u_2 \end{bmatrix} \right) dt$$

$$= \frac{1}{2}\int_0^\infty \left(0.05x_1^2 + 0.1x_2^2 + 0.01x_3^2 + x_4^2 + 0.001u_1^2 + 0.001u_2^2\right) dt.$$

The MATLAB to design the control law and simulate the system is

```
A=[-10 0 -20 0; 0 -10 -10 0; 10 5 -1 0; 0 0 1 0];
disp('eigenvalues_A'); disp(eig(A)); % Eigenvalues of matrix A
B=[10 0;0 10;0 0;0 0]; H=[0 0 0 1]; D=[0 0 0 0];
Q=[0.05 0 0 0;0 0.1 0 0;0 0 0.01 0;0 0 0 1]; G=[0.001 0;0 0.001];
[K_feedback,K,Eigenvalues]=lqr(A,B,Q,G);
% Feedback and return function coefficients, eigenvalues
disp('K_feedback'); disp(K_feedback); disp('K'); disp(K);
disp('eigenvalues A-BK_feedback'); disp(Eigenvalues);
A_closed_loop=A-B*K_feedback; % Closed-loop system
t=0:0.002:1; uu=[0*ones(max(size(t)),4)]; % Applied inputs
x0=[20 10 -10 -20]; % Initial conditions
[y,x]=lsim(A_closed_loop,B*K_feedback,H,D,uu,t,x0); plot(t,x,'LineWidth',3);
title('System Dynamics, {\itx}_1, {\itx}_2, {\itx}_3, {\itx}_4','FontSize',18);
   xlabel('Time [sec]','FontSize',18); disp('End')
```

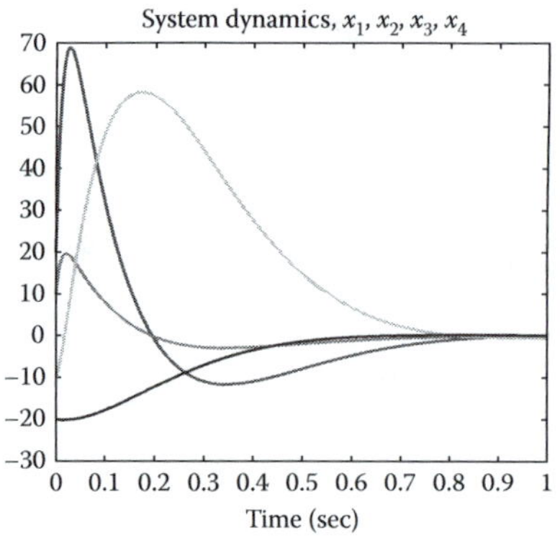

FIGURE 8.21 Dynamics of the state variables.

The return function matrix K, feedback gain matrix K_F, and the eigenvalues of the closed-loop system $\dot{x} = \left(A - BG^{-1}B^TK\right) = \left(A - BK_F\right)$ are found. The control law is

$$u = \begin{bmatrix} u_1 \\ u_2 \end{bmatrix} = -K_F x = -\begin{bmatrix} 6.78 & 0.21 & 4.77 & 29.7 \\ 0.21 & 9.12 & 1.46 & 10.9 \end{bmatrix} \begin{bmatrix} x_1 \\ x_2 \\ x_3 \\ x_4 \end{bmatrix}.$$

The dynamics of the closed-loop system states are plotted in Figure 8.21 for initial conditions $x_{10} = 20$, $x_{20} = 10$, $x_{30} = -10$, and $x_{40} = -20$. The closed-loop system is stable. ■

8.7 TRACKING CONTROL OF LINEAR SYSTEMS

Optimization and design of stabilizing control laws using the Hamilton–Jacobi theory are covered. The tracking control laws are synthesized using the tracking error $e(t) = r(t) - y(t)$. For electromechanical systems, modeled as (8.14) $\dot{x}(t) = Ax + Bu$ with the output equation $y(t) = Hx(t)$, we synthesize the tracking optimal control law by minimizing the performance functional. Using the output equation $y(t) = Hx(t)$, $e(t) = Nr(t) - y(t) = Nr(t) - Hx(t)$, $N \in \mathbb{R}^{b\times b}$.

Denoting $e(t) = \dot{x}^{ref}(t)$, consider the system dynamics

$$\dot{x}(t) = Ax + Bu, \quad y = Hx, \quad x_0(t_0) = x_0,$$
$$\dot{x}^{ref}(t) = Nr - y = Nr - Hx. \tag{8.27}$$

From (8.27), we have

$$\dot{x}_\Sigma(t) = A_\Sigma x_\Sigma + B_\Sigma u + N_\Sigma r, \quad y = Hx, \quad x_{\Sigma 0}(t_0) = x_{\Sigma 0}, \tag{8.28}$$

where $x_\Sigma = \begin{bmatrix} x \\ x^{ref} \end{bmatrix} \in \mathbb{R}^c$, $c = n + b$; $A_\Sigma = \begin{bmatrix} A & 0 \\ -H & 0 \end{bmatrix} \in \mathbb{R}^{c\times c}$; $B_\Sigma = \begin{bmatrix} B \\ 0 \end{bmatrix} \in \mathbb{R}^{c\times m}$; $N_\Sigma = \begin{bmatrix} 0 \\ N \end{bmatrix} \in \mathbb{R}^{c\times b}$.

The quadratic performance functional is

$$J = \frac{1}{2}\int_{t_0}^{t_f}\left(\begin{bmatrix} x & x^{ref}\end{bmatrix} Q \begin{bmatrix} x \\ x^{ref}\end{bmatrix} + u^T G u\right) dt. \tag{8.29}$$

From (8.28) and (8.29), the Hamiltonian function is

$$H\left(x_\Sigma, u, \frac{\partial V}{\partial x_\Sigma}\right) = \frac{1}{2}\left(x_\Sigma^T Q x_\Sigma + u^T G u\right) + \left(\frac{\partial V}{\partial x_\Sigma}\right)^T \left(A_\Sigma x_\Sigma + B_\Sigma u\right). \tag{8.30}$$

Using the first-order necessary condition for optimality (8.10), from (8.30), one finds

$$\frac{\partial H}{\partial u} = u^T G + \left(\frac{\partial V}{\partial x_\Sigma}\right)^T B_\Sigma.$$

The control law is

$$u = -G^{-1} B_\Sigma^T \frac{\partial V(x_\Sigma)}{\partial x_\Sigma} = -G^{-1}\begin{bmatrix} B \\ 0 \end{bmatrix}^T \frac{\partial V\left(\begin{bmatrix} x \\ x^{ref}\end{bmatrix}\right)}{\partial \begin{bmatrix} x \\ x^{ref}\end{bmatrix}}. \tag{8.31}$$

The solution of the Hamilton–Jacobi partial differential equation

$$-\frac{\partial V}{\partial t} = \frac{1}{2} x_\Sigma^T Q x_\Sigma + \left(\frac{\partial V}{\partial x_\Sigma}\right)^T A x_\Sigma - \frac{1}{2}\left(\frac{\partial V}{\partial x_\Sigma}\right)^T B_\Sigma G^{-1} B_\Sigma^T \frac{\partial V}{\partial x_\Sigma} \tag{8.32}$$

is satisfied by the quadratic return function

$$V(x_\Sigma) = \frac{1}{2} x_\Sigma^T K x_\Sigma. \tag{8.33}$$

From (8.32) and (8.33) the solution of the Riccati equation

$$-\dot{K} = Q + A_\Sigma^T K + K^T A_\Sigma - K^T B_\Sigma G^{-1} B_\Sigma^T K, \quad K(t_f) = K_f \tag{8.34}$$

provides the unknown symmetric matrix $K \in \mathbb{R}^{c \times c}$. The control law is found from (8.31) and (8.33) as

$$u = -G^{-1} B_\Sigma^T K x_\Sigma = -G^{-1}\begin{bmatrix} B \\ 0 \end{bmatrix}^T K \begin{bmatrix} x \\ x^{ref}\end{bmatrix}. \tag{8.35}$$

Recalling that $\dot{x}^{ref}(t) = e(t)$, one has $x^{ref}(t) = \int e(t)dt$. We have a control law with the state feedback and integral term for $e(t)$ as

$$u(t) = -G^{-1} B_\Sigma^T K x_\Sigma(t) = -G^{-1}\begin{bmatrix} B \\ 0 \end{bmatrix}^T K \begin{bmatrix} x(t) \\ \int e(t)dt \end{bmatrix}. \tag{8.36}$$

For the derived control law with state feedback and integral term $\int e(t)dt$, the pole-placement (*modal control*) problem can be solved using the state-space methods as well as the transfer function algebra, see Practice Problem 8.3.

8.8 STATE TRANSFORMATION METHOD AND TRACKING CONTROL

The tracking control problem can be solved by designing the proportional-integral control laws using the *state transformation* method. The tracking error vector is $e(t) = Nr(t) - y(t) = Nr(t) - Hx^{sys}(t)$. For linear systems

$$\dot{x}^{sys} = A^{sys}x^{sys} + B^{sys}u, \quad y(t) = Hx^{sys}(t), \tag{8.37}$$

we specify the evolution of the dynamic tracking error as

$$\dot{e}(t) = -I_E e - HA^{sys}x^{sys} - HB^{sys}u, \tag{8.38}$$

where $I_E \in \mathbb{R}^{b\times b}$ is the positive-definite diagonal matrix, $I_E > 0$. Using the identity matrix I, one may assign $I_E = I$.

From (8.37) and (8.38), applying the expanded state vector $x(t) = \begin{bmatrix} x^{sys}(t) \\ e(t) \end{bmatrix}$, one finds

$$\dot{x} = \begin{bmatrix} \dot{x}^{sys} \\ \dot{e} \end{bmatrix} = \begin{bmatrix} A^{sys} & 0 \\ -HA^{sys} & -I_E \end{bmatrix} \begin{bmatrix} x^{sys} \\ e \end{bmatrix} + \begin{bmatrix} B^{sys} \\ -HB^{sys} \end{bmatrix} u = Ax + Bu, \quad x = \begin{bmatrix} x^{sys} \\ e \end{bmatrix}. \tag{8.39}$$

Define a vector

$$z = \begin{bmatrix} x \\ u \end{bmatrix},$$

where the evolution of u is governed by $\dot{u} = -I_U u + I_V v$ with control vector $v \in \mathbb{R}^m$. Here, $I_U \in \mathbb{R}^{m\times m}$ and $I_V \in \mathbb{R}^{m\times m}$ are the positive-definite diagonal matrices, $I_U > 0$, $I_V > 0$. For example, $I_U = I_V = I$.

Consider

$$z = \begin{bmatrix} x \\ u \end{bmatrix} = \begin{bmatrix} x^{sys} \\ e \\ u \end{bmatrix}, \quad z \in \mathbb{R}^{n+b+m}, \quad \text{and} \quad v \in \mathbb{R}^m,$$

with

$$\dot{e}(t) = -I_E e - HA^{sys}x^{sys} - HB^{sys}u, \quad \dot{u} = -I_U u + I_V v. \tag{8.40}$$

Using (8.39) and (8.40), one obtains the system model as

$$\dot{z}(t) = \begin{bmatrix} A & B \\ 0 & -I_U \end{bmatrix} z + \begin{bmatrix} 0 \\ I_V \end{bmatrix} v = A_z z + B_z v, \quad y = Hx^{sys}, \quad z(t_0) = z_0, \tag{8.41}$$

where $A_z \in \mathbb{R}^{(c+m)\times(c+m)}$ and $B_z \in \mathbb{R}^{(c+m)\times m}$ are the constant-coefficient matrices.

Minimizing the quadratic functional

$$J = \frac{1}{2}\int_{t_0}^{t_f}\left(z^T Q_z z + v^T G_z v\right)dt, \quad Q_z \in \mathbb{R}^{(c+m)\times(c+m)}, \quad Q_z \geq 0, \quad G_z \in \mathbb{R}^{m\times m}, \quad G > 0, \tag{8.42}$$

we apply the first-order necessary condition for optimality (8.10) $\frac{\partial H}{\partial u} = 0$ to the Hamiltonian $H = \frac{1}{2}z^T Q_z z + \frac{1}{2}v^T G_z v + \frac{\partial V}{\partial z}^T\left(A_z z + B_z v\right)$. One finds

$$v = -G_z^{-1}B_z^T\frac{\partial V}{\partial z}. \tag{8.43}$$

The solution of the Hamilton–Jacobi differential equation

$$-\frac{\partial V}{\partial t} = \frac{1}{2}z^T Q_z z + \left(\frac{\partial V}{\partial z}\right)^T A_z z - \frac{1}{2}\left(\frac{\partial V}{\partial z}\right)^T B_z G_z^{-1}B_z^T\frac{\partial V}{\partial z} \tag{8.44}$$

is satisfied by the continuous and differentiable quadratic return function

$$V(z) = \frac{1}{2}z^T Kz. \tag{8.45}$$

Using (8.43) and (8.45), one obtains the control function

$$v = -G_z^{-1}B_z^T Kz. \tag{8.46}$$

From (8.44), the Riccati equation to find the unknown matrix $K \in \mathbb{R}^{(c+m)\times(c+m)}$ is

$$-\dot{K} = Q_z^{+} KA_z + A_z^T K - KB_z G_z^{-1}B_z^T K, \quad K(t_f) = K_f.$$

Using (8.39) $\dot{x}(t) = Ax + Bu$, we have

$$u = B^{-1}\left(\dot{x}(t) - Ax\right) = \left(B^T B\right)^{-1}B^T\left(\dot{x}(t) - Ax\right). \tag{8.47}$$

Applying (8.40), (8.46), and (8.47), one obtains

$$\begin{aligned}\dot{u}(t) &= -G_z^{-1}B_z^T Kz - I_U u = -G_z^{-1}\begin{bmatrix}0\\ I_V\end{bmatrix}^T Kz - I_U u = -G_z^{-1}\begin{bmatrix}0\\ I_V\end{bmatrix}^T\begin{bmatrix}K_{11} & K_{21}^T\\ K_{21} & K_{22}\end{bmatrix}\begin{bmatrix}x\\ u\end{bmatrix} - I_U u\\ &= -G_z^{-1}I_V K_{21}x - \left(G_z^{-1}I_V K_{22} + I_U\right)u\\ &= \left[-G_z^{-1}I_V K_{21} + \left(G_z^{-1}I_V K_{22} + I_U\right)B^T A\right]x - \left(G_z^{-1}I_V K_{22} + I_U\right)B^T\dot{x} = K_{F2}x + K_{F1}\dot{x}.\end{aligned} \tag{8.48}$$

From (8.48), the proportional-integral control law with the state feedback and tracking error feedback is

$$u=-\left(G_z^{-1}I_VK_{22}+I_U\right)B^Tx+\int\left[-G_z^{-1}I_VK_{21}+\left(G_z^{-1}I_VK_{22}+I_U\right)B^TA\right]xdt$$

$$=-\left(G_z^{-1}I_VK_{22}+I_U\right)B^T\begin{bmatrix}x^{sys}\\ e\end{bmatrix}+\left[-G_z^{-1}I_VK_{21}+\left(G_z^{-1}I_VK_{22}+I_U\right)B^TA\right]\int\begin{bmatrix}x^{sys}\\ e\end{bmatrix}dt$$

$$=K_{F1}x+\int K_{F2}xdt,\quad x=\begin{bmatrix}x^{sys}\\ e\end{bmatrix}. \tag{8.49}$$

We designed a tracking control law with state feedback, because $x(t)=\begin{bmatrix}x^{sys}(t)\\ e(t)\end{bmatrix}$. To implement control law (8.49), the states $x^{sys}(t)$ and tracking error $e(t)$ must be directly measured or observed.

Example 8.21: Controllability and Observability

Assume system (8.37) is controllable and observable. That is, the controllability $C^{sys}=\left[B^{sys}\ A^{sys}B^{sys}\ \cdots\ A^{sys(n-1)}B^{sys}\right]$ and observability $O^{sys}=\left[C^{sys}\ C^{sys}A^{sys}\ \cdots\ C^{sys}A^{sys(n-1)}\right]^T$ matrices have full rank, rank(C^{sys}) = n, and rank(O^{sys}) = n.

System (8.41) is controllable and observable because for $C=\left[B_z\ A_zB_z\ \cdots\ A_z^{(n-1)}B_z\right]$ and $O=\left[C\ CA_z\ \cdots\ CA_z^{(n-1)}\right]^T$, rank($C$) = $n+b+m$, and rank(O) = $n+b+m$. ■

For nonlinear electromechanical systems, the proposed procedure is applied to derive control laws. Consider electromechanical systems described by differential equations

$$\dot{x}^{sys}(t)=F^{sys}(x^{sys})+B^{sys}(x^{sys})u,\quad y=Hx^{sys},\quad e(t)=Nr(t)-y(t)=Nr(t)-Hx^{sys}(t), \tag{8.50}$$

where $F^{sys}(\cdot)$ and $B^{sys}(\cdot)$ are the smooth Lipschitz maps.

To guarantee stability, the tracking error and state evolutions must be bounded for bounded input. We specify the dynamic evolution of *de*/*dt* as (8.40)

$$\dot{e}(t)=-I_Ee-HF^{sys}(x^{sys})-HB^{sys}(x^{sys})u,$$

where $I_E\in\mathbb{R}^{b\times b}$ is the diagonal matrix, $I_E>0$, for example, $I_E=I$.

The dynamics of state and error vectors are

$$\dot{x}(t)=\begin{bmatrix}\dot{x}^{sys}\\ \dot{e}\end{bmatrix}=\begin{bmatrix}F^{sys}(x^{sys})\\ -HF^{sys}(x^{sys})\end{bmatrix}+\begin{bmatrix}0\\ -I_E\end{bmatrix}\begin{bmatrix}x^{sys}\\ e\end{bmatrix}+\begin{bmatrix}B^{sys}(x^{sys})\\ -HB^{sys}(x)\end{bmatrix}u=F(x)+B(x)u. \tag{8.51}$$

Denoting $x=\begin{bmatrix}x^{sys}\\ e\end{bmatrix}$, and applying the *state transformation* procedure using $z=\begin{bmatrix}x\\ u\end{bmatrix}$, where the evolution of u is governed by (8.40) $\dot{u}=-I_Uu+I_Vv$, $v\in\mathbb{R}^m$, we have

$$\dot{z}(t)=\begin{bmatrix}\dot{x}\\ \dot{u}\end{bmatrix}=\begin{bmatrix}F(x)\\ 0\end{bmatrix}+\begin{bmatrix}B(x)u\\ -I_Uu\end{bmatrix}+\begin{bmatrix}0\\ I_V\end{bmatrix}v=F_z(z)+B_zv,\quad y=Hx^{sys}. \tag{8.52}$$

The controllability of (8.52) is guaranteed because the physical systems are controllable. Using the Lie bracket operator $\left[\mathrm{ad}_{F_z}^k B_z\right]=\left[F_z \cdots [F_z, B_z]\right]$ one finds that $C=\left[B_z \cdots [\mathrm{ad}_{F_z}^k B_z]\right]$ spans $(n+b+m)$-space with the rank $(n + b + m)$.

We minimize the quadratic functional (8.42) $J=\int_{t0}^{t_f}\left(\frac{1}{2}z^T Q_z z+\frac{1}{2}v^T G_z v\right)dt,\ Q_z \geq 0,\ G_z > 0$ subject to (8.52). Applying the first-order necessary condition for optimality $\frac{\partial H}{\partial u}=0$ to Hamiltonian $H=\frac{1}{2}z^T Q_z z+\frac{1}{2}v^T G_z v+\frac{\partial V}{\partial z}^T\left[F_z(z)+B_z v\right]$, we have

$$v=-G_z^{-1}B_z^T\frac{\partial V}{\partial z}. \tag{8.53}$$

The solution of the Hamilton–Jacobi equation

$$-\frac{\partial V}{\partial t}=\frac{1}{2}z^T Q_z z+\frac{\partial V}{\partial z}^T F_z(z)-\frac{1}{2}\frac{\partial V}{\partial z}^T B_z G_z^{-1}(z)B_z^T\frac{\partial V}{\partial z} \tag{8.54}$$

is approximated by the nonquadratic return function

$$V(z)=\sum_{i=1}^{N}\frac{2\gamma+1}{2i+2\gamma}\left(z^{\frac{i+\gamma}{2\gamma+1}}\right)^T K_i z^{\frac{i+\gamma}{2\gamma+1}} \quad \text{or} \quad V(z)=\sum_{i=1}^{N}\frac{2\gamma+1}{2i+2\gamma}\left(|z|^{\frac{i+\gamma}{2\gamma+1}}\right)^T K_i |z|^{\frac{i+\gamma}{2\gamma+1}},\quad K_i>0. \tag{8.55}$$

The unknown symmetric matrices $K_i=K_i^T$ is found by solving (8.54). For the quadratic return function $V(z) = ½z^T Kz$ ($\gamma = 0$ and $N = 1$), using (8.53), we have $v=-G_z^{-1}B_z^T Kz$. From (8.51), $u=B^{-1}(x)\left[\dot{x}(t)-F(x)\right]=\left(B^T(x)B(x)\right)^{-1}B^T(x)\left[\dot{x}(t)-F(x)\right]$. Using (8.40) and (8.53) one obtains

$$\begin{aligned}\dot{u}(t)&=-G_z^{-1}B_z^T\frac{\partial V}{\partial z}-I_U u=-G_z^{-1}\begin{bmatrix}0\\ I_V\end{bmatrix}^T\frac{\partial V(x,u)}{\partial[xu]^T}-I_U u\\ &=-G_z^{-1}\begin{bmatrix}0\\ I_V\end{bmatrix}^T Kz-I_U u=-G_z^{-1}\begin{bmatrix}0\\ I_V\end{bmatrix}^T\begin{bmatrix}K_{11} & K_{21}^T\\ K_{21} & K_{22}\end{bmatrix}\begin{bmatrix}x\\ u\end{bmatrix}-I_U u\\ &=-G_z^{-1}I_V K_{21}x-\left(G_z^{-1}I_V K_{22}+I_U\right)u\\ &=\left[-G_z^{-1}I_V K_{21}x+\left(G_z^{-1}I_V K_{22}+I_U\right)B^T(x)F(x)\right]-\left(G_z^{-1}I_V K_{22}+I_U\right)B^T(x)\dot{x}.\end{aligned} \tag{8.56}$$

A proportional-integral control law with the state feedback and tracking error feedback is

$$u=-\left(G_z^{-1}I_V K_{22}+I_U\right)B^T(x)x+\int\left[-G_z^{-1}I_V K_{21}x+\left(G_z^{-1}I_V K_{22}+I_U\right)B^T(x)F(x)\right]dt,\quad x=\begin{bmatrix}x^{sys}\\ e\end{bmatrix}. \tag{8.57}$$

Example 8.22

Considering the *force control* problem, as reported in Example 8.4, a one-dimensional motion of a moving frictionless mass is described by

$$\frac{dx_1}{dt} = u, \quad \frac{dx_2}{dt} = x_1, \quad y = x_2.$$

Consider the PID control

$$u = k_p e + k_i \int e\,dt + k_d \frac{de}{dt}, \quad e = (r - y) = (r - x_2).$$

The transfer function of the closed-loop system is $G(s) = \dfrac{k_d s^2 + k_p s + k_i}{s^3 + k_d s^2 + k_p s + k_i}$.

The characteristic equation is $s^3 + k_d s^2 + k_p s + k_i = 0$.

For the PD control law $u = k_p e + k_d \dfrac{de}{dt}$, we have

$$G(s) = \frac{k_d s + k_p}{s^2 + k_d s + k_p}$$

and

$$s^2 + k_d s + k_p = 0.$$

Applying the pole-placement approach, the feedback gains can be found to ensure the specified time constant T, damping coefficient ξ, overshoot, etc. We have

$$T = \sqrt{\frac{1}{k_p}} \quad \text{and} \quad \xi = \frac{k_d}{2\sqrt{k_p}}.$$

Let $T = \sqrt{\dfrac{1}{k_p}} = 0.2$ sec, while the two damping coefficients are $\xi = 0.707$ and $\xi = 1$.
For $T = 0.2$ sec and $\xi = 0.707$, the feedback gains of $u = k_p e + k_d \dfrac{de}{dt}$ are $k_p = 25$ and $k_d = 7.07$.
For $T = 0.2$ sec and $\xi = 1$, the feedback gains are $k_p = 25$ and $k_d = 10$.
The resulting closed-loop output dynamics $y = x_2$ for two sets of feedback gains in

$$u = k_p e + k_d \frac{de}{dt}$$

are documented in Figures 8.22a and b, indicated by dashed lines if $r = \pm 1$.
We demonstrate the design the tracking control law using the *state transformation* concept. Let the evolution of the dynamic tracking error be

$$\dot{e}(t) = -e - x_1 \quad \text{and} \quad \dot{u} = -u + v.$$

We obtain

$$z = \begin{bmatrix} x_1 \\ x_2 \\ e \\ u \end{bmatrix}, \quad A_z = \begin{bmatrix} 0 & 0 & 0 & 1 \\ 1 & 0 & 0 & 0 \\ -1 & 0 & -1 & 0 \\ 0 & 0 & 0 & -1 \end{bmatrix} \quad \text{and} \quad B_z = \begin{bmatrix} 0 \\ 0 \\ 0 \\ 1 \end{bmatrix}.$$

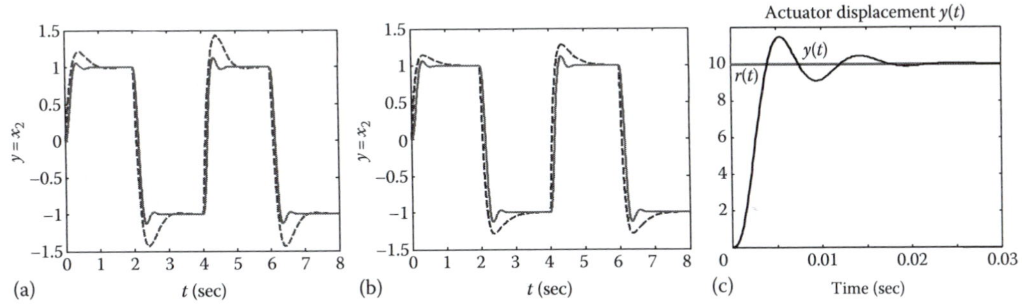

FIGURE 8.22 Output dynamics $y = x_2(t)$ of the closed-loop system:
(a) Proportional–derivative $u = k_p e + k_d\, de/dt$ control law (dashed line) with $k_p = 25$ and $k_d = 7.07$, and, the tracking control law $u = K_{F1}\begin{bmatrix} x_1 \\ x_2 \\ e \end{bmatrix} + \int K_{F2}\begin{bmatrix} x_1 \\ x_2 \\ e \end{bmatrix} dt$ (solid line);
(b) Proportional–derivative control law (dashed line) with $k_p = 25$ and, $k_d = 10$, and, the tracking control law $u = K_{F1}\begin{bmatrix} x_1 \\ x_2 \\ e \end{bmatrix} + \int K_{F2}\begin{bmatrix} x_1 \\ x_2 \\ e \end{bmatrix} dt$ (solid line);
(c) Closed-loop system dynamics if $r(t) = 10$ μm (Example 8.23).

The controllability matrix $C = \begin{bmatrix} B_z & A_z B & A_z^2 B_z & A_z^3 B_z \end{bmatrix}$ is

$$C = \begin{bmatrix} 0 & 1 & -1 & 1 \\ 0 & 0 & 1 & -1 \\ 0 & 0 & -1 & 2 \\ 1 & -1 & 1 & -1 \end{bmatrix}.$$

For the controllability C and observability $O = \begin{bmatrix} C & CA_z & CA_z^2 & CA_z^3 \end{bmatrix}^T$ matrices, rank$(C) = 4$ and rank$(O) = 4$. The number of uncontrollable and unobservable states is zero. The system with the extended vector z is controllable and observable.

For $Q_z = \begin{bmatrix} 1 & 0 & 0 & 0 \\ 0 & 1 & 0 & 0 \\ 0 & 0 & 1\times 10^7 & 0 \\ 0 & 0 & 0 & 1 \end{bmatrix}$ and $G_z = 1$,

the solution of

$$Q_z + KA_z + A_z^T K - KB_z G_z^{-1} B_z^T K = 0$$

results in

$$K = \begin{bmatrix} 8.69\times 10^3 & 2.84\times 10^1 & -8.12\times 10^4 & 4.03\times 10^2 \\ 2.84\times 10^1 & 3.16\times 10^3 & 2.76\times 10^3 & 1 \\ -8.12\times 10^4 & 2.76\times 10^3 & 1.2\times 10^6 & -2.76\times 10^3 \\ 4.03\times 10^2 & 1 & -2.76\times 10^3 & 2.74\times 10^1 \end{bmatrix}, \quad K > 0, \quad K = K^T.$$

The control law is

$$u = K_{F1}\begin{bmatrix} x_1 \\ x_2 \\ e \end{bmatrix} + K_{F2}\int \begin{bmatrix} x_1 \\ x_2 \\ e \end{bmatrix} dt, \; K_{F1} = \begin{bmatrix} -27.4 & 0 & 0 \end{bmatrix}, \; K_{F2} = \begin{bmatrix} -403 & -1 & 2758.4 \end{bmatrix}.$$

Figures 8.22a and b report simulations for the closed-loop systems if $r = \pm 1$ with a PD control law

$$u = k_p e + k_d \frac{de}{dt},$$

and designed control algorithm

$$u = K_{F1}\begin{bmatrix} x_1 \\ x_2 \\ e \end{bmatrix} + K_{F2}\int \begin{bmatrix} x_1 \\ x_2 \\ e \end{bmatrix} dt.$$

There is no derivative feedback in

$$u = K_{F1}\begin{bmatrix} x_1 \\ x_2 \\ e \end{bmatrix} + K_{F2}\int \begin{bmatrix} x_1 \\ x_2 \\ e \end{bmatrix} dt,$$

which ensures better performance than PID algorithms. ■

Example 8.23

We design the tracking control law for a system with a PZT actuator controlled by changing the applied voltage u. The second-order equation of motion is

$$m\frac{d^2y}{dt^2} + b\frac{dy}{dt} + k_e y = k_e d_e u,$$

where y is the actuator displacement. The PZT actuator parameters are $k_e = 3000$, $b = 1$, $d_e = 0.000001$, and $m = 0.02$. We have a set of two first-order differential equations

$$\frac{dy}{dt} = v,$$

$$\frac{dv}{dt} = -\frac{k_e}{m}y - \frac{b}{m}v + \frac{k_e d_e}{m}u.$$

Using the *state transformation* method, the tracking control law (8.49) is

$$u(t) = K_{F1}\begin{bmatrix} y(t) \\ v(t) \\ e(t) \end{bmatrix} + \int K_{F2}\begin{bmatrix} y(\tau) \\ v(\tau) \\ e(\tau) \end{bmatrix} d\tau.$$

The quadratic functional (8.42) is minimized with

$$Q_z = \begin{bmatrix} 1 & 0 & 0 & 0 \\ 0 & 1 & 0 & 0 \\ 0 & 0 & 1\times 10^{10} & 0 \\ 0 & 0 & 0 & 1 \end{bmatrix} \quad \text{and} \quad G_z = 10.$$

The feedback matrices K_{F1} and K_{F2} are found. The closed-loop actuator dynamics is documented in Figure 8.22c. The settling time is 0.022 sec, and the tracking error converges to zero.

The differential equations of PZT actuators may be found accounting for an inherent hysteresis. We have

$$m\frac{d^2y}{dt^2}+b\frac{dy}{dt}+k_e y=k_e\left(d_e u-z_h\right),$$

$$\dot{z}_h=\alpha d_e\dot{u}-\beta|\dot{u}|z_h-\gamma\dot{u}|z_h|,$$

where α and β are the constants.

We have the state-space nonlinear high-fidelity PZT actuator model as

$$\frac{dy}{dt}=v,$$

$$\frac{dv}{dt}=-\frac{k_e}{m}y-\frac{b}{m}v-\frac{k_e}{m}z_h+\frac{kd_e}{m}u,$$

$$\frac{dz_h}{dt}=-\beta|\dot{u}|z_h+\alpha d_e\dot{u}-\gamma|z_h|\dot{u},$$

for which a tracking control law can be designed. However, z_h cannot be directly measured. The tracking control law with directly measured velocity, displacement, and tracking error feedback will ensure tracking, stability, and near-optimal performance. The experimental results are reported in Example 8.49. ■

One may derive high-fidelity models and design corresponding control algorithms. The consistency, *implementability*, and practicality of any synthesized control laws are studied. There is no end for enhancing the model accuracy, which, in turn, may result in considerable complexity. The model used should be consistent and sufficiently accurate. The *minimal-complexity* control laws can be proportional-integral algorithms, which can be synthesized using the Lyapunov theory, search, or nonlinear optimization methods. The *minimal-complexity* control laws may not ensure optimal performance. The trade-off between hardware complexity (affected by sensors and control laws to be implemented), system performance, and capabilities must be studied. Usually, the electromechanical systems are open-loop stable. Adequate proportional or proportional-integral control laws may ensure near-optimal performance.

8.9 MINIMUM-TIME CONTROL

For dynamic systems (8.8) $\dot{x}(t)=F(x)+B(x)u$, $-1\le u\le 1$, the minimum-time control laws can be designed minimizing

$$J=\int_{t_0}^{t_f}1\,dt \quad \text{or} \quad J=\frac{1}{2}\int_{t_0}^{t_f}W(x)dt. \tag{8.58}$$

For nonlinear systems (8.8), using $J=\int_{t_0}^{t_f}1\,dt$, the Hamilton–Jacobi equation is

$$-\frac{\partial V}{\partial t}=\min_{-1\le u\le 1}\left[1+\left(\frac{\partial V}{\partial x}\right)^T\left(F(x)+B(x)u\right)\right]. \tag{8.59}$$

Using the first-order necessary condition for optimality (8.10), the relay-type control law is

$$u=-\text{sgn}\left(B^T(x)\frac{\partial V}{\partial x}\right), \quad -1\le u\le 1. \tag{8.60}$$

Control law (8.60) may not be applied due to the chattering phenomena, switching, losses, and other undesirable effects. The relay-type control laws with a dead zone

$$u = -\mathrm{sgn}\left(B(x)^T \frac{\partial V}{\partial x} \right)\Bigg|_{dead\ zone}, \quad -1 \le u \le 1, \tag{8.61}$$

may be considered as possible alternatives. The application of hard-switching relay-type control laws for electromechanical systems is limited or impractical. Furthermore, it is impossible to realize the hard-switching control even if desired.

Example 8.24

We synthesize control laws for the system described by the following differential equations

$$\dot{x}_1(t) = x_2^7 + x_1^5 u_1, \quad -1 \le u_1 \le 1,$$

$$\dot{x}_2(t) = -x_2^3 + x_1^3 x_2^5 u_2, \quad -1 \le u_2 \le 1.$$

The performance functional is $J = \int_{t_0}^{t_f} 1\, dt$. The Hamilton–Jacobi equation is

$$-\frac{\partial V}{\partial t} = \min_{u \in U} \left[1 + \left(\frac{\partial V}{\partial x} \right)^T \left(F(x) + B(x)u \right) \right]$$

$$= \min_{\substack{-1 \le u_1 \le 1 \\ -1 \le u_2 \le 1}} \left[1 + \frac{\partial V}{\partial x_1} \left(x_2^7 + x_1^5 u_1 \right) + \frac{\partial V}{\partial x_2} \left(-x_2^3 + x_1^3 x_2^5 u_2 \right) \right].$$

From the first-order necessary condition for optimality (8.10), the control law (8.60) becomes

$$u_1 = -\mathrm{sgn}\left(x_1^5 \frac{\partial V}{\partial x_1} \right),$$

$$u_2 = -\mathrm{sgn}\left(x_1^3 x_2^5 \frac{\partial V}{\partial x_2} \right).$$

Finding the return function $V(x)$ is a challenging problem. We apply the approximation

$$\mathrm{sgn}\, z \approx \frac{z}{\sqrt{z^2 + a^2}}, \quad a \ll 1.$$

Assume that the Hamilton–Jacobi equation is approximated by the quadratic function $V(x) = k_{11}x_1^2 + 2k_{12}x_1x_2 + k_{22}x_2^2$. One finds $k_{11} = 0.25$, $k_{12} = 0.5$, and $k_{22} = 0.25$. Hence,

$$u_1 = -\mathrm{sgn}[x_1^5(0.5x_1 + 0.5x_2)],$$

$$u_2 = -\mathrm{sgn}[x_1^3 x_2^5(0.5x_1 + 0.5x_2)].$$

The Simulink model is reported in Figure 8.23a. To numerically solve the nonlinear differential equations with discontinuous u_1 and u_2, robust numeric methods must be applied. The Configuration Parameters icon is displayed in Figure 8.23a. The transient dynamics is documented in Figure 8.23b for the initial conditions $x_{10} = 5$ and $x_{20} = -5$. There is a significant switching activity of $u_1(t)$ and $u_2(t)$, which affects numerics and accuracy. The relative tolerance (accuracy) should be ~1×10^{-10}. Physical systems cannot exhibit the mathematically derived hard-switching control functions and switching. Relay-type hard switching, if hypothetically exhibited, is undesirable phenomena.

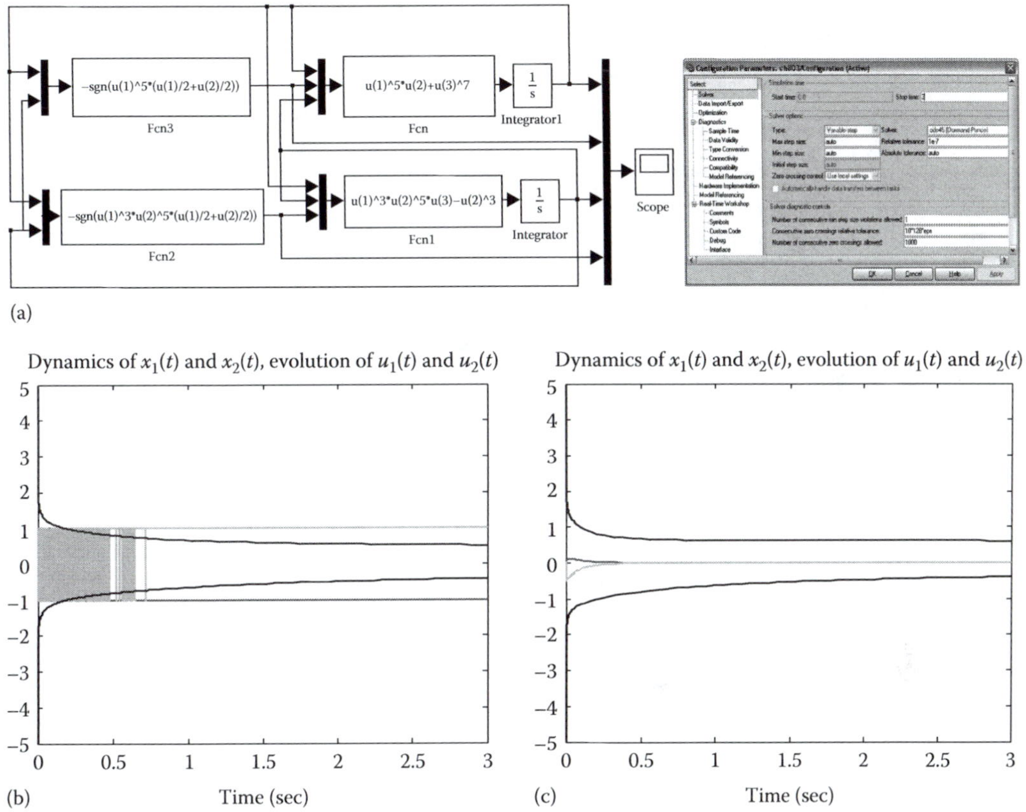

FIGURE 8.23 (a) Simulink® model and the configuration parameters icon; (b) Minimum-time control $u_1 = -\text{sgn}\left[x_1^5(0.5x_1 + 0.5x_2)\right]$, $u_2 = -\text{sgn}\left[x_1^3 x_2^5(0.5x_1 + 0.5x_2)\right]$: Dynamics of $x_1(t)$ and $x_2(t)$, and switching control $u_1(t)$ and $u_2(t)$ activity ($u_1 = \pm 1$ and $u_2 = \pm 1$) if $x_{10} = 5$ and $x_{20} = -5$; (c) Continuous control $u_1 = -x_1^5(0.5x_1 + 0.5x_2)$, $u_2 = -x_1^3 x_2^5(0.5x_1 + 0.5x_2)$: Evolution of $x_1(t)$, $x_2(t)$, $u_1(t)$, and $u_2(t)$.

Figure 8.23c reports the system evolution for a closed-loop system with a continuous control

$$u_1 = -x_1^5(0.5x_1 + 0.5x_2),$$

$$u_2 = -x_1^3 x_2^5(0.5x_1 + 0.5x_2).$$

This control law is found using the quadratic functional. The closed-loop system is stable and exhibits adequate dynamics without undesirable switching. ■

Example 8.25

We design an optimal relay-type control law for a system studied in Examples 8.4 and 8.19. The control constraints are considered, and the equations of motion are

$$\dot{x}_1(t) = x_2,$$

$$\dot{x}_2(t) = u, \quad -1 \le u \le 1.$$

The calculus of variations is applied. The control takes values $u = 1$ and $u = -1$.

If $u = 1$, from $\dot{x}_1(t) = x_2$, $\dot{x}_2(t) = 1$, one has $\dfrac{dx_2}{dx_1} = \dfrac{1}{x_2}$. The integration gives $x_2^2 = 2x_1 + c_1$.

If $u = -1$, from $\dot{x}_1(t) = x_2$, $\dot{x}_2(t) = -1$, we obtain $\frac{dx_2}{dx_1} = -\frac{1}{x_2}$. The integration yields $x_2^2 = -2x_1 + c_2$.

With the switching $u = \pm 1$, the switching curve is derived as a function of the state variables. A comparison of $x_2^2 = 2x_1 + c_1$ and $x_2^2 = -2x_1 + c_2$ gives the switching curve $-x_1 - \frac{1}{2}x_2|x_2| = 0$. The control takes the values $u = 1$ and $u = -1$. Using the switching curve $-x_1 - \frac{1}{2}x_2|x_2| = 0$, one finds the minimum-time relay control law

$$u = -\operatorname{sgn}\left(x_1 + \frac{1}{2}x_2|x_2|\right), \quad -1 \le u \le 1.$$

We express a control law using the switching curve $\upsilon(x)$, which was found using the calculus of variations

$$u = -\operatorname{sgn}(\upsilon(x)), \quad \upsilon(x) = x_1 + \frac{1}{2}x_2|x_2|, \quad -1 \le u \le 1.$$

We derived a control law using the calculus of variations by analyzing the solutions of the differential equations with the relay control switching. The Hamilton–Jacobi theory is applied. We minimize the functional $J = \int_{t_0}^{t_f} 1\,dt$. From

$$-\frac{\partial V}{\partial t} = \min_{-1 \le u \le 1}\left[1 + \frac{\partial V}{\partial x_1}x_2 + \frac{\partial V}{\partial x_2}u\right],$$

a control function is

$$u = -\operatorname{sgn}\left(\frac{\partial V}{\partial x_2}\right).$$

The solution of the Hamilton–Jacobi partial differential equation is given by the nonquadratic, positive-definite continuous, and differentiable return functions

$$V(x_1, x_2) = k_{11}x_1^2 + k_{12}x_1x_2 + k_{22}x_2^2|x_2|.$$

The control law is

$$u = -\operatorname{sgn}\left(x_1 + \frac{1}{2}x_2|x_2|\right).$$

The return function $V(x_1, x_2) = \frac{1}{2}k_{11}x_1^2 + k_{12}x_1x_2 + \frac{1}{2}k_{22}x_2^2$, results in

$$u = -\operatorname{sgn}(x_1 + x_2)$$

which also stabilizes the closed-loop system.

Denoting $\upsilon(x) = x_1 + x_2$, one has $u = \begin{cases} 1 & \text{if } \upsilon(x_1, x_2) < 0 \\ -1 & \text{if } \upsilon(x_1, x_2) > 0 \end{cases}$.

Optimization using the Hamilton–Jacobi theory matches to the calculus of variations. The transient dynamics is analyzed. The switching curve, the phase-plane evolution of the variables, and the transient behavior for different initial conditions are documented in Figure 8.24a. Figure 8.24b illustrates the Simulink model that provides different options to perform simulations. The evolutions of $x_1(t)$, $x_2(t)$, and $u(t)$ are reported.

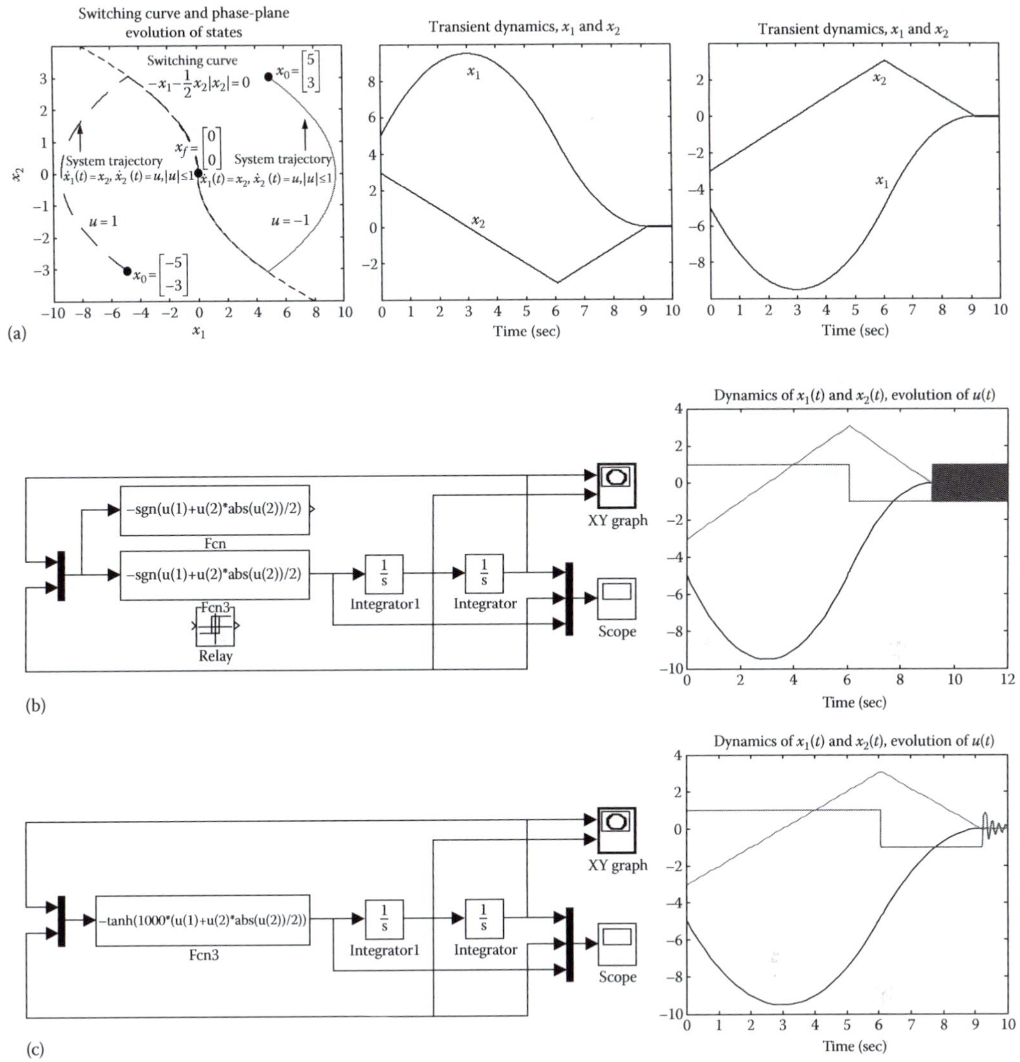

FIGURE 8.24 (a) Phase-plane evolution and system dynamics with minimum-time control law; (b) Simulink® model and evolutions of $x_1(t)$, $x_2(t)$, and $u(t)$ if $u = -\text{sgn}\left(x_1 + \frac{1}{2}x_2|x_2|\right)$; (c) Simulink model for the soft-switching control law $u = -\tanh\left[1000\left(x_1 + \frac{1}{2}x_2|x_2|\right)\right]$ and evolutions of $x_1(t)$, $x_2(t)$, and $u(t)$.

The hard switching of $u(t)$ is undesirable or unacceptable. Sections 8.9 and 8.10 report the design of soft-switching control laws using the continuous real-analytic functions. For system $\dot{x}_1(t) = x_2$, $\dot{x}_2(t) = u$, $-1 \le u \le 1$, the soft-switching control law is

$$u = -\tanh\left[1000\left(x_1 + \frac{1}{2}x_2|x_2|\right)\right]$$

with $\upsilon(x) = 1000\left(x_1 + \frac{1}{2}x_2|x_2|\right)$.

The evolutions of $x_1(t)$, $x_2(t)$, and $u(t)$, illustrated in Figure 8.24c, provide evidence of the significant advantages of soft switching. The settling time remains the same, guarantying near-minimum-time dynamics, but undesirable switching is eliminated. ■

8.10 SLIDING MODE CONTROL

Minimum-time control results in relay-type control functions with switching curves, surfaces, or manifolds. The undesirable phenomena result due to hard switching, chattering, ripple, etc. These lead to unacceptable overall performance, such as low efficiency, wearing, electromagnetic overloading, etc. Sliding mode control has an analogy to the minimum-time control. Soft- and hard-switching sliding mode control laws can be designed. Soft-switching control ensures good performance without hard switching and chattering. To design control laws, we model the states and errors dynamics as

$$\dot{x}(t) = F(x)x + B(x)u, \quad u_{\min} \le u(t,x,e) \le u_{\max}, \quad u_{\min} < 0, \quad u_{\max} > 0, \tag{8.62}$$

$$\dot{e}(t) = N\dot{r}(t) - H\dot{x}.$$

The soft-switching control law, which can be obtained using the Hamilton–Jacobi concept, is

$$u(t, x, e) = -G\phi(\upsilon),\ \upsilon(\cdot)\colon \mathbb{R}_{\ge 0} \times \mathbb{R}^n \times \mathbb{R}^b \to \mathbb{R}^m,\ u_{\min} \le u(t, x, e) \le u_{\max},\ G \in \mathbb{R}^{m\times m},\ G > 0, \tag{8.63}$$

where ϕ is the continuous real-analytic function, for example, tanh(·) and $\tan^{-1}(\cdot)$; $\upsilon(\cdot)$ is the switching manifold.

Example 8.26

Soft-switching control laws can be designed using the Hamilton–Jacobi, Lyapunov, and other methods. The *admissible* control laws with the switching manifolds $\upsilon(\cdot)$ are found using $V(x)$ and $V(t, x, e)$. Example 8.25 examines the soft-switching control

$$u = -\tanh\left[1000\left(x_1 + \frac{1}{2}x_2|x_2|\right)\right]$$

with $\upsilon(x) = 1000\left(x_1 + \frac{1}{2}x_2|x_2|\right)$.

The simulation results for

$$\dot{x}_1(t) = x_2,\ \dot{x}_2(t) = u,\ -1 \le u \le 1$$

with

$$u = -\tanh\left[1000\left(x_1 + \frac{1}{2}x_2|x_2|\right)\right]$$

are reported in Figure 8.24c. ■

The discontinuous hard-switching tracking control law is

$$u(t,x,e) = -G\,\mathrm{sgn}(\upsilon), \quad G \in \mathbb{R}^{m\times m}, \quad G > 0. \tag{8.64}$$

For example, the hard-switching tracking control law (8.64) yields

$$u(t,x,e) = \begin{cases} u_{\max},\ \forall\upsilon(t,x,e) < 0 \\ 0, \quad \forall\upsilon(t,x,e) = 0, \quad u_{\min} \le u(t,x,e) \le u_{\max}. \\ u_{\min},\ \forall\upsilon(t,x,e) > 0 \end{cases}$$

The analysis performed in Examples 8.25 through 8.27 demonstrate the advantages of soft switching.

Example 8.27

For a rigid-body mechanical system, using Newtonian dynamics, one has

$$\frac{dx_1}{dt} = ax_1 + bu,$$

$$\frac{dx_2}{dt} = x_1, \quad -1 \le u \le 1,$$

where x_1 is the velocity; x_2 is the displacement.

For the translational and rotational motion, the coefficients a and b are $a = -B_v/m$ or $a = -B_m/J$, and, $b = 1/m$ or $b = 1/J$. System parameters, such as friction coefficient, mass, moment of inertia, and others, vary. One examines system robustness and *sensitivity* to parameter variations. For varying $a(\cdot)$ and $b(\cdot)$, we have $a_{\min} \le a(\cdot) \le a_{\max}$, $a_{\min} > 0$, $a_{\max} > 0$, and, $b_{\min} \le b(\cdot) \le b_{\max}$, $b_{\min} > 0$, $b_{\max} > 0$. One may design the hard and soft switching control laws. Using a linear switching curve $\upsilon(x) = 100(x_1 + x_2)$, from (8.63) and (8.64), one finds the hard- and soft-switching control laws as

$$u = -\text{sgn}[100(x_1 + x_2)]$$

and

$$u = -\tanh[100(x_1 + x_2)].$$

The variations of $a(\cdot) \in \begin{bmatrix} a_{\min} & a_{\max} \end{bmatrix}$ and $b(\cdot) \in \begin{bmatrix} b_{\min} & b_{\max} \end{bmatrix}$ can be continuous or discontinuous. Let $a = a_0 - a_v \sin(\pi t)$ and $b = b_0 + b_v \cos(\pi t)$, where $a_0 = -1$, $a_v \in \begin{bmatrix} 0 & 0.5 \end{bmatrix}$ and $b_0 = 1$, $b_v \in \begin{bmatrix} 0 & 0.5 \end{bmatrix}$. The Simulink models to perform simulations for the closed-loop systems with the hard and soft-switching control laws are reported in Figures 8.25.

The simulation results for the initial conditions $x_{10} = -3$ and $x_{20} = -5$ are documented in Figures 8.26 for: (1) $a_v = 0$ and $b_v = 0$; (2) $a_v = 0.25$ and $b_v = 0.25$; (3) $a_v = 0.5$ and $b_v = 0.5$.

The hard-switching control law $u = -\text{sgn}(x_1 + x_2)$ results in continuous switching activity with $u = \pm 1$. The soft switching leads to a preferable continuous-time control activity guaranteeing the superior performance and capabilities. The soft-switching control law $u = -\tanh[100(x_1 + x_2)]$ is *implementable*. For the hard-switching control laws, the discontinuous switching of force and torque cannot be ensured even if desired. Furthermore, the actuator–mechanism flexible *torque limiter* couplings are used to prevent malfunctions and breakdowns. These *torque limiters* prevent the mechanical failures due to high instantaneous torques. The electronic and mechanical hardware solutions cannot ensure hard switching. ■

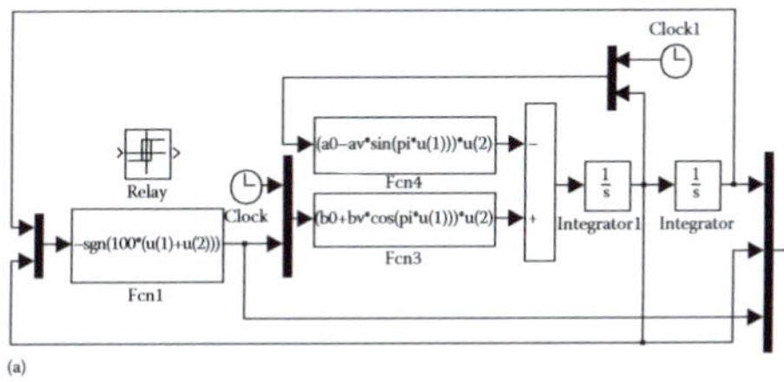

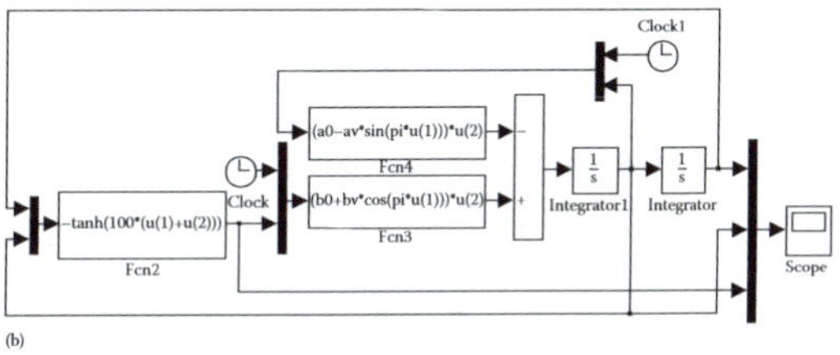

FIGURE 8.25 (a) Simulink® model: Closed-loop system with a hard-switching control law $u = -\text{sgn}[100(x_1 + x_2)]$; (b) Simulink model: Closed-loop system with a soft-switching control law $u = -\tanh[100(x_1 + x_2)]$.

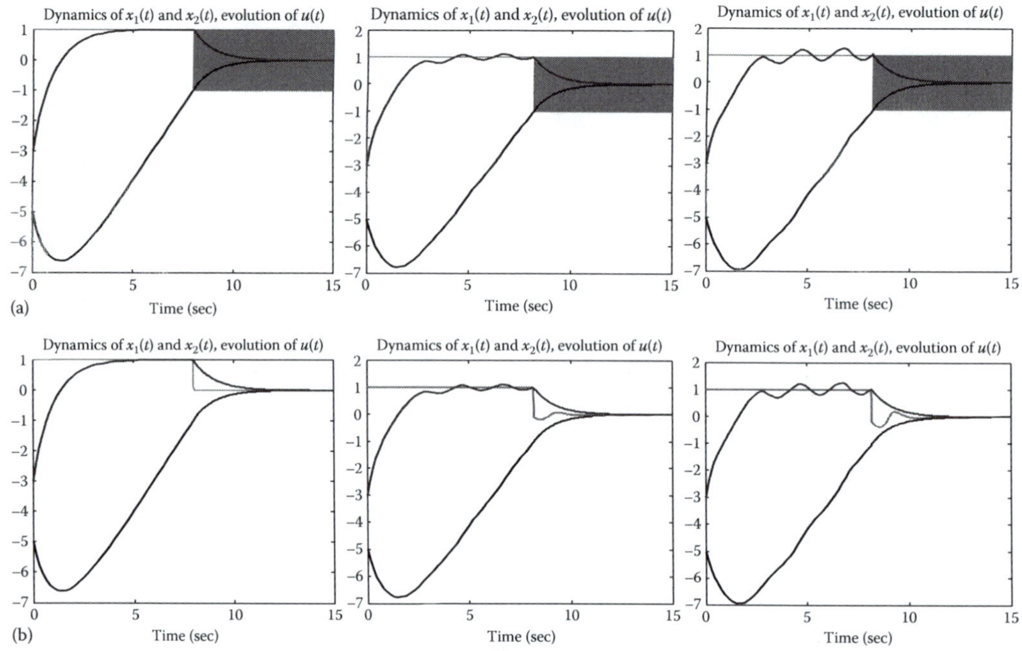

FIGURE 8.26 (a) Closed-loop system dynamics with a hard-switching control law $u = -\text{sgn}[100(x_1 + x_2)]$; (b) System dynamics with a soft-switching control law $u = -\tanh[100(x_1 + x_2)]$ if (1) $a_v = 0$, $b_v = 0$; (2) $a_v = 0.25$, $b_v = 0.25$; (3) $a_v = 0.5$, $b_v = 0.5$.

8.11 CONSTRAINED CONTROL OF ELECTROMECHANICAL SYSTEMS

We examine nonlinear control and optimization of dynamic multi-input/multi-output electromechanical systems with control bound. Electromechanical, electronic, and mechanical systems possess asymmetric or symmetric limits on states and control. There is a need to synthesize performance functionals and design control laws that account for these limits. Solution of constrained optimization problem using the calculus of variations is very complex. The designed-bounded control laws $u = \phi[\varphi(x)]$, $u(\cdot)$: $\mathbb{R}^n \to \mathbb{R}^m$, $u \in U$ should be within the closed admissible set $U = \{u \in \mathbb{R}^m | u_{\min} \le u \le u_{\max}, u_{\max} > 0, u_{\min} < 0\}$, $U \subset \mathbb{R}^m$. Control laws must be designed for symmetric and asymmetric bounds. For symmetric and asymmetric bounds, we have $|u_{\max}| = |u_{\min}|$ and $|u_{\max}| \ne |u_{\min}|$. The asymmetric limits $u \in U$, $U = \{u \in \mathbb{R}^m | u_{\min} \le u \le u_{\max}, u_{\max} > 0, u_{\min} = 0\}$ are exhibited by electronic devices (transistors, DC–DC regulators and converters, etc.), electrostatic, variable-reluctance, hydraulic, pneumatic, and thermal actuators, etc.

One finds mathematical models given by nonlinear differential equations

$$\dot{x}(t) = F(x) + B(x)u, \quad x(t_0) = x_0, \quad u = \phi[\varphi(x)], \quad u_{\min} \le u \le u_{\max}, \quad u \in U, \quad t \ge 0, \tag{8.65}$$

where $x \in X \subset \mathbb{R}^n$ is the state vector; $u \in U \subset \mathbb{R}^m$ is the control vector; $F(\cdot)$: $\mathbb{R}^n \to \mathbb{R}^n$ and $B(\cdot)$: $\mathbb{R}^n \to \mathbb{R}^{n\times m}$ are the continuous nonlinear maps, $F(0) = 0$ and $B(0) = 0$.

The continuous, piecewise continuous, or discontinuous bounded function $\phi(\cdot)$ represents the inherent limits in physical systems. For most systems, $\phi(\cdot)$ is continuous.

Our goals is to minimize the functional

$$J = \int_{t_0}^{t_f} \left[W_x(x) + W_u(u)\right] dt = \int_{t_0}^{t_f} \left[W_x(x) + \int \left(\Phi^{-1}(u)\right)^T G du \right] dt,$$

$$W_x(\cdot): \mathbb{R}^n \to \mathbb{R}_{\ge 0}, \quad W_u(\cdot): \mathbb{R}^m \to \mathbb{R}_{\ge 0}, \quad G \in \mathbb{R}^{m\times m}, \quad G > 0, \tag{8.66}$$

subject to the system dynamics (8.65) $\dot{x}(t) = F(x) + B(x)u$ with $u = \phi(\cdot)$, $u_{min} \le u \le u_{max}$, and design a constrained control $u = \Phi[\varphi(x)]$, $u \in U$ with an admissible function $\Phi[\varphi(x)]: \mathbb{R}^n \to \mathbb{R}^m$, $\Phi \in U$, $\forall x \in X$. Here, $\Phi(\cdot): \mathbb{R}^n \to \mathbb{R}^m$ is the bounded, integrable, one-to-one, real-analytic continuous function that models the system bounds ϕ. The algebraic, transcendental (exponential, hyperbolic, logarithmic and trigonometric), and other $\Phi \in U$ can be found. In (8.66), the positive-definite integrands $W_x(x)$ and $W_u(u)$ are used.

Example 8.28

Consider symmetric limits $u_{min} \le \phi \le u_{max}$, $|u_{max}| = |u_{min}|$ such as:

1. Sign, relay, and saturation bounds $-1 \le \phi \le 1$;
2. Continuous sigmoid bounds $-1 \le \phi \le 1$.

These bounds can be modeled by the odd functions. The odd function

$$\Phi = u_{max} \tanh ax, \quad u_{min} \le \Phi \le u_{max}, \quad a > 0$$

describes the saturation bounds $u_{min} \le \phi \le u_{max}$.

For

$$\Phi = \tanh ax, \quad \tanh ax = \frac{\sinh ax}{\cosh ax} = \frac{e^{ax} - e^{-ax}}{e^{ax} + e^{-ax}}, \quad \text{and} \quad \Phi^{-1} = \frac{1}{a}\tanh^{-1} x = \frac{1}{2a}\ln\left(\frac{1+x}{1-x}\right),$$

one has $\frac{d}{dx}\tanh x = \operatorname{sech}^2 x$, $\frac{d}{dx} ar\tanh x = \frac{1}{1-x^2}$, $\int \tanh ax\, dx = \frac{1}{a}\ln\cosh ax$,

and $\int \tanh^{-1}\frac{1}{b}x\, dx = x\tanh^{-1}\frac{1}{b}x + \frac{1}{2}b\ln\left(b^2 - x^2\right)$.

Consider $\Phi = \tanh^{2n+1} ax$, $-1 \le \Phi \le 1$, $a > 0$ with the horizontal asymptotes ± 1 and domain $(-\infty, +\infty)$ as illustrated in Figures 8.27a and b. For $\Phi = \tanh^{2n+1} ax$, $\Phi^{-1} = \frac{1}{a}\tanh^{-1} x^{\frac{1}{2n+1}}$, $|x| < 1$. Function $\Phi = \tanh^{2n+1} ax$, $-1 \le \Phi \le 1$, with $n = 0, 1, 2, \ldots$ and $a > 0$ describes the sign, relay, and saturation bounds $-1 \le \phi \le 1$ as shown in Figures 8.27a and b.

For $\Phi = \tanh^{2n+1}(ax)$, $n = 4$, $\Phi = \tanh^9(ax)$, $\Phi^{-1} = \frac{1}{a}\tanh^{-1}\sqrt[9]{x}$,

and

$$\frac{1}{a}\int \tanh^{-1}\sqrt[9]{x}\, dx = \frac{1}{a}\left[x\tanh^{-1}\sqrt[9]{x} + \frac{1}{2}\ln\left(1 - x^{\frac{2}{9}}\right) + \frac{1}{24}x^{\frac{2}{9}}\left(4x^{\frac{4}{9}} + 3x^{\frac{2}{3}} + 6x^{\frac{2}{9}} + 12\right)\right].$$

For a given $-1 \le \phi \le 1$, one finds Φ, which models ϕ. The arctangent, Gudermannian $\mathrm{gd}(x) = \arcsin(\tanh(x))$, and algebraic functions $\left(\Phi = \frac{x}{\sqrt{1+x^2}},\ \Phi = \frac{x}{1+|x|} \text{ and others}\right)$, approximate the piece-wise continuous saturation with $-1 \le \phi \le 1$. ■

Example 8.29: Asymmetric Bounds

Many physical systems exhibit asymmetric bounds ϕ. Typical bounds in one-quadrant DC–DC converters are $0 \le \phi \le 1$, and $u \in U$, $U = \{u \in \mathbb{R}^m | \ u_{min} \le u \le u_{max},\ u_{max} = 1,\ u_{min} = 0\}$. One may use the continuous, differentiable, and integrable functions

$$\Phi = \frac{e^{ax^2} - e^{-ax|x|}}{e^{ax^2} + e^{-ax|x|}}, \quad \Phi = \tanh\left[\frac{1}{2}a\left(|x| + x\right)\right], \quad \text{and} \quad \Phi = 1 - e^{-\left(\frac{1}{a}x\right)^{2n+1}}, \quad a > 0.$$

Figures 8.27c and d report the plots of these functions with the horizontal asymptotes 0 and +1. For $0 \le \phi \le 1$, we found $0 \le \Phi \le 1$. ■

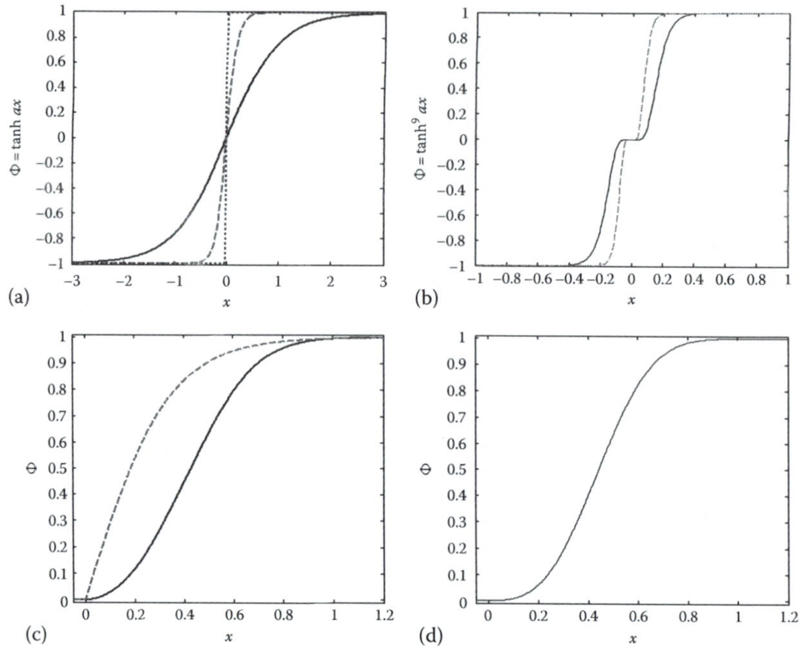

FIGURE 8.27 (a) Symmetric control bounds $-1 \le \phi \le 1$: Continuous $\Phi = \tanh^{2n+1}(ax)$ models saturation and sign bounds. Plots of $\Phi = \tanh(ax)$, $a = 1$ (solid line), $a = 5$ (dashed line), and $a = 100$ (dotted line); (b) Symmetric control bounds $-1 \le \phi \le 1$: Continuous $\Phi = \tanh^{2n+1}(ax)$ models relays with dead zone. Plots of $\Phi = \tanh^9(ax)$, $a = 10$ (solid line) and $a = 20$ (dashed line);
(c) Asymmetric control bounds $0 \le \phi \le 1$. Plots of $\Phi = \dfrac{e^{ax^2} - e^{-ax|x|}}{e^{ax^2} + e^{-ax|x|}}$, $a = 3$ (solid line) and $\Phi = \tanh\left[\dfrac{1}{2}a\left(|x| + x\right)\right]$, $a = 3$ (dashed line) for Example 8.29;
(d) Asymmetric bound $0 \le \phi \le 1$. Plots of $\Phi = 1 - e^{-\left(\frac{1}{a}x\right)^{2n+1}}$, $a = 0.5$, $n = 1$ (Example 8.29).

Consider dynamic systems (8.65) with symmetric or asymmetric bounds $\phi \in U$, modeled by $\Phi \in U$. Minimization of the positive-definite performance functional (8.66), using the necessary conditions for optimality (8.10) and (8.13) on Hamiltonian

$$H\left(x, u, \frac{\partial V}{\partial x}\right) = W_x(x) + \int \left(\Phi^{-1}(u)\right)^T G\, du + \frac{\partial V}{\partial x}^T \left[F(x) + B(x)u\right], \tag{8.67}$$

with

$$-\frac{\partial V}{\partial t} = \min_{u \in U}\left\{W_x(x) + \int \left(\Phi^{-1}(u)\right)^T G du + \frac{\partial V}{\partial x}^T \left[F(x) + B(x)u\right]\right\} \tag{8.68}$$

results in the specified-by-constraints and defined-by-integrand $W_u(u)$ constrained control law

$$u = -\Phi\left(G^{-1}B^T(x)\frac{\partial V(x)}{\partial x}\right), \quad u \in U. \tag{8.69}$$

In particular, using the first-order necessary condition for optimality $\frac{\partial H}{\partial u}=0$, the minimization of the Hamiltonian (8.67) yields

$$\left(\Phi^{-1}(u)\right)^T G+\frac{\partial V}{\partial x}^T B(x)=0.$$

One obtains the constrained *admissible* control law (8.69). The second-order necessary condition for optimality is satisfied because $\frac{\partial^2 H}{\partial u^T \times \partial u}>0$. Hence, the absolute minimum corresponds to a critical point. Control law (8.69) guarantees an absolute minimum to (8.66), and globally minimizes the Hamiltonian (8.67).

The integration by parts $\int v\,dw = vw\int w^T dv^T$ is applied. For $\int \left(\Phi^{-1}(u)\right)^T G\,du$ with control (8.69), we have

$$\int \left(\Phi^{-1}(u)\right)^T G\,du=\int \left[\Phi^{-1}\left(\Phi\left(G^{-1}B^T(x)\frac{\partial V}{\partial x}\right)\right)\right]^T Gd\left[\Phi\left(G^{-1}B^T(x)\frac{\partial V}{\partial x}\right)\right]$$

$$=\frac{\partial V}{\partial x}^T B(x)\Phi\left(G^{-1}B^T(x)\frac{\partial V}{\partial x}\right)-\int\left[\Phi\left(G^{-1}B^T(x)\frac{\partial V}{\partial x}\right)\right]^T d\left(B^T(x)\frac{\partial V}{\partial x}\right). \quad (8.70)$$

Using (8.68), (8.69), and (8.70), one finds

$$-\frac{\partial V}{\partial t}=\frac{1}{2}x^TQx+\frac{\partial V}{\partial x}^T F(x)-\int\left[\Phi\left(G^{-1}B^T(x)\frac{\partial V}{\partial x}\right)\right]^T d\left(B^T(x)\frac{\partial V}{\partial x}\right). \quad (8.71)$$

The solution of (8.71) is approximated by the real-analytic, continuous, and differentiable fractional polynomial return function

$$V(x)=\sum_{i=1}^{N}\frac{2\gamma+1}{2i+2\gamma}\left(x^{\frac{i+\gamma}{2\gamma+1}}\right)^T K_i x^{\frac{i+\gamma}{2\gamma+1}},\text{ or } V(x)=\sum_{i=1}^{N}\frac{2\gamma+1}{2i+2\gamma}\left(|x|^{\frac{i+\gamma}{2\gamma+1}}\right)^T K_i |x|^{\frac{i+\gamma}{2\gamma+1}},\ \gamma=0,1,2,\ldots,\ K_i\in\mathbb{R}^{n\times n}. \quad (8.72)$$

The matrices $K_i \in \mathbb{R}^{n\times n}$ are found by solving (8.71). The quadratic

$$V(x)=\frac{1}{2}x^T K_1 x\ (\gamma=0 \text{ and } N=1),$$

nonquadratic $V(x)=\frac{1}{2}x^T K_1 x+\frac{1}{4}\left(x^2\right)^T K_2 x^2$ ($\gamma=0$ and $N=2$),
and the rational- and high-order $V(x)$ are used depending on Taylor's series expansion.

Example 8.30: Performance Integrands for Symmetric and Asymmetric Bounds

Consider a symmetric saturation $-1\le\phi\le 1$. For

$$\Phi=\tanh x,\quad W_u(u)=\int \tanh^{-1}uG\,du=G\left[u\tanh^{-1}u+\frac{1}{2}\ln\left(1-u^2\right)\right],\quad |u|<1.$$

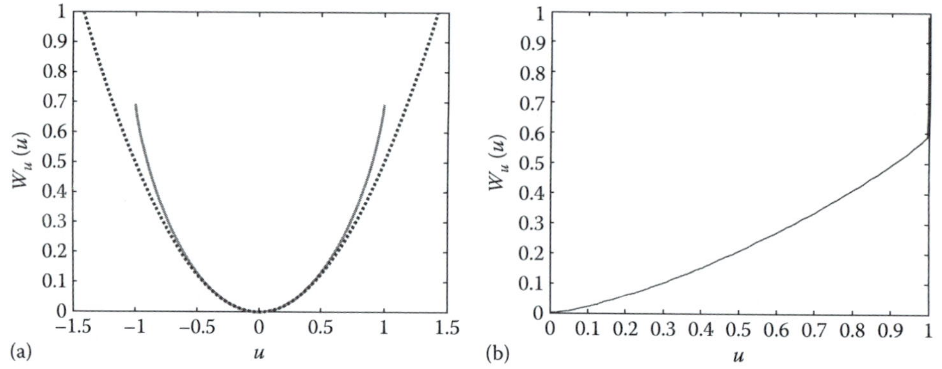

FIGURE 8.28 Plots of real-valued performance integrands $W_u(u)=\int(\Phi^{-1}(u))^T G\,du$:
(a) Saturation $-1 \le u \le 1$ with the resulting specified-by-constraint integrand $W_u(u)=\int \tanh^{-1}uG\,du=G\left[u\tanh^{-1}u+\frac{1}{2}\ln\left(1-u^2\right)\right]$ (solid line), and, quadratic integrand $W_u(u)=½u^2$ (dashed line);
(b) Asymmetric constraint $0 \le u \le 1$ with the specified-by-constraint integrand $W_u(u)=\int\left(1-e^{-\left(\frac{1}{a}u\right)^3}\right)^{-1}G^{-1}\,du=-0.303G^{-1}\sqrt[3]{\ln(1-u)}\left(3(x-1)-\frac{\Gamma\left(\frac{1}{3},-\ln(1-u)\right)}{\sqrt[3]{-\ln(1-u)}}\right), G=1.$

Figure 8.28a illustrates the specified-by-constraint $W_u(u)$ if $G = 1$. The comparison of $W_u(u)$ with the conventional quadratic integrand $½u^2$ is reported, if u is not bounded. The integrand

$$W_u(u)=\int \tanh^{-1}uG\,du$$

is relevant to the constrained optimization problem and consistent with the limits imposed $-1 \le u \le 1$.

Consider an asymmetric bound $0 \le \phi \le 1$. Example 8.29 reports $\Phi = 1-e^{-\left(\frac{1}{a}x\right)^{2n+1}}$, $a = 0.75$, and $n = 1$ to describe limits $0 \le \phi \le 1$ by $0 \le \Phi \le 1$. The integrand $W_u(u)$ is

$$W_u(u)=\int\left(1-e^{-\left(\frac{1}{a}u\right)^3}\right)^{-1}G\,du=-0.909G\int\sqrt[3]{\ln(1-u)}\,du$$

$$=-0.303G\sqrt[3]{\ln(1-u)}\left(3(x-1)-\frac{\Gamma\left(\frac{1}{3},-\ln(1-u)\right)}{\sqrt[3]{-\ln(1-u)}}\right).$$

The plot for $W_u(u)$, $0 \le u \le 1$ is documented in Figure 8.28b. ■

Example 8.31

For a system with the saturation bound

$$\dot{x}(t)=ax+bu,\quad b>0,\quad u_{\min}\le u\le u_{\max},\quad U=\{|u|\le 1\},$$

we minimize a positive-definite functional (8.66)

$$J=\int_0^\infty\left(\frac{1}{2}q_{11}x^2+\int\tanh^{-1}u\,du\right)dt.$$

Using the first-order necessary condition for the optimality for the Hamiltonian (8.67), one finds the constrained control (8.69)

$$u = -\tanh\left(b\frac{\partial V}{\partial x}\right).$$

Using (8.71), we have

$$-\frac{\partial V}{\partial t} = \frac{1}{2}q_{11}x^2 + \frac{\partial V}{\partial x}ax - \int \tanh\left(b\frac{\partial V}{\partial x}\right)d\left(b\frac{\partial V}{\partial x}\right),$$

where $\int \tanh\left(b\frac{\partial V}{\partial x}\right)d\left(b\frac{\partial V}{\partial x}\right) = \ln\cosh\left(b\frac{\partial V}{\partial x}\right)$.

One solves

$$-\frac{\partial V}{\partial t} = \frac{1}{2}q_{11}x^2 + \frac{\partial V}{\partial x}ax - \ln\cosh\left(b\frac{\partial V}{\partial x}\right).$$

From $\ln z = 2\sum_{n=0}^{\infty}\frac{1}{2n+1}\left(\frac{z-1}{z+1}\right)^{2n+1}$ and $\cosh z = 1 + \frac{1}{2!}z^2 + \frac{1}{4!}z^4 + \cdots = \sum_{n=0}^{\infty}\frac{1}{2n!}z^{2n}$, one finds

$$\ln(\cosh z) = \frac{1}{2}z^2 - \frac{1}{12}z^4 + \frac{1}{45}z^6 - \frac{17}{2520}z^8 + O(z).$$

Depending on the variations of z, the derived expression is truncated. For $|z| \le 2$, the least-squares approximation gives $\ln\cosh(z) \approx 0.35z^2$.

The solution of

$$\frac{1}{2}q_{11}x^2 + \frac{\partial V}{\partial x}ax = 0.35\left(b\frac{\partial V}{\partial x}\right)^2$$

is found by using $V(x) = \frac{1}{2}k_1x^2$.

The constrained control law is

$$u = -\tanh(bk_1x),\quad |u| \le 1,\quad k_1 > 0,\quad k_1 = \frac{-a \pm \sqrt{a^2 + 0.7b^2q_{11}}}{-0.7b^2}.$$

Using nonquadratic $W_x(x) = \frac{1}{2}x^2 + \frac{1}{4}x^4$, the solution of

$$\frac{1}{2}x^2 + \frac{1}{4}x^4 + \frac{\partial V}{\partial x}ax = 0.35\left(b\frac{\partial V}{\partial x}\right)^2$$

is found using $V(x) = \frac{1}{2}k_1x^2 + \frac{1}{4}k_2x^4$. One finds the unknown positive-definite k_1 and k_2. The constrained control law is

$$u = -\tanh(bk_1x + bk_2x^3),\quad |u| \le 1,\quad (k_1, k_2) > 0.$$

The closed-loop system $\dot{x}(t) = ax - b\tanh\left(bk_1x + bk_2x^3\right)$ is stable.

The sufficient conditions for stability are applied if $a > 0$. If $a > 0$, the sufficient conditions are satisfied if $|b\tanh(bk_1x + bk_2x^3)| > ax$, $\forall x \in X$, $\forall t \in [t_0, t_f)$. ■

Example 8.32: Control of a DC–DC Switching Converter

The high-switching-frequency PWM DC–DC converters are nonlinear systems with inherent control bounds. For different converter topologies, nonlinear differential equations were derived in Chapter 7. Consider a one-quadrant step-down *buck* PWM DC–DC converter with a MOSFET that is switching at a high frequency, as covered in Section 7.4.2. The switching activity and duty cycle are controlled by the comparator. The PWM voltage is applied to the *RL* load, formed by R_L and L_L in series. Figures 8.29 documents a controlled DC–DC converter and images of the TPS5410D regulator.

The signal-level voltage u_c is compared to the rectangular voltage u_t. By varying the duty cycle $d_D = t_{on}/(t_{on} + t_{off})$, $0 \le d_D \le 1$, one regulates the output voltage u_{RL}. For an inherent asymmetric bound $0 \le \phi \le 1$, we have $u \equiv d_D$, $u \in U$, $0 \le u \le 1$. The goal is to stabilize the output voltage u_{RL} at the specified value $u_{RL\,ref}$ if the *RL* load is applied, or R_L varies. The constrained stabilizing control law must be designed.

The differential equations are derived in Section 7.4.2 using the voltage u_C at the *LC* filter capacitor, current through the filter inductor i_L, and current in the load i_{RL}. One has

$$\frac{du_C}{dt} = \frac{1}{C}\left(i_L - i_{RL}\right),$$
$$\frac{di_L}{dt} = \frac{1}{L}\left[-u_C - \left(r_L + r_C\right)i_L + r_C i_{RL} - r_s i_L u + V_d u\right],$$
$$\frac{di_{RL}}{dt} = \frac{1}{L_L}\left[u_C + r_C i_L - \left(R_L + r_C\right)i_{RL}\right].$$

The parameters are $r_s = 0.01$ ohm, $r_L = 0.023$ ohm, $r_C = 0.05$ ohm, $C = 0.001$ F, $L = 0.001$ H, $L_L = 0.0047$ H, and $V_d = 30$ V. We design stabilizing control laws. The LQR control problem is solved by linearizing a nonlinear model. The results reported in Section 8.6 are used. Neglecting the bounds and letting $r_s = 0$, we have

$$\dot{x} = Ax + Bu = \begin{bmatrix} 0 & \frac{1}{C} & -\frac{1}{C} \\ -\frac{1}{L} & -\frac{r_L + r_C}{L} & \frac{r_C}{L} \\ \frac{1}{L_L} & \frac{r_C}{L_L} & -\frac{R_L + r_C}{L_L} \end{bmatrix} \begin{bmatrix} u_C \\ i_L \\ i_{RL} \end{bmatrix} + \begin{bmatrix} 0 \\ \frac{V_d}{L} \\ 0 \end{bmatrix} u.$$

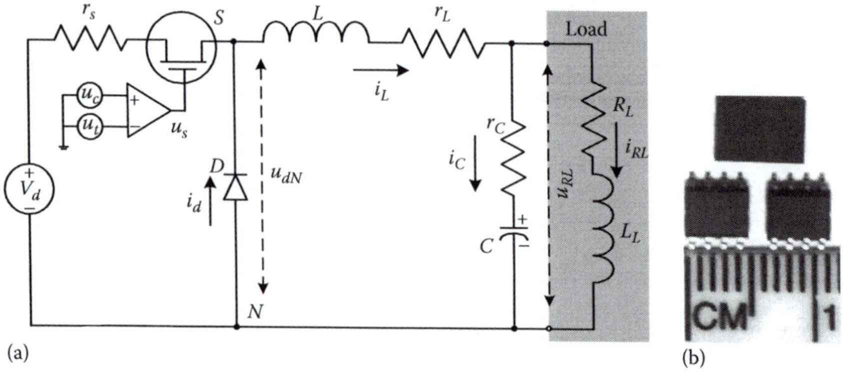

FIGURE 8.29 (a) *Step-down* switching converter;
(b) Texas instruments TPS5410D *buck* regulator, 5.5–36 V input voltage, 1.23–31 V output voltage, 1 A output current, 500 kHz.

For the open-loop stable PWM converter, with $R_L = 10$ ohm, the linearized models give the eigenvalues $\lambda_{1,2} = -90.7 \pm 1015i$ and $\lambda_3 = -2054$.

Assigning $Q \in \mathbb{R}^{3\times3}$ with $q_{11} = q_{22} = 1$, $q_{33} = 100$ and $G = 1$, we minimize the functional

$$J = \frac{1}{2}\int_0^\infty \left(q_{11}u_c^2 + q_{22}i_L^2 + q_{33}i_{RL}^2 + gu^2\right).$$

The positive-definite $K \in \mathbb{R}^{3\times3}$, $K = K^T$ is found by solving the Riccati equation, and

$$K = \begin{bmatrix} 1.22\times10^{-3} & 3.88\times10^{-5} & 1\times10^{-3} \\ 3.88\times10^{-5} & 3.45\times10^{-5} & 3.94\times10^{-5} \\ 1\times10^{-3} & 3.94\times10^{-5} & 2.25\times10^{-2} \end{bmatrix}.$$

The control law is $u = -1.16u_C - 1.04i_L - 1.18i_{RL}$.

The MATLAB file is

```
% Parameters and constant-coefficient matrices A, B, C and D
rL=0.023; rC=0.07; C=0.001; L=0.001; LL=0.0047; RL=10; Vd=30;
A=[0 1/C -1/C; -1/L -(rL+rC)/L rC/L; 1/LL rC/LL -(RL+rC)/LL]; disp('eigenvalues_A'); disp(eig(A));
B=[0 Vd/L 0]'; H=[0 0 1]; D=[0 0 0];
Q=[1 0 0; 0 1 0; 0 0 100]; G=[1];
% Weighting matrices Q and G  [K_feedback,K,Eigenvalues]=lqr(A,B,Q,G);
% Feedback and return function coefficients
disp('K_feedback'); disp(K_feedback); disp('K'); disp(K);
disp('eigenvalues A-BK_feedback'); disp(Eigenvalues);
```

For the closed-loop system, the eigenvalues are found to be $\lambda_{1,2} = -1676 \pm 463.3i$ and $\lambda_3 = -29{,}950$.

The experimental studies are conducted. The closed-loop dynamics for the varying load R_L, if the output voltage u_{RL} is specified to be 4 and 12 V, are reported in Figures 8.30a and b. The peak loads R_L are: (1) R_L is 4 ohm for $u_{RL\,ref} = 4$ V; (2) R_L is 12 ohm for $u_{RL\,ref} = 12$ V. The output voltage u_{RL} is stabilized by the LQR control law at the specified reference voltages $u_{RL\,ref} = 4$ V and $u_{RL\,ref} = 12$ V. The experiments demonstrate that bounds and nonlinearities affect the system dynamics and performance. The oscillatory dynamics with overshoots and lengthy transients are observed.

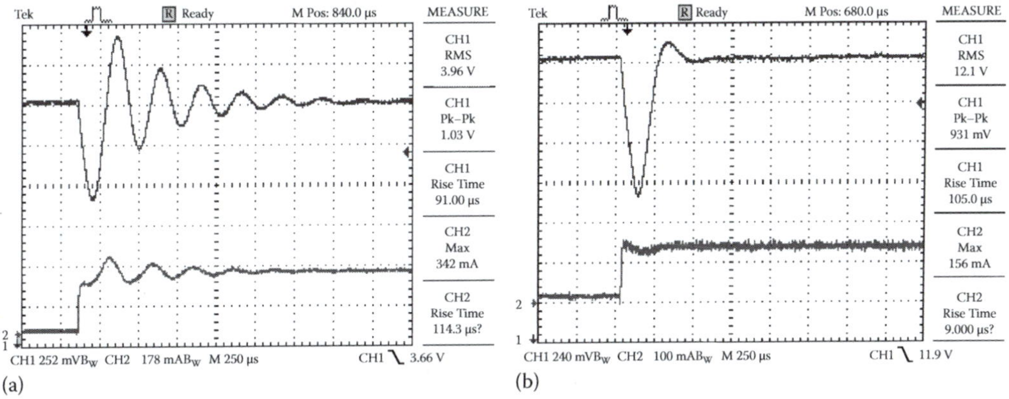

FIGURE 8.30 Closed-loop converter dynamics with the LQR control law $u = -1.16u_C - 1.04i_L - 1.18i_{RL}$. Evolutions of $u_{RL}(t)$ (top line) and $i_{RL}(t)$ (bottom line) if the load R_L is applied:
(a) The output voltage u_{RL} is stabilized for $u_{RL\,ref} = 4$ V with $i_{RL} = 0.28$, $R_L = 14$ ohm;
(b) The output voltage u_{RL} at the load is stabilized for $u_{RL\,ref} = 12.1$ V with $i_{RL} = 0.15$ A, $R_L = 80$ ohm.

The minimum-time hard-switching relay control law is designed using the maximum principle. As reported in Section 8.8, we solve the optimization problem. By minimizing $J = \int_0^{\infty} dt$, one obtains

$$u = -\operatorname{sgn}_0^1 \left[\frac{-r_s i_L + V_d}{L} \left(1\times 10^{-4} u_C + 8.85\times 10^{-5} i_L + 1.1\times 10^{-4} i_{RL} \right) \right].$$

The experimental results are reported in Figures 8.31a and b. Due to near hard-switching control activities (ideal hard switching cannot be realized because the comparator and MOSFET transients are within tens of nanoseconds), for any load and disturbance, there will be the limit cycles, voltage chattering, and current ripple. These phenomena result in low reliability, low efficiency, heating, high noise, etc, and the overall performance is inadequate.

We design a constrained control law. An asymmetric bound on the duty cycle $0 \le \phi \le 1$ is modeled by

$$\Phi = \frac{e^{a|x|} - e^{-ax}}{e^{a|x|} + e^{-ax}}, \quad a = 2, 0 \le \Phi \le 1$$

with $\Phi^{-1} = \frac{1}{a} \ln \sqrt{\frac{-u-1}{|u|-1}} = \frac{1}{2a} \ln \left(\frac{-u-1}{|u|-1} \right)$.

$$\text{Hence, } W_u(u) = \int \left(\frac{e^{a|u|} - e^{-au}}{e^{a|u|} + e^{-au}} \right)^{-1} G\, du = \int \frac{1}{a} \ln \sqrt{\frac{-u-1}{|u|-1}} G\, du$$

$$= \frac{G}{4a} (\operatorname{sgn}(u) + 1) \left[\ln(1-u) + \ln(u+1) + u \ln \left(-\frac{u+1}{u-1} \right) \right].$$

The resulting plots for $\Phi(\cdot)$, $\Phi^{-1}(\cdot)$, and $W_u(u)$ are reported in Figures 8.32.

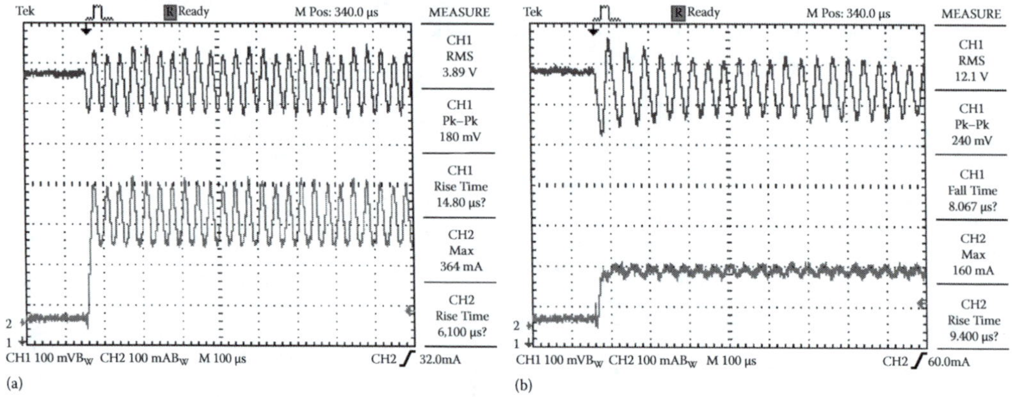

FIGURE 8.31 Closed-loop dynamics with $u = -\operatorname{sgn}_0^1 \left[\frac{-r_s i_L + V_d}{L} \left(1\times 10^{-4} u_C + 8.85\times 10^{-5} i_L + 1.1\times 10^{-4} i_{RL} \right) \right]$: Evolutions of $u_{RL}(t)$ (top line) and $i_{RL}(t)$ (bottom line) if the load R_L is applied:
(a) For $u_{RL\,ref} = 4$ V, the output voltage u_{RL} and current i_{RL} oscillate if any R_L is applied. There are the limit cycles with the voltage chattering $\Delta u_{RL} = \pm 0.09$ V and current ripple $\Delta i_{RL} = \pm 0.05$ A (for $R_L = 14$ ohm);
(b) For $u_{RL\,ref} = 12.1$ V, the output voltage u_{RL} and current i_{RL} oscillate if any R_L is applied. There are the limit cycles with the voltage chattering $\Delta u_{RL} = \pm 0.09$ V and current ripple $\Delta i_{RL} = \pm 0.025$ A (for $R_L = 80$ ohm).

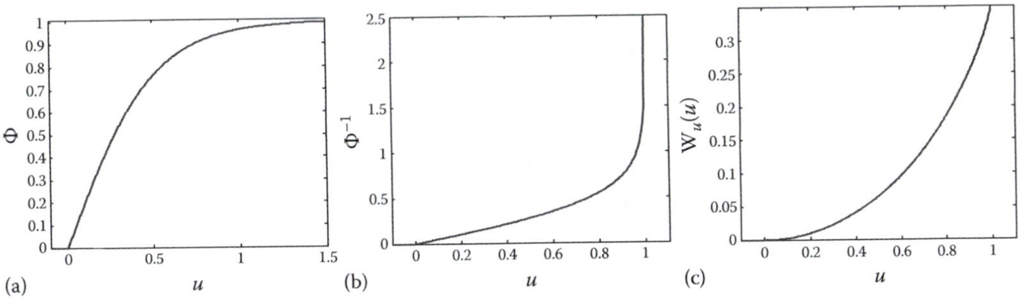

FIGURE 8.32 (a) Plot for $\Phi = \dfrac{e^{a|x|} - e^{-ax}}{e^{a|x|} + e^{-ax}}$, $0 \le \Phi \le 1$, $a = 2$ with horizontal asymptotes 0 and +1. This $0 \le \Phi \le 1$ models an asymmetric bound $0 \le \phi \le 1$;
(b) Plot for an inverse function $\Phi^{-1} = \dfrac{1}{a}\ln\sqrt{\dfrac{-u-1}{|u|-1}} = \dfrac{1}{2a}\ln\left(\dfrac{-u-1}{|u|-1}\right)$; (c) Plot for an integrand $W_u(u) = \dfrac{G}{4a}\left(\mathrm{sgn}(u)+1\right)\left[\ln(1-u)+\ln(u+1)+u\ln\left(-\dfrac{u+1}{u-1}\right)\right]$.

The constrained control law

$$u = -\Phi\left(G^{-1}B^T(x)\frac{\partial V(x)}{\partial x}\right), \quad B(x) = \frac{1}{L}\begin{bmatrix} 0 \\ -r_s i_L + V_d \\ 0 \end{bmatrix}$$

is found using (8.69) and (8.71).

Using $\displaystyle\int \frac{e^{a|x|} - e^{-ax}}{e^{a|x|} + e^{-ax}}\,dx = -\frac{1}{2a}\left(\mathrm{sgn}(x)+1\right)\left[ax - \ln\left(e^{2ax}+1\right) + \ln 2\right]$, equation (8.71) is solved using the quadratic return function (8.72) $V(x) = \frac{1}{2}x^T K_1 x$ ($\gamma = 0$ and $N = 1$), which approximates (8.71) when $\left|B^T(x)\dfrac{\partial V}{\partial x}\right|$ is ~1. The constrained control law is

$$u = -\Phi\left(G^{-1}B^T(x)\frac{\partial V}{\partial x}\right) = -\Phi\left(\frac{-r_s i_L + V_d}{L}\frac{\partial V}{\partial i_L}\right) = -\Phi\left(\frac{-r_s i_L + V_d}{L}\left(k_{21}u_C + k_{22}i_L + k_{23}i_{RL}\right)\right), \quad 0 \le \Phi \le 1,$$

$$k_{21} = 1.62 \times 10^{-5}, \quad k_{22} = 1.55 \times 10^{-5}, \quad k_{23} = 1.47 \times 10^{-5}.$$

The experimental results are reported in Figures 8.33a and b. Fast dynamics, minimal overshoot, stability, desired voltage and current evolutions, and superior capabilities are guaranteed. ■

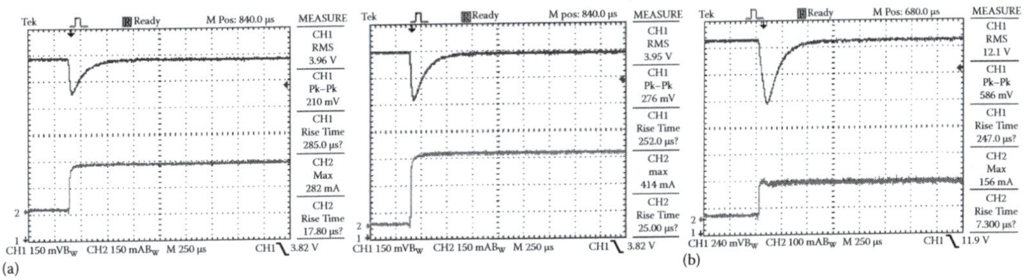

FIGURE 8.33 Converter dynamics with $u = -\Phi\left(\dfrac{-r_s i_L + V_d}{L}\left(k_{21}u_C + k_{22}i_L + k_{23}i_{RL}\right)\right)$, $0 \le \Phi \le 1$: Evolutions of $u_{RL}(t)$ (top line) and $i_{RL}(t)$ (bottom line) if the R_L is applied:
(a) The output voltage u_{RL} is stabilized for $u_{RL\,ref} = 4$ V with $i_{RL} = 0.28$ A ($R_L = 14$ ohm) and $i_{RL} = 0.414$ A ($R_L = 10$ ohm);
(b) The output voltage u_{RL} is stabilized for $u_{RL\,ref} = 12.1$ V with $i_{RL} = 0.15$ A ($R_L = 80$ ohm).

8.12 CONSTRAINED TRACKING CONTROL OF ELECTROMECHANICAL SYSTEMS

Section 8.11 outlined the design of stabilizing constrained control laws. Our goal is to analytically design a tracking control law such that $u \in U$ for a system described by (8.65)

$$\dot{x}(t) = F(x) + B(x)u,\ x(t_0) = x_0,\ y = Hx,\ u = \phi[\varphi(x)],\ u_{\min} \le u \le u_{\max},\ u \in U,\ t \ge 0.$$

With the output equation $y = Hx$, the tracking error is $e = (r - y)$.
We define the evolution of the reference vector $x^{ref}(t)$ as

$$\dot{x}^{ref}(t) = r - y = r - Hx. \tag{8.73}$$

The evolution of the dynamic tracking error is

$$\dot{e}(t) = -I_E e + \dot{r} - \dot{y} = -I_E e + \dot{r} - H\dot{x} = -I_E e + \dot{r} - HF(x) - HB(x)u, \tag{8.74}$$

where $I_E \in \mathbb{R}^{b\times b}$ is the diagonal positive-definite matrix, $I_E > 0$. For example, $I_E = I$, where I is the identity matrix.

Using (8.65), (8.73) and (8.74), the governing equation for the expanded vector $\mathbf{x} = \begin{bmatrix} x & x_{ref} & e \end{bmatrix}^T$ is

$$\dot{\mathbf{x}}(t) = \begin{bmatrix} \dot{x}(t) \\ \dot{x}^{ref}(t) \\ \dot{e}(t) \end{bmatrix} = \begin{bmatrix} F(x) \\ 0 \\ -HF(x) \end{bmatrix} + \begin{bmatrix} 0 & 0 & 0 \\ -H & 0 & 0 \\ 0 & 0 & -I_E \end{bmatrix} \begin{bmatrix} x \\ x^{ref} \\ e \end{bmatrix} + \begin{bmatrix} B(x) \\ 0 \\ -HB(x) \end{bmatrix} u + \begin{bmatrix} 0 \\ r \\ \dot{r} \end{bmatrix},\quad u \in U, \tag{8.75}$$

$$\dot{\mathbf{x}}(t) = \mathbf{F}(\mathbf{x}) + A\mathbf{x} + \mathbf{B}(\mathbf{x})u + F_r(r),\quad \mathbf{F}(\mathbf{x}) = \begin{bmatrix} F(x) \\ 0 \\ -HF(x) \end{bmatrix},\quad A = \begin{bmatrix} 0 & 0 & 0 \\ -H & 0 & 0 \\ 0 & 0 & -I_E \end{bmatrix},$$

$$\mathbf{B}(\mathbf{x}) = \begin{bmatrix} B(x) \\ 0 \\ -HB(x) \end{bmatrix},\quad F_r(r) = \begin{bmatrix} 0 \\ r \\ \dot{r} \end{bmatrix}.$$

The controllability and observability of the system (8.75) with the extended vector $\mathbf{x}$ is guaranteed because the physical systems (8.65) are controllable and observable. Mathematically, using the Lie bracket operator $\left[\mathrm{ad}_{\mathbf{F}}^k \mathbf{B}_z\right] = \left[\mathbf{F} \cdots [\mathbf{F}, \mathbf{B}]\right]$ one finds that $C = \left[\mathbf{B} \cdots [\mathrm{ad}_{\mathbf{F}}^k \mathbf{B}]\right]$ spans $(n+2b)$ space with the rank equal to $(n + 2b)$. The relevant results are documented in Example 8.33 by solving a practical problem.

Using the reported constrained optimization procedure, one minimizes the performance functional

$$J = \int_{t_0}^{\infty} \left[W_x(\mathbf{x}) + \int \left(\Phi^{-1}(u)\right)^T G\, du \right] dt,\quad G > 0. \tag{8.76}$$

The Hamiltonian is

$$H\left(\mathbf{x}, u, \frac{\partial V}{\partial \mathbf{x}}\right) = W_x(\mathbf{x}) + \int \left(\Phi^{-1}(u)\right)^T G\, du + \frac{\partial V}{\partial \mathbf{x}}^T \left[\mathbf{F}(\mathbf{x}) + A\mathbf{x} + \mathbf{B}(\mathbf{x})u\right]. \tag{8.77}$$

The first-order necessary condition for optimality (8.10) $\frac{\partial H}{\partial u} = 0$ yields the constrained control

$$u = -\Phi\left(G^{-1}\mathbf{B}^T(\mathbf{x})\frac{\partial V(\mathbf{x})}{\partial \mathbf{x}}\right), \quad u \in U. \tag{8.78}$$

The solution of the Hamilton–Jacobi–Bellman equation

$$-\frac{\partial V}{\partial t} = W_x(\mathbf{x}) + \frac{\partial V}{\partial \mathbf{x}}^T \mathbf{F}(\mathbf{x}) - \int\left[\Phi\left(G^{-1}\mathbf{B}^T(\mathbf{x})\frac{\partial V}{\partial \mathbf{x}}\right)\right]^T d\left(\mathbf{B}^T(\mathbf{x})\frac{\partial V}{\partial \mathbf{x}}\right) \tag{8.79}$$

is approximated by the return function

$$V(\mathbf{x}) = \sum_{i=1}^{N} \frac{2\gamma+1}{2i+2\gamma}\left(\mathbf{x}^{\frac{i+\gamma}{2\gamma+1}}\right)^T K_i \mathbf{x}^{\frac{i+\gamma}{2\gamma+1}}, \quad \text{or} \quad V(\mathbf{x}) = \sum_{i=1}^{N} \frac{2\gamma+1}{2i+2\gamma}\left(|\mathbf{x}|^{\frac{i+\gamma}{2\gamma+1}}\right)^T K_i |\mathbf{x}|^{\frac{i+\gamma}{2\gamma+1}}, \; K_i = K_i^T, \; K_i > 0. \tag{8.80}$$

From (8.73), one has $x^{ref} = \int e\, dt$. Using $V(\mathbf{x})$ (8.80), the tracking control law (8.78) with the state feedback, proportional error term, and integral tracking error feedback $\mathbf{x} = \begin{bmatrix} x \\ \int e\, dt \\ e \end{bmatrix}$ results.

The matrices K_i are found by solving (8.79). For example, the quadratic $V(\mathbf{x}) = \frac{1}{2}\mathbf{x}^T K_1 \mathbf{x}$ ($\gamma = 0$ and $N = 1$) or nonquadratic $V(\mathbf{x}) = \frac{1}{2}\mathbf{x}^T K_1 \mathbf{x} + \frac{1}{4}\left(\mathbf{x}^2\right)^T K_2 \mathbf{x}^2$ ($\gamma = 0$ and $N = 2$) are used applying the series expansions of

$$\int\left[\Phi\left(G^{-1}\mathbf{B}^T(\mathbf{x})\frac{\partial V}{\partial \mathbf{x}}\right)\right]^T d\left(\mathbf{B}^T(\mathbf{x})\frac{\partial V}{\partial \mathbf{x}}\right)$$

in (8.79).

Using $V(\mathbf{x}) = \frac{1}{2}\mathbf{x}^T K_1 \mathbf{x}$, the tracking control law (8.78) is

$$u = -\Phi\left(G^{-1}\mathbf{B}^T(\mathbf{x})\frac{\partial V(\mathbf{x})}{\partial \mathbf{x}}\right) = -\Phi\left(G^{-1}\mathbf{B}^T(\mathbf{x})K_1\mathbf{x}\right) = -\Phi\left(G^{-1}\mathbf{B}^T(\mathbf{x})K_1\begin{bmatrix} x \\ \int e\, dt \\ e \end{bmatrix}\right). \tag{8.81}$$

The constrained tracking control law is designed.

Example 8.33: Bounded Tracking Control Law Design for the Second-Order System

Consider a system with the saturation bound on control $|u| \le 1$

$$\frac{dx_1}{dt} = u, \quad \frac{dx_2}{dt} = x_1, \quad |u| \le 1, \quad y = x_2.$$

Our goal is to design tracking control laws. To stabilize this system, the derivative term should be used. With a PID control law

$$u = k_p e + k_i \int e\, dt + k_d \frac{de}{dt},$$

neglecting the control bounds, the transfer function is

$$G(s)=\frac{k_d s^2+k_p s+k_i}{s^3+k_d s^2+k_p s+k_i}$$

with the characteristic equation $s^3 + k_d s^2 + k_p s + k_i = 0$.

For the PD control law $u=k_p e+k_d\frac{de}{dt}$, we have $G(s)=\frac{k_d s+k_p}{s^2+k_d s+k_p}$.

The characteristic equation is $s^2 + k_d s + k_p = 0$.

Applying the pole-placement approach, the feedback gains k_p and k_d can be found to ensure the specified time constant $T=\sqrt{\frac{1}{k_p}}$ and damping coefficient $\xi=\frac{k_d}{2\sqrt{k_p}}$. Letting $T = 1$ sec and $\xi = 0.707$, we have

$u=k_p e+k_d\frac{de}{dt}$, $e = (r - x_2)$ with $k_p = 1$ and $k_d = 1.41$.

For $T = 1$ sec and $\xi = 1$, we have $k_p = 1$ and $k_d = 2$.

The high-gain control with large k_p and k_d results in an unstable system due to the bound $|u| \le 1$. Another drawback is that the derivative term $k_d de/dt$ is impractical due to noise.

We synthesize a constrained control law using the proposed concept. The expanded vector is $\mathbf{x}=\begin{bmatrix}x_1 & x_2 & x_{ref} & e\end{bmatrix}^T$.

We define

$$\frac{dx^{ref}}{dt}=r-y=r-x_2,\quad\text{and}\quad e=(r-y)=(r-x_2).$$

The dynamic tracking error evolves as

$$\frac{de}{dt}=-e+\dot{r}-\dot{x}_2=-e+\dot{r}-x_1.$$

One obtains (8.75)

$$\dot{\mathbf{x}}=\begin{bmatrix}\dot{x}_1\\ \dot{x}_2\\ \dot{x}^{ref}\\ \dot{e}\end{bmatrix}=\begin{bmatrix}0&0&0&0\\1&0&0&0\\0&-1&0&0\\-1&0&0&-1\end{bmatrix}\begin{bmatrix}x_1\\ x_2\\ x^{ref}\\ e\end{bmatrix}+\begin{bmatrix}1\\0\\0\\0\end{bmatrix}u=\mathbf{A}\mathbf{x}+\mathbf{B}u,\quad |u|\le 1,\quad y=x_2.$$

The controllability $C=\begin{bmatrix}\mathbf{B} & \mathbf{AB} & \mathbf{A}^2\mathbf{B} & \mathbf{A}^3\mathbf{B}\end{bmatrix}$, $C=\begin{bmatrix}1&0&0&0\\0&1&0&0\\0&0&-1&0\\0&-1&1&-1\end{bmatrix}$

and observability $O=\begin{bmatrix}C & C\mathbf{A} & C\mathbf{A}^2 & C\mathbf{A}^3\end{bmatrix}^T$ matrices have full ranks. That is, rank(C) = 4 and rank(O) = 4. The number of uncontrollable and unobservable states is zero. The system with the extended state vector $\mathbf{x}=\begin{bmatrix}x_1 & x_2 & x_{ref} & e\end{bmatrix}^T$ is controllable and observable.

Letting $Q=\begin{bmatrix}1&0&0&0\\0&1&0&0\\0&0&1&0\\0&0&0&1\end{bmatrix}$ and $G = 1$, we solve (8.79) using the quadratic return function (8.80) $V(\mathbf{x})=\frac{1}{2}\mathbf{x}^T K_1\mathbf{x}$ and approximating

$$\int\tanh\left(G^{-1}\mathbf{B}^T\frac{\partial V}{\partial\mathbf{x}}\right)d\left(\mathbf{B}^T\frac{\partial V}{\partial\mathbf{x}}\right)=\ln\cosh\left(\mathbf{B}^T\frac{\partial V}{\partial\mathbf{x}}\right)\approx\frac{1}{3}\left(\mathbf{B}^T\frac{\partial V}{\partial\mathbf{x}}\right)^2.$$

Solving the quadratic equation $K_1A + A^TK_1 - \frac{2}{3}K_1BG^{-1}B^TK_1 + Q = 0$, we obtain

$$K = \begin{bmatrix} 1.83 & 1.94 & -0.817 & -0.063 \\ 1.94 & 4.75 & -2.31 & 0.261 \\ -0.817 & -2.31 & 2.37 & -0.077 \\ -0.063 & 0.261 & -0.077 & 0.497 \end{bmatrix}, \quad K > 0, \quad K = K^T.$$

The constrained control law is

$$u = -\tanh\left(K_F \begin{bmatrix} x_1 \\ x_2 \\ \int e\,dt \\ e \end{bmatrix}\right), \quad K_F = \begin{bmatrix} 1.83 & 1.94 & -0.817 & -0.063 \end{bmatrix}.$$

Figures 8.34a and b report the simulation results for closed-loop systems if $r = \pm 1$ with the PD control law and constrained control. The designed optimal control $u = -\tanh\left(K_F \begin{bmatrix} x_1 \\ x_2 \\ \int e\,dt \\ e \end{bmatrix}\right) = \tanh\left(-1.83x_1 - 1.94x_2 + 0.817\int e\,dt + 0.063e\right)$ is practical, technology compliant, and ensures better performance than PID algorithms. ■

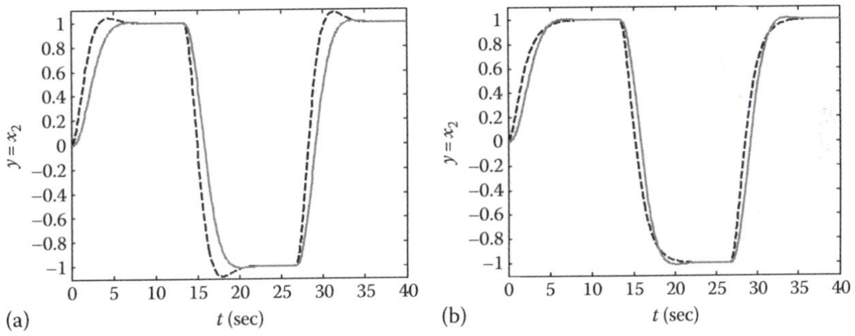

FIGURE 8.34 Output dynamics $y = x_2$ of the closed-loop systems;
(a) Proportional-derivative control law $u = k_pe + k_d(de/dt)$ with $k_p = 1$ and $k_d = 1.41$ (dashed line), and, the bounded tracking control law $u = -\tanh\left(K_F \begin{bmatrix} x_1 \\ x_2 \\ \int e\,dt \\ e \end{bmatrix}\right)$, $K_F = \begin{bmatrix} 1.83 & 1.94 & -0.817 & -0.063 \end{bmatrix}$ (solid line), $r = \pm 1$;
(b) Proportional-derivative control law with $k_p = 1$ and $k_d = 2$ (dashed line), and, the bounded tracking control law $u = -\tanh\left(K_F \begin{bmatrix} x_1 \\ x_2 \\ \int e\,dt \\ e \end{bmatrix}\right) = \tanh\left(-1.83x_1 - 1.94x_2 + 0.817\int e\,dt + 0.063e\right)$ (solid line), $r = \pm 1$.

8.13 OPTIMIZATION OF SYSTEMS USING NONQUADRATIC PERFORMANCE FUNCTIONALS

The Hamilton–Jacobi theory, maximum principle, and Lyapunov concept provide the designer with methods to solve linear and nonlinear control problems. Quadratic and nonquadratic integrands are applied to solve optimization problems. Control laws, robustness, and closed-loop performance are predefined by the integrands and functionals. The closed-loop system dynamics is optimal with respect to the minimizing functional.

The admissible bounded control laws

$$u = -\Phi\left(G^{-1}B^T(x)\frac{\partial V(x)}{\partial x}\right), \quad u \in U,$$

were designed in Sections 8.11 and 8.12 by minimizing

$$J = \int_{t_0}^{t_f}\left[W_x(x) + \int\left(\Phi^{-1}(u)\right)^T G^{-1}\,du\right]dt.$$

The performance integrands $W_u(u)$ are synthesized using the bounded, integrable, one-to-one, real-analytic continuous function $\Phi \in U$.

Consider electromechanical systems modeled by linear and nonlinear differential equations (8.8) and (8.14)

$$\dot{x}(t) = Ax + Bu, \quad x(t_0) = x_0$$

and

$$\dot{x}(t) = F(x) + B(x)u, \quad x(t_0) = x_0.$$

Consider the performance functional

$$J = \int_{t_0}^{t_f}\frac{1}{2}\left[\omega_1^T(x)Q\omega_1(x) + \dot{\omega}_2^T(x)P\dot{\omega}_2(x)\right]dt, \quad Q \geq 0, \quad P > 0, \tag{8.82}$$

where $\omega_1(\cdot):\mathbb{R}^n \to \mathbb{R}_{\geq 0}$ and $\omega_2(\cdot):\mathbb{R}^n \to \mathbb{R}_{\geq 0}$ are the differentiable real-analytic continuous odd functions; $Q \in \mathbb{R}^{n\times n}$ and $P \in \mathbb{R}^{n\times n}$ are the positive-definite diagonal weighting matrices.

Using (8.82), the system performance and stability are specified by positive-definite integrands $\omega_1^T(x)Q\omega_1(x)$ and $\dot{\omega}_2^T(x)P\dot{\omega}_2(x)$. The performance functional (8.82) depends on the states, control, control efforts, energy, etc. In general, $\omega_1(x)$ and $\omega_2(x)$ are different, $\omega_1(x) \neq \omega_2(x)$. To simplify deviations, let $\omega(x) = \omega_1(x) = \omega_2(x)$. Using $\omega(\cdot)$: $\mathbb{R}^n \to \mathbb{R}_{\geq 0}$, we minimize

$$J = \int_{t_0}^{t_f}\frac{1}{2}\left[\omega^T(x)Q\omega(x) + \dot{\omega}^T(x)P\dot{\omega}(x)\right]dt, \quad Q \geq 0, \quad P > 0, \tag{8.83}$$

subject system models (8.8) and (8.14).

For linear systems $\dot{x}(t) = Ax + Bu$, we have

$$\dot{\omega}(x) = \frac{\partial \omega}{\partial x}\dot{x} = \frac{\partial \omega}{\partial x}\left(Ax + Bu\right).$$

In $\dot{\omega}^T(x)P\dot{\omega}(x)$, the element-wise product $\circ$, similar to the Hadamard product, is used. That is,

$$\nabla\omega(x)\circ\dot{x} = \left[\nabla\omega(x_1)\circ\dot{x}_1\cdots\nabla\omega(x_n)\circ\dot{x}_n\right], \nabla\omega(x)\circ\dot{x} : \mathbb{R}^n \times \mathbb{R}^n \to \mathbb{R}^n.$$

The positive-definite integrand $\dot{\omega}^T(x)P\dot{\omega}(x)$ consistently assesses the system energy and power changes for $x \in X$ and $u \in U$. The gradient vector field of a scalar function $\omega(x)$ is used in $\dot{\omega}^T(x)P\dot{\omega}(x)$, where $\nabla\omega$ is consistent with the laws of physics, as illustrated in Example 8.34.

Example 8.34

The classical mechanics defines the force and torque as $F = -\nabla W_c$ and $T = -\nabla W_c$, where W_c is the positive-definite coenergy function. We apply the physics-consistent term

$$\frac{1}{2}\dot{\omega}^T(x)P\dot{\omega}(x) = \frac{1}{2}\dot{x}^T \circ \nabla\omega^T(x)P\nabla\omega(x)\circ\dot{x}.$$

Using $\omega(x) \equiv W_c$, one has $\nabla\omega \equiv \nabla W_c$.

The positive-definite physics-consistent energy and power terms are used. ■

Using (8.83), for linear and nonlinear systems, one minimizes

$$J = \int_{t_0}^{t_f} \frac{1}{2}\left[\omega^T(x)Q\omega(x) + \dot{x}^T \circ \frac{\partial\omega}{\partial x}^T P \frac{\partial\omega}{\partial x}\circ\dot{x}\right]dt. \tag{8.84}$$

As the differentiable and integrable real-valued continuous function $\omega(x)$, one may use:

- $\omega(x) = x$, which leads to the quadratic integrand functions;
- $\omega(x) = x^3$, $\omega(x) = \tanh(x)$, $\omega(x) = e^{|-x|}$ or $\omega(x) = e^{-|x|}$, which result in nonquadratic functionals.

These $\omega(x)$ result in positive-definite $\omega(x)^T Q\omega(x)$ and $\dot{\omega}(x)^T P\dot{\omega}(x)$.

Example 8.35

For a system described by

$$\dot{x} = u,$$

We minimize the functional

$$J = \int_{t_0}^{t_f} \frac{1}{2}\left[\omega^T(x)Q\omega(x) + \dot{x}^T \frac{\partial\omega}{\partial x}^T P \frac{\partial\omega}{\partial x}\dot{x}\right]dt = \int_{t_0}^{t_f} \frac{1}{2}\left[\omega^T(x)Q\omega(x) + u\frac{\partial\omega}{\partial x}^T P \frac{\partial\omega}{\partial x}u\right]dt.$$

Let $Q = 1$ and $P = 1$. Consider $\omega(x) = x$ and $\omega(x) = \tanh(x)$.

For $\omega(x) = x$, one has $J = \int_{t_0}^{t_f} \frac{1}{2}\left(x^2 + u^2\right)dt.$

If $\omega(x) = \tanh(x)$, from $\frac{d}{dx}\tanh x = \mathrm{sech}^2 x$, we have $J = \int_{t_0}^{t_f} \frac{1}{2}\left(\tanh^2 x + \mathrm{sech}^4 x u^2\right)dt.$

The resulting three-dimensional plots for the positive-definite performance integrands $\frac{1}{2}\left(x^2 + u^2\right)$ and $\frac{1}{2}\left(\tanh^2 x + \mathrm{sech}^4 x u^2\right)$ are reported in Figure 8.35. The MATLAB statement is

```
x=linspace(-2,2, 50); u=x; [X,U]=meshgrid(x,u);
W=(X.*X+U.*U)./2; W=(tanh(X).^2+(sech(X).^4).*(U.^2))./2;
surf(x,u,W); xlabel('{\itx}','FontSize',18');
ylabel('{\itu}','FontSize',18'); zlabel('{\itW}','FontSize',18');
```

■

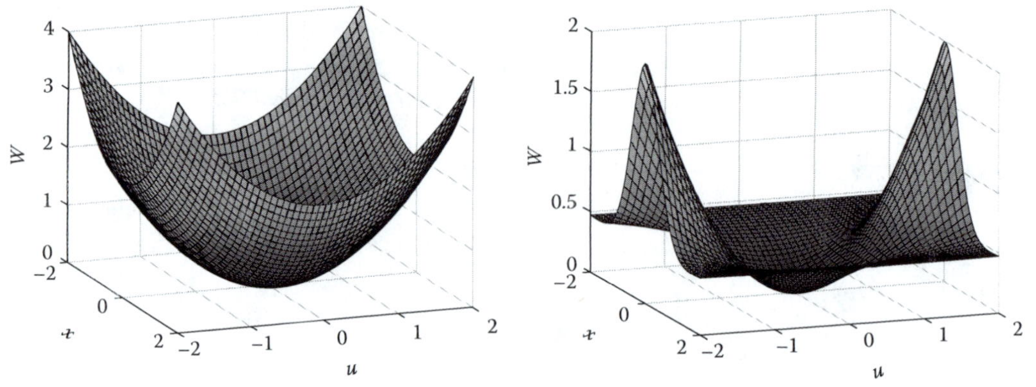

FIGURE 8.35 Surfaces for the integrands $\frac{1}{2}(x^2+u^2)$ and $\frac{1}{2}(\tanh^2 x+\operatorname{sech}^4 x u^2)$.

For linear systems (8.14) and performance functional (8.84), the Hamiltonian function is

$$H\left(x,u,\frac{\partial V}{\partial x}\right)=\frac{1}{2}\omega^T(x)Q\omega(x)+\frac{1}{2}\dot{\omega}^T(x)P\dot{\omega}(x)+\frac{\partial V}{\partial x}^T(Ax+Bu)$$

$$=\frac{1}{2}\omega^T(x)Q\omega(x)+\frac{1}{2}(Ax+Bu)^T P_\omega\left(\frac{\partial\omega}{\partial x}\right)(Ax+Bu)+\frac{\partial V}{\partial x}^T(Ax+Bu), \quad (8.85)$$

where $P_\omega(\cdot)$: $\mathbb{R}^n\to\mathbb{R}^{n\times n}$ is the diagonal map,

$$P_\omega\left(\frac{\partial\omega}{\partial x}\right)=\operatorname{diag}^T\left(\frac{\partial\omega}{\partial x}\right)P\operatorname{diag}\left(\frac{\partial\omega}{\partial x}\right)=\begin{bmatrix} p_{11}\left(\frac{\partial\omega}{\partial x_1}\right)^2 & \cdots & 0\\ \vdots & \ddots & \vdots\\ 0 & \cdots & p_{nn}\left(\frac{\partial\omega}{\partial x_n}\right)^2\end{bmatrix},\quad \forall p_{ii}>0.$$

Applying the first-order condition for optimality (8.10) to the Hamiltonian function (8.85) yields an optimal control law for linear and nonlinear systems as

$$u=-\left(B^T P_\omega\left(\frac{\partial\omega}{\partial x}\right)B\right)^{-1}B^T\left(P_\omega\left(\frac{\partial\omega}{\partial x}\right)Ax+\frac{\partial V}{\partial x}\right), \quad (8.86)$$

$$u=-\left(B^T(x)P_\omega\left(\frac{\partial\omega}{\partial x}\right)B(x)\right)^{-1}B^T(x)\left(P_\omega\left(\frac{\partial\omega}{\partial x}\right)F(x)+\frac{\partial V}{\partial x}\right).$$

The second-order condition for optimality (8.13) is guaranteed, $\frac{\partial^2 H}{\partial u\times\partial u^T}=B^T P_\omega\left(\frac{\partial\omega}{\partial x}\right)B>0.$

From (8.85) and (8.86), we have

$$-\frac{\partial V}{\partial t}=\omega^T(x)Q\omega(x)-\frac{\partial V}{\partial x}^T B\left(B^T P_\omega\left(\frac{\partial\omega}{\partial x}\right)B\right)^{-1}B^T\frac{\partial V}{\partial x}. \quad (8.87)$$

Solution of (8.87) is found by using

$$V(x)=\sum_{i=1}^{N}\frac{2\gamma+1}{2i}\left(x^{\frac{i}{2\gamma+1}}\right)^{T}K_i x^{\frac{i}{2\gamma+1}},\quad \gamma=0,1,2,\ldots \tag{8.88}$$

Example 8.36: Optimization and Linear Quadratic Regular Problem

The performance integrands $\omega^T(x)\,Q\omega(x)$ and $\dot{\omega}^T(x)P\dot{\omega}(x)$ affect the control laws, stability, etc. The integrable and differentiable function $\omega(x)$ must be consistent with the laws of physics, device physics, and control. Applying $\omega(x) = x$, from (8.84), we have the generalized quadratic functional

$$J=\int_{t_0}^{t_f}\frac{1}{2}\left[x^TQx+\left(Ax+Bu\right)^T P\left(Ax+Bu\right)\right]dt. \tag{8.89}$$

The functional equation and the Hamiltonian are

$$-\frac{\partial V}{\partial t}=\min_u\left\{\frac{1}{2}\left[x^TQx+\left(Ax+Bu\right)^T P\left(Ax+Bu\right)\right]+\frac{\partial V}{\partial x}^T\left(Ax+Bu\right)\right\}, \tag{8.90}$$

$$H\left(x,u,\frac{\partial V}{\partial x}\right)=\frac{1}{2}\left[x^TQx+\left(Ax+Bu\right)^T P\left(Ax+Bu\right)\right]+\frac{\partial V}{\partial x}^T\left(Ax+Bu\right),$$

Using the first-order necessary condition for optimality (8.10), one finds

$$u=-\left(B^TPB\right)^{-1}B^T\left(PAx+\frac{\partial V}{\partial x}\right). \tag{8.91}$$

The solution of the functional equation (8.90) is given by the quadratic return function (8.88). When $\gamma = 0$ and $N = 1$, $V=\frac{1}{2}x^TKx$. From (8.91), we have

$$u=-\left(B^TPB\right)^{-1}B^T\left(PA+K\right)x. \tag{8.92}$$

Using (8.90) and (8.92), the unknown symmetric matrix $K \in \mathbb{R}^{n\times n}$ is obtained by solving the following nonlinear differential equation

$$-\dot{K}=Q-KB\left(B^TPB\right)^{-1}B^TK,\quad K\left(t_f\right)=K_f. \tag{8.93}$$

The control law (8.92) is different compared to the conventional linear control law (8.24). The equations to compute matrix K are different, see (8.23) and (8.93). ■

Example 8.37

For the first-order system

$$\frac{dx}{dt}=ax+bu,$$

the generalized quadratic performance functional (8.89) is minimized with $\omega(x) = x$. From (8.89), we have

$$J = \int_{t_0}^{\infty} \frac{1}{2}\left[Q\omega^2(x) + P\left(\frac{\partial\omega}{\partial x}\right)^2\left(a^2x^2 + 2abxu + b^2u^2\right)\right]dt$$

$$= \int_{t_0}^{\infty} \frac{1}{2}\left(x^2 + a^2x^2 + 2abxu + b^2u^2\right)dt, \quad Q = 1, \quad P = 1.$$

Using the quadratic return function $V = \frac{1}{2}kx^2$, the linear control law (8.92) is

$$u = -\frac{1}{b}(a+k)x.$$

Solving the differential equation (8.93) $-\dot{k} = 1 - k^2$ with $t_f = \infty$, we obtain $k = 1$. The control law is

$$u = -\frac{1}{b}(a+1)x.$$

The closed-loop system is stable and evolves as

$$\frac{dx}{dt} = -x. \qquad \blacksquare$$

Example 8.38

For the system

$$\frac{dx}{dt} = ax + bu,$$

we minimize the performance functional (8.89)

$$J = \int_{t_0}^{\infty} \frac{1}{2}\left[Q\omega^2(x) + P\left(\frac{\partial\omega}{\partial x}\right)^2\left(a^2x^2 + 2abxu + b^2u^2\right)\right]dt.$$

The nonquadratic integrands are designed using $\omega(x) = \tanh(x)$. Let $Q = 1$ and $P = 1$. Hence,

$$J = \int_{t_0}^{\infty} \frac{1}{2}\left[\tanh^2 x + \operatorname{sech}^4 x\left(a^2x^2 + 2abxu + b^2u^2\right)\right]dt.$$

For $x < 1$, $\tanh^2 x \approx x^2$ and $\operatorname{sech}^4 x \approx 1$. If $x < 1$, the performance functional

$$J \approx \int_{t_0}^{\infty} \frac{1}{2}\left[x^2 + a^2x^2 + 2abxu + b^2u^2\right]dt$$

is a generalized quadratic-like functional, which was used in Example 8.37 when $\omega(x) = x$. However, the approximations are accurate only for $x < 1$.

If $x \gg 1$, $\tanh^2 x \approx 1$ and $\operatorname{sech}^4 x \approx 0$. Hence, for $x \gg 1$ (for example, $x > 2$), the performance functional is

$$J \approx \frac{1}{2}\int_{t_0}^{\infty} dt.$$

Recall that $J = \int_{t_0}^{\infty} dt$ is commonly used to solve the minimum-time control problems.

Applying the first-order necessary condition for optimality (8.10), from (8.86), we have

$$u = -\frac{a}{b}x - \frac{1}{b\text{sech}^4 x}\frac{\partial V}{\partial x}.$$

The functional equation (8.87) to be solved is

$$-\frac{\partial V}{\partial t} = \frac{1}{2}\tanh^2 x - \frac{1}{2\text{sech}^4 x}\left(\frac{\partial V}{\partial x}\right)^2.$$

In $x \in X$, one approximates continuous functions $\tanh^2 x$ and $\text{sech}^4 x$. The quadratic and nonquadratic return functions can be used. Letting $V = \frac{1}{2}kx^2$, we have

$$u = -\frac{a}{b}x - \frac{1}{b\text{sech}^4 x}kx.$$

The closed-loop system evolves as

$$\frac{dx}{dt} = -\frac{k}{\text{sech}^4 x}x.$$

If $x < 1$, $\text{sech}^4 x \approx 1$, and thus

$$u \approx -\frac{a+k}{b}x.$$

For $x > 1$, we have the high-gain nonlinear control law.
Solving

$$-\frac{\partial V}{\partial t} = \frac{1}{2}\tanh^2 x - \frac{1}{2\text{sech}^4 x}\left(\frac{\partial V}{\partial x}\right)^2, \quad V = \frac{1}{2}kx^2,$$

one finds $k = 1$.

Though the quadratic return function $V = \frac{1}{2}kx^2$ may approximate the solution of nonlinear functional equation in the specified $x \in X$, nonquadratic return functions must be used. For

$$V = \frac{1}{2}k_1x^2 + \frac{1}{4}k_2x^4,$$

one has

$$u = -\frac{a}{b}x - \frac{1}{b\text{sech}^4 x}(k_1x + k_2x^3).$$

The solution of

$$\frac{1}{2}\tanh^2 x - \frac{1}{2\text{sech}^4 x}\left(\frac{\partial V}{\partial x}\right)^2 = 0$$

gives $k_1 = 1$ (equating terms with x^2) and $k_2 = 1/3$ (equating terms with x^6).

The transient dynamics of the closed-loop system for four different initial conditions ($x_0 = 0.1$, $x_0 = 1$, $x_0 = 5$, and $x_0 = 10$) are reported in Figures 8.36. The analysis indicates that the settling time is almost the same for any initial conditions. A novel nonlinear optimal control law possesses unique features as discussed above. The control bounds will increase the settling time. ■

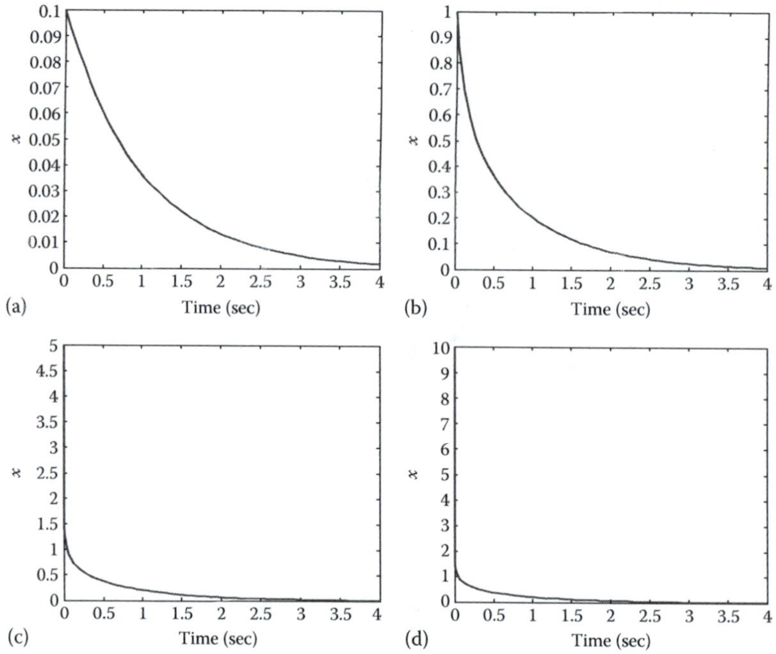

FIGURE 8.36 Transient dynamics of the closed-loop system if (a) $x_0 = 0.1$; (b) $x_0 = 1$; (c) $x_0 = 5$; (d) $x_0 = 10$.

Example 8.39

Consider a system

$$\frac{dx}{dt} = ax + bu, \quad a>0 \text{ and } b>0.$$

Our goal is to find the control law by minimizing the performance functional (8.84) with

$$\omega(x)=\frac{2}{\pi}\arctan x, \quad Q=1, \text{and } P=1.$$

We have

$$J=\int_{t_0}^{t_f}\frac{1}{2}\left[\left(\frac{2}{\pi}\arctan x\right)^2+\left(\frac{d}{dx}\left(\frac{2}{\pi}\arctan x\right)\right)^2(ax+bu)^2\right]dt, \quad \frac{d}{dx}\left(\frac{2}{\pi}\arctan x\right)=\frac{2}{\pi}\frac{1}{x^2+1}.$$

Hence, the functional is

$$J=\int_0^{\infty}\frac{1}{2}\left[\left(\frac{2}{\pi}\arctan x\right)^2+\frac{4}{\pi^2}\left(\frac{1}{x^2+1}\right)^2(ax+bu)^2\right]dt.$$

Finding the Hamiltonian

$$H\left(x,u,\frac{\partial V}{\partial x}\right)=\frac{1}{2}\left(\frac{4}{\pi^2}\arctan^2 x+\frac{4}{\pi^2}\frac{1}{(x^2+1)^2}(ax+bu)^2\right)+\frac{\partial V}{\partial x}(ax+bu),$$

we apply the first-order necessary condition for optimality (8.10). From

$$\frac{\partial H}{\partial u}=\frac{4}{\pi^2}\frac{1}{(x^2+1)^2}abx+\frac{4}{\pi^2}\frac{1}{(x^2+1)^2}b^2u+\frac{\partial V}{\partial x}b=0,$$

one finds the control law

$$u=-\frac{a}{b}x-\frac{\pi^2(x^2+1)^2}{4b}\frac{\partial V}{\partial x}.$$

From (8.87),

$$-\frac{\partial V}{\partial t}=\frac{2}{\pi^2}\arctan^2 x+\frac{2}{\pi^2}\frac{1}{(x^2+1)^2}\left(ax+bu\right)^2+\frac{\partial V}{\partial x}\left(ax+bu\right),$$

one has

$$-\frac{\partial V}{\partial t}=\frac{2}{\pi^2}\arctan^2 x+\frac{2}{\pi^2}\frac{1}{(x^2+1)^2}\left(\frac{\pi^2(x^2+1)}{4}\frac{\partial V}{\partial x}\right)^2-\frac{\partial V}{\partial x}\frac{\pi^2(x^2+1)^2}{4}\frac{\partial V}{\partial x},$$

and

$$-\frac{\partial V}{\partial t}=\frac{2}{\pi^2}\arctan^2 x-\frac{\pi^2(x^2+1)^2}{8}\left(\frac{\partial V}{\partial x}\right)^2.$$

The solution of this equation is approximated by the quadratic return function

$$V=\frac{1}{2}kx^2.$$

One solves

$$-\frac{\partial V}{\partial t}=\frac{2}{\pi^2}\arctan^2 x-\frac{\pi^2(1+x^2)^2}{8}k^2x^2,$$

where $\arctan x=x-\frac{1}{3}x^3+\frac{1}{5}x^5-\cdots\approx x$.

From

$$\frac{2}{\pi^2}x^2-\frac{\pi^2}{8}(1+x^2)^2k^2x^2=0,$$

by grouping the terms with x^2, one obtains

$$\frac{2}{\pi^2}-\frac{\pi^2}{8}k^2=0.$$

Hence, $k=\pm\sqrt{\frac{16}{\pi^4}}$ and $k=\frac{4}{\pi^2}$.

From

$$u=-\frac{a}{b}x-\frac{\pi^2(x^2+1)^2}{4b}\frac{\partial V}{\partial x}\quad\text{and}\quad V=\frac{1}{2}kx^2,$$

we have

$$u=-\frac{a}{b}x-\frac{\pi^2}{4b}k(x+2x^3+x^5).$$

Therefore,

$$u=-\frac{a}{b}x-\frac{1}{b}(x+2x^3+x^5).$$

The closed-loop system is evolves as

$$\frac{dx}{dt} = ax + bu = ax + b\left(-\frac{a}{b}x - \frac{1}{b}(x + 2x^3 + x^5)\right) = -x - 2x^3 - x^5.$$

The stability is examined by using the positive-definite quadratic Lyapunov function

$$V_L = \frac{1}{2}x^2 > 0.$$

One finds the total derivative

$$\frac{dV_L}{dt} = \frac{dV_L}{dx}\frac{dx}{dt} = -x^2 - 2x^4 - x^6 < 0.$$

Thus, the closed-loop system is asymptotically stable. ■

Example 8.40

The differential equations that describe rigid-body mechanical systems are

$$\frac{dx_1}{dt} = ax_1 + bu,$$

$$\frac{dx_2}{dt} = x_1,$$

where x_1 and x_2 are the velocity and displacement; u is the force or torque.

The performance integrands are designed with $\omega(x) = \tanh(x)$. Using the identity matrices $Q = I$ and $P = I$, functional (8.84) is

$$J = \int_{t_0}^{t_f} \frac{1}{2}\left\{\begin{bmatrix}\tanh x_1 & \tanh x_2\end{bmatrix}\begin{bmatrix}1 & 0\\ 0 & 1\end{bmatrix}\begin{bmatrix}\tanh x_1\\ \tanh x_2\end{bmatrix} + \begin{bmatrix}\dot{x}_1\text{sech}^2 x_1 & \dot{x}_2\text{sech}^2 x_2\end{bmatrix}\begin{bmatrix}1 & 0\\ 0 & 1\end{bmatrix}\begin{bmatrix}\dot{x}_1\text{sech}^2 x_1\\ \dot{x}_2\text{sech}^2 x_2\end{bmatrix}\right\}dt$$

$$= \int_{t_0}^{t_f} \frac{1}{2}\left[\tanh^2 x_1 + \tanh^2 x_2 + \text{sech}^4 x_1\left(a^2x_1^2 + 2abx_1u + b^2u^2\right) + x_1^2\text{sech}^4 x_2\right]dt.$$

The Hamiltonian is

$$H = \frac{1}{2}\left[\tanh^2 x_1 + \tanh^2 x_2 + \text{sech}^4 x_1\left(a^2x_1^2 + 2abx_1u + b^2u^2\right) + x_1^2\text{sech}^4 x_2\right]$$

$$+ \frac{\partial V}{\partial x_1}\left(ax_1 + bu\right) + \frac{\partial V}{\partial x_1}x_1.$$

The first-order necessary condition for optimality (8.10) results in a control law

$$u = -\frac{a}{b}x_1 - \frac{1}{b\text{sech}^4 x_1}\frac{\partial V}{\partial x_1}.$$

We solve (8.87)

$$-\frac{\partial V}{\partial t} = \frac{1}{2}\tanh^2 x_1 + \frac{1}{2}\tanh^2 x_2 - \frac{1}{2\text{sech}^4 x_1}\left(\frac{\partial V}{\partial x_1}\right)^2 + \frac{1}{2}x_1^2\text{sech}^4 x_2 + \frac{\partial V}{\partial x_2}x_1$$

by approximating the solution by a nonquadratic return function

$$V = \frac{1}{2}k_{11}x_1^2 + k_{12}x_1x_2 + \frac{1}{2}k_{22}x_2^2 + \frac{1}{4}k_{41}x_1^4.$$

One has

$$\tanh x = x - \frac{1}{3}x^3 + \frac{2}{15}x^5 - \frac{17}{315}x^7 + \cdots = \sum_{n=1}^{\infty} \frac{1}{(2n)!} 2^{2n}(2^{2n}-1)B_{2n}x^{2n-1}$$

and

$$\operatorname{sech} x = 1 - \frac{1}{2}x^2 + \frac{5}{24}x^4 - \frac{61}{720}x^6 + \cdots = \sum_{n=1}^{\infty} \frac{1}{(2n)!} E_{2n}x^{2n},$$

where B_n and E_n are the nth Bernoulli and Euler numbers.

Using $\tanh x \approx x - \frac{1}{3}x^3$ ($|x_1| \le 1$ and $|x_2| \le 1$) and $\operatorname{sech} x \approx 1$, and, equating terms with x_1^2, x_1x_2, x_2^2, and x_1^6, one finds $k_{11} = 1$, $k_{12} = 1$, $k_{22} = 1$, $k_{41} = 1/3$.

Having found the unknown coefficients of $V(x)$, we have

$$u = -\frac{a}{b}x_1 - \frac{1}{b\operatorname{sech}^4 x_1}\left(x_1 + x_2 + \frac{1}{3}x_1^3\right).$$

The Simulink model is reported in Figure 8.37a. Let $a = 1$ and $b = 1$. The transient dynamics of the closed-loop system with the initial conditions

$$\begin{bmatrix} x_{10} \\ x_{20} \end{bmatrix} = \begin{bmatrix} 0.1 \\ -0.1 \end{bmatrix}, \quad \begin{bmatrix} x_{10} \\ x_{20} \end{bmatrix} = \begin{bmatrix} 1 \\ -1 \end{bmatrix}, \quad \text{and} \quad \begin{bmatrix} x_{10} \\ x_{20} \end{bmatrix} = \begin{bmatrix} 2 \\ -2 \end{bmatrix}$$

are reported in Figure 8.37b.

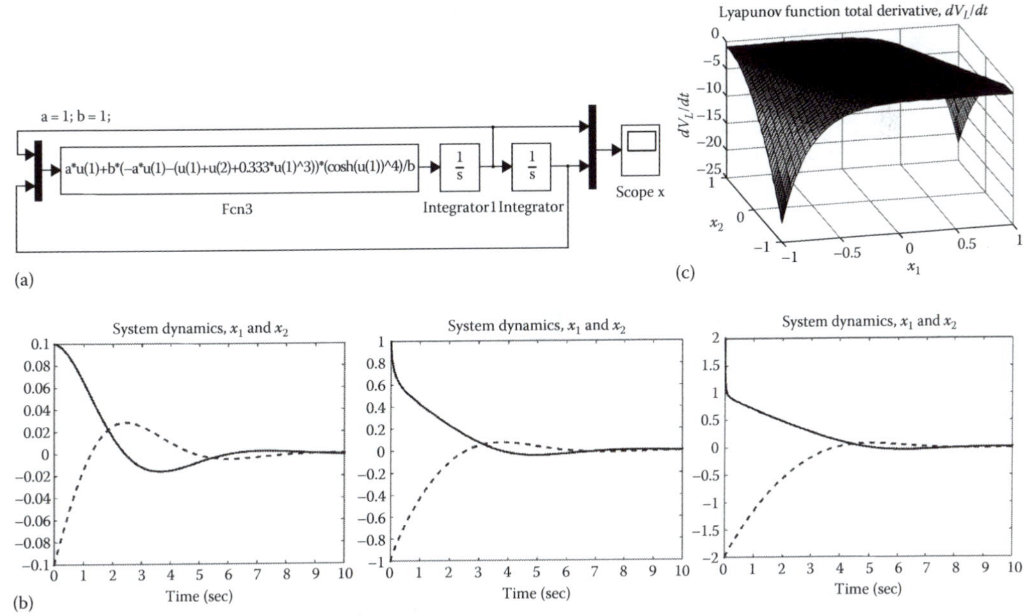

FIGURE 8.37 (a) Simulink® model to simulate the closed-loop system;
(b) Dynamics of the closed-loop system (x_1 and x_2 are plotted using the solid and dashed lines) with different initial conditions;
(c) The total derivative of the Lyapunov function.

The closed-loop system is

$$\frac{dx_1}{dt}=-\frac{1}{\operatorname{sech}^4 x_1}\frac{\partial V}{\partial x_1},\quad \frac{dx_2}{dt}=x_1.$$

The design procedure and control law

$$u=-\frac{a}{b}x_1-\frac{1}{b\operatorname{sech}^4 x_1}\frac{\partial V}{\partial x_1}$$

are simplified if one applies the quadratic return function

$$V=\frac{1}{2}k_{11}x_1^2+k_{12}x_1x_2+\frac{1}{2}k_{22}x_2^2.$$

One can examine the stability of the closed-loop system. For a positive-definite Lyapunov function

$$V_L=\frac{1}{2}x_1^2+\frac{1}{2}x_2^2,$$

the total derivative is

$$\frac{dV_L}{dt}=-\frac{1}{\operatorname{sech}^4 x_1}\left(\frac{\partial V}{\partial x_1}\right)^2+\frac{\partial V}{\partial x_2}x_1=-\frac{1}{\operatorname{sech}^4 x_1}(k_{11}x_1+k_{12}x_2)^2+x_1x_2.$$

Hence, dV_L/dt is a negative-definite. The solution of (8.90) results in $k_{11}=1$, $k_{12}=1$, and $k_{22}=1$. The total derivative of the Lyapunov function is illustrated in Figure 8.37c. The following MATLAB statement is used to perform calculations and for plotting:

```
x=linspace(-1,1,50); y=x; [x1,x2]=meshgrid(x,y); k11=1; k12=1;
dV=x1.*x2-((k11*x1+k12*x2).^2)./sech(x1).^4;
surf(x,y,dV); xlabel('{\itx}_1','FontSize',18); ylabel('{\itx}_2','FontSize',18);
zlabel('{\itdV_L/dt}','FontSize',18);
title('Lyapunov Function Total Derivative, {\itdV_L/dt}','FontSize',18);
```

The nonlinear control law designed ensures optimal system evolution with respect to the functional minimized. The closed-loop system is stable. ■

8.14 LYAPUNOV STABILITY THEORY IN ANALYSIS AND CONTROL

The electromechanical system dynamics is described by nonlinear state-space differential equations. We examine stability of time-varying open-loop and closed-loop nonlinear systems. The following theorem is formulated by applying the results of the Lyapunov stability theory.

Theorem 8.1

Consider a dynamic system described by nonlinear differential equations

$$\dot{x}(t)=F(t,x),\quad x(t_0)=x_0,\quad t\geq 0. \tag{8.94}$$

If there exists a positive-definite scalar function $V(t, x)$, $V(\cdot)$: $\mathbb{R}_{\geq 0}\times\mathbb{R}^n\to\mathbb{R}_{\geq 0}$, called the Lyapunov function, with continuous first-order partial derivatives with respect to t and x

$$\frac{dV}{dt}=\frac{\partial V}{\partial t}+\left(\frac{\partial V}{\partial x}\right)^T\frac{dx}{dt}=\frac{\partial V}{\partial t}+\left(\frac{\partial V}{\partial x}\right)^T F(t,x), \tag{8.95}$$

then

- The system and equilibrium state are stable if the total derivative of the positive-definite function $V(t, x) > 0$ is $dV(t, x)/dt \le 0$;
- The system and equilibrium state are uniformly asymptotically stable in the large if the total derivative of $V(t, x) > 0$ is negative definite, $dV(t, x)/dt < 0$;
- The system and equilibrium state are exponentially stable in the large if there exist the unbounded functions $\rho_1(\cdot)$ and $\rho_2(\cdot)$, and, continuous strictly increasing function $\rho_3(\cdot)$, $\rho_3(0) = 0$, such that

$$\rho_1\left(\|x\|\right) \le V(t,x) \le \rho_2\left(\|x\|\right) \quad \text{and} \quad \frac{dV(x)}{dt} \le -\rho_3\left(\|x\|\right).$$ ■

Example 8.41

Consider a system that is described by two nonlinear time-invariant differential equations

$$\dot{x}_1(t) = x_1 - x_1^5 - x_1^3 x_2^4,$$

$$\dot{x}_2(t) = x_2 - x_2^9, \quad t \ge 0.$$

These differential equations describe the uncontrolled or controlled (closed-loop) dynamics. For example, one may consider unstable open-loop system with control functions u_1 and u_2

$$\dot{x}_1(t) = x_1 - x_1^3 x_2^4 + u_1, \quad u_1 = -x_1^5,$$

$$\dot{x}_2(t) = x_2 + u_2, \quad u_2 = -x_2^9.$$

Let a scalar positive-definite function be

$$V\left(x_1, x_2\right) = \frac{1}{2}\left(x_1^2 + x_2^2\right).$$

The total derivative is

$$\frac{dV\left(x_1,x_2\right)}{dt} = \left(\frac{\partial V}{\partial x}\right)^T \frac{dx}{dt} = \left(\frac{\partial V}{\partial x}\right)^T F(x) = \frac{\partial V}{\partial x_1}\left(x_1 - x_1^5 - x_1^3 x_2^4\right) + \frac{\partial V}{\partial x_2}\left(x_2 - x_2^9\right) = x_1^2 - x_1^6 - x_1^4 x_2^4 + x_2^2 - x_2^{10}.$$

The total derivative of a positive-definite $V(x_1, x_2) > 0$ is

$$\frac{dV\left(x_1,x_2\right)}{dt} < 0 \quad \text{for } |x| > 1.$$

Therefore, the system is stable in the large.
For $\dot{x}_1(t) = x_1 - x_1^3 x_2^4 + u_1$, $\dot{x}_2(t) = x_2 + u_2$,
using

$$u_1 = -x_1 - x_1^5, \quad u_2 = -x_2 - x_2^9,$$

for a positive-definite

$$V\left(x_1, x_2\right) = \frac{1}{2}\left(x_1^2 + x_2^2\right),$$

one finds

$$\frac{dV\left(x_1,x_2\right)}{dt} = -x_1^6 - x_1^4 x_2^4 - x_2^{10}.$$

For a positive-definite $V(x_1, x_2) > 0$, $dV/dt < 0$. Therefore, the equilibrium state of the system is asymptotically stable in the large. ■

Example 8.42

Consider a time-varying nonlinear differential equations

$$\dot{x}_1(t) = -x_1 + x_2^3,$$

$$\dot{x}_2(t) = -e^{-10t} x_1 x_2^2 - 5x_2 - x_2^3, \quad t \geq 0.$$

A scalar positive-definite function $V(t, x_1, x_2) > 0$ is chosen as

$$V\left(t, x_1, x_2\right) = \frac{1}{2}\left(x_1^2 + e^{10t} x_2^2\right).$$

The total derivative is

$$\frac{dV\left(t, x_1, x_2\right)}{dt} = \frac{\partial V}{\partial t} + \frac{\partial V}{\partial x_1}\left(-x_1 + x_2^3\right) + \frac{\partial V}{\partial x_2}\left(-e^{-10t} x_1 x_2^2 - 5x_2 - x_2^3\right) = -x_1^2 - e^{10t} x_2^4.$$

The total derivative is negative-definite, $\frac{dV\left(x_1, x_2\right)}{dt} < 0.$

Using Theorem 8.1, one concludes that the equilibrium state is uniformly asymptotically stable. ■

Example 8.43

The system dynamics is described by the following differential equations

$$\dot{x}_1(t) = -x_1 + x_2,$$

$$\dot{x}_2(t) = -x_1 - x_2 - x_2\left|x_2\right|, \quad t \geq 0.$$

The positive-definite scalar Lyapunov candidate is

$$V(x_1, x_2) = \frac{1}{2}\left(x_1^2 + x_2^2\right).$$

Thus, $V(x_1, x_2) > 0$.
The total derivative is

$$\frac{dV(x_1, x_2)}{dt} = x_1\dot{x}_1 + x_2\dot{x}_2 = -x_1^2 - x_2^2\left(1 + \left|x_2\right|\right).$$

Therefore, $\frac{dV(x_1, x_2)}{dt} < 0.$
Hence, the equilibrium state of the system is uniformly asymptotically stable, and the quadratic function $V(x_1, x_2) = \frac{1}{2}\left(x_1^2 + x_2^2\right)$ is the Lyapunov function. ■

Example 8.44: Stability of Permanent-Magnet Synchronous Motors

Consider a permanent-magnet synchronous motor in the rotor reference frame. The mathematical model, assuming that $T_L = 0$, is given by three differential equations

$$\frac{di_{qs}^r}{dt} = -\frac{r_s}{L_{ss}} i_{qs}^r - \frac{\psi_m}{L_{ss}} \omega_r - i_{ds}^r \omega_r + \frac{1}{L_{ss}} u_{qs}^r,$$

$$\frac{di_{ds}^r}{dt} = -\frac{r_s}{L_{ss}} i_{ds}^r + i_{qs}^r \omega_r + \frac{1}{L_{ss}} u_{ds}^r,$$

$$\frac{d\omega_r}{dt} = \frac{3P^2\psi_m}{8J} i_{qs}^r - \frac{B_m}{J} \omega_r.$$

For an open-loop system,

$$u_{qs}^r = 0 \quad \text{and} \quad u_{ds}^r = 0.$$

Using the quadratic positive-definite Lyapunov function

$$V(i_{qs}^r, i_{ds}^r, \omega_r) = \frac{1}{2}\left(i_{qs}^{r2} + i_{ds}^{r2} + \omega_r^2\right),$$

the expression for the total derivative is

$$\frac{dV(i_{qs}^r, i_{ds}^r, \omega_r)}{dt} = -\frac{r_s}{L_{ss}} i_{qs}^{r2} - \frac{r_s}{L_{ss}} i_{ds}^{r2} - \frac{B_m}{J} \omega_r^2 - \frac{8J\psi_m - 3P^2 L_{ss}\psi_m}{8JL_{ss}} i_{qs}^r \omega_r.$$

Thus,

$$\frac{dV\left(i_{qs}^r, i_{ds}^r, \omega_r\right)}{dt} < 0.$$

One concludes that the equilibrium state of an open-loop drive is uniformly asymptotically stable in the large.

For a closed-loop system, one has $u_{qs}^r \neq 0$ and $u_{ds}^r = 0$.

Let

$$u_{qs}^r = -k_I i_{qs}^r - k_\omega \omega_r.$$

The following differential equations result

$$\frac{di_{qs}^r}{dt} = -\frac{r_s}{L_{ss}} i_{qs}^r - \frac{\psi_m}{L_{ss}} \omega_r - i_{ds}^r \omega_r - \frac{1}{L_{ss}}\left(k_I i_{qs}^r + k_\omega \omega_r\right),$$

$$\frac{di_{ds}^r}{dt} = -\frac{r_s}{L_{ss}} i_{ds}^r + i_{qs}^r \omega_r,$$

$$\frac{d\omega_r}{dt} = \frac{3P^2\psi_m}{8J} i_{qs}^r - \frac{B_m}{J} \omega_r.$$

Using the quadratic positive-definite Lyapunov function

$$V(i_{qs}^r, i_{ds}^r, \omega_r) = \frac{1}{2}(i_{qs}^{r2} + i_{ds}^{r2} + \omega_r^2),$$

we obtain

$$\frac{dV(i_{qs}^r, i_{ds}^r, \omega_r)}{dt} = -\frac{r_s + k_I}{L_{ss}} i_{qs}^{r2} - \frac{r_s}{L_{ss}} i_{ds}^{r2} - \frac{B_m}{J} \omega_r^2 - \frac{8J\left(\psi_m + k_\omega\right) - 3P^2 L_{ss}\psi_m}{8JL_{ss}} i_{qs}^r \omega_r.$$

Hence,

$$V\left(i_{qs}^r, i_{ds}^r, \omega_r\right) > 0 \quad \text{and} \quad \frac{dV\left(i_{qs}^r, i_{ds}^r, \omega_r\right)}{dt} < 0.$$

The conditions for asymptotic stability are guaranteed. The rate of decreasing of $V\left(i_{qs}^r, i_{ds}^r, \omega_r\right)$ affects the drive dynamics. The derived expression for *dV*/*dt* illustrates the role of the control $u_{qs}^r = -k_I i_{qs}^r - k_\omega \omega_r$. ■

Example 8.45: Lyapunov Function and Energy

The Lyapunov method provides a consistent foundation in stability analysis. Applying the basic laws of physics, one performs concurrent analyses and derives differential equations that describe the dynamics of physical systems. Using the system variable x, the system energy is found and used deriving equations of motion, finding the force and torque, etc. The scalar positive-definite Lyapunov functions $V(x)$ can be designed using the all-state-dependent energy. That is, $V(x) \equiv E(x)$. For stable systems, characterized by the positive-definite total energy $E(x)$, the energy change rate $dE(x)/dt$ decreases if the system evolves toward equilibrium. For stable systems, the total derivative of the positive-definite energy-pertinent function $V(x) > 0$ is negative definite, and, $dV(x)/dt < 0$. ■

Example 8.46

Consider a linear open-loop system

$$\dot{x}(t) = Ax,$$

or a closed-loop system

$$\dot{x}(t) = A_{open\ loop}x + Bu = (A_{open\ loop} - BK_F)x = Ax, \quad u = -K_F x.$$

Using the quadratic all-state-dependent function

$$V = \frac{1}{2}x^T Kx,$$

one finds

$$\frac{dV}{dt} = \frac{1}{2}\left(\dot{x}^T Kx + x^T K\dot{x}\right) = \frac{1}{2}x^T\left(A^T K + K^T A\right)x, \quad K = K^T.$$

The energy of the physical system $E(x)$ is positive-definite, and $V(x) > 0$, $\forall x \neq 0$. This $V(x)$ can be implicitly or explicitly derived using the total energy $E(x)$.

Using the positive-definite matrix $Q > 0$, one may set the negative-definiteness of the total derivative dV/dt or dE/dt to be

$$\frac{dV}{dt} = -\frac{1}{2}x^T Qx.$$

The resulting linear Lyapunov equation to find $K > 0$ is

$$A^T K + KA = -Q.$$

The asymptotic stability of the system is guaranteed if $K > 0$ for a given $Q > 0$. That is, $V(x) > 0$ for a given $dV(x)/dt < 0$. The specified negative definiteness of dV/dt was applied.

Alternatively, for a system characterized by $E > 0$, one applies

$$V = \frac{1}{2}x^T Kx, \quad V = E > 0$$

with the state- and parameter-dependent $K > 0$. By solving

$$A^T K + KA = -Q,$$

one obtains Q, which must be positive. This yields the rate of change as

$$\frac{dV}{dt} = -\frac{1}{2}x^T Qx$$

for a given

$$V = \frac{1}{2}x^T Kx > 0.$$

■

8.15 MINIMAL-COMPLEXITY CONTROL LAWS

To reduce hardware and software complexity, we design *minimal-complexity* control laws. Tracking control laws can be designed by applying Lyapunov's stability theory. The control laws affect the system dynamics, change the total derivative of the Lyapunov function $V(t, x, e)$, and affect stability and robustness. For $V(t, x, e) > 0$, one can derive a control function u with a goal to guarantee $dV/dt < 0$ as well as the specified rate of change of dV/dt. The feedback gains can be found by solving nonlinear matrix equations or inequalities specifying $dV(t, x, e)/dt$.

Example 8.47: Optimization and Lyapunov Theory in Control Design

We study the relationship between the Lyapunov and Hamilton–Jacobi concepts. This analysis is very important because it assesses optimization, stability, system dynamics, and system complexity. For linear systems (8.14)

$$\dot{x}(t) = Ax + Bu,$$

minimization of the positive-definite quadratic performance functional (8.15) using the first-order necessary condition for optimality (8.10) yields a linear control law (8.24)

$$u = -G^{-1}B^TKx.$$

Using the control law derived and the feedback gain matrix $K_F = G^{-1}B^TK$, we rewrite the functional (8.15) as

$$J = \frac{1}{2}\int_{t_0}^{\infty}\left(x^TQx + u^TGu\right)dt = \frac{1}{2}\int_{t_0}^{\infty}\left(x^TQx + x^TK_F^TGK_Fx\right)dt = \frac{1}{2}\int_{t_0}^{\infty} x^T\left(Q + K_F^TGK_F\right)x\,dt.$$

Using the quadratic, positive-definite function (8.20) $V(x) = \frac{1}{2}x^TKx$, for system (8.14) $\dot{x}(t) = Ax + Bu$, one finds

$$\frac{dV(x)}{dt} = \frac{1}{2}\left(\dot{x}^TKx + x^TK\dot{x}\right) = \frac{1}{2}\left[\left(Ax - BK_Fx\right)^T Kx + x^TK\left(Ax - BK_Fx\right)\right].$$

One may specify the rate of change of the Lyapunov function as

$$\frac{dV(x)}{dt} = -x^T\left(Q + K_F^TGK_F\right)x.$$

The positive-semidefinite and positive-definite constant-coefficient weighting matrices Q and G are used.

Hence, $\frac{dV(x)}{dt} < 0$. That is, the total derivative of $V(x)$ is negative-definite.

One has

$$\frac{1}{2}\left[\left(Ax - BK_Fx\right)^T Kx + x^TK\left(Ax - BK_Fx\right)\right] = -\frac{1}{2}x^T\left(Q + K_F^TGK_F\right)x.$$

Using $K_F = G^{-1}B^TK$, we find the algebraic Riccati equation, which yields the unknown matrix $K > 0$

$$-Q - A^TK - K^TA + K^TBG^{-1}B^TK = 0, \quad K = K^T.$$

This equation is in agreement with (8.23). From $V(x)=\frac{1}{2}x^TKx$ one concludes that $V(x) > 0$ and $dV(x)/dt < 0$. The closed-loop systems (8.14) with (8.24)

$$\dot{x}(t)=Ax+Bu=Ax-BG^{-1}B^TKx=\left(A-BG^{-1}B^TK\right)x=\left(A-BK_F\right)x$$

is stable.

Control laws

$$u=-G^{-1}B^TKx,\quad u=\phi(t,e,x),\text{ or } u=\phi[\varphi(t,e,x)]$$

may be defined by the designer, and the feedback gains can be derived using the requirements imposed on the Lyapunov pair,

$$V(x) > 0 \quad \text{and} \quad dV/dt < 0.$$

The use of the Hamilton–Jacobi theory and other optimization methods results in optimal control laws designed minimizing the performance functionals. To implement these control laws, all variables (e and x) must be measured or observed. Alternatively, the designer may apply the Lyapunov concept. One synthesizes the *minimal-complexity* control laws by using the directly measured variables to reduce system complexity and to simplify hardware by minimizing the number of sensors, ICs, etc. ■

The *minimal-complexity* control implies synthesis of control laws with the specified physical variables to implement u. The designer may use the tracking error $e(t)$, directly measured states $x_m(t)$, or measured performance quantities $q_s(t)$. This simplifies the hardware and reduces software complexity.

Using the stability and optimality measures, which are quantified by a Lyapunov pair, we find the *minimal-complexity* control law as a nonlinear function of error $e(t)$ and measured states $x_m(t)$

$$u=\phi(e,x_m) \tag{8.96}$$

subject to the system dynamics (8.8) $\dot{x}(t)=F(x)+B(x)u$ and constraints.

Theorem 8.2

Consider a closed-loop system (8.8)

$$\dot{x}(t)=F(x)+B(x)u$$

with control law (8.96) under the references $r \in R$ and disturbances $d \in D$. For the closed-loop systems (8.8) with (8.96)

1. Solutions are uniformly bounded;
2. Equilibrium is exponentially stable in the convex and compact state evolution set $X(X_0, U, R, D) \subset \mathbb{R}^n$;
3. Tracking is ensured and disturbance attenuation is guaranteed in the state-error evolution set $XE(X_0, E_0, U, R, D) \subset \mathbb{R}^n \times \mathbb{R}^b$

if there exists a continuous differentiable function $V(t, x, e)$, $V(\cdot)$: $\mathbb{R}_{\ge 0} \times \mathbb{R}^n \times \mathbb{R}^b \to \mathbb{R}_{\ge 0}$ in XE such that

$$V\left(t,x,e\right)>0 \quad \text{and} \quad \frac{dV(t,x,e)}{dt}\le-\rho_1\|x\|-\rho_2\|e\| \tag{8.97}$$

for all $x \in X$, $e \in E$, $u \in U$, $r \in R$, and $d \in D$ on $[t_0, \infty)$. Here, $\rho_1(\cdot)$: $\mathbb{R}_{\ge 0} \to \mathbb{R}_{\ge 0}$ and $\rho_2(\cdot)$: $\mathbb{R}_{\ge 0} \to \mathbb{R}_{\ge 0}$ are the continuous, strictly increasing functions, $\rho_1(0) = \rho_2(0) = 0$. ■

The criteria imposed on the Lyapunov pair (8.97) are examined. Using the system dynamics, the total derivative of the Lyapunov candidate $V(t, x, e)$ is obtained. The inequality

$$\frac{dV(t,x,e)}{dt} \le -\rho_1 \| x \| - \rho_2 \| e \|$$

may be solved to find the feedback coefficients. The Lyapunov candidate functions should be designed. The quadratic and nonquadratic Lyapunov candidates are applied. For example, $V(x, e)$, $V(\cdot): \mathbb{R}^n \times \mathbb{R}^b \to \mathbb{R}_{\ge 0}$ can be given as

$$V(x,e) = \sum_{i=1}^{N} \frac{1}{2i}(x^i)^T K_{xi} x^i + \sum_{i=1}^{M} \frac{1}{2i}(e^i)^T K_{ei} e^i, \; K_{xi} \in \mathbb{R}^{n \times n}, K_{ei} \in \mathbb{R}^{b \times b}, N = 1,2,3,\ldots, \text{ and } M = 1,2,3,\ldots. \tag{8.98}$$

Using the matrix-functions $K_{xi}(\cdot): \mathbb{R}_{\ge 0} \to \mathbb{R}^{n \times n}$ and $K_{ei}(\cdot): \mathbb{R}_{\ge 0} \to \mathbb{R}^{b \times b}$, the time-varying Lyapunov function $V(t, x, e)$, $V(\cdot): \mathbb{R}_{\ge 0} \times \mathbb{R}^n \times \mathbb{R}^b \to \mathbb{R}_{\ge 0}$ can be given as

$$V(t,x,e) = \sum_{i=1}^{N} \frac{1}{2i}\left(x^i\right)^T K_{xi}(t) x^i + \sum_{i=1}^{M} \frac{1}{2i}\left(e^i\right)^T K_{ei}(t) e^i.$$

The scalar Lyapunov function $V(t, x, e)$, $V(\cdot): \mathbb{R}_{\ge 0} \times \mathbb{R}^n \times \mathbb{R}^b \to \mathbb{R}_{\ge 0}$ can be expressed as

$$V(t,x,e) = \sum_{i=1}^{N} \frac{1}{2i}(x^i)^T K_{xi}(t) x^i + \sum_{i=1}^{P} \frac{1}{2i}(x^i)^T K_{xei}(t) e^i + \sum_{i=1}^{M} \frac{1}{2i}(e^i)^T K_{ei}(t) e^i,$$

$$P = 1,2,3,\ldots, \; K_{xei}(.): \mathbb{R}_{\ge 0} \to \mathbb{R}^{n \times b}. \tag{8.99}$$

The results of analytic design of control laws can be applied as reported in illustrative examples. One may derive unconstrained or constrained control laws

$$u = \varphi(t, e, x_m) \quad \text{and} \quad u = \phi(t, e, x_m)$$

using the directly measurable or observable x_m and e. For example, tracking control laws

$$u = -G_z^{-1} B_z^T(x,e) \frac{\partial V_m(x,e)}{\partial [x\ e]^T}$$

can be designed with the positive-definite $V_m(x, e)$ such that $u = \varphi(e, x_m)$. For the closed-loop system (8.8) with (8.96), criteria (8.97) must be guaranteed.

We depart from all-state control laws deriving

$$u(t) = K_{mF1} \begin{bmatrix} x_m(t) \\ e(t) \end{bmatrix} + \int K_{mF2} \begin{bmatrix} x_m(\tau) \\ e(\tau) \end{bmatrix} d\tau, \quad u(t) = \phi\left(K_{mF1} \begin{bmatrix} x_m(t) \\ e(t) \end{bmatrix} + \int K_{mF2} \begin{bmatrix} x_m(\tau) \\ e(\tau) \end{bmatrix} d\tau \right). \tag{8.100}$$

To study the closed-loop system stability, the closed-loop dynamics is examined using the Lyapunov function. The system performance and capabilities in the operating envelope XE are quantified by $V(x, e)$ and $dV(x, e)/dt$, which are relevant to the performance functionals.

Example 8.48

Consider the electric drive with a permanent-magnet DC motor (30 V, 300 rad/sec, $r_a = 2$ ohm, $k_a = 0.1$ V-sec/rad, $L_a = 0.005$ H, $B_m = 0.0001$ N-m-sec/rad, and $J = 0.0001$ kg-m^2) and a *step-down* converter as covered in Section 7.4.2. Using the Kirchhoff's law and the *averaging* concept, we have

$$\begin{bmatrix} \frac{du_C}{dt} \\ \frac{di_L}{dt} \\ \frac{di_a}{dt} \\ \frac{d\omega_r}{dt} \end{bmatrix} = \begin{bmatrix} 0 & \frac{1}{C} & -\frac{1}{C} & 0 \\ -\frac{1}{L} & -\frac{r_L + r_c}{L} & \frac{r_c}{L} & 0 \\ \frac{1}{L_a} & \frac{r_c}{L_a} & -\frac{r_a + r_c}{L_a} & -\frac{k_a}{L_a} \\ 0 & 0 & \frac{k_a}{J} & -\frac{B_m}{J} \end{bmatrix} \begin{bmatrix} u_C \\ i_L \\ i_a \\ \omega_r \end{bmatrix}$$

$$+ \begin{bmatrix} 0 \\ \frac{V_d}{L u_{t\max}} \\ 0 \\ 0 \end{bmatrix} u + \begin{bmatrix} 0 \\ -\frac{r_s}{L u_{t\max}} i_L \\ 0 \\ 0 \end{bmatrix} u - \begin{bmatrix} 0 \\ 0 \\ 0 \\ \frac{1}{J} \end{bmatrix} T_L, \quad u \in [0 \quad 10]\,\text{V}.$$

The positive-definite energy-relevant Lyapunov function is

$$V(x) = \frac{1}{2}[u_C \quad i_L \quad i_a \quad \omega_r] K_{x1} \begin{bmatrix} u_C \\ i_L \\ i_a \\ \omega_r \end{bmatrix}, \quad K_{x1} = \begin{bmatrix} C & 0 & 0 & 0 \\ 0 & L & 0 & 0 \\ 0 & 0 & L_a & 0 \\ 0 & 0 & 0 & J \end{bmatrix}, \quad K_{x1} \in \mathbb{R}^{4\times4}.$$

As reported in Section 8.11 and illustrated in Example 8.33, in the design of tracking control law, the tracking error $e = (r - y) = (r - \omega_r)$ is used.

We have

$$\frac{dx^{ref}}{dt} = r - y = r - \omega_r, \quad x^{ref} = \int e\,dt.$$

The dynamics of the tracking error is

$$\frac{de}{dt} = -100e + \dot{r} - \dot{\omega}_r.$$

The measured variables are the tracking error $e(t)$ and the angular velocity $\omega_r(t)$. For the tracking control problem, we use the positive-definite

$$V(u_C, i_L, i_a, \omega_r, x^{ref}, e), \quad x^{ref} = \int e\,dt.$$

with quadratic and quartic terms

$$V(x, x^{ref}, e) = V(x) + V(x^{ref}, e) = \frac{1}{2} x^T K_{x1} x + \frac{1}{2}\left[x^{ref} \quad e\right] \begin{bmatrix} 1 & 0 \\ 0 & 1 \end{bmatrix} \begin{bmatrix} x^{ref} \\ e \end{bmatrix} + \frac{1}{4}\left[x^{ref2} \quad e^2\right] \begin{bmatrix} 1 & 0 \\ 0 & 1 \end{bmatrix} \begin{bmatrix} x^{ref2} \\ e^2 \end{bmatrix}.$$

The *minimal-complexity* control law is synthesized as

$$u = -\text{sat}_0^{10}\left(\frac{V_d}{Lu_{t\max}}\frac{\partial V_m(x, x^{ref}, e)}{\partial i_L}\right),$$

where

$$V_m(x, x^{ref}, e) = x^T K_{x1} x + i_L\left(k_\omega \omega_r + k_{i1} x^{ref} + k_{i2} x^{ref^3} + k_{p1} e + k_{p2} e^3\right) + \begin{bmatrix} x^{ref} & e \end{bmatrix}\begin{bmatrix} 1 & 0 \\ 0 & 1 \end{bmatrix}\begin{bmatrix} x^{ref} \\ e \end{bmatrix}$$

$$+ \begin{bmatrix} x^{ref^2} & e^2 \end{bmatrix}\begin{bmatrix} 1 & 0 \\ 0 & 1 \end{bmatrix}\begin{bmatrix} x^{ref^2} \\ e^2 \end{bmatrix}.$$

The criteria (9.97) of Theorem 8.2, imposed on the Lyapunov pair, are examined. The positive-definite nonquadratic Lyapunov functions $V(u_C, i_L, i_a, \omega_r, x^{ref}, e)$ and $V_m(u_C, i_L, i_a, \omega_r, x^{ref}, e)$ are used. The feedback gains are found by solving the inequality

$$\frac{dV(e,x)}{dt} \le -\frac{1}{2}\|x\|^2 - \frac{1}{2}\|x^{ref}\|^2 - \frac{1}{4}\|x^{ref}\|^4 - \frac{1}{2}\|e\|^2 - \frac{1}{4}\|e\|^4$$

for the closed-loop system. The solution yields $k_\omega = 0.085$, $k_{p1} = 1.8$, $k_{p2} = 0.25$, $k_{i1} = 7.3$, and $k_{i2} = 0.92$. A bounded *minimal-complexity* control law is

$$u = \text{sat}_0^{10}\left(-k_\omega \omega_r + k_{p1} e + k_{p2} e^3 + k_{i1}\int e\, dt + k_{i2}\int e^3\, dt\right).$$

The $dV(x, e)/dt$ imposes the specifications on stability, efficiency, tracking error (e should converge to zero), etc. The designed *minimal-complexity* control law is experimentally verified. Different operating conditions are studied to analyze the dynamic performance. The system performance degrades at high temperatures. For T = 140°C (the maximum operating temperature), the resistance of the armature winding reaches the maximum value, and the *torque* constant k_a is minimum. The disturbance attenuation features are of interest. It is required that the angular velocity remains equal to the reference value if the load torque T_L is applied. Figure 8.38a depicts the measured dynamics for control $u(t)$ and states $u_C(t)$, $i_a(t)$, $\omega_r(t)$ when $r = \omega_{reference}$ = 255 rad/sec. Though $u_C(t)$ and $i_a(t)$ can be measured, one may simplify the system complexity avoiding the use of these variables. A motor reaches the desired (reference) angular velocity within 0.05 sec with overshoot 4.5%, and the steady-state error $e(t)$ due to the sensor error is less than 0.1%. The disturbance attenuation is studied. The load torque 0.075 N-m is applied at t = 0.07 sec. The tracking accuracy and disturbance attenuation are achieved. If the load is applied, the settling time is 0.01 sec with 2.5% deflection from $\omega_{reference}$, and the steady-state error is ~0.1%. The desired performance and capabilities are achieved.

We compare the designed *minimal-complexity* control law with the proportional-integral control law. A high frequency (36 kHz with 2.5 kHz bandwidth) PWM *servo-amplifier* 25A8 (Advanced Motion Controls) is used to implement a linear proportional–integral control

$$u = 1.8e + 7.3\int e\, dt.$$

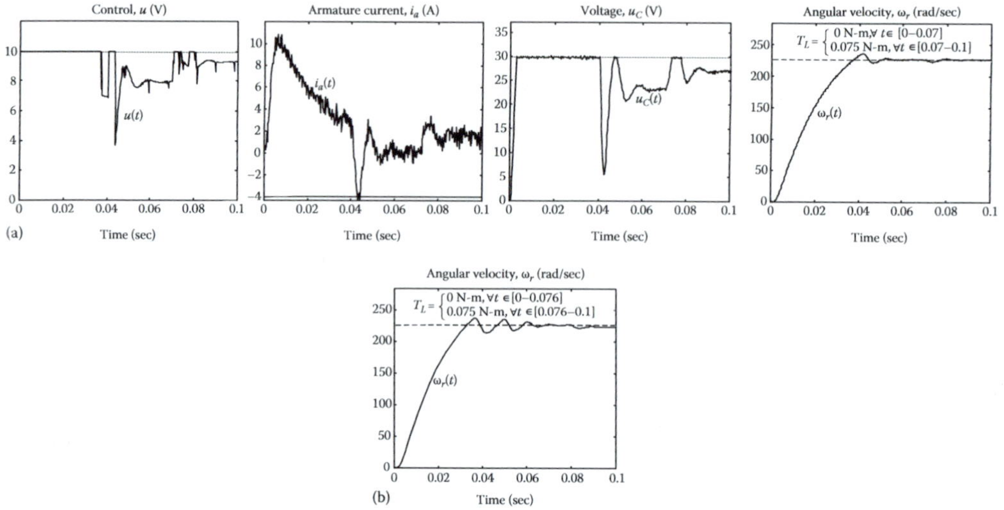

FIGURE 8.38 (a) Transient dynamics of the resulting closed-loop system; (b) Dynamics of the closed-loop system with the 25A8 *servo-amplifier.*

The experimental results are illustrated in Figure 8.38b. The motor starts from stall, and the load torque is applied at 0.076 sec. The settling time is 0.075 sec with overshoot 5.2%. As the load torque 0.075 N-m is applied at 0.076 s, the steady-state error of 1.9% results. The settling time is 0.01 sec with 3.4% maximum deflection from the assigned $\omega_{reference}$.

The analysis of two control algorithms indicates that the acceleration rate remains the same. The motor reaches the maximum angular velocity at ~0.04 sec. No matter which control laws are used, the maximum voltage is applied to the armature winding as the motor accelerates, $u_a = 30$ V and $u_{\max} = 10$ V. As the system reaches operating conditions where saturation is not a factor, the experimental results illustrate that the *admissible minimal-complexity* control law improves the closed-loop dynamic and minimizes the steady-state error. System dynamics, performance, and capabilities are improved by nonlinear feedback. The reported analytical and experimental results illustrate that nonlinear control laws guarantee better dynamics, precise tracking, disturbance attenuation, robustness, stability, etc. ■

Example 8.49

We study the eight-layered lead magnesium niobate actuator, see Figure 8.39a. A set of differential equations to model the actuator dynamics are found to be

$$\frac{dF}{dt} = -8{,}500F + 14Fu + 450u, \quad \frac{dv}{dt} = 1000F - 100000v - 2500v^3 - 2750x, \quad \frac{dx}{dt} = v.$$

The control voltage is bounded, $-100 \le u \le 100$ V. The error is the difference between the reference and actuator linear displacements. That is, $e(t) = r(t) - y(t)$, where $y(t) = x(t)$.

The tracking control law is designed using the dynamic evolutions

$$\frac{dx^{ref}}{dt} = r - y, \; x^{ref} = \int e\,dt, \quad \text{and} \quad \frac{de}{dt} = -10000e + \dot{r} - \dot{y}.$$

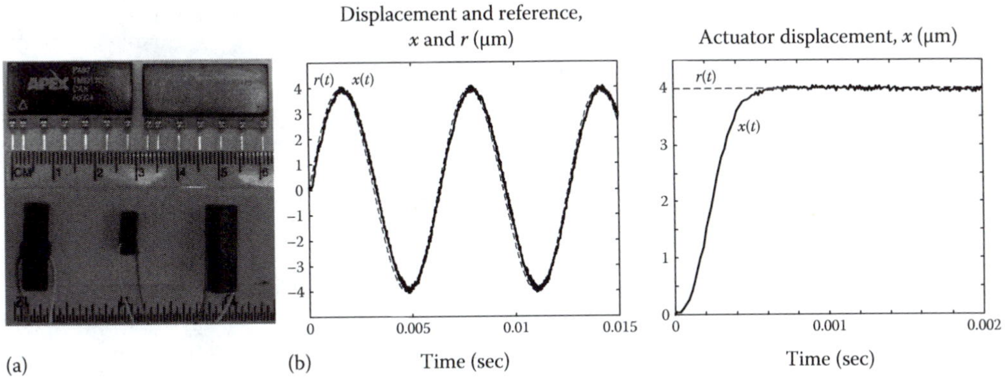

FIGURE 8.39 (a) Eight-layered lead magnesium niobate piezoelectric actuator (4 mm diameter, 0.125 mm thickness) with the controller/driver;
(b) Output dynamics if $r(t) = 4 \times 10^{-6} \sin(1000t)$ and $r(t) = 4 \times 10^{-6}$ m.

An *admissible minimal-complexity* control law is synthesized using a nonlinear error feedback. Using the criteria imposed on the Lyapunov pair (8.97), the feedback gains are found by using the displacement and error. We solve the inequality

$$\frac{dV(e,x)}{dt} \le -\|x\|^2 - \left\|x^{ref}\right\|^{4/3} - \left\|x^{ref}\right\|^2 - \|e\|^{4/3} - \|e\|^2.$$

The positive-definite Lyapunov function is

$$V(x,e) = \frac{1}{2}\begin{bmatrix} F & v & x \end{bmatrix}\begin{bmatrix} 1 & 0 & 0 \\ 0 & 1 & 0 \\ 0 & 0 & 1 \end{bmatrix}\begin{bmatrix} F \\ v \\ x \end{bmatrix} + \frac{3}{4}\left|x^{ref}\right|^{4/3} + \frac{1}{2}x^{ref\,2} + \frac{3}{4}|e|^{4/3} + \frac{1}{2}e^2.$$

The tracking control law is

$$u = \text{sat}_{-100}^{+100}\left(8.6e + 1.5\,\text{sgn}(e)|e|^{1/3} + 3.9\int e\,dt + 0.52\int \text{sgn}(e)|e|^{1/3}\,dt\right).$$

The criteria imposed on the Lyapunov pair (8.97) are satisfied. Hence, the bounded control law guarantees stability, ensures tracking, and guarantees the specified negativeness of dV/dt. The control law is experimentally substantiated. Figure 8.39b illustrates the transient dynamics for $x(t)$ if references are $r(t) = 4 \times 10^{-6} \sin(1000t)$ and $r(t) = 4 \times 10^{-6}$ m. The stability is guaranteed, desired performance is achieved, and the output precisely follows the reference $r(t)$. ■

Example 8.50

We design and implement a closed-loop system for an axial-topology hard disk drive. The requirements are accuracy, fast repositioning, minimal steady-state positional tracking error, disturbance attenuation, and, minimal settling time. The angular displacement θ_r is measured by a sensor, while the angular velocity ω_r can be estimated using an *observer*. The actuator parameters are: $N = 108$, $r_a = 6.4$ ohm, $L_a = 3.32 \times 10^{-5}$ H, $B_{max} = 0.4$ T, $B_m = 5 \times 10^{-8}$ N-m-sec/rad, and $J = 1.4 \times 10^{-6}$ kg-m^2.

Using the tracking error $e(t)$, constrained tracking control laws are designed, examined, and tested. The proportional-integral control law is

$$u = \text{sat}_{-1}^{+1}\left(k_p e + k_i \int e\, d\tau\right),\ e = r - \theta_r,\ k_p = 20,\ k_i = 21.$$

The integral tracking control law (8.36) with control bonds is

$$u = \text{sat}_{-1}^{+1}\left(-k_1 i_a - k_2\omega_r - k_3\theta_r + k_i \int e\, d\tau\right),\ k_1 = 0.47,\ k_2 = 0.02,\ k_3 = 2.8,\ k_i = 21.$$

The constrained *minimal-complexity* tracking control law with state feedback is designed using the *state transformation* method applying the Lyapunov theory. We obtain

$$u = \text{sat}_{-1}^{+1}\left(-k_\theta\theta_r + k_p e + k_i \int e\, d\tau\right),\ k_\theta = 2.8,\ k_p = 20,\ k_i = 21.$$

The designed control laws are tested and evaluated. These algorithms are discretized and implemented using a 16-bit PIC16F877A microcontroller with the clock rate of 20 MHz ensuring the sampling time T_s = 0.001 sec. The image of the experimental testbed is documented in Figure 8.40a. The dynamics of the closed-loop system with the constrained proportional-integral, integral tracking, and *minimal-complexity* proportional-integral with state feedback control laws are reported in Figure 8.40b. The reference angular displacement θ_{ref} is ~0.105 rad. Optimal performance is achieved using a *minimal-complexity* tracking control law with state feedback

$$u = \text{sat}_{-1}^{+1}\left(-k_\theta\theta_r + k_p e + k_i \int e\, d\tau\right),\quad e = r - \theta_r.$$

This control law ensures high accuracy, fast repositioning, minimal settling time, disturbance attenuation, and minimal overshoot. ■

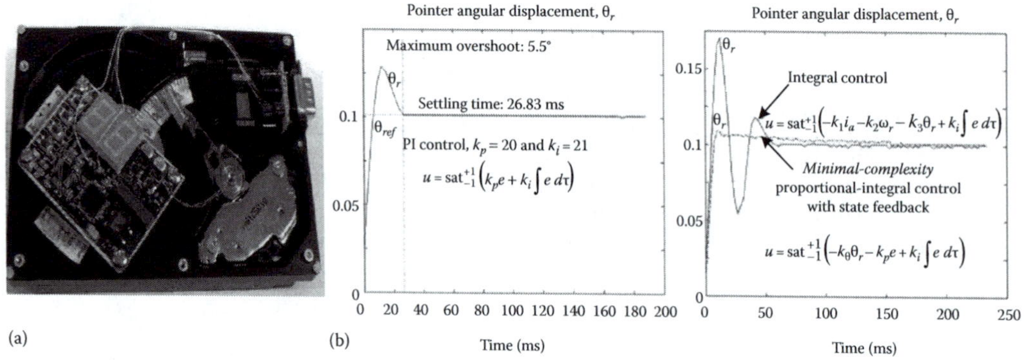

FIGURE 8.40 (a) Closed-loop servo with axial topology actuator, sensors, and microcontroller; (b) Dynamics of the closed-loop servo with the bounded proportional-integral $u = \text{sat}_{-1}^{+1}\left(k_p e + k_i \int e\, d\tau\right)$, integral $u = \text{sat}_{-1}^{+1}\left(-k_1 i_a - k_2\omega_r - k_3\theta_r + k_i \int e\, d\tau\right)$, and *minimal-complexity* control laws $u = \text{sat}_{-1}^{+1}\left(-k_\theta\theta_r + k_p e + k_i \int e\, d\tau\right)$.

8.16 CONTROL OF LINEAR DISCRETE-TIME SYSTEMS

8.16.1 Linear Discrete-Time Systems

Consider a discrete-time system described by the state-space difference equation

$$x_{n+1} = A_n x_n + B_n u_n,\ n \ge 0. \tag{8.101}$$

Different performance indexes are applied to optimize the closed-loop system dynamics. For example, the quadratic performance index to be minimized is

$$J = \sum_{n=0}^{N-1} \left[x_n^T Q_n x_n + u_n^T G_n u_n \right],\ Q_n \ge 0,\ G_n > 0. \tag{8.102}$$

We find the control law that guarantees the minimum value of the performance index (8.102). For linear dynamic systems (8.101) and quadratic performance index (8.102), the solution of the Hamilton–Jacobi equation

$$V(x_n) = \min_{u_n} \left[x_n^T Q_n x_n + u_n^T G_n u_n + V(x_{n+1}) \right] \tag{8.103}$$

is satisfied by the quadratic return function

$$V(x_n) = x_n^T K_n x_n,\ K_n > 0. \tag{8.104}$$

From (8.103), using (8.104), we have

$$\begin{aligned} V(x_n) &= \min_{u_n} \left[x_n^T Q_n x_n + u_n^T G_n u_n + \left(A_n x_n + B_n u_n \right)^T K_{n+1} \left(A_n x_n + B_n u_n \right) \right] \\ &= \min_{u_n} \left[x_n^T Q_n x_n + u_n^T G_n u_n + x_n^T A_n^T K_{n+1} A_n x_n + x_n^T A_n^T K_{n+1} B_n u_n + u_n^T B_n^T K_{n+1} A_n x_n + u_n^T B_n^T K_{n+1} B_n u_n \right]. \end{aligned} \tag{8.105}$$

The Hamiltonian is

$$H = x_n^T Q_n x_n + u_n^T G_n u_n + x_n^T A_n^T K_{n+1} A_n x_n + x_n^T A_n^T K_{n+1} B_n u_n + u_n^T B_n^T K_{n+1} A_n x_n + u_n^T B_n^T K_{n+1} B_n u_n. \tag{8.106}$$

The application of the first-order necessary condition for optimality (8.10) for Hamiltonian (8.106) yields

$$u_n^T G_n + x_n^T A_n^T K_{n+1} B_n + u_n^T B_n^T K_{n+1} B_n = 0.$$

Hence, the control law is

$$u_n = -\left(G_n + B_n^T K_{n+1} B_n \right)^{-1} B_n^T K_{n+1} A_n x_n. \tag{8.107}$$

The second-order necessary condition for optimality (8.13) is guaranteed because

$$\frac{\partial^2 H\left(x_n, u_n, V\left(x_{n+1}\right)\right)}{\partial u_n \times \partial u_n^T} = \frac{\partial^2 \left(u_n^T G_n u_n + u_n^T B_n^T K_{n+1} B_n u_n \right)}{\partial u_n \times \partial u_n^T} = 2G_n + 2B_n^T K_{n+1} B_n > 0,\ K_{n+1} > 0.$$

Using the control law (8.107), from (8.105), one finds

$$x_n^T K_n x_n = x_n^T Q_n x_n + x_n^T A_n^T K_{n+1} A_n x_n - x_n^T A_n^T K_{n+1} B_n \left(G_n + B_n^T K_{n+1} B_n\right)^{-1} B_n K_{n+1} A_n x. \quad (8.108)$$

From (8.108), the difference equation to find the unknown symmetric matrix of the quadratic return function is

$$K_n = Q_n + A_n^T K_{n+1} A_n - A_n^T K_{n+1} B_n \left(G_n + B_n^T K_{n+1} B_n\right)^{-1} B_n K_{n+1} A_n. \quad (8.109)$$

If in performance index (8.102) $N = \infty$, we have

$$J = \sum_{n=0}^{\infty} \left[x_n^T Q_n x_n + u_n^T G_n u_n \right], \; Q_n \geq 0, G_n > 0. \quad (8.110)$$

The optimal control law is

$$u_n = -\left(G_n + B_n^T K_n B_n\right)^{-1} B_n^T K_n A_n x_n, \quad (8.111)$$

where the unknown symmetric matrix K_n is found by solving

$$-K_n + Q_n + A_n^T K_n A_n - A_n^T K_n B_n \left(G_n + B_n^T K_n B_n\right)^{-1} B_n K_n A_n = 0, \quad K_n = K_n^T. \quad (8.112)$$

Matrix K_n is positive-definite. The closed-loop system (8.101) with (8.111) becomes

$$x_{n+1} = \left[A_n - B_n \left(G_n + B_n^T K_{n+1} B_n\right)^{-1} B_n^T K_{n+1} A_n \right] x_n. \quad (8.113)$$

Example 8.51

For the second-order discrete-time system

$$x_{n+1} = \begin{bmatrix} x_{1n+1} \\ x_{2n+1} \end{bmatrix} = A_n x_n + B_n u_n = \begin{bmatrix} 1 & 2 \\ 0 & 3 \end{bmatrix} \begin{bmatrix} x_{1n} \\ x_{2n} \end{bmatrix} + \begin{bmatrix} 4 & 5 \\ 6 & 7 \end{bmatrix} \begin{bmatrix} u_{1n} \\ u_{2n} \end{bmatrix},$$

we find the digital control law by minimizing the performance index

$$J = \sum_{n=0}^{\infty} \left[x_n^T Q_n x_n + u_n^T G_n u_n \right] = \sum_{n=0}^{\infty} \left[\begin{bmatrix} x_{1n} & x_{2n} \end{bmatrix} \begin{bmatrix} 10 & 0 \\ 0 & 10 \end{bmatrix} \begin{bmatrix} x_{1n} \\ x_{2n} \end{bmatrix} + \begin{bmatrix} u_{1n} & u_{2n} \end{bmatrix} \begin{bmatrix} 5 & 0 \\ 0 & 5 \end{bmatrix} \begin{bmatrix} u_{1n} \\ u_{2n} \end{bmatrix} \right]$$

$$= \sum_{n=0}^{\infty} \left(10x_{1n}^2 + 10x_{2n}^2 + 5u_{1n}^2 + 5u_{2n}^2 \right).$$

Using the `dlqr` command, we have

```
A=[1 2;0 3];B=[4 5;6 7]; Q=10*eye(size(A)); G=5*eye(size(B));
[Kfeedback,Kn,Eigenvalues]=dlqr(A,B,Q,G)
```

One obtains $K_n = \begin{bmatrix} 20 & -0.71 \\ -0.71 & 10.6 \end{bmatrix}$.

The control law is

$$u_n = -\left(G_n + B_n^T K_{n+1} B_n\right)^{-1} B_n^T K_n A_n x_n = -\begin{bmatrix} -0.285 & 0.231 \\ 0.332 & 0.222 \end{bmatrix}\begin{bmatrix} x_{1n} \\ x_{2n} \end{bmatrix},$$

$$u_{1n} = 0.285x_{1n} - 0.231x_{2n}, \quad u_{2n} = -0.332x_{1n} - 0.222x_{2n}.$$

The eigenvalues are 0.522 and 0.0159. The system is stable because the eigenvalues are within the unit circle. ■

Example 8.52

For the third-order system

$$x_{n+1} = \begin{bmatrix} x_{1n+1} \\ x_{2n+1} \\ x_{3n+1} \end{bmatrix} = A_n x_n + B_n u_n = \begin{bmatrix} 1 & 1 & 2 \\ 3 & 3 & 4 \\ 5 & 5 & 6 \end{bmatrix}\begin{bmatrix} x_{1n} \\ x_{2n} \\ x_{2n} \end{bmatrix} + \begin{bmatrix} 10 \\ 20 \\ 30 \end{bmatrix} u_n,$$

we find a control law by minimizing the quadratic performance index

$$J = \sum_{n=0}^{\infty}\left[x_n^T Q_n x_n + u_n^T G_n u_n\right] = \sum_{n=0}^{\infty}\left[\begin{bmatrix} x_{1n} & x_{2n} & x_{3n} \end{bmatrix}\begin{bmatrix} 1 & 0 & 0 \\ 0 & 10 & 0 \\ 0 & 0 & 100 \end{bmatrix}\begin{bmatrix} x_{1n} \\ x_{2n} \\ x_{3n} \end{bmatrix} + 1000u_n^2\right]$$

$$= \sum_{n=0}^{\infty}\left(x_{1n}^2 + 10x_{2n}^2 + 100x_{3n}^2 + 1000u_n^2\right).$$

The output equation is $y_n = H_n x_n + D_n u_n$, $H_n = \begin{bmatrix} 1 & 1 & 1 \end{bmatrix}$, and $D_n = [0]$. We have

```
A=[1 1 2; 3 3 4; 5 5 6]; B=[10; 20; 30];
Q=eye(size(A)); Q(2,2)=10; Q(3,3)=100; G=1000;
[Kfeedback,Kn,Eigenvalies]=dlqr(A,B,Q,G)
```

One finds $K_n = \begin{bmatrix} 48.4 & 47.4 & 31.1 \\ 47.4 & 57.4 & 31.1 \\ 31.1 & 31.1 & 140 \end{bmatrix}$.

The control law is

$$u_n = -0.155x_{1n} - 0.155x_{2n} - 0.2x_{3n}.$$

The dynamics of the closed-loop system, which is stable because the eigenvalues are within the unit circle, is simulated using the filter command. Having found the closed-loop system dynamics as $x_{n+1} = \left[A_n - B_n\left(G_n + B_n^T K_{n+1} B_n\right)^{-1} B_n^T K_n A_n\right] x_n$, one finds the numerator and denominator of the transfer function in the z-domain. The MATLAB statements are

```
A_closed=A-B*Kfeedback; H=[1 1 1]; D=[0];
[num,den]=ss2tf(A_closed,B,H,D);
x0=[10 0 -10]; k=0:1:20; u=1*[ones(1,21)];
x=filter(num,den,u,x0); plot(k,x,'-',k,x,'o','LineWidth',3);
title('System Dynamics, x(k)','FontSize',18);
xlabel('Discrete Time, k','FontSize',18);
```

The simulation results are illustrated in Figure 8.41. The stabilization problem was solved. The closed-loop system is stable. ■

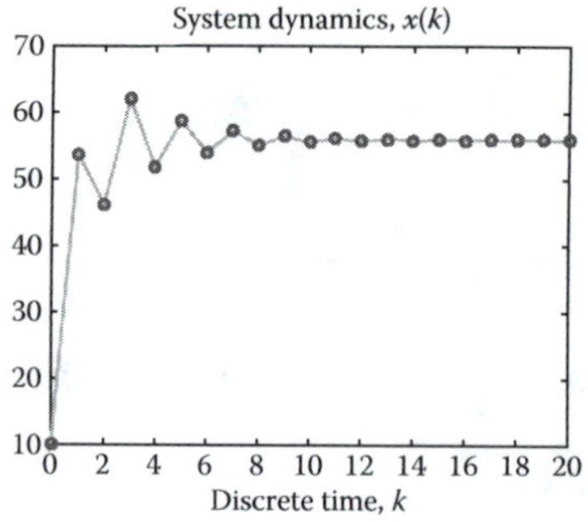

FIGURE 8.41 Output dynamics of the closed-loop system.

8.16.2 Constrained Optimization of Discrete-Time Systems

Due to the constraints imposed on controls $u_{n\,\min} \le u_n \le u_{n\,\max}$, the designer must synthesize bounded control laws. The system is described by the state-space difference equation

$$x_{n+1} = A_n x_n + B_n u_n, \quad x_{n0} \in X_0, \quad u_n \in U, \quad u_{n\min} \le u_n \le u_{n\max}, \quad n \ge 0. \tag{8.114}$$

The control constraints are described by continuous integrable one-to-one bounded functions $\Phi \in U$, for which the inverse function Φ^{-1} exists. The nonquadratic performance index to be minimized is

$$J = \sum_{n=0}^{N-1} \left[x_n^T Q_n x_n + 2\int \left(\Phi^{-1}(u_n)\right)^T G_n \, du_n - u_n^T B_n^T K_{n+1} B_n u_n \right]. \tag{8.115}$$

Performance indexes must be positive-definite. Hence,

$$\left[x_n^T Q_n x_n + 2\int \left(\Phi^{-1}(u_n)\right)^T G_n \, du_n \right] > u_n^T B_n^T K_{n+1} B_n u_n \quad \text{for all } x_n \in X \text{ and } u_n \in U. \tag{8.116}$$

The Hamilton–Jacobi recursive equation is

$$V(x_n) = \min_{u_n \in U} \left[x_n^T Q_n x_n + 2\int \left(\Phi^{-1}(u_n)\right)^T G_n \, du_n - u_n^T B_n^T K_{n+1} B_n u_n + V(x_{n+1}) \right]. \tag{8.117}$$

Using the quadratic return function $V(x_n) = x_n^T K_n x_n$ (8.104), one finds

$$x_n^T K_n x_n = \min_{u_n \in U} \left[x_n^T Q_n x_n + 2\int \left(\Phi^{-1}(u_n)\right)^T G_n \, du_n - u_n^T B_n^T K_{n+1} B_n u_n + \left(A_n x_n + B_n u_n\right)^T K_{n+1} \left(A_n x_n + B_n u_n\right) \right]. \tag{8.118}$$

Applying the first-order necessary condition for optimality (8.10) for

$$H = x_n^T Q_n x_n + 2\int \left(\Phi^{-1}(u_n)\right)^T G_n du_n - u_n^T B_n^T K_{n+1} B_n u_n + \left(A_n x_n + B_n u_n\right)^T K_{n+1}\left(A_n x_n + B_n u_n\right),$$

the bounded control law is

$$u_n = -\Phi\left(G_n^{-1} B_n^T K_{n+1} A_n x_n\right), \quad u_n \in U. \tag{8.119}$$

One-to-one integrable functions Φ and Φ^{-1} lie in the first and third quadrants, and $G_n > 0$. Hence,

$$\frac{\partial^2\left(2\int \left(\Phi^{-1}(u_n)\right)^T G_n\, du_n\right)}{\partial u_n \times \partial u_n^T} > 0.$$

One concludes that the second-order necessary condition for optimality (8.13) is satisfied. From (8.118) and (8.119), we have

$$\begin{aligned} x_n^T K_n x_n = {} & x_n^T Q_n x_n + 2\int x_n^T A_n^T K_{n+1} B_n d\left(\Phi\left(G_n^{-1} B_n^T K_{n+1} A_n x_n\right)\right) + x_n^T A_n^T K_{n+1} A_n x_n \\ & - 2 x_n^T A_n^T K_{n+1} B_n \Phi\left(G_n^{-1} B_n^T K_{n+1} A_n x_n\right). \end{aligned} \tag{8.120}$$

The second term on the right-hand side of (8.120) is

$$\begin{aligned} & 2\int x_n^T A_n^T K_{n+1} B_n d\left(\Phi\left(G_n^{-1} B_n^T K_{n+1} A_n x_n\right)\right) \\ & = 2 x_n^T A_n^T K_{n+1} B_n \Phi\left(G_n^{-1} B_n^T K_{n+1} A_n x_n\right) - 2\int \left(\Phi\left(G_n^{-1} B_n^T K_{n+1} A_n x_n\right)\right)^T d\left(B_n^T K_{n+1} A_n x_n\right). \end{aligned} \tag{8.121}$$

From (8.120), using (8.121), the unknown matrix K_{n+1} is found by solving

$$x_n^T K_n x_n = x_n^T Q_n x_n + x_n^T A_n^T K_{n+1} A_n x_n - 2\int \left(\Phi\left(G_n^{-1} B_n^T K_{n+1} A_n x_n\right)\right)^T d\left(B_n^T K_{n+1} A_n x_n\right). \tag{8.122}$$

Describing the control bounds using the continuous integrable bounded functions $\Phi \in U$, one finds the expression for $2\int \left(\Phi\left(G_n^{-1} B_n^T K_{n+1} A_n x_n\right)\right)^T d\left(B_n^T K_{n+1} A_n x_n\right)$. For example, using the hyperbolic tangent function to describe the saturation- and relay-type constraints, one obtains

$$\int \tanh z\, dz = \log\cosh z \quad \text{and} \quad \int \tanh^g z\, dz = -\frac{\tanh^{g-1} z}{g-1} + \int \tanh^{g-2} z\, dz, \quad g \neq 1.$$

Matrix K_{n+1} is found by solving the recursive equation (8.122), and the feedback gains result. Minimizing the nonquadratic performance index (8.115) for $N = \infty$, the bounded control law is found as

$$u_n = -\Phi\left(G_n^{-1} B_n^T K_n A_n x_n\right), \quad u_n \in U. \tag{8.123}$$

For open-loop unstable systems (8.114), the *admissibility* concept may be applied to verify the stability of the resulting closed-loop system in the operating envelope $x_n \in X$ and $u_n \in U$. The closed-loop system (8.114) with (8.123) evolves in X. A subset of the admissible domain of stability $S \subset \mathbb{R}^n$ is found by using the Lyapunov stability theory as

$$S = \left\{ x_n \in \mathbb{R}^n : x_{n0} \in X_0, \quad u_n \in U \middle| V(0) = 0, \quad V(x_n) > 0, \quad \Delta V(x_n) < 0, \quad \forall x_n \in X(X_0, U) \right\}.$$

The region of attraction can be studied, and S is an *invariant* domain. The quadratic Lyapunov function $V(x_n) = x_n^T K_n x_n$ is positive-definite if $K_n > 0$. Stability is guaranteed if the first difference

$$\Delta V(x_n) = V(x_{n+1}) - V(x_n) = x_n^T A_n^T K_{n+1} A_n x_n - 2 x_n^T A_n^T K_{n+1} B_n \Phi\left(G_n^{-1} B_n^T K_{n+1} A_n x_n\right)$$

$$+ \Phi\left(G_n^{-1} B_n^T K_{n+1} A_n x_n\right)^T B_n^T K_{n+1} B_n \Phi\left(G_n^{-1} B_n^T K_{n+1} A_n x_n\right) - x_n^T K_n x_n \tag{8.124}$$

is negative-definite for all $x_n \in X$. The evolution of the closed-loop system depends on the initial conditions, constraints, references, disturbances, parameter variations, etc. The *sufficiency* analysis of stability is performed studying sets $S \subset \mathbb{R}^n$ and $X(X_0, U) \subset \mathbb{R}^n$. Stability is guaranteed if $X \subseteq S$ for which $V(x_n) > 0$ and $\Delta V(x_n) < 0$.

Nonlinear discrete-time electromechanical systems are described as

$$x_{n+1} = F(x_n) + B(x_n) u_n, \quad x_{n0} \in X_0, \; u_n \in U, \quad u_{n\min} \le u_n \le u_{n\max}, \quad n \ge 0. \tag{8.125}$$

To design a nonlinear *admissible* control law $u_n \in U$, we describe the exhibited control bounds by $\Phi \in U$. The nonquadratic performance index is

$$J = \sum_{n=0}^{N-1} \left[x_n^T Q_n x_n - u_n^T B^T(x_n) K_{n+1} B(x_n) u_n + 2 \int \left(\Phi^{-1}(u_n)\right)^T G_n \, du_n \right]. \tag{8.126}$$

The integrand $2\int \left(\Phi^{-1}(u_n)\right)^T G_n \, du_n$ is positive-definite. The positive definiteness of the performance index is guaranteed if

$$x_n^T Q_n x_n + 2 \int \left(\Phi^{-1}(u_n)\right)^T G_n \, du_n > u_n^T B^T(x_n) K_{n+1} B(x_n) u_n \quad \text{for all } x_n \in X \text{ and } u_n \in U. \tag{8.127}$$

The positive definiteness of the performance index in $x_n \in X$ and $u_n \in U$ can be studied when the positive-definite symmetric matrix K_{n+1} is found. The inequality (8.127) is ensured by using Q_n and G_n. The first- and second-order necessary conditions for optimality are applied. For the quadratic return function (8.104), we have

$$V(x_{n+1}) = x_{n+1}^T K_{n+1} x_{n+1} = \left(F(x_n) + B(x_n) u_n\right)^T K_{n+1} \left(F(x_n) + B(x_n) u_n\right). \tag{8.128}$$

Therefore

$$x_n^T K_n x_n = \min_{u_n \in U} \left[x_n^T Q_n x_n - u_n^T B^T(x_n) K_{n+1} B(x_n) u_n \right.$$

$$\left. + 2 \int \left(\Phi^{-1}(u_n)\right)^T G_n \, du_n + \left(F(x_n) + B(x_n) u_n\right)^T K_{n+1} \left(F(x_n) + B(x_n) u_n\right) \right]. \tag{8.129}$$

Using the first-order necessary condition for optimality (8.10), a bounded control law is

$$u_n = -\Phi\left(G_n^{-1}B^T(x_n)K_{n+1}F(x_n)\right), \quad u_n \in U. \tag{8.130}$$

The second-order necessary condition for optimality (8.13) is satisfied because

$$\frac{\partial^2\left(2\int\left(\Phi^{-1}(u_n)\right)^T G_n\, du_n\right)}{\partial u_n \times \partial u_n^T} > 0.$$

From (8.129), using the bounded control law (8.130), we have the recursive equation

$$x_n^T K_n x_n = x_n^T Q_n x_n + 2\int F^T(x_n)K_{n+1}B(x_n)d\left(\Phi\left(G_n^{-1}B^T(x_n)K_{n+1}F(x_n)\right)\right) + F^T(x_n)K_{n+1}F(x_n)$$
$$-2F^T(x_n)K_{n+1}B(x_n)\Phi\left(G_n^{-1}B^T(x_n)K_{n+1}F(x_n)\right). \tag{8.131}$$

The integration by parts gives

$$2\int F^T(x_n)K_{n+1}B(x_n)d\left(\Phi\left(G_n^{-1}B^T(x_n)K_{n+1}F(x_n)\right)\right)$$
$$= 2F^T(x_n)K_{n+1}B(x_n)\Phi\left(G_n^{-1}B^T(x_n)K_{n+1}F(x_n)\right) - 2\int\left(\Phi\left(G_n^{-1}B^T(x_n)K_{n+1}F(x_n)\right)\right)^T d\left(B^T(x_n)K_{n+1}F(x_n)\right).$$

The equation to find the unknown symmetric matrix K_{n+1} is

$$x_n^T K_n x_n = x_n^T Q_n x_n + F^T(x_n)K_{n+1}F(x_n) - 2\int\left(\Phi\left(G_n^{-1}B^T(x_n)K_{n+1}F(x_n)\right)\right)^T d\left(B^T(x_n)K_{n+1}F(x_n)\right). \tag{8.132}$$

For a given $\Phi \in U$, one finds $2\int\left(\Phi\left(G_n^{-1}B^T(x_n)K_{n+1}F(x_n)\right)\right)^T d\left(B^T(x_n)K_{n+1}F(x_n)\right)$. Equation (8.132) can be solved. The *admissibility* concept is applied to verify the stability of the resulting closed-loop system that evolves in $X \subset \mathbb{R}^c$, $\left\{x_{n+1} = F(x_n) - B(x_n)\Phi\left(G_n^{-1}B^T(x_n)K_{n+1}F(x_n)\right),\ x_{n0} \in X_0\right\} \in X(X_0, U)$. Using the Lyapunov stability theory, the domain of stability $S \subset \mathbb{R}^n$ is found by applying the sufficient conditions under which the discrete-time closed-loop system (8.125) with (8.130) is stable. The positive-definite quadratic function (8.104) is used. To guarantee the stability, the first difference

$$\Delta V(x_n) = V(x_{n+1}) - V(x_n)$$
$$= F^T(x_n)K_{n+1}F(x_n) - 2F^T(x_n)K_{n+1}B(x_n)\Phi\left(G_n^{-1}B^T(x_n)K_{n+1}F(x_n)\right)$$
$$+\Phi\left(G_n^{-1}B^T(x_n)K_{n+1}F(x_n)\right)^T B^T(x_n)K_{n+1}B(x_n)\Phi\left(G_n^{-1}B^T(x_n)K_{n+1}F(x_n)\right) - x_n^T K_n x_n \tag{8.133}$$

must be negative-definite for all $x_n \in X$ and $u_n \in U$. One defines the set

$$S = \left\{x_n \in \mathbb{R}^n : x_{n0} \in X_0,\ u_n \in U \,\middle|\, V(0) = 0,\ V(x_n) > 0,\ \Delta V(x_n) < 0,\ \forall x_n \in X(X_0, U)\right\}.$$

The stability analysis is performed by studying S and $X(X_0, U)$. The stability of the closed-loop system (8.125) with (8.130) is guaranteed if $X \subseteq S$.

8.16.3 Tracking Control of Discrete-Time Systems

We study systems modeled by the following difference equation in the state-space form

$$x_{n+1}^{system} = A_n x_n^{system} + B_n u_n, \quad x_{n0}^{system} \in X_0, \quad u_n \in U, \quad n \geq 0. \tag{8.134}$$

The output equation is $y_n = H_n x_n^{system}$. The exogenous system is described as

$$x_n^{ref} = x_{n-1}^{ref} + r_n - y_n. \tag{8.135}$$

Using (8.134) and (8.135), one finds

$$x_{n+1}^{ref} = x_n^{ref} + r_{n+1} - y_{n+1} = x_n^{ref} + r_{n+1} - H_n\left(A_n x_n^{system} + B_n u_n\right). \tag{8.136}$$

Hence

$$x_{n+1} = \begin{bmatrix} x_{n+1}^{system} \\ x_{n+1}^{ref} \end{bmatrix} = \begin{bmatrix} A_n & 0 \\ -H_n A_n & I_n \end{bmatrix} x_n + \begin{bmatrix} B_n \\ -H_n B_n \end{bmatrix} u_n + \begin{bmatrix} 0 \\ I_n \end{bmatrix} r_{n+1}, \quad x_n = \begin{bmatrix} x_n^{system} \\ x_n^{ref} \end{bmatrix}. \tag{8.137}$$

To synthesize the bounded control law, we minimize the nonquadratic performance index

$$J = \sum_{n=0}^{N-1} \left[x_n^T Q_n x_n + 2\int \left(\Phi^{-1}(u_n)\right)^T G_n \, du_n - u_n^T \begin{bmatrix} B_n \\ -H_n B_n \end{bmatrix}^T K_{n+1} \begin{bmatrix} B_n \\ -H_n B_n \end{bmatrix} u_n \right]. \tag{8.138}$$

For the quadratic return function (8.104), the Hamilton–Jacobi equation is

$$x_n^T K_n x_n = \min_{u_n \in U} \left[x_n^T Q_n x_n + 2\int \left(\Phi^{-1}(u_n)\right)^T G_n \, du_n - u_n^T \begin{bmatrix} B_n \\ -H_n B_n \end{bmatrix}^T K_{n+1} \begin{bmatrix} B_n \\ -H_n B_n \end{bmatrix} u_n \right.$$

$$\left. + \left(\begin{bmatrix} A_n & 0 \\ -H_n A_n & I_n \end{bmatrix} x_n + \begin{bmatrix} B_n \\ -H_n B_n \end{bmatrix} u_n \right)^T K_{n+1} \left(\begin{bmatrix} A_n & 0 \\ -H_n A_n & I_n \end{bmatrix} x_n + \begin{bmatrix} B_n \\ -H_n B_n \end{bmatrix} u_n \right) \right]. \tag{8.139}$$

The resulting Hamiltonian and first-order necessary condition for optimality yield the bounded tracking control law

$$u_n = -\Phi\left(G_n^{-1} \begin{bmatrix} B_n \\ -H_n B_n \end{bmatrix}^T K_{n+1} \begin{bmatrix} A_n & 0 \\ -H_n A_n & I_n \end{bmatrix} x_n \right), \quad u_n \in U. \tag{8.140}$$

The unknown matrix K_{n+1} is found by solving

$$x_n^T K_n x_n = x_n^T Q_n x_n + x_n^T \begin{bmatrix} A_n & 0 \\ -H_n A_n & I_n \end{bmatrix}^T K_{n+1} \begin{bmatrix} A_n & 0 \\ -H_n A_n & I_n \end{bmatrix} x_n$$

$$-2\int \left(\Phi\left(G_n^{-1} \begin{bmatrix} B_n \\ -H_n B_n \end{bmatrix}^T K_{n+1} \begin{bmatrix} A_n & 0 \\ -H_n A_n & I_n \end{bmatrix} x_n \right) \right)^T d\left(\begin{bmatrix} B_n \\ -H_n B_n \end{bmatrix}^T K_{n+1} \begin{bmatrix} A_n & 0 \\ -H_n A_n & I_n \end{bmatrix} x_n \right). \tag{8.141}$$

The tracking control problem can be solved for linear and nonlinear discrete-time systems applying the *state transformation* concept reported in Sections 8.8 and 8.12. The control law with the state feedback results.

8.17 PHYSICS AND ESSENCE OF CONTROL

The basics of electromechanical motion devices, microelectronics and power electronics hardware, system optimization, and other fundamentals are covered. We discuss the essence of control laws and controllers from design and hardware standpoints.

Control laws are designed to be implemented on analog, digital, and *hybrid* hardware platforms. Optimal control laws are derived for physical systems modeled by differential and difference equations. Distinct mathematical models with different levels of hierarchy, consistency, accuracy, details, and other descriptive features can be used. In these mathematical models, u refers to the control variable or vector, which is a physical quantity to be varied to control a system. From classical mechanics, u may imply the applied force F or torque T. Using the Newtonian mechanics, we considered the *force* and *torque control* problems. To develop these F and T, actuators are used. For electromagnetic, electrostatic, piezoelectric, and other actuators, the voltage V and current i could be considered as the control variables u. Actuators are controlled by using power electronics. The power transistors are switched (controlled) by using the signal-level continuous-time voltage u_c developed by analog or digital ICs.

To control electromechanical system components, continuous-time control functions u are used. Digital control results in u_k, which is converted to analog u using the digital-to-analog converters, and the microcontrollers outputs are in the form of analog signals. Consistent PID, *admissible*, soft switching, *minimal-complexity*, and other control laws are reported. Relay and hard-switching control laws have a limited practicality due to abstract problem formulation under hypothetical assumptions. For example, from the hardware standpoint, it is impossible to develop force F and torque T with high-frequency and hard-switching activity. Even if intended, one is not able to generate high-frequency switching of F and T. The need for the force and torque *limiters* was emphasized to reduce the instantaneous force, torque, load, and disturbances. Having emphasized the mechanical features, one recalls the physics of electrostatic and electromagnetic actuators. Electromechanical motion devices are not controlled by applying the constant voltage and reversing the voltage polarity. Furthermore, the *LC* and *LCL* filters are always used to filter the PWM voltage. The switching frequency of transistors is limited, and there is a transient dynamics of switching. Due to the device physics, the output PWM voltage waveform is not an ideal train of pulses. The settling time of power transistors is within nano- or microseconds. The designer always intents to ensure soft switching of the power transistors to minimize losses, ensure efficiency, minimize loads, etc. There are *LC* output filters to supply the adequate voltage to actuators.

Our major focus was concentrated on consistent control concepts which ensure the best performance and *achievable* capabilities. Consistent and practical PID, stabilizing, tracking, soft-switching, and *minimal-complexity* control laws were designed. These control algorithms can be implemented by analog and digital controllers ensuring near-optimal overall system performance. The codesign consistency in analytic design and hardware is guaranteed.

PRACTICE PROBLEMS

8.1 Consider an electric drive with a permanent-magnet DC motor modeled by the following differential equations (Example 8.5)

$$\frac{di_a}{dt} = -\frac{r_a}{L_a} i_a - \frac{k_a}{L_a}\omega_r + \frac{1}{L_a}u_a, \quad \frac{d\omega_r}{dt} = \frac{k_a}{J} i_a - \frac{B_m}{J}\omega_r - \frac{1}{J}T_L.$$

The output is the angular velocity ω_r. That is, $y = \omega_r$. The tracking error is $e = r - y = r - \omega_r$. Consider a control law with the current, angular velocity, and proportional feedback

$$u(t) = -k_I i_a - k_\omega \omega_r + k_p e.$$

We find the feedback gains of a control law to ensure the desired eigenvalues $s_1 = -10$ and $s_2 = -100$.

The transfer function is $G_{sys}(s) = \dfrac{\Omega(s)}{U(s)} = \dfrac{k_a}{L_aJs^2 + (r_aJ + L_aB_m)s + r_aB_m + k_a^2}$.

With the current and angular velocity feedback terms $-k_I i_a$ and $-k_\omega\omega_r$, using $G_{sys}(s)$, we have

$$G^*_{sys}(s) = \frac{k_a}{L_aJs^2 + (r_aJ + k_IJ + L_aB_m)s + r_aB_m + k_IB_m + k_a^2 + k_ak_\omega}.$$

The transfer function for the closed-loop system is

$$G_\Sigma(s) = \frac{\Omega(s)}{R(s)} = \frac{G_{PID}(s)G^*_{sys}(s)}{1 + G_{PID}(s)G^*_{sys}(s)} = \frac{k_pG^*_{sys}(s)}{1 + k_pG^*_{sys}(s)}$$

$$= \frac{k_pk_a}{L_aJs^2 + (r_aJ + k_IJ + L_aB_m)s + r_aB_m + k_IB_m + k_a^2 + k_ak_\omega + k_pk_a}$$

The characteristic equation is

$$s^2 + \frac{r_aJ + k_IJ + L_aB_m}{L_aJ}s + \frac{r_aB_m + k_IB_m + k_a^2 + k_ak_\omega + k_pk_a}{L_aJ} = 0.$$

The assigned characteristic equation is

$$(s + 10)(s + 100) = s^2 + 110s + 1000 = 0.$$

The comparison of characteristic equations yields

$$\frac{r_aJ + k_IJ + L_aB_m}{L_aJ} = 110 \quad \text{and} \quad \frac{r_aB_m + k_IB_m + k_a^2 + k_ak_\omega + k_pk_a}{L_aJ} = 1000.$$

As a practice example, let all motor parameters be equal to 1 (r_a = 1 ohm, L_a = 1 H, k_a = 1 V-sec/rad, B_m = 1 N-m-sec/rad, and J = 1 kg-m^2).

Let $k_\omega = 1$. One finds the feedback gains $k_I = 108$ and $k_p = 889$.

The steady-state error is $\lim\limits_{t\to\infty} e(t) = \lim\limits_{s\to 0} \dfrac{sR(s)}{1 + G_{PID}G^*_{sys}}$.

For the unit step $r(t) = 1(t)$, $R(s) = 1/s$. Hence

$$\lim_{t\to\infty} e(t) = \lim_{s\to 0}\frac{sR(s)}{1 + G_{PID}G^*_{sys}} = \lim_{s\to 0}\frac{sR(s)}{1 + k_pG^*_{sys}}$$

$$= \lim_{s\to 0}\frac{1}{1 + k_p\dfrac{k_a}{L_aJs^2 + (r_aJ + k_IJ + L_aB_m)s + r_aB_m + k_IB_m + k_a^2 + k_ak_\omega}}$$

$$= \lim_{s\to 0}\frac{L_aJs^2 + (r_aJ + k_IJ + L_aB_m)s + r_aB_m + k_IB_m + k_a^2 + k_ak_\omega}{L_aJs^2 + (r_aJ + k_IJ + L_aB_m)s + r_aB_m + k_IB_m + k_a^2 + k_ak_\omega + k_pk_a} = \frac{111}{1000} = 0.111.$$

Therefore, $\lim_{t\to\infty} e(t) = e(\infty) = 0.111$.

Let the desired eigenvalues be $s_1 = -10$ and $s_2 = -100$, and the specified tracking error be $\lim_{t\to\infty} e(t) = e(\infty) = 0.001$. One can find k_I, k_ω and k_p in $u(t) = -k_I i_a - k_\omega \omega_r + k_p e$.

The comparison of characteristic equations yields

$$\frac{r_aJ + k_IJ + L_aB_m}{L_aJ} = 110 \quad \text{and} \quad \frac{r_aB_m + k_IB_m + k_a^2 + k_ak_\omega + k_pk_a}{L_aJ} = 1000.$$

We also have

$$\lim_{t\to\infty} e(t) = \lim_{s\to 0} \frac{L_aJs^2 + (r_aJ + k_IJ + L_aB_m)s + r_aB_m + k_IB_m + k_a^2 + k_ak_\omega}{L_aJs^2 + (r_aJ + k_IJ + L_aB_m)s + r_aB_m + k_IB_m + k_a^2 + k_ak_\omega + k_pk_a} = 0.001.$$

Thus, $\dfrac{r_aB_m + k_IB_m + k_a^2 + k_ak_\omega}{r_aB_m + k_IB_m + k_a^2 + k_ak_\omega + k_pk_a} = 0.001.$

Using the motor parameters, one solves three equations with three unknowns k_I, k_ω, and k_p. We have

$$\begin{bmatrix} J & 0 & 0 \\ B_m & k_a & k_a \\ 0.999B_m & 0.999k_a & -0.001k_a \end{bmatrix} \begin{bmatrix} k_I \\ k_\omega \\ k_p \end{bmatrix} = \begin{bmatrix} 110L_aJ - r_aJ - L_aB_m \\ 1000L_aJ - r_aB_m - k_a^2 \\ -0.999r_aB_m - 0.999k_a^2 \end{bmatrix}.$$

In $u(t) = -k_I i_a - k_\omega \omega_r + k_p e$, the feedback gains k_I, k_ω, and k_p must be positive. If the design results in a positive feedback, the eigenvalues and tracking error must be refined to find $k_I > 0$, $k_\omega > 0$ and $k_p > 0$.

8.2 Consider a permanent-magnet DC motor. For simplicity, let all motor parameters be equal to 1, that is, $r_a = 1$ ohm, $L_a = 1$ H, $k_a = 1$ V-sec/rad, $B_m = 1$ N-m-sec/rad, and $J = 1$ kg-m^2. Consider the control law

$$u(t) = -k_\omega \frac{d\omega_r}{dt} + k_p e.$$

Let us find the control feedback coefficients k_ω and k_p to ensure $s_1 = -1$ and $s_2 = -3$.

Recall that $G_{sys}(s) = \dfrac{\Omega(s)}{U(s)} = \dfrac{k_a}{L_aJs^2 + (r_aJ + L_aB_m)s + r_aB_m + k_a^2}$.

The transfer function for the closed-loop system is

$$G_\Sigma(s) = \frac{\Omega(s)}{R(s)} = \frac{G_{PID}(s)\dfrac{G_{sys}(s)}{1 + k_\omega sG_{sys}(s)}}{1 + G_{PID}(s)\dfrac{G_{sys}(s)}{1 + k_\omega sG_{sys}(s)}} = \frac{\dfrac{k_pk_a}{L_aJs^2 + (r_aJ + L_aB_m)s + r_aB_m + k_a^2}}{1 + \dfrac{(k_p + k_\omega s)k_a}{L_aJs^2 + (r_aJ + L_aB_m)s + r_aB_m + k_a^2}}$$

$$= \frac{k_pk_a}{L_aJs^2 + (r_aJ + L_aB_m + k_\omega k_a)s + r_aB_m + k_a^2 + k_pk_a}.$$

The resulting characteristic equation $s^2 + \dfrac{r_aJ + L_aB_m + k_\omega k_a}{L_aJ}s + \dfrac{r_aB_m + k_a^2 + k_pk_a}{L_aJ} = 0$ is compared with the assigned $(s + 1)(s + 3) = s^2 + 4s + 3 = 0$.

One finds the feedback coefficients $k_\omega = 2$ and $k_p = 1$.

8.3 For an electric drive with a permanent-magnet motor, find the feedback gains of a control law with the current, angular velocity, and integral feedback

$$u(t) = -k_I i_a - k_\omega \omega_r + k_r \int e\,dt$$

to guarantee the desired eigenvalues $s_1 = -1$, $s_2 = -1$, and $s_3 = -10$. Find the steady-state error if $r(t) = 1(t)$.

The transfer function for an electric drive is $G_{sys}(s) = \dfrac{\Omega(s)}{U(s)} = \dfrac{k_a}{L_a J s^2 + (r_a J + L_a B_m)s + r_a B_m + k_a^2}$.

With the current and angular velocity feedback terms $-k_I i_a$ and $-k_\omega \omega_r$, using $G_{sys}(s)$, one finds

$$G_{sys}^*(s) = \frac{k_a}{L_a J s^2 + (r_a J + k_I J + L_a B_m)s + (r_a + k_I)B_m + k_a^2 + k_a k_\omega}.$$

The transfer function for the closed-loop system is

$$G_\Sigma(s) = \frac{\Omega(s)}{R(s)} = \frac{G_{PID}(s)G_{sys}^*(s)}{1 + G_{PID}(s)G_{sys}^*(s)} = \frac{\dfrac{k_i}{s}G_{sys}^*(s)}{1 + \dfrac{k_i}{s}G_{sys}^*(s)}$$

$$= \frac{k_i k_a}{L_a J s^3 + (r_a J + k_I J + L_a B_m)s^2 + (r_a B_m + k_I B_m + k_a^2 + k_a k_\omega)s + k_i k_a}.$$

The characteristic equation is

$$L_a J s^3 + (r_a J + k_I J + L_a B_m)s^2 + (r_a B_m + k_I B_m + k_a^2 + k_a k_\omega)s + k_i k_a = 0.$$

The desired characteristic equation is $(s + 1)(s + 1)(s + 10) = s^3 + 12s^2 + 21s + 10 = 0$.
The comparison of

$$L_a J s^3 + (r_a J + k_I J + L_a B_m)s^2 + (r_a B_m + k_I B_m + k_a^2 + k_a k_\omega)s + k_i k_a = 0 \text{ and } s^3 + 12s^2 + 21s + 10 = 0$$

yields $\dfrac{r_a J + k_I J + L_a B_m}{L_a J} = 12$, $\dfrac{r_a B_m + k_I B_m + k_a^2 + k_a k_\omega}{L_a J} = 21$, and $\dfrac{k_i k_a}{L_a J} = 10$.
Letting all motor parameters be equal to 1, the feedback gains are $k_I = 10$, $k_\omega = 9$, and $k_i = 10$.
The steady-state error is

$$\lim_{t\to\infty} e(t) = \lim_{s\to 0} \frac{sR(s)}{1 + G_{PID}G_{sys}^*}.$$

For the unit step input $r(t) = 1(t)$, one has $R(s) = 1/s$. Hence

$$\lim_{t\to\infty} e(t) = \lim_{s\to 0} \frac{sR(s)}{1 + G_{PID}G_{sys}^*} = \lim_{s\to 0} \frac{sR(s)}{1 + \dfrac{k_i}{s}G_{sys}^*}$$

$$= \lim_{s\to 0} \frac{1}{1 + \dfrac{k_i}{s}\dfrac{k_a}{L_a J s^2 + (r_a J + k_I J + L_a B_m)s + (r_a + k_I)B_m + k_a^2 + k_a k_\omega}}$$

$$= \lim_{s\to 0} \frac{s\left[L_a J s^2 + (r_a J + k_I J + L_a B_m)s + (r_a + k_I)B_m + k_a^2 + k_a k_\omega\right]}{s\left[L_a J s^2 + (r_a J + k_I J + L_a B_m)s + (r_a + k_I)B_m + k_a^2 + k_a k_\omega\right] + k_i k_a} = 0.$$

Therefore, $\lim_{t\to\infty} e(t) = e(\infty) = 0$.

8.4 Consider an electric drive with a permanent-magnet DC motor. We find the feedback gains of a PID control law

$$u(t) = k_p e + k_{i1}\int e\,dt + k_{i2}\iint e\,dt + k_d\frac{de}{dt}$$

to guarantee the characteristic eigenvalues $s_1 = -1$, $s_2 = -1$, $s_3 = -1$, and $s_4 = -1$.
The transfer function for an electric drive is

$$G_{sys}(s) = \frac{\Omega(s)}{U(s)} = \frac{k_a}{L_aJs^2 + \left(r_aJ + L_aB_m\right)s + r_aB_m + k_a^2} = \frac{k_a}{as^2 + bs + c},$$

where we assume the motor parameters to be equal to 1, yielding $a = 1$, $b = 2$, and $c = 2$.
The transfer function for the closed-loop system is

$$G(s) = \frac{\Omega(s)}{R(s)} = \frac{G_{PID}(s)G_{sys}(s)}{1 + G_{PID}(s)G_{sys}(s)} = \frac{\dfrac{k_ps^2 + k_{i1}s + k_{i2} + k_ds^3}{s^2}G_{sys}(s)}{1 + \dfrac{k_ps^2 + k_{i1}s + k_{i2} + k_ds^3}{s^2}G_{sys}(s)}$$

$$= \frac{k_a\left(k_ps^2 + k_{i1}s + k_{i2} + k_ds^3\right)}{as^4 + \left(b + k_ak_d\right)s^3 + \left(c + k_ak_p\right)s^2 + k_ak_{i1}s + k_ak_{i2}}$$

The characteristic equation of the closed-loop system

$$s^4 + \frac{b + k_ak_d}{a}s^3 + \frac{c + k_ak_p}{a}s^2 + \frac{k_ak_{i1}}{a}s + \frac{k_ak_{i2}}{a} = 0$$

is compared with the desired characteristic equation

$$(s + 1)(s + 1)(s + 1)(s + 1) = s^4 + 4s^3 + 6s^2 + 4s + 1 = 0.$$

One finds the feedback gains $k_p = 4$, $k_{i1} = 4$, $k_{i2} = 2$, and $k_d = 2$.
For the unit step input $r(t) = 1(t)$, one has $R(s) = 1/s$. The steady-state error is

$$\lim_{t\to\infty} e(t) = \lim_{s\to 0}\frac{sR(s)}{1 + G_{PID}G_{sys}} = \lim_{s\to 0}\frac{s^2\left(as^2 + bs + c\right)}{s^2\left(as^2 + bs + c\right) + \left(k_ps^2 + k_{i1}s + k_{i2} + k_ds^3\right)k_a} = 0.$$

There is a zero steady-state error with the studied control law.

For the sinusoidal reference $r(t) = \sin(\omega_0 t)\,1(t)$ with $R(s) = \dfrac{\omega_0}{s^2 + \omega_0^2}$, the steady-state error is

$$\lim_{t\to\infty} e(t) = \lim_{s\to 0}\frac{sR(s)}{1 + G_{PID}G_{sys}} = \lim_{s\to 0}\frac{s\dfrac{\omega_0}{s^2 + \omega_0^2}\left(as^2 + bs + c\right)}{s^2\left(as^2 + bs + c\right) + \left(k_ps^2 + k_{i1}s + k_{i2} + k_ds^3\right)k_a} = 0.$$

Hence, $\lim_{t\to\infty} e(t) = e(\infty) = 0$.

8.5 Consider a nonlinear system $\frac{dx}{dt} = -\left|x^{2/3}\right| x$.

The positive-definite Lyapunov function is $V = \left|x^{2/3}\right|^2 > 0$.

The total derivative dV/dt is

$$\frac{dV}{dt} = \frac{dV}{dx}\frac{dx}{dt} = \frac{d}{dx}\left|x^{2/3}\right|^2 \frac{dx}{dt} = -\frac{4x}{3\left|x^{2/3}\right|}\left|x^{2/3}\right| x = -\frac{4}{3}x^2 < 0.$$

The system is asymptotically stable because for $V(x) > 0$, $dV(x)/dt < 0$.

8.6 Consider a nonlinear electric drive assuming that all coefficients are 1. A nonlinear model with a nonlinear *BH* characteristic, nonlinear friction, and other effects is

$$\frac{di}{dt} = -i - i^3 - \omega + u,$$

$$\frac{d\omega}{dt} = i - \omega - \omega^3.$$

The positive-definite total kinetic energy Lyapunov function is $V = \frac{1}{2}\left(i^2 + \omega^2\right) > 0$.

We examine the open-loop system stability and plot V and dV/dt.
The total derivative of $V(i, \omega)$ is

$$\frac{dV}{dt} = \frac{dV}{di}\frac{di}{dt} + \frac{dV}{d\omega}\frac{d\omega}{dt} = i\left(-i - i^3 - \omega\right) + \omega\left(i - \omega - \omega^3\right) = -i^2 - i^4 - \omega^2 - \omega^4 < 0.$$

The system is asymptotically stable because for $V(i, \omega) > 0$, $dV(i, \omega)/dt < 0$.
The three-dimensional plots for V and dV/dt are documented in Figures 8.42.

```
x=linspace(-1,1,50); y=x; [X,Y]=meshgrid(x,y);
V=(X.^2+Y.^2)./2 ; surf(x,y,V);
xlabel('{\iti}','FontSize',16);
ylabel('{\it\omega}','FontSize',16);
zlabel('{\itV}','FontSize',16);
title('Lyapunov Function, {\itV}','FontSize',16); pause
dV=-X.^2-X.^4-Y.^2-Y.^4; surf(x,y,dV);
zlabel('{\itdV/dt}','FontSize',16);
title('Lyapunov Function Total Derivative, {\itdV/dt}','FontSize',16);
```

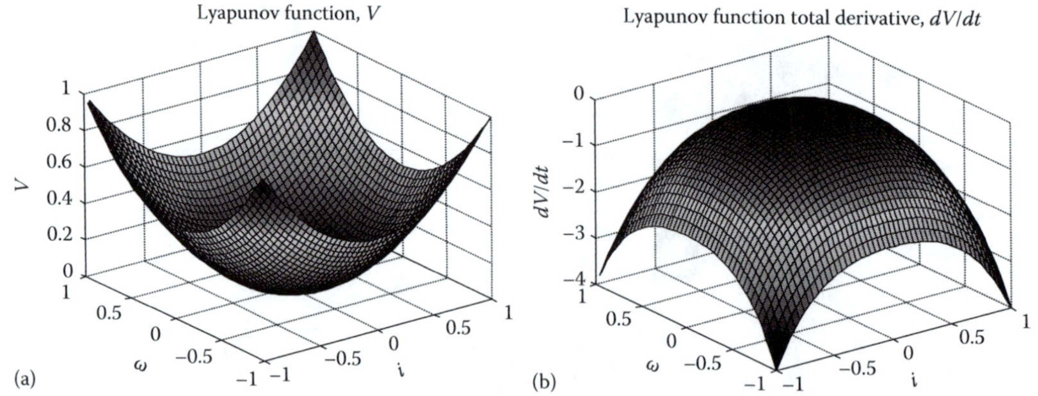

FIGURE 8.42 Three-dimensional plots for (a) $V = \frac{1}{2}(i^2 + \omega^2)$; (b) $\frac{dV}{dt} = -i^2 - i^4 - \omega^2 - \omega^4$.

8.7 Consider an axial topology hard disk drive actuator. Let a nonlinear model of a servo be

$$\frac{di}{dt} = -i - i^3 - \omega + u,$$

$$\frac{d\omega}{dt} = i - \omega - \omega^3 - \theta - \theta^3,$$

$$\frac{d\theta}{dt} = \omega.$$

We study the open-loop system stability using a positive-definite Lyapunov function

$$V = \frac{1}{2}\left(i^2 + \omega^2 + 2\omega\theta + \theta^2\right) > 0.$$

The total derivative of $V(i, \omega, \theta)$ is

$$\frac{dV}{dt} = \frac{dV}{di}\frac{di}{dt} + \frac{dV}{d\omega}\frac{d\omega}{dt} + \frac{dV}{d\theta}\frac{d\theta}{dt}$$

$$= i\left(-i - i^3 - \omega\right) + \left(\omega + \theta\right)\left(i - \omega - \omega^3 - \theta - \theta^3\right) + \left(\omega + \theta\right)\omega$$

$$= -i^2 - i^4 - \omega^4 - \omega\theta^3 + i\theta - \omega\theta - \theta\omega^3 - \theta^2 - \theta^4 < 0.$$

The open-loop system is stable because for $V(i, \omega, \theta) > 0$, we found $dV(i, \omega, \theta)/dt < 0$.

8.8 Using the Lyapunov theory, study the stability of the system

$$\dot{x}_1(t) = -x_1 + 10x_2,$$

$$\dot{x}_2(t) = -10x_1 - x_2^7, \quad t \geq 0.$$

Consider a positive-definite scalar function

$$V(x_1, x_2) = \frac{1}{2}\left(x_1^2 + x_2^2\right) > 0.$$

The total derivative is

$$\frac{dV(x_1, x_2)}{dt} = x_1\dot{x}_1 + x_2\dot{x}_2 = x_1(-x_1 + 10x_2) + x_2(-10x_1 - x_2^7) = -x_1^2 - x_2^8.$$

Therefore, $\frac{dV(x_1, x_2)}{dt} < 0.$

We conclude that the system is asymptotically stable, and the quadratic positive-definite function

$$V(x_1, x_2) = \frac{1}{2}\left(x_1^2 + x_2^2\right)$$

is the Lyapunov function.

HOMEWORK PROBLEMS

8.1 Why one should control electromechanical systems? List the specifications imposed on systems.

8.2 Explain the differences between bounded and unbounded control laws. Provide the examples. Explain how control bounds influence the system performance and capabilities. What are the challenges in the design of bounded control laws. How should a designer approach and solve control problems for electromechanical systems?

8.3 Let the control law be $u = k_{p1}e + k_{p2}\text{sgn}(e)|e^{1/7}| + k_{p3}e^3 + k_i\int e\,dt$.
Explain the use of feedback terms. Explain why one should study the bounded control law $u = \text{sat}_{u_{\min}}^{u_{\max}}\left(k_{p1}e + k_{p2}\,\text{sgn}(e)\left|e^{1.7}\right| + k_{p3}e^3 + k_i\int e\,dt\right)$. Propose the additional feedback terms to improve the system performance.

8.4 Let the performance functional be $J = \min\limits_{t,x,e}\int_0^{\infty}\left(x^2 + e^4 + t\left|e\right|\right)dt$.
Explain which performance characteristics are specified and how. Justify the results. If strict specifications should be imposed on the settling time and tracking error, propose the additional integrands in the performance functional.

8.5 Let the system be modeled as $\frac{d\omega}{dt} = -\omega + 100u$. The control is bounded as $-10 \le u \le 10$ V. Propose the bounded PID control laws and simulate the closed-loop system in Simulink. Let the desired speed be ±200 rad/sec. Find the feedback coefficients by varying the feedback gains. Report the simulation results for the closed-loop system.

8.6 Consider a pointing system actuated by a geared permanent-magnet DC motor. The angular displacement of the pointing stage is $y(t) = k_{gear}\theta_r$. The motor data are as follows: $u_{\max} = 24$ V ($-24 \le u_a \le 24$ V), $i_{a\max} = 10$ A, $\omega_{r\max} = 240$ rad/sec, $r_a = 1$ ohm, $L_a = 0.005$ H, $k_a = 0.1$ V-sec/rad (N-m/A), $J = 0.0005$ kg-m², and $B_m = 0.0005$ N-m-sec/rad. The reduction gear ratio is 10:1. Design and analyze: (1) Unbounded linear and nonlinear PID control laws; (2) Bounded PID control laws. For different control laws and feedback coefficients, study the transient dynamics for $r(t) = 0.1$ rad and $r(t) = 1$ rad. Perform simulations in MATLAB.

8.7 Consider a closed-loop system with a permanent-magnet DC motor.
The control law is $u(t) = k_pe - k_Ii$:
1. Derive and report a transfer function of a closed-loop system. Report the resulting characteristic equation;
2. Let the desired characteristic eigenvalues of the closed-loop system be −10 and −100. Derive and report the values of the feedback gains to ensure the aforementioned specified eigenvalues. For simplicity, let $r_a = k_a = L_a = B_m = J = 1$;
3. Synthesize (derive) and report the performance functional J with the terms (integrands) to evaluate and estimate the closed loop system performance on the settling time, tracking error, and power.

8.8 Consider a permanent-magnet DC motor in an electric drive application:
1. Derive a transfer function of the closed-loop electric drive system with a PID control law $u = k_pe + k_i\int e\,dt + k_d\frac{de}{dt}$. Report the characteristic equation;
2. Using the characteristic equations, derive relationships and calculate the values of k_p, k_i, and k_d to guarantee the desired eigenvalues at −10, −10, and −10. For simplicity, let $r_a = k_a = L_a = B_m = J = 1$.

8.9 Consider a system described by $\frac{d\omega_r}{dt} = -\omega_r - \theta_r + u,\ \frac{d\theta_r}{dt} = \omega_r$.
Derive and report a transfer function for a closed-loop servosystem with the PID control law. The desired eigenvalues are −1, −1, and −1. Derive the feedback gain coefficients k_p, k_i, and k_d.

8.10 Consider the second-order system $\frac{dx_1}{dt} = -x_1 + x_2,\ \frac{dx_2}{dt} = x_1 + u$.
Applying MATLAB, find a control law that minimizes the quadratic functional

$$J = \frac{1}{2}\int_0^{\infty}\left(x_1^2 + 2x_2^2 + 3u^2\right)dt.$$

Study the closed-loop system stability.

8.11 Using the Lyapunov theory, study the stability of the system described by the following differential equations:

$$\dot{x}_1(t) = -x_1 + 10x_2,$$

$$\dot{x}_2(t) = -10x_1 - x_2^7, \quad t \geq 0.$$

8.12 Consider the system

$$\frac{d\omega_r}{dt} = -\omega_r - 2\omega_r^3 - \theta_r - 2\theta_r^3,$$

$$\frac{d\theta_r}{dt} = \omega_r.$$

Applying the Lyapunov function

$$V = \frac{1}{2}\omega_r^2 + \omega_r\theta_r + \frac{1}{2}\theta_r^2,$$

using the Lyapunov stability theory, prove that the system is stable or unstable.

8.13 Consider a permanent-magnet synchronous motor in the rotor reference frame. The relevant problem was solved in Example 8.44. The nonlinear mathematical model is

$$\frac{di_{qs}^r}{dt} = -i_{qs}^r - i_{qs}^{r3} - \omega_r - i_{ds}^r\omega_r + u_{qs}^r,$$

$$\frac{di_{ds}^r}{dt} = -i_{ds}^r - i_{ds}^{r3} + i_{qs}^r\omega_r + u_{ds}^r,$$

$$\frac{d\omega_r}{dt} = i_{qs}^r - \omega_r - \omega_r^3.$$

Using the Lyspunov stability theory, examine the stability of the open-loop system.

8.14 Let the system be described by the following differential equations

$$\dot{x}_1(t) = -x_1 + 10x_2,$$

$$\dot{x}_2(t) = x_1 + u.$$

Derive (synthesize) the control law that will asymptotically stabilize this system. Using the Lyapunov stability theory, examine the stability of the closed-loop system.

REFERENCES

1. B. C. Kuo, *Automatic Control Systems*, Prentice-Hall, Englewood Cliffs, NJ, 1995.
2. S. E. Lyshevski, *Control Systems Theory with Engineering Applications*, Birkhauser, Boston, MA, 2000.
3. S. E. Lyshevski, *Electromechanical Systems and Devices*, CRC Press, Boca Raton, FL, 2008.
4. K. Ogata, *Discrete-Time Control Systems*, Prentice-Hall, Upper Saddle River, NJ, 1995.
5. K. Ogata, *Modern Control Engineering*, Prentice-Hall, Upper Saddle River, NJ, 1997.

Appendix

Derivatives
Formulas and Rules

$\frac{d}{dx}(cf(x)) = cf'(x)$, c is the constant. $(f(x) \pm g(x))' = f'(x) \pm g'(x)$

$\frac{d}{dx}(x^n) = nx^{n-1}$, n is the number. $\frac{d}{dx}(c) = 0$, c is the constant.

$(fg)' = f'g + fg'$ (Product Rule) $\left(\frac{f}{g}\right)' = \frac{f'g - fg'}{g^2}$ (Quotient Rule)

$\frac{d}{dx}(f(g(x))) = f'(g(x))g'(x)$ (Chain Rule)

$\frac{d}{dx}(\mathbf{e}^{g(x)}) = g'(x)\mathbf{e}^{g(x)}$ $\frac{d}{dx}(\ln g(x)) = \frac{g'(x)}{g(x)}$

Common Derivatives
Polynomials

$\frac{d}{dx}(c) = 0$ $\frac{d}{dx}(x) = 1$ $\frac{d}{dx}(cx) = c$ $\frac{d}{dx}(x^n) = nx^{n-1}$ $\frac{d}{dx}(cx^n) = ncx^{n-1}$

Trigonometric Functions

$\frac{d}{dx}(\sin x) = \cos x$ $\frac{d}{dx}(\cos x) = -\sin x$ $\frac{d}{dx}(\tan x) = \sec^2 x$

$\frac{d}{dx}(\sec x) = \sec x \tan x$ $\frac{d}{dx}(\csc x) = -\csc x \cot x$ $\frac{d}{dx}(\cot x) = -\csc^2 x$

Inverse Trigonometric Functions

$\frac{d}{dx}(\sin^{-1} x) = \frac{1}{\sqrt{1-x^2}}$ $\frac{d}{dx}(\cos^{-1} x) = -\frac{1}{\sqrt{1-x^2}}$ $\frac{d}{dx}(\tan^{-1} x) = \frac{1}{1+x^2}$

$\frac{d}{dx}(\sec^{-1} x) = \frac{1}{|x|\sqrt{x^2-1}}$ $\frac{d}{dx}(\csc^{-1} x) = -\frac{1}{|x|\sqrt{x^2-1}}$ $\frac{d}{dx}(\cot^{-1} x) = -\frac{1}{1+x^2}$

Exponential and Logarithm Functions

$\frac{d}{dx}(a^x) = a^x \ln(a)$ $\frac{d}{dx}(\mathbf{e}^x) = \mathbf{e}^x$

$\frac{d}{dx}(\ln(x)) = \frac{1}{x}, x > 0$ $\frac{d}{dx}(\ln|x|) = \frac{1}{x}, x \neq 0$ $\frac{d}{dx}(\log_a(x)) = \frac{1}{x \ln a}, x > 0$

$\frac{d}{dx}(e^{g(x)}) = e^{g(x)} \frac{d}{dx} g(x)$

Hyperbolic Trigonometric Functions

$$\frac{d}{dx}(\sinh x) = \cosh x \qquad \frac{d}{dx}(\cosh x) = \sinh x \qquad \frac{d}{dx}(\tanh x) = \text{sech}^2 x$$

$$\frac{d}{dx}(\text{sech}\, x) = -\text{sech}\, x \tanh x \qquad \frac{d}{dx}(\text{csch}\, x) = -\text{csch}\, x \coth x \qquad \frac{d}{dx}(\coth x) = -\text{csch}^2 x$$

Some Integrals

Formulas and Rules

$$\int cf(x)dx = c\int f(x)dx, \quad c \text{ is a constant.} \qquad \int f(x) \pm g(x)dx = \int f(x)dx \pm \int g(x)dx$$

$$\int_a^b f(x)dx = F(x)\Big|_a^b = F(b) - F(a) \text{ where } F(x) = \int f(x)dx$$

$$\int udv = uv - \int vdu \quad \int_a^b udv = uv\Big|_a^b - \int_a^b vdu \ (\text{integration by part})$$

Polynomials

$$\int dx = x + c \qquad \int k\,dx = k\,x + c \qquad \int x^n dx = \frac{1}{n+1} x^{n+1} + c, n \neq -1$$

$$\int \frac{1}{x} dx = \ln|x| + c \qquad \int x^{-1} dx = \ln|x| + c \qquad \int x^{-n} dx = \frac{1}{-n+1} x^{-n+1} + c, n \neq 1$$

$$\int \frac{1}{ax+b} dx = \frac{1}{a} \ln|ax+b| + c \qquad \int x^{\frac{p}{q}} dx = \frac{1}{\frac{p}{q}+1} x^{\frac{p}{q}+1} + c = \frac{q}{p+q} x^{\frac{p+q}{q}} + c$$

Trigonometric Functions

$$\int \cos u\,du = \sin u + c \qquad \int \sin u\,du = -\cos u + c \qquad \int \sec^2 u\,du = \tan u + c$$

$$\int \sec u \tan u\,du = \sec u + c \qquad \int \csc u \cot u\,du = -\csc u + c \qquad \int \csc^2 u\,du = -\cot u + c$$

$$\int \tan u\,du = \ln|\sec u| + c \qquad \int \cot u\,du = \ln|\sin u| + c$$

$$\int \sec u\,du = \ln|\sec u + \tan u| + c \qquad \int \sec^3 u\,du = \frac{1}{2}\left(\sec u \tan u + \ln|\sec u + \tan u|\right) + c$$

$$\int \csc u\,du = \ln|\csc u - \cot u| + c \qquad \int \csc^3 u\,du = \frac{1}{2}\left(-\csc u \cot u + \ln|\csc u - \cot u|\right) + c$$

Exponential and Logarithm Functions

$$\int \mathbf{e}^u\,du = \mathbf{e}^u + c \qquad \int a^u\,du = \frac{a^u}{\ln a} + c \qquad \int \ln u\,du = u\ln(u) - u + c$$

$$\int \mathbf{e}^{au} \sin(bu)du = \frac{\mathbf{e}^{au}}{a^2+b^2}\left(a\sin(bu) - b\cos(bu)\right) + c \qquad \int u\mathbf{e}^u du = (u-1)\mathbf{e}^u + c$$

$$\int \mathbf{e}^{au} \cos(bu)du = \frac{\mathbf{e}^{au}}{a^2+b^2}\left(a\cos(bu) + b\sin(bu)\right) + c \qquad \int \frac{1}{u\ln u} du = \ln|\ln u| + c$$

Inverse Trigonometric Functions

$$\int \frac{1}{\sqrt{a^2 - u^2}}\, du = \sin^{-1}\left(\frac{u}{a}\right) + c \qquad \int \sin^{-1} u\, du = u \sin^{-1} u + \sqrt{1-u^2} + c$$

$$\int \frac{1}{\sqrt{a^2 + u^2}}\, du = \frac{1}{a}\tan^{-1}\left(\frac{u}{a}\right) + c \qquad \int \tan^{-1} u\, du = u \tan^{-1} u - \frac{1}{2}\ln\left(1+u^2\right) + c$$

$$\int \frac{1}{u\sqrt{u^2 - a^2}}\, du = \frac{1}{a}\sec^{-1}\left(\frac{u}{a}\right) + c \qquad \int \cos^{-1} u\, du = u \cos^{-1} u - \sqrt{1-u^2} + c$$

Hyperbolic Trigonometric Functions

$$\int \sinh u\, du = \cosh u + c \qquad \int \cosh u\, du = \sinh u + c \qquad \int \operatorname{sech}^2 u\, du = \tanh u + c$$

$$\int \tanh u\, du = \ln\left(\cosh u\right) + c \qquad \int \operatorname{sech} u\, du = \tan^{-1}\left|\sinh u\right| + c$$

$$\int \operatorname{sech} \tanh u\, du = -\operatorname{sech} u + c \qquad \int \operatorname{csch} \coth u\, du = -\operatorname{csch} u + c \qquad \int \operatorname{csch}^2 u\, du = -\coth u + c$$

Index

D

E

F

R

S

T

V

W

Z